Vol. 28. **The Analytical Chemistry o**  ʋ). Edited by C. A. Streuli and Philip R.

Vol. 29. **The Analytical Chemistry of** ...ree parts). By J. H. Karchmer

Vol. 30. **Ultramicro Elemental Analysis.** By Günther Tölg

Vol. 31. **Photometric Organic Analysis** (*in two parts*). By Eugene Sawicki

Vol. 32. **Determination of Organic Compounds: Methods and Procedures.** By Frederick T. Weiss

Vol. 33. **Masking and Demasking of Chemical Reactions.** By D. D. Perrin

Vol. 34. **Neutron Activation Analysis.** By D. De Soete, R. Gijbels, and J. Hoste

Vol. 35. **Laser Raman Spectroscopy.** By Marvin C. Tobin

Vol. 36. **Emission Spectrochemical Analysis.** By Morris Slavin

Vol. 37. **Analytical Chemistry of Phosphorus Compounds.** Edited by M. Halmann

Vol. 38. **Luminescence Spectrometry in Analytical Chemistry.** By J. D. Winefordner, S. G. Schulman and T. C. O'Haver

Vol. 39. **Activation Analysis with Neutron Generators.** By Sam S. Nargolwalla and Edwin P. Przybylowicz

Vol. 40. **Determination of Gaseous Elements in Metals.** Edited by Lynn L. Lewis, Laben M. Melnick, and Ben D. Holt

Vol. 41. **Analysis of Silicones.** Edited by A. Lee Smith

Vol. 42. **Foundations of Ultracentrifugal Analysis.** By H. Fujita

Vol. 43. **Chemical Infrared Fourier Transform Spectroscopy.** By Peter R. Griffiths

Vol. 44. **Microscale Manipulations in Chemistry.** By T. S. Ma and V. Horak

Vol. 45. **Thermometric Titrations.** By J. Barthel

Vol. 46. **Trace Analysis: Spectroscopic Methods for Elements.** Edited by J. D. Winefordner

Vol. 47. **Contamination Control in Trace Element Analysis.** By Morris Zief and James W. Mitchell

Vol. 48. **Analytical Applications of NMR.** By D. E. Leyden and R. H. Cox

Vol. 49. **Measurement of Dissolved Oxygen.** By Michael L. Hitchman

Vol. 50. **Analytical Laser Spectroscopy.** Edited by Nicolo Omenetto

Vol. 51. **Trace Element Analysis of Geological Materials.** By Roger D. Reeves and Robert R. Brooks

Vol. 52. **Chemical Analysis by Microwave Rotational Spectroscopy.** By Ravi Varma and Lawrence W. Hrubesh

Vol. 53. **Information Theory As Applied to Chemical Analysis.** By Karel Eckschlager and Vladimir Štěpánek

Vol. 54. **Applied Infrared Spectroscopy: Fundamentals, Techniques, and Analytical Problem-solving.** By A. Lee Smith

Vol. 55. **Archaeological Chemistry.** By Zvi Goffer

Vol. 56. **Immobilized Enzymes in Analytical and Clinical Chemistry.** By P. W. Carr and L. D. Bowers

Vol. 57. **Photoacoustics and Photoacoustic Spectroscopy.** By Allan Rosencwaig

Vol. 58. **Analysis of Pesticide Residues.** Edited by H. Anson Moye

Vol. 59. **Affinity Chromatography.** By William H. Scouten

Vol. 60. **Quality Control in Analytical Chemistry.** By G. Kateman and F. W. Pijpers

Vol. 61. **Direct Characterization of Fineparticles.** By Brian H. Kaye

Vol. 62. **Flow Injection Analysis.** By J. Ruzicka and E. H. Hansen

(*continued on back*)

(*continued from front*)

Vol. 63. **Applied Electron Spectroscopy for Chemical Analysis.** Edited by Hassan Windawi and Floyd Ho

Vol. 64. **Analytical Aspects of Environmental Chemistry.** Edited by David F. S. Natusch and Philip K. Hopke

Vol. 65. **The Interpretation of Analytical Chemical Data by the Use of Cluster Analysis.** By D. Luc Massart and Leonard Kaufman

Vol. 66. **Solid Phase Biochemistry: Analytical and Synthetic Aspects.** Edited by William H. Scouten

Vol. 67. **An Introduction to Photoelectron Spectroscopy.** By Pradip K. Ghosh

Vol. 68. **Room Temperature Phosphorimetry for Chemical Analysis.** By Tuan Vo-Dinh

Vol. 69. **Potentiometry and Potentiometric Titrations.** By E. P. Serjeant

Vol. 70. **Design and Application of Process Analyzer Systems.** By Paul E. Mix

Vol. 71. **Analysis of Organic and Biological Surfaces.** Edited by Patrick Echlin

Vol. 72. **Small Bore Liquid Chromatography Columns: Their Properties and Uses.** Edited by Raymond P. W. Scott

Vol. 73. **Modern Methods of Particle Size Analysis.** Edited by Howard G. Barth

Vol. 74. **Auger Electron Spectroscopy.** By Michael Thompson, M. D. Baker, Alec Christie, and J. F. Tyson

Vol. 75. **Spot Test Analysis: Clinical, Environmental, Forensic and Geochemical Applications.** By Ervin Jungreis

Vol. 76. **Receptor Modeling in Environmental Chemistry.** By Philip K. Hopke

Vol. 77. **Molecular Luminescence Spectroscopy — Part 1: Methods and Applications.** Edited by Stephen G. Schulman

Vol. 78. **Inorganic Chromatographic Analysis.** Edited by John C. MacDonald

Vol. 79. **Analytical Solution Calorimetry.** Edited by J. K. Grime

Vol. 80. **Selected Methods of Trace Metal Analysis: Biological and Environmental Samples.** By Jon C. VanLoon

Vol. 81. **The Analysis of Extraterrestrial Materials.** By Isidore Adler

Vol. 82. **Chemometrics.** By Muhammad A. Sharaf, Deborah L. Illman, and Bruce R. Kowalski

Vol. 83. **Fourier Transform Infrared Spectrometry.** By Peter R. Griffiths and James A. de Haseth

Vol. 84. **Trace Analysis: Spectroscopic Methods for Molecules.** Edited by Gary Christian and James B. Callis

Vol. 85. **Ultratrace Analysis of Pharmaceuticals and Other Compounds of Interest.** Edited by S. Ahuja

Vol. 86. **Secondary Ion Mass Spectrometry: Basic Concepts, Instrumental Aspects, Applications and Trends.** By A. Benninghoven, F. G. Rüdenauer, and H. W. Werner

Vol. 87. **Analytical Applications of Lasers.** Edited by Edward H. Piepmeier

Vol. 88. **Applied Geochemical Analysis.** by C. O. Ingamells and F. F. Pitard

Applied Geochemical Analysis

CHEMICAL ANALYSIS

A SERIES OF MONOGRAPHS ON
ANALYTICAL CHEMISTRY AND ITS APPLICATIONS

Editors
P. J. ELVING, J. D. WINEFORDNER
Editor Emeritus: **I. M. KOLTHOFF**

VOLUME 88

A WILEY-INTERSCIENCE PUBLICATION

JOHN WILEY & SONS

New York / Chichester / Brisbane / Toronto / Singapore

Applied Geochemical Analysis

C. O. INGAMELLS

*Rainbow Lakes
Dunnellon, Florida*

FRANCIS F. PITARD

*Gy and Pitard Sampling Consultants
Broomfield, Colorado*

A WILEY-INTERSCIENCE PUBLICATION

JOHN WILEY & SONS

New York / Chichester / Brisbane / Toronto / Singapore

Library of Congress Cataloging in Publication Data:

Ingamells, C. O.
  Applied geochemical analysis.

  (Chemical analysis, ISSN 0069-2883 ; v. 88)
  "A Wiley-Interscience publication."
  Bibliography: p.
  Includes index.
  1. Geochemistry, Analytic. I. Pitard, Francis F.
II. Title. III. Series.

QE516.3.I54 1986   551.9   86-13277
ISBN 0-471-83279-0

Printed in the United States of America

10 9 8 7 6 5 4 3 2 1

PREFACE

This book began many years ago in the Minnesota Rock Analysis Laboratory and within the French Atomic Energy Commission. Fortunate circumstances brought us together after each of us had accumulated a wealth of experience in a dozen fields, and had been privileged to work with acknowledged experts in these fields. Atomic energy, mining exploration, steelworks quality control, geochronology, statistical analysis, sampling theory, computer programming, exploration data evaluation— all were peripheral to the application of classical and instrumental techniques to the accurate analysis of ores, rocks, minerals, and metallurgical and environmental samples.

The result, we hope, is a work that will prove of value to geologists, mining engineers, geochronologists, and geochemists—and especially to inorganic chemists and instrumentalists, both in academia and in industry. Research metallurgists may also benefit, since the chemistry they apply differs only in scale from that of the analyst. Some topics in this book reflect an audacity that could disturb old habits in some circles. Innovations are many and have been carefully selected. The difficulty in preparing the text has not been what to include, but what to leave out. Much of the material cannot be found elsewhere—it has been drawn from our lifetime experience. Imperfections in it surely exist; we will be grateful if these are brought to our attention.

Our emphasis on practical techniques and the presentation of many practical case studies should assist readers in performing a good job in a field becoming more challenging every day. Oliver Ingamells has been an active member of the Peace Corps for several years, and it is not by pure hazard that this publication is specifically oriented toward the laboratory facilities of developing countries to assist them to perform excellent quality work with limited budgets. It is our belief that analytical quality is not necessarily a function of the number of costly gadgets at the investigator's disposal.

Chapter 1 covers in detail many problems associated with the sampling operation that always precedes the analysis of geological materials. Our objective was to provide the information necessary for the reader to better appreciate the unsuspected problems often encountered during the sampling operation. Sampling is already part of the analytical process, and it should be handled with analytical care, accuracy, and precision. Sampling is too often overlooked and nearly always followed by lengthy and expensive analytical work. Everyone, from the analyst to the president of a company, should regard sampling as an "insurance policy" against losses generated by poor data. The reasons why sampling is overlooked are because it is a costly operation and an error-generating process that leaves no apparent tracks of its shortcomings. Indeed, it is easier to track down analytical errors than sampling errors. Consequently, bad sampling practices plague the entire industry and are probably responsible for the losses of many millions of dollars.

Chapters 2–4 strongly emphasize the classical analysis of rocks and minerals. With the impressive development of instrumental technology, lengthy and costly primary analytical work tends to be overlooked, if not forgotten. In some circles, primary methods are regarded as part of an old age and are replaced by fast, cheap, and easy instrumental methods of analysis. Sometimes, those actions are justified but more often lead to collection of useless data. In this volume, we look at two aspects of this problem. First, in many developing countries, laboratories do not have access to all the sophisticated and expensive technology. Nevertheless these laboratories can perform excellent analytical work. A skilled chemist is often better off using a simple and reliable machine on which he or she has control over all parameters than a complicated microprocessor optimizing all parameters that does not give any flexibility. Modern instrumentation often gives a false sense of security to the analyst. This leads to the second aspect of the problem: The skilled analyst wants to buy reliability; however, the manufacturer wants to sell gadgets; unfortunately, the more the gadgets, the less the reliability.

Chapter 5 provides the reader with a strategy to tailor rapid methods of analysis. Many carefully researched procedures are available; however, emphasis is made on the fact that they are designed for specific materials and, most often, should not be applied to other materials without thorough investigation. Six areas responsible for most of the analytical failures are presented.

Chapter 6 describes rapid and practical techniques of calculation used in basic statistics, sampling, analytical chemistry, x-ray matrix corrections, determination of mineral formulas, and so on. A special effort is made to adhere only to the essential and not to confuse the reader in developments of little interest. For many, this chapter could be an attractive, refreshing course on useful and fundamental concepts that are essential to the effective solution of many simple problems.

Chapter 7 presents general guidelines for the preparation of geochemical standards. Justifications for a company to set up its own standard laboratory are presented.

Acknowledgment is made of assistance from specialized texts, which were of great help to the writers in preparing the first edition of this book; in particular, the former accomplishments of Mr. Brent P. Fabbi from the U.S. Geological Survey, Menlo Park, California; Ms. J. C. Engels, also from the U.S. Geological Survey, Menlo Park, California; Dr. J. Visman, Manager of Western Research Laboratory, Canada Center for Mineral and Energy Technology; Dr. Pierre M. Gy, consulting Engineer and Professor at the Paris School of Physics and Chemistry; and Mr. Sydney Abbey of the Geological Survey of Canada.

Finally, we express our deepest appreciation for the patience and assistance of a long-time friend, Helene Waibel, who has been so helpful during the preparation of the manuscript.

The contents of this book are the sole responsibility of the authors.

Francis F. Pitard

Broomfield, Colorado
August 1986

C. O. Ingamells

Rainbow Lakes
Dunnellon, Florida

CONTENTS

CHAPTER 1   SAMPLING   1

CHAPTER 2   THE ART OF CHEMICAL ANALYSIS   85

CHAPTER 3   THE CLASSICAL ROCK OR MINERAL
            ANALYSIS   190

CHAPTER 4   THE ELEMENTS   311

CHAPTER 5   RAPID CHEMICAL METHODS   510

CHAPTER 6   BASIC CALCULATIONS AND
            RECOMMENDATIONS   579

CHAPTER 7   GUIDELINES FOR PREPARATION OF
            GEOCHEMICAL STANDARDS   687

            INDEX   721

Applied Geochemical Analysis

CHAPTER

1

SAMPLING

1.1. Gy's Sampling Theory
1.2. Ingamells' and Switzer's Constant K_s
1.3. Visman's Sampling Theory
1.4. Relationships among the Theories
1.5. Geostatistics
1.6. The Laboratory Sample
1.7. Calibrating Standards
1.8. Samples and Subsamples
1.9. Weight Proportions and Volume Proportions
1.10. Size and Number of Particles in a Mixture
1.11. Contribution of a Single Grain
1.12. Histograms
1.13. Standard Deviation σ and Estimated Standard Deviation s
1.14. Gaussian, or Normal, Distribution
1.15. Poisson Statistics
1.16. Relative Deviation R
1.17. Homogeneity
1.18. Reduction of Samples to Laboratory Subsamples
1.19. Gy's Sampling Slide Rule
1.20. Determination of Visman Constants A and B
1.21. Determination of Gangue Concentration L
1.22. Sampling Diagrams
1.23. Sampling Diagrams for Segregated Mixtures
1.24. Construction of Sampling and Subsampling Diagrams
1.25. Usefulness of Sampling and Subsampling Diagrams
1.26. Planning a Sampling Campaign or Exploration
1.27. Effect of Variable Sample or Subsample Weight: Weighted Average
1.28. Minimizing Exploration Costs
1.29. The Point of Diminishing Returns
1.30. Evaluation of Preliminary Data
1.31. Manipulation of Skewed Data
1.32. Data from Segregated Ore Bodies
1.33. Double Poisson Distribution
1.34. Fitting Statistical Models
1.35. Purposes of Sampling
1.36. Field Sampling Methods
1.37. Sampling for Potassium–Argon Dating

1.38. Mixing and Blending
1.39. Contamination
1.40. Preparation of the Laboratory Sample for Analysis
1.41. Screen Tests: The Tyler Standard Screen Scale
References

If you wish to sample the ocean for its salt content, several cupfuls of seawater taken in several of the world's seas would probably yield a useful distribution of *assay values* and a useful average value.

If you wish to sample the ocean for its herring content, a million cupfuls would probably yield a false distribution of assay values and an erroneous average.

This example points out the important principle that sampling procedures must be designed with a purpose in mind. Sampling may have to be quite different for different purposes. Assaying the ocean for its herring content requires large samples; cupfuls are obviously too small. In fact, nets are used; the individual sample size is the volume of ocean through which the net is drawn during a single sweep.

No one would blame the person who counts the herring in the nets for a consequently poor estimate of the herring content of the ocean; yet geologists have sometimes been led into the habit of collecting minuscule samples of "mountains" containing large chunks (nuggets) of ore sparsely distributed and attributing "anomalous" assay values to deficiencies in analytical techniques.

In these days of diminishing mineral reserves, exploitation of increasingly marginal deposits becomes necessary. As the margin narrows, accuracy in evaluation becomes ever more important. False evaluation may lead to abandonment of a viable property or to exploitation of one that is too lean to be profitable.

Success of an evaluation depends on tiny analytical subsamples having, within tolerable limits, the same concentration of the valuable element, X, as the ore surrounding the position of the corresponding field samples. If this requirement is not met, the most competent analyst cannot generate useful results. Since analysts are likely to be held responsible for confused data, they are well-advised to scrutinize the processes whereby the samples are generated.

The information chain leads from the mountain to the balance pan through sample collection and reduction; it then leads back to the mountain via physical, instrumental, or chemical analysis and data evaluation. Links in this chain have been studied by a number of investigators, and a variety of sampling theories and data evaluation disciplines have been developed.

The several sampling theories all depend on the same statistical principles but differ in emphasis and applicability. Each is based on a more
or less artificial model.

1.1. GY'S SAMPLING THEORY

Pierre Gy [1–3] has done more than anyone else to develop sampling
theory and promote its application in the real world. His monographs
should be required reading for anyone involved in geochemical analysis.

Gy proposes that the sampling characteristics of a granular material
containing a small proportion of a valuable component, X, depend on (1)
the shape of the particles, (2) the distribution of particle sizes, (3) the
mineralogical composition of each of the phases in a two-phase mixture,
and (4) the degree to which X is liberated from the gangue. Gy's sampling
constant G is thus the product of four factors: (1) f is a shape factor,
equal to 1.00 if all particles are perfect cubes: f is the ratio of particle
volume to that of a cube that will just pass the same screen; (2) g is a
particle size distribution factor, equal to 1.00 if all particles are the same
size—the larger the proportion of fine particles, the lower the value; (3)
m is a mineralogical composition factor,

$$ m = \frac{p_w d_H + q_w d_L}{q_w} \tag{1.1} $$

where d_H and d_L are the densities of ore mineral and gangue, respectively,
and $q_w = 1 - p_w$ is the weight proportion of the valuable constituent
X—derivation of equation (1.1) is a matter of some difficulty; and (4) b
is a liberation factor related to the proportion of ore mineral liberated
from gangue, approaching 1.00 as liberation becomes complete during the
reduction process. Gy describes experimental methods for estimating
each of these four factors and thereby the sampling constant G.

When $G = fgmb$ has been estimated for a specific material, the uncertainty in a w-gram sample or subsample is

$$ \pm s_R = \sqrt{\frac{Gu'^3}{w}} \tag{1.2} $$

where $s_R = s/K$ is the absolute uncertainty divided by the true X content,
K, of the mixture, and u' is the linear dimension of the largest particles,
that is, the linear dimension of the screen mesh that will pass 95% of the
particles.

From (1.2), one may calculate the sample or subsample weight required for any desired uncertainty s_R:

$$w = \frac{Gu'^3}{s_R^2} \tag{1.3}$$

The product Gu'^3 may be estimated by repetitive assay of a granular material, avoiding a need to find the individual factors f, g, m, and b:

$$Gu'^3 = ws_R^2 \tag{1.4}$$

where w is the weight of each sample used in the exercise, and s_R is the relative standard deviation of a sufficient number of assays. It must be noted that the above is by no means a complete exposition of Gy's sampling theory.

1.2. INGAMELLS' AND SWITZER'S CONSTANT K_s

A measure of the sampleability of a well-mixed granular material is simply the weight necessary for a 1% sampling uncertainty. This defines a sampling constant K_s (in grams):

$$K_s = R^2 w \quad \text{or} \quad w = \frac{K_s}{R^2} \tag{1.5}$$

where R (in percent) is the relative uncertainty in a w-gram sample or subsample, related to Gy's s_R and the standard deviation in a sufficient number of assays each using a w-gram sample:

$$R = \frac{100s}{K} = 100s_R = \sqrt{\frac{K_s}{w}} \tag{1.6}$$

Ingamells' and Switzer's approach [4] depends on the following proposition: The sampling characteristics of any mass of material may be duplicated, for a single constituent X in a hypothetical mixture of uniform cubes of two minerals of X contents H and L percent and densities d_H and d_L, respectively.

It has been shown that, for heavy metal ores,

$$K_s \approx \frac{10^4(K - L)(H - L)u^3 d_H}{K^2} \tag{1.7}$$

where K is the overall X content (in percent and u^3 is the volume of a hypothetical cube. Thus, K_s may be estimated either by repetitive determination or from physical characteristics of the material being sampled.

1.3. VISMAN'S SAMPLING THEORY

Visman's sampling equation,

$$S^2 = \frac{A}{W} + \frac{B}{N} + (SE)^2 \tag{1.8}$$

expresses the fact that variances are additive. In (1.8), S is the uncertainty in the average of N assays using sample weight $w = W/N$, SE is the uncertainty due to subsampling and analytical errors, A is a homogeneity constant, and B is a segregation constant. The variance A/W is due to the nugget effect, the variance B/N is due to large-scale segregation, and $(SE)^2$ is irrelevant variance.

The Visman sampling constants A and B are calculated from

$$A = \frac{w_1(s_1^2 - s_2^2)}{1 - 1/Q} \tag{1.9}$$

$$B = s_1^2 - \frac{A}{w_1} = s_2^2 - \frac{A}{w_2} \tag{1.10}$$

where s_1 and s_2 are the standard deviations in two sets of assays deriving from sample weights w_1 and w_2, and $Q = w_2/w_1$ [5].

1.4. RELATIONSHIPS AMONG THE THEORIES

The sampling constants G, K_s, and A are used to find (1) the weight of sample or subsample needed for a given uncertainty, and (2) the uncertainty expected for any sample or subsample weight. The three constants are related (approximately) in the following way:

$$A = GK^2u'^3 = K_sK^2 \times 10^{-4} = (K - L)(H - L)u^3D \tag{1.11}$$

where D is a function of mineral densities often approximated by the density of the ore mineral. It is also approximately true that

$$fg = \frac{u^3}{u'^3} \quad \text{and} \quad b = \frac{(K - L)^2}{K^2} \tag{1.12}$$

These relationships enable selection of the easiest and most convenient procedure for estimating a sampling constant. They also provide a means for cross-checking in cases where estimation of the several sampling parameters is difficult.

1.5. GEOSTATISTICS

Geostatistics is a data evaluation discipline that attempts prediction of mining grades and cutoffs from exploration data. It depends, among other devices, on the construction of *variograms,* which are essentially measures of the *zone of influence* in a given direction of each sample taken within a mountain [6, 7].

Suppose samples taken 100 ft apart in a straight line assay 0.62, 0.60, 0.58, 0.52, and 0.01; it is easy to decide that the ore diminishes slowly in grade through the first four positions and peters out somewhere between the fourth and fifth. Suppose, however, that the assays were 0.62, 3.98, 0.20, 0.52, and 0.36 (a more likely circumstance in many cases); then interpretation is more difficult. With large data sets, it is sometimes possible to predict changes in mining grade throughout an ore body using geostatistical principles.

To visualize the geostatistical process, it is useful to write Visman's equation (1.8) as follows:

S^2 = total variance

= variance due to small-scale inhomogeneity

+ variance due to large-scale inhomogeneity

+ variance due to reduction and subsampling error

or

S^2 = total variance

= nugget effect + geostatistical variance + irrelevant variance

Geostatistics measures and utilizes only the variance in assay data due to large-scale changes in grade throughout the ore body. The geostatistician must assume that irrelevant variance due to subsampling and analytical error is low. If it is not low, his calculations are likely to show a false nugget effect or be otherwise in error. If there is a real nugget effect—

that is, the ore is present in large pockets or crystalline masses—geostatistical calculations will reveal the relative importance of this effect.

When data show a large nugget effect, this indicates that the field samples may be too small or that reduction and subsampling errors are not adequately controlled. Given the Visman sampling constants A and B, it is possible to calculate an optimum field sample weight

$$w_{opt} = \frac{A}{B} \qquad (1.13)$$

that will yield maximum information with minimum effort. If the field sample weight is too small, data may be confusing and difficult to evaluate; if it is too large, collection, reduction, and subsampling costs will be unnecessarily high.

1.6. THE LABORATORY SAMPLE

However field samples are collected and reduced, the laboratory is likely to receive a large number of samples. Often each will weigh 1–2 kg and will have been crushed to 0.25 in. It is the responsibility of the analyst to reduce each of these samples to an analytical subsample weighing from a few milligrams (spectrographic analysis) to several grams (fire assay). This reduction must proceed in such a way that the analytical subsample contains, within acceptable limits, the same concentration of the constituent of interest as the 1–2 kg provided. Since many samples must be processed, reduction and subsampling procedures should be rapid and efficient. Sampling theory provides a guide to optimizing the process.

The subsampling of properly collected samples depends heavily on the degree of segregation of the parent material. With unsegregated materials, the properly collected sample is of the minimum size that will yield an acceptable uncertainty and should be so handled that inappreciable added uncertainty enters during subsampling. With segregated materials, of which several samples need to be taken, the requirement is that subsampling uncertainty shall be no more than one-third of the sampling uncertainty.

It is always uneconomic to diminish the subsampling uncertainty below one-third of the sampling uncertainty. Observance of this principle may save a lot of unnecessary work. For example, if a series of samples of optimum weight have been taken from a segregated mass for which $A = 50$, $B = 0.05$, the sampling uncertainty is, from (1.8) with $N = 1$ and $w = A/B$, ±0.32. If the overall X content of the mass is 1%, the relative uncertainty in each sample is 32%. Subsampling procedures cannot profitably be refined past the point where subsampling uncertainty is ± 10–11%. This means that only one-hundredth of the subsample weight calculated from (1.3) or similar relationships need be employed.

On the other hand, if properly collected samples come from an unsegregated mass, it is inexcusable to subsample them in such a way that a 10–11% uncertainty is introduced.

1.7   CALIBRATING STANDARDS

During a mineral exploration, hundreds of field samples may be collected every working day. It is desirable to report their analyses or assays as promptly as possible so as to provide a continuing guide to the exploration geologist. This usually means that rapid instrumental methods of analysis should be employed.

Rapid instrumental methods require calibration; calibration is most satisfactory if a few of the actual samples analyzed by primary methods are used to develop working calibration curves. If a sampling constant K_s can be determined for each calibrating standard, error bars can be drawn on each calibration point. If this is done, variability in repetitive instrument readings from calibrating standards will cause no concern provided the error bar limits are not exceeded.

Measurement of sampling errors during calibration of X-ray methods and others for which matrix corrections are essential is especially important. If deviations due to subsampling uncertainty are dealt with as though they were due to a matrix effect, false corrections are likely.

1.8.   SAMPLES AND SUBSAMPLES

A sample is a portion of a mass of material, but is not necessarily a representative portion of the whole. If the mass is heterogeneous, its composition with respect to the element of interest, X, may have to be estimated through analysis or assay of many samples, no one of which is expected to correspond exactly in composition to that of the whole mass. Variance in assay values from a large number of samples provides a measure of several characteristics of the parent mass—degree of segregation or nonuniformity, inhomogeneity, effective grain size of the ore mineral, banding, nugget content, and so on.

A subsample is a split of a sample removed from it in such a way that there is some confidence in its having the same concentration of the element of interest, X, as the parent sample. A most important application of sampling theory is the devising of methods of sample reduction that will ensure subsamples of the same X content as the samples from which they derive.

Certain elementary principles must be well understood by anyone who wishes to control sampling and subsampling error. These principles are as applicable to the sampling of a mountain as they are to the subsampling of a small bag or bottle of rock or mineral powder generated from the mountain. The importance of these principles cannot be overemphasized: Ignorance of them may result, and has on occasion resulted, in large expenditures without reward.

1.9. WEIGHT PROPORTIONS AND VOLUME PROPORTIONS

In a mixture of two minerals or of an ore mineral and a gangue, the weight proportions of the two components are the same as their volume proportions only if the densities of all minerals in the mixture are the same. Most often, analytical or assay values are reported in terms of weight proportions—percent, parts per million, ounces per ton, and so on. Sampling characteristics of the mixture are, however, more dependent on volume proportions: It is therefore necessary to relate these two ways of measuring concentration.

Consider a mixture of two minerals each containing the element of interest, X. The mixture is described by the following parameters:

p_v = volume proportion of mineral with X content L wt. %

q_v = volume proportion of mineral with X content H wt. %

p_w = weight proportion of mineral with X content L wt. %

q_w = weight proportion of mineral with X content H wt. %

K = overall X content of mixture, wt. %

L = X content of component of mixture present in proportion p_v by volume and p_w by weight, wt. %

H = X content of component of mixture present in proportion q_v by volume and q_w by weight, wt. %

d_L = density of mineral with X content L (%), g/cc

d_H = density of mineral with X content H (%), g/cc

10 SAMPLING

Then, identically,

$$K = p_w L + q_w H \tag{1.14}$$

$$p_w = 1 - q_w = \frac{H - K}{H - L} \tag{1.15}$$

$$q_w = 1 - p_w = \frac{K - L}{H - L} \tag{1.16}$$

$$p_v = \frac{d_H p_w}{d_H p_w + d_L q_w} \tag{1.17}$$

$$q_v = \frac{d_L q_w}{d_H p_w + d_L q_w}$$

If the overall density of the mixture is d grams per centimeter cubed,

$$d = \frac{d_H d_L}{d_H p_w + d_L q_w} = \frac{d_H d_L (H - L)}{d_H (H - K) + d_L (K - L)} \tag{1.18}$$

These relationships are all identities when applied to bimineralic mixtures: With mixtures of several minerals of different densities and X contents, the same formulas may be used if all minerals but one are grouped together and treated as a single mineral. In practice, the two constituents of a mixture are an ore mineral and a gangue [8, 9].

Some examples may illustrate the usefulness of these formulas:

Example. To investigate the validity of the formulas given above, a mixture of an amphibolite with 0.1445% K_2O and an orthoclase with 14.92% K_2O was prepared. The weight in the mixture was 4.4601 g amphibolite and 0.0429 g orthoclase. The density of orthoclase is 2.7 g/cc and that of amphibolite is 3.0 g/cc. The average of 40 determinations of K_2O on 0.1000-g samples of the mixture was 0.286% (Table 1.1).

From (1.15) and (1.16), the proportions by weight of the two minerals are

$$p_w = \frac{14.92 - 0.286}{14.92 - 0.1445} = 0.9904; \quad q_w = 1 - p_w$$

From (1.18), the density of the mineral mixture is

$$d = \frac{(2.7 \times 3.0)(14.92 - 0.1445)}{2.7(14.92 - 0.286) + 3.0(0.286 - 0.1445)}$$

$$= 2.997$$

Table 1.1. Determination of K_2O in a Mixture[a]

	Mixed on Paper Only, Spooned Out	Mixed on Mechanical Roller, Spooned Out	Remaining Material, Split to ~0.1 g
1	0.247	0.230	0.300
2	0.300	0.246	0.271
3	0.236	0.251	0.261
4	0.258	0.345	0.331
5	0.304	0.350	0.248
6	0.330	0.254	0.333
7	0.247	0.314	0.198
8	0.275	0.242	0.340
9	0.212	0.320	0.321
10	0.311	0.297	0.283
11	0.258	0.312	0.278
12	0.187	0.300	0.309
13			0.378
14		(subsample weight, 0.1000 g)	0.337
15			0.279
16			0.348
Mean value	0.2638	0.288	0.301
Standard deviation	0.04226	0.0420	0.0452
Relative deviation, R (%)	16.0	14.6	15.0
Spread	0.143	0.120	0.180

Over all 40 determinations

$\overline{X}$ = arithmetic mean	0.2860
s = standard deviation	0.0451
R = relative deviation	15.75%
Calculated R for 48 mesh (or #)	13.69%
Calculated R for 35 mesh (or #)	23.02%

[a] The mixture was prepared by sieving out the $+48$, -35 mesh portion of both the orthoclase Or-1 [S. S. Goldich, C. O. Ingamells, N. H. Suhr and D. H. Anderson, *Can. J. Earth Sci.* **4,** 747 (1967)] and an amphibolite. Portions of each were weighed (4.4601 g amphibolite and 0.0429 g orthoclase) and thoroughly blended. The K_2O content of the sized amphibolite was repetitively determined at 0.1445%. The K_2O content of the sized orthoclase was identical to that of the unsieved material [J. C. Engels and C. O. Ingamells, *Talanta* **17,** 783 (1970)]. All but about 0.3 g was used in acquiring the above values. The remainder was used to determine K_2O on twelve 0.01-g subsamples. Thus, the rules for sampling with replacement do not exactly apply. [From J. C. Engels and C. O. Ingamells, *Geochim. Cosmochim. Acta* **34,** 1007–1017 (1970).]

Example. A lateritic mud sample is collected, dried, and analyzed for its nickel content. The loss in weight on drying was 45%. The nickel-content of the dried material was determined at 1.12%. The density of the dried material was measured at 2.1 g/cc. We would like to know the weight of nickel in a cubic meter of the mud as collected.

The density of the mud as received may be found from (1.18), with $H = 1.12$, $L = 0$ (the nickel content of water), $d_H = 2.1$ g/cc, and $d_L = 1$ g/cc (the density of water). The nickel content of the mud [by (1.14)] = $0.55 \times 1.12 = 0.616$, and $d = (2.1 \times 1.0 \times 1.12)/[2.1(1.12 - 0.616) + 1.0(0.616)] = 1.404$ g/cc. Therefore, a cubic meter weighs 1.404×10^6 g and contains $1.12 \times 0.55 \times 1.404 \times 10^6/100 = 8649$ g nickel.

1.10. SIZE AND NUMBER OF PARTICLES IN A MIXTURE

In a mixture of two minerals of X contents by weight H and L and of uniform particle size, the number of particles in w grams of the mixture is

$$n = n_H + n_L = \frac{d_L w_H + d_H w_L}{d_H d_L u^3} \tag{1.19}$$

where n_H is the number of grains of the mineral with X content $H\%$, n_L is the number of grains of the mineral with X content $L\%$ by weight in w grams, and u is the side of a cube of volume equal to the volume of one of the particles in the mixture. In a real mixture, u is the *effective* linear mesh size of the mixture—that is, the mesh size of a hypothetical mixture of uniform grain size having the same sampling characteristics as the real mixture being examined.

1.11. CONTRIBUTION OF A SINGLE GRAIN

When a sample of weight w grams is taken from a mixture of two minerals of uniform grain size, each grain of each mineral contributes to the overall X content of the sample an amount that depends on the X content of that grain. Let this contribution be c_L for the grains with X content L and c_H for grains with X content H. Then the X content of a w-gram sample is

$$K = p_w L + q_w H = c_L n_L + c_H n_H$$

and

$$c_L = \frac{L d_L u^3}{w} \qquad c_H = \frac{H d_H u^3}{w} \tag{1.20}$$

Example (Table 1.1). The contribution of a single grain of orthoclase to a single analytical result for potassium using this mixture is, from (1.20),

$$c_H = \frac{14.92 \times 2.7 \times (0.032)^3}{0.1000} = 0.0132\% \text{ per grain}$$

Similarly, the contribution of a single grain of amphibolite is

$$c_L = \frac{0.1445 \times 3.0 \times (0.032)^3}{0.1000} = 0.000142\% \text{ per grain}$$

The factor $u = 0.032$ was obtained from statistical evaluation of the data in Table 1.1 in a manner to be described. Both components of the mixture were screened to $+48$, -35 mesh (0.0297–0.0420 cm); u must obviously lie between these limits.

As an approximation, since the amphibolite contains 0.1445% K_2O, a 0.1000-g sample with one grain of orthoclase contains $0.1445 + 0.0132 = 0.158\%$ K_2O; a sample with two grains of orthoclase contains $0.1445 + 2 \times 0.0132 = 0.171\%$; a sample with three grains contains $0.1445 + 3 \times 0.0132 = 0.184\%$; and so on. Examination of the data (Table 1.1) indicates that the lowest of the 40 values, 0.187% K_2O, probably derives from a subsample containing three grains of orthoclase.

The approximation in the above calculation is that the number of grains of amphibolite remains constant in 0.1000-g samples, regardless of how many grains of orthoclase are present. This approximation is good only when the number of orthoclase grains is very small relative to the number of amphibolite grains. The exact relationship between the number of grains of a high-X component and the X content of the mixture is

$$K_i = L + \frac{(H - L)d_H u^3 n_H}{w} \tag{1.21}$$

where K_i are the values corresponding to various values (0, 1, 2, . . .) of n_H.

Example (Table 1.1). Equation (1.21) for these data becomes

$$K_i = 0.1445 + (14.92 - 0.1445)(2.7)(0.032)^3 n_H/0.1000$$

$$= 0.1445 + 0.01307 n_H$$

and we may prepare a table of values with $n_H = 0, 1, 2, \ldots$:

n_H	K_i	n_H	K_i	n_H	K_i	n_H	K_i
0	0.1445	5	0.2099	10	0.2752	15	0.3406
1	0.1576	6	0.2229	11	0.2883	16	0.3536
2	0.1706	7	0.2360	12	0.3013	17	0.3667
3	0.1837	8	0.2491	13	0.3144	18	0.3798
4	0.1968	9	0.2621	14	0.3275	19	0.3928

In the real mixture, mineral grains are only approximately of uniform size; nevertheless, the pattern of K_2O values resembles that of the table above. If the sample weight were 0.01 g instead of 0.1 g, the values to be expected are, from (1.21), $K_i = 0.1445 + 0.1307 n_H$, yielding the following:

n_H	K_i
0	0.1445
1	0.2752
2	0.4059
3	0.5366
$\vdots$	$\vdots$

Twelve determinations of K_2O using subsamples of approximately 0.01 g were performed. Results follow, with n_H calculated by rearranging (1.21):

$$n_H = \frac{(K_i - 0.1445)w}{0.001307}$$

$K_i = K_2O$ (%)	w (g)	n_H (calculated)
0.327	0.00910	1.3
0.425	0.00907	2.0
0.143	0.00846	0.0
0.390	0.00935	1.8
0.387	0.01137	2.1
0.317	0.01075	1.4
0.485	0.01014	2.6
0.267	0.01053	1.0
0.297	0.01219	1.4
0.157	0.01004	0.1
0.437	0.01029	2.3
0.287	0.01021	1.1

It is evident that the wide variance in analytical values for K_2O is due to the chance distribution of orthoclase grains in the several samples.

We now define a quantity c such that the overall X content, K, of a mixture is

$$K = L + c\bar{n}_H \quad \text{or} \quad K = L + cz$$

where $\bar{n}_H$ is the average number of grains of a high-X component in a two-mineral mixture and will be given the symbol z.

The parameter c is not exactly the same as c_H: The relationship between the two is

$$c = \frac{c_H(H - L)}{H} \quad \text{or} \quad c_H = \frac{cH}{H - L}$$

When H is much larger than L, $c \approx c_H$; this is the common case in mineral exploration. The distinction should not be forgotten, however, when dealing with mixtures in which there is more than one major constituent.

While each subsample must contain an integral number n_H of minor mineral grains, the *average* subsample need not contain an integral number of grains. The overall X content of the sample being subsampled is estimated from the average of all values,

$$K = L + cz \tag{1.22}$$

where z is the average of the n_H in the several subsamples.

The average number z of grains of a minor constituent assumes importance in analysis for trace or minor constituents because this number is often small, even when samples have been put through fine screens. For example, a rock powder containing 200-mesh grains of cassiterite in which the tin content is 20 ppm (parts per million) contains about 10 grains of cassiterite per gram. If 10-mg subsamples are taken for spectrographic determination of tin, only 1 in 10 of them will contain a grain of cassiterite. The average number z of cassiterite grains per 10-mg subsample is about 0.1.

Example. The U.S. Geological Survey (USGS) granite G-1, as supplied, contains chromite grains that are fairly uniform in size and shape. More than half the chromium content of G-1 is accounted for by these chromite grains. One gram of G-1, as supplied, contains an average of about 5 chromite grains, most of which will just pass a 170-mesh screen. The grains (crystals of chromite) are somewhat elongated; in the nomenclature adopted above, their effective mesh size is $u \approx 0.01$ cm. The density of chromite is about 4.5 g/cc; one chromite grain weighs about 4.5×10^{-6} g.

A spectrographer taking a 10-mg subsample of G-1 will find about 200 ppm chromium if the 10-mg subsample happens to contain a grain of chromite, or 400 ppm if it happens to contain two grains. Most often, however, a 10-mg subsample will contain no grain of chromite, and the spectrographer will measure only the background, or gangue, concentration of chromium, about 8 ppm. He may report this and claim high precision except for an occasional "anomalous" value.

A chemist using 1-g subsamples of G-1 will find the results very variable since they depend on how many grains of chromite will appear in each of the subsamples. About 20 ppm chromium will be reported with poor precision.

While the average 1-g subsample of G-1 contains about five grains of chromite, any specific 1-g sample may contain more or less than five grains: the distribution of results will approximate the Poisson model (see below).

The validity of some of the relationships developed above may be demonstrated using chromium in G-1 as an example. We will suppose:

$K =$ overall concentration of chromium in G-1 = 0.0020% (20 ppm)

$L =$ gangue concentration of chromium in G-1 (i.e., that part of the chromium content unaffected by subsampling difficulties) = 0.0008% (8 ppm)

$H =$ concentration of chromium in chromite, $FeCr_2O_4$ = 46%

$d_L =$ density of G-1 = 2.8 g/cc

$d_H =$ density of chromite = 4.5 g/cc

$u =$ effective linear mesh size of chromite in G-1 = 0.01 cm

From these parameters, we may calculate:

$$p_w = \frac{H - K}{H - L} = 0.999974 \quad \text{[from (1.15)]}$$

$$q_w = \frac{K - L}{H - L} = 0.000026 \quad \text{[from (1.16)]}$$

$$p_v = \frac{d_H p_w}{d_H p_w + d_L q_w} = 0.999984 \quad \text{[from (1.17)]}$$

$$q_v = \frac{d_L q_w}{d_H p_w + d_L q_w} = 0.000016 \quad [\text{from (1.18)}]$$

$$K = p_w L + q_w H = 0.0020\% \text{ Cr} \quad [\text{from (1.14)}]$$

The average number z of chromite grains in a 10-mg subsample may be found by rearranging (1.22) and putting $c \approx c_H = H d_H u^3/w$, with $w = 0.01$ g:

$$z = \frac{K - L}{c} = 0.058 \text{ grains/10 mg G-1}.$$

In general, if the average number of high-X grains of ore mineral in a sample or subsample is z, the probability that a single sample or subsample will contain n grains of high-X ore mineral is given by the Poisson formula

$$P_n = \frac{z^n e^{-z}}{n!} \tag{1.23}$$

where P_n is the probability that n grains will appear in a randomly collected sample or subsample, and e is the base of the natural logarithms. For chromite in 10-mg samples of G-1, $z = 0.058$, and the Poisson distribution is:

$n_H = z_i$	P_n
0	0.943
1	0.054
2	0.002

This means that in one thousand 10-mg subsamples of G-1, 943 are likely to contain no chromite grain, 54 are likely to contain one chromite grain, and 2 are likely to contain two chromite grains; 1 subsample in 1000 may contain more than two chromite grains. Translated into assay values obtained on 10-mg subsamples, with $c \approx c_H = H d_H u^3/w = 46 \times 4.5 \times (0.01)^3/0.01 = 0.0207\%$ chromium per grain from (1.20), 943 assays of 10-mg subsamples will show the gangue value of 8 ppm chromium, 54 will show 215 ppm, 2 will show 422 ppm, and 1 will show an even higher value if 1000 samples are taken.

Application of these formulas is by no means limited to 200-mesh powders. A high-grade molybdenum ore containing 0.48% MoS_2 may have one 1-cc grain of molybdenite/kg. If 1-kg samples of 0.5-in. material are properly reduced to analytical subsamples and analyzed for MoS_2, assay values will show a variance that depends on the distribution of 1-cc grains of MoS_2 among the several 1-kg samples.

Example. Suppose a molybdenum ore has the following:

K = 0.300% Mo (overall concentration of Mo)

L = 0.100% Mo (one-third of the Mo is evenly distributed)

H = 60% Mo in MoS_2

u = 1.27 cm (0.5 in.) effective grain size

d_H = 4.8 g/cc (density of molybdenite)

d_L = 2.8 g/cc (average density of gangue minerals)

From (1.16), $q_w = (K - L)/(H - L) = 0.003339$; from (1.20), $c \approx c_H = Hd_Hu^3/w = 0.5889$ for 1-kg samples; from (1.22), $z = (K - L)/c = 0.3396$ grains/kg. Analytical values will be distributed as follows:

$n_H = z_i$	K (%)	Poisson Distribution of 1000 Assays
0	0.100	712
1	0.689	241
2	1.278	41
3	1.867	5
4 or more	2.456+	1

Note that it is impossible, in this example, for any 1-kg sample to contain the correct overall concentation, 0.300%, of molybdenum. Samples containing no molybdenite grain will show the gangue concentration, 0.100% molybdenum; samples containing one molybdenite grain will show 0.689% molybdenum.

In a real situation, of course, the molybdenite grains will not all be 0.5-in. cubes: Samples with one grain of molybdenite may assay from, say, 0.4 to 1.0%; those with two grains may assay from 1.0 to 1.5%; and so on. Nevertheless, a histogram of assay values will show a periodicity, which can be used to detect and quantify the nugget effect.

1.12. HISTOGRAMS

Histograms should be constructed using a rational interval. Use of a completely arbitrary interval may lead to a false picture of a distribution. One rational interval is that determined by the precision of the analytical method that develops the data. When there is a nugget effect, a more useful histogram interval is that corresponding to the contribution of a single grain of ore mineral to an assay value.

Example. Table 1.2 shows the results of assays for cobalt in drill samples taken from a lateritic deposit. A superficial look at these data might lead to the conclusion that there is very little cobalt in the vicinity of drill holes

Table 1.2. Cobalt Assays in a Lateritic Ore

					Drill Hole Number							Depth
1	2	3	4	5	6	7	8	9	10	11	12	(m)
0.03	0.10	1.07	0.64	0.34	0.14	0.09	0.16	0.21	0.20	0.28	0.22	1
0.07	0.20	0.16	0.24	0.20	0.24	0.25	0.36	0.73	2.42	0.81	0.53	2
0.02	0.02	0.03	0.41	0.31	0.46	0.29	0.33	0.28	0.41	0.35	0.11	3
0.09	0.04	0.04	0.03	0.09	0.08	0.09	0.12	0.50	0.28	0.09	0.47	4
0.02	0.03	0.05	0.28	0.23	0.33	1.01	0.17	0.10	0.07	0.03	0.08	5
0.11	0.22	0.21	0.24	0.21	0.20	0.20	0.20	0.21	0.18	0.14	0.13	6
0.05	0.04	0.04	0.03	0.03	0.04	0.04	0.03	0.05	0.10	0.16	0.12	7
0.02	0.02	0.01	0.03	0.01	0.02	0.06	0.05	0.08	0.17	0.35	0.28	8
0.02	0.02	0.03	0.03	0.05	0.03	0.02	0.03	0.03	0.08	0.09	0.05	9
0.02	0.02	0.03	0.02	0.08	0.14	0.12	0.30	1.34	1.04	0.50	0.27	10
0.02	0.02	0.02	0.02	0.02	0.02	0.04	0.07	0.12	0.16	0.30	0.43	11
0.20	0.26	0.17	0.12	0.12	0.10	0.22	0.23	0.27	0.29	0.22	0.18	12

Calculations: (field sample weight, 10 kg)

	Across Bed	Down Holes
s_1	0.2864	0.2864
$\overline{X}$	0.2020	0.2020
h_1	0.0585	0.0585
s_2	0.1330	0.1087
h_2	0.1249	0.1503
A	701.9	765.9
B	0.0118	0.0054
L	0.046	0.031
$w_{(z=1)}$	28,961	26,169
w_{opt}	59,301	140,784

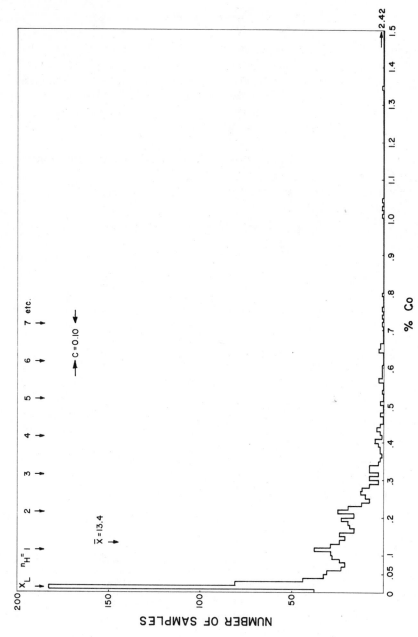

Figure 1.1. Histogram of cobalt data in lateritic ore.

1 and 2 and that mining in the vicinity of drill holes 9 and 10 would yield more cobalt. Such is not the case: There is, in fact, just as much cobalt in the vicinity of holes 1 and 2 as elsewhere. Actual mining generates ore that uniformly contains about 0.2% cobalt.

The explanation is that cobalt occurs in pockets of high-manganese material; these are only occasionally intersected by the drill. The high-manganese material may contain up to 5% cobalt. The laterite ore is very wet, of the consistency of thick mud. In drilling, it is almost impossible to prevent some of the material in every 1-m length of core from contaminating the next length of core. Consequently, a high-cobalt, high-manganese sample is often followed by one or more overestimates of cobalt content.

Figure 1.1 is a histogram of the data of Table 1.2. The contribution c of a single "grain" (pocket, nugget) of cobalt–manganese material is estimated at 0.10% cobalt, and the periodicity in the data is evident.

Methods for determining the size of grains of a constituent of interest will now be explored. These methods require application of some elementary statistical principles.

1.13. STANDARD DEVIATION σ AND ESTIMATED STANDARD DEVIATION s

The standard deviation is an index used to measure the dispersion of a number of measurements about their arithmetic mean. In dealing with analytical or assay values, there are almost always too few of them to permit an exact measure of their dispersion, and only an estimate can be made. The standard deviation σ is a theoretical quantity; the estimated standard deviation s is an experimental approximation to σ. Nevertheless, the estimate s is often referred to as the standard deviation, without mention of the fact that it is an estimate only.

Commonly, the standard deviation in a set of assay values is estimated through use of the formula

$$s^2 = \frac{1}{N-1} [(x_1 - \bar{x})^2 + (x_2 - \bar{x})^2 + \cdots + (x_N - \bar{x})^2] \quad (1.24)$$

where N is the number of assays, $x_1, x_2, \ldots, x_N$ are their respective values, $\bar{x}$ is the arithmetic mean, and the square of the estimated standard deviation, s^2, is called the variance in the results.

When the number of assay values is small, as is often the case, the standard deviation σ is poorly estimated by (1.24): The estimate is usually too low. Equation (1.24) is not the only way to estimate a standard deviation. If the process by which the values are obtained is such that a very large number of them would be symmetrically distributed about their arithmetic mean, a standard deviation may be estimated from a few values by averaging the differences between successive analytical values and multiplying the average by 0.89. For this device to be useful, the variance in assay values must be random; there must be no drift due to sampling or instrumental imperfection.

Example. The standard deviation in the first 12 values of Table 1.1 estimated from (1.24) is 0.042. Averaging the differences between successive values yields

0.247	0.053
0.300	0.064
0.236	0.022
0.258	0.046
0.304	0.026
0.330	0.083
0.247	0.028
0.275	0.063
0.212	0.099
0.311	0.053
0.258	0.071
0.187	11)0.608

$$s = 0.055 \times 0.89 = 0.049$$

A very rapid estimation of σ when only three assays are available may be made by multiplying their spread or range by 0.6. With four values, the estimate is made by multiplying the spread by 0.5. The estimate of σ obtained from a few values is poor in any case, and the trouble involved in using (1.24) is unwarranted.

With 8–16 measurements, the spread divided by 3 gives a rough, but often useful, estimate of the standard deviation. For example, the spread in the 12 values given above is $0.330 - 0.187 = 0.143$; the standard deviation estimated from this range is $0.143/3 = 0.047$.

The variance s^2 in a set of values is of importance because variances are additive. Any set of analytical values contains uncertainties due to sampling, subsampling, and analytical error. Each of these sources of uncertainty contribute to the overall uncertainty. Their contributions are additive through the corresponding variances:

$$s^2_{total} = s^2_{sampling} + s^2_{subsampling} + s^2_{analytical} \tag{1.25}$$

With a set of assay values obtained during the early stages of an exploration, the total variance can be calculated from 1.24 or one of the alternate methods suggested. The analytical variance can be found by repetitive analysis of samples with known subsampling characteristics, the
subsampling variance can be estimated through careful evaluation of the
subsampling process, and the variance due to inhomogeneity and heterogeneity of the mountain being examined can be found by difference.
Through such logic, the process of sampling and analyzing a mountain
may be optimized; methods for doing this will be suggested.

1.14. GAUSSIAN, OR NORMAL, DISTRIBUTION

When a large number of measurements are symmetrically distributed
about their mean and a histogram of their frequencies is defined by the
bell-shaped curve

$$y = ae^{-mx^2}$$

in which a and m are appropriate constants, their distribution is said to
be Gaussian, or normal. A large proportion of statistical literature is based
on the assumption of a Gaussian distribution. When real sets of values
are not symmetrically distributed, empirical devices are used to "restore"
symmetry, so that the available wealth of statistical theory may be used
in data evaluation. One of these empirical devices is to convert the values
to their logarithms: if this develops a symmetrical distribution, the original
distribution is said to be log normal.

In dealing with sets of analytical values, the mathematical approximations and the assumptions of Gaussian statistics often become invalid,
and conclusions drawn through their use may be false. While the lognormal empiricism is sometimes useful, it can easily be abused.

1.15. POISSON STATISTICS

A frequent cause of failure of Gaussian statistics to deal adequately with
real problems is the existence of a critical component in a mixture being
present in isolated or low-frequency grains randomly distributed among
a much larger number of gangue grains. If a sample or subsample contains,
on average, less than about five or six of these grains, a limited number
of repetitive assays may yield either widely variant or precise but erroneous results. In an example cited above, for example, several deter-

minations of chromium in the USGS granite G-1 using 10-mg subsamples in a spectrographic procedure may yield an average value of 8 ppm with high precision. Several determinations using 1-g subsamples in a chemical procedure may yield an average value of 20 ppm with very poor precision. The distribution of chemical results will best be described by the Poisson formula (1.23) given above.

A characteristic of the Poisson distribution is that the standard deviation in measurements of the number z of particles per average sample is the square root of that number:

$$\sigma_z = \sqrt{z} \tag{1.26}$$

Multiplying both sides of (1.26) by c, the contribution of a single grain to an assay value, gives the standard deviation of the assay values:

$$c\sigma_z = \sigma = c\sqrt{z}$$

or, in practice,

$$s = c\sqrt{z} \quad \text{or} \quad s^2 = c^2 z \tag{1.27}$$

It is very important to note that the estimation of σ becomes increasingly precarious as the number of grains per subsample decreases. For the estimated standard deviation s to have meaning, a sufficient number of measurements must be made; this number may be very large when z is small. For example, determination of chromium using 10-mg subsamples of G-1 will yield about 20 gangue values for every value that derives from a grain of chromite in a subsample. The theoretical standard deviation for chromium in G-1 using 10-mg subsamples is, from (1.27), $\sigma = 0.0050\%$ chromium. For any estimate of σ that will reliably approach the true value using (1.24), many hundreds of determinations would be necessary if 10-mg subsamples are used. If the purpose of repetitive determination is to discover the sampling characteristics of a material, a sample or subsample size should be selected such that each sample or subsample contains, on average, at least one and preferably five or six grains of the minor constituent. The expected standard deviation using smaller or larger subsamples may then be calculated using a *sampling constant* K_s and the relative deviation R:

$$K_s = R^2 w = \frac{10^4 s^2}{K^2} w \quad \text{or} \quad s = \frac{K}{100}\sqrt{\frac{K_s}{w}} \tag{1.28}$$

1.16   RELATIVE DEVIATION *R*

The relative deviation in a set of values is the standard deviation divided by the true value and multiplied by 100 to express the result in percent. An estimated relative deviation, referred to as the coefficient of variation, is found by dividing the estimated standard deviation by the arithmetic mean and multiplying by 100. The subtle distinction between the terms *relative deviation* and *coefficient of variation* should be noted. As here defined, the former is a theoretical quantity and the latter is an estimate obtained by experiment.

The relative deviation *R* is sometimes a complex quantity. It may be composed of several deviations $R_1, R_2, \ldots, R_N$, which add statistically as follows:

$$R^2 = R_1^2 + R_2^2 + \cdots + R_N^2 \qquad (1.29)$$

One of the components of the relative error *R* is the relative error due to sampling or subsampling, and this may be referred to as the relative sampling error, or the relative error due to sampling. The relative sampling error may itself be complex, including several errors introduced during the several steps of the sampling and subsampling process. Since this chapter concerns sampling error, the symbol *R* may be used to designate relative sampling error where it is understood that other errors do not come into question.

1.17   HOMOGENEITY

Completely homogeneous materials are so rare that they may be considered nonexistent. Even natural and synthetic glasses, optically pure minerals, and natural liquids seldom qualify as completely homogeneous. Natural waters are almost always in a state of disequilibrium or dynamic equilibrium and begin to change in composition at the instant samples are collected. Single mineral crystals often show gradations in composition from the inside to the outside or from one end to the other. Glasses may be structured and very variable in elemental composition; their surfaces may be quite different from their interior. Inclusions of foreign mineral, which may be solid, liquid, or gas, are the rule rather than the exception. Such inclusions may be so small that they are not visible under the microscope. Submicroscopic inclusions may be thermodynamically stable in a two-phase system and may make up an appreciable part of the composition of the mixture [10]. Barnes [11] and Barnes and Clarke [12] have remarked on difficulties met in the sampling of natural waters. Kistler [13] found wide variations in the magnesium content of an optically homogeneous mica. Cruft et al. [14] found that substantially all the carbonate and silica in a single crystal of apatite were present in inclusions. Ingamells and Gittins [15] found submicroscopic calcite and halite in near-gem quality scapolite. Weed and Leonard [16] reported variations in composition and X-ray properties within single flakes of hydrobiotite–vermiculite. Engels [17] showed that carefully separated hornblendes contain enough altered biotite in the form of thin veneers and inclusions to yield a false potassium–argon age.

The word *homogeneity* is used here to mean degree of homogeneity in a sense that presumes that the substance under investigation is thoroughly mixed—that is, all its particles are randomly distributed among themselves. A sample that is, in practice, homogeneous at the 1-g subsampling level (with respect to a specific constituent) may not be, in practice, homogeneous at the 10-mg subsampling level. To say that a sample is homogeneous at the 1-g subsampling level means only that several 1-g subsamples will each yield the same answer, within the capabilities of the analytical method, or within the requirements of the investigation. Thus, samples may be adequately homogeneous for a routine control operation in which two significant figures are all that is required and inhomogeneous in an investigation requiring four-figure accuracy. The degree of homogeneity of a material is expressed in terms of the sampling constant K_s (1.28), which may be estimated for any constituent of a mixture in several ways.

1.18. REDUCTION OF SAMPLES TO LABORATORY SUBSAMPLES

Given a sample of weight W that must be reduced to weight w, to find the subsample weight demanded by an analytical method for determining the constituent of interest X, one may follow a number of procedures. If these procedures are rationally designed, they will take into account the following parameters: U, the effective grain size of the as-submitted material, usually that of a minor mineral that contains a major proportion of the element of interest in the mixture; u_{anal}, the grain size to which the material must be reduced to provide an analytical subsample that contains, within the requirements of the investigation, the same concentration, K, of the element of interest X as the original sample; W, the weight of sample supplied; and w, the weight of the analytical subsample. If the material is submitted as a crushed pulp, it should be screened through a mesh that will retain a few particles; these should be examined for their

X content. If the largest particles are all gangue minerals, they may be crushed and added to the rest of the sample. If the largest particles contain ore mineral, they establish the effective mesh size U of the as-submitted material. If the material is submitted as lumps or chunks, these should be examined closely for included grains of ore mineral; the largest grain of ore mineral, whether liberated or not, determines the effective grain size U.

Every submitted sample carries a sampling error. This error may be estimated from the effective grain size, the composition of the pure ore mineral, and the composition of the gangue. Expressed as a standard deviation, this error is, for unsegregated (well-mixed) materials,

$$S = \sqrt{\frac{A}{W}} = \sqrt{\frac{(H - L)(K - L)U^3 d_H}{W}} \qquad (1.30)$$

For unsegregated materials, $B = 0$.

The value for S, the uncertainty existing in the sample as supplied, must govern further operations. If S is very large, attention to this should be drawn before proceeding; those supplying the sample will have to be concerned. In any case, further reduction procedures need not be designed to do more than retain the degree of meaningfulness that exists in the as-supplied material. Efforts to produce an exact value from a meaningless sample are futile. The rule to follow is that if additional variances introduced during subsampling are less than one-tenth of the original variance, they have no significance; this principle has been most clearly stated by Youden [18].

If S is very small, a useful rule of thumb is that a sample may be split in two (using sound blending and splitting procedure) if it is first reduced to pass 100% through the next finer screen on the Tyler $\sqrt{2}$ screen scale. For example, if $U = 0.168$ cm (10 mesh on the Tyler scale), the material may be split in two if it is first reduced to pass 100% through a 14-mesh screen; it may be split in four if it is first passed 100% through a 20-mesh screen; and so on.

If S is large, such a procedure involves unnecessary work. It is only necessary to grind fine enough that the subsampling variance s^2 is less than one-tenth the existing variance S^2. The mesh size, u, to which the material should be reduced before splitting $1:1$ is estimated by rearranging (1.30):

$$u^3 = \frac{s^2 w}{(H - L)(K - L)d_H} \qquad (1.31)$$

where $s^2 = \frac{1}{10}S^2$; w is the weight of the generated subsample—for a $1:1$ split, $w = \frac{1}{2}W$; d_H is the density of the ore mineral; K is the overall grade (break-even grade may be used for calculation); L is the gangue concentration of X; and H is the X content of the ore mineral.

Example. A 2-kg sample of molybdenum ore is received for analysis. Screening through a 10-mesh screen leaves a few pieces of molybdenite, MoS_2, on the screen. The largest of these pieces weighs 2.2 g. The density of molybdenite is 4.8, so that this largest piece corresponds to a cube of MoS_2 of side $\sqrt[3]{2.2/4.8} = 0.76$ cm. If the overall grade is estimated as $K = 0.3$ and the molybdenum concentration in the gangue is 0.1%, since $H = 60\%$ Mo in MoS_2, the uncertainty existing in the sample on receipt is, from (1.30),

$$S = \sqrt{\frac{(60 - 0.1)(0.3 - 0.1)}{2000}} (0.76)^3 (4.8) = \pm 0.113$$

and the mesh size to which the material should be reduced before splitting $1:1$ is

$$u^3 = \frac{10 \times (0.0113)^2 \times 1000}{(60 - 0.1)(0.3 - 0.1)4.8} = 0.0222 \quad \text{or}$$

$$u = 0.28 \text{ cm} \quad \text{(between 7 and 8 mesh)}$$

If, however, the sample as supplied is passed through a 10-mesh screen (linear opening 0.168 cm) and a few particles of MoS_2 remain on the 12 mesh (0.141 cm), $S = 0.0117$, and the permissible mesh size for $1:1$ splitting is $u^3 = 0.000238$, and $u = 0.062$, or about 28 mesh.

It remains to decide how finely the material must be ground to provide an analytical subsample of a size demanded by the method to be employed:

$$w = \frac{(H - L)(K - L)d_H u^3}{s^2} \tag{1.32}$$

If the minimum variance required in subsampling is one-tenth the variance existing in the sample as received, $s^2 = \frac{1}{10}S^2$. The maximum useful subsample weight at any final mesh size u is therefore

$$w_{anal} = \frac{10(H - L)(K - L)d_H u^3}{S^2} \tag{1.33}$$

where S is the uncertainty existing in the as-received sample. The maximum useful subsample is

$$w_{anal} = \frac{10 W u_{anal}^3}{U^3} \tag{1.34}$$

where W is the total weight of sample received, U is the effective linear grain size of the as-received material, and u_{anal} is the linear mesh size of the laboratory sample from which the analytical subsample is removed.

Example. Two kilograms of molybdenum ore containing pieces of MoS_2 up to 4 mesh (0.476 cm) are reduced to pass a 150-mesh (0.0105-cm) screen using the following scheme: Two kilograms of 4-mesh maximum ore mineral particle size are crushed and passed through 14 mesh, split 1:1 to 1 kg minus 14 mesh, crushed and passed through 35 mesh, split 1:1 three times to 125 g 35 mesh, and crushed and passed through 150 mesh.

The maximum useful weight of the analytical subsample is

$$w_{anal} = 10 \left(\frac{0.0105}{0.476}\right)^3 \times 2000 = 0.215 \text{ g}$$

If a method using a 1-g analytical subsample were employed, it would be unnecessary to reduce this material to as fine as 150 mesh; if it were reduced only to 80 mesh (0.0177 cm),

$$w_{anal} = 10 \left(\frac{0.0177}{0.476}\right)^3 \times 2000 = 1.03 \text{ g}$$

and 80-mesh material is adequate, in this instance, if 1-g subsamples are taken, but only because the original sample poorly represents the bulk material from which it was removed. If the original 2-kg sample had been 100% minus 20 mesh (0.084 cm), the useful subsample weight at 150 mesh is, from (1.34),

$$w_{anal} = 10 \left(\frac{0.0105}{0.0840}\right)^3 \times 2000 = 39 \text{ g}$$

At 80 mesh, the useful subsample weight is

$$w_{anal} = 10 \left(\frac{0.0177}{0.0840}\right)^3 \times 2000 = 187 \text{ g}$$

and the integrity of the original sample is not preserved if 1-g subsamples of 80-mesh material are taken for analysis when the original sample has an effective grain size of 20 mesh.

The original sample as received may have been oversampled; in such a case, the value S used in the above equations should be the tolerable error. For instance, if one receives 2 kg of minus 100 mesh material, there may be no need to do any grinding before splitting out a 1-g subsample for analysis; the only requirement may be thorough blending. However, there are sure to be cases in which even 2 kg of a minus 100 mesh sample may not be representative. Crushing and screening can be expensive and time-consuming, especially when the number of samples is large, as in mining exploration. Inadequate reduction procedures yield useless results; oversampling is wasted effort; a rational means for striking a balance is essential.

1.19.   GY'S SAMPLING SLIDE RULE

To permit convenient and rapid application of Gy's sampling theory, he has designed an ingenious slide rule, which permits quick calculations based on sampling theory. The slide rule carries a table giving corresponding values of l and D/L, the ratio of biggest particle size to liberation size. Any one of the parameters of Gy's basic formula can be found instantly when the others are known. For example, a single setting of the slide rule will tell to what mesh a sample must be reduced before splitting out a subsample of any specific weight. A brief description of Gy's sampling theory and of the proper use of Gy's sampling slide rule, with examples, has been prepared by Ottley [19]. The slide rule is an indispensable tool for anyone continually faced with the problems of sampling broken ores.

1.20.   DETERMINATION OF VISMAN CONSTANTS A AND B

As applied to unknown materials that are both inhomogeneous and segregated, Visman's sampling theory requires the collection of two series of samples, one series of large samples and one series of small samples. The members of each series are reduced and analyzed for the constituent of interest, X, care being taken that variance in the results does not originate from errors of sample reduction, from laboratory subsampling error, or from analytical error. The importance of controlling reduction error using Gy's principles or their equivalent and of minimizing laboratory subsampling errors and analytical errors is obvious. If any of these are out of control, variances essential to the application of Visman's theory may be obscured. Visman sampling constants are calculated from equations (1.9) and (1.10).

The two constants having been determined, the uncertainty in the average of N results, each derived from a sample of w grams, is

$$S = \sqrt{\frac{A}{W} + \frac{B}{N}} \qquad (1.35)$$

where $W = Nw$, the total weight of samples each of weight w. Note that as the segregation constant B increases, the number N of samples taken assumes greater importance. If the ore body, shipment, or mountain is not at all segregated with respect to the element of interest, B is zero, and it makes no difference into how many increments the total weight of samples is divided. With segregated materials, there is an optimum individual sample weight. As data accumulate during an exploration, increasingly accurate estimates of A and B will permit adjustment of field sample weights so as to obtain the most information at the lowest cost.

Visman's sampling theory was developed semiempirically. Duncan [20], Visman et al. [21], Visman [22], and Switzer [23] have examined it critically, and its validity in several respects has been established. However, some of its approximations may not be acceptable when segregation contributes heavily to the observed variance. The problems of segregation cannot always be summarized in a single constant; the science of geostatistics applies itself to the accurate control of segregation variance.

An important practical fact that may influence selection of individual field sample weights is that subsampling errors (as distinct from sampling errors) are likely to be small for small samples and maximum for medium samples and are likely to diminish when samples become very large (always supposing sound reduction procedures are followed). For example, a 1-kg sample can easily be reduced in its entirety to 100 mesh and will often exhibit negligible subsampling error in comparison to a large sampling error. With very large field samples (say over 100 kg), reduction errors and subsampling aberrations are likely to cancel in a well-devised reduction process. It is with medium-sized (say, 10-kg) samples that accidents such as the loss or gain of "nuggets" is most likely. Thus, practical considerations may override a calculation that established the optimum field sample weight at an intermediate level. Most often, the choice is between small manageable samples and very large samples. In practical terms, an ore body is best investigated either by drilling or by trenching or its equivalent according to its sampling characteristics.

1.21. DETERMINATION OF GANGUE CONCENTRATION L

It may be noted that many of the formuls given above rely on an estimate of the X content of the two minerals in a mixture. Very often, the X content of the high-X mineral—the ore mineral—is known, but the X content of the low-X mineral, or gangue, is difficult to estimate. In the case of ores, the gangue concentration may be regarded as the portion of the overall grade that is perfectly sampleable. An ore that contains a friable mineral like, for example, malachite, $Cu_2CO_3(OH)_2$, is easy to subsample because the constituent containing the element of interest, copper, is readily reduced to fine powder. The major part of the copper in such an ore will be included in L when sampling problems are under consideration. On the other hand, an ore in which the constituent of interest resides in a hard and intractable mineral like chromite, $FeCr_2O_4$, will likely have a low gangue content of chromium. Most of the chromium will remain in resistant grains of chromite, which will often retain the mesh size of the screen through which the material is passed. Native metals (gold, copper, etc.) are extremely difficult to reduce in particle size, and L remains near zero.

Several methods for estimating the gangue concentration of an element are available.

1. With closely sized material (e.g., mineral separates), analysis of a series of small subsamples gives a probability

$$P_{(z_i = 0)} = 1 - (1 - e^{-z})^N \qquad (1.36)$$

where least one of the N subsamples is free from contaminating grains of X content L. The average number z of contaminating grains per subsample may be estimated as

$$z = \frac{(\overline{K} - L)^2}{s^2} = \frac{\overline{K} - L}{H - L} \frac{w}{u^3 d_H} \qquad (1.37)$$

where w is the subsample weight in grams, s^2 is the standard deviation in a series of N values, u is the effective grain size in centimeters, and d_H is the density of the contaminating mineral in grams per centimeter cubed. Solution of (1.37) for L gives

$$L = \frac{H + \overline{K}}{2} \pm \frac{1}{2} \sqrt{(H - \overline{K})^2 + \frac{4ws^2}{u^3 d_H}} \qquad (1.38)$$

If $P_{(z_i = 0)}$, calculated using a value for z obtained from (1.37), is 0.98 or higher, it is reasonably certain that the extreme low value in the N determinations represents the true value of L.

Example (Table 1.1). Using (1.38), with $H = 14.92$, $K = 0.286$, $u = 0.032$, $s = 0.0451$, and $d_H = 2.7$, L calculates to 0.131% K_2O (compared to the true value, 0.145). From (1.37), $z = (0.286 - 0.131)^2/(0.0451)^2 = 11.8$ grains per 0.1-g sample. Substitution of this value for z in (1.36), with $N = 40$, gives $P_{(z_i = 0)} = 1 - (1 - e^{-11.8})^N = 0.0$. There is almost no probability that any one of the 40 values represents the composition of the major constituent. With 10-mg samples of the same material, however, the probability $P_{(z_i = 0)}$ is $1 - (1 - e^{-1.18})^N = 0.99$ for 12 determinations, and, in fact, the lowest of 12 values using 10-mg subsamples was 0.143% K_2O, compared to the true value, 0.145%.

2. Analysis of two series of subsamples of differing weights w_1 and w_2 will yield two skew distributions if the weights are carefully selected. The two series of subsamples used in calculating the Visman constants A and B may be utilized. If the modes of these two series are Y_1 and Y_2, and the weighted mean of all values is taken as an estimate of K, L is estimated from

$$Y = K - \frac{K - L}{2z + 1} \quad \text{i.e.,} \quad 2z = \frac{L - Y}{Y - K} \tag{1.39}$$

For two subsamples of different weights w_1 and w_2,

$$2z_1 = \frac{L - Y_1}{Y_1 - K} \quad \text{and} \quad 2z_2 = \frac{L - Y_2}{Y_2 - K}$$

or

$$\frac{z_1(L - Y_2)}{Y_2 - K} = \frac{z_2(L - Y_1)}{Y_1 - K} \tag{1.40}$$

Substituting (1.37) in (1.40) yields

$$L = \frac{Y_1 w_2(K - Y_2) - Y_2 w_1(K - Y_1)}{w_2(K - Y_2) - w_1(K - Y_1)} \tag{1.41}$$

The rationale of this device is explained in Section 1.22.

3. With closely sized mixtures, or those in which a minor component occurs in grains of uniform size (e.g., lateritic ore containing chromite), L may sometimes be found from a series of determinations even when z calculated from (1.37) is too large to yield a high value for $P_{(z_i = 0)}$ when substituted in (1.36). The method depends on detecting discrete differences between successive determinations x_i due to the varying number of grains of high-X mineral that are distributed, following the laws of chance, among the several subsamples taken for analysis. Each analytical value generates a series of possible values of L for each x_i:

$$L = \tfrac{1}{2}(x_i + K) \pm \tfrac{1}{2}\sqrt{(x_i - K)^2 + 4s^2 Z_i} \qquad (1.42)$$

By comparing these values among all K_i, the value for L most nearly the same for all K_i can be found, in some cases by inspection. A computer program has been written to make the comparison in cases where the result is not obvious [24]. Application of this program to the 40 K_2O determinations of Table 1.1 quickly yielded the correct K_2O content of the amphibole.

4. The composition of a single-mineral phase in a mixture can sometimes be determined using the electron microprobe. Electron microprobe analyses are highly dependent on analyzed samples used for calibration, especially when low-X gangue minerals are being examined. The reference samples must be analyzed by other methods using large samples; this means that 100% pure materials must be available. For example, a thoroughly competent bulk analysis of the orthoclase–amphibole mixture of Table 1.1 would yield 0.286% K_2O; the microprobe looks only at the amphibole grains, with 0.145% K_2O. Despite its 99% purity, this amphibole is worthless as a microprobe standard for K_2O (unless L has been determined by one of the methods suggested here).

Probably a combination of these methods may be used in any specific case. It may be a complaint that the effort and trouble needed to apply these devices is exorbitant. The reply to this complaint is that the effort and trouble must be balanced against the alternatives, which invite worthless conclusions or lead to excessive sampling and subsampling costs. A large exploration may generate many thousands of samples at a cost of millions of dollars. It is folly to reduce and analyze these without giving attention to sampling and subsampling principles, avoiding gross error on the one hand and unnecessary effort on the other.

1.22. SAMPLING DIAGRAMS

A sampling diagram for any mixture (a rock powder, an ore pile, a purified mineral separate, a shipment, a mountain) can, in principle, be prepared by repetitively measuring the X content in samples or subsamples of different weights and plotting averages ($\overline{X}$), standard deviations ($\pm s$), ranges (error bars), and some other parameters against sample or subsample weight. As sample weight diminishes, the deviations increase, and there is an increasing probability that minor constituents of the mixture will be excluded from one or more samples or subsamples. If the minor constituents are rich in X, there will be a tendency toward lower values as sample weight diminishes. If the minor constituents are low in X, there will be a tendency toward higher values as sample weight diminishes.

It is very necessary to clearly recognize that the measured X content is *not* independent of sample or subsample weight.

Given the sampling constants of a mixture, uncertainty in the measurement of X content at any sample or subsample weight can be estimated from Visman's equation (1.8), and the need for repetitive determination disappears. It is, therefore, desirable to calculate the sampling constants from known physical characteristics of the mixture. A two-mineral model permits such calculation.

Figure 1.2 is a sampling diagram for a hypothetical mixture of two minerals of uniform grain size in which there is no segregation. The diagram was constructed from the following parameters:

K = true overall X content of mixture, $= 0.500\%$

L = X content of low-X mineral, $= 0.100\%$

H = X content of high-X mineral, $= 100\%$

u = linear grain size, $= 1$ cm

d_H = density of high-X mineral, $= 5.00$ g/cc

From these, we may calculate many secondary characteristics of the mixture, such as the most probable result, Y, for a given sample weight.

The number of grains, z, of a high-X mineral in a sample or subsample of weight w grams is, from (1.22) with $\overline{X} = K$,

$$z = \frac{K - L}{c} \tag{1.43}$$

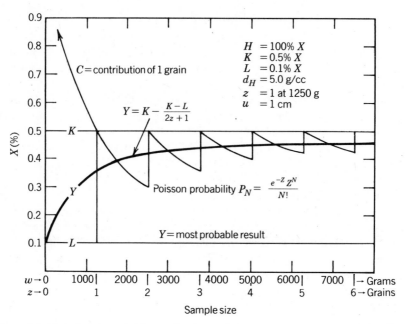

Figure 1.2. Plot of most probable result versus sample weight for a mixture containing 1-cc grains of ore mineral of density 5 in 1250 g of a substrate with 0.1% of the element of interest.

From (1.20),

$$c = \frac{(H - L)d_H u^3}{w} \tag{1.44}$$

Substituting (1.44) in (1.43) gives

$$z = \frac{K - L}{H - L} \frac{w}{u^3 d_H} \tag{1.45}$$

Putting $z = 1$ and solving (1.45) for w,

$$w_{(z=1)} = \frac{(H - L)d_H u^3}{K - L} \tag{1.46}$$

When $K = 0.500$, $L = 0.100$, $H = 100.0$, $d_H = 5.00$, and $u = 1$, $w_{(z=1)}$ = 1250 g. If samples weighing 1250 g are taken and properly reduced to analytical subsamples and analyzed for their X content, the distribution of results will be described by the Poisson model (1.23). At this sample weight, there is an equal probability of taking one grain or zero grains in a single sample. The curve showing the most probable result, Y, as a function of sample weight w is a vertical line at $w = 1250$ g drawn from $Y = L$ to $Y = K$. At every sample weight corresponding to a whole number of grains, that is, at 1250, 1250 × 2, 1250 × 3, . . . grams, the curve $Y = f(X)$ shows such a discontinuity. Figure 1.2 shows $Y = f(X)$ corresponding to the example given.

With samples weighing less than 1250 g, the most probable result is $Y = L$. If samples are very small—say, less than $\frac{1}{10}w_{(z=1)}$, in this case, less than about 100 g—it is very likely that none but a few determinations of K will show anything but the X content of the low-X mineral.

In real circumstances, mineral grains are seldom all of exactly the same size; also, the gangue composition is not exactly defined. As a useful approximation, therefore, a smooth curve may be drawn through the "teeth" of the Poisson curve of Figure 1.2. The equation of this curve is (1.39), $Y = K - (K - L)/(2z + 1)$, or, substituting (1.45) for z,

$$Y = \frac{2K(K - L)w + L(H - L)u^3 d_H}{2(K - L)w + (H - L)u^3 d_H} \tag{1.47}$$

The most probable value, Y, may also be expressed in terms of the Visman sampling constant A, by solving (1.11) for $u^3 d_H$ and substituting in (1.47),

$$Y = \frac{2(K - L)^2 Kw + AL}{2(K - L)^2 w + A} \tag{1.48}$$

1.23. SAMPLING DIAGRAMS FOR SEGREGATED MIXTURES

With segregated mixtures (i.e., mixtures in which all particles are not randomly distributed), it is necessary to estimate not only a homogeneity constant A but also a segregation constant B. With segregated mixtures, the uncertainty S in the average of many results depends not only on the total weight of samples taken but also on the number of samples, that is, on the number of increments into which the total weight of samples is divided. From (1.8), if the total weight of samples could be divided into an infinite number of increments, $S = \sqrt{A/W}$. For a single sample, $S = s' = \sqrt{A/w + B}$; for samples of optimum individual weight A/B, $S = s_v = \sqrt{2A/w}$, where s_v is the standard error of the mean in a set of values derived from samples of weight A/B. Note that while values for s_v generate a continuous error curve on the sampling diagram, they are only meaningful at weights that are multiples of $w_{opt} = A/B$.

1.24 CONSTRUCTION OF SAMPLING AND SUBSAMPLING DIAGRAMS

To construct a sampling diagram, it is first desirable to calculate the Visman sampling constants A and B either by the method of repetitive determination using two series of samples of different individual weight or by estimating K, L, H, u, and d_H (also d_L if the element of interest, X, is present in two major constituents) using (1.8), (1.9), and (1.11).

If A can be found through use of these formulae, a single series of determinations can be used to find B from (1.10) provided the sample weight w_2 is sufficiently large to include in each sample, on average, at least five or six grains of the relevant mineral.

Putting various values for w in the equations

$$s = \sqrt{\frac{A}{w}} \qquad s' = \sqrt{\frac{A}{w} + B} \qquad s_v = \sqrt{\frac{2A}{w}} \qquad (1.49)$$

yields three pairs of error curves (Fig. 1.4) for segregated materials. When there is no segregation, $B = 0$, $s = s'$, s_v is meaningless, and only one pair of error curves is obtained.

Note that at individual sample or subsample weights at which z is less than about 5 or 6, the error estimates s, s', and s_v are no longer reasonable because of the increasing assymetry of the distribution of values.

Example (Table 1.1). The sampling constant A may be found in two ways:

1. From the standard deviation in 40 values; from (1.6) and (1.11),

$$A = K_s K^2 \times 10^{-4} = 0.00020$$

2. From the known composition of the mixture; from (1.30),

$$A = (H - L)(K - L)u^3 d_H$$
$$= (14.92 - 0.145)(0.286 - 0.1445)(0.032)^3(2.7)$$
$$= 0.00019$$

Prepare a table of values using (1.48) and (1.49):

w (g)	Y (%)	$s = \sqrt{A/w}$
10^{-4}	0.147	1.42
10^{-3}	0.168	0.450
10^{-2}	0.238	0.142
10^{-1}	0.279	0.045
1	0.285	0.014
10	0.286	0.005

$$w_{(z=1)} = 0.010$$

from (1.46) or from

$$w_{(z=1)} = \frac{A}{(K - L)^2}$$

$$K_s \approx 25 \text{ g}$$

Plot s, Y, K, and L against w. Mark $w_{(z=1)}$ and K_s on the y axis (Fig. 1.3).

In this example, the segregation constant B is assumed to be zero; that is, it is assumed that the mineral particles are randomly distributed—the

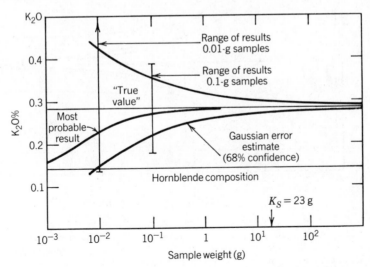

Figure 1.3. Subsampling diagram for an amphibole–orthoclase mixture.

sample is well mixed. With segregated mixtures, s' and s_v should also be calculated from (1.49) and plotted. Figure 1.4 is a sampling diagram for an ore body that is both nonuniform and segregated. Constants A and B were found by analyzing series of samples of different weights using (1.9) and (1.10).

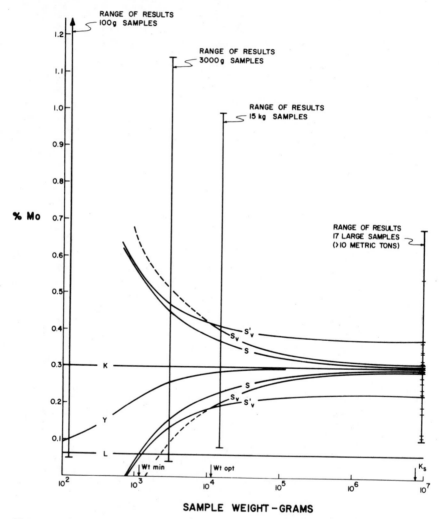

Figure 1.4. Sampling diagram for an ore body that is both nonuniform and segregated. A total sample weight of 10^6 g (1 tonne), if collected in appropriately sized increments, will yield an overall grade estimate within about 5% of the true value. At this sample weight, the point of diminishing returns has been passed. Even 5–10 times as many samples will not appreciably improve the grade estimate.

1.25.   USEFULNESS OF SAMPLING AND SUBSAMPLING DIAGRAMS

Once a sampling diagram has been prepared, one can tell at a glance how large a sample or subsample should be taken to achieve the desired sampling precision. For segregated materials, the total weight of sample does not establish sampling precision; the number of samples taken is equally important. This is shown in Figure 1.4. If samples of $w_{opt} = A/B$ grams are taken, the relative sampling error in the average of N values is given by the s_v curve.

Note that as the total weight of samples, each of $w_{opt} = A/B$, increases, a point of diminishing returns is reached. In Figure 1.4, this point is reached at a total weight of samples of about 10^6 g (1 tonne). At this sample weight, even 10 times as many samples will not appreciably improve the overall grade estimate. The s'_v curve shows the uncertainty that exists in a single sample of weight w grams: Of 17 samples each weighing more than 10 tonnes (and each properly reduced to analytical subsamples!), 10 yielded values enclosed by the s'_v curve; in theory, 68% of the 17 should have fallen within these limits.

In practice, of course, one cannot know at the start of an exploration what sample sizes are most useful. However, a preliminary sampling diagram can be prepared as a guide using either a geologist's estimates of K, L, H, d_H, and u in (1.11) or the first assay values accumulated. Sampling practice can then be revised during the exploration as the several estimates improve. Table 1.2 shows preliminary values for cobalt in a lateritic ore from which 20-kg core samples were taken. These preliminary values show that 20-kg samples are much too small to optimize exploration costs and give the required information.

After a field sample has been collected, it is necessary to reduce it to an analytical subsample. The reduction process usually takes several stages. A subsampling diagram may be prepared at each stage and provides a certainty that the analytical subsample will have the same X content, within acceptable limits, as the field sample from which it derives. Without some such control, it is very possible to introduce an intolerable sampling error at one or more stages in the reduction.

Gy's sampling slide rule is an invaluable aid in designing reduction procedures. Subsampling diagrams utilize the same principles and may, in some circumstances, be easier to use and understand. Where possible, both devices should be used.

Example (Figure 1.5). Seven hundred grams of tungsten ore were ground to pass 10 mesh (0.168 cm) and submitted for WO_3 determination. A

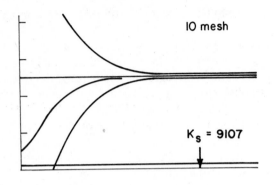

10 mesh

$K_s = 9107$

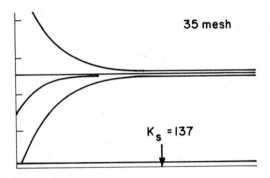

35 mesh

$K_s = 137$

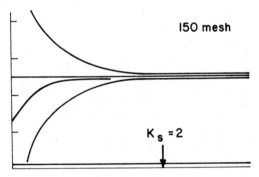

150 mesh

$K_s = 2$

Figure 1.5. Sample reduction scheme of a tungsten ore.

42

sampling constant for this material was calculated from (1.7),

$$K_s = \frac{10^4 \times (2.4 - 0.1)(80 - 0.1)(0.168)^3(6.02)}{(2.4)^2} = 9107 \text{ g}$$

by estimating $K = 2.4$, $H = 80$, $L = 0.1$, and $d_H = 6.02$. For a relative sampling error of 1% or less, 9000 g, not 700 g, are required. The error already present in the as-received sample is, from (1.28),

$$R = \sqrt{\frac{K_s}{w}} = \sqrt{\frac{9107}{700}} = 3.6\%$$

No amount of care during further sample preparation and analysis can diminish this uncertainty; however, care must be taken that it is not increased. Splitting the sample without further grinding is, in this case, impermissible. The entire 700 g were ground to pass a 35-mesh screen, and a sampling diagram was prepared for the 35-mesh material. The K_s for the 35-mesh material is 137 g. Since the original sampling error was 3.6%, a subsampling error of 1% is acceptable [18], and no significant error will be incurred by splitting out 137 g of the 35-mesh material. The 700-g sample was split into four portions of 175 g each. One of these portions was ground to pass 150 mesh, and a new $K_s = 2.0$ g was calculated. Since the analytical method uses 2-g subsamples, this 150-mesh pulp was mixed and submitted to the analyst. If the analytical method were to use a smaller subsample (e.g., 100 mg in a colorimetric procedure), grinding to 150 mesh would be inadequate. The sampling error thereby introduced would be $R = \sqrt{K_s/w} = \sqrt{2.0/0.1} = 4.5\%$, which is unacceptable. Figure 1.5 shows the three sampling diagrams for this example [24].

If these principles are not followed during sample reduction, large errors may be (and have often been) introduced, errors so large, sometimes, that a mining exploration costing millions of dollars may yield worthless conclusions. Pierre Gy [1] has rightly pointed out that "sampling operators should be perfectly conscious that sampling is a control operation belonging in the analytical chain, that a sample should be respected as if it were pure gold (in many cases it is worth considerably more than its weight in gold), and that sampling cannot be improvised."

The chain of events leading from the mountain to the balance pan is no stronger than its weakest link. Efforts to strengthen any other link are wasted, for one weak link will cause the strongest chain to break. It is mandatory that the geochemical analyst examine each link carefully and make certain that it is adequately strong; otherwise, bad data may be generated for which the analyst may be held responsible, whoever is at fault.

1.26. PLANNING A SAMPLING CAMPAIGN OR EXPLORATION

Procedures for sampling and analysis of unknown masses of material— ore bodies, shipments, mountains, rock piles, slag dumps, and so on— should be designed to minimize costs, using available sampling theory to avoid misdirected effort, and applying analytical techniques in the most efficient manner. Any evaluation of the unknown mass must take place in several stages or steps:

1. Pattern sample collection.
2. Sample collection.
3. Reduce samples to analytical subsamples.
4. Assay for the element(s) of interest, X.
5. Analyze and interpret data.
6. Loop back to 1.
7. Continue until information is as adequate as cost permits.

This chain of steps is no stronger than its weakest link. Each link should be strengthened optimally, avoiding unnecessary effort and gaining all possible information from accumulating data. Since all data originate at the analytical link, this link should be strengthened first.

Loop I.   Visually inspect a few preliminary samples (expertly collected hand samples may be adequate), and guess or estimate the following:

u = effective in situ grain size of ore mineral or other constituent of interest, cm

K = overall grade (break-even grade may be used), % of X

H = X content of ore mineral or other constituent of interest

d_H = density of ore mineral, g/cc

d_L = density of gangue or host material (unnecessary when constituent of interest makes up a minor proportion of the whole)

Calculate a preliminary estimate of the minimum field sample weight from (1.46) or its equivalent. The minimum field sample weight is that weight that contains, on average, one "grain" of the constituent of interest. A grain in this sense may be a pocket, lens, striation, or any other concentration of X. When more than one constituent is of interest, a minimum sample weight may be calculated for each. The largest of these estimated weights should be accepted as w_{min}, or plans should be made to group samples appropriately during their reduction to laboratory pulps, or to group the analytical values obtained from too small field samples before evaluating them. Assays of samples for a component of which there is less than one grain per average sample are certain to yield confusing results unless the causes of their anomalous means and variances are recognized and taken into proper account.

From the preliminary samples, prepare a series of laboratory standards of varying X content to be employed throughout the exploration to calibrate and control routine methods of analysis. Establish a sampling constant K_s for each relevant constituent of each laboratory standard, and make sure that the subsampling characteristics of the standards are adequate for the analytical methods to be employed.

Develop rapid assay methods of adequate precision and accuracy using the laboratory standards as controls. Because of differing mineralogies, routine methods that are satisfactory for one ore deposit may fail when applied to another. For example, one ore might yield all nickel into solution on boiling with acid—another might require an alkaline fusion. A fine-grained ore may yield satisfactory assays using emission spectroscopy; a coarse-grained ore may present severe subsampling problems, so that the small subsamples used by the spectroscopist cannot suffice.

Analyze each laboratory standard by primary methods. For hydrous materials, establish an exact sampling weight base line. Preliminary metallurgical tests and mineralogical examination may profitably be conducted during establishment of laboratory standards. Use the knowledge gained to design reduction procedures to be applied to field samples using Gy's principles or their equivalent. Failure to carefully design all steps in the reduction procedure, from field to balance pan, will threaten the success of the exploration. Several duplicate reductions should be made to ensure the adequacy of the process.

Loop II. Collect, reduce, and analyze two sets of 10–100 field samples—one set of individual weight at least $w_{(z=1)}$, and one set of individual weight five or more times the individual weight of the first set. As an alternative, collect a sufficient number of field samples each of weight at least $w_{(z=1)}$, and group these in sets of 8 or 10 (or more); the averages of assays on each of these sets provides a set of values corresponding to large field samples. This alternative assumes that reduction, subsampling, and analytical errors are well controlled and that the grouping is done in a knowledgeable manner. Calculate Visman constants A and B [equations (1.9) and (1.10)] and $w_{opt} = A/B$.

Example (Table 1.2). These preliminary samples yield values that show clearly that the field sample weight is much too small. Further exploration using this field sample weight (20 kg) will be unnecessarily expensive and will yield confusing data. Visman sampling constants were calculated by synthesizing large samples in two different ways—one by taking the average cobalt content of each of the drill holes, and one by averaging cobalt values from the same level. The results show (poorly because the field samples are too small but nevertheless clearly in a qualitative sense) that there is little segregation, either in depth or in expanse, in this ore body. There are large, randomly distributed concentrations of cobalt that cannot be satisfactorily evaluated using 20-kg field samples.

Construct a preliminary sampling diagram for the ore body, slag pile, or mountain.

Loop III. Pattern further sample collection on the basis of accumulated information. Preliminary estimates of optimum field sample weight are likely to be poor. The second and succeeding rounds of sample collection must take into account the information gained during preliminary sampling and analysis. In some cases, preliminary field sample weights may be found to be orders of magnitude larger or smaller than the optimum. The expense of altering the field sampling strategy may be high. Costs must be rationally balanced against reality.

It should be obvious that the processes of sample reduction, subsampling, and analysis must be very carefully controlled if bad decisions are to be avoided. If variance in assay values originates during the treatment of the field samples, and this variance is attributed to field relationships, resulting error of decision may be very expensive.

If w_{opt} turns out to be small (i.e., A is small or B is large), the Visman approach to practical sampling loses its usefulness. Highly segregated mountains are difficult to evaluate; Matheron's geostatistics provide the best approach possible. The relationship between Visman's theory and geostatistics is expressed in

$$S^2 = \frac{A}{W} + \text{geostatistics} + SE^2 \qquad (1.50)$$

or

$$S^2 = \text{nugget effect} + \frac{B}{N} + SE^2 \qquad (1.51)$$

If w_{opt} turns out to be large (i.e., A is large and B is small), the ore body or mountain is homogeneous on a large scale, and Matheron's geostatistics are of limited use. If w_{opt} is so large that practical considerations exclude it, it becomes economically important to make an accurate estimation of L and to calculate $w_{(z=1)} = A/(K - L)^2$.

Construct new sampling diagrams, reestimate optimum and minimum field sample weights as exploration proceeds, and adjust field operations accordingly. Analytical operations should be governed, meanwhile, by developing knowledge of the mass of material being examined.

Loops IV, V, VI, and so on.   While continuing to loop through the information-gathering process, develop an increasingly accurate picture of the distribution of the constituent of interest throughout the ore body, the mountain, and so on. Discontinue the exploration at the point of diminishing returns. The analyst should always remember that his operation stands at the very center of the exploration, and that if he or she does not keep close watch on it, it is likely to fail in its purpose. The analyst must not blindly accept samples that depend for their integrity on the intuition of uninformed technicians or of persons whose expertise does not include a full knowledge of the theory of sampling.

The largest sampling difficulties occur when the constituent of interest is a minor or trace amount of native metal (gold, copper, etc.). Landy [25] sampled copper-bearing greenstones from South Mountain, Pennsylvania, finally reducing his samples to 200 mesh for spectrographic analysis. Copper in these samples varied from traces to over 10,000 ppm. It is interesting to calculate the sample weight required for accurate determination of copper in rocks of this kind. The copper occurs as 1-mm flakes evenly distributed throughout the rock mass, and the overall concentration of copper is 1%. The dimensions of a copper flake are about $1 \times 1 \times 0.1$ mm; its volume is $u^3 = 0.1 \times 0.1 \times 0.01 = 0.0001$ cc. The density of copper is 8.9 g/cc. A sampling constant for an ore containing 1-mm flakes of copper is given by (1.11). Supposing the gangue carries no copper,

$$K_s = (H - L)(K - L)u^3 d_H \frac{10^4}{K^2} = 8900 \text{ g}$$

To ensure a sampling error of 1% or less (68% confidence), about 9 kg of sample must be taken. Reduction of this large sample to the 10 mg that is loaded into a spectrographic electrode requires the reduction of the copper flakes to a linear dimension; from (1.31),

$$u = \sqrt[3]{\frac{S^2 w}{(H - L)(K - L)d_H}} = 0.001 \text{ cm}$$

for a sampling error of 1% at the 68% confidence level. The copper particles must be reduced to cubes of side 10 μm or to flakes of 100 μm, that is, to 150 mesh, if spectrographic analysis is to be meaningful. If the copper-bearing sample is simply pulverized without screening, the probability is great that the metallic flecks will be substantially unaltered. It is extremely difficult to reduce metallic particles to fine-mesh size.

With gold ores, the sampling problem becomes even more difficult, since 1 ppm (1 g/tonne) is a significant concentration. The mesh size to which a gold ore must be reduced to ensure a sampling error no greater than 1% with 68% confidence is, from (1.31), 0.0025 cm when 28-g samples are taken if 1 ppm is significant. The sampling constant for a gold ore with 1 ppm of native gold ground to pass a 200-mesh screen is $K_s = 15$ kg.

Plainly, heavy metals should not be sought using methods (e.g., dc arc spectroscopy) that demand small analytical subsamples. The hard facts of sampling provide the best reason for retaining the fire assay, which uses large samples, in preference to chemical methods that are usually satisfied with much smaller samples. With very low grade gold ores, extraction methods using several hundred grams of finely ground and screened sample are sometimes necessary.

Exploration programs in which heavy metals are sought by rapid spectrographic methods are almost sure to fail; the small flecks of precious metal will most often be missed entirely. When one does appear, by chance, in a spectrographic sample, it is probable that the signal it develops will overwhelm the detecting system. A repeat will likely show none of the element sought.

Example. A gold ore carrying 100 ppm gold in quartz has been ground to pass a 200-mesh (0.0074-cm) screen. The sampling constant K_s is estimated to be 150 g. If 28-g subsamples are used in a fire assay method, the relative sampling error is $\sqrt{150/28} = 2.4\%$. If 10-mg subsamples are used in a spectrographic method, $R = \sqrt{150/0.01} = 120\%$. If the ore carried only 1 ppm gold, K_s is 100 times larger; the method using 28-g subsamples would show $R = \sqrt{15000/28} = 23\%$, which may be marginally acceptable. The method using 10-mg subsamples would show $R = 1200\%$ (!), which shows only the method is worthless.

1.27.   EFFECT OF VARIABLE SAMPLE OR SUBSAMPLE WEIGHT: WEIGHTED AVERAGE

It often happens that available data is derived from samples or subsamples that are not all of the same weight. Such data makes the estimation of sampling constants difficult, especially when there is appreciable segregation. The Visman formulas (1.9) and (1.10) utilize constant sample weights to separate variance due to segregation and variance due to inhomogeneity.

When subsample weights vary in a series of N determinations, a weighted average is estimated from

$$\overline{X} = \frac{1}{\overline{w}N} \sum (w_i X_i) \tag{1.52}$$

where w_i is the individual weight of the subsample with X content X_i, and

$$\overline{w} = \sum \frac{w_i}{N} \tag{1.53}$$

(i.e., the average subsample weight). The relative deviation in $\overline{X}$, the average of all X_i, is given by

$$R = \sqrt{\frac{K_s}{\overline{w}N}} \tag{1.54}$$

where K_s, the sampling constant, is estimated from

$$K_s = \frac{10^4 \sum w_i(X_i - \overline{X})^2}{(N - 1)\overline{X}^2}$$

The Visman homogeneity constant A may be found from (1.9). Determination of the segregation constant B is likely to be precarious when assay values derive from samples of widely different weight.

The relationship between observed variance and effective sample weight has not been sufficiently appreciated; neither has the fact that the most probable result of an analysis is a function of sample or subsample weight. As a consequence, field sample weights and the subsample weights at various stages of reduction have sometimes not been accurately measured, and the calculation of realistic sampling constants is not possible.

1.28. MINIMIZING EXPLORATION COSTS

The preoccupation of Matheron's geostatistics with variance due to segregation has led him to deny the importance of effective field sample weight in data evaluation. He states [6], "Geostatistics actually show that accuracy is the same with pieces (of drill core) of 5 mm and 50 cm, as every miner understands instinctively." This is, of course, true if all variance is segregation variance. When segregation is extreme, it is the number N of samples taken that assumes importance and not their individual weight. This is clear from the Visman formula (1.8). Unfortunately, the application of geostatistical theory is often attempted when the data originates from an essentially unsegregated ore body. The geostatistician discovers that, in his or her frame of reference, there is a large "nugget" effect.

Also, there has been little appreciation among miners and geologists of the difference between reduction and subsampling error and analytical error. Miesch has said [26], "Analytical errors include errors due to crushing, splitting, homogenization, laboratory subsampling, procedures at bench or instrument, and even typing and recording the lab report. The analyst has to be concerned with these sources of analytical error, but from the geologist's standpoint, it's the total analytical error that's important!" Miesch's attitude is representative of that of many geologists and miners. If the geochemical analyst accepts the load of responsibility implicit in Miesch's statement, he or she must assume some sort of control over the sampling, reduction, and subsampling processes or be found in an untenable position. If exploration costs are to be minimized, the whole sampling exercise must be controlled; the geochemical analyst must be a part of that control.

Any system of data evaluation assumes a priori that the data points are sufficiently accurate to warrant evaluation. It is assumed that the assay value developed from a field sample reflects to a measurable degree the composition of the ore surrounding the place from which the field sample was taken. It is also assumed that the assay value is, within acceptable limits, representative of the composition of the field sample itself. For these assumptions to be true, the field sample must be

1. large enough to represent the surrounding ore and
2. reduced to an analytical subsample in such a way that the gram or so weighed out for analysis has the same composition as the whole field sample.

Two opposing factors decide the appropriate field sample weight: If the material being examined is highly segregated, many small samples need to be taken; if the ore body is coarse grained, samples should be large enough that each field sample contains, on average, at least one, and preferably five or six, "grains" of the constituent of interest. When the ore is both segregated and coarse grained, there is evidently an optimum field sample weight, which will yield the most information at the lowest cost:

$$C = PW + QN + F = PNw + QN + F \qquad (1.55)$$

where P is the cost per gram of sample, Q is the cost per sample, and F is the fixed cost. Solving Visman's equation (1.8) for N and substituting in (1.55),

$$C = \frac{PA}{S^2} + \frac{PBw}{S^2} + \frac{AQ}{wS^2} + \frac{BQ}{S^2} + F \qquad (1.56)$$

Differentiating (1.56) with respect to w,

$$\frac{dC}{dw} = \frac{PB}{S^2} - \frac{AQ}{w^2 S^2} \qquad (1.57)$$

Minimum cost is achieved when $dC/dw = 0$, that is, when

$$\frac{PB}{S^2} = \frac{AQ}{w^2 S^2} \quad \text{or} \quad w = \sqrt{\frac{A}{B}\frac{Q}{P}} \qquad (1.58)$$

Substituting (1.58) and (1.8) in (1.56) gives the minimum cost,

$$C_{\min} = \frac{AP}{S^2} + \frac{BQ}{S^2} + F + \frac{2}{S^2}\sqrt{\frac{A}{B}\frac{Q}{P}} \qquad (1.59)$$

for any desired variance S^2.

The optimum field sampling weight from the standpoint of sampling alone (without consideration of costs) is A/B. If a core diameter can be chosen such that $Q/P = A/B$, the exploration will be optimized by taking core lengths such that $w = A/B$. Under this ideal condition, $Q = AP/B$, and (1.59) becomes

$$C_{\min} = \frac{4AP}{S^2} = \frac{2APN}{B} \qquad \left(w = \frac{A}{B} = \frac{Q}{P}\right) \qquad (1.60)$$

If we call the cost per foot of drilling E, the drill radius r, and the density of the ore d (E is in dollars per foot, r is in centimeters, and d is in grams per centimeter cubed),

$$P = \frac{E}{2.54 \times 12(\pi r^2 d)} \qquad \text{(dollars/g)} \qquad (1.61)$$

If the cost per sample, Q, can be estimated (it will include transportation, storage, cost per sample for reduction and analysis, and so on), a value for Q/P for any specific drill rig can be calculated. Neither Q nor P are exactly constants, but only a rough estimate is needed. When a large number of small samples is handled, Q will be high; with a small number of large samples, P will be high.

If the estimate of Q/P is very different from A/B, as determined from (1.9) and (1.10), revision of the sampling plan should be considered.

The important conclusion is that the cost of an exploration that aims at a specified uncertainty S in its estimate of ore grade (of the whole ore body or a block of ore within it) is directly dependent on the cost per foot of drilling only if the individual field sample weight is optimized. Estimation of this optimum field sample weight would therefore be made as early in the exploration as possible to avoid misguided effort.

Any preliminary estimate of the optimum field sample weight may be used to determine the type of equipment to be used in field sampling. If, for example, the optimum weight is estimated to be 100 g, the smallest available drill is indicated. If it calculates to 100 kg, a large drill is needed. If it calculates to several tons, drifting or trenching should be considered. When more information is available, the optimum field sample weight can be attained by taking longer or shorter lengths of core per sample. If it turns out that preliminary estimates of A, B and A/B are poor, decision to change the drilling program accordingly must be made. If field samples are too small or too large, data derived from them are not of much value; it is folly to continue to collect them.

The geochemical analyst must remember that all these processes depend on the accurate analysis of properly reduced subsamples.

1.29. THE POINT OF DIMINISHING RETURNS

When field samples of weight $w_{opt} = A/B$ are taken, the uncertainty in a grade estimate for a mountain or for any block of ore within the mountain is related to the number of samples analyzed:

$$S^2 = \frac{2B}{N} \quad \text{(when samples of } w_{opt} \text{ are taken)} \qquad (1.62)$$

If this is written $S = \sqrt{2B/N}$, any grade estimate may be followed by an error statement $\pm \sqrt{2B/N}$. The greater the number of samples that have been taken, the less the improvement in the grade estimate that follows from the collection of more samples. It requires four times as many samples to cut the error estimate in half and a hundred times as many to improve it by an order of magnitude. Taking a few more samples after many have been taken is a waste of time, money, and effort. One must always double or quadruple the number of samples taken within any block of ore to achieve meaningful improvement in the grade estimate. This is true when samples of optimum size are collected. Of course, if samples are much larger or much smaller than the optimum, no rule can be made! Almost any action may improve a situation that is totally inadequate!

1.30. EVALUATION OF PRELIMINARY DATA

If we regard the 12 drill holes of Table 1.2, from each of which twelve 1-m samples each weighing 10 kg were taken as the preliminary to a large exploration, it is desirable to calculate from the assay values some estimates of the sampling characteristics of the ore. These estimates may then be used to plan the details of further exploration.

One notices immediately that the standard deviation of these values is larger than the mean. This may occur for one or two or more reasons: The ore body may be badly segregated, the field samples may be too small, or errors in reduction or analysis may have been committed. We shall suppose that duplicate reductions and analyses have excluded the last possibility and decide whether the high variance is due to segregation or to large grains of cobalt-bearing material in the ore.

We first synthesize "large" samples by averaging the available values in groups of 12. This may be done in two ways, averaging down the drill holes or across the bed. We then find the standard deviation in the set of 12 large samples and calculate Visman constants A and B from (1.9) and (1.10). The calculation shows that there is little segregation, either down the holes or across the bed. The harmonic means h_1 and h_2 are used as estimators of the most probable results of the two series of samples, and the gangue concentration L is calculated from (1.41). The optimum sample weight is 59 kg from samples taken across the bed and 141 kg for samples taken down the holes. The calculated minimum field sample weights are 29 and 26 kg, respectively.

The only certain conclusion from this data is that the 10-kg field samples are too small. It is also fairly certain that the ore body is not segregated to a significant extent, either vertically or horizontally, and that the manipulations of geostatistics are probably inapplicable. The best approach to further sampling and exploration will be to take a relatively few very large samples, the assays of which (supposing they are properly reduced!) will differ little from one part of the ore body to another.

Casual inspection of the data of Table 1.2 may not yield such a conclusion. It is more likely that the presence of isolated pockets of high-grade ore would be suspected, and the intuitive approach might be to drill large numbers of small holes to find these pockets and their distribution. This would, of course, be an expensive mistake.

In this particular case, further drilling confirmed the evaluation beyond much doubt. It also showed that the first estimates of optimum field sample weight were too low by at least an order of magnitude. The initial samples were so small that most of them measured only the gangue concentration and contained little or no information. Only the chance collection of a few cobalt-rich samples saved the exercise and provided a useful clue to the broad sampling problem.

1.31. MANIPULATION OF SKEWED DATA

Exploration data are often highly skewed. It is important to discover the underlying causes of this skewness in any particular case. If it is due to segregation, field samples should be small and numerous, and Matherson's geostatistics should be employed in data evaluation. If it is due to coarse-grained ore or randomly distributed pockets of ore, larger field samples should be taken, and the use of geostatistics is counterindicated.

When field samples are unavoidably too small, sample reduction must be done with great care; even then, reduction errors are apt to be appreciable. Methods exist whereby the quality of a data set can be improved by manipulation. There is no general method, but any specific case offers this possibility. The following is an example of such manipulation.

Let us construct an imaginary ore body that contains 1-cc grains of MoS_2 randomly scattered throughout. The overall grade of this imaginary deposit is 0.400% MoS_2. The field samples, each weighing 10 kg, are split in two, and one-half is put through a 0.5-in. jaw crusher. The resulting 5 kg are blended, and 500 g are split out for analysis. Most of the original 1-cm grains survive the crushing; we will suppose that 25% of them are ground up and mixed with the gangue, giving an average gangue concentration L of 0.100% MoS_2. The average grade due to intact grains of MoS_2 is then 0.300% MoS_2, and the overall grade is 0.400% MoS_2. The density of MoS_2 is 4.8 g/cc. One grain of MoS_2 contributes $c = (H - L)d_H u^3 / w = 0.97\%$ MoS_2 to a single 500-g field sample (1.44). A single field sample containing one 1-cm grain of MoS_2 will show a grade of $0.100 + 0.970 = 1.070\%$ MoS_2.

Note that it is impossible, in this example, for any 500-g field sample to show the correct overall concentration of MoS_2. Those field samples that carry no grain of MoS_2 will show the gangue concentration, 0.100%; those that carry one grain will show 1.07%; those that carry two grains will show 2.04%; and so on. The assay values follow a Poisson distribution (1.23) with interval equal to the average number of grains per field sample. This average number is given by (1.45). In the present example, $z = 0.312$ grains of MoS_2 per average field sample. The Poisson probability distribution is given in Table 1.3.

If we have 100 or more values from field samples all collected within the same uniform block of ore, the average of all these values would probably come close to the true value, 0.400% MoS_2. However, if we would like to look at just a few data points, neither their individual values nor their average would be reliable. Suppose the large block of ore with more than 100 data points is a 400-ft cube, and we would like to look at

Table 1.3. Poisson Distribution, $z = 0.312$, $c = 0.97\%$

z_i	Percentage of MoS_2	Distribution of 1000 Values
0	0.100	732
1	1.070	228
2	2.040	36
3	3.010	4
≥ 4	≥ 3.980	< 1

Table 1.4. Possible Distributions (% MoS₂) of Seven Assay Values, y_i, $z = 0.312$, $c = 0.97$

0.100	0.100	0.100	0.100	0.100	0.100	
0.100	0.100	0.100	0.100	0.100	0.100	
0.100	0.100	0.100	0.100	0.100	0.100	
0.100	0.100	0.100	0.100	0.100	0.100	
0.100	1.070	1.070	1.070	0.100	0.100	
0.100	1.070	1.070	1.070	1.070	1.070	
0.100	1.070	2.040	3.010	2.040	1.070	
0.100	0.516	0.654	0.793	0.516	0.376	$\bar{y}$

one of the eight 200-ft blocks contained in this 400-ft block. Suppose that this 200-ft block contains only seven data points; these may be distributed in a number of different ways, like cards in a poker hand. Table 1.4 shows some very possible distributions.

This example is, of course, somewhat artificial. In the real case, the amount of ore mineral reduced to gangue during crushing will not always amount to 0.100% MoS₂. Some MoS₂ may be in the gangue category before crushing. The in situ grains of ore mineral will not always be exactly cubes. Some much larger grains or flakes of MoS₂ may be present in the ore body and will occasionally end up in a 500-g field sample. The net effect of these and other easily imaginable realities will be a spread of results in each of the categories of Table 1.4. In a real set of values, those samples that lack an intact grain of MoS₂ will vary in MoS₂ content from about 0.05% to about 0.5% MoS₂, with most values clustering about 0.100% MoS₂. Those samples with one intact grain of MoS₂ may vary from about 0.5 to about 1.5%. Those samples with two intact grains may vary from about 1.5 to about 2.5% MoS₂, and so on. In each case, the real values will cluster about the ideal values. Table 1.5 has been invented

Table 1.5. Possible Distributions (% MoS₂) of Seven Assay Values, y_i, $z = 0.312$, $c = 0.97$, Nonuniform Grains

0.050	0.302	0.070	0.213	0.085	0.101	
0.090	0.092	0.096	0.094	0.105	0.109	y_i
0.092	0.100	0.122	0.120	0.130	0.117	
0.108	0.118	0.140	0.099	0.192	0.232	
0.131	1.008	1.021	1.141	0.410	0.150	
0.145	1.060	1.132	0.800	0.976	0.899	
0.200	1.382	2.020	2.980	2.020	1.032	
0.117	0.580	0.657	0.778	0.560	0.371	$\bar{y}$

with these simple principles in mind; this table should be regarded as representing the real case corresponding to the artificial case of Table 1.4.

Note that the distribution of data points, y_i, in Table 1.5 is skewed. Neither the averages nor the individual values come close to the true grade, 0.400% MoS_2. We would like to manipulate this data so as to obtain values more closely representing the true grade of the ore in the vicinity of these too small samples. This may be done by utilizing a knowledge of the sampling characteristics of the ore body as a whole. This knowledge is implicit in the Visman sampling constants A and B. The homogeneity constant A is a measure of small-scale inhomogeneity, and the segregation constant B is a measure of large-scale inhomogeneity. Since our artificial ore body is completely uniform, that is, all constituents are randomly distributed, the segregation constant $B = 0$. The homogeneity constant A is found from (1.11): $A = (K - L)(H - L)u^3 d_H = (0.400 - 0.100)(100 - 0.100)(1)^3(4.8) = 143.9$. To perform the desired manipulation, we use (1.11), (1.39), and (1.45) and obtain

$$K_{i(200)} = \frac{y_{i(200)} + L_{200}}{2} + \frac{1}{2}\sqrt{(y_{i(200)} - L_{200})^2 + \frac{2A_{400}(y_{i(200)} - L_{200})}{wK_{400} - wL_{400}}}$$

$$(1.63)$$

where $K_{i(200)}$ is the manipulated grade for the data point $y_{i(200)}$, L_{200} is the estimate of the gangue concentration of MoS_2 in the 200-ft block, A_{400} is the Visman homogeneity constant estimated from data available from the 400-ft block, K_{400} is the weighted average of all data from the 400-ft block, L_{400} is the estimated gangue value for the 400-ft block, and w is the effective sample weight. Results of this manipulation on the data of Table 1.5 are shown in Table 1.6.

Of these manipulated values, those deriving from y_i that are not most

Table 1.6. Manipulated Data (% MoS_2) from Table 1.5

0.050	0.527	0.070	0.397	0.085	0.122	$K_{i(200)}$
0.200	0.092	0.190	0.094	0.191	0.170	
0.204	0.157	0.249	0.218	0.253	0.199	
0.236	0.216	0.283	0.145	0.366	0.426	
0.277	1.349	1.346	1.493	0.666	0.282	
0.301	1.405	1.463	1.123	1.309	1.236	
0.384	1.746	2.387	3.392	2.402	1.381	
0.236	0.786	0.856	0.980	0.753	0.545	$\overline{K}_{i(200)}$

probable are falsely high. A criterion for rejecting these too high values is that those higher than the average K_i, $\overline{K}_i$, are not the most probable y_i. To calculate the best estimate $\overline{K}$ of the MoS_2 content of each 200-ft block, those y_i that are greater than the corresponding $\overline{K}_i$ are subtracted in the following manner, using column 2 of Tables 1.5 and 1.6 as an example (the asterisks indicate $y_i > \overline{K}_i$):

$\overline{K}_i$		y_i
0.527		0.302
0.092		0.092
0.157		0.100
0.216		0.118
1.349*		1.008*
1.405*		1.060*
1.746*		1.382*
5.492	(sum)	
0.785	(average)	

$$K = \frac{5.492 - 1.008 - 1.060 - 1.382}{7 - 3}$$

$$= \frac{2.042}{4} = 0.511$$

What we are trying to do is break up the MoS_2 grains in the field samples with high y_i mathematically and distribute the excess MoS_2 in these samples among the samples with low y_i in a manner such that the manipulated values will more nearly approach the true value. The differences ($K_i - y_i$) for samples with $y_i > \overline{K}_i$ represent the gangue concentration. In this way, samples that do not contain a grain of MoS_2 have their assay values incremented according to the number of field samples that happen to contain one or more grains of MoS_2. It does not matter, in calculation, to which of the gangue y_i one adds the increment contributed by y_i derived from samples that do contain a grain of ore mineral. To demonstrate, we have added the increments $K_i - y_i$ for the ore-mineral-bearing samples to the y_i derived from non-ore-mineral-bearing samples (Table 1.7). Samples with y_i greater than $\overline{K}_i$ are omitted.

It should be obvious that the average $\overline{K}$ of the censored manipulated values is a better estimate of K than the average $\bar{y}$ of all y_i. In this artificial example, we know that K should be 0.4% or slightly higher. Inclusion of the value 2.980, representing a sample with three grains of MoS_2, results

Table 1.7. Completed Manipulation, (% MoS$_2$)

0.050	0.527	0.437	0.397	0.467	0.471	
0.200	0.456	0.522	0.506	0.524	0.507	
0.204	0.502	0.574	0.218	0.253	0.199	
0.236	0.557	0.283	0.497	0.366	0.426	
0.277	—	—	—	0.666	0.282	
0.301	—	—	1.123	—	—	
0.384	—	—	—	—	—	
0.236	0.510	0.454	0.548	0.455	0.377	$\overline{K}$

in an average somewhat higher than 0.400. The chances of including a sample with three grains of MoS$_2$ are only 4 in 1000, or 0.17 in 42. Table 1.8 compares the values of $\overline{y}$, $\overline{K_i}$, $\overline{K}$ for the six data sets from the hypothetical 200-ft block.

Let us now suppose that all the 42 data points of Table 1.5 originate in a single 200-ft block and that, as before, we have determined $A_{400} = 143.9$, $K_{400} = 0.400$, and $L_{400} = 0.100$ by examining the much larger data set from the 400-ft block. We will, however, substitute a gangue value, 0.050, for the aberrant 2.980 to remove the bias introduced by this unlikely value. The differences $(K_i - y_i)$ for those samples that have $y_i > \overline{K_i}$ are added, as in Table 1.7, to the lowest K_i, and the high K_i are then excluded. The results of this manipulation are shown in Table 1.9. The estimate of K, $\overline{K}$, comes to 0.397, very close to the actual value, 0.400. A histogram of arbitrary interval 0.05% MoS$_2$ is shown for each of these sets of values in Figure 1.6.

If the unlikely value 2.980 were included in these calculations, $\overline{K}$ would come to 0.430% MoS$_2$. This may draw attention to the precarious nature of operations based on too few and too small field samples.

Note that the number of grains of MoS$_2$ in the 42 field samples averages $\frac{14}{42} = 0.333$. We have included in our artificial array somewhat too many high-MoS$_2$ values, since the correct value for z is 0.312. This is reflected in the high $\overline{y}$. If the average had included the unlikely value 2.980, the average of the y_i would be $\overline{y} = 0.492$.

In general, the manipulation will nearly always bring the observed values closer to their true value however the laws of chance distribute grains

Table 1.8. Comparison of $\overline{y}$, $\overline{K_i}$, and $\overline{K}$

$\overline{y}$	0.117	0.580	0.657	0.778	0.560	0.371
$\overline{K_i}$	0.236	0.786	0.856	0.980	0.753	0.545
$\overline{K}$	0.236	0.510	0.454	0.548	0.455	0.377

Table 1.9. Manipulation (% MoS$_2$) of 42 Data Points $y_{i(200)}$ [a]

y_i	K_i	Less K_i	With $y_i > \overline{K}_i$
0.050	0.050		0.399*
0.090	0.200		0.518*
0.092	0.204		0.518*
0.108	0.236		0.236
0.131	0.277		0.277
0.145	0.301		0.301
0.200	0.384		0.384
0.302	0.522		0.522
0.092	0.204		0.516*
0.100	0.221		0.524*
0.118	0.255		0.255
1.008	1.319	(0.311)	0.397
1.060	1.374	(0.314)	0.397
1.382	1.712	(0.330)	0.397
0.070	0.151		0.481*
0.096	0.212		0.521*
0.122	0.262		0.262
0.140	0.293		0.293
1.021	1.332	(0.311)	0.397
1.132	1.450	(0.318)	0.397
2.020	2.369	(0.349)	0.397
0.213	0.403		0.403
0.094	0.208		0.523*
0.120	0.258		0.258
0.099	0.219		0.527*
1.141	1.459	(0.318)	0.397
0.800	1.095	(0.295)	0.397
0.050	0.050		0.399*
0.085	0.189		0.507*
0.105	0.230		0.230
0.130	0.276		0.276
0.192	0.373		0.373
0.410	0.655		0.655
0.976	1.284	(0.308)	0.397
2.020	2.369	(0.349)	0.397
0.101	0.223		0.518*
0.109	0.238		0.238
0.117	0.253		0.253
0.232	0.429		0.429
0.150	0.309		0.309
0.899	1.202	(0.303)	0.397
1.032	1.344	(0.312)	0.397
$\overline{y} = 0.442$	$\overline{K}_i = 0.628$		$\overline{K} = 0.397$

[a] The differences in parentheses have been added to the lowest K_i in reverse order. The resulting values are marked with an asterisk.

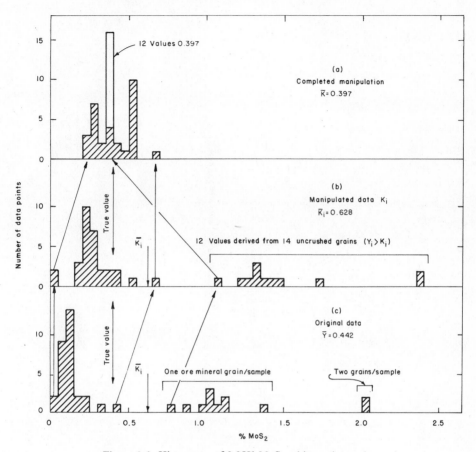

Figure 1.6. Histograms of 0.05% MoS$_2$ arbitrary intervals.

of ore mineral among the several field samples—*provided the skewness this device essays to correct is due to inhomogeneity, not segregation.*

Whether skewness is due to segregation (large-scale inhomogeneity) or to inhomogeneity on a small scale is determined by the Visman constants.

Figure 1.7 shows the result of manipulating a real set of 52 data points from a 200-ft block of ore. Visman constants A and B were estimated from about 350 data points within a 400-ft parent block. The values used in the manipulation are $A_{400} = 92$ and $K_{400} = 0.352$. The average of the raw data is $\bar{y} = 0.506$. If one removes the 2 data points at 3.5% MoS$_2$, the average of the remaining 50 data points is $\bar{y} = 0.352$. The best estimate of the true overall MoS$_2$ content of the 200-ft block is given by $\bar{K} = 0.426\%$ MoS$_2$.

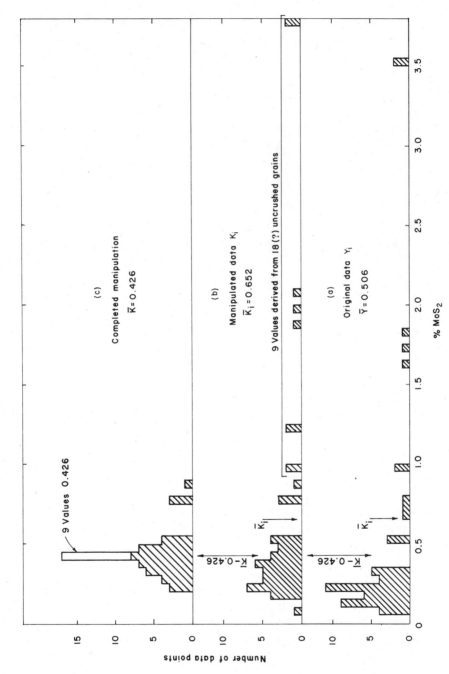

Figure 1.7. Manipulation of a real set of data from a 200-ft block of ore.

Attention is drawn to the similarity of this real data set to the invented data set of Figure 1.6. Note especially that no data point corresponds with the estimate $\overline{K}$ = 0.426% MoS_2 of the true grade of the 200-ft block.

There are nine data points with values higher than $\overline{K}_i$. One may easily suppose that four of these came from samples that contained one uncrushed grain of ore mineral, three came from samples that contained two uncrushed grains, and two came from samples that contained four uncrushed grains. The gods of chance chose to omit samples containing three uncrushed grains. In an adjacent 200-ft block of the same ore body, there are no data points corresponding to four grains of uncrushed mineral and two data points corresponding to three uncrushed grains.

1.32. DATA FROM SEGREGATED ORE BODIES

One must be cautious in applying manipulations of the kind suggested above lest variance due to segregation be misinterpreted. The assumption that skewness is due to randomly scattered grains of ore mineral in too small field samples must be proven before applying calculations based on that assumption. Skewed distributions may arise from segregation; in this case, the described manipulation is invalid. If the Visman segregation constant B is not zero or very low, the manipulation of data must be modified to take this into account; otherwise, false smoothing of the data will result. Suppose that the 400-ft assays of real ore develop a calculated value for B of 0.49. The uncertainty in each data point is then given by the Visman formula (1.8), and $S = \sqrt{A/w + B} = \sqrt{92/500 + 0.49} = 0.82$. The uncertainty due to segregation is $\sqrt{B} = 0.70\%$ MoS_2. The cutoff $\overline{K}_i$, used to decide which data points should be excluded after manipulation, must be increased by 0.70% MoS_2. This would leave only five data points which the constant A tells us are unnaturally high. As a result, fewer of the low y_i would be incremented, and fewer of the high y_i would be excluded.

Introduction of this very necessary modification of the manipulation described above makes this manipulation generally applicable to any data set. If either A is very low or B is very high, the manipulation will not change the original data appreciably.

1.33. DOUBLE POISSON DISTRIBUTION

When samples taken from a mountain or other mass of material that contains the constituent of interest in discrete grains are subsampled in such a way that the subsamples also contain discrete grains of reduced size, a double Poisson distribution of the assay values is likely.

When field samples each contain, on average, a limited number Z of ore mineral grains, and the subsamples they generate also contain a limited number z of ore mineral grains, the distribution of assay values is doubly Poisson. The probability π_r of r grains of ore mineral appearing in any subsample is determined by the sum of the probabilities of r grains being generated from samples with n grains:

$$\pi_r = \frac{e^{-Zf} f^r}{r!} \sum \frac{Z^n e^{-n} n^r}{n!} \qquad (n = 0, 1, 2, 3, \ldots) \qquad (1.64)$$

where $f = C/c = z/Z$ is the ratio of the number of grains in the average subsample to the number of grains in the average sample. When $r = 0$, $\pi_r = \pi_0$ is the probability that no grains of ore mineral will appear in a specific subsample:

$$\pi_0 = e^{-Zf} \sum \frac{Z^n e^{-n}}{n!} \qquad (n = 0, 1, 2, 3, \ldots) \qquad (1.65)$$

We also have the following relationships [equations (1.27) and (1.22)]:

$$s^2 = C^2 Z + c^2 z \qquad (1.66)$$

$$K = L + CZ = L + cz \qquad (1.67)$$

where C is the contribution of a single ore mineral grain to the X content of a field sample of weight W, c is the contribution of a single ore mineral grain to the X content of a subsample of weight w, Z is the number of ore mineral grains in the average field sample, and z is the number of ore mineral grains in the average laboratory subsample. From 1.66 and the identifies $f = C/c = z/Z$,

$$f = \frac{(K - L)C}{s^2 - (K - L)C} = \frac{s^2 - (K - L)c}{(K - L)c} \qquad (1.68)$$

$$Z = \frac{K - L}{C} \quad \text{and} \quad z = \frac{K - L}{c} \qquad (1.69)$$

In evaluating a badly skewed data set, estimates of K, L, C, and s^2 may be used to find preliminary values for Z and f; substitution in (1.64) or (1.65) will show how well the double Poisson model fits the data. Improved estimates of L and C may follow, and reiteration may result in a better fit.

Example. An ore contains one grain of ore mineral per 6.394 kg ore; each ore mineral grain weighs 19.1 g. The concentration of ore mineral is 19.1/ 6394 × 100 = 0.3%. In addition, the ore contains 0.1% of finely divided ore mineral evenly distributed throughout the mass. Samples each weighing 10 kg are taken. The average sample contains 1.564 grains of ore mineral. The distribution of the true grades of 1000 samples is Poisson, supposing the ore mineral grains are randomly distributed:

True Grade (wt. %)	Number of Grains of Ore Mineral	Poisson Probability
0.100	0	209
0.291	1	327
0.482	2	256
0.673	3	133
0.864	4	52
1.055	5	16
1.246	6	4
1.437	7	1
1.628	8	0

Each of the samples are reduced by crushing in such a way that each ore mineral grain is broken into 10 pieces. The crushed material is mixed and one-twentieth of it is split out. The subsample weight is 500 g. The average number of grains of ore mineral per subsample is 0.782. Of 1000 samples, the 209 that contained no grain of ore mineral yield 209 subsamples containing no grain of ore mineral. The 327 samples that contained one grain of ore mineral yield subsamples containing 1, 2, 3, . . . grains, as do the samples containing 2, 3, 4, . . . grains. The overall probabilities are shown in Table 1.10.

Note that in this artificial example it is impossible for the final assay value to show the true grade for samples that originally contained an odd number of ore mineral grains. The distribution of assay values is highly skewed, with too many low values and a few unrealistically high values. Such distributions are common in mining exploration data and are often due to the combined effect of too small samples and poor reduction practice.

To give a real example, Table 1.11 presents 67 assay values obtained from a single 200-ft block of an ore body during an exploratory drilling program.

Table 1.10. 1000 Samples Subsampled

Grade	Number of Grains	Poisson Probability	Probabilities in Each Subsample									
			0	1	2	3	4	5	6	7	8	9
0.100	0	209	209	—	—	—	—	—	—	—	—	—
0.291	1	327	198	99	25	4	1	0	—	—	—	—
0.482	2	256	94	94	47	16	4	1	0	—	—	—
0.673	3	133	30	45	33	17	6	2	0	—	—	—
0.864	4	52	7	14	14	9	5	2	1	0	0	—
1.055	5	16	1	3	4	3	2	1	1	1	—	0
1.246	6	4	0	1	1	2	0	—	—	—	—	—
1.437	7	1	—	—	0	1	0	—	—	—	—	—
1.628	8	0	—	—	—	—	—	—	—	—	—	—
Total probabilities			539	256	124	52	18	6	2	1	0	0
Measured grade			0.100	0.482	0.864	1.246	1.628	2.010	2.392	2.674	—	—

66

Table 1.11. Assay Values from Single 200-ft Ore Body

Percentage of Molybdenum	Number of Assay Values[a]
0.0–0.2	32 (36)
0.2–0.6	24 (17)
0.6–1.0	7 (8)
1.0–1.4	0 (3)
1.4–1.8	1 (1)
1.8–2.2	2 (1)
2.2–2.6	0
2.6–3.0	0 (1)
3.0–3.4	0
3.4–3.8	1

[a] Numbers in parentheses are the frequencies in Table 1.10 reduced to a base of 67. The artificial example was designed to simulate this real data set using the processes described in the text.

1.34. FITTING STATISTICAL MODELS

The double Poisson distribution described is one example of a model that may be used in attempts to evaluate a data set. At one time, scientists became convinced that the Gaussian distribution was universally applicable, and an overwhelmingly large majority of practical applications of statistical theory are based on this distribution. A common error has been to reject "outliers" that cannot be made to fit the Gaussian model or some modification of it (e.g., the lognormal model). The tendency has been to make the data fit a preconceived model instead of searching for a model that fits the data.

It is now apparent that outliers are often the most important data points in a given data set. Any successful evaluation must account for them, and in a rational way. The Gaussian distribution approximates a real distribution only when the items (grains) contributing to a data point are large in number. The lognormal model is an empiricism designed to obscure lack of "normality" in the data. This does not mean that it should not be used—only that its empiricism should be recognized and that it should be abandoned if it cannot account for outliers. There are many instances in which the double Poisson model is of no value. For example, data derived from a highly segregated ore body may exhibit a skew that, at first sight, may lead one to suspect a double Poisson distribution; appropriate testing of this distribution will lead to an understanding of the seg-regation-generated skewness and to the use of a more appropriate statistical device (e.g., the geostatistics of Matheron).

In data sets derived from ore bodies and in trace element data, the Gaussian approximation is seldom valid; its uninformed application is likely to lead to erroneous conclusions.

1.35. PURPOSES OF SAMPLING

In the preceding, emphasis has been on the gross composition of materials being sampled and subsampled. Sampling has other purposes, some of which are worth mentioning. In geochronology, for example, whole-rock rubidium–strontium or potassium–argon ages depend more on the character of the portions taken for analysis than they do on the sample being chemically representative of the whole-rock mass. Age determination on specific minerals from the mass depend more on the rejection of altered material and on clean mineral separations from geologically homogeneous rock than they do on gross sampling procedure.

A variation of trace element content in specific minerals in a rock may be more meaningful than variations in the same trace elements in the whole rock. Thermodynamic (phase equilibrium) studies may require determination of elemental ratios, or the ratios of valence states of the same element, without close regard for total rock composition; ferrous–ferric ratios may be significant, for example.

In sedimentary rocks, organic carbon may need to be distinguished from carbonate carbon. Pollen grains or humic acids may be the constituents of interest, and their gross concentration in the rock may be of small importance. Efforts to obtain a representative portion of a sedimentary rock may not require any consideration of the sampling statistics outlined above.

In paleomagnetic work, the orientation of the specimens taken is of supreme importance. The location of samples is sometimes of more consequence than their exact elemental composition.

Sampling cannot be considered adequately without reference to its purpose. An important corollary is that samples collected for one purpose may be worthless for another purpose. It is essential that the geochemical analyst understand the purpose for which a sample is collected, and that analytical techniques are devised accordingly.

Because sample collection is usually expensive, appropriate efforts should be made to sample in such a way that the product is useful in as many anticipated applications as possible.

Fortunately, it is seldom that the sampling of a completely unknown mass of material has to be faced. Always, one should make full use of all available information before beginning the sampling operation. This is best done by the geologist or engineer who fully appreciates innumerable subtleties within the problem that are inamenable to objective treatment, supposing the geologist or engineer is also aware of sampling theory. If there is a close consulting relationship among all those who apply their disciplines to the same problem, and if this relationship extends to the geochemical analyst, a small expenditure of time may result in a much greater reward for the effort. While the geochemical analyst should, of course, leave field sampling to those who understand the problem most thoroughly, he or she must be prepared to give advice on the requirements and the limitations of the analytical procedure to be followed.

Knowing the requirements of a specific investigation, the geochemical analyst may be able to decide that rough and rapid analytical procedures will be adequate or that only the most sophisticated analysis will meet the needs of the moment.

Geologists and mining engineers are prone to underestimate the enormity of their sampling problems. Too often, the results of what should be regarded as a preliminary survey are published as the final word, and conclusions are drawn from very raw data. While the first responsibility for such mistakes is the geologist's or engineer's, it is the geochemical analyst who produces the data, whether he or she likes it or not. It is up to the geochemical analyst to explain the limitations of the methods used and to evaluate the work in the context of the overall investigation. When unwarranted conclusions are drawn from the data produced, he or she must complain or risk being later criticized.

1.36. FIELD SAMPLING METHODS

Diamond drilling is a most effective way to take trivial samples of large rock masses. As light and portable equipment becomes more generally available, the practice of collecting "hand" specimens will hopefully diminish.

Hand specimens are usually of qualitative interest only. They are sometimes altered, contaminated, and quantitatively useless. The geochemical analyst should avoid wasting time on them. Any good mineralogist can give, by simple inspection, as good an analysis of the average hand specimen as the analyst can in two weeks' work, supposing the analysis is intended to represent the rock mass from which the specimen was taken.

Those who take small drill cores (or hand samples) should consider the statistics of sampling. Baird et al. [27] have done so in sampling granites. They concluded that a core 2 cm in diameter and 30 cm long was adequate for the coarsest rock in their collection. They did not make a clear distinction of the rock components that were thus adequately sampled. Sampling designed for major rock constituents may be totally inadequate if minor or trace constituents are sought. Even major constituents may be subject to gross sampling error. Consider, for example, two extreme cases: first, a pegmatite, and second, a fine-grained basalt. In the first case, a 30-cm length of 2-cm core may conceivably consist of a single piece of feldspar. Further reduction of this sample presents no problem; one small portion of it will differ inappreciably from another; however, its composition is plainly not representative of the whole-rock mass. In the second case, a 30-cm length of 2-cm core may well come close to representing the overall composition—but a randomly selected portion of it may not be representative either of the core or of the whole-rock mass.

In analyzing a subsample of a core sample, it is obviously desirable to consider the purpose of the analysis within the investigation. That the sample of pegmatite is homogenous does not mean that analysis by sophisticated methods is necessary; the most trivial analysis, optical inspection, may yield all necessary information. With the fine-grained basalt, the situation is different. An estimate of grain size will make it possible to decide how well the 2 × 30 cm core represents the overall composition of the rock and provide a guide to further treatment and methods of analysis to be applied. Analytical methods should be selected to match sampling precision and to meet the requirements of the investigation. Very often, a rapid rock analysis, using instrumental or slop methods, is all that is warranted. Sometimes, however, the purposes of the investigation and the integrity of the sample may call for application of sophisticated chemical methods. Geologists and mining engineers are often unaware of the difference between a rapid and a primary analysis and must be informed when they ask for one and need the other.

1.37. SAMPLING FOR POTASSIUM–ARGON DATING

An extreme example of the need for careful and knowledgeable sampling procedures is afforded by the K–Ar dating of low-potassium minerals when these minerals are cogenetic with high-potassium minerals that have suffered diffusional or other loss of either potassium or of radiogenic argon. The effect of minute quantities of such high-potassium minerals in low-potassium mineral separates is astonishingly large, and a general failure among geochronologists to recognize it has led to the publication of many erroneous K–Ar ages.

Some definitions are necessary. The *derived age* of a low-potassium mineral separate is the K–Ar age calculated from determinations of radiogenic argon and of potassium. The *true age* of a low-potassium mineral separate is the K–Ar age that would be derived if the separate were 100% pure—that is, free from any contaminant of a different K–Ar age. Both the derived age and the true age are calculated ages based on accepted values for decay constants and on certain fundamental assumptions that need not be elaborated here.

The general K–Ar age equation may be written

$$e^{a/\lambda} = 1 + \frac{A^*}{K\lambda\lambda_e} \qquad (1.70)$$

where a is a K–Ar age, λ is the reciprocal of the sum of two decay constants [numerical value ≈ 1885 my], λ_e is the ^{40}K–^{40}Ar decay constant, and A^* and K are the ^{40}Ar (radiogenic) and ^{40}K concentrations expressed in the same units (moles per gram). In bimineralic mixtures of potassium contents K_L and K_H, radiogenic argon contents A_L^* and A_H^*, and K–Ar ages a_L and a_H, from (1.14), we have the identities

$$p_w K_L + q_w K_H = K \quad \text{and} \quad p_w A_L^* + q_w A_H^* = A^* \qquad (1.71)$$

Multiplying (1.70) through by K and substituting the identities (1.71),

$$
\begin{aligned}
K e^{a/\lambda} &= K + \frac{A^*}{\lambda\lambda_e} = K + \frac{p_w A_L^* + q_w A_H^*}{\lambda\lambda_e} \\
&= p_w K_L + q_w K_H + \frac{p_w A_L^* K_L}{K_L \lambda\lambda_e} + \frac{q_w A_H^* K_H}{K_H \lambda\lambda_e} \\
&= p_w K_L \left(1 + \frac{A_L^*}{K_L \lambda\lambda_e}\right) + q_w K_H \left(1 + \frac{A_H^*}{K_H \lambda\lambda_e}\right) \\
&= p_w K_L e^{a_L/\lambda} + q_w K_H e^{a_H/\lambda}
\end{aligned}
\qquad (1.72)
$$

From (1.71), $p_w = (K_H - K)/(K_H - K_L)$ and $q_w = (K - K_L)/(K_H - K_L)$. Substituting these values in (1.72) gives (1.73),

$$K(K_H - K_L)e^{a/\lambda} = K_L(K_H - K)e^{a_L/\lambda} + K_H(K - K_L)e^{a_H/\lambda} \qquad (1.73)$$

The sampling error incurred during the removal of an analytical subsample for either potassium or argon determination is given by (1.30) and (1.28):

$$s = \sqrt{\frac{(K - K_L)(K_H - K_L)d_H u^3}{w}} = \frac{K}{100}\sqrt{\frac{K_s}{w}} \qquad (1.74)$$

where w is the analytical subsample weight in grams and K_s is the laboratory sampling constant; K is the measured K_2O content of the low-potassium mineral of K_2O content K_L contaminated with a proportion q_w of a high-potassium mineral of K_2O content K_H. The sampling error is measured by a standard deviation s. The high-potassium contaminant has an effective mesh size u centimeters and a density d_H grams per centimeter cubed.

The two sources of error in K–Ar age determinations on low-potassium mineral separates must be controlled if accurate ages are to be obtained. The error due to contamination by a high-potassium mineral of a different K–Ar age may be very large. For example, a hornblende with 0.200% K_2O and a K–Ar age a_L of 170 my contaminated with 0.5% of a biotite with 10.0% K_2O and a K–Ar age a_H of 70 my will develop a derived age a of 150 my from (1.73). The error due to the subsampling of the impure low-potassium mineral separate may also be large if K_2O and Ar are determined on separate subsamples; refer to the amphibole–orthoclase mixture of Table 1.1.

To circumvent these errors, an understanding of sampling theory is essential. To find the true K–Ar age of the low-potassium mineral in a contaminated separate, determination of K, K_L, K_H, A^*, and A_H^* is necessary. The K_H and A_H^* may usually be accurately determined by analysis of a carefully prepared separate of the high-potassium mineral. A *small* proportion of a low-potassium contaminant in a high-potassium mineral separate has a (usually) minor effect. An exception occurs when the low-potassium mineral (e.g., beryl) may have a large content of extraneous argon. Determination of K and A^* requires careful attention to subsampling difficulties. Determination of K_L is most difficult but can be achieved through use of one or another of the devices to be described. It is fortunate that direct determination of A_L^* is usually unnecessary.

If K and A^* can be determined both on the same subsample, using, for example, the $LiBO_2$ fusion method [28] or the ^{39}Ar–^{40}Ar method [29], the K–Ar age so derived may be geochronologically meaningless and not at all reproducible (because of subsampling difficulties), but each pair of values (a, K) so obtained can be used in (1.73) to calculate a true value for a_L.

If K and A^* cannot be determined on the same subsample, it is necessary to eliminate subsampling errors.

Two methods of diminishing subsampling error are available: The subsample weight may be increased or the impure low-potassium mineral separate may be ground finer. The questions how much to grind and how fine to grind it deserve special attention. It is, of course, necessary to screen out the fines during the sample reduction, so as to keep air–argon contamination under control. Unless grinding is carefully and skillfully performed, the process may lead to intolerable losses of sample.

The amphibole–orthoclase mixture of Table 1.1 has a sampling constant for potassium of $K_s \approx 25$, and 25 g must be taken to achieve a relative sampling error of 1% or less (68% confidence). Evidently, a 25-g subsample is impracticable; if K and A^* are to be determined on different subsamples, it will be necessary to grind to a finer mesh size before splitting out subsamples for the K and A^* determinations. Since the derived or calculated age of the mixture is not of consequence, it is not necessary to take 25 g for grinding; only enough for as many analyses as are contemplated should be taken, say, 1–2 g. If this should turn out to be insufficient, and another 1–2 g are taken for grinding, the second lot will likely show a different derived age and a different K_2O content. This is of no consequence in finding the true K–Ar age of the low-potassium mineral, but may be disconcerting if the reasons for the difference are not appreciated.

The principles that should govern the proper reduction of samples to analytical subsamples have been outlined above. To make full use of these principles in planning the reduction and analysis of a low-potassium mineral separate, an estimate of K_L, the true K_2O content of the low-potassium mineral is necessary. Several methods for making such an estimate are available: At the risk of repetition, a summary of the relevant methods follows.

1. While separation of enough 100.00% pure low-potassium mineral for A^* determination may be impracticably difficult, it is sometimes possible to hand pick a few milligrams under a binocular microscope. Accurate K_2O values may easily be obtained using only 5–10 mg of material even when the K_2O content is 0.1% or less.

2. Analysis of a series of small subsamples of a low-potassium mineral separate gives a probability (1.36) that at least one of N subsamples is free from high-potassium contaminating grains and represents the composition of the low-potassium mineral.

3. In a closely sized mineral separate, the number of high-potassium

mineral grains per subsample may be estimated from

$$z = \frac{(K - K_L)^2}{s^2} = \frac{w(K - K_L)}{(K_H - K_L)u^3 d_H} \qquad (1.75)$$

where w is the subsample weight in grams, s^2 is the square of the standard deviation in a series of determinations, u is the effective grain size in centimeters, and d_H is the density of the high-potassium mineral in grams per centimeter cubed. Rearrangement of (1.75) gives

$$K_L = \frac{K_H + K}{2} - \frac{1}{2}\sqrt{(K_H - K)^2 + \frac{4ws^2}{u^3 d_H}} \qquad (1.76)$$

Despite a difficulty in estimating u^3, (1.76) often gives a very good estimate of K_L, certainly good enough to use in (1.36) to find the probability that the lowest value in a series represents the composition of the pure low-potassium mineral.

4. Analyses of two series of subsamples of different weights yields an estimate of K_L through (1.41) when the contaminate exists in discrete particles.

5. With closely sized mixtures, K_L may sometimes be found through use of (1.44).

6. The electron microprobe may sometimes be used to advantage.

1.38. MIXING AND BLENDING

Most of the sampling theory outlined above is based on the assumption that during reduction the material being sampled is well mixed; that is, all mineral grains are randomly distributed among themselves. It is not easy, in practice, to maintain thorough mixing throughout the reduction process. Various widely employed mixing devices should be regarded with suspicion; some of them actually segregate minerals of different particle shape and density. An example is the V-blendor found in almost every sample preparation room. Anyone who has faith in this device should place in it a mixture of two dissimilar materials of different color (e.g., red glass beads and quartz sand of the same particle size; black biotite flakes and crushed peridotite) and watch what happens. In general, machines that use mechanical violence and look and sound as though they are efficient are most likely to cause segregation of heavy and light, large and small, and flat and round particles.

Possibly the best general method for mixing is to pour the sample through a splitter repeatedly, combining the two halves by pouring them into a cone between passes. With small samples, slow tumbling on a sheet of cloth or paper is probably the most rapid and effective method. It is necessary to make sure that the cloth or paper does not contain any element being sought in the sample; this sometimes excludes glazed papers, which may have high concentrations of, for example, alkali metals, barium sulfate, and TiO_2, on their surface. Plastics are prone to develop static charges and selectively attract certain minerals.

Some combinations of minerals resist mixing. Rock powders containing magnetite afford a common example. The magnetite is prone to collect in aggregates, especially if the sample has been treated magnetically at some stage in its preparation. Sometimes mixing turns out to be impossible: in such cases, division of the sample into two weighed portions followed by separate analyses is necessary. Stony meteorites are best handled in this way when they contain more than a small proportion of metallic material. The division of the sample into two parts may be accomplished physically, but it is sometimes easier to make the separation chemically by dissolving out a portion with a suitable reagent (e.g., meteorites may be treated with bromine and water to dissolve metallics and sulfides).

1.39. CONTAMINATION

It is, of course, impossible to collect, reduce, grind, screen, and mix rock or mineral samples without introducing some contamination from the equipment and the environment. The best that can be done is to make sure that critical contaminants are excluded. Which contaminants can be tolerated and which cannot depends on the purpose at hand. The person who designs the procedure must understand this purpose so as to avoid equipment and devices that will introduce intolerable contamination.

Contamination originates in every material with which the sample comes in contact, including air, skin, paper, cloth, bags, bottles, shovels, drills, mules, picks, hammers, sampling machinery, and so on; in the laboratory, this includes glass and plastic ware, stirring rods, covers, wipers, and the laboratory atmosphere (which may be very different from that in the sampling area). The increasing interest in trace elements makes it mandatory for the analyst to scrutinize every possible source of contamination and try to take them all into account. Sometimes it is difficult to decide whether a constituent is a contaminant or not. For example, should chlorine found in tourmalines collected on a seacoast be regarded as part of the sample? What is the significance of ammonium ion in stilp-nomelane collected in a cow pasture? Should one report a few parts per million of copper in a sample that has been passed through brass screens? What is the meaning of a total carbon determination on a rock sample if the original specimen had lichens growing on it? How much meaning is there in a ferrous iron determination if the sample was ground between metal plates in a plate mill? Is a K–Ar age determination on a plagioclase (with about 0.02% K_2O) valid when potassium feldspars and biotites are regularly put through the same sampling equipment? Should specimens labeled with white paint containing zinc oxide be analyzed for traces of zinc? Should rocks collected in the wilderness and carried in cloth bags on pack mules be examined for traces of chlorine or sodium? Should a sample ground in a "tungsten carbide" mortar be examined for cobalt when it is known that tungsten carbide is made by cementation with cobalt?

The geochemical analyst should be careful not to waste time performing useless analyses. When he or she becomes aware of a contamination, the originator of the sample should be consulted and an understanding attempted. Geologists and miners, like chemists, do not fall into traps on purpose; they do it inadvertently. An analyst who proceeds blindly with an analysis of a hopelessly contaminated sample is reduced from the rank of scientist to that of technician.

Because contamination is a very important and difficult problem for all concerned, it is useful to discuss some of the common sources of this plague from the point of view of the working analyst.

Sodium Chloride. Most geochemical studies of the chlorine content of rocks are probably invalid: The chloride measured derives from perspiration.

Tramp Iron. Samples prepared in mining laboratories, where iron contamination is often of little consequence and rapid reduction of large samples is necessary, are apt to contain large amounts of free iron: the CAAS syenite-1 contains about 0.3% iron as metal. In a conventional analysis, iron is reported as FeO and Fe_2O_3; when free iron is present, this convention leads to a high value for FeO and a high total. Of the several reference rock samples available in 1976, the USGS samples G-1, W-1, G-2, BCR-1, AGV-1, PCC-1, GSP-1 and DTS-1 are reasonably free from tramp iron contamination. The Tanganyika (Tanzania) tonalite T-1 contains about 0.1% free iron. A difficulty with tramp iron contamination is that it is prone to segregate; iron determinations on small subsamples are likely to exhibit a high sampling variance. In trace element analysis, the minor constituents of the tramp iron (Mn, Si, Cr, Mo, Ni, etc.) may contribute appreciably to the values obtained and also introduce subsampling variance due to segregation of the tramp iron particles.

Glass Containers. New glass bottles commonly used to contain samples may contaminate their contents with alkali and alkaline earth oxides (and silica). New glass surfaces carry submicroscopic blebs of extruded low-melting glass, which are contributed to materials enclosed. Adams [30] presents a clear picture of this source of contamination. Fabbi [31] has recognized that success in the X-ray spectrometric determination of light elements depends on removing this source of contamination by scrubbing the glass plates against which pelletized samples are pressed. All new glass should be scrubbed with a brush in soap and water, then washed with concentrated sulfuric acid containing chromic acid (not potassium dichromate), leached in an acid solution containing hydrogen peroxide, rinsed with distilled water, and oven dried.

Detergents, Dish-Washing Compounds, and so on. No responsible analyst should ever use a cleaning agent of which the composition is unknown. Many proprietary cleaners contain phosphates, silicates, and complex organics. Some leave glassware "sparkling clean" in appearance, with a film of silicate, phosphate, and organic material that may destroy an analytical procedure, especially in analyses for trace elements. The cleaning of glassware should be done with its application in mind, using materials of known composition and with known effect on the procedure to be employed.

Paper Towels and Wiping Cloths. Paper towels may contain one or more of a long list of elements and should never be used to wipe glass surfaces that may come in contact with samples or sample solutions. Even analytical-grade filter paper may contain enough zinc, potassium, chloride, or fluoride to spoil a procedure in which traces of these are to be determined. Glassine weighing papers carry large amounts of sodium. If it is necessary to use cloth or paper in an analytical procedure, it is advisable to ash a quantity of it and find out what elements it contains. Paper often contains large amounts of clay, titanium oxide, or barium sulfate.

Mixing Cloths and Papers. Thick, smooth glazed papers may appear to be ideal for mixing rock and mineral powders. They are not. The glaze may consist of a mineral layer that grossly contaminates any sample it touches. Rubberized mixing cloths may contribute elements such as antimony and sulfur, which are used as vulcanizing agents. Rough papers (e.g., brown Kraft paper) may contain almost any element in the periodic table in amount sufficient to disturb trace element concentration. Probably the best paper to use for mixing and splitting samples is the 100% rag tracing paper (Albanene) used by draftsmen. This may be obtained in sheets and rolls of convenient size. Its cost should not be a consideration; each batch should be sampled, and the sample should be ashed and analyzed spectrographically.

Crushing and Grinding Equipment. Contamination during grinding cannot be avoided. The composition of each and every implement used in crushing and grinding laboratory samples should be known. Steel mortars are, in general, undesirable; if they are used, they should be made of hardened plain-carbon steel. The design of steel mortars for laboratory use was the subject of a study at the University of Minnesota by Goldich and Sandell [32]. This study led to the design of Figure 1.8. Essential features of a

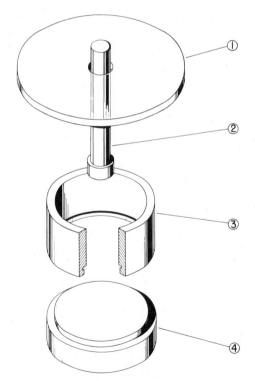

Figure 1.8. Steel Mortar for laboratory use: (1) Cover. (2) Pestle (SAE 1045 or 1055 steel). (3) Ring (SAE 1095 steel). (4) Base (SAE 1015 or 1020 steel).

useful steel mortar are that high-alloy steels be avoided, that there be no abrading surfaces (this rejects the common Plattner mortar), and that working surfaces must be as hard as possible. Construction of such a mortar is expensive. The pestle, made of SAE 1045 or 1055 steel, is hardened maximally at the working end and annealed at the other end (so that chips are not broken off during use). The ring should be of SAE 1095 steel and hardened by quenching, but not to the point where it will break on impact. The base is of SAE 1015 or 1020 steel, with the working surface deeply carburized and hardened maximally. After fabrication and heat treatment, the ring and the base must be precision ground to fit. The ring must not ride up on the base during use or sample will be lost. This requires very careful machining.

For silicates, agate mortars are best: The small amount of silica introduced from the mortar is often of no consequence. Boron carbide mortars are valuable when hard materials such as ilmenite or chromite are to be ground, and silica contamination is inadmissible. Composition mortars of various kinds are now offered in place of agate, which is in short supply: These must be used with caution. If their composition is unknown, chips should be examined spectrographically: One common variety contains appreciable chromium, which may invalidate trace determinations of this element.

In considering the innumerable grinding devices available, one must look not only at their grinding efficiency but also at the ease with which they may be cleaned after use and the extent to which sample is lost during and after grinding. A machine that grinds a sample in a few minutes without effort is attractive; if it retains a tenth of the sample in its internal parts and takes an hour to clean, it is useless.

Screens. Screens of brass or stainless steel contribute surprizingly little copper, zinc, or iron to samples shaken in them. The largest contamination is often from the solder used to fix the screen in its ring. It is probably best to always use silk or nylon screens whenever practicable if trace metals are to be determined. Unfortunately, during the early stages of reduction of large samples, it is difficult to keep track of possible metallic contamination from large screens.

Cleaning Equipment. A major source of contamination is the difficulty in cleaning sampling equipment. Machinery is difficult to clean in proportion to its complexity: this supplies the best reason to avoid mechanical sampling devices where possible. The mortar illustrated in Figure 1.8 may require 15 min of operation to reduce 100 g of a 1-in. material to 10 mesh, while a mechanical mill may require only a minute or two; but the mortar can be cleaned ready for the next sample in a minute, while the mill may require a half hour or more. If one considers also that the mortar will add negligible contamination, while contamination from the mill is uncontrollable, the advantage of the mortar is evident. All mechanical sampling devices should be disassembled after each use, and all the separate parts cleaned.

A motorized wire brush mounted over a sink is invaluable for cleaning mortars and pestles as well as parts of sampling machinery.

A supersonic bath is almost an essential in a laboratory where screens are in continuous use. The bath should be large enough to fully submerge the screens.

Nylon screens can be washed in soap and water; however, the dollar value of samples is often so high that justification of a new nylon screen for each sample is not difficult.

An air blast is often used to blow dust from sampling equipment, often without consideration of the final resting place of the dust. It is better to use a vacuum brush; most of the dust ends up in a receptacle instead of being randomly distributed throughout the working area. If an air blast is used, it should be in a hood that will take away the dust.

Cross-Contamination. Besides the obvious possibility of cross-contamination due to imperfect cleaning of sampling equipment, there are sources of cross-contamination that are not immediately obvious. Glass or plastic vessels that have contained or have been washed with materials containing fluoride, phosphate, or high metal concentrations may contribute these to solid or liquid samples. Plastic and glass vessels are capable of absorbing or adsorbing metals from solutions; reuse of such vessels may result in gross contamination. For example, Pyrex flasks used to contain fluoride-bearing solutions should not be used in any method that is sensitive to fluoride ion; polythene vessels that have contained chromate solutions will deliver chromium to any material stored in them; trace determination of mercury in natural waters is dependent for success on preventing the mercury from depositing on container walls, and reuse of such containers may induce error.

Plastic and other soft material may be impregnated with hard, sharp particles of materials being stored or treated and may deliver particles of these materials to other samples. Minerals may be electrostatically attracted to plastic surfaces and held so firmly that ordinary cleaning routines are ineffective. A molybdenite ore, for example, may turn its plastic container black; the sample is depleted in molybdenum, and a new sample placed in the same container may deliver a falsely high assay. Those engaged in sample preparation are sometimes insufficiently aware of the importance of such phenomena and process the samples they receive in such a way that the final assay values are badly distorted. The geochemical analyst must be sure that the reasons for bad values are well understood and should be in a position to track down and correct cross-contamination errors of the kind described. Reuse of containers is probably the greatest difficulty to surmount. Containers are sometimes expensive; if they look clean, it seems wasteful to discard them. One must balance the high cost of making sure they are clean enough for the purpose at hand against the cost of replacing them. This balance is often difficult to strike.

1.40. PREPARATION OF THE LABORATORY SAMPLE FOR ANALYSIS

When a mass of material has been efficiently sampled, mineral separations have been completed, unavoidable contaminations have been measured, and the analyst has received a small vial containing the results of all these efforts, he or she must decide on the preliminary steps to be taken prior to analysis for the constituents of interest.

The first step should usually be a microscopic examination. This will often give much useful information in a very short time. Under a binocular microscope, the presence of a relatively few grains of a minor mineral rich in the constituent of interest will warn of subsampling difficulties. A judgment of the need for further grinding, screening, or purification can be made. Simple chemical tests can be made in a few moments: CO_2, P_2O_5, and S^{2-} may be quickly detected and estimated. Organic matter or graphite may be noted. A magnet will show the presence of iron or magnetite, and a test with copper sulfate solution will distinguish between the two. Metallic flecks may be detected and identified. Most important, isolated grains of minor constituents (columbite in carbonatite, chromite in dunite, etc.) will be noted; this may save much confusion by avoiding irreproducibility in analysis for trace and minor elements.

1.41. SCREEN TESTS: THE TYLER STANDARD SCREEN SCALE

The Tyler standard screen scale has as its base an opening of 0.0029 in. (0.0103 cm): Other openings increase or decrease in the ratio of $\sqrt{2}$. For finer sizing, intermediate sieve openings increase in the ratio $\sqrt[4]{2}$. The Tyler scale is not the same as the U.S. scale, but the difference is largely one of nomenclature. The Tyler Handbook 53, *Testing Sieves and Their Uses*, gives full information on various standard sieve series [33].

Whenever possible, samples submitted for analysis will be first passed through an appropriate screen. The material must be fine enough for adequate subsampling but should not be ground any finer than is necessary. Routine application of the sampling constant principle serves as a guide. Sample treatment depends on several factors, among which are the accuracy required, the integrity of the sample as received, the subsample weight to be used for the analysis, the concentration(s) of the element(s) sought, and the feasibility of replicate determinations. When replicate determinations are not practicable, it is mandatory that sample preparation be carried out as though the sample were a worst case, that is, as though many ore mineral particles retain the mesh size of the screens through which they are passed.

REFERENCES

1. P. Gy, L'Echantillonage des Minerais en Vrac, *Rev. Ind. Min.*, **1** (Special issue; January 15, 1967).
2. P. Gy, *Rev. Ind. Min.* 15 Septembre, 1971. A pamphlet "The sampling of broken ores—a review of principles" has been compiled by Pierre Gy, consulting engineer, Cannes, France, summarizing, in English, some of his works.
3. P. Gy, "Contribution a l'etude de l'heterogeneite d'un lot de matiere morcelee," These de Docteur-Ingenieur, Université de Nancy, 1972.
4. C. O. Ingamells and P. Switzer, "A Proposed Sampling Constant for Use in Geochemical Analysis," *Talanta* **20,** 547–568 (1973).
5. J. Visman, "A General Sampling Theory," *Mat. Res. Stds.* **9**(11), 8 (1969).
6. G. Matheron, "Principles of Geostatistics," *Econ. Geol.* **58,** 1246–1266 (1963). Reference is made to Matheron's *Treatise of Applied Geostatistics* in 3 volumes, Editions Technip., Paris. 1962
7. D. C. Krige, *S.A. Inst. Min. Met. J.* **61,** 231 (1960). Krige's original work is in Afrikaans; synopses of it occur throughout the geostatistical literature.
8. A. D. Wilson, "The Sampling of Silicate Rock Powders for Chemical Analysis," *Analyst* **89,** 18–30 (1964). Errors in silicate analysis due to the heterogeneity of the powdered laboratory sample are dependent on the distribution of the various mineral particles in the powder. In a mathematical appendix, expressions are derived that define sampling errors. Practical aspects of the problem are emphasized. For major constituents of silicate rocks, a 1-g sample of 72-mesh powder is usually adequate, but for minor constituents a larger sample, or a more finely ground sample, is required. In extreme cases, e.g., gold in quartz, very large samples may be necessary if sampling error is to be reduced to a tolerable level.

9. A. W. Kleeman, "Sampling Error in the Chemical Analysis of Rocks," *J. Geol. Soc. Australia* **14**, 43–48 (1967). The size of a representative sample of rock powder may vary considerably for each constituents. Collecting a sample of rock from its environment may be a larger problem than its sub-sampling for analysis. The work of Wilson is simplified to the point where it may be comprehensible to geologists.

10. H.-P. Aubauer, "Two-phase Systems with Dispersed Particles of a Stable Size," *Acta Metall.* **20**, 165–180 (1972).

11. I. Barnes, "Geochemistry of Birch Creek, Inyo County, California—a Travertine Depositing Creek in an Arid Climate," *Geochim. Cosmochim. Acta* **29**, 85–112 (1963).

12. I. Barnes and F. E. Clarke, Chemical Properties of Ground Water and Their Corrosion and Encrustation Effects on Wells, U.S. Geol. Surv. Prof. Paper 498-D, D1-D58, 1969. "Of interest is the demonstration that simple stable equilibrium models fail in nearly every case to predict composition of water yielded by the wells studied. Only one stable phase (calcite) was found to exhibit behavior approximately predictable from equilibrium consideration."

13. R. W. Kistler, U.S. Geological Survey, personal communication. Repetitive K_2O determinations on a white mica, using 100-mg subsamples, varied from 7.59 to 8.90 in seven determinations. Microprobe determinations of magnesium varied by a factor of 2 within the same crystal.

14. E. F. Cruft, C. O. Ingamells, and J. Muysson, "Chemical Analysis and the Stoichiometry of Apatite," *Geochim. Cosmochim. Acta* **29**, 581-597 (1965).

15. C. O. Ingamells and J. Gittins, "The Stoichiometry of Scapolite," *Can. Mineral.* **9**, 214–236 (1967).

16. S. B. Weed and R. A. Leonard, "Variation in Properties within Single Flakes of Hydrobiotite-Vermiculite," *Soil* **104**, 416–421 (1967).

17. J. C. Engels, "Effects of Sample Purity on Discordant Mineral Ages Found in K–Ar Dating," *J. Geol.* **79**, 609–616 (1971).

18. W. J. Youden, "The Role of Statistics in Regulatory Work," *J.A.O.A.C.* **50**, 1007–1013 (1967).

19. D. J. Ottley, "Gy's Sampling Slide Rule," *World Mining* **19**(9), 40 (1966). A most valuable summary of Gy's sampling theory and the use of the slide rule, supplied with American orders for the slide rule, is an expanded version of this *World Mining* article.

20. A. J. Duncan, *Mat. Res. Stds.* **11**(1), 25 (1971).

21. J. Visman, A. J. Duncan, and M. Lerner, *Mat. Res. Stds.* **11**(8), 32 (1971).

22. J. Visman, *J. Mat.* **7**(3), 345 (1972).

23. P. Switzer, Stanford University, personal communication.

24. P. K. Samland and C. O. Ingamells, "New Approaches to Geostatistical Evaluations," Proc. 12th Comp. Appl. Conf. Denver, Colo, 1974, Vol. II.

25. R. A. Landy, "Variation in Chemical Composition of Rock Bodies: Metabasalts in the Iron Springs Quadrangle, South Mountain, Pennsylvania. Ph.D. Dissertation, The Pennsylvania State University, University Park, 1961.

26. A. T. Miesch, Theory of Error in Geochemical Data, U.S. Geological Survey, Paper 574-A-A1-17, 1967.

27. A. K. Baird, D. B. McIntyre, E. E. Welday, and D. M. Morton. A Test of Chemical Variability and Field Sampling Methods, Lakeview Mountain Tonalite, Lakeview Mountains, Southern California Batholith, Short contributions to *California Geol.*, Spec. Rept. 92, California Div. Mines and Geol., 11–19 (1967).

28. C. O. Ingamells, "Lithium Metaborate Flux in Silicate Analysis," *Anal. Chem. Acta,* **52,** 323–334 (1970).

29. J. C. Engels and C. O. Ingamells, "Effect of Sample Inhomogeneity in K–Ar Dating," *Geochim. Cosmochim. Acta* **34,** 1007–1017, Pergamon Press (1970).

30. P. B. Adams, "The biology of Glass," *New Scientist* **41,** 25, 316 (1969).

31. B. P. Fabbi, *Appl. Spectry.* 25 (1971).

32. E. B. Sandell, "Contamination of Silicate Samples Crushed in Steel Mortars," *Anal. Chem.* **19,** 652–653 (1947).

33. *The Tyler Handbook 53,* "Testing Sieves and Their Uses," W. S. Tyler, Incorporated Screening Division, Mentor, Ohio.

CHAPTER

2

THE ART OF CHEMICAL ANALYSIS

2.1. Tools of the Art
    2.1.1. Analytical Balance
           Single-Deflection Weighing
           Single-Swing Weighing
           Equality of Beam Arms
           Calibration of Weights
           Ultimate Weighing
           Location of Balance
           Calibration of Weights
           Determination of Air Density
    2.1.2. Laboratory Ware
           Volumetric Apparatus
           Beakers and Other Glassware
           Platinum Ware
           Gold and Silver Ware
           Zirconium, Nickel, and Iron Crucibles
           Graphite and Vitrified Carbon Crucibles
           Porcelain Ware
           Silica Glass: Vitreous Silica
           Desiccators and Desiccants
2.2. Reagents
    2.2.1. Distilled Water
    2.2.2. Acids
    2.2.3. Ammonia
    2.2.4. Organic Reagents
           Aluminon
           Purification of Aluminon
    2.2.5. Fluxes
           Lithium Metaborate, $LiBO_2$
           Purification of Lithium Metaborate
           Procedures Using Lithium Metaborate
           Sodium Carbonate
           Potassium Pyrosulfate
           Sodium Peroxide
           Potassium Carbonate
2.3. Instrumentation
    2.3.1. Emission Spectroscopy

2.3.2. Emission Spectrometry

2.3.3. The Internal Standard Principle

2.3.4. Inductively Coupled Plasma Used as Source in Emission Spectrometry

Physical Description

Analytical Characteristics

Interferences Encountered in Plasma Spectroscopy

Chemical Interferences

Spectral Interferences

How To Minimize Spectral Interferences

Advantages and Disadvantages of ICP

2.3.5. X-ray Methods

2.3.6. A Pelletizing Die

2.3.7. Atomic Absorption Spectroscopy

Principle

Nebulization

Atomization

Drift and Background Noise

Ratio of Signal to Background Noise

Detection Limit

Flame Type and Stoichiometry

Lamp Current

Slit Width

Flame Optimization

Burner Position

Reproducibility

Calibration Curve and Limit of Linearity

Acid Concentration

Concentration of Other Elements

Choice of Wavelength

Interferences

Solutions to Interference Problems

2.4. Dissolution Procedures

2.4.1.   Lithium Metaborate Fusion, Solution in Dilute Nitric Acid

2.4.2.   Sulfuric–Hydrofluoric Acid Attack

2.4.3.   Perchloric–Hydrofluoric Acid Attack

2.4.4.   Attack with Sodium Peroxide, Solution in Acid

2.4.5.   Sodium Hydroxide Fusion, Solution in Acid

2.4.6.   Solution in Phosphoric Acid

2.4.7.   Sodium and Potassium Pyrosulfate, Solution in Sulfuric Acid

2.4.8.   Sodium Carbonate Sinter, Solution in Hydrochloric Acid

2.4.9.   Sodium Peroxide Sinter, Solution in Hydrochloric Acid

2.4.10. Fusions Using Ammonium Salts

2.4.11. Solution Procedures for Specific Minerals in Rocks

2.4.12. Sodium Carbonate Fusion

References

2.1. TOOLS OF THE ART

2.1.1. Analytical Balance

The analytical balance is the central and indispensable implement in all quantitative analytical work. No operation can be performed without its aid, and the accuracy of all analytical operations is ultimately dependent on its proper use.

A good analytical balance is capable of making measurements with a precision of 1 ppm and in this respect is almost unique among instruments. Full use of this remarkable capability requires, however, an awareness of numerous factors that may affect the weighing operation: Some of these factors are not always given sufficient attention, and as a consequence, inadvertent errors in weighing are all too frequent.

In ordinary weighing, the actual downward force exerted by the object being weighed is

$$F = G \frac{M(m - b)}{D^2} \tag{2.1}$$

where G is the gravitational constant, M is the effective mass of the earth, D is the distance between the earth's center of gravity and that of the object being weighed, m is the mass of the latter, and b is the diminution of that mass due to the buoyant effect of the air it displaces.

The fact that D and b are not strictly constant is often ignored: This can lead to error. The wide acceptance of single-pan balances has helped obscure the source of such errors.

Forces other than those implicit in (2.1) may affect the accuracy of a weighing. Of these, the analyst must be aware. A force may exist between magnetic materials and ferrous structural components of the balance. Static electric charges must be dissipated if accurate weighing is to be achieved (as much as a 30-mg error has been observed in weighing a 20-g glass tube that has become statically charged by drying with a paper towel). Air currents within the balance case may create an upward or downward draft of air. Such air currents have three common causes: The object being weighed has not come to temperature equilibrium with the air in the balance case; the case may be exposed to unbalanced radiant energy from a light source or a radiator; the balance may be mounted in an area having a temperature gradient, for example, in the corner of an uninsulated room. The means for overcoming such effects are obvious, except perhaps for eliminating the effects of static charge.

Static charges are easily induced on glass and plastic by wiping or rubbing with cloth or paper. They need to be run to ground. Several devices for doing this are effective. A piece of radioactive mineral (e.g., carnotite) may be kept in the balance case: This acts too slowly (through ionization of the air) to have much effect on the weighing of a charged piece of apparatus but effectively prevents the accumulation of static charges on the balance parts (e.g., the glass plate that forms the floor of the case). If a piece of chamois leather is kept in a desiccator containing water, wiping a piece of glass to be weighed with this damp chamois will usually remove static charge. Whenever a piece of apparatus made of nonconducting material is to be weighed, it is advisable to check the weighing after making an effort to remove any static charge it may carry.

A weighing error that is usually additive to the convection effect mentioned above is that due to the expansion of gases with temperature. A covered crucible containing air that is warmer than that in the balance case has an apparent weight less than the correct apparent weight. The error amounts to about 0.1 mg/°C for a 25-ml crucible. It is only necessary to hold a covered platinum crucible between thumb and forefinger for a few seconds and note the apparent loss in weight to be convinced of the reality of this effect. It may take several minutes for a slightly warm crucible to come to temperature equilibrium in a balance case.

The existence of weighing errors due to poor balance location is best detected by repeatedly weighing an object (preferably a 25-ml platinum crucible) over a period of several days. The object should be kept in a desiccator near the balance during the test. The greatest permissible variation in apparent weight should be no greater than can be explained by changes in air density (for a 25-ml platinum crucible, it should be no more than 0.1 mg). A badly located balance in perfect working order may show a variation in apparent weight of a 25-ml platinum crucible approaching a milligram or more. Plainly, accurate work is impossible under such circumstances.

Changes in apparent weight occur with changes in air density due to fluctuations in barometric pressure and humidity. The metal in a 25-ml platinum crucible displaces somewhat more than 1 cc of air, and between extremes of air density, the mass of this volume of air may change by almost 0.1 mg. A porcelain crucible of the same capacity has at least 10 times the physical volume, and its apparent weight may change by nearly a milligram under the same circumstances. The air density effect on the counterbalancing weights must, of course, be considered as well: If the object to be weighed is made of a material with the same specific gravity as the weights, the effect will exactly cancel. If the object to be weighed is more dense than the weights (e.g., platinum, sp. gr. ~22), its apparent weight increases with increasing air density; if it is less dense than the weights (e.g., porcelain, sp. gr. ~2.4), its apparent weight decreases with increasing air density. The specific gravity of ordinary analytical weights is about 8.

In practice, all this means is that a 25- or 30-ml platinum crucible is about the largest object, in point of physical size, that may be safely weighed to the nearest 0.1 mg without in some way compensating for the air buoyancy effect [1].

When physically large pieces of equipment are used—for example, the absorption bulb in a gravimetric carbon determination, a pycnometer, an alkalimeter, or a weight buret—air buoyancy effects can introduce a major error if atmospheric conditions change between successive weighings (see Table 2.1). With single-pan balances, it is difficult to counter such effects: They are therefore most useful in air-conditioned environments held at constant humidity, in regions where barometric pressure does not change drastically, and in analytical operations that do not require the weighing of the same article over an extended period of time [2].

The sources of weighing error discussed above are all more or less independent of the type of balance used. Other errors may occur as a consequence of peculiarities in balance design. The many types of balance

Table 2.1. Air Buoyancy Effects with Major Error in Weighing Light Bulb if Atmospheric Conditions Change between Successive Weighings with Single-Pan Electronic Balance

Measurement Number	Date	Weight	Weather Conditions
1	3-15-83	23.1803	End of snow storm
2	3-18-83	23.1814	Light snow
3	3-22-83	23.1803	Cold and cloudy
4	3-28-83	23.1819	Cloudy and mild
5	4-07-83	23.1769	Cool and sunny
6	4-11-83	23.1835	Cloudy, warm, and windy
7	5-17-83	23.1799	Snow storm
8	12-28-84	23.1838	Cool, dry, and sunny
9	1-04-85	23.1796	Very cold, dry, and sunny
10	1-09-85	23.1824	Very cold and cloudy
11	1-11-85	23.1791	Cold and light snow
12	1-16-85	23.1818	Cold, dry, and sunny
13	1-21-85	23.1810	Very cold and cloudy
14	1-22-85	23.1805	Cold and sunny
15	1-25-85	23.1789	Cold and sunny
Mean		23.1808	
Standard deviation		0.0018	

now in common use differ markedly in their susceptibility to the various causes of error. There can be little doubt that a simple equal-arm balance, uncluttered with damping and automatic weight-changing devices, is least likely to suffer from error-inducing defects.

Disadvantages in the use of a simple balance are sometimes overemphasized, especially by those interested in selling a more complex and much more expensive device. A simple balance in first-class condition (this requirement is so obvious that we apologize for mentioning it!) and equipped with a 100-mg chain weight as the only concession to automation can perform at least as quickly as any of the more complex instruments when a sensitivity of 0.1 mg is adequate. To be sure, some skill is needed, but this is easily acquired.

It is quite unnecessary to take the time to "count swings" in order to attain 0.1 mg accuracy with such a balance. A single-deflection weighing is more than adequate; the precision of the chain weight is unlikely to be better than 0.05 mg, and the single-deflection weighing is easily capable of this precision.

The symmetrical equal-arm balance offers the possibility of easily overcoming two of the most common weighing difficulties—the temperature effect and the air buoyancy effect—by the simple expedient of using a tare, which should be as nearly as possible the same shape, size, and material as the object weighed and should be carried through all heating operations to which the latter is subjected. With single-pan balances, the use of a tare requires two separate weighings, and the temperature effect may not be effectively canceled.

Damping devices on a high-quality balance are usually undesirable. Magnetic dampers are capable of interfering with the weighing of ferrous materials; air dampers are prone to trap dust between piston and cylinder and are very sensitive to temperature changes. They do not result in any appreciable gain in the speed of an accurate weighing. They may be of some advantage when a balance is used in a specific routine operation in which many weighings of moderate accuracy are required with the greatest possible speed, especially when these must be performed by semiskilled personnel.

Many balances use the *aperiodic* principle (especially single-pan balances). The final figures in the weight are derived by measuring the deflection of the beam from its central position, the object being weighed conterbalanced only partially. Such balances always have damping devices. Many of them work well in controlled atmosphere environments, but they are chronically sensitive to temperature changes because of the near impossibility of designing a beam and pan system (especially a nonsymmetrical system) with a center of gravity unaffected by temperature.

The *constant-load* device commonly employed in such balances (weights are taken off rather than added) overcomes a major difficulty of the aperiodic principle, that of change of sensitivity with load, but the larger mass of the moving parts results in slower response.

Automatic weight-changing devices may appear to be a convenience, but they have several disadvantages. First among these is the difficulty associated with keeping the weights and supporting mechanism clean. With these devices, calibration of weights is difficult—use of the convenient substitution method is in some cases a practical impossibility. The unavoidable lack of symmetry makes the balance very susceptible to humidity effects, and changes in air density may lead to a shifting rest point.

In all of what follows it will be presumed that weighings can be made on a simple two-pan balance equipped with a 100-mg chain weight: Only where gross error may develop from the use of other types will mention be made of them.

Single-Deflection Weighing

A two-pan balance with separate beam and pan arrests is required for proper application of this technique. The balance sensitivity should be set so that one division on the pointer scale is equivalent to 0.1 mg when the balance is not under load. The rest point should be set two to four divisions off center so that when the beam and pans are released, the pointer will swing to about the fourth division on the pointer scale. The pan release mechanism must be adjusted so that no push is given to either pan when the button is depressed. The point to which the needle swings is taken as the rest point of the balance, and subsequent weighings are made by adding weights until just such a swing occurs. With a chain weight, a weighing need not take more than a minute or so—less if (as is often the case) the gross weight is known within a few milligrams. The chain weight is manipulated with one hand and the pan release button with the other; after a little practice, it will be found unnecessary to wait for the pointer to complete its single swing before the need to add or remove weight becomes apparent. A final check of the rest point should be made after lifting and relowering the beam to be sure that the knife-edges have not moved on their bearings during the operation.

Single-Swing Weighing

Presuming always that only balances in perfect operating condition are to be considered, the single-swing weighing is less desirable than the single-deflection weighing: It takes more time and is probably no more precise. It may be used when the balance does not have separate beam and pan arrests or when the pan arrest cannot be adjusted adequately for a single-deflection weighing.

For a single-swing weighing using a chain weight, the beam is started swinging after adding weights to within a milligram or so of the correct weight, and the chain is adjusted until the needle swings as many divisions to the right as to the left. Often it may be convenient to reach a preliminary balance using the single-swing method and to make a final adjustment using the single-deflection method.

With practice, it is possible to use a combination of these methods and achieve greater weighing speed than is possible with the best single-pan balance.

Equality of Beam Arms

It is impossible to manufacture a balance in which the two arm lengths are exactly the same: It is only a question of how closely this ideal is approached. If the same balance is used in all operations, a difference in arm length is immaterial because all weighings are proportionately affected. However, if other balances take part in the operation—and this is the usual situation because those used in factory calibration of, for example, glassware must be included—error may be introduced. It is therefore desirable to determine the beam arm ratio if only to be sure that no error can be introduced on this account. To do this, two identical weights are required. One is placed on each pan, and the change in rest point, or the apparent difference in weight, noted. To establish the identity of the two weights, their positions are interchanged. Two 10-g weights from a calibrated set are convenient for this test. With asymmetrical balances, the test should be made with both 10-g and 100-g weights.

PROCEDURE. Determine the rest point of the balance. Put identical 10-g weights on each pan and adjust to the same rest point using the chain weight. If x is the chain reading, all future weighings should be multiplied by a factor of $10/(10 + x)$. Unless the two 10-g weights are known to be identical, the procedure should be repeated with the weights interchanged, and the average of the two values of x so obtained should be used. Note that this procedure involves an approximation: It is assumed that x will be so small in relation to the total weight involved that no correction on its account is necessary. Note also that x may have a negative value (the right arm may be the longer): In such a case, the balance should be brought to a rest point with a few milligrams showing on the chain weight scale, and x should be measured from this setting.

Most analytical balances worth the name will not show a sufficient difference in beam arm lengths to require any special attention: The situation may be different with larger, heavier, and less expensive balances that may be used for the calibration of volumetric glassware and for similar operations. Such balances should always be tested in the manner described.

It is important to recognize that the beam arm ratio is affected by other factors than the obvious structural one. A temperature gradient within the balance case or a condition where more light strikes one side of the balance than the other may affect the ratio appreciably. It is very likely that the effective ratio will change with load under such conditions, especially with asymmetrical balances. So little time is required to make sure of the absence of undesirable effects of this kind that there can be no excuse for tolerating the possibility that they exist.

Calibration of Weights

Calibration of weights is best done on a simple two-pan balance without a chain weight. The chain weight is not capable of a precision better than about ±0.05 mg, and a good analytical balance with a 5-mg rider traveling the full length of the beam will easily attain ±0.02 mg. However, a set of weights used routinely with a chain balance should be checked periodically, and a quick routine for doing this is desirable. It is best to keep a set of calibrated weights that are used only for this purpose: Weighing each of these in turn, using the working weights, and using various combinations of the calibrated weights to check chain and rider takes only a few minutes and immediately shows any deterioration of the working weights. Balances with automatic weight placement devices, single-pan balances, and aperiodic balances of all kinds should be frequently checked using weights calibrated as described below.

Most balances with a chain weight are equipped with a notched beam carrying a rider with an effective weight of 1 g so that only the 1-g and heavier weights of a set need be used. The chain weight and the rider are most subject to deterioration and need frequent checking against a reference set. They should always be checked before attempting the calibration of larger weights using the chain balance. The following procedure is recommended: It assumes that the beam rider has an effective weight of 1.00000 g, that the chain weight is without gross error, and that the balance arms are essentially equal.

PROCEDURE. Begin using the 5-g; 2'-g, 2''-g; and 1 g weights. Put a small object (about 10 mg) on the left pan and leave it there during the following operations:

Its purpose is to make sure that readings will always be possible on the chain weight scale.

1.　Using the chain, bring the balance to its rest point. Call this initial chain reading R.

2.　Put the 1-g weight on the left pan and the rider in the 1.0 notch. Bring the balance to its rest point using the chain. Call this chain reading W_1. Calculate $D_1 = W_1 - R$.

3.　Return the rider to the 0.0 notch and obtain a chain reading W_2 with the 2'-g, weight on the left pan and the 2"-g on the right pan. Calculate $D_2 = W_2 - R$.

4.　With the rider in the 1.0 notch, the 1-g weight on the right pan, and the 2'-g weight on the left pan, obtain a chain reading W_3 and calculate $D_3 = W_3 - R$.

5.　With the rider in the 1.0 notch, the 2'-g and 2"-g weights on the right pan, and the 5-g weight on the left pan, obtain a chain reading W_4 and calculate $D_4 = W_4 - R$.

The actual weights (referred to the rider as 1.00000) are then found as follows:

$$1\text{-g weight} = 1.00000 - D_1$$

$$2'\text{-g weight} = 2.00000 + D_3 - D_1$$

$$2''\text{-g weight} = 2.00000 + D_3 - D_2 - D_1$$

$$5\text{-g weight} = 5.00000 - D_4 + 2D_3 - D_2 - 2D_1$$

Weigh the 10-g weights against the 5-g; 2'-g; 2"-g; and 1-g weights, and treat the 20- and 50-g weights similarly.

A refinement of the procedure is to make two weighings for each counterbalance of weights, interchanging them so as to take into account the possibility of inequality in the balance arms. However, the rider and the chain weight cannot be included in such an operation so that it is usually not worthwhile. This simple procedure does not constitute a complete calibration: It is used as a check only. It requires that the rider and chain be in good condition. Checking the rider and chain may be done very simply by weighing calibrated weights or by a substitution method:

PROCEDURE. Place an object weighing about 1.1 g on the left pan. Counterbalance it with calibrated weights ranging from 10 to 500 mg. Remove these weights one at a time in various combinations, substituting them with suitable portions of the chain and rider. In this test, the chain should vary from its ideal weight by no more than 0.1 mg along its whole length, and the rider should give reproducible readings when it is turned in its notch and weighed repetitively.

Sometimes, after long use, the beam notches become worn, and reproducible weighings are impossible. In such cases, it is best to discontinue the use of the rider, and resort to a set of calibrated fractional weights (500, 200, 200, and 100 mg). These fractional weights should be checked from time to time using the method described above for the 5-, 2-, 2-, and 1-gram weights.

The ordinary analytical balance equipped with a chain weight is not, in practice, capable of weighing accurately and reliably to much better than ± 0.1 mg. If it is necessary to extend weighing accuracy to ± 0.02 mg, the chain device must be dispensed with. However, in most "macro" work—even of the highest caliber—weighing to better than ± 0.1 mg is not meaningful. Besides the numerous difficulties associated with the weighing operation itself (air buoyancy effects, temperature effects, humidity effects, etc.), which begin to have significance at the ± 0.1-mg level, the limited accuracy of chemical operations and the inexact stoichiometry and impurity of even the most carefully prepared weighing forms seldom justifies extension beyond the ± 0.1-mg level of precision. In many instances, the increased speed of weighing with a chain weight more than compensates for its limitations; for example, an ignited ammonia precipitate containing a large proportion of Al_2O_3 avidly gathers moisture and, even in a closed crucible, must be weighed quickly. Weighing to ± 0.02 mg using an analytical balance is generally a slow operation and cannot in ordinary practice be usefully performed when the weighing form is at all hygroscopic.

In microchemical work, the situation is of course quite different: Here, the whole scale of operation is diminished, the various effects mentioned above are reduced proportionately, and special techniques are used—many of which are not available on the macro scale.

Ultimate Weighing

Weighing with maximum accuracy is beset with difficulties. Besides the common difficulties mentioned above, extension of accuracy beyond the 0.1-mg level requires consideration of factors that are not relevant at that level. Stoichiometry of weighing forms is often exact to 1 part in 1000 but may not be exact to 1 part in 10,000. A good set of Class S weights may be presumed exact to 0.1 mg but not to 0.01 mg. Air buoyancy and temperature effects are relatively easy to control to 0.1 mg but not to 0.01 mg. The differing densities of sample and weighing form are nearly always quite immaterial when 1:1000 accuracy is adequate, but not necessarily at the 1:10,000 accuracy level. If absolute mass instead of relative weight is required, difficulties are compounded.

The first essentials to highly accurate weighing are a balance of superior quality in perfect working order located in an ideal position; a set of accurately calibrated weights; and equipment for measuring air density, or, at least, changes in air density. These essentials are dealt with in turn.

Location of Balance

The balance requires a table or other support that is level and free from vibration. Temperature gradients in the vicinity of the balance must not exist. Temperature changes must not be excessive: Balances differ greatly in their susceptibility to error on this account. Illumination must be symmetrical, and artificial lights must not be close enough to the balance case to cause any marked rise in temperature therein. Humidity must be controlled, especially if the balance construction is not exactly symmetric. Balances with lacquered parts are especially sensitive to humidity changes.

Undesirable vibration may be detected by placing a shallow dish of mercury on the balance support and reflecting a beam of light from it onto the ceiling. Temperature gradients may be found by blackening the bulbs of identical thermometers and hanging them at opposite ends of the balance case; uneven illumination may also be detected in this way. Effects due to humidity changes may be discovered by noting the change in rest point when a container of water (or a desiccant in humid climates) is left in the closed balance case for a few hours.

The final test of a balance location is made by repetitively weighing an object (preferably of stainless steel or brass, to avoid air buoyancy complications) at different times during several days. If the weight is not reproducible, the cause almost certainly involves one or more of the factors mentioned above. In an actual test of a balance intended for weighing to ±0.05 mg, a platinum crucible varied in apparent weight by as much as 2.5 mg, always showing the greater weight in the early afternoon. The difficulty was traced to a shaft of sunlight, very diffuse and not easily noticed, that struck one side of the balance.

Calibration of Weights

The rough weight calibration described above, while practical for routine checking of a chain balance, is inadequate for weighing of high accuracy. It is limited mostly by intrinsic inaccuracies of the chain device and, as written, does not adequately consider differences in arm length. For accurate calibration, a rider balance without a chain should be used. The rider should preferably weigh 5 mg and travel from end to end of a beam graduated to 0.05 mg. Its use during weight calibration should be avoided as much as possible.

The principle used in the method to be described is similar to that used in the rapid practical method. Differences between nominally identical weights are determined by observing changes in the rest point of the balance. Effect of beam arm inequality is canceled by making each comparison of weights twice, with the weights interchanged between the pans. All weights are measured in terms of a standard weight—which may be one of the set being calibrated if only internal consistency is required or one certified by an agency (e.g., the National Bureau of Standards) with access to primary standard weights.

The rest point of the balance is obtained during each reading by the method of "counting swings." Readings on the pointer scale are estimated to the nearest 0.1 division. It is convenient to mentally divide the scale from left to right into 200 divisions numbered from 0 to 200. The beam is released in such a way that it swings with an amplitude of at least 100 of these divisions. Swinging may be increased or diminished by waving the fingers at a good distance from the appropriate balance pan, creating a small air current. After closing the door and after four or five preliminary swings, three swings to one side and four swings to the other are observed and recorded as in the following example. Obviously, trivial variations of this scheme are possible.

Swings left	Swings right
66	172
68	170
71	167
74	
4)279	3)510
70	170
	− 70
	100

In this example, the rest point is exactly at the middle of the pointer scale but is recorded as 100; the advantage of this procedure is that there is no chance of confusing signs when calculating the results.

PROCEDURE:

1. Determine the rest point of the empty balance.
2. Put nominally equal weights on each pan and determine the rest point.
3. Add a 1-mg (or 2-mg) weight to one pan or the other, and determine

the rest point. Calculate the sensitivity of the balance in divisions per milligram at each load.

Important: Sensitivity must be assumed to change with load and must therefore be redetermined during each double weighing.

4. Remove the small weight, and proceed to compare pairs of weights and combinations of weights, in every case interchanging pans and recording two rest points.
5. Calculate the differences in mass between the several pairs of weights or combinations of weights.

When all the necessary weight differences have been accumulated, calculation of the true mass of each piece relative to one special weight taken as a standard or reference is possible. It is easiest to perform the calibration in several steps and to use auxiliary weights where necessary, as indicated below. Divide the set of weights into groups in the following manner, or in any of several possible alternate ways.

Group I: 50   20′ $\dfrac{10'\ \ 10''}{20''}$ $\dfrac{5\ \ 2'\ \ 2''\ \ 1}{10'}$ 10″

(i.e., group 10′ and 10″ together, also 5, 2′, 2″, 1: Keep these groupings intact throughout this step)

Group II: 5   2′   2″   1′   1″

(note the need for an auxiliary 1-g weight)

Group III: 0.5   0.2′   0.2″   0.1′   0.1″

Group IV: 0.05   0.02′   0.02″   0.01′   0.01″

Group V: 0.005   0.002′   0.002″   0.001′   0.001″

Procedure for each group is essentially the same as in the rapid practical method, with the refinements mentioned. Four differences are obtained, and hence the relative weights in each group in terms of 10′, 1′, 0.1′, 0.01′, and 0.001′. The several groups are then compared and each is multiplied by a factor to bring the whole set into internal consistency. Comparison of any one of the weights with an externally derived standard and further multiplication by a factor completes the calibration. A factor for correction of error due to unequal beam arms may also be applied when the calibrated weights are always used on the same balance.

It has been assumed that all the weights in the set are close enough to their nominal value that the differences can be found without use of the 5-mg rider. Calibration of the rider against a 5-mg weight is done without difficulty. An adjustment of the calculations may be required if it turns out that the 1-mg weight used in sensitivity determination differs appreciably from its face value. It may therefore be desirable to calibrate the smaller weights first, unless there is some assurance that the 1-mg (or 2-mg) weight is accurate.

Determination of Air Density

The only apparatus needed for this determination is a (preferably evacuated) sealed glass bulb of known volume (about 100–200 cc) and known mass. A perfectly adequate makeshift bulb can be obtained by removing the metallic parts of an ordinary light bulb, retaining one of the sealed-in copper wires so that it can be fashioned into a hook for suspending the bulb above the balance pan. The volume of the bulb can most easily be found by measuring the volume of water it displaces from a suitable vessel. Determination of its mass is more difficult: It may be weighed in vacuo if equipment is available or the true mass may be calculated from the observed mass and an independent air density measurement.

In many common applications, correction of weighings of any vessel at different times does not require knowledge of the absolute mass of the bulb. The apparent weight of the bulb and of any object being weighed is its true weight less the weight of the air it displaces plus the weight of the air displaced by the balance weights:

$$W_a = W - Vd_a + vd \qquad (2.2)$$

$$= W - \frac{Wd_a}{d} + \frac{Wd_a}{d_w}$$

where W_a is the apparent weight of the bulb or other object, W is its true weight, V is the volume of air it displaces, d_a is air density, and v is the volume of the weights that counterbalance the bulb.

The difference in apparent weight at two air densities d'_a and d''_a is

$$W'_a - W''_a = W(d'_a - d''_a)\left(\frac{1}{d_w} - \frac{1}{d}\right) \qquad (2.3)$$

Air density depends on the composition of the atmosphere (the greatest variable being its water vapor content), on pressure and temperature, and on the local value of the gravitational constant. Thus, for any given object at a single location, there is a linear relationship between changes in apparent weight and changes in air density. This makes it easy to correct for atmospheric changes between weighings of crucibles, bulbs, and other objects. It is only necessary to weigh the calibration bulb every time an unknown is weighed. The ratio between the changes in the apparent weight of the bulb and of the unknown (due to air density change) is a constant that depends only on the densities of the several objects involved—the unknown, the calibration bulb, and the weights.

It is very often not necessary to find an absolute weight for objects being weighed in the course of an analytical procedure; but it is essential that changes in observed weight due to air density changes be kept in consideration.

The easiest and most practical way to avoid error due to such changes is to carry a tare as close in form and composition as possible to the object to be weighed throughout all operations. It will then seldom be necessary to apply air density corrections in the manner suggested above. A notable exception is the determination of mineral or rock density using a pycnometer; in this instance, such corrections are mandatory.

2.1.2. Laboratory Ware

Volumetric Apparatus

Volumetric apparatus is second only to the balance in importance. Each piece should be calibrated; even if calibrated glassware is purchased, the analyst should confirm the calibration with his own balance and weights under the prevailing conditions of use. Balance and weights may be internally consistent but nonetheless may deviate from the absolute to an extent that error is introduced by accepting a calibration made in another locality and under different conditions.

Volumetric glassware should be of the highest possible quality: The saving effected by purchase of less than the best is not worthwhile, except in the case of large numbers of pieces used in rapid or slop procedures. Even then, it is worth the effort to check the calibration of each using a simple and rapid procedure. Weighing a volumetric flask empty and then full of water and using a trip scale takes very little time and avoids the possibility of a grossly miscalibrated flask [3, 4].

Burettes, in particular, should be of first quality: They should be fabricated of precision-bore tubing, and the markings should be easily readable. Stopcocks with Teflon plugs are a convenience but not a necessity. Such stopcocks should be carefully tested to be sure they do not leak, and they should be tightened if necessary even to the point where they are inconveniently difficult to operate. Glass plugs in stopcocks should be greased with a material that is easily removed; the use of silicones to grease glass stopcocks is undesirable because this material is difficult to remove when it finds its way into the burette proper.

Five burette types are of value: a 50-ml straight burette, a 50-ml burette with an offset tip (a so-called titration burette) for use in titrating hot solutions, a 10-ml burette graduated to 0.02 ml, and two 5-ml burettes graduated to 0.01 ml—one with a regular tip and one with a removable platinum tip for micro- or semi-microtitrations, Automatic burettes should be avoided except in routine repetitive operations and in some procedures using solutions that must be protected from the air (e.g., titanous salts and alkali hydroxide solutions, Kark Fischer reagent).

It is very convenient to have a 5-ml micrometer burette, which is used for delivering very small volumes with high accuracy and for rapid calibration of ordinary burettes (see below). A micrometer burette must be of high quality to be of much value: The barrel must be of precision bore tubing and the glass plunger must fit perfectly; the micrometer part must be a finely machined tool, essentially the same as a micrometer caliper, and not a plastic toy.

Any burette can most easily be calibrated by weighing the water it delivers. A calibrated micrometer burette can be used to calibrate other pieces of volumetric apparatus using a simple and rapid procedure. For burettes, this is easily done by attaching the tip of the micrometer burette to that of the piece to be calibrated with a short length of small-bore heavy-wall tubing. After removal of all air bubbles, water (or other liquid) is drawn into the micrometer burette from the piece being calibrated, and corresponding scale readings are recorded. Using this device, it is possible to calibrate pipettes and burettes up to 5 ml in total volume and larger burettes by emptying the micrometer burette between readings.

Calibration of a burette by weighing the water it delivers is accomplished using a weighing bottle of suitable size with a hole drilled in the cover that is just large enough to admit the burette tip.

PROCEDURE. Weigh the weighing bottle against a similar bottle used as a tare. Thoroughly clean the burette, and fill it with water at a known temperature. Have a small beaker of water handy. Immerse the burette tip to a depth of 0.5 cm in the water in the beaker, and immediately replace the beaker with the weighing bottle. Deliver a suitable volume of water from the burette and weigh. Repeat the procedure, wetting the burette tip each time in the same manner until the contents of the burette has been transferred to the weighing bottle. Record room temperature and measure air density (either by weighing a calibrated bulb or by weighing a vessel first evacuated, and then filled with air).

Calculate the delivered volumes of water, using the relationships

$$M = m + md_a \left(\frac{1}{d_m} - \frac{1}{d_w} \right) \quad \text{and} \quad V = \frac{M}{d_m} \qquad (2.4)$$

where M is the true mass of water, m is its apparent mass when weighed against weights of density d_w, d_m is the density of water at the temperature of the experiment, and d_a is the density of air determined at the time of the experiment. Here V is the volume delivered: The volume delivered at another temperature depends on the coefficient of expansion of the glass of which the burette is made; it may be assumed equal to 25×10^{-6}. If laboratory temperature is reasonably constant, expansion of the glass is usually negligible. The coefficient of expansion of volumetric solutions, however, is about 10 times that of the glass and must not be neglected in accurate work.

It is most convenient to base all temperature corrections in volumetric work to an idealized 20 °C or some other convenient temperature, depending on local conditions. If all solutions to be used in volumetric work are made to volume at the same temperature, the whole problem of temperature correction is much simplified.

The draining characteristic of burettes and pipettes must be considered. In general, the faster a burette or pipette is drained, the less the volume it will deliver even if it is allowed to drain for the same length of time. Good burettes have tip orifices such that only a small difference in delivered volume exists when the burette is drained at a maximum rate. The calibration should be repeated under a condition of maximum flow rate, delivering the whole capacity into the weighing bottle to determine the extent of the draining effect. After delivery, drainage from the walls of the burette will take place, and the time required for drainage to make up the discrepancy between the volume delivered in parts and the volume delivered at one stroke may be determined. Drainage time depends on the nature of the solution being dispensed and may vary from one burette to another. These considerations must be kept in mind if the maximum accuracy is to be attained in volumetric work.

In calibrating 5- and 10-ml burettes, the advantage of using a micrometer burette as a calibrating device is plain. Liquid can be withdrawn from the burette being calibrated at different rates and the effects observed at once without time-consuming weighings and calculations. This is of special importance with small-bore burettes because the internal surface is proportionately greater than larger burettes.

Several mistakes are common in the use of burettes. The most serious of these is a failure to control evaporation, which results in concentration of the active substance. A 50-ml burette can very easily be read to the nearest 0.02 ml and a 5-ml burette to the nearest 0.002 ml. Obviously, such precise reading becomes meaningless if much more than these small volumes should evaporate from the solution during transfer from reagent bottle to burette. The sophomoric practice of pouring solution from the storage container into a beaker and then into the burette can lead to evaporation losses far greater than 0.02 ml. In a dry atmosphere, 10 times this amount may evaporate in a very few minutes. This can easily be demonstrated by following the evaporation loss from a small beaker of solution placed on the pan of a trip balance.

Losses through evaporation can easily be avoided, and at the same time the process of filling a burette can be speeded up by drawing the solution directly from the bottle into the burette. Burettes of 5 and 10 ml are commonly fabricated with a funnel top, intended for insertion of a suction tube so that the burette may be filled from the tip. This funnel should be cut off. Filling is accomplished by attaching a rubber bulb to the burette tip, inverting the burette, dipping the top just under the surface of the liquid in the reagent bottle, and opening the stopcock. This method has several advantages besides the obvious one of avoiding evaporation: The solution always passes through the burette in the same direction, avoiding transfer of stopcock grease to the body of the burette; solution remaining in the burette from a previous titration, and which may have become concentrated through evaporation, is drawn off and discarded; filling is extremely rapid—almost as rapid as when an automatic burette is used and without any of the disadvantages of the latter; and accidental spillage is almost impossible. Larger burettes can be filled in the same manner if the top is drawn down to a narrow bore so that solution does not pour out on inversion; however, a 50-ml burette can easily be filled by direct pouring from the reagent bottle, especially if the top is flared slightly. All burettes should be covered with a small glass or plastic tube immediately after filling. A small test tube with a plug of glass wool at the bottom is convenient.

Error due to evaporation from the solution in the burette is small if all precautions are taken, but such evaporation does occur. It is minimized if the inside glass surfaces are clean all the way to the top of the burette. If the top is slightly greasy, so that droplets of solution adhere to the upper part of the burette, these may evaporate rapidly to near dryness and be redissolved when fresh solution is poured into the burette. For this reason, it is desirable to run the burette completely empty before each refilling and rinse it with a few milliliters of the fresh solution. A solution left standing in a burette for an hour or so should never be used in a titration that pretends to accuracy.

A further common source of error in the use of burettes and pipettes is the failure to thoroughly rinse with the solution being dispensed. Any thoroughly clean burette or pipette carries a film of water on its inner surface that is not as readily removed as one might believe. If the burette is not thoroughly and repeatedly rinsed with the standard solution, it may be diluted very appreciably; the effect is very noticeable with 5- and 10-ml burettes and small pipettes, which have more surface area per unit volume than larger pieces. It is necessary to pass several (at least six) portions of the solution through the burette, allowing it to drain well each time, before it is certain that no dilution effect will be observed. Fewer rinsings are required for 50-ml burettes, but at least four are recommended, with thorough drainage after each rinsing.

Drainage errors are sometimes a problem. A procedure may call for addition of the titrant as quickly as possible (as in the titration of iron in hot hydrochloric acid solution after reduction with stannous chloride). A 50-ml burette drained slowly (the titrating solution dispensed dropwise) over a period of, say, 5 min, may deliver appreciably more solution between 0 and 50 on the scale than the same burette run quickly from 0 to 50, allowed to stand for 5 min, and then drained to the 50 mark. The draining characteristics of a particular burette using a specific solution should be observed experimentally, and suitable steps should be taken to correct the observed volumes. Unfortunately, different solutions behave differently; there is no simple routine that can be followed to remove this difficulty.

In principle, weight burettes are superior to those that measure volume; but they are so much more inconvenient to use that they are reserved for special applications, for example, in dispensing concentrated solutions, which deviate markedly in their behavior from the water used to calibrate a burette. Careful attention to all the factors mentioned above make it possible to control titrant volumes to 1:1000 or better, and such precision is almost always adequate. It must be added that failure to observe all precautions may lead to errors of 1:100 or more.

The use of 5- and 10-ml burettes should always be preferred to the more common 50-ml burette when less than 10 ml of titrant is required. The expedient described for filling the small burettes actually makes them more convenient and less troublesome than the larger burettes, and the 10-fold increase in reading precision practically shifts the cause of any analytical error from the readout to the chemistry of the solution being titrated.

Of course, all solutions dispensed from burettes must be prepared with care, filtered when appropriate, sterilized when necessary, and free from minute specks of foreign material that will plug the fine tip of a good burette. The time spent by careless and sloppy chemists poking pieces of wire into burette tips would be much better spent in preparing clean standard solutions.

Most of what has been said concerning burettes applies generally to transfer pipettes. Pipettes should be long and slim and finely made; they should always be calibrated by the user. In the dispensing of solutions, the following technique is recommended. It differs from that commonly described but is equally effective and has the advantage that it can be used whether the vessel from which the solution is drawn is of glass, plastic, or metal.

PROCEDURE. After thorough cleaning, draw a little of the solution into the pipette from the reagent container. Tip and rotate the pipette until the solution wets the whole inner surface. Drain, discarding the liquid. Repeat this at least four times—more often with small pipettes. Draw the pipette full of liquid to a point about an inch above the mark. Wipe off the outside with a piece of clean hardened filter paper or suitable equivalent and lower the tip into the reagent bottle until it just touches the surface of the liquid. Release finger pressure until the meniscus moves just to the mark. Gently remove the filled pipette from the bottle to the receiving vessel, and release the liquid, holding the pipette vertical. Ten seconds after the flow stops, touch the tip to the surface of the delivered liquid and remove the pipette without shaking.

A good transfer pipette has a long slender stem; it can be easily used to extract liquid from a 10- or 25-ml volumetric flask; it is shaped internally so that it will drain efficiently; there should be no internal shoulders or lumps that inhibit free drainage.

The so-called lambda pipettes, with delivery volumes of 1 ml or less, are very useful in certain circumstances. They are, however, very difficult to calibrate and are seldom capable of high accuracy. The exact volume dispensed, especially with the smaller sizes, depends heavily on the technique used. Probably, when volumes less than 1 ml are to be delivered, it is better to use a micrometer burette.

It should be remembered that pipette calibration applies only to relatively dilute solutions; delivery of liquids even slightly more or less viscous than water will not be accurate.

Automatic pipetting devices and dilutors are extremely useful in routine and repetitive operations but seldom find application in accurate primary analyses.

Volumetric flasks are indispensable. It is particularly important to have one or more carefully calibrated 2-l flasks in which standard solutions are prepared. One or two of each of an assortment of smaller flasks should be calibrated. These can be used to quickly check the calibration of any number of other flasks. It is only necessary to weigh the flasks dry and then full of water and compare the apparent weight of water contained with that in the calibrated flask. The volume correction in milliliters is, to a good approximation, equal to the difference in apparent weight of water contained in grams. Most class A volumetric flasks will be found to meet specifications; occasionally, however, a gross error may be found. The time required to check the accuracy of flasks destined for routine use is small and may prevent a rogue flask from introducing an error every time it is used.

Beakers and Other Glassware

New glass always carries on its surface a layer of submicroscopic blebs of composition different from the body of the glass. This impurity is usually high in alkali metals and may contain a concentration of trace elements higher than that of the glass proper. All new glassware should be scrubbed in soap (not phosphate detergent) with a brush and then cleaned with acids or given some other treatment appropriate to the intended use. Bottles may be shaken with a strong solution of sodium or potassium carbonate and sand prior to rinsing and treatment with acids.

Some analytical procedures leave a film on glassware or alter the glass surface in such a way that large errors may occur if apparatus is used in other procedures. Solutions containing phosphates and fluorides are especially apt to permanently contaminate glassware. For example, the determination of aluminum with aluminon is practically impossible in flasks that have been employed in determinations using hydrofluoric acid. Colorimetric phosphorus determinations are certain to fail if flasks have been previously used to contain phosphoric acid. No reasonable amount of washing can remove sodium from glasses that have contained strong sodium hydroxide solution.

Platinum Ware

The platinum available in a laboratory is probably more often determined by financial considerations than anything else. It is very convenient and conducive to good analyses to have a plentiful supply of dishes, crucibles, and accessories of platinum, but it is certainly not essential. The only platinum essential to a good silicate analysis is a 25-ml crucible and cover and a pair of platinum-tipped tongs. Obviously, it is desirable to have several such crucibles so that several analyses can be started simultaneously.

Properly cared for, the life of a platinum crucible is almost indefinite: Crucibles have been maintained in continuous use for many years without any deterioration except the loss in weight occasioned by frequent cleaning. On the other hand, improperly used platinum can be ruined in a single operation. It is therefore important to take all precautions in the use of platinum ware not only because of the high cost but also because of the nuisance of replacement. Those who are limited to a very small collection of platinum need to take extra care; they may be rewarded with an ability to outperform those who use platinum prodigally.

Possibly the commonest difficulty in using platinum is its contamination with base metals, usually iron. In an ordinary silicate analysis, if improper heating conditions prevail during the various fusions and ignitions, several milligrams of iron may be taken up by a platinum crucible. All or part of this may be released in a subsequent analysis if it is not removed in the meantime. Plainly, such contamination cannot be tolerated and must be avoided.

In the analysis of iron-bearing minerals, ores containing copper, nickel, lead, and so on, and rocks containing sulfides, it is sometimes very difficult to avoid contamination of the platinum, and steps are necessary to remove the contamination and return it to the analysis.

Despite statements made to the contrary, it is quite possible to carry out sodium carbonate fusions of even highly ferrous materials (e.g., iron ore and ferrous amphibole) in platinum crucibles without serious difficulty, even without the addition of oxidants to the flux (the addition of an oxidant is nevertheless a wise precaution). For example, two separate analyses of the CAAS Syenite-1 were carried out in a platinum crucible without mishap. This sample contains 33.5% iron as Fe_2O_3, 1.92% NiO, and 1% CuO; the nickel and copper are present in the form of sulfides. The sulfur content of the rock is 12%. The sample was given a preliminary roasting to oxidize most of the sulfur, and about 0.1 g sodium peroxide was added with the sodium carbonate before fusion. A small part of the copper deposited on the platinum during the fusion but was easily removed by treatment with hydrofluoric and nitric acids. The total weight loss of the crucible during fusion and subsequent acid treatment was less than 1 mg. The fusion was accomplished in a gas flame using first a Tirrill and then a Meker burner.

The manner in which platinum is heated during a sodium carbonate fusion is of utmost importance. Fusion in a muffle furnace or over an electric burner results in considerable loss of platinum, whereas fusion over a properly controlled gas burner does not. It is difficult to make general statements concerning gas flames because gas pressure and composition vary widely, and many types of burners are available. One essential requirement must be met, however: The flame must contain excess air. This is the requirement that most commonly leads to difficulty. Some combinations of burner and gas supply are incapable of providing a sufficiently hot oxidizing flame. A reducing flame simply must not be used to heat platinum, especially in contact with ferrous materials.

A simple test of any burner and its gas supply can be made by heating a nickel crucible of the same size in exactly the same position as that taken by the platinum crucible. A film of oxide will appear on the nickel crucible if the flame is oxidizing. It may be found that a very slight adjustment of the flame will change it from an excess-air condition to an excess-gas condition; this change is shown by the alternate appearance and disappearance of the oxide film on the bright nickel surface. With some experience, visual inspection of the flame will reveal its character: The reducing flame is more opaque. A good oxidizing flame over a Meker burner is almost invisible in a bright room, and the crucible heated by it will glow white-red.

Protracted heating of a platinum crucible in a reducing flame results in embrittlement and deterioration even when base metals are absent. Because of the clear superiority of a gas flame as a heat source during sodium carbonate fusions (and other operations), every effort should be made to obtain an adequate gas supply and burners that can be used efficiently with it.

Contamination in platinum cannot always be detected visually. Very appreciable amounts of iron may impart only a slight dullness; however, on heating the empty crucible in an oxidizing flame or in a furnace at 800 °C, a purple stain of iron oxides will usually develop within an hour or so. This can be removed (if it has not soaked too deeply into the crucible) by leaching in hot hydrochloric acid. The crucible should then be reheated to discover if contamination remains.

If iron contamination is not removed at once, it will gradually work its way deeper into the platinum, and its removal may be impossible. In such a case, the crucible may be considered a loss.

After each using, when it is certain that any contamination has been removed, crucibles should be burnished with moist round-grained sand, not to abrade the surface but to refine the grain structure of the metal. After sanding, the crucible should be washed thoroughly with soap and water, thoroughly dried, and then heated strongly in an oxidizing flame.

Platinum alloyed with 3.5% rhodium has some advantages over pure platinum for crucibles. However, if it is possible to have only one type of crucible, pure platinum must be preferred. The chief advantage of the alloy is that it is stiffer and easier to handle; the soft platinum is easily deformed, especially when hot. Lids of pure platinum are prone to stick to the crucibles unless cleaning and sanding are very carefully performed.

Rhodium alloy crucibles acquire a dark stain when heated with alkaline oxides like CaO, SrO, and so on. A similar stain results from sodium carbonate fusions, but it represents a very small weight change and may usually be ignored. In the attack of samples that result in unusually large losses of platinum from the crucible (chromites, manganese minerals, and so on) or when fusions must be done in a muffle or over an electric burner, rhodium alloy crucibles are undesirable because the dissolved rhodium complicates the subsequent analysis. For example, it interferes in the iron determination and in the colorimetric determination of titanium.

Each piece of platinum ware, whether alloy or not, should be provided with a wooden reshaping form, which should be frequently used. A jewelers' burnishing tool is invaluable for working out dents, rough spots, and scratches, all of which may interfere in the best use of the ware. The burnishing tool should slide smoothly over the surface of the metal; if it sticks and scratches, the platinum is embrittled through carbonization or other contamination and needs repeated heating and working to restore it.

Many of the difficulties encountered by amateurs, such as lids sticking to crucibles, crucibles sticking to triangles, pieces of silica triangle attached to the outside of the crucible, an inability to obtain the same weight repeatedly throughout the analysis, fusion cakes that cannot be easily removed, large weight losses during fusion, and so on, are due primarily to failure to properly clean and maintain the ware. No crucible should be used to start the analysis of a rock or mineral until it is perfectly smooth and clean, free of any visible graininess, and thoroughly washed with soap, dried, and ignited. Once ignited, it should be handled only with clean platinum-tipped tongs when hot and with well-made rubber-tipped tongs when cold and should never come in contact with any surface that is not perfectly free from loose particles or alkali salts (e.g., sodium chloride from the fingers). Silica triangles used to support the crucible during heating must be thoroughly clean; finger marks on them make a tiny fusion at the point where crucible touches triangle, and pieces of silica adhere to the crucible.

Maintenance of platinum dishes, which are not often heated to high temperature, is not difficult; but regular sanding and shaping is very necessary. Flat-bottom dishes, invaluable in operations where rapid evaporation on a hot plate is required, must be carefully worked to maintain flatness. Neglect of this soon results in dishes that have all the disadvantages of both flat and round-bottom dishes.

Gold and Silver Ware

Gold crucibles and dishes are not widely used. They have, however, characteristics that make them desirable in certain applications. Gold is more inert and less liable to contamination than platinum. It is, however, softer and has a much lower melting point. Gold crucibles must be heated very cautiously over flames. In the furnace, temperature should not exceed 900 °C, and heating must be done in such a way that two gold items, or one of gold and one of platinum, do not touch; it is easy to weld the items together. Gold crucibles should be cooled in the furnace before removal with platinum-tipped tongs.

Gold crucibles are useful for sodium hydroxide fusions, in the analysis of certain minerals that resist attack by sodium carbonate, or rocks and ores containing copper, lead, and other metals that might alloy with platinum at the higher temperature of a carbonate fusion. Fusions with mixed sodium carbonate and peroxide are best accomplished in gold crucibles.

The curcumin procedure for determining traces of boron calls for a sodium hydroxide fusion, which is best performed in a gold crucible.

Silver crucibles may also be used for sodium hydroxide fusions. Silver dishes may replace platinum in many procedures and are relatively easy to maintain in good condition. They are best cleaned with sodium bicarbonate (mild abrasive action) and aqueous ammonia. Small amounts of silver, which may dissolve in some procedures, are easily removed as chloride or as metal precipitated on zinc.

Zirconium, Nickel, and Iron Crucibles

Iron crucibles of high purity are cheap and very useful. They are used for sodium peroxide fusions in every case where introduction of iron into the analysis is of no consequence. Properly employed, they are far superior to nickel for peroxide fusions.

Large iron crucibles (100 ml) are very convenient when a peroxide fusion is to be leached with water and the precipitated oxides are not of interest, as in determinations of chlorine and sulfur. In the attack of shales, coal, and other carbon-bearing rocks, an improvised bomb can be formed by two large iron crucibles closely fitted inside an iron pipe or band. This simple device is very effective and much more convenient than a conventional peroxide bomb.

Nickel crucibles are indicated for carbonate fusions when platinum is unavailable or would be adversely affected. Usually mixed sodium and potassium carbonates are used because of a lowered melting point of the mixture. An example of this application is the zinc oxide—mixed carbonate fusion used in attacking minerals for fluorine determination.

Zirconium crucibles are widely used for peroxide fusions over a flame. In a furnace, zirconium oxidizes rapidly above about 600 °C. A major disadvantage of zirconium is that small amounts dissolved during the fusion may upset analytical procedures. For example, the stability of zirconium peroxy compounds makes determination of chromium by the carbazide method of doubtful reliability.

Graphite and Vitrified Carbon Crucibles

These have become invaluable for borate fusions, indispensable to many rapid analytical methods. They are preferred to their chief competitor, platinum–gold alloy, for their relatively low cost and their freedom from contaminants.

Fusions in graphite are always made in furnaces, with a controlled air supply to prevent reduction of certain elements to metal, which may volatilize. Too much air must be avoided, or the crucibles will burn away rapidly.

Despite the real aesthetic appeal of vitreous carbon crucibles, it is doubtful that they give enough of an improvement over graphite to compensate for their much higher cost.

Boron nitride, which looks and feels like white graphite, may sometimes be used as a crucible material in borate fusions. It gives the possibility of fusion in an oxidizing atmosphere, which may be an advantage when ores of copper, lead, and so on are to be fused.

Porcelain Ware

Porcelain crucibles may be extensively used in laboratories with a limited supply of platinum. This does not constitute much of an economy over a long period because porcelain deteriorates rapidly in most procedures even with the best of care. In using porcelain crucibles for weighing, the smallest possible should be used to minimize weighing errors due to absorption and loss of moisture, changes in air density, and actual loss in weight of the porcelain itself. Porcelain crucibles are satisfactory for pyrosulfate fusions, ignition loss determination, roasting of sulfides and arsenides, gravimetric determination of calcium as carbonate, magnesium as pyrophosphate, and the ammonia precipitate in ordinary rock analysis. If a crucible weighing more than about 10 g is used, it is most important to carry a second crucible through all heating operations to use as a tare; otherwise weighing errors may be serious.

Porcelain crucibles with porous bottoms are very useful. Those with a fused-in sintered disk are less satisfactory than those made in one piece: Royal Berlin type 1A2 crucibles have proven most satisfactory in many operations. An empty crucible should always be carried along as a tare. It is desirable to reserve a set of crucibles for every specific routine because it is sometimes impossible to completely remove an ignited and weighed precipitate, and residuals from one procedure should not be allowed to contaminate another.

Every laboratory should have a number of glazed porcelain plates on which to place platinum and other crucibles during various stages of analysis. Porcelain desiccator plates are very useful for storing and carrying crucibles between operations.

Porcelain casseroles are very satisfactory for hydrochloric acid dehydrations of silica in samples that do not contain fluorine. They are preferred over platinum in many cases where much iron or manganese is present and may be used for all ordinary rocks when platinum is not available. Numerous parallel analyses in platinum and porcelain show that the error arising from use of the latter need not exceed 0.1% if all precautions are taken. The value for silica using porcelain for dehydration is usually low, not high as might be expected, probably because it is very difficult to scrub out the last traces of precipitated silica.

Silica Glass: Vitreous Silica

Clear silica glass crucibles are invaluable in the analysis of samples that yield large ammonia group precipitates containing much iron and easily reducible elements such as arsenic. They should be equipped with flat lids, and a tare should be carried through all operations because accurate weighing is impossible otherwise. Ignition of a complex ammonia group in platinum risks damage to the latter. In addition, the subsequent pyrosulfate fusion is much more easily accomplished in silica crucibles.

Silica crucibles must be very thoroughly cleaned before use, using chromic acid cleaning solution. Once cleaned, they should not be so much as touched by the fingers for traces of alkali cause their rapid deterioration. If ammonia group precipitates are not thoroughly washed, traces of alkali therein will cause rapid deterioration of the crucible. Triple precipitations of large ammonia groups should be the rule; an etched silica crucible is an indication of less-than-perfect washing of the ammonia group precipitate.

Large fused silica dishes are invaluable during the separation of the alkali metals by the J. L. Smith method.

Silica glass stirring rods are preferable to those of Pyrex during the main portion of a silicate analysis. The ends should be fire polished in an oxy-hydrogen flame to a perfect hemisphere to avoid the danger of chips being broken off.

Vycor glass, which is nearly pure silica, may be substituted for it in many applications. Vycor evaporating dishes may be used in analyses of fluorine-bearing materials in place of platinum when the latter is not available; attack on the Vycor is less of consequence than is the case if porcelain casseroles are used. Vycor crucibles are generally heavier than silica glass crucibles and less desirable on this account.

Desiccators and Desiccants

Many of the weighing forms used in quantitative analysis are highly surface active and rapidly absorb or adsorb water vapor and other gases from the air. Oxides (especially the alkaline earths) rapidly absorb carbon dioxide. It is essential to have an evacuable desiccator charged with a highly effective desiccant. Of the available desiccants, magnesium perchlorate and barium oxide are by far the most useful. Calcium sulfate and calcium chloride are inadequate and should not be used. Indicating silica gel is convenient but has a very low capacity. Any really good desiccant cannot be regenerated by any simple procedure, and attempts to regenerate magnesium perchlorate or barium oxide are futile.

A vacuum desiccator should be used in conjunction with a drying tower containing soda lime and magnesium perchlorate. The crucible containing the ignited weighing form is placed in the desiccator while still as hot as practicable, the air is immediately pumped out, and then dry CO_2-free air is admitted slowly through the drying tower. With extremely active ignition products, only one crucible should be allowed per desiccator (together with its tare if one is used). The smaller the desiccator, the better.

Tubulatures and stopcocks should be greased with care using heavy silicone grease. The flanges of the desiccator lid should be greased with the lightest possible coating of stopcock grease.

Desiccator plates should be of porcelain and must be kept scrupulously clean. Desiccators and plates of metal or plastic are not recommended.

Weighing oxides such as Al_2O_3, CaO, La_2O_3 is impossible if desiccator and desiccant are not in first-class condition.

2.2. REAGENTS

2.2.1. Distilled Water

Water is the most essential of reagents, and its quality is of first importance in almost all analytical operations—even those that do not use water directly. Ion exchange water should not be used in analytical work except in specific applications where experiment has shown that it is adequate. The chief reason for this is that complete exclusion of dissolved or colloidal resins and nonionic impurities such as SiO_2 is almost impossible in any ion exchange system of reasonable cost. The effect of these contaminants in analytical operations can be disastrous to their outcome. Glass and plastic ware adsorbs them on surfaces, where they perform their intended function of exchanging ions. In solution, they may act as chelating agents and cause unwanted reactions that spoil various analyses, especially those involving colorimetric determinations of trace elements.

The type of still that should be used depends on the quality of the water supply as well as the intended use. Ordinary laboratory stills are not always capable of producing water of sufficient purity. Claims that water with 1 ppm or less of dissolved solids can be produced are often justified, but water with 1 ppm solids is quite inadequate for many ordinary procedures. One part per million is 1 mg/l; some lengthy analyses require several liters of water, the salt content of which is likely to be included in the results. Several milligrams of contaminant so introduced leads to an intolerable error. In regions where the water supply contains much lime, a double distillation is nearly always essential.

Conductivity tests and conductivity meters are not very good guides to the suitability of water for analytical work. It is better to test the water by determining silica, calcium, magnesium, sodium, potassium, and other critical elements by flame photometry or flame atomic absorption spectrophotometry. Alternatively, several liters of water may be evaporated to dryness in platinum (by making as many measured additions as necessary during the evaporation) and the residue examined spectrographically after weighing.

Since the success of an analytical operation depends heavily on the quality of the water available, it is worth taking some pains to ensure its purity.

2.2.2. Acids

The ordinary mineral acids are obtainable in grades sufficiently pure for most purposes, with the exception of phosphoric acid, which often contains organic matter and sulfate. Probably most of the impurity in reagent-grade acids originates in the glass of the container; thus, it is advantageous to seek a supply source where long storage is not likely. In some applications, redistillation of nitric, hydrochloric, and hydrofluoric acids is necessary.

Purification of phosphoric acid is not feasible in the ordinary laboratory, and there is no alternative to testing each batch in the procedures to which it is applied. For example, determination of ferrous iron by the pyrophosphate method depends on the absence or removal of organic and other reducing material it may contain. Some batches have proven worthless in this procedure—no reasonable pretreatment will remove interfering reducing substances.

Because the source of impurity in acids is most often their container, it is necessary that each bottle be examined for contamination if one is to be completely certain of its adequacy. Usually, judicious inclusion of blank determinations in any suite of analyses will detect any real difficulty. In atomic absorption spectrophotometry especially, where dilutions of the analytical solutions may be extreme, it is necessary to be certain that acid-contributed blanks are kept under control.

2.2.3. Ammonia

Ammonium hydroxide is a nearly saturated solution of ammonia gas in water. Ammonium hydroxide supplied in glass bottles should never be used in accurate analytical work because the alkaline solution dissolves glass and forms insoluble metal hydroxides, which may or may not precipitate visibly. Ammonium hydroxide should be prepared in small batches by saturating pure distilled water with ammonia gas. Only in rough routine work is the use of the reagent, as ordinarily supplied, permissible.

2.2.4. Organic Reagents

Possibly one of the largest causes of difficulty in practical analysis is the unreliability of organic reagents. Many are so grossly impure that their reactions do not even closely resemble those of the pure material. Further, published procedures involving organic reagents were often worked out using impure reagents, and the reported results are not reproducible by those using pure reagents or reagents containing different impurities. These statements apply, to some degree, to all the commonly used organic reagents with few exceptions. It would be impossible to deal with the effects of impurity in organic reagents with any degree of completeness. Some examples must suffice. Those using procedures that depend on reactions between organic compounds and metallic elements should ascertain the effectiveness of their reagent under the conditions of their particular procedure [5–9].

It is probably best to carry through any procedure using organic reagents from several sources. If each of two or more reagents behave in essentially the same manner, there is probably no difficulty. More often than not, reagents from different sources will give different results, sometimes so different as to be in startling contrast.

One experience with some of these reagents is outlined below. The implication is that most organic reagents are subject to similar aberrations, and their use must take this into account.

Aluminon

Aluminon, aurin tricarboxylic acid, ammonium salt, has been used for many years as a reagent for aluminum, and it probably remains the best of several such. However, reagent quality varies widely, to the extent that some aluminons are almost worthless as colorimetric reagents for aluminum. In experiments with 10 reagents from different sources, a rather astonishing difference in their behavior was observed.

Plainly, the use of aluminon as a reagent for aluminum will succeed only if the reagent is carefully selected or synthesized. It seems probable that many of the statements found in the literature concerning this reagent are of little value because few analysts have actually worked with the pure reagent.

The 10 reagents were tested by reacting in acetate–borate buffer solution at pH 3.8, using 12 μg Al_2O_3 in the form of ammonium alum solution and 2 mg aluminon in water solution in a total volume of 25 ml. Blanks were identically prepared but without the aluminum addition. Over the 10 reagents, absorbances in 2-cm cells due to the colored complex varied from 0.059 to 0.278. The blank absorbances varied from 0.053 to 0.438.

Numerous attempts to purify these reagents met with little success. Of several methods of purification, the following seems the most promising.

Purification of Aluminon

PROCEDURE. Dissolve 2 g aluminon in 100 ml water. Stir in a separatory flask with 5 ml concentrated hydrochloric acid and 50 ml amyl alcohol. Precipitated aurin tricarboxylic acid and some other acids dissolve in the alcohol layer. Remove and discard the acid layer, and wash three times with 30-ml portions of water. Discard the washings. Extract the reagent from the organic layer by adding successive 25-ml portions of 0.4% aqueous ammonia, agitating, and drawing off the aqueous extract. The performance of the successive extracts was quite different with the different reagents, indicating that they contain several different compounds.

2.2.5. Fluxes

Lithium Metaborate, LiBO₂

Lithium metaborate fusions of most silicates and some nonsilicates are readily soluble in dilute acid [10]. This provides a convenient method for preparing whole-rock solutions suitable for many different analytical procedures. Probably no other reagent is as universally effective in its action on rock-forming minerals. Zircon, sphene, tourmaline, corundum, and many other refractory minerals yield readily to $LiBO_2$ fusion. If the fusion is performed in graphite, the molten glass that results may be poured from the crucible without loss. Pouring into dilute acid results in shattering of the glass, and solution is accomplished in a few minutes without heating, simply by stirring the suspension. Some nonsilicates (e.g., magnetite) require the addition of silica to the melt. Ferrous minerals in general and chromite in particular sometimes are not rapidly attacked in graphite but react readily with the flux in an oxidizing atmosphere when fusion is accomplished in platinum, gold, palau, or boron nitride crucibles. In vacuum systems, molybdenum crucibles may be used. Various additions may be made to the flux for special purposes. Cerium or lanthanum oxide may serve as "heavy absorbers" in X-ray fluorescence analysis or as releasing agents when solutions of the melt are to be used in atomic absorption spectrometry. Cobalt, lutecium, or strontium may be added to act as internal standards in emission spectrometry or spectrography. Standard solutions of various elements may be added in known and varied amounts to lithium metaborate solutions of the samples when the method of standard additions is applied or when standard samples are lacking and the preparation of synthetic standards is necessary.

When additions are made to the flux, the mixture should be fused, ground, and screened before use. Mixtures not so treated are prone to segregation and lead to error when the additive concentration is critical to the success of the procedure.

Whatever the source of the reagent, its stoichiometry and purity should always be proven before use. Commercial reagents may contain large amounts of impurity, and this may be present in the form of relatively large grains and particles. Before analysis of the reagent, it should be crushed to pass 16 mesh or finer and thoroughly mixed. Then several portions should be taken for analysis to discover its effective homogeneity.

Analysis of the reagent is best accomplished by weighing 1.000 g into a platinum crucible, fusing for 10–15 minutes at 950 °C, observing the weight loss, and treating the fused residue with hydrofluoric and sulfuric acids, evaporating, and reigniting. The residue of lithium sulfate should come within a milligram of theoretical. Spectrographic examination of several 10-mg portions will reveal the presence of trace metals and also indicate the homogeneity of the reagent. If it is to be used for mass spectrometric work, especially in geochronology, mass spectrometric determinations of, for example, potassium, rubidium, lead, strontium, and uranium should not be omitted.

Preparation of lithium metaborate is most conveniently accomplished starting with the carbonate and reacting it simply by mixing it intimately with finely powdered (impalpable powder) boric acid and heating. The reaction

$$Li_2CO_3 + 2 H_3BO_3 = 2LiBO_2 + 3H_2O + CO_2$$

occurs without any need for water addition or fusion. If the product of a preliminary heating to about 300 °C is crushed and reheated to 625 °C, the reaction will be 95% or more complete. Procedures involving the aqueous reaction of lithium hydroxide and boric acid have not proven as successful: Other lithium borates may be formed, solutions are apt to become supersaturated, and attack on reaction vessels introduces unwanted impurity.

If purification of the reagent is necessary (and this is most often the case, even when reagent-grade chemicals are used), the following procedure is recommended.

Purification of Lithium Metaborate

PROCEDURE. Add 500 g $LiBO_2$ to 3000 ml hot water, and stir mechanically while heating to 90 °C. Allow to cool slowly while stirring until the temperature falls to 85 °C or slightly lower. Some crystallization of the dihydrate should occur. Allow the sludge of undissolved metaborate, precipitated dihydrate, and impurities to settle; then filter as quickly as possible. A fluted 40-cm No. 42 Whatman paper may be used, or a pad of hardened paper in a Büchner funnel. Allow the filtrate to stand undisturbed for 48 h. The octahydrate will separate in large crystals.

Pour off and discard the supernate, and blot the crystals as dry as possible with analytical-grade filter paper or other material known to be free from relevant trace impurities. The liquid should be removed as completely as possible or the crystals may melt in the next operation.

Transfer the octahydrate to a silver, platinum, or porcelain dish, and place in a *ventilated* oven at 40 °C until the crystals crumble to a fine powder of dihydrate. A vacuum oven may be used. Drying may take 3–4 days.

Slowly increase the oven temperature to 200 °C, and then transfer small portions of the product to a platinum dish, cover, and heat to 625 °C. A large volume increase may occur. Hold at 625 °C for 1 h, cool, crush, screen through 16 mesh or finer, and mix the product thoroughly before analyzing it for purity and stoichiometry.

The dihydrate loses water much below 625 °C, but the product is unmanageably fluffy and light. At 625 °C, incipient sintering occurs, crushing is easy, and a reasonably heavy powder results. Lithium metaborate undergoes a polymorphic transition at 785 °C. Heating above this temperature but below the melting point may be desirable because an appreciably denser product is so obtained.

If the material is to be used for X-ray work and a "heavy absorber" is to be added, or if vacuum fusions are intended, or if the bulk density of the powder is inconveniently low, fusion of the product should be carried out in pure platinum utensils. The fused material contains large amounts of dissolved atmospheric gases and must be treated in a vacuum furnace if this is of consequence.

The fused material cools to a crystalline mass that is relatively easy to crush and screen. It yields needle-shaped particles that are rather difficult to handle during weighing since they show a strong tendency to mat together. This is especially true when fusion is accomplished under vacuum and cooling is slow.

As the molten material cools and freezes, there is evolution of heat and an increase in volume. This makes it very necessary to carry out the fusion in a covered vessel to avoid splatter of its contents. Heat transfer in small quantities, such as used in an analysis of a rock or mineral, is rapid enough that this effect does not occur during ordinary use of the material. It exhibits itself only with large preparatory batches.

Procedures Using Lithium Metaborate

PELLET PREPARATION FOR X-RAY FLUORESCENCE ANALYSIS. Mix 0.1–0.5 g of sample with 2–7 times its weight of $LiBO_2$ in a porcelain crucible, and transfer the mixture to a high-purity graphite crucible. Heat in a 900–1000 °C muffle for 25–20 min. The higher temperature should be used with lower flux–sample ratios or when refractory minerals such as chromite are present. Loss of alkali metals should be suspected at the higher temperature: If this is of consequence, the temperature should be lowered, and a higher flux–sample ratio used.

Allow the fusion to cool completely in the crucible, remove the resulting glass bead, weigh it, and transfer it to shaker mill capsule with tungsten carbide ends. Crush the bead using a suitable pestle or plunger, add a tungsten carbide ball and sufficient cellulose powder to bring the total weight to a constant amount, and shake for 10 min or long enough to reduce particle size to the point where particle size effects are negligible. For the lighter elements, reduction to minus 400 mesh or finer is necessary. Transfer the mixture to a boron carbide (or agate) mortar, and mix and grind thoroughly to ensure comminution and homogeneity. Then press into a pellet using a die containing a working surface that will not contaminate the pellet and about 5 g methyl cellulose as a backing material. If the die is so constructed that the working surface is protected by a rim of compressed backing material, pellets so prepared may be used and reused indefinitely if kept in a desiccator. A suitable die has been described by Fabbi [11].

DISSOLUTION PROCEDURE FOR RAPID ROCK ANALYSIS. Mix 0.1–0.2 g of sample with 0.5–0.7 g $LiBO_2$ and transfer to a graphite crucible. Fire at 900 °C for 15 min. Pour the molten material into a plastic beaker containing a stirring bar and 100 ml of 4% nitric acid. Stir to solution, which should take less than 5 min.

DISSOLUTION PROCEDURE FOR SMALL SAMPLES. Weigh a 40-ml platinum crucible, add 20 mg or less of sample, and reweigh. If necessary, determine loss or gain on ignition or at any appropriate temperature. For nonsilicates, an addition of silica may be desirable. Add seven times as much $LiBO_2$ as sample (weighing it accurately), support the crucible in a hole in an asbestos board so that the upper half is above the board, cover, and place a second crucible on the lid to act as a condenser. Dip two tubes into the condenser, one to supply cold water and the other to remove it through a suction pump. Heat the lower crucible with an oxidizing Meker flame for 10–15 min. Cool. Add an exactly measured volume of dilute acid (usually 1 ml of 4% nitric acid for each milligram of sample), add a small stirring bar, and stir to solution.

ION EXCHANGE TECHNIQUE. After fusion in graphite, pour the melted material into a water suspension of a strongly acid cation exchange resin (Dowex 50). Solution will be slower than when a mineral acid solution is used; but if there is sufficient resin, it will eventually be complete. The resin may be filtered off, transferred to a column, and appropriately eluted. In general, the divalent metals and the alkalies will be found on the resin and silica, lithium, boron, and much of the iron in the sample in the solution. This procedure is of value in isotope dilution work, when the "spike" is added to the aqueous mixture during solution. Quantitative recovery of specific metals is then not essential. This procedure is due to Govindaraju [12], who gives relevant details. It must be modified to suit the problem at hand.

Sodium Carbonate

Sodium carbonate is the most essential of fluxes in the gravimetric analysis of silicates. Since gravimetric analyses are almost always made with the intent of maximum accuracy, the sodium carbonate must be free from metallic contaminants. Calcium and magnesium are the most common of these. The best analytical reagent should be acquired and in a quantity that will last for some time. The whole batch should be mixed and tested by the following procedure.

TEST FOR PURITY OF SODIUM CARBONATE. Transfer 25 g of the reagent to a 1000-ml beaker and add 500 ml water. Stir to solution while heating to boiling and allow to cool. Examine the solution carefully for any precipitate or turbidity. None should appear. Add 10 g diammonium phosphate in the form of a freshly prepared and filtered solution and 60 ml aqueous ammonia prepared by saturating pure water with ammonia gas. Allow to stand with occasional stirring for at least 4 days. If any visible precipitate appears, the reagent should be rejected. If no precipitate is visible, filter the solution, wash the filter very thoroughly with 1:20 aqueous ammonia, ignite the paper in platinum, and weigh the residue. It should not exceed 0.1–0.2 mg. Even when no phosphate precipitate is visible appreciable magnesium may be present in the form of finely crystalline material, hence the need for filtration and weighing even if no visible precipitate is seen.

A total blank for a gravimetric silicate analysis may be determined using a similar procedure. If this is intended, the various additions of hydrochloric acid, ammonium chloride, ammonium persulfate, and so on may be made to the sodium carbonate solution before neutralizing with ammonia and adding a 10% excess. No gravimetric analysis should be attempted without such a blank run. Batches of sodium carbonate have been found that will contribute as much as 1.00% to the total of a gravimetric analysis. Plainly such errors should be avoided.

Potassium Pyrosulfate

Potassium pyrosulfate is very often contaminated with bisulfate. This results in "noisy" fusions, with the nuisance of splattering and foaming of the melt. A good reagent should melt quietly in a platinum or silica crucible without any splattering.

The bisulfate may be converted to pyrosulfate by cautious heating, but it is very difficult to judge the point at which excess H_2SO_4 is completely removed and SO_3 starts to come off from decomposition of pyrosulfate.

Impurities in reagent-grade $K_2S_2O_7$ are not often a problem. Nevertheless, when the reagent is used in trace element work, blank determinations should always be made.

Sodium Peroxide

Sodium peroxide would be a much more useful flux, and much more widely used, if a pure reagent were easily obtainable. It is most often contaminated with hydroxide, carbonate, and small flakes and pieces of metallic oxides, usually of iron or nickel. Calorimetric-grade Na_2O_2 specially prepared for use in Parr peroxide bombs has proven to be most satisfactory. Since contamination is most often present in the form of discrete particles, blank determinations are not reliable. For this reason, sodium peroxide should be used with caution, and appropriate tests for adequate purity and homogeneity should be made with the intended use in mind.

Sodium peroxide fusions are most useful in the attack of chromite and other refractory minerals, which are not easily attacked by carbonate fusion. Most peroxide fusions should be made in thick-walled iron crucibles. Only in cases where iron is to be determined or will interfere seriously in a procedure should other crucibles be used. In such cases, sintering in platinum crucibles at a temperature not exceeding 500 °C, at which temperature platinum is not appreciably attacked by sodium peroxide, is called for. In exceptional cases, zirconium crucibles may prove best. A difficulty with zirconium crucibles is the stability of zirconium peroxy compounds: zirconium dissolved from the crucible may make it impossible to completely remove peroxide from the solution of the melt, and this may cause difficulty, for example, in the determination of chromium using symdiphenylcarbazide.

A fusion with peroxide in an iron crucible should be performed over a Tirrill or Bunsen burner. The gas flame should be so adjusted that its inner cone (the cool part of the flame) strikes the bottom of the crucible. The relatively cool reducing gases in this part of the flame effectively prevent burnout of the crucible bottom. The mixture within the crucible is heated from the sides, with consequently gentler and more even heating.

Sodium peroxide is used in small quantities as an additive to sodium carbonate during fusion with that salt, usually to make certain that ferrous compounds and sulfides are oxidized and cannot react with platinum crucibles to form metallic iron. The use of much sodium peroxide in a carbonate fusion may result in heavy attack on a platinum crucible so that it should be used sparingly in this application.

Potassium Carbonate

Potassium carbonate is used alone or in eutectic mixture with sodium carbonate in some special cases. Some analytical schemes for niobium- and tantalum-bearing minerals benefit from its use. The lower melting point of the eutectic mixture is advantageous in cases where there is danger of loss by volatilization during fusion of samples in which halogens are to be determined. Thus, fluorine-bearing samples are fused with mixed sodium–potassium carbonate.

2.3. INSTRUMENTATION

Much of ordinary silicate analysis can be accomplished with very little instrumentation: a balance, a Duboscq colorimeter or its equivalent (a good set of Nessler tubes can be made to suffice), a potentiometer of the simplest design, and a miscellany of ordinary items (thermometers, hydrometers, stopwatch, etc.) are all that are really necessary to a minimal laboratory. Those with a limited budget should not feel hopelessly deprived: it is entirely possible to turn out superlative work, albeit limited somewhat in scope, using very little expensive equipment. It may be remarked that the most expensive instrumentation cannot produce good results if preliminary sample preparation is not expertly done. The advantages of extensive instrumentation are in the direction of speed and convenience rather than excellence. It is entirely possible for a laboratory with an unlimited budget to produce less useful results than one less endowed, partly because of a tendency to neglect fundamental skills in an affluent atmosphere and partly because too great a reliance on indirect instrumental methods of analysis increases the possibility of undetected errors of many kinds. The maintenance of a large number of complex instruments is a vexing problem that tends to draw attention away from more mundane matters like sample preparation and homogeneity. Often an automated analytical system will accept samples so fast that their preparation is hurried to the point where results do not approach the capabilities of the instruments used.

It is suggested that an impoverished laboratory should apply itself to problems in which careful and accurate work are essential and not attempt to compete with larger facilities on their own ground. At the same time, those fortunate enough to possess one or more major analytical instruments should fully recognize the limitations of each and apply them in areas where they will do the best work.

There are several instances in which some instrumentation is almost essential—for example, in the determination of alkali metals. Gravimetric sodium and potassium determinations in silicates are lengthy and can be corrected for the effects of lithium, rubidium, and cesium only through prohibitively complex and difficult procedures. Determinations of rubidium and cesium by classical methods are possible but require extreme skill and much time. Flame emission and absorption equipment so greatly improves the capability of determining these elements that one or the other may well be considered a necessity. Probably atomic absorption spectrometry is to be preferred if a choice has to be made between the two because of its capabilities for other elements. Flame emission remains the technique of choice if the alkalies alone are to be considered.

Flame emission equipment should be purchased with the alkali metals first in mind. This means that the monochromator must have sufficient resolving power to isolate the rubidium 780-μm line from potassium 767-μm line. A low-temperature (air–natural gas) flame is preferred for low background and stability. Unfortunately, this practically limits use of the equipment to determine the alkalies and alkaline earths. A less sophisticated flame photometer capable of rapidly determining sodium and potassium (and perhaps lithium) but not rubidium or cesium is a convenience that is probably not essential unless large numbers of samples are to be routinely run for these elements.

For the alkali metals, flame absorption is less desirable than flame emission, largely because it is necessary to work in a much narrower concentration range, and interelement effects are less easily controlled. Adequate results are, however, quite possible, and the capability of determining many other elements with the same equipment make it most attractive. A deciding factor might be the availability of a direct-reading emission spectrometer, which performs comparably with atomic absorption in many circumstances.

Atomic fluorescence spectrometry may well become a major determinative technique. Obviously, an apparatus that can readily be used in all three modes—emission, absorption, and fluorescence—would be most desirable. It is worth emphasizing again that in a silicate analysis laboratory the first consideration must be the ability to determine the alkali metals with speed and assurance. Other capabilities are hardly less desirable but are secondary because alternatives do exist for most other elements.

In a list of near-essential instruments, a photometer should be placed after flame equipment, a simple filter instrument or a spectrophotometer of almost any degree of sophistication. In the absorptiometric procedures commonly applied in silicate analysis, the quality of the results depends less on the complexity of the readout instrument than on the skill with which the analytical solution is prepared. Only the most highly refined chemical techniques will produce a colored species in solution so accurately that a sophisticated instrument is necessary.

Aside from the cost of purchase and upkeep, simple filter photometers have one considerable advantage over more complex instruments; the quality of the light used in the determination is exactly reproducible, a condition difficult to attain when slit and wavelength must be adjusted, as on an ordinary spectrophotometer. This is important when readings are made on the shoulder of a transmittance–concentration curve or when the absorption of a second species overlaps that of the one being measured. Simple photometers are also very much easier and faster to read; seldom does the extra reading precision of a sophisticated instrument have much meaning in routine analysis.

Possibly the most serious limitation of simple photometers and some low-priced spectrophotometers is their narrow wavelength range, commonly 400–700 nm. Numerous useful determinations are accomplished outside this range.

Some instruments may be fitted with direct-reading attachments and printout devices. These must be considered expensive luxuries, except in cases when very many routine determinations are to be made. A more generally useful attachment permits recording the results on a strip chart. This may serve several purposes. It is particularly valuable in flame work because it enables integration of a noisy signal which may be most difficult to interpret by observing a galvanometer needle. It also makes detection of drift and decay of the signal easily possible.

After flame equipment and a photometer, a pH meter is probably most important. Many would regard it as a necessity, and with some justification, particularly if its use in other than simple pH determination is considered. A pH meter is essentially a sensitive voltmeter, one that draws negligible current. In its most common application, it measures the potential difference between a glass and a calomel electrode immersed in an aqueous solution at or near room temperature and reads out in pH units. Most pH meters are readily usable as millivoltmeters and as such can be used in electrometric titrations. There is, of course, nothing to prevent the use of a meter graduated only in pH units as an indicating instrument in titrations of various kinds.

The development of specific ion electrodes has given the pH meter new importance in the analytical laboratory. The ordinary glass electrode is an electrode that is specific for the hydrogen ion—or approximately so. Thus, there is no difference in principle between it and the other specific ion electrodes, and the same instrument is conveniently used to measure the potential of them all. Whatever the ion being measured, there is ideally a logarithmic relationship between its concentration and the potential of the appropriate electrode. In practice, the logarithmic relationship does not always apply. Some pH electrodes—glass electrodes—operate in approximately the theoretical way only between pH about 2 and 10. The behavior of the other specific ion electrodes is likewise always short of the ideal. Plainly the substitution of an expensive meter for a low-priced one will not change the output of the electrodes. As in other instances, the high-priced instrument does not necessarily result in a noticeable improvement in performance.

Taking everything into consideration, there are two sensible alternatives in the purchase of a pH meter: (1) a simple, rugged, battery-operated meter and (2) a sophisticated, multipurpose line-operated meter that can be used as a potentiometer; can readily be connected to a strip chart recorder; has the means for scale expansion, which enables switching from one range to another without recalibrating; is easily serviced; and is not subject to deterioration in long use. Some intermediate instruments are less than satisfactory; and one must be careful not to be led astray by the excessive claims of their manufacturers.

Some strip chart recorders are capable of accepting the signal from an electrode pair directly or through a resistance box. If one of these is available, it may be used to good advantage with specific ion electrodes, in electrometric titrations, as well as in other applications; and a minimum pH meter will be adequate in other applications.

Completing the list of near essentials is a microscope. This may be of almost any type, depending on the nature of the work to be accomplished and the knowledgeability of those involved. Those with no training in microscopy or petrography may obtain as much benefit from a powerful hand lens as from a research microscope. However, chemical microscopy is such a powerful tool that anyone engaged in mineral analysis should attempt to discover some of its possibilities. A petrographic microscope will be preferred by one trained in its use. Otherwise, more can probably be accomplished with an instrument designed primarily for microchemical manipulations. In the latter case, a useful characteristic is a large working distance, including as much space between objective lens and specimen as possible. Turret nosepieces are apt to get in the way. A binocular has its advantages but is often less flexible than a single-tube simple microscope with a flat stage. Ideally, it should be possible to put a 250-ml beaker under the microscope and examine crystalline material on the bottom of it at 50–100 × magnification. At the same time, the capability of determining the refractive index on small mineral grains and determining optical properties should be present. The ability to examine mineral specimens in detail under the microscope before performing any chemical work is a most valuable asset, which frequently avoids troublesome difficulties. For example, a mineral containing 1% or so of pyrite as an impurity will not be subjected to phosphoric acid attack in a ferrous iron determination; the presence of acid-soluble carbonate can be detected in a moment without expending more than a milligram or so of sample; a rock containing many zircons will be analyzed for zirconium, which might otherwise be reported as aluminum; organic material and graphite will be noted and a combustion carbon determination considered; the purity of mineral samples can be estimated and a decision made concerning further purification prior to analysis; and so on. Even a single quick look at a sample may reveal characteristics that will have a bearing on subsequent treatment. Admittedly, what is gained by preliminary microscopic examination depends strongly on knowledgeability; this is gained by experience.

An ability to make microscopic examinations of samples for analysis is particularly important when emission spectrographic facilities are not available. If the minerals in the sample can be identified, it is usually possible to find in the literature what, if any, unusual elements are likely to be present and to look for them chemically during subsequent operations. Several elements, notably zirconium, thorium, and the rare earths, will be undetected and reported chiefly as aluminum in the ordinary classical analysis. These elements occur commonly in easily recognizable minerals and should be suspected present as a result of the microscopic examination before analysis is begun. Boron and fluorine also occur in easily recognizable minerals.

Chemical microscopy is not as widely used as it once was for quantitative analytical work, largely because of the development of several instrumental techniques, but it should not be entirely avoided—especially in laboratories lacking major instrumentation. The relatively new field of ultramicroanalysis has borrowed many of the tricks and techniques of the chemical microscopist and, by combining them with sophisticated instrumentation, has made possible the quantitative analysis of minute samples. An awareness of the possibilities in this direction should be possessed by every first-rate analyst so that use may be made of them when occasion arises. Most often, the simplest expedient combines a separation under the microscope followed by a determination using colorimetry or photometry. For example, a few micrograms of a mineral suspected to contain boron may be quickly picked out by a steady hand or a micromanipulator and fused with a milligram or so of sodium peroxide and then analyzed for boron with curcumin. The size of the sample can be estimated from its volume and specific gravity if a microbalance is not available.

2.3.1. Emission Spectroscopy

Of all the major instrumental techniques, emission spectroscopy is without question the most universally valuable even though it gives results that are seldom as precise as is possible with other techniques. The reason it is so valuable is its ability to yield qualitative information quickly and sensitively with very small samples. The information is permanently recorded on a photographic plate, and quantitative work involves interpretation of this record. No other technique yields so much information about an inorganic sample so quickly [13, 14].

Plates may be examined visually under magnification to obtain qualitative and semiquantitative values for most metals and some nonmetals. Quantitative work requires the use of a microphotometer or a densitometer.

While the accuracy and precision of spectrographic results is not usually high (by the use of the most refined techniques, precision seldom reaches 2% of the amount present, and an accuracy of 5% is considered good), it is adequate in many applications, and in the case of many trace elements, it is better than can be obtained by many other commonly used techniques. The most serious limitations of accuracy are due to sampling difficulties and to inherent nonreproducibility in the photographic process. A very great advantage is that interferences are almost always detectable by inspection of the plate. Emission spectrometry (see below) eliminates the photographic source of error but is much more subject to undetected interference effects.

The sampling problem can be avoided in various ways, notably by fusing a relatively large sample into a glass with a suitable flux, crushing to a powder, and proceeding with the diluted sample. A solution may also be prepared and excited. Both these expedients dilute the sample and lower sensitivity; they are thus less desirable when traces are sought.

Among instrumental techniques, spectroscopy requires a higher degree of skill and more experience than most. Selection of analytical lines that will give good results in a particular matrix is one of numerous problems.

2.3.2. Emission Spectrometry

With an emission spectrometer, exit slits are mounted in the focal plane of the instrument, and the light of specific wavelengths is allowed to fall on photomultipliers. The signal so produced is integrated over a predetermined period of time and activates a readout of some description. Most often, a ratio of two signals, one from a line in the spectrum of the unknown element and the other from an added internal standard, is recorded. The spectrometer thus avoids the errors due to the photographic process and also the time-consuming process of reading the plates. Selection of analytical lines is of great importance and must of course be done before the sample is burned. If interferences exist, there is no way of detecting them. As a result, the process is most useful in the analysis of large numbers of very similar samples. In contrast to emission spectrography, it is better applied to the determination of major constituents than trace constituents because dilution of the sample and the resulting simplification of the spectrum makes it easier to select lines that will be free from interference. Under ideal conditions, precision (and accuracy if good standards are available) on the order of 1% of the amount present can be attained without very much difficulty. The technique is extremely valuable in the routine analysis of silicate rocks for major constituents, yielding results for many elements as good or better than can be obtained by any other rapid technique. If a spark solution excitation is used, the same solution can also be used for the determination of the alkali metals using the flame photometer, as well as several trace constituents using atomic absorption. Another successful form of excitation uses a "tape machine":

The sample is fused into a glass with a suitable flux (usually alkali borate), and the glass is crushed to a powder and fed into a spark on a tape. This device, like the spark solution arrangement, effectively dilutes the sample so that major elements that are relatively free from matrix effects and interferences may be determined.

The use of direct-reading instruments with a dc arc source in the determination of trace elements does not generally lead to highly precise and accurate results, but they may be very adequate when interest is in unusual concentrations of specific elements. The statement applies, of course, to samples that vary widely in composition, as randomly collected rocks. It does not apply to trace element determination in samples of the same material (limestone, cement, etc.), where the matrix is always essentially the same.

Direct-reading spectrometers have been in use for many years in the metal working industries, particularly in steel plants; and routine determinations of not only manganese, silicon, copper, chromium, molybdenum, nickel, and other metals but also of nonmetals such as carbon, sulfur, and phosphorus are made with great speed and high accuracy. Analyzed standards are used for calibration. Vacuum spectrometers are required for the nonmetals. Methods for steel analysis have been well standardized. This has been possible because of the uniform matrix. Development of spectrometric methods applicable to the determination of traces in minerals, with the infinite variety of compositions they exhibit, is a much more difficult assignment. Nevertheless, the very great speed of the technique makes it highly attractive in the analysis of large numbers of exploration samples: Those that show high concentrations of the elements sought can always be examined by slower, but more reliable, methods.

2.3.3. The Internal Standard Principle

Use of an internal standard involves the addition of a known amount of an element not present in the sample. A ratio between its concentration and that of some other element is then measured. The added element unavoidably has different analytical characteristics from the unknown element. This fact has been cited as a disadvantage of the internal standard method.

In isotope dilution mass spectrometry, the internal standard is an isotope of the element being determined, and differences in behavior between unknown and standard are small. Isotope dilution mass spectrometry thus approaches an ideal application of the internal standard principle. In most techniques, for example, emission spectroscopy, an element other than the unknown has to be used. As a result, it is necessary to pay some attention to the characteristics of the two elements, such as excitation potential, type of spectral radiation, volatility, chemical reactivity, and so on. There exists a ready source of error due to differences in such characteristics, which must be avoided as far as possible.

Despite these considerations, the internal standard principle makes it possible to improve precision very considerably, even as much as an order of magnitude. The reason is that, aside from its primary purpose, an internal standard enables simple control of dilution effects. Part of the success of isotope dilution mass spectrometry arises from the fact that once the sample is "spiked," it is no longer necessary to retain all of it in order to complete the analysis. In the flame photometry of sodium and potassium, if no internal standard is used, extreme care is necessary to retain all the sample throughout chemical preparation, and dilutions prior to flaming must be made with high accuracy for good results. If, on the other hand, lithium is added in known amounts during the solution of the sample, exact dilution before flaming is not necessary; an automatic dilutor precise to 1 or 2% is adequate even when results good to 0.1% are desired. The fact that lithium is spectrographically not an ideal internal standard element for sodium and potassium is irrelevant provided a stable and reproducible set of analytical conditions can be maintained.

When an internal standard is used in a secondary method, that is, both unknown samples and known standards are spiked and put through the analytical process, the advantages of the technique are enhanced. Two sets of ratios are then measured: that between internal standard and the element being determined in both unknowns and standards and that between the ratios so obtained. Very often, even gross interferences may be minimized to an acceptable degree through such a process.

The commonly employed method of standard additions may be regarded as an internal standard method that uses the same element as the internal standard. In this method, a signal is generated from the unspiked sample and from other similar samples spiked with known amounts of the element in question. Extrapolation of the results to zero addition yields the required result. Again, if the method is used in secondary mode, its advantages are enhanced. The method of standard additions is not as frequently used as might be expected, probably because it may appear to require much more work than some other procedures. However, for some purposes, it provides the only certain means of avoiding serious errors. A good example is its application in atomic absorption spectrophotometry. This technique has the characteristic of sometimes yielding a calibration curve that (probably because of self-absorption effects) exhibits a maximum: Increase of concentration eventually results in a decrease of signal. The standard addition technique quickly draws attention to the existence of the effect and makes it possible to avoid otherwise undetectable gross error.

A modification of the method of standard additions that deserves more attention is to mix unknowns and standards in varying proportions either before sample attack or after solution. If the preparatory operations are reasonably rapid, the extra time and trouble required in the weighing and preparation of the samples may be well worthwhile. In cases where samples may be retained and used repeatedly as standards (e.g., X-ray fluorescence analysis), accumulation of a bank of mixed samples may be of great value. In other cases (e.g., flame absorption spectrophotometry), standard and sample may be aspirated into the flame at the same time using a controlled flow from each solution. In this manner, the need to prepare extra samples is partly eliminated.

Whenever any internal standard or standard addition method is used, it is essential to include an appropriate number of known samples among the unknowns if full advantage of the technique is to be taken. Failure to do this requires the assumption that calibration curves are straight lines and also that interferences and matrix effects are inconsequential. Neither assumption can be justified. Sometimes it is possible to adjust the procedure or the readout instrument so as to correct such effects, but they must, of course, be discovered and interpreted first. This can only be done by using known samples, of grossly similar composition to the unknowns.

The internal standard principle applied to the determination of potassium by flame photometry: So-called internal standards have been used for many years in emission spectroscopy and spectrometry and in the flame photometry of the alkali metals. In these applications, the term *internal standard* is something of a misnomer; *internal reference* would be better.

There appears to be some rather widespread misconceptions concerning the internal standard principle, and evidence of these appears in the literature from time to time.

The simplest application of the internal standard technique is found in the accurate determination of potassium by flame photometry in a low-temperature flame using lithium as the reference element. Precision of $\pm 0.1\%$ of the amount present is routinely achieved by geochronologists using this technique in K–Ar age determination. In a common procedure, the potassium-bearing sample is fused with exactly seven times its weight of lithium metaborate and the melt is dissolved in dilute acid. The signals generated by potassium and by lithium in an air–propane flame are measured and ratioed. Under favorable circumstances, the ratio is proportional to the potassium content of the sample. Note, however, that this ratio does not provide a measure of potassium unless similar ratio measurements from known samples (standards) are made. The output of the flame photometer must be plotted against the potassium content of known samples to achieve a calibration (see Fig. 2.1).

Plotting the signal due to potassium and that due to lithium separately ideally yields straight-line graphs through a well-established zero (see Fig. 2.2).

Under these ideal conditions, the volume to which the samples are diluted is not of consequence because the K-signal–Li-signal ratio stays the same for each sample. In practice, the plot of signal or concentration for each element is rarely linear. In determining the K-signal–Li-signal ratio in a low-temperature flame under carefully proper conditions, dilutions for high-potassium samples are not at all critical, and a 10%, or even 20%, dilution error has very little effect on the observed ratio; this is because the signal-versus-concentration plots for both potassium and lithium are very nearly linear. This is easily demonstrated by diluting a sample with water and noting the very small change in the instrument readout.

The real advantage in using the internal standard technique in accurate potassium determination does not, however, lie in the possibility of making sloppy dilutions but in the control afforded over variability in the atomizer. Only a very sophisticated atomizing system is capable of delivering aerosol to the flame in a sufficiently constant manner, and without

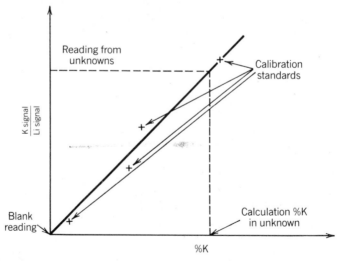

Figure 2.1. Plot of output of flame photometer versus potassium content of known samples.

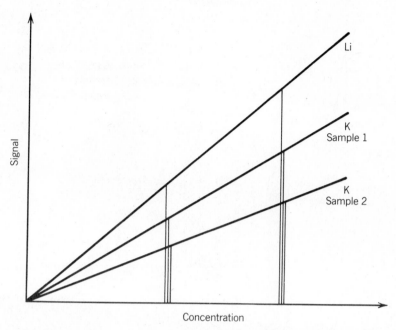

Figure 2.2. Plot of signal due to potassium and that due to lithium separately ideally yields straight-line graphs through a well-established zero.

an internal "standard" (more correctly, an internal reference), it is difficult to attain the 0.1% relative deviation required in K–Ar work. While the amount of sample excited during a readout period may vary, the K-signal–Li-signal ratio remains essentially constant, and an accurate instantaneous readout is possible (see Fig. 2.3). Signal integration is avoided, and small changes in atomization rate are nullified.

Several authors have complained that lithium is an unsuitable internal standard for potassium determination because the excitation characteristics of the two elements are very different. This complaint has validity when the variable factor is the temperature of the flame; if flame temperature varies, the K-signal–Li-signal ratio does not remain constant, and the internal standard principle loses its usefulness. Flame temperature is dependent on solution composition; a "radiation buffer" may be used in relatively high concentration. Lithium borate serves this function well. The air–gas ratio in the flame must be very tightly controlled to avoid changes in flame composition and temperature.

Difficulties are encountered in determining low-level potassium, and the causes of these difficulties may usefully be explored because they become major factors in AA determinations using the internal standard

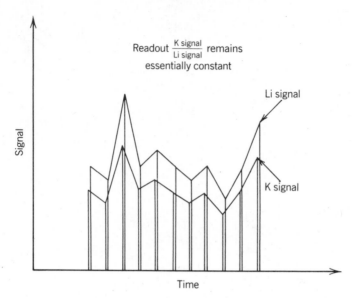

Figure 2.3. Illustration showing K-signal-Li-signal ratio remains essentially constant.

mode. A major problem lies in establishing the "blank" reading when the reagent-grade lithium metaborate contains potassium impurity. It is almost impossible to obtain reagents completely free from potassium. A plot of signal versus concentration in which the blank is ignored will look like the plot shown in Figure 2.4.

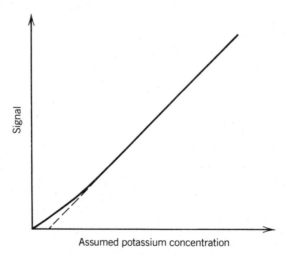

Figure 2.4. Illustration of problems encountered when determining very low levels of potassium.

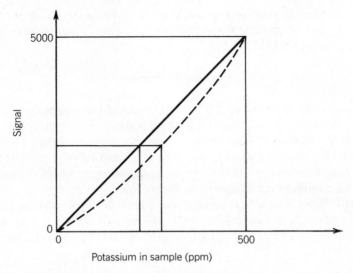

Figure 2.5. Illustration of problems encountered when determining low levels of potassium with appreciable amount of potassium present in blank.

Suppose the true blank is 100 ppm potassium, and we set this at zero; also suppose we have a calibrating standard with 500 ppm potassium, and we set the sensitivity of the instrument to read 5000 with this sample. The calibration graph looks like that shown in Figure 2.5, with the true curve shown as a dashed line. All samples with less than 500 ppm will give too low a result.

Blank readings may not always be due to contamination in reagents; other constituents in the sample may generate a blank. In potassium determinations on high-iron, low-potassium minerals, iron contributes an appreciable signal that must be considered.

Another difficulty lies in the fact that the calibration curves are not exactly linear. This is easily overcome in flame potassium determinations by making appropriate dilutions and by using calibrating standards as near as possible to the unknowns. In a well-setup procedure using a low-temperature flame, approximate linearity extends at least over an order of magnitude with a deviation of no more than about 1%. With high-temperature flames, linearity is not nearly as good.

2.3.4. Inductively Coupled Plasma Used as Source in Emission Spectrometry

This chapter summarizes, in general the terms, theory, instrumentation, and application of inductively coupled plasma (ICP) technology. The use of this method is widening in such areas as the analysis of ores, mill products, and geochemical materials and in environmental monitoring, among others [15–17].

Physical Description

The device shown in Figure 2.6 produces excited free atoms from a liquid sample sprayed into a high-temperature plasma (e.g., high-temperature gas that is a mixture of ions and electrons in a magnetic field). The wavelength of the resulting atomic emission is a function of a particular element, and the intensity is a function of concentration. A typical ICP spectrometer configuration is shown in Figure 2.7.

When radio frequency current energizes the coil surrounding the plasma, coupling between ions in the plasma and the current in the coil transfer kinetic energy to the ionized gas and thus sustains the high tem-

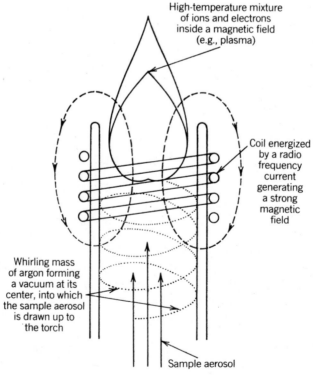

High-temperature mixture of ions and electrons inside a magnetic field (e.g., plasma)

Coil energized by a radio frequency current generating a strong magnetic field

Whirling mass of argon forming a vacuum at its center, into which the sample aerosol is drawn up to the torch

Sample aerosol

Figure 2.6. Illustration of typical inductively coupled plasma torch used as source in emission spectrometry.

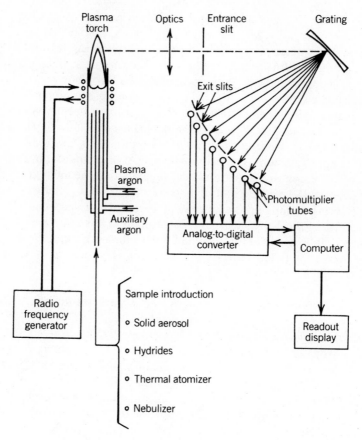

Figure 2.7. Illustration of typical ICP configuration.

perature. Because argon gas generally maintains the plasma, simplified background spectra and lower plasma power requirements result [18].

ICP nebulizers are designed to operate at very low flow rates since too much turbulence may shut down the torch; thus, ICP nebulizers generally contain much finer capillaries. As a result, small amounts of solids in suspension in solutions to be analyzed may generate serious problems to the analytical chemist. It is recommended to use nebulizing systems that generate extremely fine aerosol that will not disturb the plasma torch, thus producing a more stable signal with less background noise; also, it is critical to prevent large droplets or even large aerosols from reaching the plasma torch to minimize chemical interference mechanisms.

Analytical Characteristics

Analytical characteristics include a wide linear working range, low limits of detection, and freedom from chemical interferences. Because the aerosol has a relatively long residence time in a very hot region prior to spectroscopic observation, the sample compounds are largely destroyed (e.g., the sample is completely vaporized into free atoms), which is not always the case in cooler combustion flames used for atomic absorption.

Unlike most analytical techniques, simultaneous analysis of up to greater than 50 elements can occur with proper instrumentation; however, it should be emphasized that it is very unlikely that the analytical chemist will successfully succeed putting into solution so many elements without encountering serious problems or even incompatibilities.

Because of the hotter source temperature and large number of spectral lines available, elements not normally determined by atomic absorption could be readily determined by ICP spectrometry (e.g., P, S, U, Th, Re, W, Ge, Nb, Ta, Sn, and rare earths).

Interferences Encountered in Plasma Spectroscopy

Physical interferences also called influences: Because the capillary tube used in front of the nebulizing system is very small (much smaller than in atomic absorption), a slight change in solution viscosity or surface tension may affect the nebulization flow rate, the size of the aerosol, and consequently the fraction of aerosol reaching the plasma torch. The presence of high concentrations of salts or acids in solution may alter the behavior of the analyte inside the plasma torch and greatly affect the total signal.

Chemical Interferences

Ionization interferences are likely to be a problem. The presence of several elements adding free electrons to the plasma might displace the ionization equilibrium toward more free atoms:

$$\text{Free atoms} \rightleftarrows \text{ions} + e^-$$

The concentration of free electrons from the argon is so high inside the plasma that it reacts as a buffer in the ionization reaction. However, addition of an excess of an easily ionized element to both standards and samples is recommended to compensate for ionization equilibrium problems.

Spectral Interferences

The biggest problem facing the user of an ICP spectrometer is the problem of spectral interferences. These are line interference, light scattering, line broadening [19], and recombination of electrons with ionized atoms, which produces light background. As discussed in Section 2.3.7, these interferences are additive in nature, and a technique such as standard addition cannot minimize them and may not even detect them.

How To Minimize Spectral Interferences

Several things can be done to minimize spectral interferences. The most effective solution is to choose an interference-free wavelength; however, this is not always possible with spectrometers having fixed exit slits. In

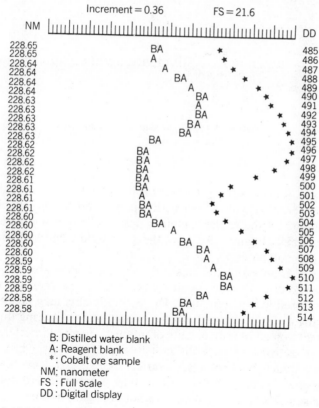

B: Distilled water blank
A: Reagent blank
*: Cobalt ore sample
NM: nanometer
FS : Full scale
DD: Digital display

Figure 2.8. Initial plot showing reagent blank, *A*, distilled water blank, *B*, and sample, *.

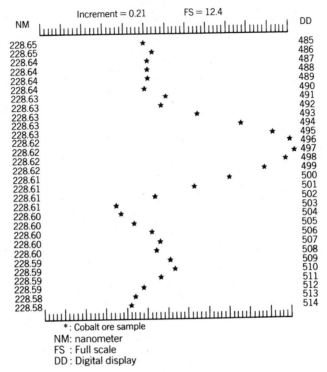

Figure 2.9. Final plot with blank subtracted.

many cases, it is possible to correct for both background and line-broadening interferences with a subtraction performed by the attached computer. Figure 2.8 shows the computer initial scan of a sample containing 0.0123% cobalt. Figure 2.9 shows the scan following blank subtraction done by the computer.

If line overlap problems exist and alternate lines are not available, correction coefficients must be studied carefully, and when obtained, they must be stored in the computer. These coefficients may be obtained by introducing a single element into the plasma and measuring the effect of that element on all other elements of interest. Spectral interferences render ICP spectrometry very difficult to use effectively when analyzing different products made of a large variety of matrices. Nevertheless, ICP spectrometry is a powerful way to analyze geochemical routine samples or water samples where matrices are not changing frequently.

Advantages and Disadvantages of ICP

Advantages. ICP spectrometry is a multielement technique with a large linear dynamic range. It has attractive detection limits for elements such as phosphorus, sulfur, boron, germanium, niobium, tantalum, uranium, thorium, silicon, tin, and rare earths. Chemical interferences are very few.

Disadvantages. Spectral interferences are a serious problem when analyzing a wide variety of products made of very different matrices. The nebulizing system is extremely delicate and the source of many interruptions. Its initial cost is very high, and its operating cost is high (e.g., argon gas, changing nebulizing systems, problems with vacuum system since each photomultiplier is a source of leaks, etc.).

2.3.5. X-ray Methods

Analysis by X-ray diffraction, which detects specific crystalline compounds in a solid sample by recording their X-ray "fingerprints," is widely used and extremely important. No attempt to deal with it is made here, except to mention that some equipment used for diffraction may be modified for use in the fluorescent mode and is useful for elemental analysis.

X-ray fluorescence (XRF) spectrometry has been employed extensively in the rapid analysis of silicate rocks and minerals and with very considerable success. Unfortunately, more has been sometimes expected of it than its capabilities warrant. Interelement interferences and matrix effects are major difficulties. These are capable of correction but only through lengthy calculations that must in practice be done with the aid of a computer. In general, the elements immediately preceding the one being determined in the periodic table have the greatest interfering effects. Also, interferences tend to be of larger magnitude among the light elements; the elements of lower atomic number also pose difficult problems because of the low energy of their fluorescent X-rays [20]. Measurements of elements as light as beryllium have been accomplished, but in actual practice, it is not sensible to attempt determinations of any elements lighter than magnesium except in special circumstances. Certainly, if an emission spectrometer is available, it will be seldom indeed that magnesium will be determined by XRF, except in magnesium minerals where a large number of determinations of this element alone need to be turned out.

Probably more than any other technique for analysis of silicates, XRF should be regarded as strictly secondary because it requires analyzed standards for calibration. The standards should be as close in general nature to the unknowns as possible. In this way, acceptable results can be obtained for many elements without resorting to difficult or impossible corrections for absorption and enhancement effects.

It is somewhat unfortunate that the limitations of the XRF technique are not at all apparent in results obtained after expert sample preparation. Precision is extremely good for most elements: It is relatively easy to obtain a precision of $\pm 1\%$ of the amount present. In no other case is the difference between precision and accuracy so clearly apparent, for these precise results may be (in extreme cases) inaccurate by a factor of 2 or even more.

Accuracy for major constituents is much improved, as a rule, by dilution of the sample or by adding a "heavy absorber" (a device that accomplishes essentially the same thing without as grossly reducing the concentration of the element sought) [21]. However, such dilution reduces the possibility of determining minor and trace constituents. This is a particularly serious defect in silicate analysis because magnesium is a very common minor constituent, and its determination by XRF is frequently marginal even without dilution.

An important advantage of X-ray over emission spectrometry is the much lesser skill required to obtain acceptable results. Selection of analytical lines is no great problem because there are relatively few to choose from, and the actual physical manipulation of the machine is something that anyone can learn quickly. As long as the inherent deficiencies of the method are recognized, there is no great difficulty in acquiring the necessary manipulative skills [22].

Sample preparation has been amply demonstrated to be the largest source of controllable error in X-ray spectrometry. Several methods of preparing samples of silicate rocks and minerals have been proposed in the literature [23, 24], and probably many more have been tried with varying degrees of success. The simplest is to pour the rock powder into a sample holder, and proceed to read the signal. This works well in a few isolated cases (e.g., determining iron in glass sand) but is quite inadequate in most instances. The rock powder may be pressed into a pellet, with a backing of boric acid, methyl cellulose, or other material containing only light elements, and examined directly. Heavy absorbers (La_2O_3, CeO_2, or Bi_2O_3) may be mixed with the powder before pelleting. The powder may be fused with lithium metaborate and the resulting glass poured into a hot graphite mould or onto an aluminum slab to form a "window" or "button." For this purpose, the metaborate, $Li_2O \cdot B_2O_3$, is much preferred to the tetraborate, $Li_2O \cdot 2B_2O_3$, because it attacks almost all refractory minerals and incorporates them into the glass. Unfortunately, it is extremely difficult to obtain a homogenous glass in this way; even when the original button is ground to powder and refused, it may still show inhomogeneity. A better approach and one that is now quite universal is to fuse with either metaborate or tetraborate, grind the resulting glass, and press it into a pellet. Heavy absorber, if used, should be added before fusion.

A pellet preparation technique that is of particular value with oxide mixtures containing no silica is to fuse with potassium pyrosulfate in a platinum crucible and pour the homogenous melt into a glass ring supported on a glass plate. The glass pieces should be heated to 400 °C or so on a hot plate before pouring to prevent their cracking. Using this device, it is possible to determine the several constituents of mixed oxides of, for example, iron, aluminum, titanium, tantalum, niobium with excellent precision. Accuracy can be made as good as desired by preparing synthetic mixed oxides as near as necessary to the unknowns and using them for calibration.

Solutions of metals can be analyzed by absorbing them in a suitable volume of cellulose powder or other X-ray inert material, drying, and pressing the mixture onto a backing of cellulose. Using this device, extremely small samples can be examined for many elements that give great difficulty in other techniques—the rare earth metals, for example. Again, the desired accuracy can be attained by preparing known mixtures and treating them in exactly the same fashion as the unknowns.

It is worth noting that the chemical behavior of the various elements has very little effect on the X-ray determinative technique except to the extent that it affects sample preparation. Thus, sulfur, chlorine, phosphorus, and other nonmetals may be determined with essentially the same precision and accuracy as metals near them on the periodic table. This is in contrast to emission spectroscopy, which determines most nonmetals with difficulty.

The electron microprobe is increasingly used as a tool for elemental analysis. It differs in principle from XRF only by the fact that excitation of the sample is accomplished by means of an electron beam [25, 26]. Fluorescent X-rays are collimated and those of specific energy used to activate a detector and readout device in essentially the same manner as with X-ray excitation. The difficulties in the method are essentially the same; even sample preparation involves similar problems. The difference is in the scale of operation: While the ordinary pellet used in X-ray work may approach an inch in diameter, the sample examined in the electron probe microanalyzer may be several micrometers across. The same problems of finding suitable standards for calibration or of calculating out absorption and enhancement and other matrix effects still exist. In the electron probe the few difficulties derive from the high energy of the electron beam. For example, potassium in micas migrates away from the beam (in effect, volatilizes) and must be determined with dispatch and with this effect in mind. Arsenic and other volatile elements give difficulty for a similar reason.

Methods of sample preparation for the electron probe microanalyzer depend to a great extent on the peculiarities of the instrument that happens to be available, but in general, it is necessary to have a finely polished surface, which must be rendered a conductor (usually by depositing a light element such as carbon on the surface). Some probes have the ability to accept polished specimens as large as an ordinary thin section. These are obviously a convenience in petrographic work. However, analysis must still be of a very small area; there is little possibility of obtaining the gross composition of a large sample.

The greatest drawback to the use of the electron microprobe as an analytical tool is the frequent difficulty of obtaining adequate standards. In most cases, it is necessary to compare an unknown with a known of quite different composition. The accuracy of the final analysis depends to an extreme degree on the ability of the operator to take into account the effect of matrix differences between sample and standard. A great deal of work has been done on this problem, and a considerable degree of success has been achieved, but it remains a serious one [27, 28]. Quantitative elemental analysis is not, of course, the major use of the electron microprobe. Much of its enormous usefulness stems from its ability to compare concentrations of specific elements in adjacent areas of a sample. In such an application, the exact weight or volume percent of the element in question is of lesser consequence. In mineral analysis, its ability to detect and analyze, even semiquantitatively, minute inclusions and impurities in mineral grains is often of supreme value in the solution of mineralogical problems.

2.3.6. A Pelletizing Die

A pelletizing die that produces extremely durable pellets with plane mirrorlike reproducible surfaces has been specifically designed by Fabbi to fit the ARL hydraulic press [11]. Pellets having a strong backing and edge are mandatory if standards and samples are to be used for any length of time. Smooth, plane reproducible pellet surfaces are an important requirement for XRF analysis and especially so for the elements sodium and titanium. The pressure exerted on the edge of the samples by the

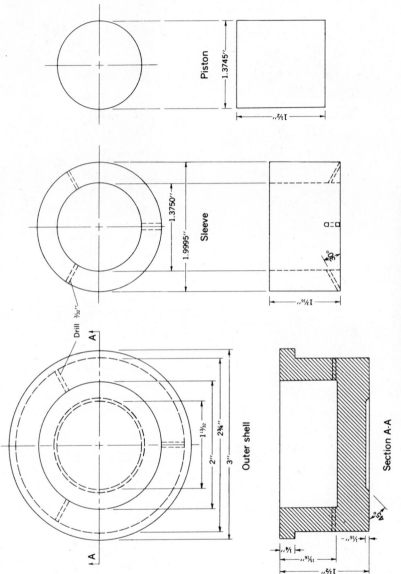

Figure 2.10. Illustration of pelletizing die.

145

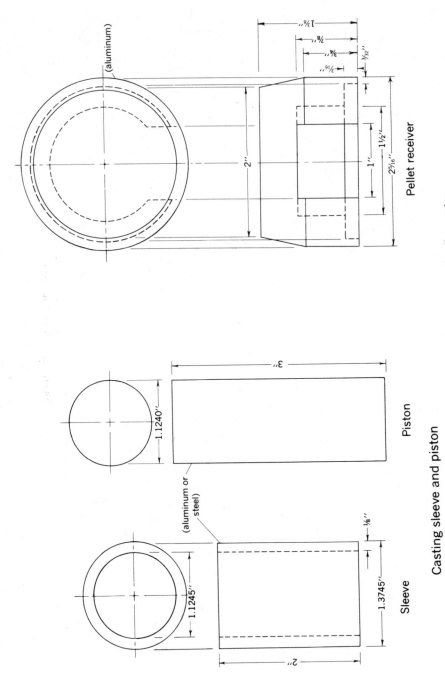

(aluminum)

1 3/8"

7/8"

3/4"

9/16"

3/32"

2"

1 1/2"

1"

2 5/16"

Pellet receiver

3"

1.1240"

Piston

(aluminum or steel)

1.1245"

Sleeve

1.3745"

1/8"

2"

Casting sleeve and piston

Figure 2.11. Drawing of casting sleeve-piston and pellet receiver.

146

sample holders in the G.E. spectograph causes the edges to crumble. The pellets tilt or are pushed forward and the resulting X-ray intensities emitted from the samples are high and nonproportional. Fused pellets having no supporting edge are usable for a very short period due to edge crumbling and generally deteriorate before a complete analysis can be performed.

Figure 2.10 is an illustration of the pelletizing die (which should be of carbon tool steel). Figure 2.11 is a drawing of the casting sleeve piston and the pellet receiver. A cleaned flashlight lens is placed into the die outer shell and the die sleeve is then inserted. The casting sleeve is placed inside the die sleeve flush on the lens and the sample poured in. Moderate pressure exerted by hand on the casting piston forms a semirigid pancake out of the sample. After slowly withdrawing the casting sleeve and piston together, 20–30 ml methyl cellulose powder is poured evenly over the sample in the die. The die piston is inserted and the pellet is compacted in the press. The piston and die assembly is removed from the die outer shell and replaced in the press inverted (so that the pellet is at the top of the die sleeve), and minimal pressure is applied to extrude the pellet. The window in the receiver allows one to observe when the pellet is extruded. Figure 2.12 is a photograph of a finished pellet and a section cut through a pellet.

One inch must be cut off of the ARL piston to accommodate the die. If the inner surface of the die outer shell is not completely flat, the glass lenses will crack. A 5–20-ml-thick piece of plastic cut to 1.35 in. diameter and placed between the bottom of the die and the glass lens will extend the usable life of the lens.

Four replicate (1:1 sample to cellulose) pellets of BCR-1 were prepared to determine sample homogeneity and pellet reproducibility. The samples were counted for 1,000,000 counts, the theoretical counting error being 0.31% at 3σ. The relative standard deviation is 0.40% for the four repli-

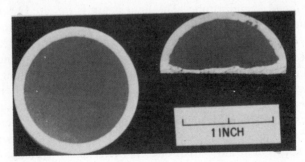

Figure 2.12. Photograph of finished pellet and section cut through pellet.

cates. Hence, at an Fe_2O_3 value of 13.29% with 95% confidence one could expect $13.29 \times 2(0.40)/100 = \pm 0.11\%$, or weight percent Fe_2O_3 = 13.29% ± 0.11%, whereas the theoretical value obtainable is 13.29 ± 0.08%. The figure 0.40% includes errors due to counting, specimen inhomogeneities, uneven specimen surface, placement errors, and grain size.

2.3.7. Atomic Absorption Spectrometry

Exponential growth in application of the atomic absorption technique during the last two decades and its resultant use by persons not fully aware of underlying principles has led some analytical chemists to hold it in low regard. Proper understanding and application of fundamentals, however, show it to be one of the most powerful, rapid, and cost-effective determinative techniques available to the practicing analyst [29–31].

Following is a brief survey of several factors not thoroughly treated in manufacturers' manuals in which the competent analyst should find some interest.

Attention should be brought to the fact that expensive automated spectrometers are not necessarily essential. Instead, it is recommended to have good basic components (e.g., monochromator, power supply, nebulizing system) and stay in control of all parameters that are a key to quality and in many instances to productivity. In other words, stay away from gadgets and put emphasis on simplicity, reliability, and understandability.

Principle

In AA spectrometry, a beam of light is generated by excitation of a small portion of the element to be determined. This is accomplished in a specially designed vacuum tube. The light beam is directed at a phototube and its intensity is measured electronically. Atoms of the element in question are generated in a flame or in a furnace or other device and placed in the path of the light beam. The signal reaching the phototube is diminished in proportion to the number of atoms of the element in question that absorb light quanta from the beam. In principle, the only interference in this process occurs when the spectral line of the element sought happens to exactly correspond in wavelength to that of another element in the sample. Such direct interferences are rare and easily dealt with by measuring absorption at more than one characteristic wavelength.

Nebulization

A sample in the form of a solution is aspirated through a capillary into a chamber using a stream of oxidant gas (oxygen, air, nitrous oxide). The aerosol is mixed with a fuel gas (acetylene, propane, methane) and swept into the base of a flame. Excess large droplets of sample solution run down the side of the chamber and are rejected (Fig. 2.13 summarizes this process).

This process has always been the vulnerable feature of atomic absorption spectrometry, contributing a large proportion of difficulties and annoyances encountered by the working analyst. It probably limits the precision and accuracy of atomic absorption data to ±1% relative (i.e.,

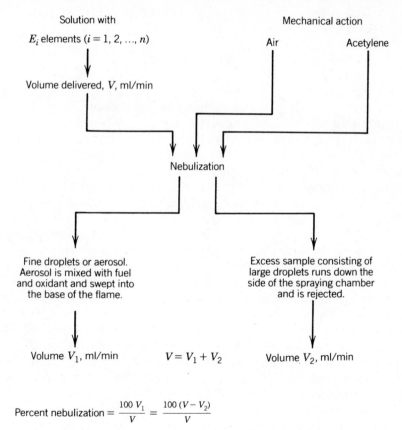

Figure 2.13. Nebulization process in AA spectrometry.

R = 1% except under unusual circumstances not likely to hold in routine analytical situations).

Some rules of thumb may help minimize the difficulty: Solutions to be analyzed should not contain more than 2% (w/v) salts (e.g., derived from fusion procedures) and no more than 5% (v/v) mineral acid (hydrochloric, sulfuric, nitric, perchloric).

Nebulizer materials (Teflon, stainless steel, platinum, gold plate) must be compatible with the solutions to be atomized. Not only should cleaning of the nebulizer be frequent and thorough, but all parts should be examined for wear and deterioration, preferably under a lens or microscope. The tiny hydromechanical parts in a nebulizer wear rapidly and need frequent replacement or restoration. Many AA instruments deliver solution to the flame through a flexible plastic capillary; this tube should be as short as possible to minimize fluctuations due to viscosity and surface tension effects.

Atomization

When the aerosol developed by the nebulizer enters the flame, it is first desiccated, and then salts and oxides are thermally decomposed [32]. The composition of the flame may be varied so as to provide either an oxidizing or a reducing environment in the working portion of the flame. The analyst must adjust the fuel oxidant ratio according to the element sought and the type of interference expected.

These adjustments are designed to generate a population of free metal atoms in exactly that portion of the flame through which a light beam of appropriate wavelength passes on its way to a detector. Obviously, very careful adjustment of flame conditions is necessary for optimum results; the transient population of free atoms must be at a maximum in just that part of the flame being examined. At the same time, interferences of all kinds must be minimized (Figure 2-14 summarizes this process).

Atomization may be accomplished by direct electrothermal heating of the sample in a tiny furnace, usually a graphite tube through which the analyzing light beam is passed. The liquid sample is first dried and ashed within the furnace at the highest temperature that will not volatilize the element sought; then temperature is rapidly increased and the ashed sample is pyrolyzed. Sometimes, additives or preliminary chemicals are required to obtain the desired results. For example, nickel salts may be added to control the volatilization of selenium; sodium salts may improve determination of tungsten; aluminum salts may enhance a molybdenum signal but also increase interference effects; and so on.

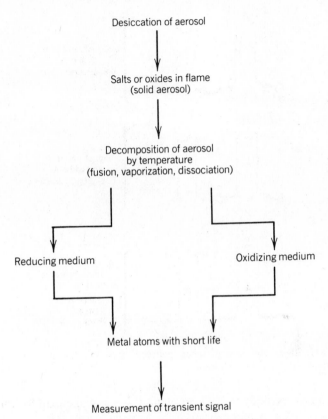

Figure 2.14. Atomization process in AA spectrometry.

Drift and Background Noise

In practice, development of an absolutely constant signal from the tube is almost never achieved; electrical background noise due to imperfect photomultiplier response and electronic multiplication as well as inherent instability in the lamp and its power supply make it mandatory to take these factors into account. In addition, pneumatic background noise due to nebulization irregularity and instability of the atomizing device (flame or furnace) must be considered and controlled or counteracted. Figure 2.15 shows the difference between drift and background noise. Drift is the low frequency of a permanent and disorderly phenomenon that may affect both the baseline and the signal. Background noise is the high frequency of a permanent phenomenon.

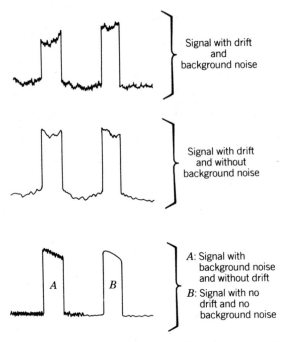

Figure 2.15. Illustration of difference between drift and background noise in AA spectrometry.

When drift is likely to be a problem, it is recommended for the analyst to use a strip chart recorder.

Ratio of Signal to Background Noise

Recommendations of the manufacturer as to wavelength and operating conditions should not be accepted without trial; it is sometimes useful to choose conditions that yield lower sensitivity but a better signal-to-background-noise ratio. This ratio depends not only on instrument characteristics but also on the nature of the solutions being examined; thus the manufacturer's recommendations should be taken as a starting point in setting up a procedure and not as the final and only word.

Detection Limit

Detection limit is defined as the concentration of an element of interest that yields a ratio of signal to background noise not greater than 2 after optimization of all parameters (see Fig. 2.16).

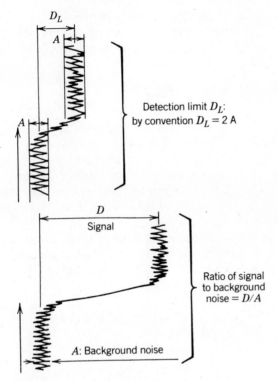

Figure 2.16. Definition of detection limit and ratio of signal to background noise.

Flame Type and Stoichiometry

Selection of a flame type for a specific determination is important. Should one use air–propane, air–butane, air–acetylene, nitrous oxide–acetylene, air–hydrogen, or oxy-hydrogen? Stoichiometry of the flame is a parameter that demands very careful control. It depends on the competence of the fuel and gas regulating mechanisms. It is easy to generate a fuel-rich or an oxidant-rich flame, but an exactly stoichiometric flame is very difficult to achieve; almost always, one works with a slightly fuel-rich or oxidant-rich flame. With the air–acetylene flame, a yellow color indicates an excess of acetylene, a blue color, an excess of air. In a stoichiometric air–acetylene flame, there is a small but observable yellow zone above the blue cone. Generally, a flame slightly on the side of excess fuel (i.e., slightly reducing) is most effective because elements must be present as free atoms for detection. High flame temperatures (such as are generated in oxy-hydrogen or oxy-acetylene flames) generate higher concentrations of free atoms, but the higher flow rate counteracts this effect: each element is optimally determined by selection of the most appropriate gas–oxidant combination. Interferences of several kinds are usually minimized in high-temperature flames.

Lamp Current

The lamp should be allowed one-half hour warm-up at the desired current. The range shown in Figure 2.17 is from low values to the maximum advised by the manufacturer. Sensitivity changes slowly, passes through a maximum, and then diminishes when the intensity is increasing.

The ratio of signal to background noise (BGN) can change quickly, particularly for small values of intensities. The background noise increases as the lamp ages.

To maintain the same background noise level, it may be necessary to raise the lamp current progressively as the lamp is aging. The shape of the curve in Figure 2.18 is a function of instrumental parameters: apparatus, monochromator, slit, and so on.

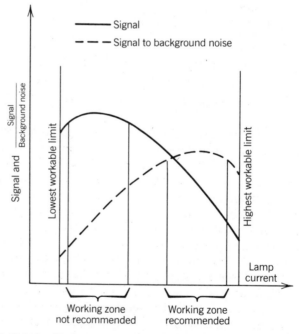

Figure 2.17. Relationship between lamp current and signal-to-background-noise ratio.

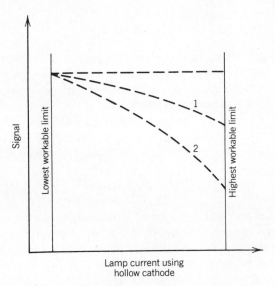

Figure 2.18. Variation of signal with lamp current obtained using various instruments for a particular element: (1) obtained using a good instrument; (2) obtained using a bad instrument. Both curves were obtained using the same hollow cathode.

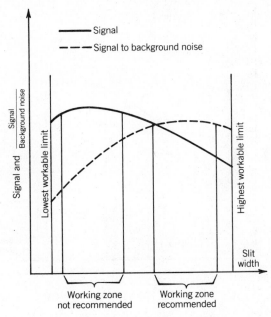

Figure 2.19. Effect of slit width on total signal and on ratio of signal to background noise.

155

Slit Width

Use of too large a slit may cause excessive curvature of absorbance curves while use of too narrow a slit may allow too small an amount of light to hit the photomultiplier: As slit width decreases, sensitivity increases; as slit width decreases, the background noise of the lamp decreases but the electronic background noise increases; also, as slit width decreases, spectral interferences decrease. Figure 2.19 shows the effect of slit width on total signal and on the ratio of signal to background noise.

Flame Optimization

There are two basic facets of flame optimization, that is, correct burner positioning and proper adjustment of fuel–oxidant mixture. Figure 2.20 shows the sensitivity as a function of the nebulizing pressure of the oxidant. For a better understanding of combustion and flame phenomena it is recommended to read Anderson [33] and Rasmuson, Fassel, and Kniseley [34].

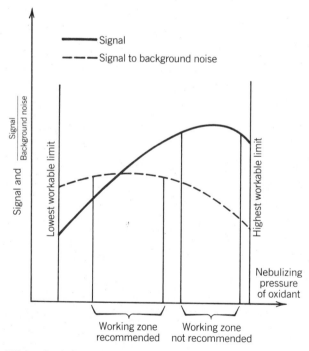

Figure 2.20. Effect of nebulizing oxidant pressure on total signal and on ratio of signal to background noise.

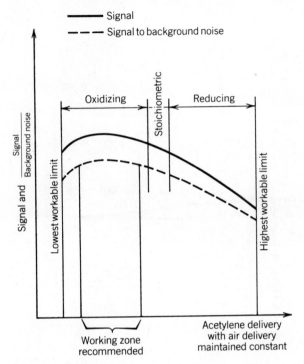

Figure 2.21. Effect of flame stoichiometry on total signal and on ratio of signal to background noise for easily reduced elements.

Stoichiometry. Adjustment of flame stoichiometry is important for elements that are easily reduced such as lead, copper, silver, cadmium, bismuth, and so on. As shown in Figure 2.21, the maximum of sensitivity of the signal observed in the oxidizing zone coincides with the maximum of sensitivity observed for the ratio of signal to background noise. As shown in Figure 2.22, adjustment of flame stoichiometry could be even more critical for elements that are easily oxidized such as molybdenum, chromium, and silicon, where the maximum sensitivity is observed inside the reducing zone.

Total Delivery of Flame. This factor is not as important as stoichiometry. Total delivery of the flame has an effect on the pneumatic background noise, which increases when the delivery is too high, and also an effect on the sensitivity. (If the delivery is too low, the flame is shorter and the maximum atomic density is lower; thus, the sensitivity decreases. In such a case, the positioning of the optical beam in the flame becomes critical. See Fig. 2.23.).

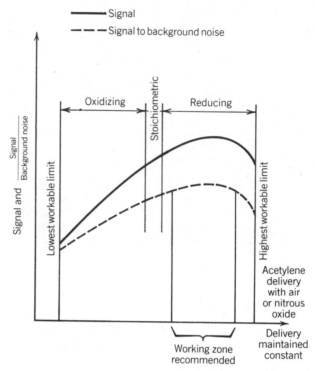

Figure 2.22. Effect of flame stoichiometry on total signal and on ratio of signal to background noise for easily oxidized elements.

Burner Position

Usually there are four burner positioning controls (vertical position, parallelism, eccentricity, and focal point, as shown in Fig. 2.24).

The correct burner position is the position of the burner that yields maximum sensitivity. For some apparatus, the optical beam is a single disk at the focal point; for others, there is a more diffused halo, as shown in Figure 2.25.

Positioning of the focal point in the flame may be critical for many elements and should be studied in many cases, as indicated in Figure 2.26.

Reproducibility

It is recommended to use a strip chart recorder and pass the same solution through the spectrometer at least 15 times with alternate introduction of

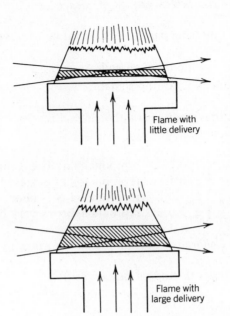

Figure 2.23. Populated atomic zones (hatched zones) obtained with different flame delivery.

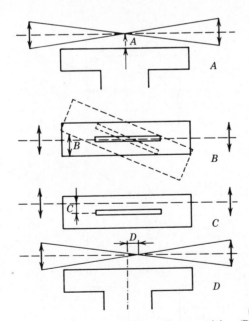

Figure 2.24. (A) Vertical position; (B) parallelism; (C) eccentricity; (D) position of burner toward focal point.

159

pure water. The following parameters would be observed:

1. the stability of the baseline and
2. the reproducibility of the signal.

Relative standard deviation (σ) (with a 65% confidence interval) relate to reproducibility as indicated:

$$\sigma = \begin{cases} \pm\ 0.2\%, & \text{reproducibility is excellent} \\ \pm\ 0.5\%, & \text{reproducibility is good} \\ \pm\ 1.0\%, & \text{reproducibility is poor} \\ \pm\ 2.0\%, & \text{reproducibility is very bad} \end{cases}$$

Calibration Curve and Limit of Linearity

There are three types of calibration curves:

1. always straight;

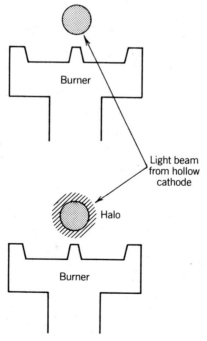

Figure 2.25. Illustration of diffused halo around focal point in some instruments.

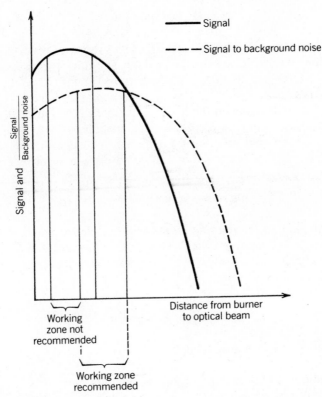

Figure 2.26. Effect of vertical position of burner on total signal and on ratio of signal to background noise.

2. straight in the low values, curved after a particular upper limit; and

3. always curved (e.g., iron, lead, bismuth).

Limit of linearity and curvature depends mainly on monochromator performance. There are several ways to improve the linearity of the calibration curve:

1. change wavelength;
2. use narrow slit width;
3. use an electronic curve corrector;
4. change burner position;
5. and change the pressure support.

Acid Concentration

The acid concentration may affect the signal observed. For instance, during the nebulization process, as the acid concentration increases, the viscosity also increases; then the sensitivity is decreasing. Inside the flame, various effects may also be observed, such as the influence of the concentration of some acids on the aerosol size, which will affect the atomization rate. The optimization of the signal stability versus the acid concentration is generally overlooked by analysts; nevertheless, it is an essential parameter to optimize in atomic absorption spectrometry.

Concentration of Other Elements

The preparation of standard solutions with an element at high concentration may introduce impurities. Also, high concentrations of one element may change the viscosity of the solutions, the flow rate, the aerosol size, and thus the sensitivity.

Choice of Wavelength

The optimum wavelength is a function of:

1. diminution of sensitivity (e.g., for elements such as nickel and iron it may be convenient to use a secondary wavelength instead of doing time-consuming dilutions),
2. optimizing the ratio of signal to background noise (e.g., some wavelengths are more stable than others, and the most sensitive wavelength is not necessarily the most stable), and
3. obtaining more linearity (e.g., some secondary wavelengths for nickel and iron are more linear than the most sensitive wavelength).

Interferences

Atomic absorption spectrometry has the justified reputation to be a technique with relatively few interference problems; however, these problems do exist and must be prevented. Interference mechanisms generally are a combination of influences (e.g., variation of temperature, atmospheric pressure, lamp current, and flame stoichiometry) and interferences originated by the variation of other constituents and their concentration (see Fig. 2.27). The literature often refers to nonspecific interferences and interferences produced by matrix effects.

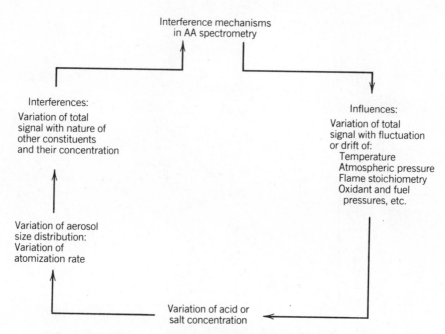

Figure 2.27. Illustration of interference mechanisms in AA spectrometry.

Nonspecific or Additive Interferences. Nonspecific interferences are always additive. They appear when some constituents of the solution respond on the instrument almost exactly in the same way as the analyte (e.g., superposition of wavelength, often observed in emission spectroscopy but rarely in atomic absorption spectroscopy). Additive interferences also appear when some constituents of the solution respond in a very different way than the analyte (e.g., formation of a diffusing fog inside the flame or the tiny furnace, known as a scattering effect, which is frequent in atomic absorption spectroscopy). See the illustration of additive interferences in Figure 2.28.

Matrix Effect of Proportional Interferences. Interferences produced by matrix effects are always proportional and extremely frequent in atomic absorption spectroscopy. Fortunately, these interferences are easily detected and sometimes corrected by the method of standard additions. See the illustration of proportional interferences in Figure 2.28.

Solutions to Interference Problems

Sample Must Resemble Standard. As much as practically possible, standards must be prepared similar in composition to the samples. To achieve

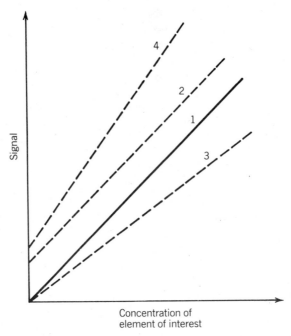

Figure 2.28. Illustration of additive and proportional interferences. Signal versus concentration of the element of interest: (1) with no interference; (2) with an additive interference; (3) with a proportional interference; (4) with a combination of additive and proportional interferences.

this, it is necessary to collect as much information as possible for each sample and each standard. This process is often time-consuming, expensive, difficult, and unrealistic.

Modification of Operating Parameters. Several operating parameters may be changed to solve interference phenomena, such as wavelength, flame stoichiometry, nature of the flame, or even the atomization technique.

Saturation of Interference. Effect of some elements tends toward an asymptote as their concentration is increasing. In such a case, the addition of a sufficient and constant quantity of these elements until the asymptote is nearby is the solution to the interference problem (see Fig. 2.29). An example is shown in Figure 2.30, where the interference of sodium on the determination of tungsten has been stabilized. Making measurements outside the working zone is not recommended; results are likely to be extremely erratic.

Addition of an Element To Reduce Interference. Addition of lanthanum chloride reduces interferences of phosphorus and aluminum when determining calcium. Addition of aluminum chloride reduces interferences of some elements when determining molybdenum. Addition of strontium chloride reduces interferences of some elements when determining magnesium with an air–acetylene flame [35].

Methods Suppressing Additive Interferences. Additive interferences may often be corrected by quantifying the interfering element (e.g., measurement of the total signal on two different wavelengths). There are also what are called background correction methods. Presently, three methods of background correction are in use:

1. measurement of the background using an adjacent nonabsorbing line,
2. the continuum method using mainly deuterium arc, and
3. various methods using Zeeman line splitting [36, 37].

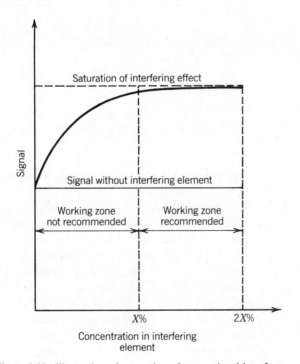

Figure 2.29. Illustration of saturation of proportional interference.

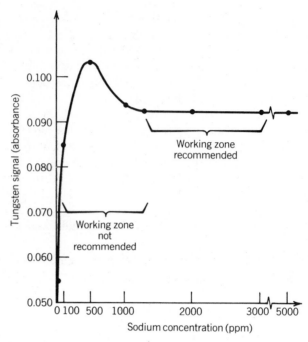

Figure 2.30. Determination of tungsten by AA spectrometry. Illustration of effect of sodium on total signal.

Background correction methods are not essential for flame analysis; however, they are recommended. They become essential with electrothermal atomization.

Method Detecting Proportional Interferences: Standard Addition Method. In a first step, measure the signal from the solution and then the signal from the solution to which we have added known quantities of analyte. In a second step, measure the signal from a blank representing as closely as possible the matrix of the solution and then the signal from the blank to which we have added the same known quantities of analyte as for the solution. Draw on a graph paper the two regression curves after plotting results. If the two regression lines are parallel, conclusion may be made that there is no proportional interferences (see Figs. 2.31 and 2.32). This method is very powerful in detecting proportional interferences. However, sometimes this method cannot correct them and works properly only inside a zone of good linearity.

Method Using Extrapolation to Infinite Dilution. When assuming that the sensitivity is not sufficient, the analyst might not be able to dilute the sample enough to eliminate the effects of the matrix. However, it may be possible to proceed to a series of different dilutions on the sample and on a standard to allow a reading of sufficient sensitivity. Then a graph is drawn with ordinate R, the signal from the sample to the signal from the standard for each dilution d, and abscissa α, the value of the dilution factor, which is equal to $1/d$. Extrapolate the curve until it intercepts the ordinate axis at a point R_0, which is assumed to be the correct ratio of signal from the sample to signal from the standard (see Fig. 2.33). This method is lengthy but very useful in some difficult cases. Knowing the value of the standard it is then possible to deduct the value of the unknown sample.

Method Using Internal Standard. Use of internal standards in AA spectrometry requires a full understanding of the principle if difficulties peculiar to this technique are to be overcome (see Section 2.3.3). In AA

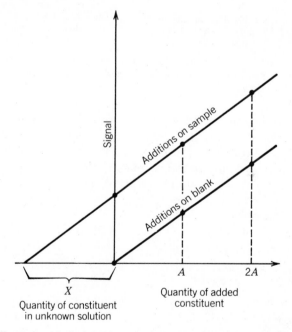

Figure 2.31. Illustration of principle of standard addition method; case where no proportional interference has been detected because two regression lines have the same slope.

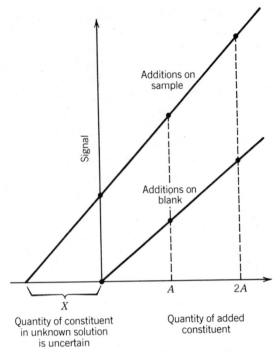

Figure 2.32. Illustration of principle of standard addition method; case where proportional interference is detected because two regression lines have different slopes.

spectrometry, the internal standard principle remains the same, but some difficulties that are minor in flame photometry become serious. There is difficulty in establishing a true zero, and the range of linear response is small, especially for the internal standards. Wide variations in flame temperature and atomization rate cannot be tolerated. The essential requirements are:

1. Linearization of calibration curves for both internal standard element and element sought.
2. Exactly the same concentration of internal standard element in all analyte solutions.
3. Maintenance of constant atomizer and burner conditions.
4. A plot of readout concentrations against true concentrations using standards. This plot need not be linear even if the separate calibration curves for the two elements are linearized.

The procedure should be as follows (e.g., using nickel against cadmium as internal standards):

1. Optimize flame conditions for nickel, using A channel.
2. Plot nickel signal versus nickel content using standard nickel solutions of the same nature as unknown solutions.
3. Generate corrected nickel signals via a linearization process.
4. Using optimum flame conditions for nickel, plot cadmium signal versus cadmium concentration using standard cadmium solutions of the same nature as the unknown solutions and the B channel.
5. Generate corrected cadmium signals via a linearization process.
6. Choose an optimum cadmium concentration. Introduce this concentration of cadmium into all solutions, standards and unknowns.
7. Go to double-channel internal standard mode using nickel standard solutions, all with the same optimum cadmium content, and read off concentration of nickel.
8. Plot readout concentrations against actual concentrations.

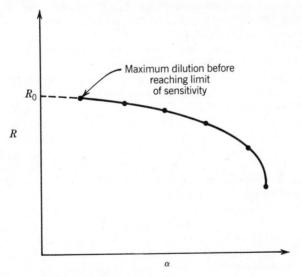

α : Dilution factor, $= 1/D$
D : Dilution
R : Ratio of signal from sample to signal from standard
R_0: Correct ratio of sample to standard

Figure 2.33. Illustration of principle of extrapolation to an infinite dilution.

9. Add the same optimum concentration of cadmium to unknown, and obtain nickel content from graph prepared in previous step.

During all these operations, flame conditions must remain unchanged. If any adjustment is made, the whole process must be repeated.

Solutions should all be diluted to bring their nickel content into an optimum range, but always in such a way that the cadmium content is absolutely constant.

2.4. DISSOLUTION PROCEDURES

2.4.1. Lithium Metaborate Fusion, Solution in Dilute Nitric Acid

REAGENTS: Lithium Metaborate. Anhydrous $LiBO_2$ or $LI_2O \cdot B_2O_3$ is somewhat difficult to prepare free from impurity. As a result, commercially available reagents are sometimes inadequate and must be purified before use. Alternatively, $LiBO_2$ may be prepared by reacting lithium carbonate with boric acid [38, 39]. The molten salt dissolves air gases, including H_2O, HCl, SO_2, CO_2, Ar, and so on, in appreciable quantities, so that vacuum fusion may be desirable in some applications. Lithium metaborate forms a dihydrate above 37 °C and an octahydrate below that temperature. A saturated aqueous solution contains about 1% $LiBO_2$ at 0 °C, about 3% at room temperature, and about 13% at 100 °C. Supersaturated solutions are easily prepared. Other lithium borates, for example, $Li_2B_4O_7$, may be formed.

Preparation of $LiBO_2$ from lithium hydroxide and boric acid may result in a product that is not stoichiometrically $LiBO_2$: Preparation is best accomplished using powdered Li_2CO_3 and H_3BO_3. Weigh equivalent amounts of finely powdered lithium carbonate and "impalpable powder" boric acid (1 mol Li_2CO_3 to 2 mol H_3BO_3) and mix very thoroughly, using a mechanical mixing device. Transfer to a platinum or silver dish and heat slowly to 250–300 °C. Crush and grind the resulting cake, and reheat to 625 °C. The reaction will be about 95% complete if the reactants were thoroughly mixed and finely powdered, and the product may be used in this condition for some purposes. Often, however, further purification is desirable.

To purify lithium metaborate, add 500 g to 3 l of hot (70–80 °C) water while stirring mechanically. Continue to stir and heat to 90 °C. Cool slowly, with stirring, until the temperature falls to 85 °C. Some crystallization (of the dihydrate) should occur. Allow the sludge of undissolved lithium metaborate and impurities to settle, and filter as rapidly as possible through a fluted 40-cm Whatman No. 42 paper or use a Büchner funnel. Allow the filtrate to stand undisturbed for 48 h. $LiBO_2 \cdot 8H_2O$ will crystallize. Pour off and discard the supernate and blot the crystals as dry as possible with analytical-grade filter paper.

Transfer the octahydrate to a silver, platinum, or porcelain dish (depending on the intended application) and place in a *ventilated* oven at 40 °C or slightly lower temperature until the crystals crumble to a fine powder of dihydrate. Drying may take 3–4 days or longer with large batches. Too high a drying temperature will result in a cake or puddle that is very difficult to handle. After dehydration is substantially complete, increase the oven temperature to 180–200 °C for several hours to partially decompose the dihydrate.

Transfer small portions of the dihydrate to a platinum dish, and heat to 625 °C. If the dihydrate was not partially decomposed by raising the oven temperature to 200 °C, the second heating will result in a large volume increase (as much as 10-fold). After heating for at least an hour at 625 °C, cool, crush, and screen through 16 mesh or finer using a brass, steel, or nylon screen depending on the intended application. Mix the product very thoroughly.

The dihydrate loses water much below the 625 °C temperature recommended, but the product is unmanageably light and fluffy. Heating small portions of the reagent to 800 °C and cooling rapidly will result in a much heavier powder: $LiBO_2$ undergoes a polymorphic transition at 785 °C. Fusion or vacuum fusion may be desirable for some purposes. Vacuum fusion should be accomplished in capacious vessels using small quantities because a large heat loss during the polymorphic transition at 785 °C may result in violent splattering of the still molten portion of the cooling mass.

Whatever its source, the purity, homogeneity, and stoichiometry of the reagent should be determined before use. Spectrographic analysis of several subsamples of the mixture is advisable in case impurities occur in isolated particles. Stoichiometry is best determined by fusing a 1-g sample in a platinum crucible, determining the loss on fusion, and then treating the residue with hydrofluoric and sulfuric acids and evaporating. Drying and ignition will yield lithium sulfate, the weight of which should come within a milligram of theoretical.

Evaporation of the lithium sulfate solution to dryness and ignition is somewhat difficult. The following procedure is recommended: Use a 60-ml crucible. Treat the fused $LiBO_2$ with 2 ml concentrated sulfuric acid in 5 ml water, and add 30 ml hydrofluoric acid. Heat on a steam bath until most of the water and hydrofluoric acid has evaporated. Repeat the treatment with hydrofluoric acid and again evaporate. Pack the crucible half full of analytical filter paper pulp, and add 5 ml concentrated ammonia solution. Dry in the oven, transfer to a cold muffle furnace, and heat slowly, gradually raising the temperature over a period of several hours to 450 °C. Hold at this temperature until carbon is removed. Heat to 600 °C, cool, and weigh Li_2SO_4. The procedure is somewhat less difficult if it is carried out in a flat-bottomed platinum dish. Heating on a temperature-controlled hot plate is then feasible.

Lithium metaborate fusions using pure $LiBO_2$ are much to be preferred to alternatives in which mixtures of $Li_2B_4O_7$ and Li_2CO_3 or Li_2CO_3 and H_3BO_3 are used, largely because of the homogeneity and purity of the flux. The internal standard principle is inapplicable when mixtures are used because their composition may vary from sample to sample. The evolution of gases during fusion results in appreciable losses of volatile metals and in spattering. Borate glasses are so viscous that even prefusion of mixed fluxes may not result in a homogeneous material. If mixed fluxes are used, they should be prepared in large batches, fused, crushed, screened, and mixed before use. They may then be satisfactory for routine applications.

When it is desirable to include extra alkali or extra boric acid in the flux, it is best to use pure $LiBO_2$ and add an appropriate weighed amount of boric oxide or lithium carbonate to each sample rather than to use a mixed flux. Sometimes, it is advantageous to add silica to the fusion mixture (analysis of iron ores, chromite, and spinels). This is best added as purified, powdered quartz, and the appropriate amount should be weighed and added to each sample.

APPARATUS: GRAPHITE CRUCIBLES. These are obtainable as stock items from manufacturers of graphite electrodes. They may be fabricated by machining graphite rods. Vitreous carbon crucibles may be substituted. If graphite crucibles are carefully machined free of ridges and other imperfections and if they are preignited and the powdery inside surface is left undisturbed, $LiBO_2$ melts may be poured from them without appreciable loss. The graphite must, of course, be of high purity in most analytical applications. If fusions show a tendency to stick, a different grade of graphite should be tried. In preparing crucibles from graphite rod, the cavity should have a hemispherical bottom. It may be desirable to prepare a reaming tool to touch up inside surfaces of used crucibles and remove the small beads of molten material that sometimes remain after fusion.

The internal standard principle is important to the use of rapid methods involving pour-out of the molten mixture of flux and sample. If either lithium or boron from the flux is used as an internal standard, and if flux and sample are thoroughly mixed before fusion, the very small amount of molten material that remains in the crucible after pour-out will usually have little effect on the results. It is, of course, necessary to be sure that none of the material from one fusion finds its way into the next, hence the suggestion that a reaming tool be used to touch up crucibles before reuse.

In ordinary practice, graphite crucibles are good for a dozen or so fusions before they become too thin to be serviceable. Their life depends to a large extent on the atmosphere within the furnace and on temperature. If air circulation is too great, crucibles may burn out very rapidly.

PLASTIC WARE AND STIRRING APPARATUS. Beakers of polyethylene are probably best to use in the dissolution of lithium metaborate melts. Stirring bars, used over a magnetic stirring unit, are most convenient. They should be nonporous; some Teflon-coated stirring bars do not meet this requirement. A variety in which the iron slug is wrapped in paper and enclosed in transparent plastic is very desirable because evidence of failure is immediately apparent. Much difficulty can result from the use of defective stirring bars, which carry material from one determination to the next; they can develop a leak in the plastic coating and slowly deliver iron originating from the enclosed magnet into analytical solutions.

If the as-prepared solution is to be diluted before use, this is best accomplished by means of a diluter. In trace element work, the 1% accuracy of such a device is nearly always adequate. If the internal standard principle is used, no error is introduced in the dilution step. The continuous use of conventional pipettes and volumetric flasks with silicate solutions in analytical procedures using lithium metaborate–nitric acid solutions (or any other solutions containing silicic acid) leads to difficulty because of the deposition of silica gel inside the apparatus, which cannot be removed in ordinary cleaning processes. The deposited silica gel adsorbs or absorbs ions from solution and may be the largest single source of contamination in an otherwise well-conceived procedure.

GLASSWARE. Before each using, glassware should be scrubbed with a brush, using soap if the glass is new. New glassware carries submicroscopic beads and blebs of alkali-rich contamination that can only be removed by rather drastic treatment. After scrubbing, cleaning with chromic-sulfuric acid cleaning solution followed by treatment with dilute nitric or hydrochloric acid containing peroxide and thorough washing with distilled water is desirable. If volumetric flasks are used, they may be scoured by shaking a suspension of sand in sodium hydroxide solution in them prior to the acid cleaning treatment.

Since almost all rapid procedures with lithium metaborate–nitric acid solutions require that the element of interest be diluted extensively before analysis, it is essential that the contribution of the glass surfaces to the solution composition be minimized in every way possible. Probably much of the difficulty encountered in atomic absorption analysis, where dilutions are sometimes extreme, is due to contamination from inadequately clean glassware.

Glassware used for solutions containing silicate should never be allowed to dry without thorough cleaning. Deposited siliceous matter is extremely difficult to remove.

PROCEDURE. Weigh 0.1000 g of sample into a porcelain or platinum crucible, and add 0.700 g lithium metaborate. Mix thoroughly with a metal spatula or rod. Black-glazed porcelain crucibles are usually the best for this purpose because the white metaborate can easily be seen. Black-glazed crucibles contain uranium and should not be used when a trace of this element is of interest.

Transfer the mixture quantitatively to a preignited graphite crucible. In most procedures a blank and a standard, or several standards, should be weighed up at the same time as the sample or samples. The blank must contain silica; fusions without silica do not pour well. If the samples are deficient in silica (iron ores, chromite, phosphate rock, etc.), up to 0.1 g pure powdered quartz should be included in the mixture.

If a brush is used in transferring the mixture to the graphite crucible, it should be employed in such a way that particles of sample and flux are not forced into the bristle. Sable artist brushes seem best for this purpose. The flux and sample should not be weighed into the graphite crucible and then mixed because touching the graphite surface with the stirring rod or spatula causes the melt to stick to the crucible. Sample and flux must be thoroughly mixed for best results.

Place the crucible and charge in a furnace held at 950 °C for 15 min. In 200- or 250-ml plastic beakers, put 100 ml of 4% (v/v) nitric acid and a plastic-coated stirring bar. Place these over magnetic stirring units. Pour the melt into the dilute nitric acid, and stir to solution, which should be complete within a few minutes. Immediately make any dilutions that will be required in the analytical procedures to follow. Also add any other reagents that may be required, including lanthanum solution (as a releasing agent in AA spectrometry), fluoride (to complex silica), or ferric nitrate (used in rapid procedures for phosphorus).

Following is the solution procedure for small samples.

PROCEDURE. Tare a 40- or 60-ml platinum (or gold-lined platinum) crucible, and add 20 mg or less of sample. Reweigh. If necessary, determine loss in weight at 105 °C or other appropriate temperature or roast at 500 °C to oxidize sulfur. If the sample contains little or no silica, add up to 20 mg pure powdered quartz. Add exactly seven times the weight of sample of lithium metaborate, and mix by tapping the crucible. Mixing should be as thorough as possible. Finally tap the crucible so that the charge is collected in one pile toward the side of the crucible. Cover with a tight-fitting cover, and support the crucible in a hole in an asbestos board so that the lower half or third is below the board. Place a second crucible containing water on the lid to act as a condenser if the alkali metals, chlorine, or other volatile element is to be determined. Dip two tubes into the condenser, one to deliver cooling water and the other to remove it through a suction pump. Heat the lower crucible with an oxidizing Meker flame for 10–15 min. It is essential that the flame be oxidizing. Heating in a reducing flame will result in rapid transfer of metals such as iron, manganese, and cobalt into the platinum. To test for an oxidizing flame, set up a nickel crucible in exactly the same position as the platinum, and observe its surface during 5 or 10 min of heating. The nickel must be completely filmed with oxide at all times. If the flame is reducing, bright surfaces will appear on the nickel as oxide is reduced to metal. Failure to observe such precaution will result in contamination of the platinum ware, intersample contamination, and hopelessly confusing results.

At the end of the fusion, allow the whole assembly to cool, turn off the condenser water, and dismantle. From a burette, add exactly 1 ml of 4% (v/v) nitric acid for every 10 mg of sample to the residue in the crucible, washing down the lid in the process. Cover, add a small stirring bar, and stir to solution over a magnetic stirring unit. The uncertainty in the exact volume of solution is not usually of consequence if standards are used to calibrate and the procedure is exactly the same each time. If lithium or boron is used as internal standards, the exact final volume is irrelevant. Experience has shown that the errors introduced through solution of the melt and subsequent dilution to volume in a volumetric flask are generally larger than those due to the slight uncertainty in the recommended procedure. Analytical imprecision in the analysis of 10–20-mg silicate samples is seldom perceptible in comparison to sampling imprecision at this level.

The solution obtained in the above procedure may be treated in exactly the same manner as that obtained from 100-mg samples fused in graphite. It has the advantage that chlorine and fluorine and possibly sulfur are more likely to be quantitatively retained in the solution.

When larger samples are to be fused with lithium metaborate, it is best to fuse and crush the flux to reduce its volume, which might otherwise overflow even a larger graphite crucible. Samples up to 500 mg or even 1 g may sometimes be successfully dissolved by fusion with four to seven times their weight of fused $LiBO_2$, followed by pouring the melt into 100–250 ml of dilute nitric acid and stirring. Thus, 0.5 g sphene, very difficult to dissolve by other procedures, especially when it contains contaminating zircons, is quickly put into solution in this way: For sphene, the best flux–sample ratio is $7:1$, and 50 ml of 4% nitric acid per 100 mg sample is sufficient.

Fusion in platinum crucibles is not advised for samples larger than 100–200 mg if solution in dilute acid other than hydrofluoric is contemplated.

2.4.2. Sulfuric–Hydrofluoric Acid Attack

APPARATUS. PLATINUM DISHES. Flat-bottomed, 55-ml platinum–rhodium dishes are best for the purpose. These should be equipped with well-made hardwood forms, permitting careful reshaping after each use. If the bottoms are permitted to become rounded, the advantages of this shape of dish are lost. Frequent working is necessary. Teflon vessels are less desirable, unless it can be demonstrated that they do not exhibit any memory effect due to microporosity. Teflon cannot, of course, be used in ignition procedures, which are sometimes desirable following this attack.

STIRRING RODS. Stirring rods of stiff platinum–rhodium wire with machined round ends to prevent scratching are best. They should be of a length that will permit them to remain in the dishes during heating. Teflon rods may be used but are something of a nuisance when they must be removed before ignition procedures. Platinum tubes, with the ends welded shut, are very convenient but cause much difficulty if they develop leaks. Stirring bars of some kind are almost always essential; efforts to carry through the procedure without them are sometimes successful; most often the sample becomes gelatinized into a siliceous mass that dissolves very slowly if swirling the dish is the only means of mixing its contents.

COVERS. Teflon "watch glasses" should be used in the early stages of the procedure when long digestion with hydrofluoric acid is necessary. After fluoride has been removed, silica watch glasses may be used. Glass covers are undesirable.

PROCEDURE. Transfer 0.5000 g sample to a 55-ml platinum dish, and roast, if necessary, at 500 °C for several hours to oxidize iron and remove sulfides. Cool, cover the dish with a silica watch glass, add a little water, and swirl until the sample is thoroughly wetted. Using a pipette, add 5–10 ml of 1:1 nitric acid. Heat on the steam bath or on a low-temperature plate until any effervescence ceases. Remove and wash off the silica cover. Introduce a platinum rod and add 10 ml hydrofluoric acid. While heating on the steam bath or plate (below the point where bubbles appear), stir constantly until there is no danger that the sample will gelatinize. Heat, uncovered, as strongly as possible without any effervescence until most of the liquid has evaporated. Cool, wash down the sides of the dish with a little water, add 5 ml of 1:1 sulfuric acid and 10 ml hydrofluoric acid, mix thoroughly, cover with a Teflon cover glass, and heat at steam bath temperature for an hour or more or overnight. If attack is incomplete, repeat the treatment with hydrofluoric acid. Evaporate slowly to fumes of sulfuric acid. Wash down the sides of the dish with water and again evaporate.

Complete removal of fluoride, necessary in some subsequent procedures, is extremely difficult, especially when much aluminum or magnesium is present. Dilution and reevaporation to fumes of sulfuric acid may be repeated several times without complete success. Insoluble aluminum and magnesium minerals, not present in the original sample, may form, sometimes in crystals that can be mistaken for unattacked sample. Some minerals (e.g., zircon, tourmaline, sphene, some garnets, ilmenite, and kyanite) are not attacked rapidly and frequently remain among the inadvertently synthesized minerals.

When removal of fluoride is judged completed, wash down the dish with water, cover with a silica watch glass, and digest until all soluble sulfates are dissolved. Filter through a tight paper into an appropriate volumetric flask, and wash thoroughly with hot water. Scrub the dish with a policeman, and transfer its contents completely to the filter.

Ideally, the solution in the volumetric flask will contain all the nonvolatile elements in the sample that form soluble sulfates. Of the common elements, only barium and some strontium should remain in the insoluble residue. In practice, almost any element may be present in the residue, and the solution may contain fluoride and silica despite every effort to avoid this. Often some of the sample remains undecomposed.

As a result of these complications and before there existed any useful alternative to the hydrofluoric acid attack, various devices were developed to improve the method. Among these, the Teflon-lined "bomb" is probably the most successful. This device, described in various modifications, results in much more complete attack of refractory minerals, and reduces the time and trouble attached to the procedure; it does not reduce the difficulty in completely removing fluoride from the sample solution. Generally, the application of determinative methods to solutions prepared in this way should be limited to those in which fluoride does not interfere, and the insoluble residues should always be examined for the element sought.

The most useful modification of the sulfuric–hydrofluoric acid dissolution procedure is that developed by Abbey and Maxwell for the determination of the alkali metals. A somewhat modified version of this method follows.

PROCEDURE. After complete solution of the sample as described above but before the filtration, add 100 mg MgO as sulfate, and evaporate the solution to complete dryness, fuming off all the free sulfuric acid. Increase the heat, first on the hot plate and then over a burner, until bisulfate has been decomposed, and then ignite to red heat. Cool and add 10–20 ml water. Digest on the steam bath for an hour, breaking up the solid matter with the platinum rod. Filter into a 150-ml beaker through a tight filter, and wash thoroughly with hot water. To the filtrate, add 100 mg calcium carbonate, and digest on the steam bath, with occasional stirring, until the solution is clear and the precipitate has settled. Filter through the same paper into a 50- or 100-ml volumetric flask, and wash thoroughly with hot water. The solution contains substantially all the alkali metals in the sample plus an indefinite amount of calcium and magnesium. Dilute to volume, and remove an aliquot for a proximate determination of calcium and magnesium by EDTA titration. Prepare standard sodium and potassium solutions, matching the unknown in calcium, magnesium, sodium, and potassium content and starting with pure alkali sulfates. Compare standards and unknowns using a flame photometer. This provides one of the best possible primary methods for alkali determination. Lithium, however, is seldom completely recovered; determinations of this element by this method are always low. There is also the possibility of loss of rubidium and cesium if the ignition step is not cut as short as possible. As in all other procedures of this type, it is advisable to carry through several known samples to check the performance of the method.

2.4.3. Perchloric–Hydrofluoric Acid Attack

The procedure is essentially the same as with sulfuric acid except that it is much more difficult, as a rule, to completely remove fluoride by evaporation with perchloric acid. After decomposition of the sample is complete, evaporation of excess perchloric acid should be followed by a new addition of perchloric acid and reevaporation almost to dryness. Perchloric acid has the advantage that most metal perchlorates are more water soluble than the corresponding sulfates. The outstanding exception is titanium. Sulfuric acid solutions of titanium are stable; when this element is fumed with perchloric acid, titanic acid is likely to separate as an intractable precipitate, which may carry down other elements such as aluminum and phosphorus. Clear solutions obtained through the perchloric–hydrofluoric attack are very likely to contain fluoride, which complexes titanium and holds it in solution.

Perchloric acid solutions should not be evaporated to complete dryness; an excess of perchloric acid should always remain.

The danger of violent reaction or explosion on evaporating perchloric acid solutions containing organic matter or reducing materials (especially antimony compounds) is entirely removed if sufficient nitric acid is used, as directed in the procedure.

Perchloric acid solutions of rocks and minerals are most useful in procedures in which extraction with organic solvents is anticipated. They are also useful in atomic absorption procedures because perchlorate has a generally enhancing effect on the signals obtained from many elements. The addition of alcohol, instead of water, to perchloric acid residues often enhances the signal obtained in flame emission procedures. For example, flame emission spectrometry of the rare earths becomes practicable in alcoholic perchloric acid solution.

2.4.4. Attack with Sodium Peroxide, Solution in Acid

This procedure would be very useful if pure sodium peroxide were more easily obtainable. Possibly the best reagent to use is that sold by the Parr Instrument Company for use in their peroxide bomb calorimeters. Most reagent-grade sodium peroxides are insufficiently pure for good analytical work. Every batch of reagent should be tested in the procedure in which it is to be used. If found satisfactory, it can be of very great value, for sodium peroxide decomposes many minerals that resist most other reagents. Unfortunately, impurities are likely to be present in discrete particles—for example, of oxidized alloy steel from the vessels in which it is prepared. Duplicate determinations should be the rule when metallic elements are sought in procedures using Na_2O_2.

Sodium peroxide fusions may be accomplished in platinum, iron, nickel, silver, gold, or zirconium crucibles. Each has advantages. Zirconium crucibles are probably less desirable than the others despite the fact that they are very resistant to the flux. Zirconium forms peroxy compounds, and even the small amount that finds its way into solution of an analytical sample may cause major difficulty. For example, in chromium determinations, complete removal of peroxide, essential to the success of most methods, is difficult when zirconium is present. Iron or platinum crucibles are most often used. Iron crucibles of high purity are easily available and, if used properly, permit rapid solution of almost all minerals in a minimum of time, with iron the only contaminant. Platinum crucibles can be used if heating conditions are closely controlled. In this application, pure platinum, not platinum–rhodium, crucibles are recommended. Procedures for Na_2O_2 fusion follow:

FUSION IN PLATINUM. Transfer 0.1–1.0 g of the sample to a 30–50-ml platinum crucible, and add five times its weight of Na_2O_2. Mix thoroughly with a glass or platinum rod. Weigh 1–2 g anhydrous sodium carbonate onto a piece of hard-surfaced paper (albanene). Rinse off the stirring rod in this. Transfer the sodium carbonate to the crucible, covering the peroxide–sample mixture as completely as possible. Cover and transfer the crucible and contents to a furnace held at 500 °C, and heat for 1 h or slightly longer. At a slightly higher temperature (about 525 °C), sodium peroxide begins to attack platinum rapidly. At 500 °C, attack is minimal. The temperature of the furnace is critical: If crucibles used in the procedure become coated with a yellowish film of platinum oxides, the temperature is too high. In a well-controlled procedure, with most mineral samples, no more than a fraction of a milligram of platinum should be introduced into the analytical sample.

FUSION IN IRON. Load the crucible as above (fusion in platinum). Mount the crucible (which need not be covered) on a triangle, and heat with a Bunsen or Tirrill flame. The flame must be adjusted so that the inner cone of unburned gases envelops the lower part of the crucible. This has the effect of heating the charge around the edges rather than from the bottom. If the flame is adjusted properly, fusion will occur quietly. If the flame is adjusted to overheat the bottom of the crucible, the iron may ignite, and the charge may break through the crucible. The key to a successful, smooth, and trouble-free fusion is close control of the flame. Fusion should be complete in a few minutes and should be continued until the melt is homogeneous and dull red. In the later stages, it is desirable to pick up the crucible in a pair of tongs and swirl it so that the bottom is momentarily exposed: There should be no particles of sample visible. As soon as fusion is complete, the crucible should be placed in a shallow dish of water to cool as rapidly as possible.

BOMB FUSIONS. The Parr peroxide bomb is an extremely useful device. Detailed directions for the fusion of various sorts of samples are given by the Parr Instrument Company. The Parr bomb is mostly used for the decomposition of organic materials such as coal and oil. Probably the possibilities of this attack have been underdeveloped. A fusion resulting from a determination of the calorific value of coal or oil in a Parr peroxide bomb calorimeter should be usable for the determination of almost every constituent of the coal or oil since it is produced in a closed system. By substituting pure carbon for the coal sample, adding a weighed amount of any silicate mineral, and proceeding as for a determination of calorific value, a fusion of any sample can be obtained in a few minutes without loss of any constituent. Sulfur, chlorine, bromine, iodine, fluorine, boron, carbon, and other volatile elements can be completely fixed for their determination without possibility of loss.

A device for attack of samples containing organic matter can be improvised from two iron crucibles and an iron tube. The crucibles should be large and should just fit within the tube. The charge—sample + sodium peroxide + accelerators, if necessary—is placed in one crucible, the other is inverted over it, and both are enclosed in the closely fitting tube. Heating the crucible containing the charge starts the reaction. The covering crucible effectively stops spatter.

For the dissolution of peroxide fusions, several procedures are available. Commonly, the crucible containing the melt is submerged in a relatively large volume of water in a beaker or a platinum or silver dish and rapidly goes into solution on gentle heating. Elements of the sodium hydroxide group are of course precipitated. If an iron crucible was used, there will be a large residue of hydrated iron oxides. In a few cases, notably methods for determining chromium, the residue may be filtered off through a double paper (Whatman No. 41H or 541 inside Whatman No. 44 or S and S blue ribbon) and washed thoroughly with dilute $NaOH$–Na_2CO_3 solution. In this filtration, all solutions should be at room temperature.

If the entire sample is to be dissolved in acid, the melt should be treated with water as above. The crucible should be removed, thoroughly scrubbed, and rinsed, if necessary, with a little of the dissolving acid. The required amount of acid should be added, at a single stroke, to the suspension of oxides and hydroxides. The mixture should be mixed immediately to prevent the precipitation of silicic acid. Flecks of iron oxide scale from an iron crucible may be filtered off and discarded.

An effective way to dissolve peroxide melts, valuable when volatile elements (fluorine, chlorine) are of interest, is to place the crucible containing the melt in a large platinum or silver dish, cover with a watch glass, and direct a stream of water from a wash bottle in such a way that it runs down the inside convex surface of the watch glass and drips into the crucible. When enough water has been added to dissolve the peroxide, acid (phosphoric or perchloric) is added in the same way from a pipette.

2.4.5. Sodium Hydroxide Fusion, Solution in Acid

REAGENTS: SODIUM HYDROXIDE. The only real advantage to using sodium hydroxide instead of sodium peroxide in attacking silicate rocks and minerals is that the former is usually of higher purity [40]. It is, nevertheless, not obtainable in as pure a form as sodium carbonate, and some reagents are of poor quality. Large amounts of water are released during the fusion, and this leads to procedural difficulties. If the reagent is moist, from having absorbed water from the air, such difficulties become extreme, and spattering is very difficult to avoid. Many procedures therefore suggest preliminary fusion of the sodium hydroxide in the crucible before adding the sample. This may result in conversion of an appreciable part of it to carbonate, especially when only small amounts are involved.

PROCEDURE. Transfer an appropriate weight of sodium hydroxide pellets to a silver or gold crucible (nickel, platinum, or zirconium crucibles may also be used). The weight of flux should be 4–10 times that of sample. Heat on a hot plate or over a small flame until the hydroxide melts and water is removed. Cool, add the sample, and reheat slowly, preferably on a hot plate if a high enough temperature can be so attained; otherwise, heat over a small flame until the attack is complete. It is very difficult to determine completeness of attack, and this must be classed as a serious disadvantage of the method. Many common silicate minerals are dissolved at once, while some minor minerals may require 1 h heating or more at low red heat for complete attack.

Fusion in a furnace at 500 °C is sometimes recommended. When applied to materials of which the behavior is known, such heating is convenient. The rate of heating and the final temperature can be adjusted to give acceptable results. When the method is applied to samples of varying composition or grain size, it is impossible to predict behavior in a tightly written procedure, and close attention must be given to each fusion to avoid spattering and to make sure attack is complete.

Decompose the fusion by heating crucible and contents in a large volume of water. Then add a calculated amount of acid (usually hydrochloric) all at once and stir.

2.4.6. Solution in Phosphoric Acid

REAGENTS. Phosphoric acid, reagent grade, is approximately 85% H_3PO_4 [41]. It very frequently contains reducing substances and is the most difficult of the ordinary mineral acids to obtain in highly purified form. When used for the dissolution of silicates, the excess water must be removed, and the organic or other reducing constituents should be oxidized. These requirements are best met by heating with the addition of a little chromic acid or potassium dichromate. The excess of chromic acid is destroyed by raising the temperature until the yellow chromate color disappears.

Anhydrous phosphoric acid may also be prepared by adding the calculated amount of P_2O_5 to the 85% acid. For dissolving refractory silicates, the adition of sodium pyrophosphate is desirable.

2.4.7. Sodium and Potassium Pyrosulfate, Solution in Sulfuric Acid

REAGENTS. Potassium pyrosulfate, rather than sodium pyrosulfate, is almost always used. An exception is the analysis of tantalum and columbium minerals where occasionally these metals are required in sodium, not potassium, solution [42]. The analytical-grade reagents are seldom stoichiometric $K_2S_2O_7$. Bisulfate and/or free sulfuric acid is often present and results in a melt that spatters. If a trial fusion indicates that bisulfate is present, the flux should be preheated until it gives off slight fumes of SO_3 without any boiling or spattering.

APPARATUS. Pyrosulfate fusions in platinum always result in some attack on the crucible, which under extreme conditions may amount to several milligrams. The presence of platinum in analytical solutions causes difficulty, and it is therefore better to make pyrosulfate fusions in fused silica crucibles (Vycor may be substituted in many procedures). If the sample does not contain fluoride, attack on the silica glass is usually negligible. In some procedures, it is convenient to make the fusion in small Pyrex beakers. Small amounts of silica are dissolved, but this is often of no consequence.

Pyrosulfate fusions are most often performed on residues obtained during the course of a systematic analysis but occasionally offer the best means of a preliminary attack. They are almost essential in the analysis of tantalum and columbium minerals and are also convenient with titanium minerals. If the fusion is carefully performed in an open silica crucible, fusion is continued until most of the SO_3 is driven off. Then extra pyrosulfate is added, and any fluorine in the sample will be removed. This is necessary in the analysis of microlite and similar minerals.

During a pyrosulfate fusion, the melt frequently becomes opaque, so that it is difficult to see whether attack is complete. Removal of the source of heat and swirling the crucible as it cools will often permit observation of unattacked particles because just before the melt freezes, it loses its opacity.

The use of electric burners—parabolic aluminum mirrors with a heating element mounted within them—is advantageous for pyrosulfate fusions. The wire-wound element must always be covered with a thimble-shaped shield of silica glass to prevent contamination of the crucible by oxides of chromium and iron, which evaporate from the element.

Pyrosulfate fusions should never be attempted within a furnace. At temperatures where the flux is effective, there is a considerable P_{SO_3}. The condensing effect of a lid on a bottom-heated crucible is of some importance. Only when the expulsion of fluorine from a sample is desired or when attack on the sample is known to be ready and fast should pyrosulfate fusions be carried out in open crucibles.

The most frequent use of pyrosulfate fusions is in the dissolution of weighed ammonia group oxide precipitates in the systematic analysis of silicate rocks and minerals. In accurate work, refinements of a simple procedure are always necessary. In analyses that begin with a pyrosulfate fusion, these refinements are of even more importance. The fusion always leaves some unattacked material. This may consist of minerals originally present in the material being attacked or may be synthesized during the fusion—inadvertently. In the systematic analysis, silica remaining in solution after two or three dehydrations with hydrochloric or perchloric acid and present in the mixed ammonia group oxides forms compounds with iron, nickel, and other elements that remain undissolved on treatment with dilute sulfuric acid. These compounds appear as a small insoluble residue that must be filtered off, washed with diluted sulfuric acid, ignited, and treated with hydrofluoric and sulfuric acids. After evaporation to dryness to be sure of removal of fluoride, a small pyrosulfate fusion makes it possible to recover the metals involved and add them to the main solution.

Solution of pyrosulfate fusions, except for the silicates they contain, is not difficult. The melt is simply soaked in water or dilute sulfuric acid until it goes into solution. When tantalum, columbium, or much titanium is present, it is better to leach the fusion cold. Heating, even to steam bath temperature, is apt to result in precipitation of intractable residues. If hydrogen peroxide is added to the leaching solution, and the leaching process is allowed to proceed slowly without heating, columbium and tantalum are easily obtained in a clear solution, from which they may be precipitated by thermal decomposition of the peroxide. Peroxide should never be added before removal of the crucible from the leaching solution.

2.4.8. Sodium Carbonate Sinter: Solution in Hydrochloric Acid

REAGENTS: SODIUM CARBONATE. Highly pure sodium carbonate is commercially obtainable. This remains the main reason for the use of this flux in silicate analysis. The purity of reagent-grade Na_2CO_3 should not be taken on faith. It should be acquired in large lots, and each lot should be tested in the following manner:

Weigh 25–50 g of the reagent into a 1000-ml beaker, and add 500 ml water. Cover and slowly add 100 ml hydrochloric acid. Dissolve 25 g diammonium phosphate in water, and filter the solution. Add it to the solution of the reagent. Then add ammonia solution prepared by saturating water with tank ammonia gas until the solution is alkaline, and add one-tenth of its volume of saturated ammonia solution. Let stand, with occasional stirring, for several days. If any precipitate appears, the reagent is useless for accurate work. If no precipitate is apparent, filter through a Whatman No. 44 paper, scrub, and wash thoroughly with 5% ammonia solution and ignite the paper and weigh the residue. More than a few tenths of a milligram of residue should exclude the reagent from use. It is unlikely that any residue originates in the hydrochloric acid; if this is considered possible, it may be tested by evaporating several liters (diluted 1:1 with distilled water) in a platinum dish and weighing the insoluble residue.

Of a dozen or more batches of reagent-grade sodium carbonate tested in this way, only four or five proved satisfactory.

The sodium carbonate sinter is most often used in the rapid analysis of low-grade limestones, dolomites, shales, iron ores, and nonsilicates that contain siliceous impurities.

PROCEDURE. Transfer 1 g of sample to a platinum crucible, and heat at 450 °C, covered, for 1 h. Remove the cover, and hold at 450–500 °C overnight or long enough to completely remove organic material. Then raise the temperature to 750–800 °C for 1 h. Cool. Add 1 g sodium carbonate, mix thoroughly, and heat at 800–1000 °C for 1 h. Cool and drop the crucible and contents into a beaker containing 50–1000 ml of 1:1 hydrochloric acid. Solution is usually more complete if the leaching is continued at room temperature with occasional stirring, but boiling in dilute acid will sometimes be necessary. Leaching with water alone is apt to be very slow.

2.4.9. Sodium Peroxide Sinter, Solution in Hydrochloric Acid

A large quantity of silicate minerals may be decomposed by sintering at low temperature with sodium peroxide in zirconium crucibles. This technique is also particularly effective for the decomposition of lateritic and garnieritic materials for the determination of nickel, cobalt, manganese, chromium, iron, aluminum, magnesium, copper, zinc, and to some extent silica. This technique is suitable for the simultaneous decomposition of many samples during geochemical exploration. Resistant minerals such as chromite and cassiterite are successfully attacked if ground first to 100% minus 325 mesh.

PROCEDURE. Transfer 0.25–1.00 g of the sample to a 30–50-ml zirconium crucible (note 1), and add four times its weight of Na_2O_2 (note 2). Mix thoroughly with a glass rod. Weigh 1 g Na_2O_2 onto a piece of hard-surfaced paper (Albanene). Rinse off the stirring rod in this. Transfer the Na_2O_2 to the crucible, covering the peroxide–sample mixture as completely as possible.

Note 1: Erosion rate of zirconium crucibles at the sintering temperature is extremely small. If conducted properly, a decomposition by sintering performed at 480 °C in a zirconium crucible may be repeated 1000 times before the zirconium crucible will be out of service [43].

Note 2: For the efficiency of the decomposition, the sodium peroxide must be fresh, very dry, and as fine as possible [44].

Cover, and transfer the crucible and contents in a furnace regulated at 200 °C for about 15 min. Then sinter at 480 ± 5 °C for 1 h (note 3).

Note 3: For the solid phase reaction to be most effective, fusion temperature must be approached; but it should not be reached at any time [45].

When sintering is completed, the crucible should cool for a maximum of 2 min. Put the crucible sideways in a 400-ml Pyrex beaker and place a watch glass on the beaker. Add, at once, 150 ml distilled water preheated at 60 °C (reaction may be violent, eye protection is a must). When the reaction is complete, wash the crucible once with distilled water and twice with 25 ml concentrated hydrochloric acid. Heat near boiling (note 4) until the solution is clear. Cool and add two drops of hydrogen peroxide (note 5).

Note 4. Excess boiling is not necessary and will only generate losses.

Note 5. Addition of a small quantity of hydrogen peroxide has been found to keep silica into solution for a longer period of time.

When solution is at room temperature, transfer to a 250-ml volumetric flask for instrumental determination of the elements listed above.

2.4.10.   Fusions Using Ammonium Salts

The J. L. Smith method for separating alkali metals from silicates provides a procedure for the attack of carbonate rocks that is extremely useful but seems to have been seldom used. When ammonia chloride is admixed with calcium carbonate and heated, calcium chloride is formed, and this, together with the excess of lime, attacks silicates readily. Even low-grade limestones can be obtained in solution without fluxing with alkali salts by simply mixing them with ammonium chloride and heating.

PROCEDURE. Thoroughly mix 5 g limestone sample with 0.5 g NH_4Cl. Transfer the mixture to a platinum crucible (A J. L. Smith crucible may be used but is not essential). Heat slowly to minimize loss of ammonium chloride by volatilization, and gradually increase the temperature to a full red heat. An electric furnace or a burner may be used, but the top of the crucible should be kept relatively cool. Use of an asbestos board with a hole in which the crucible is supported over a flame is probably the best device to use. A special condenser arrangement, as used in the determination of the alkali metals, may be used but is unnecessary for the prupose at hand unless it is desired to quantitatively retain the sodium and potassium in the sample. When the sinter is complete, dissolve the residue in dilute acid.

Ammonium bisulfate can be used as a flux for several minerals. The sample and flux are mixed and heated cautiously to minimize loss of ammonia. Usually any substance that can be attacked by ammonium bisulfate is also soluble in concentrated sulfuric acid, so that this flux finds limited use.

2.4.11.   Solution Procedures for Specific Minerals and Rocks

A procedure for iron ore, magnetite, hematite, rocks containing large amounts of ferrous minerals, sulfide ore minerals, and many rocks low in silica follows.

PROCEDURE. Weigh 1 g or more of sample into a 400-ml Pyrex beaker. Add 50 ml water and 75 ml concentrated hydrochloric acid. Heat on the steam bath overnight. Filter through a Whatman No. 44 paper, and wash thoroughly with 5% hydrochloric acid and then with water until no free acid remains on the paper. Transfer paper and residue to a platinum crucible, place in a cold furnace, and gradually heat to 450 °C. Hold at this temperature until paper is removed. Cool. Add 5 g (less if the residue is small) Na_2CO_3, and mix with a small rod. Rinse the rod in 1 g Na_2CO_3 by rotating it on a piece of smooth paper. Transfer the extra flux to the crucible to act as a cover. Heat over a small flame, gradually increasing temperature until fusion is complete. Be certain that the flame is oxidizing at all times. A reducing flame will cause alloying of iron in the sample with the platinum. Cool. When cold, heat the crucible for a minute or so in an oxidizing flame until it just begins to show a dull red color. Cool rapidly on an aluminum plate. Add 10 ml water, and heat gently until the cake loosens. Wash the contents of the crucible into the main solution, being careful that the resulting effervescence does not result in loss. heat and stir to solution. If a residue remains, filter it off and repeat the fusion.

2.4.12. Sodium Carbonate Fusion

PROCEDURE. Weigh a 25-ml platinum crucible, placing it on top of its lid on the balance pan. Add to the crucible 0.5–0.8 g of rock or mineral powder, and reweigh. Whether the sample weight shall be exactly 0.5000 or exactly 0.8000 g is optional. The size of the sample depends to some extent on its nature. Materials with large percentages of carbonate or hydrogen (limestone, vermiculite) demand the larger sample weight. In cases where the sample rapidly exchanges water or carbon dioxide with ambient air, variations in weighing procedure may be necessary. Record the weight of the platinum crucible plus lid and the weight with the sample added.

Transfer crucible and sample to an oven, and heat at 110 °C (or other appropriate temperature) for 1 h. Cool in an evacuated desiccator containing anhydrone or barium oxide. Reweigh. Return the crucible and sample to the oven, and hold at 110 °C for 10–18 h (overnight). Cool in an evacuated desiccator containing anhydrone or barium oxide as before, and reweigh. If appreciable weight loss occurred during the second heating, the H_2O determination is probably meaningless. Sometimes a weight gain occurs. This indicates that constituents of the sample are slowly oxidized in air, and the "moisture" determination was not worthwhile.

Heat the crucible and sample at 450 °C in a furnace for 10–18 h (overnight). If sulfides are known to be present and their removal is advantageous, the first hour of heating should take place with the crucible covered. The cover should then be removed for the remaining time. Cool the crucible and contents in an evacuated desiccator containing anhydrone or barium oxide, and reweigh. The loss or gain in weight may or may not be significant. It does no harm to record it.

Weigh 4.5 g sodium carbonate (anhydrous reagent powder, which has been tested for relevant impurities) and add this to the rock or mineral powder in the crucible. Mix sample and flux thoroughly using a platinum rod. Weigh 0.5 g sodium carbonate, and rinse the rod with it by rotating the latter in it. Cover the mixture in the crucible with the extra sodium carbonate.

Support the crucible on a silica triangle, and heat with a small oxidizing flame in such a way that the sample is attacked without excessive spattering. It is essential that the flame be oxidizing; that is, the gas and air supply to the burner must be such that there is always excess air in the flame. If the local gas supply or the construction of the burner is such that this condition cannot be achieved, it is not possible to accomplish a sodium carbonate fusion of silicates in sodium carbonate without iron, cobalt, nickel, zinc, and other elements being taken up by the platinum during fusion. A good test for an oxidizing flame is accomplished by heating a nickel crucible of the same size as the platinum crucible under identical conditions. If the nickel crucible shows any sign of a bright nickel surface during such heating, the flame is reducing and must not be used to heat platinum crucibles, especially when alkaline fusions are conducted in them.

Using appropriate burners, increase the heat until the sample is completely attacked, bearing in mind that the flame must be always oxidizing. In the final stages of the heating, the melt should be fluid and quiet. Some minerals are resistant to attack, and it is sometimes difficult to decide whether or not reaction is complete.

REFERENCES

1. I. M. Kolthoff, E. B. Sandell, E. J. Meehan and S. Bruckenstein. *Quantitative Chemical Analysis. Errors Due to Buoyancy of Air*, 4th ed., Macmillan, New York, ed. (1969), p. 495.

2. H. E. Almer, "Response of Microchemical Balances to Changes in Relative Humidity," *J. Res. Nat. Bur. of Stds. C. Engin. Instrument.* **64C**, No. 4, June 9 (1960).

3. F. R. Bacon, "The Chemical Durability of Silicate Glass," *Glass Industry*, **9**, August/September/October (1968).

4. P. B. Adams, "The biology of glass," *New Scient.*, **2**, 25–27 (1969).

5. R. Pribil and V. Vesely, "Determination of EDTA, DPTA, and TTHA in Their Mixtures," *Lab. of Anal. Chem.*, J. Heyrovsky Polarographic Inst., Czechoslovak Academy of Sciences, Prague.

6. A. J. Barnard, Jr., E. F. Joy, K. Little, and J. D. Brooks, "Practical Analysis of High-Purity Chemicals—I. Preparation and Characterization of High-Purity EDTA," *Talanta, 17*, 785–799. (1970).

7. R. C. Mehrotra and K. N. Tandon, "Adsorption Indicators in Precipitation Titrations," *Talanta, 11*, 1093–1111. (1984).

8. B. Budesinsky, "Xylenol Orange and Methylthymol Blue as Chromogenic Reagents," Nuclear Research Institute, Czechoslovak Academy of Sciences, Rzhezh, Czechoslovakia. 1966.

9. H. G. C. King and G. Pruden, "The Component of Commercial Titan Yellow most Reactive Towards Magnesium: Its Isolation and Use in Determining Magnesium in Silicate Minerals". *Analyst, 92*, 83–90 (1967).

10. C. O. Ingamells, "Absorptiometric Methods in Rapid Silicate Analysis," *Anal. Chem.* 38, 1228–1234 (1966).

11. B. P. Fabbi, "A Die for Pelletizing Samples for X-Ray Fluorescence Analysis," *X-Ray Spectrom.* **1**, 39–41 (1972).

12. K. Govindaraju and R. Montanari, *X-Ray Spectrom.* **7**, 148 (1978).

13. B. F. Scribner and M. Margoshes, "Emission Spectroscopy," in I. M. Kolthoff and P. J. Elving, eds., *A Treatise on Analytical Chemistry,* Vol. 6, Part 1, Wiley-Interscience, New York, 1965, p. 3347.

14. A. L. Gray, "Optical Emission Spectroscopy," in T. Mulvey and R. K. Webster, eds., *Modern Physical Techniques in Material Technology,* Oxford University Press, London, 1974, p. 232.

15. M. A. Floyd, V. A. Fassel, R. K. Winge, J. M. Katzenberger, and A. P. D'Silva, "Inductively Coupled Plasma—Atomic Emission Spectroscopy: A Computer Controlled, Scanning Monochrometer System for the Rapid Sequential Determination of the Elements," *Anal. Chem.* **52**, 431–438 (1980).

16. V. A. Fassel and R. N. Kniseley, "Inductively Coupled Plasma, Optical Emission Spectroscopy," *Anal. Chem.* **46**, 1104 (1974).

17. R. H. Scott and M. L. Kokot, "Application of Inductively Coupled Plasmas to the Analysis of Geochemical Samples," *Anal. Chem. Acta,* **75**, 257–270 (1975).

18. R. M. Barnes, "Short Course on Modern Emission Spectroscopy,"—Society for Applied Spectroscopy. BOSTON FACSS 1978

19. G. F. Larson and V. A. Fassel, "Line Broadening and Radiative Recombination Background Interferences in ICP-AES," *App. Spectro.* **33**, 592–600 (1979).

20. H. J. Rose, Jr., I. Adler, and F. J. Flanagan, "X-Ray Fluorescence Analysis of the Light Elements in Rocks and Minerals," *U.S. Geol. Sur.,* **17**(4), (1963).

21. G. K. Czamanske, J. Hower, and R. C. Millard, "Non-proportional, Non-linear Results from X-ray Emission Techniques Involving Moderate-dilution Rock Fusion," *Geochim. Cosmochi. Acta,* **30**, 745–756 (1966).

22. C. O. Ingamells and J. J. Fox, "Deconvolution of Energy-dispersive X-ray Peaks Using the Poisson Probability Function," *X-Ray Spectrom.* **8**(2), 79–84 1979.

23. B. P. Fabbi, "A Refined Fusion X-Ray Fluorescence Technique, and Determination of Major and Minor Elements in Silicate Standards," *Am. Mineral.* **57**, 237–245 (1972).

24. K. Govindaraju, "X-Ray Spectrometric Determination of Major Elements in Silicate Rock Samples, Using a Thin Film Technique Based on Ion Exchange Dissolution," *X-Ray Spectrom.* **2**, 57–62 (1973).

25. J. V. Smith, "X-Ray-Emission Microanalysis of Rock-Forming Minerals—I. Experimental Techniques," *J. Geol.* **73**(6), 830–864 (1965).

26. P. Mihalik, "The Electron Microprobe X-Ray Analyzer," National Institute for Metallurgy, Johannesburg, *Chem. Process.,* April–May (1967).

27. C. A. Friskney and C. W. Haworth, "Electron-Probe Micro-analysis of Metal

Oxides: Comparison of Correction Methods," *Brit. J. Appl. Phys.* (*J. Phys. D*), Ser. 2. **1**, 873–879 (1968).

28. S. J. B. Reed, "Probe Current Stability in Electronprobe Micro-analysis," *J. Sci. Instru.* (*J. Phys. E*) Ser. 2, **1**, 136–139 (1968).

29. W. Slavin, *Atomic Absorption Spectroscopy,* Interscience, New York, 1968.

30. J. A. Dean and T. C. Rains, eds, *Flame Emission and Atomic Absorption Spectrometry,* Vol. 1, *Theory,* Dekker, New York, 1969.

31. B. V. L'Vov, *Atomic Absorption Spectrochemical Analysis,* American Elsevier, New York, 1970.

32. D. J. Halls, "The Formation of Atoms in the Air-Acetylene Flame," *Spectrochim. Acta,* **32B**, 221–230 (1977).

33. R. C. Anderson, "Combustion and Flame," *J. Chem. Ed.* **44**, 248–260 (1967).

34. J. O. Rasmuson, V. A. Fassel, and R. N. Kniseley, "An Experimental and Theoretical Evaluation of the Nitrous Oxide-Acetylene Flame as an Atomization Cell for Flame Spectroscopy," *Spectrochim. Acta* **28B**, 365–406 (1973).

35. K. Govindaraju, "Use of Complexing Agents as Solvents for Atomic Absorption Determination of Calcium and Magnesium in Silicates," *Appl. Spectrosc.* **24**(1), 81–85 (1970).

36. R. A. Newstead and P. J. Whiteside, "Background Correction for Atomic Absorption Spectrophotometers," *Europ. Spectrosc. New* **11**, 8–10 (1977).

37. R. A. Newstead, W. J. Price, and P. J. Whiteside, "Background Correction in Atomic Absorption Analysis," *Prog. Analyt. Atom. Spectrosc.* **1**, 267–298 (1978).

38. C. O. Ingamells, *Anal. Chem. Acta* **52**, 323, (1970).

39. N. H. Suhr and C. O. Ingamells, *Anal. Chem.* **38**, 730, (1966).

40. C. B. Belcher, *Talanta* **10**, 75, (1963).

41. C. O. Ingamells, *Talanta* **2**, 171, (1959).

42. C. W. Sill, Accuracy and Trace Analysis, National Bureau of Standards, Spec. Publ. 422, 1976.

43. C. B. Belcher, "Sodium Peroxide as a Flux in Refractory and Mineral Analysis," *Talanta* **10**, 75–81 (1963).

44. J. A. Corbett, W. C. Godbeer, and N. C. Watson, "The Application of Sodium Peroxide Sintering Techniques to the Analysis of Minerals and Rocks by Atomic Absorption Spectroscopy, *Australas. Inst. Min. Metall.,* No. 250, June 1974.

45. Jan Dolezal, "Decomposition Techniques in Inorganic Analysis, Pavel Povondra and Zdenek Sulcek, 1966 [Trans Ota Sofr, 1966; English edition, 1968, Iliffe Books; published in USA American Elsevier, New York, 1968.]

CHAPTER

3

THE CLASSICAL ROCK OR MINERAL ANALYSIS

3.1. Classical Silicate Analysis
    3.1.1. Determination of H_2O^- and Loss on Ignition
    3.1.2. Fusion
    3.1.3. Determination of Silica, SiO_2
    3.1.4. Separation and Analysis of the Ammonia Group
    3.1.5. Interferences, Refinements of General Procedure
    3.1.6. Interferences, Modifications, Difficulties
    3.1.7. Procedure When Sample Attack Is Incomplete
    3.1.8. Other Members of Acid Group
    3.1.9. Separation of Hydrogen Sulfide Group
    3.1.10. Behavior of H_2S Group Elements in Regular Procedure
    3.1.11. Ammonium Sulfide Group
    3.1.12. Problems with Phosphate, Sulfate, and so on
    3.1.13. Extended Analysis of Ammonia Group Oxides
    3.1.14. Vanadium and Uranium
    3.1.15. Strontium and Barium
3.2. Classical Analysis of Typical Minerals
    3.2.1. Biotite
    3.2.2. Lepidolite
    3.2.3. Alkali Feldspars
    3.2.4. Fluorophlogopite
    3.2.5. Amphiboles
    3.2.6. Tourmaline
    3.2.7. Sodalite
    3.2.8. Microlite (Pyrochlore)
    3.2.9. Ardennite: An Arsenic Mineral
    3.2.10. Scapolite
    3.2.11. Catapleiite
    3.2.12. Tellurium Minerals: Spiroffite and Denningite
    3.2.13. Chromite
    3.2.14. An Impure Mineral Called Nepheline
    3.2.15. Olivine
    3.2.16. Sphene
    3.2.17. Spencite
    3.2.18. Ulvospinel
    3.2.19. Leucophosphite
    3.2.20. Triphylite

3.2.21. Triphylite–Lithiophilite
3.2.22. Garnet
3.2.23. Actinolite
3.2.24. Pyroxenes
3.2.25. Diopside
3.2.26. Andalusite
3.2.27. Manganese Zinc Ferrite
3.2.28. Vermiculite and Hydrobiotite
3.2.29. Apatite
3.2.30. Manganese Dolomite
3.2.31. Chloritoid
3.2.32. Muscovite and Other "White Micas"
References

The term *classical analysis* has become something of a misnomer. It is used here to signify not the way things were done in ancient times but the hard core of well-tried methods of maximum accuracy. Those who talk of the "superiority" of new methods are missing the point. Whenever a primary method is developed and is proven more accurate than the old, it is incorporated in the "classical" scheme. Thus, by definition, the classical analysis is more accurate than any other. The fact that the classical procedures often require more skill and more knowledgeability than those that involve little more than putting the sample in a machine and reading a signal is irrelevant.

A common error is to confuse accuracy, which cannot be objectively measured, with precision, which can. There are numerous examples of highly precise methods from which systematic error can be eliminated only by means too tedious to warrant the effort. Those who promote such methods on the basis of their precision alone anarchise the art and science of geochemical analysis.

Besides drawing the distinction between *precision* and *accuracy*, it is necessary to observe the difference between *analysis* and *determination*. Analysis is the opposite of synthesis; it is the separation—partial or complete—of a material into its constituents. How these constituents are determined, after their separation, is an entirely different matter.

Many instrumental methods reduce almost to zero the amount of analysis necessary prior to determination of one or another constituent: Probably this is a cause of the common failure to distinguish between the two concepts.

The classical silicate analysis, skilfully performed, provides high accuracy, but without consideration of time and cost. Instrumentation is used to the extent that it can provide greater accuracy. There is no dependence on samples analyzed by someone else.

3.1.   CLASSICAL SILICATE ANALYSIS

In the "main portion" of the classical silicate analysis, it is initially assumed that a rock or mineral consists of SiO_2, Al_2O_3, TiO_2, Fe_2O_3, FeO, MnO, MgO, CaO, Na_2O, K_2O, H_2O, and P_2O_5. Standard procedure is designed to separate these as completely as is necessary for their accurate determination. Numerous other constituents are dealt with as variations from the simple ideal. This may seem unrealistic but has proven to be a most practical artifice.

The alkalies, Na_2O and K_2O, ferrous iron (i.e., "oxygen deficiency"), H_2O (divided into H_2O^- and H_2O^+), and P_2O_5 are determined on separate samples preferably before beginning the main portion of the analysis, which deals with the separation and determination of SiO_2, Al_2O_3, total iron as Fe_2O_3, MnO, MgO, and CaO.

The behavior of other constituents is taken into account through refinements of the basic procedure.

Note that a total approaching 100% is not necessarily an indication of a good analysis. The analyses in the Actual columns both total significantly lower than 100%, although the values given are highly accurate.

Detailed procedures that follow should be mastered thoroughly by anyone anxious to attain geoanalytical proficiency and applied to simple rock or mineral samples before any attempt is made to analyze more complex materials. As written, these procedures can give highly satisfying results when applied to properly prepared samples of ordinary granitic rocks, which do not contain appreciable amounts of "extra" elements. It may be of interest to compare analyses of the standard granite G-1 made using the basic procedure with and without corrections for "extra" elements. This comparison has been made in Table 3.1. A similar comparison using the CAAS syenite Syenite-1 illustrates the inadequacy of the basic scheme of analysis when larger concentrations of extra elements are present.

As an outline of the basic classical scheme of analysis is given in Table 3.2. Although this scheme seems simple, one should remember that its simplicity was only developed through lifetime efforts of a large number of dedicated analysts—Washington, Hillebrand, Lundell, Ellestad, to name a few of the Americans who contributed importantly. Their work, of course, was developed from that of Berzelius, Smith, Benedetti-Pichler, and many other European scientists.

Table 3.1. Comparison of "Basic" and Actual Analyses of Silicates[a]

	USGS G-1 (%)		CAAS Syenite-1 (%)	
	Basic	Actual	Basic	Actual
SiO_2	72.54	72.52	59.77	59.78
Al_2O_3	14.18	14.08	10.21	9.01
TiO_2	0.26	0.26	0.48	0.48
Fe_2O_3	0.85	0.85	2.24	2.22
FeO	0.94	0.94	5.43	5.45
MnO	0.026	0.026	0.40	0.41
MgO	0.36	0.34	4.11	4.06
CaO	1.37	1.36	10.15	10.09
Na_2O	3.29	3.29	3.38	3.38
K_2O	5.55	5.52	2.62	2.60
H_2O^+	0.25	0.25	0.32	0.32
H_2O^-	0.02	0.02	0.14	0.14
P_2O_5	0.09	0.09	0.23	0.23
Total	99.72	99.54	99.48	98.17

[a] The "basic" analysis is one that assumes that a rock or mineral consists only of the 13 constituents listed. The actual analysis gives the true values for these constituents. Note that the largest error is in the alumina: Failure to take minor constituents into account leads to only a small error in the case of the granite (USGS G-1); in the case of syenite (CAAS Syenite-1), the error is large, amounting to 1.20%, much of which is rare earth elements.

The sample is weighed and dried: Loss in weight is recorded as H_2O^- After sodium carbonate fusion, silica is removed by dehydration using hydrochloric acid and determined by observing the loss in weight following treatment with hydrofluoric acid. Alumina, lime, and magnesia are successively precipitated with ammonia, oxalate, and phosphate and ignited to oxide, carbonate, and pyrophosphate. Determination is gravimetric. Iron, titanium, and phosphorus remain and are weighed with the alumina: They are determined, and Fe_2O_3, TiO_2, and P_2O_5 are subtracted from the impure Al_2O_3 to obtain a correct value. Manganese is weighed with the magnesium as pyrophosphate: It is determined colorimetrically in the weighed $Mg_2P_2O_7$ and the calculated MgO value corrected accordingly. Ferrous iron (oxygen deficiency), total hydrogen, and the alkali metals are determined on separate subsamples of the rock or mineral powder, making sure that subsampling problems and sampling weight baseline problems are kept under control [1].

Table 3.2.　The Basic "Classical" Scheme of Analysis

Weigh platinum crucible.

Add sample, and reweigh.

Dry at 105 °C, weigh, and report H_2O^-.

Roast at 450 °C, and weigh. Note loss on ignition.

Add sodium carbonate, plus oxidant if necessary. Fuse.

Leach with water, add hydrochloric acid, and evaporate to dryness.

Take up with hydrochloric acid and water. Filter. Wash free of sodium salts.

Reevaporate filtrate, take up with acid and water, filter, and wash.

Combine and ignite precipitates in original crucible. Weight SiO_2^+.

Treat impure SiO_2 with HF, H_2SO_4. Ignite and weigh residue. Calculate percentage of SiO_2.

Precipitate ammonia group at pH 6.4. Dissolve and reprecipitate.

Ignite ammonia group oxides in crucible containing residue from SiO_2.

Weigh ammonia group oxides. Fuse with $K_2S_2O_7$, and leach with dilute H_2SO_4. Add H_2O_2, and compare peroxytitanate color with standards. Remove peroxide. Remove platinum. Titrate iron after $SnCl_2$ reduction using dichromate.

Add oxalate to combined filtrates. Neutralize with NH_3. Filter off CaC_2O_4, reprecipitate, ignite, weigh $CaCO_3$ or CaO. Dissolve $CaCO_3$ or CaO in HNO_3, determine MnO contamination colorimetrically after oxidation to MnO_4^-.

Add diammonium phosphate and ammonia to combined filtrates. Collect precipitated magnesium and manganese ammonium phosphates, redissolve, and reprecipitate. Ignite, weigh $Mg_2P_2O_7 + Mn_2P_2O_7$. Dissolve precipitate, determine Mn colorimetrically. Determine residual calcium in $Mg_2P_2O_7$ as oxalate after separation as sulfate from 80% alcohol solution.

Determine H_2O (total), Na_2O, K_2O, and P_2O_5 on separate samples.

3.1.1.　Determination of H_2O^- and Loss on Ignition

WEIGH SAMPLE IN PLATINUM CRUCIBLE. Clean, burnish, wash, polish, ignite, cool, and weigh a platinum crucible and lid. A crucible of 25 ml capacity is best: Smaller crucibles cannot contain subsequent reactions, and larger crucibles are subject to weighing errors.

　　Without removing the crucible from the balance pan, add to it 0.7000 or 0.8000 g of the carefully prepared and thoroughly mixed sample.

It is worth the little extra trouble to make the subsample weight exactly 0.7 or 0.8 g, but the process of transferring the material must be guided by an awareness of the danger of segregation during the transfer. With some samples, the analytical subsample must be split out to avoid segregation of heavy mineral particles.

The crucible is best weighed while standing on its lid on the balance pan: This makes it unnecessary to handle it during weighing.

PERCENTAGE OF H_2O^- AND LOSS ON IGNITION. Transfer the crucible and contents to an oven and heat uncovered at 105 °C for 2 h. Transfer to a desiccator containing anhydrous magnesium perchlorate, cool, and weigh [2]. This gives the percentage of H_2O^-.

Cover the crucible and heat over a small oxidizing flame, gradually increasing the flame over a period of a half hour or more until the bottom of the crucible is perceptibly dull red. Remove the lid, and inspect it for condensed volatiles: In the ordinary case there will be none. Replace the lid a little to one side, so as to permit access of air, and continue heating (at dull red heat) for 20–30 min. If there is a deposit on the lid, heat it in a free (oxidizing!) flame for a few seconds.

Alternatively, roasting may be performed in a furnace: In this case it is best to put the covered crucible in a cold muffle, bring the temperature slowly to 450–500 °C, and remove the lid and heat for a further 30 min. A disadvantage in using a furnace is that no observation of condensing volatiles may be made.

The preliminary heating is done with the crucible closed because sulfur, the most common volatile, is more completely removed in this way. The final heating in an open crucible results in oxidation of at least some of the ferrous minerals; iron is less likely to cause subsequent difficulty when it is in the ferric condition.

Overheating of the crucible during roasting must be avoided to prevent attack of the platinum, for example by iron-bearing compounds and sulfides or arsenides.

With samples containing no sulfide, little water, and no oxidizable material or carbonates, roasting may be omitted. Roasting *should* be omitted if the sample is likely to fuse or sinter at the low temperature employed. If a volatile deposit on the lid of the crucible is observed, an effort should be made to identify it; this may be done on a separate sample. Among constituents that have been observed to volatilize from the sample and condense on the crucible lid are metallic chlorides, arsenic compounds, thallium compounds (in practice usually encountered in mineral analysis when the minerals have been separated using Clerici solution), sulfur and selenium compounds, ammonium salts (usually the sulfate), silica (because of fluoride in the sample), boron compounds, and mercury compounds.

Such important clues are lost if roasting is performed in a furnace.

Cool the crucible containing the roasted sample in a desiccator, and weigh.

The loss in weight seldom has any quantitative significance but may provide useful qualitative information.

3.1.2. Fusion

CONDUCTING THE FUSION. Place the crucible on a clean porcelain plate. Weigh 4 g anhydrous sodium carbonate (which has been tested for its magnesium, calcium, and heavy metal content), and transfer about 3 g of it to the crucible. Mix thoroughly with a small clean rod, preferably of platinum or silica (glass is less desirable).

With ordinary rocks, it is unnecessary to add an oxidant to the flux; however, if much ferrous iron is present despite the roasting (as in ferrous amphiboles), add about 0.1 g sodium peroxide of established purity and mix thoroughly. The use of nitrates as fusion oxidants is undesirable because they preclude the use of platinum dishes in dehydrating silica and lead to imperfect separation of manganese from the ammonia group later in the procedure.

After mixing flux and sample, rinse the mixing rod by rotating it in the remaining gram of sodium carbonate, and transfer the latter to the crucible, covering the mixture therein as completely as possible.

Cover the crucible and heat very slowly over a gas burner, gradually increasing the flame over a period of about 20 min until the mixture begins to sinter.

It is extremely important to have an oxidizing flame at all times: Failure to observe this precaution will lead to reduction of iron (also cobalt, nickel, and some other elements) and its deposition in the platinum as an alloy. It is impossible to make any general statement concerning the type of burner to be used in this operation because gas supplies vary from one laboratory to another, and burners that are eminently satisfactory in one location may be useless in another.

Finally increase the temperature to a bright red heat for 5–10 min or until inspection of the fused mass shows no sign of continuing reaction.

A Meker burner is best used during the final heating, but emphasis must remain on the necessity for an oxidizing flame. A simple test for an oxidizing flame is made by replacing the platinum crucible with a nickel crucible of the same shape and size. If the red hot nickel crucible becomes filmed with oxide, it is reasonably certain that the flame is sufficiently fuel lean; if, on the other hand, the nickel crucible shows flickers of bright metallic surface in the flame, the burner needs adjustment.

Fusions have been accomplished in furnaces or over electric burners, but these are less satisfactory than good gas burners because attack on the platinum crucible is usually greater (amounting to two milligrams or more as against a fraction of a milligram noted during skillful use of gas flames). Fusion in a muffle furnace makes the use of a larger crucible almost mandatory to prevent boiling over and leads to increased weighing errors throughout the rest of the analysis. If fusion must be performed in a furnace, the charged crucible should be placed in the cold furnace, and the temperature should be gradually raised to 900–1000 °C, slowly enough that boil-over does not occur: since every sample reacts differently, and since there is no possibility of observing the progress of the fusion in a furnace, it is very difficult to establish the optimum heating rate or period. Electric burners are more flexible in this respect and are recommended over closed muffles if suitable gas burners are unavailable or if gas supply is inadequate.

Heating over open wire-wound elements should never be considered: Their use will inevitably lead to base metal contamination of the platinum.

When the melt is quiet and reaction is judged complete, heat the top of the crucible and the lid for a minute or so with a free Bunsen or Tirrill (oxidizing!) flame to melt the small spatters of flux that will have accumulated on the inside of the lid and on the sides of the crucible. Remove the lid, invert it, and lay it on the porcelain plate. Pick up the crucible and swirl or tip it so that the melt solidifies in a cup-shaped layer on the walls.

With some high-iron or high-magnesium samples, this may not be possible, but in general it may be assumed that fusion temperature was too low if the melt is too viscous to be thus treated. Some melts are thixotropic and require a quick motion to liquefy them.

COOLING THE MELT. Cover the crucible and allow it to cool on the porcelain plate or on a clean aluminum slab. Reheat briefly over an oxidizing Meker flame just until the bottom of the crucible begins to show a dull red color. This should require about 30 s. Cool. The melt is now ready to be removed from the crucible.

If fusion and brief reheating have been skillfully performed, the melt will easily come free from the crucible with the addition of a little water. Should this not be so, it is probable that fusion is incomplete or that alloying of iron (cobalt, nickel, and so on) with the platinum has occurred.

Statements have appeared from time to time to the effect that contamination of platinum crucibles with iron from the sample is almost inevitable during sodium carbonate fusions. These are, to say the least, exaggerations. It is admittedly somewhat difficult to prevent reduction of iron during fusions in platinum over gas burners, and in laboratories with an inadequate gas supply or a lack of awareness of the need for oxidizing flames, it may indeed be impossible. However, given the proper conditions, iron ores, garnets, ferrous olivines, and other iron-bearing materials can be and have been successfully treated even without the precaution of adding an oxidant such as sodium peroxide to the flux. If reduction of iron and contamination of the platinum does occur, it is an indication of poor technique or inadequate facilities.

Solution of the sodium carbonate melt and the subsequent dehydration of silica may be accomplished in either platinum dishes or porcelain casseroles. Evaporating dishes of silica or porcelain, Vycor dishes, and Teflon vessels have been tried in this application but do not perform well. Platinum dishes undoubtedly result in the most complete recovery of silica, but the use of platinum with high-iron samples and with samples containing chromium or much manganese results in complications because of the relatively large amount of platinum that dissolves. Platinum is nevertheless preferred, despite these complications, when the highest accuracy is sought, even though corrections for dissolved platinum may be difficult. The error introduced through the use of porcelain casseroles is not usually greater than about 0.1% SiO_2 except with fluorine-bearing samples, with which gross error may result through attack of the casserole.

Since procedures differ appreciably depending on whether platinum dishes or porcelain casseroles are used for dehydrating silica, two versions of the next operation will be described.

SOLUTION OF MELT. When silica is dehydrated in platinum dishes: Add to the crucible containing the fused sample about 5 ml water and allow it to stand for several minutes. Stir gently with a silica or platinum rod to loosen the melt. Transfer the contents of the crucible as completely as possible to a cleaned, burnished, and polished 300-ml platinum dish, scrubbing the crucible with a policeman and using a total of about 30 ml water. Put about 5 ml concentrated hydrochloric acid in the crucible, cover, and stand it on the porcelain plate until the following operations are completed.

Allow the melt to disintegrate thoroughly. If any manganate (evidenced by a green color) is present, add a drop of ethanol. The liquid volume should be no more than about 60 ml.

Heating to hasten decomposition of the melt is not desirable; it is better to allow it to proceed at room temperature—overnight if necessary.

Crush the residue in the platinum dish with a silica-glass rod, and look carefully for gritty particles, which may indicate incomplete decomposition of the sample.

If unattacked material is present, it may be best to start over, perhaps grinding the sample to a finer mesh size or fusing for a longer time—unless it is decided that the offending unattacked minerals are not likely to succumb to sodium carbonate–sodium peroxide fusion and do not contain elements that will lead to error. Among common minerals that may remain intact after sodium carbonate fusion are zircon, chromite and other spinels, sillimanite, magnetite, cassiterite, microcline, and ilmenite. In cases where sample is limited and sodium carbonate attack is not completely successful, some ingenuity is often needed to rescue the analysis.

With the platinum dish containing the fusion of the sample thoroughly disintegrated by soaking in water and covered by a watch glass, add a total of about 15 ml concentrated hydrochloric acid diluted with at least an equal volume of water using a special funnel or a pipette to avoid loss by spattering. Allow the mixture in the dish to stand undisturbed until effervescence almost ceases. Add the contents of the platinum crucible, scrubbing it and rinsing thoroughly.

Heat the covered dish on the steam bath until bubbles of carbon dioxide can no longer be observed. Wash down the sides of the disk with 1:20 hydrochloric acid, raise the cover glass on a silica-glass triangle (unless fluoride is present in the sample, in which case the evaporation should proceed in an uncovered dish), and permit the solution to evaporate.

The use of glass triangles instead of hooks or ridged watch glasses is advised because they lead to a faster evaporation. Triangles of soft glass should not be used unless it can be established that hydrochloric acid does not leach calcium from them. Glass triangles and covers should not be used with samples containing fluorine.

DEHYDRATION OF SILICA. As dryness approaches, stir at intervals so that large crystals of sodium chloride do not form. As the residue dries, gently crush it to a powder with a silica-glass rod.

Silica glass is preferable to Pyrex for stirring rods, but the ends of the rod must be well rounded (in an oxy-hydrogen flame) or chips of silica may be spalled off into the analysis. Silica glass (fused quartz) is not attacked by fluosilicic acid and is more resistant to alkaline solutions than glass. It is a convenience to carry the rod used at this stage throughout the analysis; it should therefore be long enough to use in an 800-ml beaker and short enough that it does not tip out of the dish in which the silica is dehydrated. Rods and cover glasses of plastic are not easy to work with because solutions do not wet them, but they may have to be used with fluorine-bearing samples.

> When silica is dehydrated in porcelain casseroles: Clean a porcelain casserole in sulfuric–chromic acid cleaning solution, rinse with water and then with dilute hydrochloric acid to which a drop of 6% hydrogen peroxide is added. Transfer the fused sample to the casserole as described above, scrubbing the crucible and rinsing it with water. Use a total of not more than 30–40 ml water. Put 5 ml concentrated hydrochloric acid in the crucible and allow it to stand, covered.
>
> Cover the casserole with a cover glass, add a silica glass stirring rod, and slowly add 10 ml of hydrochloric acid that has been diluted with an equal volume of water. Use a pipette so that the cover glass need not be disturbed during this operation. Allow the casserole and contents to stand undisturbed for several hours or until the fusion cake disintegrates. Wash down the cover glass and the sides of the casserole, and then add the contents of the crucible, scrubbing and washing it thoroughly. If the sample is suspected of containing more than about 1% TiO_2, add an additional 10 ml concentrated hydrochloric acid.
>
> Heat, completely covered, on the steam bath until carbon dioxide is no longer evolved. Crush soft lumps of hydrous silica with the rod, raise the cover on a glass triangle, and evaporate as described before, stirring frequently. It is important to prevent formation of large crystals and crusts of salt.

Treatment of Platinum Crucible. The crucible used to fuse the sample with sodium carbonate will be carried throughout the analysis. It should be ignited and weighed at every stage to keep track of the platinum introduced into the analysis and to make sure that elements from the sample do not become alloyed with the platinum. In an ordinary analysis, the crucible should lose not more than a milligram during the whole operation. Losses greater than this usually indicate poor technique or a badly maintained crucible.

When the fusion has been removed from the crucible in the manner described, a brown or black stain may be observed, especially with Pt–Rh alloy crucibles. If this dissolves during the soaking in hydrochloric acid, it is probably of no consequence. If the stain persists after this treatment, some alloying of iron (cobalt, nickel, copper, etc.) with the platinum should be suspected. A small iron contamination may be removed by repeatedly heating the crucible to 800 °C in a furnace or over a *strongly oxidizing* flame and treating with concentrated hydrochloric acid between ignitions. Copper contamination is removed with nitric acid. Under no circumstances add both nitric and hydrochloric acids!

Carefully clean the outside of the crucible, ignite it for 1 h over an oxidizing flame, cool in a desiccator, and weigh. The crucible should be free of stain or discoloration of any kind and should weigh 0.1–0.3 mg less than before fusion.

If the loss is larger than this, special procedures will have to be applied later in the analysis. Common reasons for excessive loss of platinum are (1) use of too much sodium peroxide in the flux, (2) a poorly adjusted flame during fusion, (3) fusion in a muffle furnace—crucibles used in this way develop a ring of attack at the level of the fusion, (4) unusual concentrations of chromium or manganese in the sample, (5) alloying of base metals with the platinum and subsequent solution of the alloy in acid, and (6) inadequate pretreatment (burnishing, polishing, etc.) of the crucible.

3.1.3. Determination of Silica, SiO_2

Continue to heat the platinum dish or porcelain casserole on the steam bath until the odor of hydrogen chloride can no longer be detected after closely covering the dish for a minute and then raising the cover.

The dehydration may take several hours and may be continued overnight. It is essential to continue the dehydration for a sufficient time; otherwise, filtration will be slow, recovery will be very incomplete, and high values for alumina will ensue. Heating the dish and residue in an oven has been advocated from time to time. If this is done, the oven temperature must not exceed 100 °C, and forced air circulation is necessary. Oven drying is not recommended. Baking on a hot plate is impermissible—it results in the formation of soluble silicates, especially of magnesium, and in the formation of iron compounds which do not dissolve readily in hydrochloric acid. Imperfect dehydration of silica is a major cause of difficulty in the classical analysis. Low values for SiO_2 in peridotites and dunites, which contain much magnesium, may follow overheating during the dehydration. Since such samples do not contain much alumina, classical analysis yields poor values for Al_2O_3 unless all precautions are taken.

SOLUTION OF SALTS. Cool the platinum dish or porcelain casserole and its thoroughly dried contents, and add carefully 6–15 ml concentrated hydrochloric acid. Allow to stand for 10 min. If the titanium content of the sample is known to be low, use the lesser volume of acid. Wash down the cover and the sides of the dish with a little 1:20 hydrochloric acid, add water to a volume of 50–

70 ml, and heat on the steam bath with stirring until all the sodium chloride is dissolved. Add more water only if necessary.

FILTRATION OF SILICA. Filter as soon as possible through a 9-cm S and S black ribbon paper, catching the filtrate in a thoroughly clean 400-ml beaker. Wash with cold 1:20 hydrochloric acid, scrubbing the dish and transferring as much as possible of the silicic acid to the paper. Finally wash free of acid with *hot* water. The progress of the washing can be followed by adding a drop of very dilute (0.005%) methyl orange solution to the precipitate from time to time. As long as it shows a pink color, washing should be continued.

If the filtration of the silica is skillfully performed, the whole operation should not take more than a half hour even if eight analyses are being run concurrently. The operation must be performed rapidly because the dehydrated silicic acid gradually absorbs water and becomes more and more gelatinous.

Once the filtration slows down, the day is lost. A beginner may spend several hours waiting for filters to drain: Meanwhile, the silica is slowly becoming solubilized, and the purpose of the game is defeated.

The difference between a quick and efficient separation of silica and a long and discouraging operation is made up of a large number of trivial skills. A most common mistake is to use too large volumes of liquids during rinsing, washing, and scrubbing. The volume of the filtrate from the silica, after all washing is complete, should usually be less than 80–100 ml, but if care is not exercised throughout, the silica may still test acid after 200–300 ml have been used.

Wash bottles should be carefully prepared to deliver just the right stream of liquid, and they should be filled and at hand at the right temperature when needed. Plastic squeeze bottles should not be used; not only are they difficult to control but they also contribute organics to the analysis, and these cause difficulties.

Funnels, beakers, burners, filter papers, and so on should be ready, cleaned, and immediately available when needed. Papers should be folded into filters a half hour before use, so that the paper fibers will become water saturated. Funnel stems should be full and should remain full during the filtration. From the time that acid and water are added to the dehydrated mass in the dish and the sodium chloride has been dissolved, no more than 5 min should elapse before most of the solution has been poured through the filter. Beginners have been known to carefully stir the salts into solution and then start looking for a funnel, a beaker, and a filter paper. Having collected these, they must make up some 1:20 acid, and heat a bottle of water. By the time everything is ready for the filtration, the silicic acid is unfilterable. If the dish has been left on the steam bath, sodium chloride has recrystallized.

In the first filtration of the silica, the aim should be to wash completely free of sodium salts. The small amount of silica that goes into solution will be caught in the second dehydration or during the analysis of the ammonia group. One should be careful to churn up the precipitate with every portion of wash water and to carefully wash the upper edge of the filter. Testing with dilute methyl orange provides an indication of the thoroughness of the washing: The presumption is that if all the acid is removed, the sodium chloride is removed also. Sodium left in the silica will first be weighed as chloride or silicate and then later as sulfate, causing errors (a negative one in the SiO_2 and a positive one in the Al_2O_3). Actual tests on silica precipitates washed free of acid in the manner described show that much less than 0.1 mg sodium salts is retained. As much as 1–2 mg sodium salts may be left in precipitates when no special effort is made to test the completeness of washing. When there is difficulty in igniting silica to constant weight, this is usually due to failure to wash out all alkali salts.

If the stem of the filter funnel does not remain full, washing the edges of the filter paper will result in silicic acid being carried mechanically over the edges and into the filtrate.

SECOND SILICA DEHYDRATION. Reserve the filter paper containing the silica. Return the filtrate to the platinum dish or the casserole, and evaporate to dryness as before, stirring to prevent the formation of large crystals and crusts of sodium chloride. Dehydrate thoroughly. Finally take up the residue with 6–15 ml concentrated hydrochloric acid, allow to stand for 10 min, add just enough water to dissolve the sodium chloride, and warm to solution. Add a little ashless paper pulp, stir thoroughly, and filter through a Munktell 00 9-cm paper (or, less desirably, a Whatman No. 40 paper). Wash thoroughly with 1:20 hydrochloric acid, scrubbing the dish very carefully with the aid of small pieces of hardened filter paper (which are added to the filter). Wash free of acid with *hot* water using very dilute methyl orange solution to indicate complete washing.

Very frequently, the filtrate from the second silica dehydration will be cloudy. This is of no consequence; it is due to the hydrolysis of titanium compounds. The more of the titanium that can be washed into the filtrate, the better. If the sample contains a high concentration of titanium, it is best to use 15 ml or more of concentrated hydrochloric acid to take up the dehydrated residue. If little titanium is present, there is no purpose in using so much: The more hydrochloric acid used, the larger the volume of water necessary to solubilize the sodium salts.

If the washing of the silica is carefully done, the volume of the filtrate should not exceed 100 ml even after the numerous washings required to remove all acid from the filter. Poor technique can easily lead to a filtrate volume of 200 ml or more. Much time can be saved and better results obtained by developing the ability to wash thoroughly with small volumes of liquid. Solubility losses are always finite and can be substantially diminished by skillful washing techniques.

BURNING OFF FILTER PAPERS CONTAINING SILICA. Transfer the papers containing the silica to the same crucible that was used for the fusion (after it has been ignited and accurately weighed). Burn off the paper. This may be done over a flame or in a muffle furnace. Using a flame, support the crucible in a silica triangle and place a Bunsen or Tirrill burner with a very small flame (0.5 in. high) about 2–3 in. below the crucible. To begin with, the crucible should be tightly covered. A shield may be necessary to keep the flame from blowing to one side. Heat until all volatile matter has been removed, and then move the lid to one side to permit entrance of air and continue heating, gradually increasing the flame to its maximum, always being sure the flame is oxidizing, that is, contains excess air. Finally, heat the crucible at the maximum temperature of a good Meker burner, always maintaining an oxidizing flame.

Using a furnace, put the open crucible in the cold furnace, and gradually increase temperature to 1000 °C over a period of several hours.

In burning paper in which a residue is contained, whether the combustion is performed in a furnace or over a flame, certain conditions must be maintained. Moisture must be driven off under conditions that do not lead to physical loss (not until the paper is thoroughly charred should the temperature be raised much above 200 °C). In burning off the carbonaceous residue from the paper, the atmosphere within the crucible must be kept strongly oxidizing. Not until all carbon is gone should the temperature rise much above 450 °C. Final ignition should be at the minimum temperature for complete dehydration of the ignition product.

If ignition is performed in a furnace, there must be free air circulation but without drafts, which can cause mechanical loss.

Failure to meet these essential (and obvious) requirements will result in error. Laboratory furnaces are often poorly designed for these purposes and may require modification to make them satisfactory.

It may prove difficult to completely dehydrate silica over a Meker burner, but most often the difficulty does not originate with the burner but with the analyst who fails to completely remove alkali salts during the washing of the precipitate.

Platinum crucibles lose weight at high temperatures in an oxidizing atmosphere, but such weight loss is negligible at 1000 °C unless the platinum is alloyed with iridium or heating is long-continued. If continual weight loss during many hours of ignition to constant weight is observed, it is advisable to heat an empty crucible of the same manufacture for use as a tare. In this way, one can be certain that the weight loss is due to the crucible contents and not to slow removal of iridium or other alloy from the platinum.

Temperature settings and readings on furnace controllers are very often unreliable. Gas supplies vary widely. Empirical tests should be made to determine the exact conditions under which precipitates should be heated. In our experience, a final temperature of 1050 °C in a muffle furnace is adequate to thoroughly dehydrate precipitated silica without any measurable weight loss from well-maintained platinum or platinum–rhodium crucibles, and 2 or 3 h heating over well-adjusted Meker burners gives equally good results.

If difficulty is experienced in igniting silica to constant weight, the use of a blast lamp may be indicated; but tests made with empty crucibles should first establish the conditions that will not lead to appreciable loss of the crucible material.

The final weighing should always be made after cooling in an evacuated desiccator containing anhydrous magnesium perchlorate. Crucibles must always be cooled to less than 1 °C above room temperature before weighing. Dry air should be admitted cautiously to the desiccator through a tower containing magnesium perchlorate and soda lime before opening. Crucibles should remain covered during the weighing operation.

EVAPORATION OF SiO_2, HF, AND SO_3. When constant weight has been attained, add to the crucible enough water (about 0.5 ml) to moisten the silica, 1–10 drops of 1:1 sulfuric acid (the greater amount when much titanium is expected), and then 15 ml hydrofluoric acid. Place the crucible on an aluminum hot plate with a surface temperature slightly greater than 100 °C and leave until all silica and excess hydrofluoric acid have evaporated. Cool. Add a few drops of water, pack the crucible half full of ashless paper pulp, add several drops of ammonium hydroxide, and put the crucible in a cold furnace. Slowly raise the temperature to 450 °C over a period of several hours. Finally, ignite strongly either in the furnace or over an oxidizing Meker flame, cool in a desiccator, and weigh. If a furnace is not available, evaporate the excess sulfuric acid over a flame, holding the crucible in a pair of platinum-tipped tongs and using a swirling motion such that only the upper part of the crucible is heated by the flame until no fumes of SO_3 appear on heating to dull redness. Then heat to a bright red in a Meker flame. Heating over a flame must be done with skill and care, keeping the crucible in motion so that sulfates of titanium, iron, and so on are deposited in a thin layer on the bottom and sides and easily decompose without decrepitation.

Cool the crucible in a desiccator and weigh. Weight loss is SiO_2.

The loss in weight during the hydrofluoric–sulfuric acid treatment does not represent all the silica in the sample. Despite the double dehydration, a little always remains in solution. This is usually found almost entirely in the ammonia precipitate, from which it may be recovered. Whether or not a silica recovery from the ammonia group is worthwhile depends on the nature of the sample and the purpose of the analysis. An empirical correction of 0.1–0.2% added to the SiO_2 and subtracted from the Al_2O_3 may be adequate. Silica recoveries are of doubtful significance when porcelain casseroles are used in the early stages of the analysis. In rock analysis, routine sample preparation is seldom precise enough to give meaning to a small difference in SiO_2 and Al_2O_3 determinations.

Most often, the purpose of a classical analysis is to find the exact composition of a well-prepared sample, and the silica recovery should be included, especially with minerals in crystal structure work or with rocks or minerals that are to be set up as reference materials.

Fortunately, interfering elements in the silica determination are few. The chief ones are fluorine, tungsten, molybdenum, and boron. All but fluorine need seldom be considered. Methods for dealing with these interferences are given below.

The residue in the crucible after the silica determination is presumed to consist of TiO_2, Fe_2O_3, Al_2O_3, and P_2O_5. Unless columbium or tantalum is to be sought, the residue is left in the crucible and the main part of the ammonia group oxides added to it in the next stage.

Among other materials that may be found in the residue are unattacked minerals such as chromite, tourmaline, cassiterite, and zircon and insoluble salts such as barium sulfate and zirconium phosphate. Special operations are necessary if these materials are present (this will be described later). In the ideal case, one may proceed with the separation of the ammonia group while the determination of silica is in progress.

3.1.4. Separation and Analysis of the Ammonia Group

Addition of ammonia to the filtrate from the silica results in the precipitation of iron, aluminum, titanium, and (in the presence of excess iron and aluminum) phosphorus. If conditions are carefully controlled, precipitation of these elements is quantitative, and no manganese, calcium, or magnesium precipitates. A double precipitation is necessary to remove all sodium salts from the precipitate. A host of elements may accompany the four mentioned, some quantitatively. These are dealt with as special cases in a variety of ways and need not distract us from considering the basic scheme.

The aim during the ammonia precipitation should be to completely separate aluminum from the solution without precipitating any magnesium or manganese. Iron and titanium are completely precipitated under conditions that are optimum for aluminum. Phosphate precipitates only if it is present in minor amount as a phosphate of iron or aluminum. Thus, the procedures to be described are inapplicable to phosphate rocks.

Iron must be in the ferric condition: Its oxidation must be complete without there being any oxidation of manganese. If solutions containing filter paper are boiled excessively, or if iron has been reduced during evaporations in platinum, difficulties may arise. Titanium is difficult to precipitate completely in the absence of iron.

To prevent precipitation of manganese, pH should not be allowed to exceed 6.5, and oxidizing agents (dissolved oxygen from the air included) must be absent. Once oxidation of manganese has occurred, its reduction is difficult, especially in the presence of phosphate, which stabilizes higher valences of manganese. For this reason, the acid filtrate from the silica is boiled until methyl red retains its color in the boiling solution, and the first precipitation of the hydrated oxides is washed with ammonium chloride solution, not ammonium nitrate as sometimes recommended. A combination of nitrate and chloride makes it almost impossible to effect a complete reduction of manganese to the divalent state. Further, the presence of nitrate interferes later in the analysis with the separation of manganese as persulfate.

There may be some advantage in the use of an ammonium nitrate wash of the second ammonia precipitate, but the advantage is small: The slight possibility that iron may be lost during ignition of the precipitate (as volatile chloride) can be removed by making a final wash with water to remove a substantial part of the ammonium chloride and by properly controlling the ignition.

Methyl red is the indicator commonly used to measure pH during the ammonia precipitation but is less than satisfactory, especially when much iron is present or when the presence of manganese makes pH control a critical factor. Bromocresol purple is a better choice. However, any colored indicator in the solution removes the important possibility of observing its natural color, which sometimes provides important clues concerning unusual elements (e.g., copper, cobalt, chromium, nickel). A pH meter cannot, in practice, be used: Electrodes do not tolerate the hot solutions, contamination is difficult to avoid, washing traces of the sample from the electrodes is awkward, and readings cannot be made fast enough to be useful. Of several devices that have been tried, the use of small squares ($\frac{1}{8}$ in.) of nitrazine paper is recommended. These are dropped, one by one, into the solution as required. They float and assume a color varying from bright yellow in acid solution through green and slate between pH 6 and pH 7 to a bright blue at pH 7.5. Contamination of the solution is negligible.

Ordinary ammonium hydroxide must not be used in separating the ammonia group: It contains involatile impurities, organic matter, and dissolved gases that cannot be tolerated. To prepare a reagent, freshly boiled water should be saturated with ammonia gas derived from a cylinder or from a boiling flask containing the impure ammonium hydroxide. The receiver must be cooled in ice water, and the solution must be shielded from the atmosphere, from which carbon dioxide may be absorbed.

It is necessary to have in the solution from which the ammonia group oxides are precipitated a considerable excess of ammonium chloride to prevent precipitation of magnesium and manganese hydroxides. The ammonium chloride may be added in the form of a solution of the salt or as hydrochloric acid that is subsequently neutralized with ammonia. If the salt is used, its solution should be prepared in advance, digested hot at pH 7, and filtered to remove precipitated impurities, which may be appreciable.

The procedure to be described results in complete precipitation of iron, aluminum, titanium, zirconium, chromium, uranium, columbium, tantalum, and small amounts of phosphorus. Numerous other elements may be precipitated completely or in part. In the basic procedure, all except iron, titanium and phosphorus are reported as alumina. Some elements commonly considered to belong to the ammonia group may be precipitated very incompletely—for example, the rare earths and beryllium; probably the small amounts of these normally found in silicates will be almost quantitatively retained in a large iron and aluminum precipitate. The most seriously interfering element is fluorine, which may remain as fluoaluminate throughout the silica dehydration and prevent complete recovery of alumina in the ammonia group. If the separation of silica was imperfect, metallic silicates may be caught in the ammonia precipitate, notably calcium and zinc silicates.

PREPARATION OF FILTRATE FROM SILICA DETERMINATION. Heat the filtrate from the silica, which should have a volume no greater than 200 ml and usually contains 5–20 ml hydrochloric acid, to boiling. Boil for 5–10 min to remove dissolved oxygen, carbon dioxide, and chlorine, and reduce manganese entirely to the divalent state.

Occasionally it is desirable to add a few drops of bromine water to be sure of complete oxidation of the iron. This has proven necessary in some mineral analysis when much iron is present and silica dehydration is performed in platinum dishes. In these cases, several milligrams of platinum were dissolved from the dish, and the whole analysis became difficult as a result. In the usual case, addition of bromine to oxidize iron is undesirable and unnecessary.

Boiling should proceed over a Bunsen or Tirrill burner, with the 400-ml beaker containing the solution supported on an asbestos-filled wire gauze and with the burner placed directly under the end of the silica-glass stirring rod in the beaker. A 0.25-in. square of hardened filter paper under the end of the rod leads to smooth boiling. If properly prepared, the solution is very clean and will superheat badly if heated on a hot plate, with consequent bumping and perhaps loss. The ability to develop a slow steady boil in a clean solution is part of the art of refined analysis.

When all chlorine (and bromine, if it was added) has been removed by boiling (as evidenced by the lack of a characteristic odor that has been compared to the smell of musty books or by the persistence of methyl red in the solution), remove the burner from under the beaker and add up to 12 ml concentrated hydrochloric acid or its equivalent of ammonium chloride solution.

The amount of chloride to be added depends on the amount of hydrochloric acid present and on the amount of magnesium expected. Samples containing 50% MgO (dunite, peridotite) require at least 15 g ammonium chloride to be sure that no magnesium will precipitate with the ammonia group of elements. A normal addition is 50–100 ml of 15% (w/v) ammonium chloride solution. Less will be used when it is known that the magnesium content is low.

First Precipitation of Ammonia Group Oxides. To the still hot solution add pure ammonia (prepared by absorbing NH_3 in boiled water) slowly with gentle stirring until iron just begins to precipitate, being careful to avoid an excess at this stage. Heat the solution almost to boiling, and add ammonia dropwise with stirring until a piece of nitrazine paper dropped on the surface of the solution assumes a green color, indicating a pH of about 6.0. Heat just to boiling, but do not boil for more than a few seconds. Check the pH with another small square of nitrazine paper, and adjust by adding a drop of ammonia if necessary, finally bringing the pH as close to 6.4 as possible. Stir, wash down the sides of the beaker with a little water, allow to stand for not more than a minute, and filter as quickly as possible through an 11-cm Whatman No. 41 paper folded inside on 11-cm S and S No. 589 white ribbon paper. Catch the filtrate in an 800-ml beaker.

If three precipitations of the ammonia group oxides are contemplated (instead of the usual two), an S and S No. 589 black ribbon paper folded inside a Whatman No. 44 provides a more satisfactory combination.

An ammonia precipitation requires considerable skill and practice for success. The operation must be carried out quickly and exactly. Filtration should be rapid and the filtrate should be crystal clear. The double filters should be prepared an hour or so before the filtration and moistened with a little of the ammonium chloride wash solution; in this way, the paper fibers become saturated with liquid before the filtration. This improves filtering ability and prevents diminishing pore size during filtration, which leads to slowing of filtration rate. Even with double papers, omission of this pretreatment is likely to lead to a little of the precipitate going through the filter. Refiltration becomes a necessity, and this is a time-consuming and frustrating operation, during which precipitation of calcium carbonate or oxidized manganese compounds is very likely.

FILTRATION AND WASHING. During the filtration, keep the papers from running dry until the whole of the solution has been added. Then permit the filter to drain, and wash with hot 2% (w/v) ammonium chloride solution that has been carefully adjusted to pH 6.5 by addition of about 1 drop of concentrated ammonia solution per 500 ml. There is no need to transfer all the precipitate to the filter, but the beaker should be rinsed three or four times with a few milliliters of ammonium chloride wash solution.

Washing of the precipitate in the filter must be very thorough and is an art. The gelatinous mass should be broken up by a stream of liquid from the wash bottle, and the upper edges of the paper should be washed with particular care. About 100 ml of wash solution are normally used.

If washing is skillful, the precipitate will be all accumulated in the apex of the filter cone, the funnel stem will remain full of liquid, and the filtrate will be crystal clear.

When washing is completed and the filter has drained thoroughly, remove the inner soft paper containing the precipitate, being careful not to disturb the outer paper or break the column of liquid in the funnel stem, and spread the paper on the inner wall of the precipitation beaker. Wash the precipitate from the paper with a jet of water, fold the paper, and reserve it. (If a third ammonia precipitation is contemplated, return the paper to the filter.)

SECOND PRECIPITATION OF AMMONIA GROUP OXIDES. Wash down the sides of the beaker with hot 1:20 hydrochloric acid, add 5–10 ml concentrated hydrochloric acid (more if manganese minerals are present in the sample in unusually high concentration), and heat to boiling. Add water to a minimum of 150 ml, and boil as before to remove oxidizing substances (until methyl red is not decolorized in the boiling solution). Precipitate with ammonia as before, except that the reserved paper from the first precipitation should be torn into small pieces and macerated in the solution by vigorous stirring after the preliminary neutralization of most of the acid and before precipitation is complete.

Any further addition of ammonium chloride is not usually necessary in this second precipitation since most of the magnesium has now been removed.

Heat just to boiling, wash down the sides of the beaker with water, make a final check of pH using nitrazine paper, and filter through the reserved paper. Wash the paper pulp and adhering precipitate by decantation using hot 2% (w/v) ammonium chloride solution of pH 6.5–7.0, squeezing the pulp with the stirring rod and pouring the washings through the filter. Repeat this two or three times, and then transfer everything to the filter with the aid of a policeman. Wash the filter very carefully, especially the top edges of the paper, and churn up the pulp and precipitate with the hot wash liquid.

Because it is impossible to scrub the beaker completely free from metal hydroxides, these are recovered by adding a little hot 1:1 hydrochloric acid to the beaker, digesting for a few moments, adding a little paper pulp and enough ammonia to neutralize, and adding to the filter containing the bulk of the precipitate—or filtering through a small separate paper if this is more convenient.

When the precipitate has been thoroughly washed, churned up, and accumulated in the apex of the filter, make a final wash with 10–20 ml cold water without disturbing the precipitate to remove much of the ammonium chloride. Drain the filter, break the column of liquid in the stem, and wash off the end of the funnel and reserve the precipitate and filter until the crucible containing the residue from the silica determination is ready to receive it.

It is advisable to cover the funnel containing the precipitate with a piece of filter paper through which moisture may escape until it is dry enough to be handled conveniently. An ammonia group recovery from the filtrate is not usually necessary in rock analysis and in most mineral analyses if double filters were used as directed unless the sample contains beryllium or rare earth elements, which are not completely precipitated at pH 6.4.

A common problem when platinum dishes are used for the silica dehydration is the separation of platinum ammines in the filtrate from the ammonia group precipitate. These may cause concern if their identity is not known. They appear as a fine off-white or buff-colored precipitate that seems to accumulate slowly, especially if the ammonia group filtrate is evaporated. Attempts to filter them out of the solution fail. They either pass through the filter or dissolve and reprecipitate during washing operations. They indicate the possibility of appreciable platinum in the ammonia group precipitate; they do not affect the subsequent operations when these are properly carried out.

In the ordinary case, the ammonia group precipitate is ignited in the crucible containing the residue from the silica determination. Transfer of the bulky precipitate and accompanying papers is sometimes a little difficult. The following device has proven effective:

IGNITION OF AMMONIA GROUP OXIDES. Tear an 11-cm filter paper in half and roll the half paper into a cone with the point at the center of the torn edge. Put the point of the cone in the crucible, making a sort of funnel. Turn the paper containing the precipitate upside down, and compress the top edges into the improvised paper funnel. Using the other half of the torn filter paper, press the mass down into the crucible. Using a blunt pair of tweezers, pack the whole mass into the crucible, pressing it away from the sides as much as possible.

In this way the bulkiest of ammonia precipitates can be contained in a 25-ml crucible without difficulty and will collect into a single fluffy mass in the middle of the crucible during ignition, with none of the oxides adhering to the sides of the crucible.

Transfer the ammonia group precipitate to the crucible containing the residue from the silica determination and put into a cold furnace. Slowly raise the temperature to about 425 °C and hold at this temperature for several hours (preferably overnight). Then increase temperature to 900 °C, cool in a desiccator, and weigh. Repeat the ignition to constant weight, increasing the temperature to as high as 1100 °C if much alumina is present.

Samples containing much iron should not be heated to as high a temperature because of the possibility of forming magnetite or other spinels. If both alumina and iron are present, a short period at 1050 °C followed by an extended heating at 800 °C is recommended.

Weighing the ammonia group oxides is a matter of some difficulty, largely because ignited alumina avidly takes up water from the air and loses all its moisture content only at very high temperatures. Vacuum desiccators containing anhydrous magnesium perchlorate are a necessity. Crucibles containing the ammonia group precipitate should be taken from the furnace and placed red hot in the desiccator, covered immediately, and the desiccator evacuated using a vacuum pump or a water pump. Dry air should then be admitted to the desiccator (slowly, so that precipitates are not disturbed!) through a magnesium perchlorate tower. The crucibles must be within 1 °C of room temperature when they are removed for weighing. This means a 2–3-h cooling period. Weighing must be accomplished quickly. The balance should be set at the approximate weight before removing the crucible from the desiccator so that the final weighing takes only a few seconds.

Reserve the filtrate from the ammonia precipitate for the determination of calcium, magnesium, and manganese.
Analyze the ignited and weighed ammonia precipitate.

If the ammonia group precipitate contains only iron, titanium, aluminum, and minor amounts of phosphorus, its weight can be relied on to represent a mixture of Fe_2O_3, TiO_2, Al_2O_3, and P_2O_5. When other elements are present (as is usually the case in practice), this ideal situation is modified. Complications arising from the presence of these other elements will be dealt with later. In the absence of these complications, proceed as follows:

RECOVERY OF SMALL AMOUNTS OF SILICA. Add to the weighed mixed oxides in the platinum crucible 20 times their weight (or at least a gram in any case) of potassium pyrosulfate, and heat over a small flame (or, more conveniently, over an electric burner), gradually increasing the temperature until solution is substantially complete.

Some batches of pyrosulfate contain bisulfate, which causes difficulty by splattering at low temperature. If this is a problem, the salt should be fused in a silica dish before use.

For the fusion to be most effective, a compromise must be made between a high temperature, which leads to rapid solution of the oxides but also rapid loss of SO_3, and increased attack on the platinum crucible. Too strong heating will result in loss of SO_3 before attack is complete.

There is always a temptation to swirl the crucible during fusion. This should be resisted; it leads to lumps of unattacked oxides on the walls of the crucible, which are difficult to coax back into the reaction mixture. To observe the progress of the fusion, remove the crucible from the heat and watch it under a good light as it cools. Just before it freezes, the melt becomes lighter in color and transparent, and unattacked material at the bottom of the crucible can be clearly seen.

When the melt is quiet, remove the cover and permit most of the excess SO_3 to fume off. This will result in easier solution of the melt.
When fusion is complete, cool the crucible on an aluminum plate, and add to it about 10 ml of water. Allow to stand for a few minutes, and transfer to a clean 250-ml beaker, scrubbing and washing the crucible thoroughly. Ignite and weigh the empty crucible.

The weight of the empty crucible at this stage is important because it tells how much platinum has been introduced to the analysis from the crucible. If the pyrosulfate fusion is prolonged, as much as 2–3 mg platinum may be dissolved. If porcelain casseroles were used during the dehydration of silica, this is all the platinum in the analysis. If platinum dishes were used, there may be much more. We exclude the unlikely case in which platinum is present in the sample.

Heat the mixture in the 250-ml beaker on the steam bath until the cake has completely disintegrated. Add a drop of 30% hydrogen peroxide, and make a qualitative estimate of the titanium present from the intensity of the color produced. Add 10 ml of 1:1 (v/v) sulfuric acid and heat to boiling. If a silica recovery is intended, evaporate to fumes of sulfuric acid to dehydrate the silica.

The dehydration of silica in sulfuric acid solution is difficult. It is very easy to produce jarosites (alkali iron and aluminum sulfates) that are insoluble. Evaporation should proceed on a hot plate with good temperature control at a rapid a rate as possible without boiling and should be discontinued at the moment water is no longer evolved. The evaporation should proceed in open beakers, and a cold watch glass should be placed over the beaker from time to time to observe condensation of evaporating water. As soon as no immediate condensation is observed, the solution should be removed from the plate. Water should be added as soon as possible without violent reaction. The solution should then be boiled vigorously until iron and aluminum are in solution and filtered at once.

Cool the fumed mixture slightly, cover with a watch glass, and add water cautiously but rapidly to a volume of not more than 60 ml. Mix, and boil the solution until salts dissolve. Filter while hot through a 9-cm Whatman No. 40 paper, and wash thoroughly and carefully with a minimum volume of 1% (v/v) sulfuric acid. Catch the filtrate in a 100-ml volumetric flask (unless TiO_2 is over about 1% of the sample, in which case a larger flask may be used).

After thorough washing of the precipitate, the volume of liquid in the receiving vessel should be 80–90 ml.

Transfer the paper to the ignited and weighed crucible and burn off the paper. Ignite and weigh crucible plus silica plus occluded oxides. Add a drop of 1:1 sulfuric acid to the residue in the crucible and several drops of hydrofluoric acid. Evaporate to dryness. Ignite and weigh. Calculate the weight of recovered SiO_2; add it to the weight of SiO_2 obtained earlier, and subtract it from the weight of ammonia group oxides.

TiO_2 DETERMINATION. Fuse the residue in the crucible with a little potassium pyrosulfate, dissolve the melt in 1–2 ml of 10% sulfuric acid, and add the solution to the main solution in the volumetric flask. Wipe off the crucible, ignite it, and weigh. A further small loss in weight may be expected, but it should not exceed 0.1–0.3 mg. Reserve the crucible for the determination of calcium (or of aluminum if this element is to be weighed as phosphate).

If a silica recovery is not considered necessary, procedure is almost the same as that described, except that the fuming with sulfuric acid is omitted. The necessity to treat the insoluble residue with hydrofluoric acid remains because insoluble silicates are produced during the pyrosulfate fusion; these must be decomposed. The pyrosulfate never yields a perfect solution—there is always a residue that must not be ignored.

Titanium is determined colorimetrically with peroxide. The use of a Dubosq colorimeter is advantageous because it offers the possibility of detecting any irregularity in the color produced by peroxide. It also avoids the extensive dilution of the solution, which is necessary if a photoelectric instrument is used, and permits an easy recovery of all the solution for use in the determination of iron (and other elements if present). Results using a photoelectric instrument are certainly not better and may be much worse than those obtained by visual comparison. If uranium or vanadium are present, the eye immediately detects the mismatch of hue when comparing to pure titanium standards. Iron develops a color the trained eye can detect and compensate for; a photoelectric measurement cannot be easily compensated for iron interference. The addition of phosphate to bleach the iron color is impermissible because it interferes in subsequent procedures.

It is important to prepare a standard titanium solution with the same concentration of salts and sulfuric acid as the sample solution. It is not easy to prepare a standard titanium solution of accurately known concentration that contains no other species than sulfuric acid, potassium sulfate, and unhydrolyzed titanium salts. Measurements must be made with unknown solutions at the same temperature as the standards. All solutions should be as cool as possible because iron (which is always present in the solution derived from a rock or mineral analysis) catalyzes the decomposition of peroxide in acid solution. Peroxide should not be added to the unknown solutions until a few moments before observing them in the colorimeter, especially when much iron is present.

Iron in the unknowns leads to a visual comparison that yields a value for TiO_2 about 0.02% TiO_2 too high for each 10% Fe_2O_3 in the sample. Of course, it is possible to prepare titanium standards containing iron in known concentration, but this is an unnecessary nuisance once experience has been gained. Such experience enables one to ignore the dullness iron imparts to the peroxydized solution and read the correct value against pure titanium solutions, which exhibit a brighter color.

A convenient standard titanium solution is prepared to have 0.01 g

TiO_2/100 ml solution. Peroxydized titanium solutions are perfectly stable if they contain no iron.

> To the solution of the ammonia group oxides (in a 100-ml volumetric flask or a larger flask if TiO_2 exceeds 1% of the sample), which should contain about 5 ml sulfuric acid, add 1 ml of 30% hydrogen peroxide. Dilute to the mark, mix, and compare immediately with similarly prepared solutions of known titanium content using a Dubosq colorimeter. Do not discard any of the sample solution.

It is convenient to return all portions used (e.g., in rinsing the colorimeter cups) to the same 250-ml beaker employed in the solution of the pyrosulfate melt.

> Calculate the proportion of TiO_2 in the sample. Report percentage of TiO_2. Cover the beaker containing the whole of the peroxydized solution, and heat on the steam bath until peroxide has been destroyed. Evaporate the solution to about 50 ml. (If iron is absent in the sample, it may be necessary to evaporate to fumes of sulfuric acid to remove peroxide.)
>
> Cool the solution, add water if necessary to a volume of 50–60 ml, and with the beaker standing on a cold metal plate, add just 3 g granulated zinc. Allow to stand until reaction subsides, evaporate to a volume of 10–15 ml on the steam bath, dilute to about 50 ml, and filter off the precipitated platinum. Wash thoroughly with 1% (v/v) sulfuric acid. Ignite paper and precipitate in a weighed platinum crucible (logistics do not usually permit use of the same crucible previously employed, but it may be used), and reweigh.
>
> Add 10 ml hydrofluoric acid and a drop of sulfuric acid to the residue, evaporate to dryness, ignite, and weigh. This step is necessary because granulated zinc often contains silica. Calculate the weight of platinum, which was weighed with the ammonia group oxides, and correct accordingly.

If platinum dishes were used in dehydrating the silica, the weight of platinum found here may be very appreciable—sometimes as much as 10–20 mg. If porcelain casseroles were used, the weight of platinum recovered should correspond closely with the loss in weight of the crucible used during the analysis.

An alternative (and more often recommended) method for removing platinum from the solution is its precipitation with hydrogen sulfide. This method is adequate except in cases where much platinum is present and the iron content of the sample is high. In these cases, some iron precipitates with the platinum and the sulfur formed during reduction of iron by H_2S is a nuisance. Generally, the procedure with granulated zinc is much more convenient, especially if a reagent free from silica can be obtained.

Removal of platinum from the solution, however it is accomplished, is an essential preliminary to the determination of iron.

IRON DETERMINATION. To the combined solutions from the determination of titanium, after removal of platinum, add 30–35 ml concentrated hydrochloric acid. Heat to the boiling point, and reduce iron to the ferrous state by dropwise addition of a strong (5% in 1:4 hydrochloric acid) solution of stannous chloride. Titrate potentiometrically to remove excess stannous ion, and then oxidize the iron, using $N/10$ or (for low iron concentrations) $N/50$ potassium dichromate solution.

The titration of iron after reduction with stannous chloride has proven superior in all respects; certain critical difficulties in the procedure are easily overcome. The indicating electrodes are a bright platinum wire and a Hildebrand normal calomel electrode. Ordinary commercial calomel electrodes do not survive long in hot strong hydrochloric acid solutions and should not be used. Commercial platinum electrodes are expensive and difficult to regenerate when they become poisoned. A piece of pure platinum wire sealed in a glass tube and connected to a copper lead through a small pool of mercury in the tube is quickly prepared and easily cleaned with a piece of crocus cloth: It serves as well or better than a commercial platinum electrode. The glassware for a Hildebrand electrode can be purchased or made by any glassblower. An hour or so loads it and prepares it for use; it will last through many hundreds of determinations and can be regenerated quickly and easily. The indicating instrument is most conveniently a sensitive recorder, which will draw a titration curve as the titration proceeds. Any instrument (e.g., a pH meter) that registers millivolt signals can be used. A student potentiometer and associated galvanometer is somewhat inconvenient but is nevertheless perfectly adequate.

Directions for preparing a Hildebrand electrode or its equivalent are given in several texts on electrochemistry. Possibly the best description is that of Kolthoff and Lingane (*pH and Electrotitrations*, Wiley, Chapman and Hall, 1948).

The filtrate from the ammonia precipitate may be treated in several alternative ways: Manganese may be removed with sulfide or persulfate, nickel with dimethylglyoxime, or residual rare earths or beryllium by adding more ammonia. Carbonate may be used to separate calcium, strontium, and barium. Removal of ammonium salts may be desirable. In the ideal case, only calcium, magnesium, and manganese are taken into account. Calcium is separated as oxalate and determined as carbonate or oxide; manganese and magnesium are precipitated together as ammonium phosphates, ignited to pyrophosphates, and weighed. Manganese in the calcium and the magnesium precipitates is determined colorimetrically. Residual calcium is removed from the manganese and magnesium pyrophosphates and determined. Corrections are applied as necessary.

CALCIUM DETERMINATION. Combine the filtrates from the ammonia group precipitations and evaporate on the steam bath to 400–600 ml. If previous operations were skillfully performed, very little evaporation should be necessary.

Add 40 ml of 10% (w/v) oxalic acid to the hot solution. Bring to a slow steady boil. A piece of hardened paper under the end of the stirring rod will make boiling easier. To the boiling solution slowly add 1:1 ammonia until the solution is just alkaline to methyl orange. Then add 3 ml ammonia in excess and allow to cool completely (usually overnight).

It is necessary to have the solution at or near the boiling point during the precipitation because at low temperature the calcium oxalate precipitates as the mono-hydrate, which is undesirable.

Filter, and wash beaker and filter with 0.1% ammonium oxalate solution. A Whatman No. 42 paper should be used, and it should be prepared a half hour before use for maximum retentivity. Catch the filtrate in a 1000-ml beaker. Rinse the precipitate from the filter into the precipitation beaker, removing it as completely as possible from the paper, and dissolve it in hot 5% hydrochloric acid.

When carefully precipitated, the oxalate will be heavy and granular and will not creep during filtration and washing. It does not dissolve readily in dilute acid; it is therefore necessary to sluice it from the paper as described.

Wash the paper by dropping 6 N hydrochloric acid on it, covering, and allowing it to stand during the solution of the main part of the precipitate. The 400-ml beaker used for precipitating the ammonia group may conveniently be used to catch the calcium oxalate solution.

When the calcium oxalate is completely dissolved in the hot 5% hydrochloric acid, pour the solution through the filter. This ensures that all the calcium will be recovered and also removes traces of, for example, silica and platinum that may be present.

Wash the paper very thoroughly with dilute hydrochloric acid and then with water until a drop of methyl orange indicator placed on the paper no longer shows a red color. Reserve the paper.

In preparation for precipitating magnesium and manganese, dissolve 7 g diammonium phosphate in 25–30 ml water, and filter the solution through the paper from which the calcium oxalate was dissolved. Add the phosphate solution to the main filtrate.

Reprecipitate the calcium oxalate after diluting the solution so that it contains no more than 2 mg CaO/ml. Do not add more oxalic acid until precipitation is substantially complete, and then add only about 1 ml of 10% oxalic acid. The reprecipitation must be done in boiling solution as before, by the slow addition of 1:1 ammonia.

Cool, and filter after 4 h.

In the first precipitation of the calcium, the large amount of ammonium chloride helps prevent precipitation of magnesium. However, enough magnesium may coprecipitate to cause error if the filtering of the second precipitation is too long delayed. If the dissolving of the first precipitate was accomplished without using more acid than necessary, the calcium will precipitate almost quantitatively the second time even if only 4 h elapse. With high-magnesium rocks and minerals, it is essential that the calcium be precipitated and reprecipitated with care. With low-magnesium rocks and minerals and when only a little calcium is present, longer standing may do no harm and may improve recovery.

Filter, and transfer all the precipitate to the paper. The paper (Whatman No. 42) should be no larger than is necessary. A 9-cm paper is usually adequate. Wash thoroughly with 0.1% ammonium oxalate solution.

Be very careful to scrub and rinse the precipitation beaker thoroughly. Crystalline calcium oxalate adheres rather stubbornly to glass surfaces and is not easily visible.

Transfer paper and precipitate to a small platinum, gold, or (less desirable) porcelain crucible that has been heated to 475 °C and accurately weighed. Place in a cold muffle furnace and bring the temperature up to 475–525 °C gradually; then hold at this temperature for several hours (overnight). Weigh as $CaCO_3$. If the furnace temperature is raised too fast or air supply is insufficient, it is likely that some carbon will remain unburned at 500 °C. The residue will appear off white or grey and will weigh more than it should.

This grey color must be distinguished from the brown color often observed, which is due to traces of manganese.

If a platinum crucible was used, reignite to 950 °C, cool in a good desiccator, and weigh calcium oxide.

In the absence of interfering elements, the weights of $CaCO_3$ and CaO should correspond exactly. If strontium is present, the weight of oxide will appear too high: this provides a useful device for detecting appreciable amounts of strontium when a spectrographic examination of the sample is not feasible. Acid gases (SO_2, Cl, HCl) must be excluded from the atmosphere in the muffle. In the ideal case, the only element that may contaminate the calcium oxide or carbonate is manganese. If the calcium carbonate is pure white, no manganese is present; if it is slightly off white, manganese contamination is small and may be neglected. If the calcium carbonate is brown in color, appreciable manganese has precipitated with the calcium, and a correction needs to be made.

CORRECTION FOR MANGANESE CONTENT IN CaO. Dissolve the calcium carbonate or oxide in dilute nitric acid (1:1) containing *a little* hydrogen peroxide (to reduce the manganese oxide). Make the solution about 10% in HNO_3, and add 10–20 ml of a 1% solution of periodic acid. Heat near the boiling point for several hours. Compare the MnO_4^- color with standard solutions of permanganate using a Dubosq colorimeter, Nessler tubes, or a spectrophotometer, and determine its manganese content. Assume that the manganese was weighed as Mn_2O_3 if the calcium carbonate weight was used or as MnO_2 if the oxide weight was used.

When periodic acid is added to a solution containing peroxide, the two reagents compete, and manganese is not oxidized until all the peroxide is destroyed. A brown color may form. Unless an excessive amount of peroxide was used, this will dissipate after a time, and the permanganate pink will appear normally. Boiling the solution to remove peroxide is not very effective unless ferric iron is added.

MAGNESIUM AND MANGANESE DETERMINATIONS. To precipitate magnesium and manganese, combine the filtrates from the calcium oxalate and add to them 1 g diammonium phosphate (in the form of a filtered solution) for every 100 ml of solution. This is conveniently done as described above, using the filter on which the calcium oxalate was collected.

Add aqueous ammonia amounting to 10% of the volume of the solution containing the manganese and magnesium. Allow to stand for at least 48 h (more if only a little magnesium is present) with frequent stirring. Rub (without scratching) the sides of the beaker with the glass stirring rod to promote crystallization.

Filter through a large (11- or 12.5-cm) S and S No. 589 Blue Ribbon paper that has been soaked for a half hour in 5% ammonia. Wash superficially with 5% ammonia (25 ml concentrated ammonia solution diluted to 500 ml). Put the filtrate aside.

Dissolve the precipitate by pouring hot 5% hydrochloric acid through the filter, rinsing the precipitation beaker with several portions of the 5% acid. The volume of 5% acid to be used depends on the amount of magnesium present. If MgO in the sample is less than about 5%, not more than 50 ml dilute acid should be used. If MgO in the sample is as high as 50% (dunites, peridotites), not less than 200 ml dilute acid should be used. Catch the solution in a 250- or 400-ml beaker.

Add 1 ml of 15% (w/v) ammonium chloride solution for every 10 mg MgO expected, and dilute to 50–250 ml with water. Add a little methyl red indicator, and then add ammonia solution dropwise just to the neutral color of the indicator, stirring constantly. Magnesium ammonium phosphate should appear slowly as a microcrystalline precipitate. If each drop of ammonia solution produces a milky cloud, there is not enough ammonium chloride in the solution. Hydrochloric acid should be added, and the precipitation repeated.

With samples of high magnesium content, most of the magnesium will be precipitated without making the solution strongly alkaline. After the addition of a drop of ammonia solution, the methyl red indicator will revert to its red color as the magnesium precipitates and adds hydrogen ion to the solution:

$$Mg^{2+} + NH_4^+ + HPO_4^{2-} \rightarrow MgNH_4PO_4(\downarrow) + H^+$$

If ammonia is added too rapidly or if insufficient ammonium chloride is present (as a consequence of which the Mg^{2+} concentration is too high), the reaction

$$3Mg^{2+} + 2HPO_4^{2-} + 2OH^- \rightarrow Mg_3(PO_4)_2(\downarrow) + 2H_2O$$

may proceed. The consequence is a precipitate that contains a larger magnesium–phosphorus ratio; analytical results based on the supposition that the weighing form is $Mg_2P_2O_7$ will turn out too low. Numerous other insoluble magnesium phosphates may be formed, with a variety of magnesium–phosphorus ratios. Although it may appear that these errant compounds will usually contain a deficiency of phosphorous and lead to low calculated values for magnesium, the fact is that most determinations of magnesium by the pyrophosphate method are too high.

It is also true that careful adherence to the procedure recommended will yield exact values if impurities (tricalcium phosphate, tristrontium phosphate, manganese ammonium phosphate and its ignition product, etc.) are properly taken into account.

Add ammonia solution dropwise, with stirring, until precipitation is substantially complete. Make the solution 10% in concentrated ammonia solution, stir frequently for an hour or more, and then add 1 ml of 25% (w/v) diammonium phosphate solution. Stir, and let stand overnight.

Each time the solution is stirred, it is good policy to wash down the sides of the beaker with 5% ammonia solution, and leave a layer of dilute ammonia floating on top of the solution. For reasons we do not understand, this hastens precipitation. With samples containing only a little magnesium, precipitation may not initiate if this device is not employed, and one may be faced with the annoyance of further precipitation in the filtrate.

> Filter through a small (9-cm) Whatman No. 42 paper. Wash thoroughly with 5% ammonia solution, transferring all the precipitate to the paper. It is important that all excess diammonium phosphate be washed from the precipitate.

Magnesium (and manganese) ammonium phosphate is, for practical purposes, completely insoluble in 5% ammonia. Washing may be confidently continued until it is certain that no excess phosphate remains. The filter should be prepared so that the funnel stem remains full. The filter paper should not be torn to make a seal. It is very easy to lose some of the precipitate during washing if the filter is not completely sound.

> Transfer paper and precipitate to a weighed platinum (or porcelain, which is less desirable) crucible, and burn off the paper at as low a temperature as possible. Then ignite to 1100 °C until all carbon is removed. Weigh as $Mg_2P_2O_7$. Ignition over a good Meker burner (oxidizing flame!) is satisfactory.

There is often some difficulty in removing all the carbon during the ignition. Commonly, a part of the ignition product remains stubbornly black, even after many hours of ignition at high temperature, unless the conditions of ignition are carefully controlled. Presumably carbon becomes graphitized and is protected from the air by the precipitate. This difficulty can be avoided by first charring and then burning off the paper at as low a temperature as possible with a good air supply. The use of a well-ventilated muffle is essential, especially if platinum crucibles are used. If a flame is used, it must be oxidizing; otherwise, phosphate may decompose and attack the platinum. After paper has been destroyed, ignition to 1100 °C should proceed as rapidly as possible.

After weighing, the pyrophosphates should be broken up in the crucible with a small, sharp spatula to be sure that no carbonaceous matter remains. If any black material is seen, the ignition should be repeated, at a higher temperature if necessary. Magnesium pyrophosphate retains its composition up to at least 1200 °C. At higher temperatures, loss of P_2O_5 is possible.

The weighed residue may contain (in the ideal case being considered) $Mg_2P_2O_7$, $Mn_2P_2O_7$, and $Ca_3(PO_4)_2$.

> To the crucible containing the ignited and weighed pyrophosphates, add 1 ml of 1:1 sulfuric acid and a little water. Warm gently, and transfer the contents

of the crucible to a beaker (usually a 250-ml beaker) using a policeman if necessary. Heat on the steam bath until solution is complete, adding a drop of 6% hydrogen peroxide if necessary and more sulfuric acid (up to a total of 5 ml of 1:1 H_2SO_4 for 1 g $Mg_2P_2O_7$). Finally add water to make the solution 5% (v/v) in sulfuric acid. Then add as nearly as possible three times its volume of ethanol. The final concentration of ethanol should be between 75 and 80% by volume.

Do not use denatured alcohol. Absolute alcohol is preferable; if 96% alcohol is used, take its water content into account.

Warm on the steam bath until the precipitate of calcium sulfate becomes crystalline (no more than about 35 °C). Remove from the heat, and allow the solution to stand at room temperature for several hours (overnight) for complete precipitation of calcium sulfate. Filter through a large (12.5-cm) Whatman No. 44 paper, and wash superficially with 80% (v/v) ethanol (prepared by putting 200 ml water in a 1000-ml graduated cylinder and filling to 1000 ml with absolute ethanol).

The beaker and the filter should be kept covered as much as possible during this filtration to help prevent evaporation of the alcohol, which would lead to solution of the calcium sulfate.

Drain the filter and beaker thoroughly, and allow them to stand uncovered until all the alcohol has evaporated. Dissolve the calcium sulfate by washing the filter with 5% hydrochloric acid, and catch the solution in a 50-ml beaker. Add 1 ml of 10% oxalic acid, heat to boiling, and slowly add an excess of ammonia solution. Digest hot, cool, allow to stand for four hours (or longer), filter through a small Whatman No. 42 paper, wash with 0.1% ammonium oxalate solution, transfer to a crucible, burn off the paper, and ignite at 500–525 °C. Weigh $CaCO_3$.

The weight of calcium carbonate may be subtracted directly from the weight of magnesium pyrophosphate to correct the magnesium determination because the equivalent weights of $CaCO_3$ and $Ca_3(PO_4)_2$ are almost the same.

Evaporate the filtrate from the calcium recovery on the steam bath, and heat the residue to incipient fumes of sulfuric acid on a hot plate or over a burner. Dilute with a little water, and filter into a 100-ml volumetric flask or a 150-ml beaker. Make the solution 10% in nitric acid, add 10 ml of a 1% solution of periodic acid or about 0.1 g potassium periodate, and heat near the boiling point until oxidation of manganese is complete. Dilute to exactly 100 ml (or a larger volume if necessary), and measure the MnO_4^- color colorimetrically or photometrically.

Calculate the manganese found to $Mn_2P_2O_7$, and correct the weight of magnesium pyrophosphate accordingly.

The sum of the manganese found in the calcium carbonate or oxide and that found in the magnesium pyrophosphate should come very close to the total manganese in the sample. It is advisable to determine manganese separately, however, to be sure of this, especially when the MnO content of the sample amounts to more than a few tenths of a percent. Manganese not recovered in the calcium and magnesium precipitates is almost sure to have been retained in the ammonia precipitate. This may occur for several reasons, among which are the following:

Manganese is likely to be retained in the ammonia precipitate when:

1. The sample contains much phosphorus.
2. A large ammonia precipitate makes washing difficult.
3. Chromium or vanadium are present in the sample (chromate and vanadate formed during fusion with sodium carbonate lead to formation of oxidants (e.g., chlorate) in solution; these cause oxidation and precipitation of manganese during the ammonia precipitation.
4. Poor pH control exists during ammonia precipitation.
5. A beginner tries the ammonia precipitation for the first time (understandable delays and imperfections in the procedure are introduced inadvertently; only practice can yield a good ammonia precipitate!).
6. Silica remains in the solution because of poor technique.

3.1.5. Interferences, Refinements of General Procedure

Numerous common constituents of rocks and minerals cause trouble in the simple procedure described as the classical silicate analysis (see the first part of Section 3.1). Rocks and minerals other than silicate rocks and minerals present different problems, which, however, are often resolved using the same principles applied in silicate rock analysis.

In most cases, difficulties may be overcome by including extra steps in the analytical procedure or by modification of the basic classical scheme. Sometimes, however, modifications must be extensive, and the basic scheme may have to be abandoned altogether. Considerable ingenuity may be needed, and a thorough knowledge of solution chemistry may be challenged. It is impossible to outline procedures that are universally applicable: Several examples of unusual problems will be given, and these may serve to indicate the possibilities of solution of an analytical problem.

A common problem is incomplete attack of the sample during sodium carbonate fusion. Among many rock-forming minerals that may resist attack are chromite and other spinels, kyanite, magnetite, ilmenite, tourmaline, beryl, zircon, and sillimanite. As a rule, these will make their presence known as a gritty residue noticed during the leaching of the fusion with water or acid. If there is sufficient material available, it is often easier to begin the analysis again rather than to correct the difficulty. Finer grinding of the sample, or a longer or hotter fusion, or the addition of more oxidant to the flux may result in complete attack.

In some cases, the addition of extra oxidant (e.g., sodium peroxide) is undesirable. If the sample contains much chromium or manganese, attack on the platinum crucible may become intolerably great. It may be best, in some cases, to sinter the sample with sodium peroxide using a sodium carbonate cover at just 500 °C in a muffle furnace. If temperature is very carefully controlled, this will result in no greater attack on the crucible than fusion with sodium carbonate, and spinels like chromite will usually be completely decomposed. Above 500 °C, platinum is rapidly attacked by sodium peroxide, and sodium peroxide fusions in platinum should not be attempted over a flame.

A major disadvantage in using sodium peroxide as a flux is the difficulty in obtaining a pure reagent. Common impurities are silica, iron, manganese, and zinc. Some batches have contained pieces of metal, presumably from the manufacturing vessel. Blank determinations are unreliable because each portion is likely to contain a different amount of the contaminant. Sodium hydroxide has been used as a flux for silicates and other materials. Fusions with sodium hydroxide may be accomplished over a flame in silver, gold, zirconium, or nickel crucibles.

Fusions with sodium peroxide or sodium hydroxide may be treated in a similar manner to fusions with sodium carbonate. After leaching with water and the addition of hydrochloric acid, the problems become the same.

Because sintering with peroxide is a most useful alternative to the sodium carbonate fusion, provided a reagent of adequate purity can be obtained, directions are given in detail:

Tare the platinum crucible containing the weighed sample on a rough (e.g., top-loading) balance, add 3–4 g sodium peroxide, and mix the contents of the crucible as quickly as possible using a silica-glass or a platinum rod. Weigh 1 g sodium carbonate onto a weighing paper, and rinse the rod by rotating it in the sodium carbonate. Cover the mixture of sample and sodium peroxide with the sodium carbonate, and put the covered crucible in a muffle furnace held at 480–500 °C (not higher!) for an hour. Remove the crucible from the furnace and allow it to cool to room temperature. Place the crucible in the platinum dish or the porcelain casserole that is to be used for silica dehydration, cover the dish or casserole with a watch glass, and add 50 ml water.

Warm the dish or casserole very cautiously if solution is slow. Once solution starts, it may proceed violently, with boiling and evolution of oxygen.

If a porcelain casserole is used, add acid as soon as practicable to limit attack on the porcelain by the hot alkaline solution.

Remove the crucible as soon as the melt has been disintegrated, scrubbing and washing it thoroughly.

After fusion, leaching, and acidification, the procedure is the same as when sodium carbonate is used as a flux.

Peroxide sintering is indicated when rock samples contain ore minerals of such metals as copper, lead, silver, and arsenic, which might alloy with platinum during a carbonate fusion.

Sodium hydroxide fusions are sometimes useful. These are best accomplished in gold crucibles (although platinum, silver, or even silica crucibles may be used).

Put 5–6 g sodium hydroxide pellets in a gold crucible. Add a weighed portion of the sample. Heat over a small flame at such a rate that spattering does not occur. Continue until the sample is completely decomposed. Cool, and proceed as with a carbonate or peroxide fusion.

A procedure that is sometimes useful with samples containing large amounts of magnetite, pyrite, or other iron-bearing minerals involves a preliminary leaching with hydrochloric acid. This procedure has been successfully applied to lateritic ores and to low-grade iron ores. After roasting in the usual way, the sample is washed into a porcelain casserole and treated for several hours on the steam bath in 1:1 hydrochloric acid. Insolubles are filtered off, the filter paper is washed with dilute hydrochloric acid, and the residue is burned in the original crucible. The leached sample is fused with sodium carbonate in the usual way.

After treatment of the fusion with acid, the iron-bearing solution is added, and the separation of silica and the rest of the analysis proceeds as in the basic procedure. Manganese minerals and rocks containing much pyrolusite or related minerals are advantageously treated in a similar manner but with the addition of hydrogen peroxide to the hydrochloric acid leach solution.

Other similar pretreatments are sometimes useful. Most sulfide minerals are rapidly attacked by bromine. If a rock or mineral containing sulfides is treated with an excess liquid bromine and water, a proportion of the sulfur dissolves in the bromine (presumably as a sulfur bromide). A part of the sulfur may be oxidized to sulfate. Addition of a little hydrochloric (or other) acid to the mixture containing excess liquid bromine leads to a separation of the metals from sulfur without heating. This is of consequence when metals such as arsenic or mercury are sought and roasting is impermissible.

Molybdenum sulfide is unattacked by bromine but can be dissolved by treatment with alkaline hypochlorite if hydrochloric acid is slowly added to a well-agitated cool mixture of alkaline hypochlorite and the sample.

In the analysis of meteorites (chondrites), separation of metallic and silicate phases is almost essential for meaningful analysis. Such a separation may be made using bromine, which dissolves metals and sulfides, or chlorine. Use of the latter involves heating the sample in a chlorine atmosphere and collecting the volatile metal chlorides.

In general, a preliminary separation of nonsilicate minerals (metals, sulfides, etc.) is often useful. The separated metals may or may not be added to the analysis at an appropriate stage. With such materials as chondritic meteorites, sampling considerations are such that separate analysis of the silicate fraction and the metal sulfide fraction probably carries as much information as any overall analysis.

The selective dissolution of minerals is an important technique that no analyst should ignore. Sometimes it is relatively easy to accomplish a separation of two types of minerals in a rock and relatively easy to analyze each accurately. Analysis of the whole rock, without separation, may be a matter of extreme difficulty and may be not worth the trouble because of sampling considerations.

Sometimes, as in the analysis of sedimentary rocks, it is desirable to start with a large analytical sample (up to 10 g or more) in order to obtain accurate values for constituents present in relatively small amounts. A crude procedure is simply to leach the large sample with a suitable acid and analyze the insoluble residue. This does not take into account the appreciable solubility of many silicates. Soluble silicates may account for a large part of the silica in the sample, especially since such silicates are usually finely divided in sedimentary rocks.

In the crude procedure, organic matter may cause difficulty: It results in foaming, difficulty during filtration, and probably anomalous behavior of several elements, notably iron and aluminum, through formation of organic complexes.

A more sophisticated approach to the separation of sedimentary rock samples into soluble and insoluble fractions is as follows:

Heat the sample (which may weigh 10 g or more) at 450 °C to constant weight. Add water and hydrochloric acid to dissolve carbonates, soluble silicates, and so on. Filter, and wash with dilute hydrochloric acid. Ignite and weigh the insolubles. Evaporate the acid solution of the solubles to dryness, dehydrate the silica, and separate it. Precipitate the ammonia group of elements in the filtrate from the "soluble" silica, reprecipitate to remove traces of calcium and magnesium, and ignite and weigh. Analyze the acid-insoluble portion of the sample by the classical procedure, with or without the addition of the silica and ammonia group oxides obtained from the soluble portion.

Determine CaO, MgO, K_2O, Na_2O, and so on on separate portions of the sample or use aliquot parts of the original leach solution from which SiO_2 and the ammonia group elements have been removed.

3.1.6. Interferences, Modifications, Difficulties

Elements That Interfere in Separation of Silica. Fortunately, interferences in the silica determination are few. The most common is the interference of fluorine, which may cause low values because of the volatility of fluosilicic acid. In rock analysis, the error is usually of no account. In mineral analysis, especially of minerals containing little alumina, it may be large, and it must not be ignored. The problem of determining silica accurately in fluorine-bearing materials is greatly simplified by the fact that a large excess of aluminum chloride added to the hydrochloric acid used to dissolve the carbonate fusion can completely prevent the loss of silica by volatilization.

PROCEDURE. Determine "moisture" (loss in weight at 105 °C). Omit the roasting process with fluorine-bearing silicates. Fuse 0.6–1.0 g of the sample with sodium carbonate. Leach with water in a platinum dish. When the melt is completely disintegrated, acidify by adding 50 ml of 10% (w/v) aluminum chloride ($AlCl_3 \cdot 6H_2O$) in 1:3 hydrochloric acid. Dehydrate the silica as usual, except that the evaporation should take place in the uncovered vessel. The large amount of aluminum salt decreases the efficiency of the dehydration, and a third dehydration may be desirable. Otherwise, the process is exactly that of the "ideal" procedure.

The filtrate from the silica is usually discarded. A second sample is started in the usual way, but the value obtained for silica is rejected. If the rare earth elements are to be determined, the filtrate from the silica may be conveniently treated to recover these elements by precipitation with oxalic acid.

Unfortunately, in analyzing silicate minerals containing fluorine, there is often too little sample to warrant the use of half a gram or more for silica determination alone. In such cases, the following procedure is used:

PROCEDURE. Fuse 0.5–0.6 g of the sample with sodium carbonate in the usual way, and leach with water in a platinum dish. Filter when disintegration is complete using a Whatman 541 (hardened) paper folded inside a Whatman No. 42 paper. Wash with 2% sodium carbonate solution.

Return the residue to the platinum dish, add 50 ml of 2% sodium carbonate solution, and boil for a few minutes. Filter through the same papers, and wash thoroughly with hot water. Again return the residue to the platinum dish, washing it from the paper with water. Burn the paper in the original crucible, fuse the small residue with a little (about 0.5 g) sodium carbonate, leach with water, and add the contents of the crucible to the material in the platinum dish. Proceed with the analysis of this portion of the sample in the usual manner, dehydrating with, for example, hydrochloric acid.

Treat the sodium carbonate filtrate as follows:

To the combined filtrates in a 600-ml beaker, add water to a volume of about 300 ml and then a solution of zinc nitrate prepared by stirring an excess of zinc oxide in 20 ml of 10% (v/v) nitric acid and filter. Boil for 1 min, filter through a 12.5-cm Whatman No. 40 paper, and wash thoroughly with hot water. The precipitate may contain nearly all the silica and much of the alumina in the sample and may be extremely bulky. It must be thoroughly washed.

Add methyl red to the filtrate and nearly neutralize with nitric acid (do not acidify; the indicator should show its yellow alkaline color).

Evaporate to 200 ml, adding a drop of ammonia solution if necessary to keep the solution yellow in color.

Transfer to a platinum dish, add very dilute nitric acid just to the orange neutral point of the indicator, and add a solution prepared by dissolving 1 g zinc oxide, 2 g ammonium carbonate, and 2 ml concentrated ammonia solution in 20 ml water.

Cover the platinum dish and boil until the odor of ammonia is no longer detectable. The volume of the solution at this stage will be 50–100 ml.

Dilute to about 100 ml with warm water, stir, allow to stand for a few minutes, and filter through a Whatman No. 40 paper. Wash the precipitate with cold water. Reserve the filtrate for the determination of fluorine.

Transfer the zinc oxide precipitate to the platinum dish together with the previous zinc oxide precipitate (which contains the major part of the silica). Wash both precipitates from the paper as completely as possible using hot water. Burn the two papers in a platinum crucible at 500 °C in a muffle furnace, fuse the residue with a little sodium carbonate, leach with water, and add to the suspension in the dish.

Add hydrochloric acid and proceed with the dehydration of silica as in the ideal procedure and then with the separation of the ammonia group. A triple ammonia precipitation is advisable unless the ammonia precipitate is very small because zinc is not easily removed.

In many cases, only silica and a portion of the alumina are found in the sodium carbonate filtrate. Thus, although the net effect of the above procedure is to divide the sample into two parts that are handled separately, the duplication of effort is not as extensive as it may at first appear. Once the two separate ammonia precipitates have been weighed, they may be combined for the determination of iron, titanium, or other elements. Little, if any, of these elements will find their way into the sodium–carbonate-soluble portion of the sample.

Interference of fluorine with the silica determination may also be overcome, in some samples, through the use of boric acid. In silicate analysis, this approach is unattractive because the removal of borate, while not very difficult, is time-consuming, and the ammonia group oxides must be examined for their B_2O_3 content. Recent development of a rapid and reliable colorimetric method for boron has much improved the possibilities of the boric acid approach because boron can now be rapidly and conveniently determined in the ammonia group oxides: Extreme pains to remove it before precipitating the ammonia group are not as necessary.

As applied to the analysis of tourmaline, the procedure is as follows:

PROCEDURE. Fuse the sample in the usual way with sodium carbonate, leach in a platinum dish with water, and add 10 ml of 5% (w/v) boric acid before the hydrochloric acid addition. Evaporate *uncovered* to dryness on the steam bath, and then add alternately 2–3 ml concentrated hydrochloric acid and 2–3 ml methanol, cooling the dish before each addition and evaporating to dryness afterward, always with the dish uncovered. Care must be taken not to overheat, especially after a methanol addition, to avoid loss by spattering.

After several treatments with alcohol and acid, continue with the dehydration of silica and the precipitation of the ammonia group as usual, except that the ammonia group oxides should be ignited in a silica or Vycor crucible, separate from the residue from the silica determination.

When the ammonia group oxides have been weighed, mix them thoroughly with a platinum rod or spatula, and remove and weigh approximately 10 mg into a gold crucible. Add a pellet (about 1.3 g) of sodium hydroxide, and heat over a small flame until the hydroxide melts and disperses the oxides. Cool, add about 3 ml water, and leach. Evaporate on the steam bath to near dryness, and proceed with the determination of boron by the method of Heyes and Metcalfe. Correct the weight of ammonia group oxides for the B_2O_3 found.

The above procedure is attractive in the analysis of tourmaline, which contains boron. The examination of the ammonia group oxides for boron is necessary in any case so that no time is lost through addition of boric acid. The attractiveness of the method disappears when fluorine-bearing minerals without boron are being analyzed.

After removal of a few milligrams of the ammonia group oxides for boron determination, the remainder is treated as usual, making a correction for the portion removed.

A method for determining silica in fluorspar (fluorite) involves heating the sample with perchloric and boric acids. The volatility of boric acid from perchloric acid is utilized.

PROCEDURE. Prepare a special boric–perchloric acid mixture by adding an excess of boric acid powder to a mixture of 25 ml of 70% perchloric acid and 70 ml water, heating, with stirring, to 80 °C so that the hot diluted perchloric acid is saturated with H_3BO_3.

Weigh 0.5000 g fluorspar into a 400-ml Pyrex or silica-glass beaker, and add 15 ml of the hot acid mixture. Digest below the boiling point with the beaker uncovered until most of the water is expelled, and gradually increase the heat until the perchloric acid begins to fume. Continue heating for 5 min after strong fumes are evolved. Cool, wash down the sides of the beaker with water, and repeat the evaporation and fuming.

Dilute to 50–75 ml with water, heat to boiling, and filter. Wash with dilute hydrochloric acid and then with water, transferring all insolubles to the filter. Washing should be thorough, so that no perchlorate remains in the precipitate.

Transfer to a platinum crucible, add a drop of sulfuric acid and a few drops of water, pack the crucible half full of ashless paper pulp, add a few drops of ammonia solution, burn off the paper at the lowest possible temperature, ignite strongly, cool, and weigh. Determine SiO_2 as usual by volatilization with hydrofluoric acid, and continue with the separation of the ammonia group, calcium, and magnesium as in the ideal procedure. If the ammonia group is large, it is advisable to examine it for boron contamination as described above.

For the determination of silica in phosphatic rocks and minerals (phosphate rock, apatite, etc.), a variety of procedures are useful. Such materials invariably contain fluoride, and no analysis of them is valid unless their fluorine content is taken into account.

3.1.7.   Procedure When Sample Attack Is Incomplete

In the analysis of mineral separates, in which the impurity is often a single species, incomplete attack is most often due to that impurity. In the analysis of rocks, the unattacked impurity most often makes up a very small proportion of the sample and may remain undetected.

Unattacked residues are often to blame for a large weight of material remaining after the determination of silica by volatilization with hydrofluoric acid.

When incomplete attack is noted, there is a choice—whether to abandon the analysis and start over or to try to rescue it. When there is an abundance of sample, the first choice is the best; when the supply of sample is limited, as is often the case in analyzing mineral separates, it is necessary to deal with the unattacked residue. Gross error may result if it is ignored.

Procedures in such emergencies must be cleverly devised to suit the circumstances, but some common ones may be described: If the sodium carbonate fusion is leached in platinum, and no large insoluble residue is observed, there being only a few milligrams of unattacked mineral, this may be filtered off, heated to remove paper, and refused, either with sodium carbonate at a higher temperature or with peroxide. It may then be returned to the main solution and the analysis continued. Such an approach should not be attempted if there is a large insoluble residue on leaching the original sodium carbonate melt because of procedural difficulties.

Most often, it is best to carry the unattacked material through the separation of the silica and to treat the ignited impure silica by fusion with sodium carbonate or peroxide. The leach of this fusion may then be combined with the original leach and the analysis continued normally.

If unattacked material is carried through the silica determination, the residue therefrom should be treated separately from the rest of the analysis so that necessary corrections can be made: in particular, minerals containing alkalies and alkaline earths cause error in the silica (they are weighed first as oxides and then as sulfates, leading to too low a SiO_2 weight) and errors in the rest of the analysis (e.g., K_2SO_4 will be weighed and reported as Al_2O_3; calcium and magnesium values will be too low if these elements are present in the unattacked material). Common minerals that are apt to be imperfectly attacked during carbonate fusion and the subsequent evaporation with hydrochloric acid are chromite and zircon. The latter does not lead to error because its silica content is (usually) volatilized and reported properly as SiO_2, and ZrO_2 belongs in the ammonia group. Chromite, and especially magnesium chromite, causes gross error, as do many other spinels.

Many minerals that escape sodium carbonate attack may be treated with sodium peroxide or potassium pyrosulfate. Cassiterite and chromite are resistant to both at the temperatures attainable without excessive attack on the platinum crucible. Residues containing these minerals, if they weigh more than a milligram or so, should be scrubbed from the platinum crucible prior to further treatment. Pyrochlore may be fused in platinum with pyrosulfate over a small flame.

Since a separate analysis of the residue from the silica determination is time-consuming and may be difficult, it is best, when possible, to redesign the analytical approach and start over.

Occasionally, poor technique or a misapprehension concerning the nature of the sample may lead to extensive contamination of the platinum crucible. Such contamination may nearly always be avoided if the precaution of preroasting is observed and if the flame used to heat the crucible is always fuel lean (strongly oxidizing). The most common contaminant is iron; this is easily detected by the violet or purple stain it leaves on oxidation in a muffle furnace or an oxidizing flame. Unless alloyed iron is removed at once, it will work its way into the crucible, resulting eventually in embrittlement and cracking of the platinum—not to mention erroneous values for iron distributed randomly over any analyses in which the crucible is used. Repeated ignition at 850–900 °C alternated with hydrochloric acid treatment or pyrosulfate fusion appears to be the only feasible way to recover alloyed iron. Introduction of platinum into the analysis during such treatment is inevitable.

Lead and copper contamination of platinum crucibles is best removed by repeated treatment with mixed nitric and hydrofluoric acids. Arsenic and phosphorus are often mentioned as troublemakers. If fusion conditions are always strongly oxidizing, neither of these elements give any difficulty. Arsenic is quantitatively oxidized to arsenate, and most of it remains in this form during hydrochloric acid evaporation in porcelain.

The brown stain often observed in platinum–rhodium crucibles after alkali fusion is most often of no consequence: It is removed during the first treatment of the crucible with hydrochloric acid, and does not weigh more than 0.1–0.2 mg.

A large number of common minerals are not rapidly attacked during a sodium carbonate fusion unless heated at the maximum temperature attainable by a good Meker burner. Some amphiboles need to be ground to pass 100% through 115-mesh to be sure of their complete attack, and heated for 10–15 minutes at bright red heat. The use of sodium peroxide to aid in such fusions leads to more rapid attack, but if too much peroxide is used, attack on the platinum may be excessive.

3.1.8. Other Members of Acid Group

Elements that commonly finish with the impure silica include the acid group elements tantalum, columbium, tungsten, and tin. Titanium is partly precipitated. Zirconium may precipitate as phosphate or may remain unattacked as zircon (zirconium silicate). Silver chloride or barium sulfate may be present. Of all of these, only tantalum and columbium can be assumed quantitatively present in the insolubles following a sodium carbonate fusion and a hydrochloric acid dehydration and then only if phosphates or fluorides are absent or taken into account.

The presence of a large residue of acid group elements and titanium after volatilization of the silica with hydrofluoric acid makes it difficult to drive off sulfuric acid and to ignite the residue without loss. The difficulty is best overcome by using the device described above—by packing the crucible half full of paper pulp before ignition. Error in the silica determination can only arise if elements are weighed in one form with the impure silica and in another form after the silica is volatilized. Such errors are uncommon, provided the silicic acid was thoroughly washed to remove alkali. Boric acid added in the first stages of the analysis to volatilize fluorine will contaminate the silica if it is not removed. The boron will then volatilize with hydrofluoric acid and be reported as SiO_2. Tungsten and molybdenum oxides are volatile at high temperatures but may be incompletely removed during ignition of the impure SiO_2 and then reported as SiO_2. The difficulty can be minimized by making the final ignition at as low a temperature as necessary to decompose metallic sulfates (<800 °C) and keeping the crucible covered.

When the sample contains appreciable amounts of tantalum and columbium, the residue from the silica should always be analyzed separately because there is no easier way to obtain good values for these elements. The residue should be fused with potassium pyrosulfate. The fusion may be examined by X-ray spectroscopy or chemically. In a chemical analysis of this residue, several alternatives present themselves: If the earth acids (i.e., Ta_2O_5, Cb_2O_5, TiO_2, etc.) are of interest, the pyrosulfate melt is leached with dilute sulfuric acid containing tannin. If the melt is leached with water, barium sulfate is precipitated and may be removed; an ammonia precipitation will then leave calcium, magnesium, and so on in the filtrate, which may be added to the main part of the analysis. The ammonia precipitate may be fused with peroxide. Solution of the melt in water will yield chromate, which may be determined colorimetrically; the precipitated iron and titanium may be combined with the main part of the analysis. Determination of chromium in unattacked chromite is more desirable than returning the chromium to the main analysis, where it causes some difficulties.

3.1.9. Separation of Hydrogen Sulfide Group

In the classical scheme of analysis, dehydration and separation of the acid group is followed by treatment with hydrogen sulfide. In ordinary rock analysis and in the analysis of most silicate minerals, use of H_2S is confined to secondary operations such as the removal of incidental platinum prior to determining iron. The hydrogen sulfide group of elements—silver, mercury, lead, tin, bismuth, copper, cadmium, arsenic, antimony, germanium, molybdenum, selenium, tellurium, and the noble metals—are seldom present in more than traces and may be handled adequately without resorting to the hydrogen sulfide precipitation. Occasionally, however, one or more of these elements is present in an appreciable amount, and it may be desirable to treat the filtrate from the silica with hydrogen sulfide before continuing. Each of the elements in question presents its own problems, however, and the following must be taken as a general guide. A general procedure follows.

PROCEDURE. Dilute the filtrate from the silica with water until it is close to 1.0 N in hydrochloric acid. Into the cold solution, bubble hydrogen sulfide from a cylinder or a Kipp generator at a rate of about one bubble per second using an 8-mm o.d. glass tube fire polished at both ends and of a convenient length to be used as a stirring rod. Each solution should have its own delivery tube. As precipitation proceeds, slowly heat the solution to near the boiling point at such a rate that precipitation is nearly complete as the solution begins to boil. Then add hot water to reduce the acid concentration to 0.3 N. Continue the flow of gas without interruption while allowing the solution to cool to room temperature or to about 30 °C. Filter through a Whatman No. 40 paper (11 cm or smaller), and wash thoroughly with water containing a little hydrogen sulfide.

Evaporate the filtrate on the steam bath, uncovered, until the volume has been reduced to about 200 ml or until no sulfide remains. Add an excess of bromine water (or 1 ml bromine if much sulfur has precipitated). Boil to remove all excess bromine before continuing with the ammonia group precipitation.

It is impossible to give any general procedure for analysis of the sulfide precipitate. Usually it will be small, and spectrographic analysis will suffice. Of all the hydrogen sulfide group elements, only cadmium can be relied upon to be quantitatively present in the precipitate under the conditions of a sodium carbonate fusion after roasting followed by hydrochloric acid dehydration of silica. Much of the silver will have been removed as the insoluble chloride; mercury will have volatilized together with some or all of the selenium, bismuth, arsenic, and antimony. At least part of the germanium will have precipitated with the silica or volatilized as chloride. Some of the tin may be lost during hydrochloric acid evaporation. Platinum, and sometimes iridium or rhodium, originates in the crucibles and dishes. Thallium, if present, probably came from the Clerici solution used in mineral separation. Palladium, osmium, and gold are unlikely to be found quantitatively in the sulfide precipitate. They may alloy with the platinum crucible or volatilize in part. Copper and lead may be lost through alloying with the crucible.

Other methods of attack and other acid treatments may be used when a specific member of the hydrogen sulfide group is of interest. For example, while arsenic is usually removed during roasting, it may be quantitatively retained through the fusion and acid evaporation if roasting is omitted and a strongly oxidizing flux is used.

Since the hydrogen sulfide group is not normally removed before precipitation of the ammonia group, it is instructive to consider the behavior of each of its members during the regular procedure and to briefly indicate devices that may be used to overcome their interference with the separation and determination of the more common elements.

3.1.10.   Behavior of H_2S Group Elements in Regular Procedure

Silver. Silver is largely converted to metal and alloyed with the crucible during fusion. Alloying is superficial, and small amounts are converted to chloride during acid treatment of the crucible or scrubbed from the crucible and converted to chloride during evaporation. Silver-bearing silicates should be fused with sodium peroxide (in zirconium crucibles) or sintered with peroxide in platinum. Dehydration of silica should be accomplished with nitric acid instead of hydrochloric. The filtrate from the silica may be treated with hydrogen sulfide as described above, or silver may be precipitated as chloride. Hydrogen sulfide should not be used in solutions containing both nitrate and chloride. After removal of silver, the solution should be evaporated to dryness and the nitrate removed by repeated evaporation with hydrochloric acid before continuing with the ammonia group precipitation.

Mercury. Mercury is completely volatilized during preliminary operations.

Lead. Lead may alloy with the platinum crucible; is partly precipitated as chloride and weighed, probably as oxide, with the silica; is converted to sulfate and weighed as such with the residue from the hydrofluoric acid evaporation; is partly precipitated and weighed with the ammonia group elements, probably as PbO. Interference is overcome by fusion with peroxide instead of carbonate as for silver and dehydration of the silica with nitric instead of hydrochloric acid. Lead is precipitated as sulfate after evaporation of a sulfuric acid solution to remove all other acids followed by dilution with water, and it is separated from barium by treatment with sodium acetate solution.

Bismuth. Bismuth is partly precipitated as oxychloride and weighed with the silica, possibly as Bi_2O_3, and is finally collected and weighed with the ammonia group. There are probable losses of bismuth during repeated hydrochloric acid evaporation. There does not seem to be any satisfactory way of dealing with bismuth in the course of a regular silicate anlaysis; it should be determined on a separate sample. When but little is present, its equivalent in Bi_2O_3 may be subtracted from the weight of ammonia group oxides.

Copper. Copper may alloy superficially with the crucible during fusion if conditions are not kept strongly oxidizing and partly precipitates with the ammonia group of elements and sometimes with magnesium ammonium phosphate. The copper contamination of the ammonia group is easily determined by collecting it on zinc during removal of platinum prior to the iron determination. The precipitate is dissolved in aqua regia and the cuprammonium blue color developed by addition of aqueous ammonia under carefully controlled conditions. Visual color comparison is adequate when the only purpose is a small correction of the total weight of oxides.

When more than a few tenths of a percent of copper is present in a sample, an acid hydrogen sulfide separation is advisable since it gives less trouble than a search for copper throughout the rest of the analysis.

Most of the copper in the sample will go through the whole of the ordinary analysis and finish in the filtrate from the magnesium ammonium phosphate, where the blue ammonia complex will easily be seen if more than about 1% CuO is present in the sample.

Cadmium. The small amounts of cadmium usually present in silicate rocks and minerals are without appreciable effect on the course of the analysis. Most of the cadmium will go through the analysis and end up in the filtrate from the magnesium ammonium phosphate together with most of the zinc and copper in the sample. Samples with more than a few tenths of a percent CdO should always undergo a hydrogen sulfide separation after removal of the silica because this provides the most convenient way to determine the cadmium. There does not seem to be any simple way to determine cadmium contamination of the ammonia group oxides (contrast to copper, see above).

Arsenic. Arsenic is usually present in compounds from which it is volatilized under the conditions of roasting called for in the general procedure. If a sodium carbonate fusion of the unroasted sample is very carefully carried out under oxidizing conditions, all the arsenic will be converted to arsenate, analogous to phosphate, and will remain in that condition through the hydrochloric acid evaporations (in porcelain, tests have not been made with platinum dishes). If sufficient iron or aluminum is present in the sample, arsenic will be caught in the ammonia group precipitate (like phosphate) as aluminum or iron arsenate. Careful ignition of the precipitate will leave the arsenates intact (at temperatures below 600 °C). On heating to higher temperatures, the arsenic will be completely volatilized. The only noticeable difficulty will be in bringing the ammonia group oxides to constant weight.

If fusion conditions are not strongly oxidizing, however, the crucible may be attacked and gain weight; trivalent arsenic will be lost during the dehydration of the silica; some arsenic will be retained in the silica; the precipitation of the ammonia group will be disrupted because of the reduction of iron to the ferrous state.

To be sure of the quantitative recovery of arsenic after an oxidizing fusion, the silica should be dehydrated with nitric acid (see "Silver", "Lead"), and a hydrogen sulfide precipitation should follow the separation of silica.

Arsenic is best determined gravimetrically as magnesium ammonium arsenate or titrimetrically by permanganate oxidation from As(III) to As(V).

Antimony and Tin. In ordinary analyses, antimony and tin may be assumed to be quantitatively retained by the ammonia group and weighed as SnO_2 and Sb_2O_5. Losses through volatilization do occur, however, if hydrochloric acid is used for dehydration of silica. If appreciable amounts are present, a nitric acid dehydration is preferred. In this case, most of the tin and much of the antimony will be found in the residue from the hydrofluoric acid evaporation of the silica. This residue may be fused with pyrosulfate and added to the filtrate from the silica prior to an acid hydrogen sulfide separation.

Germanium. Germanium is probably volatilized as chloride during a regular silicate analysis. In a modification using nitric instead of hydrochloric acid, germanium will probably be counted as silica. Germanium sulfide is difficult to precipitate, and it is doubtful that any useful determination of the element can be made in the course of a regular analysis. It should be separated as the volatile chloride or bromide by distillation from a separate sample.

Molybdenum. Molybdenum oxide is volatile at high temperature, and appreciable amounts of the element in a silicate sample may be expected to give trouble. Some may be retained with the silica, from which it will partially evaporate during ignition. Most will be caught in the ammonia precipitate when a large excess of iron is present and will cause difficulty in attempts to ignite to constant weight. Sometimes much of the molybdenum will escape the ammonia precipitation. Calcium molybdate is insoluble and may precipitate with the ammonia group, finally causing high results for alumina and low results for calcium. It is not easy to precipitate molybdenum completely with hydrogen sulfide, and the general procedure given above may be inadequate with molybdenum-bearing materials.

The recommended procedure for silicates containing molybdenum follows.

PROCEDURE. Roast the sample thoroughly (18 h, or overnight) in a 500 °C muffle furnace. Leach the residue with aqueous ammonia, dilute 1:10 with water, and filter. Evaporate the filtrate in platinum to dryness. Add paper pulp, and ignite strongly to remove MoO_3. Combine the residue with the main part of the sample, fuse with carbonate, and proceed as usual. Determine the molybdenum content of the sample on a separate portion.

Selenium and Tellurium. These elements finish as selenate and tellurate after an oxidizing alkaline fusion. Selenate resembles sulfate in that it forms insoluble salts with lead and barium. Selenate is slowly reduced to selenite by hot concentrated hydrochloric acid (with partial volatilization of $SeOCl_4$) from which it liberates chlorine. Tellurate is not easily reduced by hydrochloric acid. The thermodynamically stable forms in acid solution are selenite and tellurite, but in the absence of halide, selenate and tellurate stubbornly evade reduction. Evaporation of selenium and tellurium compounds with aqua regia yields selenites and tellurites; however, in the absence of halide, selenates and tellurates are stable on evaporation with nitric acid.

Selenites and tellurites are quantitatively retained in ammonia precipitates containing iron. Selenates and tellurates are not. On ignition, any selenium present will be volatilized, but volatilization of tellurium does not occur.

Precipitation of selenium and tellurium by hydrogen sulfide is not very satisfactory when these elements are present alone and may not occur at all if they are in the 6-valent form. They are better precipitated with sulfur dioxide from 4 N hydrochloric acid. Control of acid concentration permits separation of selenium and tellurium from their 4-valent solutions.

The behavior of selenium and tellurium in an ordinary silicate analysis may best be described as capricious. Materials containing more than traces of these elements cannot be analyzed routinely. Special procedures must be devised to suit each case.

The Noble Metals. Of the noble metals, gold alloys superficially with the platinum of the crucible and is effectively removed from the analysis;

platinum, rhodium, and sometimes iridium usually originate in the laboratory ware; palladium tends to behave like nickel; osmium and ruthenium, both of which form volatile compounds under the proper conditions, are unpredictable. The analysis of samples containing appreciable amounts of the noble metals involves problems unrelated to those of ordinary silicate analysis.

The behavior of platinum is of importance because several milligrams of this metal may be introduced during preliminary operations and during the evaporation of chloride solutions in platinum dishes. Platinum metal is attacked, with the formation of chloroplatinate, by hydrochloric acid solutions of oxidizing agents. When samples containing iron are evaporated with hydrochloric acid in platinum, iron is reduced from ferric to ferrous and a stoichiometric amount of platinum is dissolved. The reaction is slow but continues during an evaporation as the ferrous iron produced is reoxidized by the air. In the case of samples containing much iron (20–50%), several milligrams of platinum may dissolve from the dish in which the evaporation is proceeding.

If not taken into account, the dissolved platinum may lead to gross error. Part of it will be carried by the dehydrated silica and will finish in the residue from the hydrofluoric acid evaporation. Part will coprecipitate with the ammonia group of elements. Part will form platinum ammines and will be weighed with the calcium if precautions are not taken. Platinum has also been found in milligram quantities in ignited magnesium pyrophosphate. If platinum is not removed from the solution before titration of iron, erroneous values for iron will ensue.

In the analysis of ferrous minerals, when platinum dishes are used for dehydrating silica, it may be advantageous to include a hydrogen sulfide precipitation before the ammonia group separation: however, there are severe disadvantages. A large hydrogen sulfide precipitate often carries down unacceptable amounts of iron, cobalt, nickel, and some other elements not considered part of the hydrogen sulfide group. Such a precipitate should always be examined for its iron content—and not merely discarded as might be supposed.

Thallium. Thallium is an analytical disaster. It most often occurs in mineral samples that have been separated using thallium malonate and other thallium salts and thus is usually irrelevant. As much as 2% thallium has been found in a "pure" mineral separated by such means. Obviously, no meaningful analysis is possible unless this contamination is taken into account.

Thallium has two valence states, I and III. In the monovalent state, it acts like an alkali metal. In the trivalent state, it acts like ferric iron.

It is prone to change from one state to another rather capriciously. The result, from the point of view of one who does not expect thallium to be present, is confusion. Thallium metal behaves like lead and can alloy with platinum.

Mineral samples contaminated with Clerici solution, when gently roasted over a small flame in a covered platinum crucible, yield a dull deposit on the lid of the crucible. Some of the thallium may be volatilized during roasting.

Thallous chloride is not very soluble and may precipitate with the silica. In this case, some of it may be volatilized during the ignition of the acid group.

If thallium is reduced by adding a little sulfurous acid early in the procedure, it will remain in the thallous state and cause no further difficulty; but if bromine is added to oxidize iron (or if much manganese is present), thallic ion will form and precipitate with iron in the ammonia precipitation. During ignitions, thallium may reduce to metal and either volatilize or alloy with the platinum.

Platinum ware suspected of thallium contamination should be boiled first in concentrated nitric acid and then in water and strongly ignited.

3.1.11. Ammonium Sulfide Group

In the classical scheme of analysis, the filtrate from the ammonia precipitate is treated with ammonium sulfide. Elements thus precipitated include manganese, cobalt, nickel, zinc and Tl(III). While an ammonium sulfide precipitation may occasionally be useful, for example, in the analysis of a zinc silicate mineral, practice has shown that the ammonium sulfide group is best handled in other ways.

Besides the elements listed as belonging to this group, the filtrate from the ammonia precipitation may contain part or all of several other elements: the rare earth elements, yttrium, scandium, uranium, vanadium, beryllium, tungsten, molybdenum, and so on. If a hydrogen sulfide precipitation was not included earlier in the procedure (as is usually the case), some of the hydrogen sulfide group elements—notably copper—may be present.

The rare earths, if present, should be removed from the solution by adding an excess of ammonia: a pH of 10 is required for the complete precipitation of these elements.

PROCEDURE. Heat the filtrate from the ammonia precipitation nearly to boiling, and add one-fifth its volume of concentrated aqueous ammonia. Cover and allow to stand until cold (overnight). Stir in some paper pulp and filter (Whatman No. 41). Wash superficially with 5% (v/v) aqueous ammonia. The precipitate will be contaminated with manganese (MnO·OH) and probably with alkaline earth carbonates. It must be redissolved in hydrochloric acid with the addition of a drop of hydrogen peroxide to reduce Mn(III) and Ce(IV). Before reprecipitating, the peroxide must be completely removed by boiling and then adding a small amount of sulfurous acid.

Reprecipitate with ammonia in large excess, cool, add paper pulp, and continue as before.

If much manganese is present, several reprecipitations may be necessary to remove it completely; in such cases, it is best to reprecipitate only once and plan on determining the manganese contamination of the ammonia group.

Ignite and weigh the precipitate and combine it with the main ammonia group precipitate.

Combine all filtrates and evaporate to remove most of the excess ammonia before proceeding with the separation of the calcium and magnesium.

A discussion of the behavior of the ammonium sulfide group elements during the regular procedure and methods for accommodating them in the regular scheme follows.

Nickel (and Palladium). Nickel is a common constituent of silicate rocks and minerals. If no special attention is paid to it, part will be weighed with the ammonia group and part will be precipitated with the magnesium. In cases where the oxalate precipitation is omitted and much magnesium is present (analysis of dunite, peridotite, etc.), all the nickel not retained in the ammonia group may be found in the phosphate precipitate. After an oxalate separation, nickel is not recovered completely with the magnesium. In either case, it saves considerable trouble to remove nickel from the ammonia group filtrate before proceeding with the oxalate precipitation.

PROCEDURE. Add to the combined filtrates from the ammonia precipitation about 0.2 g dimethylglyoxime. Cover, and heat on the steam bath for several hours, adding a little ammonia from time to time to keep the solution alkaline (pH > 6.5) and stirring frequently.

Cool. Filter through an 11-cm Whatman No. 40 paper, and wash superficially with water. Dissolve the precipitate by washing the filter alternately with hot 5% (v/v) hydrochloric acid and 5% (v/v) aqueous ammonia. Catch the filtrate in a 250-ml beaker. Add about 5 ml of a 1% solution of dimethylglyoxime in dilute ammonia, heat on the steam bath, and adjust pH to 6–7 (nitrazine paper) with ammonia or hydrochloric acid. Cool, filter through a tared porcelain or glass frit crucible. Wash with a minimum of water, dry at 120 °C, and weigh nickel dimethylglyoxime. The value so obtained may be used either in conjunction with a determination of total nickel to correct the weight of ammonia group oxides or in conjunction with the nickel found in the ammonia group to obtain the total nickel in the sample.

Palladium precipitates as dimethylglyoximate with the nickel. The combined filtrates from the nickel may be treated with persulfate to remove manganese, with oxalate to precipitate calcium, or with phosphate to separate calcium and magnesium together.

For recovery of nickel from the ammonia group oxides, see below.

Manganese. Under ideal conditions, no manganese will be found in the ammonia precipitate. However, it is difficult to prevent some oxidation and precipitation of manganese, and this is weighed (probably as Mn_3O_4) with the ammonia group oxides. To determine manganese contamination in the ammonia group oxides, proceed as follows:

PROCEDURE. After determining titanium with peroxide (see above) in the sulfate solution of the ammonia group oxides and heating to remove peroxide, evaporate or dilute the solution to a volume of about 75 ml, and add 3 ml of 0.2 N silver nitrate. With the solution at room temperature, add 10 ml of freshly prepared, nearly saturated solution of potassium persulfate. Stir, and heat on a hot plate just until small bubbles of oxygen begin to rise through the liquid. Cool in a cold water bath, transfer to a 100-ml volumetric flask, dilute to volume, and compare (in a Dubosq comparator or Nessler tubes) the MnO_4^- color with standard permanganate solutions. Save all the test solution.

Evaporate the solution to remove excess persulfate, adding a drop of peroxide if manganese should precipitate, and proceed with the removal of platinum with zinc as previously described or otherwise as described below. The silver used to catalyze the oxidation of manganese is removed by the treatment with zinc. If the weight of platinum is of consequence, the addition of silver nitrate should have been exact, so that the weight of silver may be calculated and subtracted.

Most or all of the manganese in the sample should be found in the filtrate from the ammonia group. It may be separated therefrom by an ammonium sulfide precipitation or by oxidation to insoluble $MnO \cdot OH$ with persulfate. The ammonium sulfide precipitation is now seldom used but finds occasional application, especially with samples containing much cobalt, nickel, or zinc.

PROCEDURE. Transfer the filtrates from the ammonia precipitate to an Erlenmeyer flask. Add fresh colorless ammonium sulfide solution (prepared by saturating aqueous ammonia with hydrogen sulfide). From 5 to 25 ml may be used, depending on the elements present. Fill the flask to the neck with freshly boiled and cooled water, stopper, mix, and allow to stand for several hours (overnight). Filter (S and S No. 589 white ribbon, preferably doubled) and wash with water containing a little ammonium chloride and ammonium sulfide.

If the sulfide precipitate is large, it must be dissolved in hydrochloric acid and reprecipitated to be sure that no alkaline earths or magnesium remain.

The filtrates from the sulfide precipitate should be made slightly acid with hydrochloric acid and evaporated to remove hydrogen sulfide before proceeding with the separation of calcium and magnesium. The ammonium sulfide precipitation is inelegant and troublesome. Most often, manganese is best separated using persulfate as follows:

PROCEDURE. Evaporate the combined filtrates from the ammonia group precipitate to about 300 ml, and heat to boiling. Put a piece of hardened filter paper under the end of the stirring rod so that a slow steady boil can be maintained. This is best done over a burner, with the beaker supported on an asbestos-filled gauze.

The operation is conveniently performed in the same 400-ml beaker that was used in the ammonia group precipitation.

If barium is present, the addition of a gram of ammonium sulfate may be advantageous.

To the boiling solution add dropwise a freshly prepared 25% (w/v) ammonium persulfate solution and, alternately with the ammonium persulfate, drops of 1:1 ammonia in such a way that the apparent pH of the solution varies between 5 and 7.

Ammonium persulfate decomposes in the boiling solution with the production of free acid, and the judicious addition of ammonia is necessary because Mn(III) is reduced in hydrochloric acid solution at lower pH. Acidity of the solution is measured by dropping a small square of nitrazine paper into it from time to time and observing the color when it touches the solution. Some practice is needed for success in this operation: It takes about 20–30 min gentle boiling, with the alternate additions of persulfate and ammonia, to completely precipitate the manganese.

The precipitation can also be accomplished by adding ammonia and persulfate alternately to the solutions at steam bath temperature; but in this case precipitation may not be complete for several hours.

The manganese should always be reprecipitated. This may be done from nitrate or from sulfate solution. If manganese is to be determined colorimetrically and the separation of barium is not required, nitrate is to be preferred. If the manganese is to be converted to pyrophosphate and weighed (preferred procedure when manganese exceeds 1–2% of the sample), it is better to reprecipitate from sulfate solution because nitrate makes it difficult to obtain $Mn_2P_2O_7$ free from oxidized compounds of manganese. When barium is to be recovered, a sulfate solution is necessary.

Collect the first manganese precipitate on an 11-cm Whatman No. 40 paper, and wash with 2% (w/v) ammonium nitrate or 1% (w/v) ammonium sulfate solution. Wash the precipitate from the paper into the precipitation beaker with water, and dissolve the $MnO \cdot OH$ by adding either 2% nitric acid containing hydrogen peroxide or diluted sulfuric acid. Minimum amounts of peroxide or sulfite should be used. Avoid the use of nitrate if manganese is to be separated as phosphate.

Pour the solution through the filter to dissolve the manganese remaining on it, and wash the paper thoroughly with water. Make sure the paper is free of sulfite or peroxide, and reserve it to catch the second precipitation.

Put a piece of hardened filter paper under the stirring rod in the beaker, and bring the solution to a slow steady boil. When most of the peroxide or sulfite is gone, add several drops of 25% (w/v) ammonium persulfate and boil until precipitation is substantially complete. Then slowly add diluted ammonia until the acid is neutralized. Continue to boil, with small alternate additions of persulfate and ammonia, until precipitation is complete. Adjust to pH 6–7, and filter. Wash with ammonium nitrate or sulfate as before.

Evaporate the combined filtrates to about 400 ml for the precipitation of calcium as oxalate.

If the manganese precipitate is small, dissolve it in dilute nitric acid with the aid of a little peroxide, boil to remove excess peroxide, make the solution 10% by volume in nitric acid, add 0.1 g periodic acid, and heat at steam bath temperature until the manganese is all oxidized to permanganate. Determine manganese colorimetrically.

If insoluble oxides of manganese precipitate during this operation, dissolve them with a minimum of peroxide, add a further volume of 10% nitric acid, and develop the color in a more dilute solution.

If the manganese precipitate is large or if it contains barium sulfate, return it, together with the filter paper, to the precipitation beaker, and put the beaker in a muffle furnace. Slowly raise the temperature to 500 °C to burn off the paper. Cool the beaker, and add 20 ml of 1% sulfuric acid and several drops of 30% hydrogen peroxide. Allow to stand for several hours without heating. Manganese will dissolve, leaving a residue that may contain barium sulfate, platinum, silica, and numerous other oxides. Rare earths, which may be present if no special precautions were taken earlier, will dissolve with the manganese. They may be separated from the solution by repeated precipitation at pH 10 using sulfite to help keep manganese in the divalent state.

Spectrographic examination of residues obtained in the procedure given above has shown an astonishing variety of elements, including bismuth, antimony, lead, beryllium, cerium, and the other rare earths, molybdenum, tungsten, cobalt, and vanadium, besides manganese and silica. Traces of iron and aluminum are always present. Copper and nickel are commonly found. The persulfate precipitation thus acts as a scavenging operation, removing numerous trace elements that otherwise might contaminate the calcium and magnesium precipitates. Calcium and magnesium do not appear in weighable amounts in persulfate precipitates when they are developed in the manner described. However, if care is not taken to keep pH below 7 during the precipitation, losses of calcium may occur.

The recovery of barium in the procedure is not complete; about 1 mg BaO is lost. Recovery of barium can be improved by removing ammonium salts (by evaporation with a large volume of nitric acid) from the ammonia group filtrate before proceeding with the manganese separation. In the analysis of feldspars, which frequently contain barium, magnesium is low, and the ammonia precipitation can be made with a minimum addition of ammonium chloride (2–5 g instead of the usual 15–25 plus). Recovery of barium may then be almost quantitative, even if ammonium salts are not removed.

If but little manganese is present in the sample, it may be advantageous to add 5–10 mg (as chloride) before the persulfate precipitation to take advantage of its scavenging properties.

There are some circumstances in which it is impossible to avoid precipitation of manganese with the ammonia group. When chromium or vanadium is present in the sample, these elements are oxidized during the carbonate fusion. If hydrochloric acid is used to dehydrate silica, chromate will be reduced and causes no difficulty. However, if the SiO_2 dehydration is accomplished using nitric, perchloric, or sulfuric acids, the chromate remaining in the solution will oxidize Mn(II) during the ammonia precipitation, and MnO·OH will precipitate. Vanadium behaves similarly.

Phosphate stabilizes higher valences of manganese; also, manganese ammonium phosphate is insoluble: with samples containing little iron and much phosphate, manganese may be found in the ammonia group of elements.

Cobalt. In rather limited experience with cobalt, it has always been collected nearly quantitatively in the ammonia precipitate. During ignition, it probably converts to Co_3O_4.

Zinc. Zinc silicate and zinc phosphates are insoluble at pH 7. Thus, any silica remaining in the solution after double dehydration with hydrochloric acid will bring down an equivalent amount of zinc in the ammonia group, and phosphate present in the sample may result in precipitation of zinc.

Most of the zinc is carried throughout the regular procedure and is lost. It does not remain with either the calcium or the magnesium if double precipitations are the rule. Careful neutralization of the filtrate from the magnesium ammonium phosphate will result in precipitation of zinc ammonium phosphate, but the recovery is not complete unless ammonium salts are first destroyed by nitric acid evaporation. Since recovery of elements in the phosphate filtrate is sometimes useful, the procedure is described here:

PROCEDURE. Evaporate the filtrate from the magnesium ammonium phosphate to near dryness on the steam bath. The more liquid can be removed by evaporation at steam bath temperature, the better. Add 50–100 ml concentrated nitric acid, cover the beaker (usually a 1000–1500-ml beaker) and heat on the steam bath until reaction begins. Remove from the heat; have a pan of cold water ready in case the reaction becomes disturbingly violent. Heat cautiously until reaction subsides (an hour or more). Finally evaporate to near dryness on the steam bath. Add 5–10 ml of concentrated hydrochloric acid.

Add 200–300 ml water, heat to boiling, and remove silica by filtration.

It may be useful to ignite this residue and examine it spectrographically.

Add a little methyl red indicator to the filtrate, and neutralize with ammonia. Zinc ammonium phosphate precipitates. Heat on the steam bath until the precipitate becomes crystalline in appearance, and adjust pH to 6.5–7.0 using nitrazine paper. Allow the solution to cool to room temperature, stirring occasionally, and filter (S and S Blue Ribbon No. 589). Wash with water containing a little ammonium chloride and carefully adjusted to pH 7 with ammonia or hydrochloric acid. Ignite and weigh $Zn_2P_2O_7$. Examine the weighed precipitate spectrographically for traces of copper, cadmium, and so on.

If no acid hydrogen sulfide precipitation was made in the early stages of the analysis, much of the copper in the sample will be found in the filtrate from the magnesium ammonium phosphate. In this case, after destroying ammonium salts, an acid hydrogen sulfide precipitation will remove copper (and cadmium). Zinc may then be precipitated without interference from these elements.

3.1.12. Problems with Phosphate, Sulfate, and so on

Phosphate and Arsenate. Samples that contain much phosphate (more than about 2% P_2O_5) or do not contain a large excess of iron or aluminum should be treated like phosphate rocks or minerals. The bulk of the phosphate should be removed with zirconium before continuing with the precipitation of the ammonia group. This requires a preliminary phosphate determination in order to calculate the proper addition of zirconium salt.

When much phosphate (or arsenate) is present, several errors may ensue: Calcium, strontium, barium, and some other elements that do not belong in the ammonia group may precipitate as insoluble phosphates; retention of manganese in the divalent state may be impossible, and it will precipitate as $MnO \cdot OH$ in the ammonia group. The net result will be high values for alumina and low values for many other elements.

If an addition of zirconium oxychloride is made before precipitating the ammonia group, and if this addition is carefully measured, the added ZrO_2 can be subtracted from the weight of ammonia group oxides, and there is little chance of incurring the errors mentioned above. With phosphate rocks, which require a large zirconium addition, the zirconium phosphate is filtered off and discarded before proceeding with the ammonia precipitation. With rocks containing a few percent of P_2O_5, the zirconium addition may be carried through the regular procedure with no extra effort.

Arsenate, produced during fusion with sodium carbonate, behaves much like phosphate, except that As_2O_5 is volatilized completely during strong ignition.

Sulfate. In the initial stages of a silicate rock analysis, careful roasting removes much of the sulfide sulfur in a sample. Some, however, is converted to sulfate. Sulfate minerals may be present that are unaffected by roasting. The effect of sulfate in the sample is to precipitate insoluble sulfates ($PbSO_4$, $SrSO_4$, $BaSO_4$) during the course of the analysis. A small amount of barium sulfate is likely to be undetected during the dehydration of the silica and finish in the insoluble residue from the hydrofluoric acid evaporation, being eventually counted as alumina.

When the potassium pyrosulfate fusion of the ammonia group oxides is dissolved, any residue should be carefully examined. It may contain barium (lead, strontium) sulfate. If this is the case, the residue should be filtered off and weighed. Spectrographic examination is desirable. The weight should be subtracted from the weight of ammonia group oxides.

PROCEDURE. After the silica recovery from the ammonia group (or the treatment of the insoluble residue from the first leach of the pyrosulfate fusion to remove silica), dilute the solution to 150–200 ml. Stir, and allow to stand cold for several hours. The precipitate may contain $BaSO_4$, $SrSO_4$, $PbSO_4$, $Zr(OH)PO_4$, and so on. Filter through a 9-cm Whatman No. 44 paper, and wash thoroughly with 2% (v/v) sulfuric acid. Ignite the precipitate at 500–550 °C and weigh. Evaporate the filtrate to an appropriate volume, and proceed with the determination of titanium or with the removal of zirconium as follows:

Add to the solution 1 g diammonium phosphate, stir into solution, and heat on the steam bath for 4–5 h. The precipitate of zirconyl phosphate is not easily visible when but little zirconium is present. Cool, and filter through a small S and S No. 589 Blue Ribbon paper containing paper pulp. Wash with water and then with 2% ammonium nitrate solution, discarding the washings containing nitrate. Ignite and weight ZrP_2O_7. Evaporate the filtrate from the zirconium phosphate to a suitable volume, and proceed with the determination of titanium as usual.

Other elements that may be found in the insoluble residue from the pyrosulfate fusion and sulfuric acid solution include tantalum, columbium, and tungsten. Antimony, tin, and germanium may sometimes be found. It is therefore important to have a spectrographic examination of this residue.

In the regular scheme, the aim is to obtain a true value for alumina; when many unusual elements are present, the numerous corrections become difficult, and alumina is best determined directly. Other elements in the ammonia group may best be determined on separate samples using other procedures. Extended analysis of the ammonia group oxides must be conducted with efficient knowledge of all other possibilities. Such extended analysis is most often necessary in mineral analysis when the supply of material is limited. If less than a gram or so is all that is available, every trick at the analyst's command must be exercised.

3.1.13.   Extended Analysis of Ammonia Group Oxides

After pyrosulfate fusion of the ammonia group oxides and investigation of any insoluble material and after titanium has been determined, it is sometimes desirable to determine aluminum directly instead of by difference. This circumstance arises when the corrections mentioned above become intolerably numerous and complicated and when Al_2O_3 is low.

The direct determination of aluminum is best accomplished after a sodium hydroxide separation, which also serves other purposes. Precipitation with sodium hydroxide is a clean and simple procedure when properly and skillfully performed. In unskilled hands and without attention to essential details, it becomes an abomination. The directions given below should be followed carefully. The solution should be perfectly clear to start with.

PROCEDURE. Add to the cold sulfate solution of the ammonia group oxides, which should have a volume of about 100 ml, enough of a 50% (w/v) sodium hydroxide solution to nearly, but not quite, neutralize the acid.

If a careful accounting of acid additions has been kept during previous operations, it will be possible to calculate just how much sodium hydroxide to add. It is important that this preliminary addition of alkali nearly neutralize the free acid in the solution without the permanent precipitation of iron.

Heat the solution nearly to boiling, and pour it into enough 5% (w/v) sodium hydroxide solution contained in a large silver or platinum dish to give a final concentration of 1% NaOH or slightly more than 1%.

The sample solution must be poured into the sodium hydroxide solution rapidly (i.e., all at once). The sodium hydroxide solution should not be added to the sample solution.

Quickly rinse the beaker with water and add the rinsings to the alkaline solution. Stir, and heat as quickly as possible to boiling. Use a silica-glass stirring rod or a platinum rod. Digest on the steam bath for an hour or boil for a few minutes (boiling in a metal dish is difficult).

If metal dishes are not available, alkali-resistant glass beakers may be used. Ordinary borosilicate glass beakers must not be used in this operation.

Allow to stand until cold (overnight). Filter through a double paper (an 11- or 12.5-cm Whatman No. 541 folded inside an S and S No. 589 Blue Ribbon is the best combination). Wash with a solution that is 1% (w/v) in NaOH and 1% (w/v) in Na_2SO_4. Avoid excessive volumes of wash liquid.

Remove the inner hardened paper without disturbing the other, leaving the funnel stem full of liquid. Sluice the precipitate from the paper into the precipitation vessel with cold water. Use as little water as possible. Add 5 ml concentrated hydrochloric acid to the dish, refold the paper, and return it to the funnel. Wash the papers thoroughly with hot 5% (v/v) hydrochloric acid, catching the washings in the dish but delaying the washing until the precipitate is mostly dissolved. Keep an accounting of the acid used so that the later addition of sodium hydroxide can be accurately calculated. Pour the solution in the dish through the papers if this will minimize the amount of acid used. Finally wash the papers free of acid with hot water. Collect solution and washings in the original beaker. Reserve the filter papers for the second precipitation.

During this operation, the aim should be to redissolve the precipitate in as small a volume and using as little acid as possible.

The alkaline filtrate should be reserved in an 800-ml beaker.

Nearly neutralize the hydrochloric acid solution of the sodium hydroxide precipitate with sodium hydroxide. If careful track has been kept of acid additions, the correct amount of alkali can be calculated and added at once.

Iron hydroxide should not precipitate at this stage, but the acid should be very nearly all neutralized.

Add to a platinum or silver dish (or a beaker of alkali-resistant glass), conveniently the same one used previously, enough 20% (w/v) sodium hydroxide solution to yield a final NaOH concentration of 3–5% (w/v), 1 g sodium carbonate, and 3 ml of 30% hydrogen peroxide. Provide a cover glass ready for use. Heat the hydrochloric acid solution of the ammonia group oxides nearly to boiling, and pour it all at once into the mixture in the dish. Cover immediately, wait until immediate reaction subsides, and rinse the beaker thoroughly, adding the rinsings to the dish. Heat the dish over a flame to the boiling point, and then digest on the steam bath until peroxide is destroyed. Cool to room temperature (overnight). Filter through the same papers as before. Wash with dilute sodium hydroxide containing carbonate and chloride (1% each of NaOH, Na_2CO_3, and NaCl).

Keep the filtrate separate from the first filtrate until chromium has been determined.

(If no chromium is present in the sample, and zirconium is to be determined in the procedure to follow, the addition of hydrogen peroxide is undesirable and should be omitted. Zirconium forms strong peroxy complexes that are difficult to decompose.)

Evaporate the second filtrate from the sodium hydroxide precipitations to about 70 ml in a platinum or silver dish, and filter through a paper that has been prewashed with alkali solution into a 100-ml volumetric flask. Wash with NaOH, Na_2CO_3, and NaCl solution (1% of each). Determine chromate visually using Nessler tubes or spectrophotometrically, preferably the former.

The small precipitate obtained during evaporation of the alkaline solution may contain iron, titanium, manganese, or zirconium. It should be reserved and combined with the main sodium hydroxide precipitate.

After the chromium has been determined colorimetrically, add its solution to the main solution containing most of the aluminum, and acidify the combined solutions with hydrochloric acid, adding just enough acid to give a red color with methyl orange. Proceed with the determination of aluminum as phosphate (see below).

The procedures outlined above may have to be revised to suit the purpose of the moment. For example, if copper is present in the sample and its presence is suspected in the ammonia group, the first sodium hydroxide precipitate will be dissolved in acid and the iron and titanium precipitated with a large excess of ammonia. Filtration will yield a blue solution that may be compared colorimetrically with standard copper solutions containing the same concentration of ammonia and ammonium salts. Nickel may be determined in the same solution after precipitation with dimethylglyoxime. The precipitate containing the iron, titanium, and so on can then be dissolved and examined for its chromium content as described.

Chromium, copper, and nickel are completely precipitated by sodium hydroxide if the concentration of sodium hydroxide does not exceed 1% (w/v) and no ammonium ion is present. If the concentration of sodium hydroxide much exceeds 1%, chromium is not completely precipitated even if it is retained in the trivalent form. When precipitating chromium with sodium hydroxide, all solutions should be prepared with boiled water; even dissolved oxygen from the air will cause oxidation to soluble chromate.

The determination of aluminum as phosphate is a thoroughly sound procedure, but details must be closely adhered to. Interferences are few when the aluminum derives from a sodium hydroxide separation such as that described. Chromate does not interfere, but there is danger that some of the chromium may be reduced to Cr(III), which precipitates with the aluminum, partly as oxide and partly as phosphate. Thus, care must be taken that the alkaline solution containing aluminum and chromium contains no peroxide or reducing agents at the time it is acidified. Determination of aluminum as phosphate proceeds as follows:

PROCEDURE. To the combined filtrates from the sodium hydroxide precipitations, which have been made just acid to methyl orange by addition of hydrochloric acid, add 1–2 g diammonium phosphate and then 1–2 ml in excess of hydrochloric acid. Stir in a quantity of ashless filter pulp (how much depends on how much aluminum is expected—more is better than less), heat to boiling, and add to the boiling solution, which should have a volume of 300–400 ml or more, 10 g ammonium acetate dissolved in a minimum of water. Boil for a few minutes, and filter hot through an 11-cm Whatman No. 40 filter paper. Wash superficially with hot 2% ammonium chloride solution. Do not let the paper run completely dry during the washing, and do not overwash.

Sluice the paper pulp and precipitate into the precipitation beaker with water, add enough hydrochloric acid to dissolve the aluminum phosphate, dilute to about the same volume as before, add not more than 0.5 g diammonium phosphate, and reprecipitate as before.

Filter through the same paper as before, and wash with hot 5% (w/v) ammonium nitrate solution until the washings give only the faintest opalescence with acidified silver nitrate solution.

Ignite the precipitate and paper in a weighed platinum crucible, and weigh $AlPO_4$.

Fuse the weighed residue with a very little sodium carbonate, take up the melt with water, and filter. Determine traces of chromium in the filtrate and iron in the precipitate if present. Correct the weight of $AlPO_4$ accordingly. Alternatively, determine impurities in the $AlPO_4$ spectrographically and make appropriate corrections.

If alumina is not to be determined on the filtrates from the sodium hydroxide precipitation, they may be used for the determination of vanadium; or the filtrates may be diluted to volume and appropriate portions used for determining aluminum and vanadium.

Vanadium can be separated with tannin or cupferron or determined colorimetrically with peroxide after acidification of the solutions without any separation. Which procedure to follow will depend on what other elements are present.

The sodium hydroxide precipitate contains substantially all the iron, titanium, zirconium, columbium, and tantalum in the sample as well as all the rare earths, thorium, copper, and nickel, that were caught in the ammonia precipitate. Nickel and copper may be separated and determined as described above, following an ammonia precipitation; zirconium can be precipitated as phosphate and determined as ZrP_2O_7. After such operations, the iron remains available for titration as in the regular scheme.

3.1.14. Vanadium and Uranium

Vanadium and uranium are usually found quantitatively in the ammonia group precipitate. On ignition, vanadium probably finishes as V_2O_5 and uranium as U_3O_8. There are circumstances, however, under which these elements may escape precipitation.

Vanadium (like phosphorus and arsenic) requires the presence of excess iron for its precipitation. Uranium requires that carbonate be absent. When the two elements are present together, their behavior is capricious.

The presence of vanadium will be noted during the colorimetric determination of titanium: V_2O_5 gives an orange-yellow color with peroxide in acid solution. The difficulty is circumvented by delaying the determination of titanium until after a sodium hydroxide separation, collecting the various precipitates containing iron in a sulfuric acid solution, and adding peroxide. The same device is used when chromium is present and the peroxidized solution of the ammonia group elements shows a green color due to Cr(III).

Both vanadium and uranium are quantitatively separated from iron and titanium during the sodium hydroxide separation. If aluminum is determined as phosphate, the ignited $AlPO_4$ should be examined for these elements as well as for chromium.

3.1.15. Strontium and Barium

In the regular procedure, neither strontium nor barium are quantitatively retained. If sulfur is present in the sample, some barium will be included in the ammonia precipitate as $BaSO_4$ and will cause error if not taken into account. Strontium sulfate is sufficiently soluble that its precipitation in the ammonia group is unlikely.

Barium is not precipitated by oxalate, and strontium is precipitated only partly. Barium does not precipitate as phosphate with the magnesium and, in the unmodified procedure, goes through the analysis and is lost.

Strontium should always be determined in the weighed calcium carbonate or calcium oxide—most conveniently by flame photometry or atomic absorption photometry. If much strontium is present, it can be detected by weighing the calcium precipitate first as carbonate and then as oxide. If the weight of oxide is too high, the presence of strontium should be suspected. It is very difficult to ignite strontium carbonate to oxide: An ignition temperature of at least 1050 °C is required, and the hot crucible must be placed in a vacuum desiccator containing magnesium perchlorate and/or barium oxide. The desiccator should be evacuated at once and filled with air through a magnesium perchlorate and soda-lime drying train. If weighings of the carbonates and the oxides are made with great care, the proportion of strontium in the mixture can be calculated with fair accuracy. Note that ignitions of alkaline earth carbonates and oxides must be protected from acid gases, especially sulfur dioxide and H_2S.

In the analysis of silicates that contain little magnesium, a fairly good recovery of calcium, strontium, and barium can be accomplished if ammonium salts are destroyed in the filtrate from the ammonia precipitation. Barium and most of the strontium may then be separated as sulfates and calcium as oxalate, which will bring down the rest of the strontium. The following procedure has been used in the analysis of alkali feldspars when the available sample was too small to afford the luxury of separate portions for determining barium and strontium.

PROCEDURE. Combine the filtrates from the ammonia precipitations and evaporate to dryness on the steam bath. Cool, and add 50 ml concentrated nitric acid. Heat cautiously until reaction begins, and remove from the heat, cooling in an ice water bath if necessary. Evaporate to dryness on the steam bath. Make repeated small additions of hydrochloric acid to remove all nitrate, and finally take up the residue in about 5 ml of 1:1 hydrochloric acid. Dilute to about 50 ml and filter. Discard the residue, which consists mostly of silica (spectrographic examination will confirm the absence of the alkaline earths).

Precipitate the barium as chromate from acetate-buffered solution, recover calcium and strontium as carbonates, convert the carbonates to nitrates, weigh, and determine calcium and strontium by flame photometry in neutral nitrate solution or by atomic absorption spectrometry.

After weighing the barium chromate, it is best to convert to sulfate and reweigh; there is a possibility that manganese or rare earth elements may lead to an erroneous chromate weight.

Another method for recovering calcium, strontium, and barium during the main portion of the analysis involves their separation as carbonates:

PROCEDURE. Evaporate the filtrate from the ammonia precipitates to 300–400 ml. Add a quarter its volume of saturated ammonium carbonate and a fifth as much concentrated aqueous ammonia. Allow to stand, stirring frequently, for 24 h or more. Filter, and wash with 5% (v/v) ammonia solution containing ammonium carbonate. Dissolve the precipitated carbonates in a minimum of hydrochloric acid diluted with water. Remove any manganese that may be present by precipitation with ammonia and bromine water.

Continue with the separation of barium as chromate and the collection of strontium and calcium as carbonates, as in the previous procedure.

The various filtrates from the calcium, strontium, and barium should be examined for magnesium.

The filtrate from the mixed carbonates is treated in the usual way with phosphate to recover the magnesium. The calcium content of the magnesium pyrophosphate is apt to be somewhat higher than in the usual procedure using oxalate so that its recovery and determination is mandatory.

The chief disadvantage of this procedure is the uncertainty of complete precipitation of calcium. Calcium carbonate separates very slowly from the solution and adheres strongly to the sides and bottom of the precipitation beaker. Possibly the addition of oxalate as well as carbonate may improve the method; but this requires that all oxalate be removed before attempting the separation of barium as chromate.

Generally, barium and strontium will be determined on separate samples by other methods; it is only necessary to determine strontium in the weighed calcium carbonate so as to correct the weight of the latter.

3.2. CLASSICAL ANALYSIS OF TYPICAL MINERALS

3.2.1. Biotite

To determine whether or not a biotite is pure and homogeneous, after passing optical examination, repetitive determinations of potassium are convenient. If subsamples of 20–50 mg are used, there will almost certainly be some variance in the results from 10 or 12 determinations. Of numerous biotite samples examined, only one has shown no perceptible variance at this subsampling level (PSU 5-110, with 10.00% K_2O). Even the 99.9 + % biotite LP-6 Bio 40-60# shows perceptible subsampling variance at the 50-mg level [3].

Before continuing with a biotite analysis, it should be decided whether the observed inhomogeneity with respect to potassium is acceptable or whether the sample should be further purified. Sometimes, as with LP-6 Bio 40-60#, further purification is impracticable, and the observed variance in results for potassium is acceptable. In other cases (e.g., the MIT Biotite B-3203), the variance may not be acceptable, and further purification or finer grinding may be necessary. Finer grinding as an expedient to improve subsampling characteristics is acceptable if there is no need for the analysis to represent exactly the composition of a biotite.

Unfortunately, the grinding of biotites alters their composition; some iron is oxidized, and combined (hydroxyl) water is lost or is converted to a form that escapes during drying at 110 °C. For example, MIT B-3203 showed the following values for H_2O^-, H_2O^+, and FeO before and after grinding from the as-supplied condition to pass 100% through 115 mesh:

	As Supplied	After Grinding
H_2O^-	0.08	0.24
H_2O^+	3.26	3.11
FeO	17.18	16.89

If the sample is to be used for a purpose (e.g., K–Ar age determination) in which air contamination is of consequence or when volatile constituents are to be estimated, grinding should, of course, be kept to a minimum, and in some cases fines should be discarded. It is necessary to balance the need for homogeneity against the need to maintain the sample composition. The use of sampling diagrams and their associated arguments is useful; see equations (1.48)–(1.51).

The H_2O content of biotites alters with atmospheric humidity and with temperature. Determinations of H_2O^- (oven drying at 105–110 °C) differ when the humidity of the laboratory air changes. Iron is slowly oxidized at these temperatures, possibly by the following mechanism:

$$4FeOH^+ + O_2 \rightarrow 4FeO^+ + 2H_2O$$

in which the superscript + indicates that the position of the iron in the biotite lattice remains essentially unchanged.

Note that the overall reaction involves a weight change on oven drying equivalent to only two molecules of hydrogen for every four atoms of iron. The loss in weight is only one-eighth that of the water evolved. This may demonstrate the limited worth of "moisture" determinations. It also shows the necessity of determining ferrous iron both on the undried and the dried sample or on the ground and the unground sample if the FeO content is of consequence.

A determination of total hydrogen should be made on a subsample removed from the sample at the same time as the subsample for moisture determination.

Biotites almost always contain some fluorine, and many contain chlorine. Losses of these elements are possible during drying.

A "loss on ignition" as a substitute for a total hydrogen determination is obviously bad practice.

Before beginning the main part of the analysis, determinations of fluorine and chlorine should be made. While the fluorine content of biotites is seldom great enough to interfere seriously with the analysis, some contain 1% or more; these should be treated differently.

The alkali metals should be determined after a J. L. Smith separation. Often, the sodium content is low, and the major alkali (potassium) may be derived from the total weight of chlorides or (better) sulfates by subtracting the other alkali chlorides or sulfates determined by flame photometry in neutral sulfate solution. If adequate standards are available, determination of the alkalies on a small sample by flame photometry is permissible provided the intent of the analysis is not the development of a standard sample. Potassium and rubidium are commonly determined by isotope dilution mass spectrometry, which can be used as a primary method. In this case, accuracy depends on an ability to accurately calibrate and store spike solutions and on careful measurement of mass discrimination effects. Combined neutron activation and mass spectrometry are now used widely in geochronology (the $^{39}Ar-^{40}Ar$ method) for determining K–Ar ages; this procedure yields, in principle, a potassium value. However, the method is essentially a secondary one, requiring analyzed standards to calibrate. It also has the disadvantage of limited sample size (commonly 10 mg). Few biotites are homogeneous enough to yield reliable subsamples of this size. For example, the 99.9 + % pure LP-6 Bio 40-60# is geochronologically inhomogeneous and also perceptibly inhomogeneous chemically at the 10-mg subsampling level with respect to potassium.

The main portion of the analysis usually gives little trouble. Minor concentrations of, for example, chromium and vanadium are best determined spectrographically on separate samples, and appropriate corrections to the alumina are best made on the basis of the spectrographic values. Usually the minor constituents are present as impurities, perhaps as minute inclusions in the biotite grains; sampleability with respect to these elements is often poor, and efforts to determine them exactly are not worthwhile.

Most difficulties with biotite analyses occur when the available sample is small and unusual constituents (either as impurities or in the biotite itself) are encountered. Small impure samples should not be analyzed by primary methods—the expenditure of time and talent is not warranted. A majority of mineral separates (and biotite separates in particular) deserve no more than a rapid rock analysis. When, however, the very considerable effort required to produce a pure biotite has been made, analysis by rapid slop methods is a disgrace.

The electron microprobe is often used to analyze biotites. This technique cannot distinguish between ferrous and ferric iron and cannot determine total hydrogen. It requires unusually pure (and seldom obtainable) calibrating standards or the use of complex matrix corrections that, at this writing, are not entirely satisfactory.

A typical biotite analysis is shown in Table 3.3.

Work sheets for the analysis of the MIT biotite are reproduced in Table 3.4. This analysis is of a sample ground in agate to pass 100% through a 115-mesh screen. Potassium determinations on four separate bottles of B-3203, each ground on receipt to pass a 115-mesh screen and thoroughly mixed, differed by no more than 0.02% K_2O. Potassium determinations on 50-mg subsamples of the as-received material differed by as much as 0.5% K_2O. (See Table 3.5 for analysis of MIT biotite B-3203 as supplied.)

Analysis of a chlorine-bearing biotite is shown in Table 3.6.

3.2.2. Lepidolite

The feature of lepidolites that causes most difficulty is the high fluorine content. If the sample is taken through the basic procedure, the SiO_2 value will not be much lower because the high aluminum content will complex much of the fluorine as AlF_6^{3-}. Manganese will probably be converted to Mn(III). The presence of fluoride in the filtrate from the silica will prevent normal precipitation of the ammonia group elements; attack on the glassware will introduce more silica. Without an awareness of the problem, the analyst will be puzzled by a variety of murkinesses and strange colors. Reported values will be erroneous and probably low.

The manganese in the sample of Table 3.7 causes some difficulty. Attack on the fusion crucible is greater than usual, and use of a platinum dish to dehydrate the silica—necessary because of the fluorine—adds more platinum to the solution. Platinum recoveries must be made.

If sample is plentiful, silica should be determined on a separate portion, adding extra aluminum chloride to make sure fluoride is complexed but otherwise proceeding in the usual manner. Three (instead of two) silica dehydrations are advisable; after two dehydrations, the silica remaining in solution may be determined colorimetrically. The filtrate from the silica may be discarded or used to determine lithium by flame photometry using the method of standard additions to overcome matrix effects. Rare earth elements (not usually present) may be conveniently separated from the filtrate.

Table 3.3. Analysis of Purified Biotite, PSU 5-110 ($\%$)

SiO_2	38.63
Al_2O_3	13.08
TiO_2	1.55
Cr_2O_3	0.23
Fe_2O_3	2.50
FeO	8.75
NiO	0.02
MnO	0.14
MgO	19.94
CaO	0.18
SrO	0.005
BaO	0.45
Na_2O	0.26
K_2O	10.00
Rb_2O	0.03
H_2O^+	3.52
H_2O^-	0.30
P_2O_5	0.06[a]
F	0.30
	99.95
Less O for F	0.13
Total	99.82

[a] Derived partly from minor apatite impurity.

Table 3.4. Work Sheet, Analysis of MIT Biotite B-3203 (-115 mesh)

H_2O^+
115#

Subsample weight	0.6995 g
Penfield tube + water	23.8142
Less water	23.7890, pH 4

June 20, 1960
Blank 0.0000

Percentage of H_2O (total)	0.0252
	3.60%
Less H_2O^-	0.49
	3.11

SiO_2
0.7000 g
Crucible 2
115#
May 10, 1960

Crucible before fusion	28.3152 g
Plus sample	29.0152
After fusion	28.3151
Plus SiO_2^+	28.5795
	28.5776
	28.57745
After HF	28.5774
	28.3236
	.2538

SiO_2 recovered from R_2O_3	0.0015
Plus SiO_2	0.2538
	0.2553
Correction for F, etc.	0.00105
Total weight SiO_2	.25635
Percentage of SiO_2	36.62%

Al_2O_3

Triple NH_3
Precipitation

Crucible plus R_2O_3	28.6126
	28.6122
	28.6126
	28.6120
	28.6120
	28.3151
	0.2969 − 0.0015 SiO_2 = 0.2954

SiO_2 recovery from R_2O_3	28.3157 g
	28.3155
After HF	28.3140
Weight SiO_2 recovered	0.0015

$Al_2O_3^+$	42.20%
TiO_2	3.00 Colorimetric, hydrogen peroxide

NaOH precipitation
for Cr, V, Fe, Ti

V$_2$O$_5$ 0.08 Colorimetric, hydrogen peroxide
Cr$_2$O$_3$ 0.07 Colorimetric, as chromate
Total Fe$_2$O$_3$ 19.70 Potentiometric titration, K$_2$Cr$_2$O$_7$
Al$_2$O$_3$ 19.35

Mn removed with S$_2$O$_8$$^{2-}$

MnO   0.044 Colorimetric as MnO$_4$$^-$
Plus recoveries   .002 Colorimetric as MnO$_4$$^-$
Total MnO   .046

CaO   Crucible   17.2668 g   Ca recovery from Mg$_2$P$_2$O$_7$ 0.05%
Precipitate with oxalate   Plus CaO   17.2671   Plus   0.03
                                        0.0003   Total CaO   0.08

MgO
Precipitate with PO$_4$$^{2-}$ Crucible   28.3135 g
                         Plus Mg$_2$P$_2$O$_7$ 28.4995
                                                  28.4995
                                Weight Mg$_2$P$_2$O$_7$$^+$ 0.1860
Uncorrected percentage of MgO   9.62%
                                Less CaO   0.03
                                     MgO   9.59

	Alkalies			
	Chlorides	Chloroplatinates	Na sulfates	K sulfates
J. L. Smith	28.7808	29.0159	28.7851	28.8779
Double leach	28.8567	28.7811	28.7811	28.4226
Weighed 0.5000 g	28.8568	—	—	28.3388
Calculated to	28.7811	—	—	—
dry weight				
−115#				
0.4986 g	.0757	0.2348	0.0040	0.0838

261

Table 3.4. (*continued*)

Flame	In K sulfates	In Na sulfates
	Na$_2$O 0.014%	Na$_2$O 0.29%
	Rb$_2$O 0.043	K$_2$O 0.024
	Cs$_2$O 0.004	Li$_2$O 0.018

Total Fe$_2$O$_3$ titrate

$$\frac{17.24 \times 0.10000 \times 159.7 \times 100}{1000 \times 2 \times 0.7000} = 19.67$$

0.10000 Cr(VI) after NaOH precipitation

plus 0.003 in NaOH filtrate $\dfrac{0.03}{19.70}$

Ferrous iron, 0.5000 g
HF, H$_2$SO$_4$
Titrate, 0.0997 N KMnO$_4$

$$\frac{11.79 \times .0997 \times 71.85 \times 100}{1000 \times 0.5000} = 16.89$$

FeO, pyrophosphate method 115#, 16.89; 40#, 17.03; as supplied 17.17

BaO, 1 g
Weigh BaCrO$_4$

Crucible + BaCrO$_4$ 9.1991 g
Crucible 9.2022
.0031 = 0.20% BaO

Repeat

crucible 9.1929
9.1960

0.0031

F, 0.2 g 0.25% Thorium nitrate titration, repeat 0.25%

Li$_2$O, PbCO$_3$ separation, flame 0.03%
0.25 g

262

**Table 3.5.   Analysis of MIT Biotite B-3203
(As Supplied[a]) (%)**

SiO_2	36.64
Al_2O_3	19.36
TiO_2	3.00
Cr_2O_3	0.07
V_2O_5	0.08
Fe_2O_3	0.62
FeO	17.18
MnO	0.05
MgO	9.59
Li_2O	0.03
CaO	0.09
BaO	0.20
Na_2O	0.31
K_2O	9.03
Rb_2O	0.05
Cs_2O	0.005
H_2O^+	3.26
H_2O^-	0.08
F	0.25
Cl	0.05
Less O for F, Cl	0.12
Total	99.83

[a] Analysis performed on a sample ground to pass 100% through 115 mesh; recalculated to as-received basis, using H_2O^+ and FeO determination on as-received material. Analysis of material in as-supplied condition is subject to severe subsampling errors (e.g., K_2O determinations on 50-mg subsamples of as-received material vary from 8.8 to 9.4% K_2O).

If available sample is limited (the usual case), boric acid should be added to the water leach of the sodium carbonate fusion, and the dried residue after dehydrating the silica should be treated alternately several times with methyl alcohol and concentrated hydrochloric acid to volatilize the excess boron. The filtrate from the silica should then be treated with 50–60 ml perchloric acid (after complete removal of all alcohol!) and evaporated (in a Pyrex beaker) to strong fumes of perchloric acid and then boiled for 5–10 min. After dilution and filtration to remove a small amount of silica, precipitation with ammonia proceeds as usual; ammonium perchlorate is as effective as ammonium chloride in controlling the precipitation.

Table 3.6.   Analysis of Chlorine-Bearing Biotite[a] (%)

SiO_2	33.09
Al_2O_3	17.65
TiO_2	1.30
Fe_2O_3	2.42
FeO	29.22
MnO	0.04
MgO	2.83
CaO	0.10
Na_2O	0.13
K_2O	9.04
Rb_2O	0.10
H_2O^+	2.92
H_2O^-	0.04
F	0.23
Cl	1.11
BaO	0.09
Less O for F, Cl	0.34
Total	99.97

[a] Minnesota Rock Analysis Laboratory R2208. Analysts: E. H. Oslund and D. Thaemlitz.

The manganese presents a problem through its oxidation during the perchloric acid evaporation: Addition of a little hydroxylamine hydrochloride after dilution and addition of hydrochloric acid reduces the manganese to the divalent state.

After precipitation and reprecipitation of the ammonia group (mostly aluminum), manganese may be recovered with persulfate, converted to pyrophosphate, and weighed. An oxalate precipitation is seldom worthwhile: Calcium and magnesium are precipitated with phosphate and separated after solution in dilute sulfuric acid by precipitating the calcium as sulfate.

Determination of the alkali metals requires a full-scale procedure: Lithium is not completely recovered in the J. L. Smith separation from samples containing much aluminum, and must be determined on a separate sample. Table 3.8 shows the work sheet for determining the alkali metals in the lepidolite of Table 3.7. Note the low recovery for lithium. The reported value, 5.37% Li_2O, was obtained by flame photometry on the filtrate from the silica. An identical value was obtained using the lead carbonate method of Ellestad and Horstman.

3.2.3.   Alkali Feldspars

Pure alkali feldspars present little difficulty when adequate sample is available. The major constituents are silica, alumina, the alkalies and the alkaline earths, and pure feldspars contain little else (Table 3.9).

Difficulties arise from the presence of impurities and from mixtures of feldspars that exhibit subsampling nonuniformity. Individual grains of calcium feldspar are frequently very nonuniform; this nonuniformity may be reflected in poor reproducibility in alkali and alkaline earth determinations. It is recommended that all feldspar samples be ground to pass a 115-mesh screen and thoroughly mixed before analysis.

When available sample is limited, it may be desirable to determine calcium, strontium, and barium during the main part of the analysis. The three elements may be separated together after the ammonia precipitation and recovered almost quantitatively if the addition of ammonium salts

Table 3.7.   **Analysis of Purified Lepidolite PSU 60-1252[a] (%)**

SiO_2	50.05
Al_2O_3	24.56
TiO_2	0.07
Fe_2O_3	0.08
FeO	0.03
MnO	2.19
MgO	0.02
CaO	Trace
SrO	Trace
BaO	0.01
Li_2O	5.37
Na_2O	0.32
K_2O	9.88
Rb_2O	1.99
Cs_2O	0.08
H_2O^+	0.74
H_2O^-	0.05
F	7.88
	103.32
Less O for F	3.32
Total	100.00

[a] Sample from Brown Derby Mine, near Ohio City, Gunnison County, Colorado.

Table 3.8. Determination of Alkalies in Lepidolite 60-1252

Chlorides	Chloro-platinates	Sodium Sulfates	Potassium Sulfates
28.2594	28.3793	28.1907	28.2628
28.2590	28.3792	28.1903	28.3700
28.2586	28.0904	28.1904	28.3703
28.2586	0.2888	28.0907	28.3703
28.0902	−0.00099 Li	0.0997	28.2635
0.1684	−0.00154 Na	−0.00453 Na	0.1068
−0.0780 K	−0.03096 Rb	−0.0003 K	−0.00026 Li
−0.00025 K	−0.0086 Cs	0.0949	−0.00048 Na
−0.00372 Na	0.2544		−0.01428 Rb
−0.00040 Na		5.16 Li$_2$O	−0.00046 Cs
−0.01294 Rb	9.86 K$_2$O	+0.01 in K	0.0993
−0.00042 Cs	−0.04 blank		
0.0727	−0.02 blank	5.17 Li$_2$O	9.87 K$_2$O
	+0.03 in Na		−0.04 blank
			−0.02 blank
			+0.03 in Na
5.12 Li$_2$O	9.83 K$_2$O		9.84 K$_2$O

Flame Determinations		Cl$^-$	SO$_4^{2-}$	PtCl$_6$
In Na$_2$SO$_4$:	5.06 Li$_2$O			
	0.395 Na$_2$O	0.00372	0.00453	—
	0.032 K$_2$O	0.00025	0.00030	—
In K$_2$SO$_4$:	9.9 K$_2$O			
	0.014 Li$_2$O	0.00020	0.00026	0.00099
	0.042 Na$_2$O	0.00040	0.00048	0.00154
	2.00 Rb$_2$O	0.01294	0.01428	0.03096
	0.072 Cs$_2$O	0.00042	0.00046	0.00086

Lithium determined on separate sample: 5.37% Li$_2$O

has been kept to a minimum. Large amounts of ammonium chloride are not necessary because magnesium is low or absent.

It is probably best to destroy ammonium salts in the filtrate from the ammonia precipitate. This yields a solution of small volume from which (after removal of extraneous silica) the alkaline earths may be precipitated under ideal conditions.

3.2.4. Fluorophlogopite

Problems with phlogopite micas are similar to those encountered with lepidolites. In addition, the water content is unstable, and a determination

Table 3.9. Analysis of Pure Potassium Feldspar PSU Or-1[a] (%)

SiO_2	64.39
Al_2O_3	18.58
SrO	0.035
BaO	0.82
Na_2O	1.14
K_2O	14.92
Rb_2O	0.03
H_2O^+	0.08
Total	99.99

[a] TiO_2, MnO, MgO, CaO, P_2O_5, BeO, and Cr_2O_3 absent. The sample contains about 0.02% tramp iron, introduced during preparation.

of total water should be made on a subsample weighed out at the same time as subsamples for other determinations to avoid sampling weight baseline error.

The analysis of Table 3.10 is of a synthetic phlogopite prepared by H. R. Shell. The analysis followed essentially the same scheme outlined for lepidolite; the difference in procedure lies mostly in determining total hydrogen. Only by radically increasing the amount of calcium carbonate in the Penfield tube and heating slowly and carefully was it possible to obtain the collected water free from fluorine. Analysis for alkali metals is, of course, relatively simple.

Table 3.10. Analysis of Synthetic Fluorophlogopite

Constituent	Original Analysis by H. R. Shell (%)	Penn State Analysis[a] (%)
SiO2	42.63	42.60
Al_2O_3	12.40	12.38
MgO	28.51	28.31
K_2O	11.11	11.04
Na_2O	0.10	0.10
F	9.08	8.86
H_2O total		0.34
	103.83	103.63
Less O for F	3.81	3.72
	100.02	99.91

[a] Analysts: E. Martinec, M. Bailey, and C. O. Ingamells.

An original analysis of this material by H. R. Shell showed no water and totaled 100.02%. Values for SiO_2, Al_2O_3, MgO, and K_2O were all higher than those reported in Table 3.10. The implication is that the sample accumulated 0.34% H_2O during transport and storage. This demonstrates the necessity to pay attention to the sampling weight baseline.

Difficulties with the water determination show that it is very necessary to examine the water collected during the Penfield procedure for its acid content.

3.2.5. Amphiboles

Amphiboles vary widely in composition (Table 3.11). A major difficulty in their analysis is encountered in determining total hydrogen. An oxidizing flux is essential, and the mineral must usually be ground to pass 150 or 200 mesh before it is possible to release all the water.

The ammonia precipitate is usually large and bulky; it is advisable to make three precipitations instead of the usual two and to ignite the ammonia group oxides in silica crucibles in a muffle furnace. It is necessary

Table 3.11.   Analyses of Amphiboles and a Protoamphibole

Constituent	Engels Amphibole[a] (%)	Fibrous Amphibole[b] (%)	Proamphibole[c] (%)
SiO_2	42.14	58.92	60.60
Al_2O_3	12.09	0	0.26
TiO_2	0.94	0	0
Fe_2O_3	6.19	0	0.04
FeO	13.48	0.80	0
MnO	0.63	0.03	0.01
MgO	8.67	24.36	33.39
CaO	11.56	13.38	0
Li_2O	—	0.01	2.31
Na_2O	1.63	0.12	0.12
K_2O	0.91	0.02	0.04
H_2O^+	1.66	2.19	0.18
F	—	0.05	5.23
Less O for F	—	0.02	2.19
Total	99.90	99.88	99.98

[a] A calibrating standard for low-potassium determination, utilized at USGS laboratories, Menlo Park, California.
[b] PSU 62-1717: separated by H. Kaufman.
[c] From the University of Minnesota Rock Analysis Laboratory.

to carry a blank crucible along with each analysis so as to avoid weighing errors. Amphiboles are likely to contain appreciable amounts of zinc, nickel, and other divalent metals; these should be looked for in the ammonia precipitate as well as in all filtrates.

During the sodium carbonate fusion, a temperature high enough to completely decompose the mineral is difficult to attain in platinum crucibles over a burner unless the latter is very efficiently adjusted and the gas supply is good. The use of a blast lamp may be necessary. An oxidizing flame must be maintained at all times.

Fluorine is sometimes present; all amphiboles should be analyzed for fluorine before beginning the main part of the analysis. If more than about 0.1% is present, steps should be taken to prevent its interference with the ammonia precipitation.

The alkalies usually amount to not more than 2–3%; lithium is very likely to be present in appreciable concentration and should always be suspected. Potassium is invariably minor—always less than 1% and most often much lower. It is very probable that amphiboles showing much potassium are contaminated with biotite.

Potassium in amphiboles is important because of the wide use of these minerals in K–Ar dating. Amphiboles are retentive of radiogenic argon and very often show a higher K–Ar age than coexisting biotite, which loses argon more easily during geothermal events. It is extremely difficult to completely remove all biotite from a hornblende separate, and many K–Ar ages reported for amphiboles are probably mixed ages. A hornblende with a true K–Ar age of 170 my and a K_2O content of 0.20% will yield a measured K–Ar age of 89 my if it is contaminated by 8% biotite of K–Ar age 70 my. The biotite contamination supplies most of the potassium.

3.2.6. Tourmaline

Tourmalines contain boron and fluorine. Boron might be expected to interfere with a fluorine determination, but this seems not to be the case if a separation from silica using zinc oxide is made, and the fluorine is isolated by distillation from sulfuric acid solution [4].

The regular procedure may be followed without modification, except that the ignited and weighed ammonia group oxides must be examined for boron. This is best done by thoroughly mixing the weighed oxides and removing about 10 mg for a colorimetric boron determination using the curcumin method of Heyes and Metcalf [4]. In the analysis of the tourmaline of Table 3.12, the equivalent of 0.57% B_2O_3 was found in the ammonia precipitate. The remainder of the weighed oxides is carried through the regular procedure for TiO_2, Fe_2O_3, and so on, making a correction for the small amount removed for boron determination. The residue from the SiO_2 determination must be carried through separately. It is usually small and consists mostly of TiO_2.

A difficulty is encountered in the J. L. Smith separation of the alkali metals. The weighed chlorides are contaminated with boric acid; this is easily removed by alternate evaporation with methyl alcohol and hydrochloric acid. Lithium is recovered very incompletely and must be separately determined after HF attack and a lead carbonate separation.

Determination of ferrous iron is best accomplished on a sample ground to pass 200 mesh using the pyrophosphate method. Tourmaline is not very soluble in the sulfuric–hydrofluoric acid mixture most often used in FeO determinations.

Boron is best determined by fusing 0.2 g with Na_2O_2 and Na_2CO_3 in platinum at 500 °C, leaching in a silver dish with water, filtering, boiling with $CaCO_3$, filtering, and titrating boron electrometrically with sodium hydroxide after addition of mannitol. All residues should be examined for boron using curcumin, and appropriate corrections should be made.

3.2.7. Sodalite

Sodalite presents problems only because of its unusual concentration of sodium and chlorine. It is of interest because it is the only silicate mineral that contains large amounts of chlorine and is homogeneous with respect to this element at the microprobe sampling level; it thus makes an excellent microprobe standard when such materials as scapolite are to be examined.

The J. L. Smith separation of sodium from sodalite is likely to be incomplete even when a double-leach procedure is used. For this reason, it is probably best to separate the sodium after hydrofluoric acid attack, fuming with sulfuric acid, and ignition of the sulfates. Sodium sulfate is quantitatively leached from the residue with water.

Calcium and carbon dioxide, commonly reported in sodalite analyses, are probably most often due to calcite contamination.

Determination of total hydrogen must be carefully performed, or HCl, not H_2O, may be evolved. The use of an extra quantity of calcium carbonate in the Penfield tube is necessary. The water collected must be tested for its acid content before the result can be accepted. If no water is collected, it is possible that all hydrogen evolved as HCl and was lost.

The analysis of Table 3.13 was performed by J. Muysson, McMaster University.

Table 3.12. Analysis of Tourmalines PSU 4-188 and 4-206 (%)

Constituent	4-188	4-206
SiO_2	35.88	33.86
Al_2O_3	25.29	30.79
B_2O_3	10.49	10.86
TiO_2	0.74	0.55
Fe_2O_3	3.16	17.62
FeO	5.78	1.27
MnO	0.02	0.13
MgO	10.43	0.13
CaO	3.07	0.69
SrO	0.19	0.00
BaO	0.01	—
Na_2O	1.39	2.46
K_2O	0.06	0.07
H_2O^+	2.76	0.40
H_2O^-	0.00	0.00
F	0.98	1.86
Less O for F	0.41	0.78
Total	99.84	99.91

Table 3.13. Analysis of Sodalite, McMaster University (%)

SiO_2	36.7
Al_2O_3	31.5
TiO_2	0.48
Fe_2O_3	0.14
FeO	0.54
MnO	0.11
MgO	0.17
CaO	0.26
Na_2O	23.9
K_2O	0.08
H_2O^+	0.29
H_2O^-	0.10
CO_2	0.31
F	0.00
Cl	6.84
SO_3	0.10
Less O for Cl	1.5
Total	100.0

271

Chlorine in sodalite is determined after sodium carbonate fusion: The melt is leached thoroughly with water, silica, and so on, is filtered off, the filtrate is made acid with nitric acid, and chlorine is precipitated with silver nitrate. The silver chloride is dissolved in ammonia, reprecipitated, and finally weighed in a glass-frit crucible.

3.2.8. Microlite (Pyrochlore)

Analysis of columbium and tantalum minerals is difficult. Procedures are complicated because most samples are impure. For example, the SiO_2 in the microlite of Table 3.14 as well as some of the other minor constituents probably represent impurities.

Fluorine is best separated by steam distillation from phosphoric acid containing a little perchloric acid after fusion in sodium peroxide. A still with an extremely effective splash trap must be used so that no phosphate distils. Extra silica in the form of Potters's flint (pulverized quartz) must be added to the distillation flask. Use of a fluoride ion electrode somewhat

Table 3.14. Harding Microlite, PSU 5-010[a]
(%)

SiO_2	0.60
Ta_2O_5	72.35
Nb_2O_5	6.67
ThO_2	0.01
ZrO_2	0.03
UO_2	0.32
UO_3	1.63
TiO_2	0.16
Al_2O_3	0.12
FeO	0.14
MnO	0.01
MgO	0.00
CaO	10.45
SrO	0.00
BaO	0.00
Na_2O	4.60
K_2O	0.09
H_2O^+	0.78
F	3.04
Less O for F	1.28
Total	99.71

[a] Also present: Tl (probably from Clerici solution), Cl, and Sn (<0.1%).

diminishes the effect of traces of phosphate in the distillate and may be desirable if the procedure can be reliably calibrated.

Silica is best determined on a separate sample by fusing with sodium carbonate and peroxide, leaching with water, filtering to remove most of the tantalum and columbium (as sodium salts), adding boric acid to the filtrate, acidifying with hydrochloric acid, and proceeding as usual. For the small amounts of silica usually present, it is unlikely that much of it will remain in the tantalum–columbium precipitate.

The main analysis is conducted as follows: Fuse 0.5 g with potassium pyrosulfate in an uncovered silica crucible. Heat gently until all excess SO_3 is removed. This leads to volatilization of the fluorine in the sample. Add a further 5 g potassium pyrosulfate, cover, and heat until attack is complete. Cool, transfer to a 600-ml beaker with water, add 30 ml of 30% H_2O_2, dilute to 300 ml, and stir constantly. (Use a magnetic stirrer.) Add 5 ml concentrated sulfuric acid. Continue to stir until the fusion is disintegrated. Add ammonium hydroxide just to the neutral point, and then add 100 ml concentrated nitric acid. Add 0.1 g SeO_2. Heat on the steam bath until oxygen is removed. Filter. Wash with 2% nitric acid. Ignite and weigh the earth acids. Evaporate the filtrate to a small volume on the steam bath; if a further precipitate separates, filter it off with the aid of paper pulp, wash with dilute nitric acid, ignite, and weigh. Examine this small residue spectrographically.

Make the filtrate from the earth acids barely alkaline with ammonia, and filter (the usual ammonia group precipitation). Continue with the separation of calcium and magnesium as usual.

Separate the earth acids using the tannin procedures of Schoeller and Powell. This difficult separation can be abbreviated by dividing the earth acids into two parts using tannin—one part enriched in tantalum and the other in columbium. After weighing, each can be fused with pyrosulfate and poured into a button suitable for examination by X-ray spectrometry. Standards can be prepared from pure Nb_2O_5 and Ta_2O_5 in various mixtures.

The tantalum and columbium will contain variable amounts of titanium, zirconium, and thorium. These are seldom detectable by X-ray spectrometry in the mixture; the emission spectrograph should be used.

Ferrous iron—more correctly, *oxygen deficiency*—is best determined by the pyrophosphate method. In the analysis of Table 3.14, the assumption is made that all the iron is ferrous, and the oxygen deficiency has been divided between the iron and the uranium.

In the example, the major element in the ammonia group is uranium: The ammonia group was fused with pyrosulfate, a silica recovery was made, and this was subtracted from the weight of oxides. Uranium was determined titrimetrically after sodium hydroxide separation, and iron and titanium were determined colorimetrically. In analyzing other microlites and tantalites, a different approach had to be engineered for each. For example, tantalite yields a large ammonia precipitate containing only traces of uranium.

Combined water is determined without difficulty in the usual Penfield procedure if precautions to contain the fluorine are observed.

3.2.9. Ardennite: An Arsenic Mineral

The interesting feature of the silicate mineral ardennite is its high arsenic and manganese content [5]. Despite a very small supply of the material and a consequent need to determine most elements on the main portion, very little difficulty was experienced. The analysis is given in Table 3.15.

The sample was fused with sodium carbonate as usual with a small addition of sodium peroxide. Dehydration of the silica was carried out in a porcelain casserole, and silica was determined as usual. The ammonia group oxides were ignited in a silica crucible, with an empty silica crucible carried throughout all operations as a tare. The oxides were first ignited

Table 3.15.   Ardennite, an Arsenic–
Manganese Mineral PSU 5-144 (%)

SiO_2	28.14
Al_2O_3	23.22
TiO_2	0.00
V_2O_5	0.82
Fe_2O_3	1.50
FeO	0.00
CoO	0.01
Cr_2O_3	0.12
SnO_2	0.03
NiO	0.02
ZnO	0.04
CuO	0.46
MnO	25.33
MgO	3.83
CaO	1.50
Na_2O	0.01
As_2O_5	9.85
F	0.14
H_2O^+	5.04
Less O for F	0.06
Total	100.00

to nearly constant weight at 600–650 °C and then at a higher temperature. The loss in weight above 600 °C corresponded almost exactly to the As_2O_5 in the sample. Apparently, the arsenic was oxidized to As(V) during the fusion and was not reduced during the double dehydration with hydrochloric acid. It precipitated with the aluminum as aluminum arsenate, which decomposed only on ignition above 650 °C, arsenic volatilizing at the higher temperature. Arsenic was determined titrimetrically after distillation as bromide from a separate sample.

The ammonia group oxides contained only a trace of manganese despite the large amount present. The manganese was precipitated in the filtrate with persulfate and finally determined by weighing $Mn_2P_2O_7$.

Vanadium and chromium were determined colorimetrically with peroxide after a sodium hydroxide separation. The other trace elements in the ammonia group were determined spectrographically by N. H. Suhr.

Some difficulty was experienced during the H_2O^+ determination due to a tendency of arsenic to volatilize and collect in the water.

The analysis was used by G. Donnay in a structure determination: Donnay remarks, "it is fair to state that starting with the (previously) available chemical analyses one could hardly have arrived at the correct structural formula given above" [5].

3.2.10. Scapolite

Scapolites may contain sulfur in more than one valence state, chlorine and carbon dioxide. They seldom contain fluorine except in the form of contaminating minerals. Common impurities include halite (sodium chloride), fluid inclusions, and calcite; these are often present in minute inclusions and are therefore not separable by ordinary means. Where possible, analysis should be corrected for these impurities. Carbon in scapolite is not liberated by acid treatment; it is necessary to use a combustion method. Acid-soluble carbonate, separately determined, can be attributed to contaminating calcite [6].

A determination of oxygen deficiency may reveal the condition of the sulfur in scapolite.

Most of the oxygen deficiency in the analysis of Table 3.16 may be attributed to the iron being in the ferrous condition instead of Fe_2O_3 as reported. There is probably a little pyrite in the form of minute inclusions.

3.2.11. Catapleiite

Catapleiite is a zirconium mineral. Its interest lies in the fact that if one were not advised of its zirconium content, it would all be reported as

Table 3.16.   Scapolite PSU 62-1703[a] (%)

SiO_2	54.06
Al_2O_3	21.62
TiO_2	0.01
Fe_2O_3	0.27
MnO	0.04
MgO	0.10
CaO	9.02
Li_2O	0.01
Na_2O	8.78
K_2O	1.04
H_2O^+	0.16
P_2O_5	0.02
Cl	2.57
CO_2	2.12
SO_3	0.66
Less O for F, Cl	0.61
Oxygen deficiency	0.050
Total	99.82

[a] Analyst: E. Martinec, Pennsylvania State University.

aluminum in the regular classical procedure. Analysis was not difficult: Zirconium precipitates much like aluminum and behaves about the same. The only differences are a slight difference in appearance and precipitation at a lower pH. Only a very observant analyst is likely to distinguish the precipitate from that produced by aluminum unless a sodium hydroxide separation is attempted.

Aluminum could not be detected chemically; only traces were found spectrographically.

The J. L. Smith separation of the alkalies proceeded abnormally; there was difficulty in filtering the water leach of the calcium carbonate–ammonium chloride fusion. However, after a double leach, only traces of sodium were found in the residues by HF treatment and flame photometry.

The ZrO_2 reported is the total ammonia group less its measured Al_2O_3, TiO_2, SiO_2, and Fe_2O_3 content. The analysis is shown in Table 3.17.

3.2.12. Tellurium Minerals: Spiroffite and Denningite

Analyses of tellurium minerals (Table 3.18) show what can be done with small samples. Each of the minerals spiroffite and denningite were pro-

Table 3.17. Analysis of Catapleiite PSU 5-002 (%)

ZrO_2	31.47
TiO_2	0.01
SiO_2	43.36
Al_2O_3	Trace
Fe_2O_3	0.19
MnO	0.01
MgO	0.03
CaO	0.63
Na_2O	15.40
K_2O	0.01
H_2O^+	8.56
H_2O^-	0.28
Total	99.95

Table 3.18. Tellurium Minerals: Spiroffite and Denningite (%)

Constituent	Spiroffite	Denningite
TeO_2	75.93	82.34
SeO_2	—	<0.1
MnO	14.13	10.28
ZnO	9.32	2.63
CdO	0.07	0.02
PbO	0.05	0.01
Bi_2O_3	0.08	0.00
Sb_2O_3	0.02	0.00
Al_2O_3	0.02	—
Insoluble	0.22	0.02
TiO_2	0.00	0.00
Fe_2O_3	0.00	0.00
V_2O_5	0.00	0.00
CoO	0.02	0.04
NiO	0.00	0.00
MoO_3	0.00	0.00
CaO	0.15	4.23
SrO	—	<0.1
MgO	0.02	0.20
Excess O_2	0.00	0.09
H_2O	0.03	0.03
Total	100.06	99.89

vided at less than half a gram. Qualitative examination by X-ray spectrometry provided a clue to their nature and enabled a rational approach.

All but a few milligrams of the two minerals were ground to pass 100 mesh, packed in holders, and examined with the X-ray spectrometer. The major constituents were thus identified. Synthetic mixtures were prepared containing manganese, calcium, zinc, and tellurium oxides, and these were compared with the unknowns.

About 30 mg of the unknown samples and the synthetic mixtures were examined by emission spectrography by N. H. Suhr. Quantitative values for trace elements and approximate values for the majors were thus obtained.

The possibility that the minerals might be hydrated was investigated. About 100 mg of each was placed in the end of a closed Pyrex tube. The other end was drawn out; the tube was evacuated and sealed off. The end containing the mineral was heated until the mineral was plainly decomposed. The other end of the tube was immersed in an ice bath. The tube was heated and pulled in two. The end containing the water was opened and a glass rod inserted (to remove CO_2). The tube and rod were weighed, dried, and weighed again. Only a little water was found.

Excess oxygen [i.e., the possibility that Te(VI) or Mn(III) was present] was determined on 20-mg portions using the pyrophosphate method and running known salts of Te(IV) and Te(VI) as controls. A little excess oxygen was found, indicating that some of the manganese may be present as Mn(III).

Alkali metals were sought using a few milligrams of sample; there is less than 0.1% Na_2O or K_2O present. Lithium is absent.

The main part of the analysis proceeded as follows: Dry 300 mg samples at 105 °C, and weigh; dry at 160 °C, and weigh. Dissolve in hydrochloric acid, filter off insoluble material (mostly contaminating silicates), and weigh. Precipitate selenium with hydroxylamine, and weigh. Precipitate tellurium with SO_2 and hydrazine, and weigh. Examine the precipitate for bismuth. Precipitate the sulfide group, precipitate ammonia group, and examine precipitates, which were small. Triple oxalate precipitation, weigh $CaO + Mn_3O_4 + ZnO$. Determine manganese and zinc contamination of the calcium oxide and make corrections. Double phosphate precipitation, weigh $Mn_2P_2O_7 + Mg_2P_2O_7 + Ca_3(PO_4)_2$. Determine magnesium and calcium contamination of the manganese pyrophosphate and make appropriate corrections. Destroy ammonium salts in the filtrate from the manganese, and precipitate zinc as the ammonium phosphate. Ignite and weigh $Zn_2P_2O_7$. Determine the manganese contamination of the zinc phosphate and make appropriate corrections.

The X-ray spectrographic work was done with the help of C. Randall, and the trace analyses were performed by N. H. Suhr.

3.2.13. Chromite

Chromite analysis is difficult. Several different approaches are available; each finds application, depending on the purpose at hand and the composition of the mineral. Magnesium-rich chromites behave differently from iron-rich chromites [7].

A major difficulty is in the determination of ferrous iron (oxygen deficiency). Possibly the most universally applicable method is that of Seil, who decomposes the chromite in a phosphoric–sulfuric acid solution in a stream of carbon dioxide under reduced pressure. Phosphine and sulfur dioxide produced during oxidation of the ferrous iron are led through a measured volume of standard dichromate solution, and the excess dichromate is later titrated.

Most chromites are decomposed below 360 °C in a mixture of phosphoric acid and pyrophosphate containing Mn(III), and this provides the most convenient method. The excess Mn(III) is titrated with ferrous solution using diphenylamine as indicator. Before any attempt to determine ferrous iron by this or the Seil method, the sample (or a portion of it) should be ground to pass 100% through a 325-mesh screen. Oxidation of iron during such grinding does not occur if the fines are frequently removed by screening.

Ferrous chromites are readily attacked by fuming perchloric acid, leaving the chromium in the form of chromate or dichromate. Small losses of chromium occur, however, due to the volatility of chromyl chloride. Magnesian chromites are less readily attacked by perchloric acid.

The easiest way to put chromite in solution is by fusion with sodium peroxide. Leaching of the melt with water followed by filtration yields nearly all the chromium in the filtrate and the iron, manganese, calcium, and magnesium in the precipitate. Aluminum and vanadium are partially retained in the precipitate.

Analysis of the chromites reported in Table 3.19 were made by the following methods:

MAIN PORTION. Treat 0.5000 g of −325-mesh chromite with 50–60 ml 70% perchloric acid and 5 ml nitric acid in a 400-ml beaker. Cover, and place a stirring rod so that acid distilling from the cover does not drip into the solution directly but runs down the rod. Boil until the sample is decomposed. Cool, dilute with 100 ml water, and filter. Wash the filter thoroughly with water to

**Table 3.19. Analysis of Chromites PSU
61-1437, PSU 4-228 (%)**

Constituent	61-1437	4-228
SiO_2	0.18	0.50
Cr_2O_3	42.97	54.10
Al_2O_3	16.01	12.61
TiO_2	0.70	0.22
Fe_2O_3	8.91	4.95
FeO	21.78	16.16
MgO	8.69	10.64
CaO	0.04	0.22
MnO	0.24	0.30
V_2O_5	0.32	—
NiO	0.14	—
Total	99.98	99.70

remove all perchloric acid. Ignite and weigh SiO_2^+. Treat with HF and H_2SO_4, evaporate, ignite, and reweigh. The loss in weight is SiO_2. Fuse the residue with a very little potassium pyrosulfate, leach with water, and return to the main solution.

Perform a triple ammonia precipitation, washing the precipitate thoroughly each time with 5% (w/v) ammonium chloride. Ignite and weigh the ammonia group oxides. Perform a double sodium hydroxide precipitation. Determine the iron, manganese, titanium, and so on in the precipitate and the aluminum and vanadium in the filtrate. If chromium is to be determined on the main portion, that in the ammonia group must be found.

Omit an oxalate separation. Precipitate calcium and magnesium as phosphates, separate the calcium as sulfate, and determine. Most of the chromium goes into the filtrate from the magnesium and calcium and is discarded. If sample is in short supply, it may be determined in this solution; results will be slightly low because of losses during the decomposition of the sample (usually about 0.2% low).

DETERMINATION OF CHROMIUM. Fuse 0.5000 g with sodium peroxide in an iron crucible. Leach with water, and filter. Wash with $NaOH$–Na_2CO_3 solution. Boil the filtrate and acidify. Titrate chromium plus vanadium with ferrous solution using diphenylamine as indicator. Manganese and/or titanium may be determined in the precipitate. Alternatively, run chromium on the residue from a Seil FeO determination.

DETERMINATION OF FERROUS IRON. Use the pyrophosphate method or the Seil method. To illustrate the effectiveness of separating chromium by ammonia precipitation, the ammonia group for 4-228 contained the equivalent of 0.38%

Cr_2O_3 and 0.04% MnO. Chromites contain no essential water but are likely to be contaminated with serpentine. A regular Penfield determination gives no trouble even though the chromite is incompletely decomposed

3.2.14. An Impure Mineral Called Nepheline

One of the frustrations of the geochemical analyst is a frequent inability to take part in the preparation of the samples he or she analyzes. Table 3.20 shows an analysis of a mineral sample that was supposed to be a purified nepheline. Nepheline is a sodium potassium alumino-silicate, $(Na,K)AlSiO_4$, that should present little difficulty in a regular classical analysis.

The analysis shows that the sample was grossly contaminated with a phosphate mineral, probably apatite; but at the start of the analysis there was no suspicion of this. Because of the alleged importance of the sample (of which there was only about 3 g), aluminum was separated from the ammonia group and determined as phosphate. Of course, the result was much lower than expected. There was no warning that phosphate was present in the sample, and efforts were made to find rare earth elements, beryllium, and so on.

Table 3.20.   Analysis of Impure Nepheline
PSU 5-194[a] (%)

SiO_2	34.04
Al_2O_3	26.27
TiO_2	0.085
Fe_2O_3	0.47
FeO	0.39
MnO	0.014
MgO	0.44
CaO	12.64
SrO	0.23
BaO	0.095
Na_2O	10.52
K_2O	4.87
Rb_2O	0.0086
H_2O^+	1.55
H_2O^-	0.24
P_2O_5	7.59
F	0.51
Less O for F	0.21
Total	99.75

[a] Also present, small amounts of Cl, CO_2, and SO_3.

A further difficulty appeared: When 100-mg subsamples were taken for rapid colorimetric procedures, values for the major constituents did not agree with the classical analysis. It is now obvious that subsampling difficulties were the cause of these discrepancies (every 100-mg subsample contained a different proportion of silicate mineral and phosphate mineral).

Only after much time had been spent was a phosphorus determination run. The cause of the several difficulties became obvious. The analysis was completed without further difficulty, but with the feeling that it was not worthwhile.

The geochemical analyst should always keep the possibility in mind that the geologist or mineralogist may have erred and regard with suspicion any statements implying that a mineral sample has been carefully purified.

The example given is an extreme one; most often, mineral separates are pure enough so that the impurities do not cause any obvious difficulties. The difficulties are nevertheless still present. One has only to look at the "standard" micas distributed within the geochronological community to see how much wasted effort an impure mineral separate can induce. For example, the USGS muscovite P-207 contains about 5% of impurity in the form of fine quartz inclusions and relatively coarse grains of apatite, epidote, calcite, and so on. As a result, potassium determinations on this mica are not reproducible when small subsamples are taken. See Figure 2.1. The first widely circulated biotite, MIT B-3203, was the subject of extensive investigations in numerous laboratories of high competence. An incalculable amount of effort was expended trying to achieve agreement between laboratories and methods using different subsample weights. It is now known that all this effort was wasted because impurities in the biotite made the material unsampleable at the analytical level.

Calculation of mineral formulae from analyses of impure samples is, of course, an exercise in futility.

Those commissioned to analyze mineral samples should always look carefully for impurities before beginning the work.

3.2.15. Olivine

Olivines may be ferrous (fayalite) or magnesian (forsterite). Analyses of ferrous and magnesian olivines are given in Table 3.21.

Ferrous olivines require great care during the sodium carbonate fusion to prevent iron from alloying with the crucible. Magnesian olivines give high results for magnesium and high totals unless the precipitation of the

Table 3.21. Analysis of Magnesian and Ferrous Olivine (%)

Constituent	Magnesian	Ferrous
SiO_2	44.16	33.58
Al_2O_3	0.83	0.18
TiO_2	0.02	0.00
Fe_2O_3	0.55	0.46
FeO	7.51	60.40
MnO	0.12	1.48
MgO	45.23	3.94
CaO	0.69	0.18
Na_2O	0.02	0.01
K_2O	0.02	0.01
H_2O	0.05	0.04
NiO	0.33	—
Cr_2O_3	0.44	—
Total	99.97	100.28

magnesium ammonium phosphate is done with great care from a clean solution. If little manganese is present in the sample, 5–10 mg should be added to the filtrate from the ammonia group and precipitated with persulfate to take advantage of the scavenging properties of $MnO \cdot OH$.

Aluminum should be determined spectrographically on a separate sample or separated and weighed as phosphate. To determine small amounts of aluminum by difference from the weight of a large ammonia group precipitate is unsound procedure.

Chromium in olivine is usually present as chromite impurity. Determinations of chromium are unreliable because of subsampling difficulties.

The high total for the ferrous olivine is probably due to excess oxygen in the ignited ammonia group oxides. Since in this case iron, not alumina, was determined by difference, the error is probably in the Fe_2O_3 value. It may be that Fe_2O_3 in this mineral is closer to 0.00 than 0.46, as reported.

3.2.16. Sphene

Sphene is a titanium mineral commonly containing rare earths and zirconium. The mineral formula is $CaTiSiO_5$ (Table 3.22).

After fusion with sodium peroxide, titanium is put into solution with sulfuric acid, and the dehydration of silica follows. The ammonia group is separated as usual, with special precautions to recover all the rare earths (excess ammonia added to the regular ammonia group filtrate).

Table 3.22. Analysis of Sphene Menlo Park
MP-617 (%)

SiO_2	31.47
Al_2O_3	2.56
Fe_2O_3	1.42
FeO	0.56
MgO	0.11
CaO	26.85
Na_2O	0.09
K_2O	0.11
TiO_2	34.57
Y_2O_3	0.18
La_2O_3	0.07
CeO_2	0.30
Pr_6O_{11}	0.09
Nd_2O_3	0.25
Sm_2O_3	0.06
Eu_2O_3	0.008
Gd_2O_3	0.05
Tb_4O_7	0.007
Dy_2O_3	0.04
Ho_2O_3	0.006
Er_2O_3	0.015
Tm_2O_3	0.003
Yb_2O_3	0.01
Lu_2O_3	0.00
P_2O_5	0.31
MnO	0.13
ZrO_2	0.38
F	0.40
V_2O_5	0.17
Less O for F	0.17
Total	100.04

The collected rare earth oxides are dissolved in hydrochloric acid and precipitated successively with oxalic acid (which leaves iron and titanium in solution) and ammonia. After weighing, they are analyzed spectrographically by the method of Joensuu and Suhr using scandium as an internal standard.

Manganese, calcium, and magnesium are determined in the filtrate from the ammonia group in the usual manner.

The alkali metals are determined after solution of the sample by dissolving a lithium metaborate fusion in nitric acid.

3.2.17. Spencite

The original analysis of spencite reported the rare earth elements in groups, as follows:

$$Y_2O_3 \text{ group} \qquad 28.20\%$$
$$La_2O_3 \text{ group} \qquad 4.16\%$$

Later, O. I. Joensuu analyzed a separate sample of the rare earth elements using a spectrographic technique; his results are incorporated in the analysis of Table 3.23.

The original sample was submitted under the name *thalenite*, a uranium-bearing mineral. Initial efforts at analyzing the material were based on the assumption that uranium was present. This led to several false starts.

The final analysis proceeded as follows:

A 1-g sample was treated with hydrochloric acid and evaporated to dryness. Boron was removed by alternate evaporation with hydrochloric acid and methyl alcohol. Chlorides were converted to nitrates by repeated evaporation to dryness with nitric acid. Silica was recovered by double dehydration with nitric acid. Thorium was precipitated as peroxynitrate, ignited to ThO_2, and weighed; the cerium content of the precipitate was determined colorimetrically, and corrections were made.

Manganese was precipitated with persulfate in nitric acid solution and determined colorimetrically. Cerium precipitates with the manganese; it was recovered and determined.

Aluminum, iron, titanium, and part of the rare earths were precipitated with ammonia. After a sodium hydroxide separation, the rare earths in the ammonia precipitate were recovered and added to the main solution. The rare earths were then separated by ammonia precipitation at pH 10, converted to oxalates, and ignited to oxides. Calcium and magnesium were determined in the filtrate as usual.

Boron was determined by Chapin's method, as described by Hillebrand and Lundell.

The weighed rare earth oxides were analyzed spectrographically by the procedure of Joensuu and Suhr, using scandium as an internal standard. (See Frondel [8].)

3.2.18. Ulvospinel

The impure ulvospinel of Table 3.24 (nominally $2FeO.TiO_2$) proved very difficult to analyze. Several approaches were less than successful. Sodium

Table 3.23.   Analysis of Spencite, Minnesota R-2029 (%)

Na_2O	0.11
K_2O	0.01
MgO	0.50
CaO	7.81
SrO	0.05
Fe_2O_3	3.22
FeO	0.00
MnO	0.60
Y_2O_3	17.77
La_2O_3	0.73
CeO_2	2.49
Pr_6O_{11}	0.54
Nd_2O_3	1.84
Sm_2O_3	1.07
Eu_2O_3	0.14
Gd_2O_3	1.61
Tb_4O_7	0.34
Dy_2O_3	1.92
Ho_2O_3	0.50
Er_2O_3	1.99
Tm_2O_3	0.31
Yb_2O_3	2.88
Lu_2O_3	0.27
ThO_2	1.84
Al_2O_3	3.87
TiO_2	0.27
B_2O_3	10.04
SiO_2	24.89
P_2O_5	0.02
Cl	0.45
F	0.44
H_2O^+	9.82
H_2O^-	1.93
Less O for F, Cl	0.28
Total	99.99

carbonate fusion left a large unattacked residue. Fusion with sodium peroxide was somewhat more successful. Fusion with potassium pyrosulfate left a large siliceous residue, but iron and titanium were thus obtained in solution.

The most effective approach is to fuse a 0.5-g sample in a silica crucible with potassium pyrosulfate, leach the melt with water, add sulfuric acid,

**Table 3.24.   Analysis of Ulvospinel
PSU 5-202 (%)**

SiO_2	2.34
Al_2O_3	2.84
TiO_2	19.86
ZrO_2	0.14
Cr_2O_3	0.37
Fe_2O_3	20.54
FeO	47.13
MnO	0.84
MgO	3.46
CaO	1.45
SrO	0.017
BaO	0.015
Na_2O	0.18
K_2O	0.09
H_2O^+	0.41
H_2O^-	0.07
P_2O_5	0.13
F	0.01
Total	99.89

and evaporate to incipient fumes. The residue of, for example, unattacked silicates is then filtered off, ignited, and weighed. Two separate analyses follow–one of the sulfuric acid solution and the other of the insoluble residue, which is fused with sodium carbonate and treated as usual. Finally, the two analyses are added.

In the analysis of Table 3.24, iron and titanium were separated using hydrogen peroxide and sodium hydroxide. This is an effective but difficult procedure: It is probably better to use cupferron to precipitate titanium prior to its gravimetric determination.

This material, like most spinels, is readily attacked by lithium metaborate. Probably a useful scheme could be developed based on this attack. One would have to take care that lithium was not retained in the various precipitates; boron would have to be removed either by evaporation with perchloric acid or by volatilization as methyl borate.

3.2.19.   Leucophosphite

Table 3.25 gives an example of the analysis of a new rare mineral of which the sample supply was extremely limited. The analysis proceeded as follows:

**Table 3.25.   Analysis of "Leucophosphite"
PSU 5-096 (%)**

SiO_2	Trace
Al_2O_3	0.00
MnO	0.07
MgO	0.05
CaO	0.11
NiO	0.22
FeO	0.24
Fe_2O_3	40.67
H_2O	10.76
P_2O_5	35.71
Li_2O	0.00
Na_2O	0.20
K_2O	11.80
Rb_2O	0.04
F	0.01
TiO_2	0.00
Total	99.88

Sodium, potassium, magnesium, calcium, manganese, nickel, and titanium were determined on a 20-mg subsample; the alkalies were determined by flame photometry and the others by spark solution emission spectrometry. The sample was attacked with lithium metaborate and dissolved in nitric acid containing cobalt internal standard.

Fluorine was expected, and it was determined on 20 mg after steam distillation from phosphoric acid solution by titration with thorium nitrate.

Phosphorus was determined on a 50-mg subsample after solution in nitric acid by precipitating ammonium phosphomolybdate under closely controlled conditions and weighing. A known phosphate mixture was taken through the whole procedure, and an empirical gravimetric factor was used.

Ferrous iron was determined by the pyrophosphate method using 20 mg sample.

Direct current arc emission spectrography was used to determine silicon, aluminum, barium, yttrium, ytterbium, zirconium, and beryllium using 20 mg sample.

The main part of the analysis used 0.2000 g. It was dissolved in hydrochloric acid, and phosphate was removed with zirconyl chloride before precipitating the ammonia group. Lithium was determined on the main portion by evaporating all residues with nitric acid to remove ammonium salts and using the flame photometer. Standard additions of lithium sulfate were made to calibrate.

Water was run by the Penfield method using 0.1887 g sample. If cou-
lometric apparatus had been available, a good water determination could
have been made on 5–10 mg.

It should be noted that the spectrographic analysis was a very necessary
preliminary. Without any knowledge of the composition of this sample,
the analyst would have to suppose that rare earth elements, uranium,
beryllium, and so on, were present and arrange an approach to take these
into account.

3.2.20. Triphylite

The analysis of triphylite (Table 3.26) was conducted at the same time as
that of leucophosphite (Table 3.25) and using essentially the same approach. The slightly high total may possibly be due to too high a value
for Li_2O, which was determined on the main portion of the sample after
the analysis proper had been completed. it may also be due to nonsto-
ichiometry of the weighing form for manganese, $Mn_2P_2O_7$. Lack of sample
offered no possibility for checking out the obvious error.

3.2.21. Triphylite–Lithiophilite

Possibly the best analysis of this class is that of Table 3.27. This sample
was the first to which the zirconium separation of phosphate was applied.
Attempts to use excess ferric iron to remove phosphate prior to separation

Table 3.26. Triphylite Analysis PSU 5-094
(%)

SiO_2	Trace
Al_2O_3	0.00
MnO	10.82
MgO	0.37
CaO	0.12
NiO	0.26
FeO	33.17
Fe_2O_3	0.91
H_2O	0.24
P_2O_5	44.83
Li_2O	9.20
Na_2O	0.18
K_2O	0.01
F	0.00
TiO_2	0.00
Total	100.11

Table 3.27.   Analysis of Triphylite–
Lithiophylite[a] (%)

SiO_2	0.27
Al_2O_3	0.24
Fe_2O_3	0.29
FeO	24.36
MnO	20.76
MgO	0.00
CaO	0.00
Na_2O	0.08
K_2O	0.04
Li_2O	9.32
H_2O^+	0.09
H_2O^-	0.07
F	0.00
P_2O_5	44.27
Total	99.79

[a] University of Minnesota Rock Analysis Laboratory R 2315.

of the ammonia group elements led to retention of about 20% of the manganese in the ammonia precipitate even after three careful precipitations, indicating that ferric iron is ineffective in removing phosphate when much manganese is present.

In the absence of manganese, ferric iron addition has proven useful in removing phosphate. To illustrate this, work sheets for a phosphate rock analysis in which iron was used are presented in Table 3.28.

3.2.22.   Garnet

A chief difficulty in the accurate analysis of garnets is the large proportion of ammonia group elements and the difficulty in completely separating the usually large amount of manganese. A triple ammonia precipitation performed with great care is required.

Garnets are attacked with difficulty by sodium carbonate fusion, and the maximum temperature of a Meker burner is necessary. At the same time, the danger of introducing iron into the platinum crucible is great; sodium carbonate fusions must be carried out carefully and skilfully. The likely presence of much manganese makes it undesirable to include more sodium peroxide in the flux than just enough to oxidize the iron.

Table 3.29 shows some garnet analyses. They all proceeded without difficulty through the regular procedure when the precautions mentioned

Table 3.28. Phosphate Rock Analysis, Work Sheet R2195

Constituent	Weights		Percent	Remarks
H_2O^+, 0.5 g	21.9229		1.34	H_2O^+ may not be
	21.9162		0.06	very accurate
	0.0067		1.28	because of organic
				matter in sample
H_2O^-, 0.8 g,	29.6816	Before fusion	0.06	
120–130 °C	30.4816	+ sample		
	30.4811	− H_2O^-		
Loss on ignition	30.3678	0.4811	1.444	Loss on ignition
(roasting)	30.3657	0.3656		indicates the
	30.3656	0.1155		presence of about
	30.3656			0.3% volatile
	29.6810	After fusion		material other than
				CO_2; this appeared
				to be largely organic
				material
SiO_2, 0.8 g (main	30.5003	Crucible	4.83	SiO_2 value obtained in
portion)	30.5391	+ SiO_2		main portion
	30.5391			analysis is too low
	30.5005	HF		because of presence
	0.0386			of F; note
				difference when
				$AlCl_3$ is used
SiO_2, 0.8 g, using	29.6804	Crucible	5.58	Small residue from
$AlCl_3$	29.7256	+ SiO_2		HF fused and added
	29.7256			to main solution
	29.6810			before ammonia
	0.0446			group
Al_2O_3, 0.8 g,	30.4749	Crucible	1.70	Triple ammonia
Fe_2O_3 added to	30.5080			precipitation,
↓ P_2O_5	30.5075	+ $AlPO_4$		double NaOH
	30.5075			precipitation, then
	0.0326			Al as $AlPO_4$
CaO, 0.8 g	30.8830	SrO determined	47.52	Solution evaporated to
	30.8820	as $Sr(NO_3)_2$,	0.13	small volume before
	30.8819+	0.13		precipitation of Ca
	30.5005			to recover as much
	0.3814			of the Sr as possible
	0.3812	(weight		and prevent it going
		correction)		through into the Mg
				determination

291

Table 3.28. (*continued*)

Constituent	Weights		Percent	Remarks
MgO, 0.8 g	14.6720			
	14.6720			
	14.6575	MnO in Mg, 0.02		
	0.0145	(MnO), CaO,		
	0.0004	0.01	0.64	
	0.0141			
Alkalies, 0.5 g	21.5832			
	21.5808	Flame, Na_2O,	0.11	
	0.0024	K_2O	0.06	Blanks subtracted
P_2O_5, 0.5 g	14.6163	27.1180		P_2O_5 determination
	14.6162	27.1173		after fusion,
	14.6155	27.1173	23.19	precipitation with
	14.6152	26.9358		molybdate, double
	14.6152	0.1815		precipiation with
	14.4332	0.1818	(weight	Mg as $Mg_2P_2O_7$;
	0.1820		correction)	duplicate
				determinations
				agree well
SO_3, 1.0 g	14.3120	$BaSO_4$	0.43	Condition of sulfur is
	14.2994			not known, so it is
	0.0126			reported as SO_3
CO_2, 1.0 g	1.6964	1.8366	14.02	CO_2 by evolution from
	0.1990	0.1990		H_3PO_4 solution and
	1.8954	2.0356		absorbtion in
		1.8954		ascarite
		0.1402		

Constituent	Weights	Percent	Remarks
F, 0.5 g	$1.35 \dfrac{2 \times 0.000339}{1.75} \dfrac{50}{2} \dfrac{100}{0.5} = 2.62$		Direct distillation from $HClO_4$ solution after fusion with Na_2CO_3, Na_2O_2: Th^{4+} titration
Total Fe_2O_3, 2.0 g	$13.71 \times 0.005036 \dfrac{100}{2} = 3.45$		Total iron only is determined because organic matter
	$13.71 \times 0.005036 \dfrac{100}{2} = 3.45$		precludes accurate FeO determination

292

Table 3.28. (*continued*)

Constituent	Weights	Percent	Remarks
TiO_2, 1.0 g			Titanium determined on a separate
	$\dfrac{2.3}{30}\ 0.00828\ \dfrac{100}{100}\ 100 = .06$		sample because of large amount of Fe added to main portion

Final Reported Figures

SiO_2	5.58
Al_2O_3	1.70
TiO_2	0.06
Fe_2O_3	3.45
FeO	n.d.
MgO	0.63
SrO	0.13
CaO	47.52
Na_2O	0.11
K_2O	0.06
H_2O^-	0.06
H_2O^+	1.28
P_2O_5	23.19
CO_2	14.02
SO_3	0.43
F	2.62
MnO	0.02
	100.86
Less O for F	1.10
	99.76

Organic matter n.d. (about 0.3%)

above were observed. In each case, manganese was separated by double precipitation with persulfate in the filtrate from the ammonia group. In some cases (low manganese), the manganese was determined colorimetrically; in most cases, it was double precipitated as the ammonium phosphate and determined gravimetrically as $Mn_2P_2O_7$.

The slightly high totals characteristic of garnet analyses are probably due to excess oxygen in the ignited Fe_2O_3. Ignition of ammonia group precipitates containing much alumina and iron presents problems: At temperatures high enough to completely dehydrate the alumina, iron is likely to be partly reduced to Fe_3O_4. Prolonged heating at 750–800 °C after

Table 3.29. Analyses of Garnets PSU 62-1640,
Two Manganese Garnets (%)

	62-1640	Manganese Garnets	
SiO_2	38.43	36.50	36.96
Al_2O_3	20.86	20.82	20.53
V_2O_5	0.10	—	—
TiO_2	0.57	0.07	0.07
Fe_2O_3	0.13	1.48	2.61
FeO	29.53	21.92	27.70
MnO	0.82	14.56	10.70
MgO	5.22	1.04	0.64
CaO	3.88	3.56	0.53
Na_2O	0.35	—	0.05
K_2O	0.02	—	0.07
H_2O^+	0.14	0.04	0.16
H_2O^-	0.06	0.00	0.00
P_2O_5	0.02	—	—
Total	100.13	99.99	100.02

preliminary ignition to 1050 °C is recommended, although this usually leads to a slighly high value for Al_2O_3 when it is calculated by difference.

Ferrous iron in garnets is best determined by the pyrophosphate method after grinding to pass 100% through a 200-mesh screen.

3.2.23. Actinolite

Analysis of actinolite presents no special problems. The high proportion of CaO and MgO demands some care in the separation of these elements. There is always the possibility that appreciable barium and strontium are present, and these elements should be taken into account. The analysis of Table 3.30 was performed by J. A. Maxwell at the University of Minnesota.

This sample has been extensively used as a standard for spark solution emission spectrographic work at the Pennyslvania State University by N. H. Suhr. Its subsampling homogeneity with respect to most elements, including chromium and nickel, is unusually good.

Actinolite is a variety of amphibole.

3.2.24. Pyroxenes

Pyroxenes are common rock-forming minerals that vary in composition from $Mg_2Si_2O_6$ to $CaFeSi_2O_6$ with sodium and aluminum often substituting for magnesium.

Table 3.30. Analysis of Actinolite (%)

SiO_2	55.26
Al_2O_3	2.23
TiO_2	0.04
Fe_2O_3	1.19
FeO	5.12
MnO	0.31
MgO	20.41
CaO	12.07
BaO	0.16
Na_2O	0.59
K_2O	0.10
H_2O^+	1.81
F	0.31
NiO	0.16
Cr_2O_3	0.32
Less O for F	0.13
Total	99.95

Pyroxenes present little difficulty in the classical analysis scheme except that aluminum can seldom be accurately determined by difference from the weight of the ammonia group oxides and should be separated using sodium hydroxide and determined as $AlPO_4$. Triple ammonia precipitations are usually desirable unless the double precipitations are washed with great care.

A variety of pyroxene analyses are presented in Table 3.31. Some details of the analysis of PSU 63-1784 are worth recounting: Most of the Cr_2O_3 reported is present in chromite impurity. The normal ferrous iron determination does not include ferrous iron in chromite; therefore, some of the chromite from the sample was isolated (by treatment with hydrofluoric acid) and analyzed. The regular FeO determination was corrected accordingly; of the 3.91% FeO reported, 0.20% is due to FeO in chromite. In addition, the residue from the regular FeO determination was analyzed for its FeO content; an additional 0.20% FeO was recovered, confirming that the chromite remained unattacked.

Protracted fusion necessary to attack the mineral with sodium carbonate led to solution of appreciable platinum from the crucible. In addition, some platinum was introduced during the dehydration of silica. Platinum was therefore determined in the ammonia group oxides; 0.0008 g was found. This would have led to an error of +0.11% in the Al_2O_3. The silica recovery from the ammonia group weighed 0.0023 g; this would have led to a further error in the Al_2O_3 of 0.33%.

Table 3.31.   Analyses of Pyroxenes (%)

	63-1784	61-1493	62-1639	Px-1
SiO_2	50.57	55.07	56.99	53.94
Al_2O_3	6.75	14.54	17.06	0.66
TiO_2	0.70	1.80	0.61	0.26
Fe_2O_3	1.74	3.83	3.61	1.13
FeO	3.91	1.95	0.98	1.91
V_2O_5	0.05	0.11	—	—
Cr_2O_3	0.44	0.03	—	.21
NiO	0.16	0.017	—	—
MnO	0.11	0.026	0.06	0.07
MgO	15.12	4.92	3.89	16.93
CaO	18.95	7.56	5.38	24.55
SrO	—	—	—	0.035
BaO	—	0.015	—	0.006
Li_2O	0.00	0.02	0.03	—
Na_2O	1.40	9.56	11.10	0.24
K_2O	0.02	0.01	0.01	—
H_2O	0.03	0.44	0.30	0.03
P_2O_5	0.03	0.00	—	0.00
Total	99.98	99.90	100.02	99.97

Zirconium, vanadium, chromium, nickel, cooper, beryllium, and titanium were determined by N. H. Suhr using the spectrograph. Chemical determinations of nickel in the ammonia group oxides and in the magnesium ammonium phosphate added to 0.16% NiO, exactly the spectrographic result. Chromium, calculated from the weight of chromite separated from a 2-g sample and from a separate analysis of this chromite, showed 0.45% Cr_2O_3, compared to the 0.44% obtained spectrographically.

Manganese was separated from the filtrate from the ammonia precipitate using persulfate. No manganese was found in either the ammonia group or in the weighed magnesium pyrophosphate.

Total iron was determined during the main portion analysis after a sodium hydroxide separation and also on the residues from the ferrous iron determination (including the residue from the pyrophosphate treatment of the chromite). Values were 6.09% Fe_2O_3 and 6.08% Fe_2O_3, respectively.

The alkali metals were determined gravimetrically after a J. L. Smith attack, a double leach. The equivalent of 0.013% Na_2O was found in the weighed potassium sulfates and 0.014% K_2O in the weighed sodium sulfates. No lithium was present in the sodium sulfates. In addition, sodium and potassium were run on a separate sample after $LiBO_2$ fusion; results were 0.03% K_2O and 1.40% Na_2O, agreeing almost exactly with the results from the more sophisticated procedure.

The calcium recovery from the magnesium pyrophosphate amounted to 0.0008 g $CaCO_3$. The equivalent of 0.01% SrO was found in the calcium oxide. SrO was not reported because the collection of SrO with the calcium oxalate is unreliable; it is sometimes complete and at other times very incomplete. The SrO content of the sample is probably not greater than 0.02% SrO.

3.2.25. Diopside

Diopside is a variety of pyroxene. The analysis of Table 3.32 is unusual because it is evidently of a very pure mineral separate, the only contamination being a little apatite. Al_2O_3, V_2O_5, ZrO_2, Cr_2O_3, NiO, CuO, BeO, Li_2O, K_2O, and H_2O are all absent.

It seems probable that if more attention were given to the purification of minerals submitted for analysis, more mineral analyses would look like this one. Calculating the reported values to a base of Si 2.00, the formula of this mineral is simply $CaMgSi_2O_6$. There is a slight excess of magnesium over calcium ($Mg_{1.01}$, $Ca_{0.99}$). Whether this discrepancy is real or not depends on the skill of the analyst and especially on the recovery of calcium from the ignited $Mg_2P_2O_7$, which in this case amounted to 0.0013 g $CaCO_3$.

Spectrographic analysis for traces was performed by N. H. Suhr.

3.2.26. Andalusite

Andalusite is an aluminum silicate that is chemically the same as kyanite and sillimanite, Al_2SiO_5.

Table 3.32. Analysis of Diopside PSU 63-1827 (%)

SiO_2	55.36
TiO_2	0.01
FeO	0.09
MnO	0.005
MgO	18.77
CaO	25.70
Na_2O	0.02
P_2O_5	0.06
Total	100.01

The chief difficulty in analyzing andalusite lies in handling the large precipitate of aluminum hydroxide and especially in washing it free from sodium salts. It is advisable to use only 0.5-g samples, filter through 12.5-cm papers, and perform a triple precipitation.

Alkali determinations are made after attack by lithium metaborate by flame photometry.

Among the trace elements likely to be present and that should be determined are chromium, copper, gallium, and vanadium. The andalusite of Table 3.33 was examined spectrographically by R. E. Mays, USGS.

The purpose of a classical analysis of simple minerals such as andalusite, kyanite, and sillimanite will usually be to ascertain the exact stoichiometry or departures from it; the analysis should therefore be performed with great care.

3.2.27. Manganese Zinc Ferrite

Manganese zinc ferrite is not a mineral but provides the best available example of a material in which manganese, zinc, and iron exist together. Considerable time was spent developing methods for separating these elements; the same procedures may presumably be applied to zinc manganese minerals. The methods used in analyzing spiroffite and denningite (Table 3.18) did not work well with the ferrite of Table 3.34.

Oxygen excess or deficiency is easily determined by the pyrophosphate method. If the sample has an oxygen deficiency, it is more convenient to

Table 3.33. **Analysis of Andalusite Menlo Park FD-21 (%)**

SiO_2	36.96
Al_2O_3	61.59
Fe_2O_3	0.64
MgO	0.16
CaO	0.21
Na_2O	0.00
K_2O	0.04
H_2O	0.06
TiO_2	0.08
P_2O_5	0.03
Cr_2O_3	0.02
Ga_2O_3	0.01
MnO	0.01
V_2O_5	0.02
Total	99.83

Table 3.34. Analysis of Manganese Zinc
Ferrite (%)

Al_2O_3	0.00
SiO_2	0.20
TiO_2	0.01
MnO	18.08
MgO	0.02
CaO	0.00
ZnO	9.97
Fe_2O_3	68.63
FeO	3.02
Na_2O	0.06
K_2O	0.02
Total	100.01

attack it with hydrofluoric and sulfuric acids and titrate with permanganate after dilution and boric acid addition because total iron may then be determined on the same sample after addition of hydrochloric acid, reduction with stannous chloride, and titration between two end points potentiometrically with dichromate.

Manganese is best determined by the pyrophosphate method of Lingane and Karplus. The nitric acid–pyrophosphate method is adequate once analyzed ferrite standards have been accumulated; this method is best used as a secondary method since oxidation of the manganese is not quite complete and an empirical standardization factor is desirable.

The following procedure was developed after trial of many others, and once mastered, it presents little difficulty.

PROCEDURE. Treat 1.000 g sample with 50 ml 1:1 hydrochloric acid in a porcelain dish. Add just enough nitric acid to oxidize the iron. Evaporate to dryness on the steam bath, take up the residue with hydrochloric acid, and filter. Wash thoroughly with dilute hydrochloric acid and water. Reserve the paper and residue. Evaporate the filtrate to near dryness in the porcelain dish, add 10 ml 8 N hydrochloric acid, and extract iron with chloroethyl ether (bis-dichloroethyl ether, 2,2' chloroethyl ether). Careful technique using a separatory flask permits complete separation of the iron from divalent metals. Iron may be determined in the ether extract, but it is more convenient to determine it on a separate sample. If there is a suspicion that some aqueous solution is entrained in the ether extract, shake it with a little water, make the water extract 8 N in hydrochloric acid, and repeat the extraction.

Evaporate the water layer to dryness on the steam bath, add a little nitric acid and 10 ml of 1:1 sulfuric acid, and evaporate to fumes. Cool, dilute, and precipitate with ammonia at pH 6.5. Repeat the precipitation. A double precipitation is necessary despite the small amount of ammonia group oxides to be sure of complete separation of zinc.

Wash the hydrated oxides with 1% ammonium sulfate solution. Combine the paper and precipitate with that containing the silica and insoluble material, and burn off the paper in the muffle furnace at 500 °C. Ignite to constant weight at 800 °C. Treat the residue with 10 drops sulfuric acid and 5 ml hydrofluoric acid, and evaporate most of the latter. Add a little water, pack the crucible half full of ashless paper pulp, add 2–3 ml concentrated aqueous ammonia, and burn off in the muffle at 500 °C, placing the crucible in the cold furnace and bringing to temperature slowly. Weigh the residue, and repeat the ignition at 800 °C to constant weight. The loss in weight is SiO_2. Analyze the weighed residue spectrographically, or chemically as follows: Fuse with 1 g sodium carbonate, leach with hot water containing 0.1% sodium hydroxide and a drop of 30% hydrogen peroxide, and determine chromium colorimetrically as chromate after filtration. Determine the titanium, iron, and manganese in the residue by standard colorimetric procedures, and add the small amounts found here to the major part of these elements found elsewhere. Remove manganese from the main solution by triple precipitation with persulfate from 1% sulfuric acid solution. The hydrated manganese oxide may be ignited and weighed as Mn_3O_4 or as Mn_2O_3 (or both). Ignition to constant weight at 900 °C or above yields Mn_3O_4 if the crucible and contents are cooled rapidly to prevent conversion to Mn_2O_3. Complete conversion to Mn_2O_3 may be achieved by holding the ignited oxide at 600 °C for several hours or until no further weight gain occurs. It is desirable to examine the ignited manganese oxide for its zinc content; however, spectrographic examination of several such residues showed only traces of zinc after a triple ammonium persulfate precipitation of manganese. After *two* precipitations in three trials, 0.8, 0.6, and 1.0 mg ZnO was found in the manganese oxide.

The filtrate from the manganese contains substantially all the zinc in the sample plus any calcium and magnesium and also most of the nickel and copper.

Reduce the volume of the combined filtrates by evaporation to about 400 ml, and make the solution 10% by volume in concentrated aqueous ammonia. Dissolve 5 g diammonium phosphate in water, and stir it into the ammonia solution. Allow to stand for several days with occasional stirring. Filter, wash with 5% aqueous ammonia, and dissolve and reprecipitate the mixed phosphates. Ignite at 550 °C and weight $Mg_2P_2O_7$ plus $Ca_3(PO_4)_2$. Separate the calcium from alcoholic sulfate solution as in the regular procedure, and convert to oxalate and then to carbonate for weighing. Test the alcoholic filtrate for manganese in case separation of this element was incomplete.

Double precipitate the zinc from the filtrate after removal of ammonium salts with nitric acid and filtration to remove incidental silica introduced from the glassware. Finally weigh $Zn_2P_2O_7$. Spectrographic examination of the ignited zinc phosphate for nickel and copper is desirable.

In the analysis of Table 3.34, the ammonia group oxides weighed 0.0015 g, consisting of 0.0009 g ZnO, 0.0001 g SiO_2, 0.0004 g Fe_2O_3, and 0.0001 g TiO_2.

The small amount of iron shows the effectiveness of the chloroethyl ether extraction.

3.2.28. Vermiculite and Hydrobiotite

The chief difficulty with the minerals vermiculite and hydrobiotite arises from their high water content. The sampling weight baseline must be established with care. It is desirable to spread the sample on a piece of Albanene paper, cover with another piece of paper, and expose to the laboratory air for 24 h; then several portions likely to be needed are weighed out into suitable crucibles or dishes.

These minerals are likely to contain absorbed gases in appreciable amounts, for example, nitrogen, oxygen, and helium. If they have been separated using organic liquids, some of these will almost certainly be present. Determinations of H_2O^- by drying at 105–110 °C are not very meaningful but should be reported. A determination of total hydrogen should be made at once after weighing out the several portions for analysis. A loss on ignition is sometimes useful: It may reveal volatiles other than water by comparison with the total hydrogen determination.

Usually, it is best to take a 1.0000-g sample for the main portion of the analysis instead of the usual 0.7–0.8 g.

Except for the sampling weight baseline difficulty, analysis of hydrobiotites and vermiculites present no greater problem than that of a biotite analysis.

A typical vermiculite and hydrobiotite analysis is shown in Table 3.35.

3.2.29 Apatite

The apatite analysis of Table 3.36 is of a single large crystal, which has been extensively investigated by Cruft [9] for its trace element content. Despite the fact that a single crystal was taken for analysis, most of the trace and minor constituents are not part of the apatite but derive from inclusions, some of them submicroscopic. This analysis is possibly the best available example of a false tradition among mineralogists that ignores the possibility of irrelevant impurities contributing to mineral analyses.

In most analyses of apatite and other phosphate minerals and rocks, the rare earth elements are not reported as such; they appear in the form of erroneous values for calcium, magnesium, and aluminum. In older analyses, the effects of fluorine and phosphorus on the classical scheme have not been taken into account; the usual result is a report of appreciable magnesium, which does not exist in apatites except as an impurity.

Table 3.35. Analyses of Vermiculite, Hydrobiotite PSU 5-104, 5-106 (%)

	Vermiculite 5-104	Hydrobiotite 5-106
SiO_2	35.43	35.60
Al_2O_3	11.30	11.85
TiO_2	0.91	1.13
Cr_2O_3	0.26	0.03
Fe_2O_3	6.65	10.28
FeO	0.27	0.81
NiO	0.01	—
MnO	0.05	0.08
MgO	23.56	20.17
CaO	0.39	1.44
SrO	0.01	0.005
BaO	0.03	0.17
Na_2O	0.00	0.16
K_2O	0.14	3.17
Rb_2O	0.00	0.0129
H_2O^+	9.47	7.56
H_2O^- (105 °C)	11.26	7.20
P_2O_5	0.05	0.07
F	0.02	0.21
Less O for F	—	0.09
Total	99.79	99.86

Without special procedures, elements such as lithium, strontium, and fluorine are likely to be collected and weighed as something else; fluorine causes low values for SiO_2 (which is an almost universal impurity in apatites) or interferes with the ammonia group precipitation by forming the aluminum complex or by forming insoluble fluorides with calcium or strontium.

Determination of fluorine in phosphates gives trouble in many common procedures, and many published values are grossly incorrect. No entirely satisfactory method for total hydrogen in phosphate rocks and minerals has ever been devised; the method given here yields good results, but only with the expenditure of considerable time and the acquisition of some skill.

Standard separations using alkali carbonates and hydroxides are not efficient in the presence of phosphate and fluoride and lead to erroneous values for alumina, the alkaline earths, and the rare earths.

The use of "rapid" chemical methods for phosphate analysis must be judicious; while such methods are essential for routine control analysis, they should always depend on carefully analyzed standards for calibration.

In the analysis of Table 3.36, moisture was determined by loss on drying at 105, 160, and 450 °C. This will give some measure of organic impurity, often present in phosphates.

Total hydrogen was determined by a refined Penfield method. Apatite is stable in the system $CaO-P_2O_5-H_2O$ up to 1400 °C, and the use of a flux is essential. Without taking proper precautions, hydrogen may volatilize from the fluxed sample as, for example, HF, HCl, and H_2SiF_6 as well as H_2O. The collected water should always be examined for these contaminants. Fusion with a mixture of ignited SiO_2, lead oxide, lead chromate, and calcium carbonate with some extra calcium carbonate in the collection portion of the Penfield tube was necessary to collect all the water free from contaminating acids.

Before starting the main part of the analysis, phosphorus was determined on a separate 0.5-g sample by sintering with 0.1 g silica and 1 g sodium carbonate, leaching with water, separating silica by nitric acid dehydration, recovering phosphorus from the residue after volatilizing

Table 3.36. Analysis of Apatite[a] (%)

CaO	53.98
SrO	0.27
MnO	0.01
MgO	0.02
Fe_2O_3	0.04
Ce_2O_3	0.96
La_2O_3	0.35
Y_2O_3	0.18
$Nd_2O_3{}^+$	0.44
CO_2	2.10
P_2O_5	38.37
SiO_2	1.07
Al_2O_3	0.03
H_2O^-	0.02
H_2O^+	0.12
F	3.30
Cl	0.02
UO_3	0.10
Oxygen deficiency	0.037
Less O for F, Cl	1.39
Total	99.95

[a] From Cruft et al., *Geochim. Cosmochim. Acta*, **29**, 581–597 (1965).

SiO_2 with hydrofluoric acid, and precipitating with ammonium molybdate. The phosphomolybdate precipitate was converted to magnesium ammonium phosphate, which was ignited to pyrophosphate and weighed.

Fluoride was also determined before starting the main part of the analysis by distilling from sulfuric acid solution (after sintering with sodium carbonate) and titrating with thorium nitrate.

Silica and the rare earths were determined after fusion with sodium carbonate and peroxide. The melt was leached with water, and a large excess of aluminum chloride was added to complex the fluoride. A triple dehydration recovered substantially all the silica in the sample. The residue from a hydrofluoric acid determination was fused with potassium pyrosulfate and returned to the main solution.

The filtrates from the silica were evaporated and the residue dissolved in water. A large volume of saturated oxalic acid was added, which precipitated much of the calcium and all the rare earths. (In analyzing phosphates that do not contain calcium, 100–200 mg of calcium as chloride should be added at this stage to be sure of complete precipitation of the rare earths.)

The oxalates were converted to chlorides, and the rare earths were double precipitated with ammonia at pH 10. The precipitate was ignited and weighed, taking into account the extreme difficulty with which rare earth oxides are freed from water and carbon dioxide. Excess oxygen in the ignited oxides was determined using the pyrophosphate method. Cerium was determined by the nitric acid–persulfate method. Spectrographic analysis for La_2O_3 and Y_2O_3 yielded the values shown. The Nd_2O_3+ reports the total of the other rare earths plus ThO_2. Obviously, a more extensive spectrographic examination is possible, and in this respect the analysis is incomplete.

The main part of the analysis proceeded as follows:

PROCEDURE. Determine moisture on a 1.000-g sample, add 0.2000 g specpure silica and 4 g sodium carbonate containing a little sodium peroxide, and fuse in a platinum crucible, being very careful to maintain oxidizing conditions. Treat with hydrochloric acid and dehydrate silica as usual, except that the platinum dish should remain uncovered during the evaporations to promote volatilization of fluoride and silicofluoride.

The filtrate should be evaporated a second time to remove all possible silica. If there is any doubt about the complete removal of silica and fluorine, a third evaporation to fumes after adding 50 ml perchloric acid is advised.

Alternatively, boric acid may be added to the hydrochloric acid solution of the sample (the silica addition then becomes unnecessary) and boron removed by repeated alternate evaporation with methanol and hydrochloric acid. Another alternative is to treat the sample without fusion with a mixture of boric and perchloric acids and evaporate to dryness with perchloric acid to remove boron and render silicic acid insoluble. This alternative is useful when the supply of sample is limited because silica may be determined with fair accuracy on the main portion.

In any case, silica and fluorine (and any added boron) must be removed before proceeding.

Make the solution from which silicon and fluorine have been removed 1–2 N in hydrochloric or perchloric acid in a volume of about 200 ml, and add a solution of zirconium oxychloride of known strength so that there is an excess of a few milligrams of ZrO_2 above that needed to react with the phosphorus in the sample. Digest at the boiling point for 20–30 min, and filter the bulky precipitate through a 12.5-cm Whatman No. 541 paper folded inside an S and S No. 589 Blue Ribbon paper. Wash superficially with 5% (v/v) hydrochloric acid. Return the precipitate to the beaker using a minimum of water, and add 20 ml concentrated hydrochloric acid. Digest hot for an hour. Dilute to 200 ml with water, and repeat the filtration (through the same papers) and washing. A third treatment is desirable to be sure that all the calcium and rare earth metals are removed from the precipitate.

Ignite the zirconium acid phosphate to ZrP_2O_7 in a large platinum crucible. Mix the residue thoroughly, and examine it spectrographically for impurities—or fuse it with 10 g sodium carbonate, leach with water, filter, and wash with 2% sodium carbonate solution. Dissolve the precipitate in 1:20 sulfuric acid and determine TiO_2 colorimetrically.

Evaporate the filtrates from the zirconium phosphate to remove excess acid, and conduct a double or triple ammonia precipitation. Analyze the weighed ammonia group oxides spectrographically or chemically by standard methods. It contains indeterminate amounts of zirconium and phosphate so that aluminum must be determined directly and not by difference.

Perform a double or triple persulfate precipitation on the filtrate from the ammonia group oxides, previously adding a measured amount of manganese (5–10 mg) in the form of a chloride solution if no manganese is present in the sample. Proceed with the separation of calcium and magnesium as usual. The weighed calcium carbonate or oxide should be examined spectrographically for rare earth oxides and corrections made.

The alkali metals may be determined by standard procedures, except that lithium, when present, is best determined on the main portion of the sample, in the filtrate from the magnesium ammonium phosphate, by flame photometry.

Total carbon should be determined by combustion and soluble carbonate by volatilization from phosphoric acid solution.

Oxygen excess or deficiency is best determined by the pyrophosphate method.

3.2.30. Manganese Dolomite

The analyses of Table 3.37 demonstrates the precision that is possible when pure mineral samples are analyzed by competent analysts using different methods.

Table 3.37.   Analyses of a Manganese
Dolomite[a] (%)

H_2O	0.00	0.00	—
SiO_2	0.00	0.00	—
FeO	0.39	0.39	0.39
MnO	23.43	23.17	23.31
CaO	33.80	33.75	—
MgO	0.75	0.78	—
CO_2	41.64	41.68	—

Concensus Analysis

FeO	0.39
MnO	23.31
CaO	33.78
MgO	0.76
CO_2	41.66
	99.90

[a] Minnesota Rock Analysis Laboratory R 2027.

This unusually pure mineral was analyzed several times during the exploration of the various methods for determining manganese.

During a regular analysis, in which manganese was precipitated with persulfate in the filtrate from the ammonia group, conversion of the $MnO \cdot OH$ to $Mn_2P_2O_7$ yielded 23.43% MnO. Only 0.002% MnO was found in the calcium oxalate precipitate and 0.006% MnO in the magnesium ammonium phosphate precipitate. Potassium permanganate titration by the method of Lingane and Karplus yielded 23.17% MnO.

Duplicate determinations of calcium, magnesium, and iron showed very satisfactory agreement. Carbon dioxide was determined by volatilization from hydrochloric acid and from phosphoric acid solution; comparative values were 41.64% and 41.68% CO_2.

The concensus of all analyses agrees almost exactly with the formula (Ca, Mn, Fe, Mg) CO_3.

3.2.31.   Chloritoid

Chloritoid is a brittle mica; the analysis of Table 3.38 is interesting because of the manganese content and because of the high concentration of alumina. After a double precipitation, the ammonia group of elements still contained 0.16% MnO (0.17% Mn_3O_4). The manganese was separated from the ammonia group filtrate as $MnO \cdot OH$, converted to $Mn_2P_2O_7$, and weighed. No manganese was found in the calcium oxalate precipitate,

Table 3.38. Analysis of Chloritoid, Minnesota Rock Lab No R2255 (%)

SiO_2	24.84
Al_2O_3	38.54
Fe_2O_3	2.42
FeO	20.51
MgO	1.41
CaO	0.06
Na_2O	0.02
K_2O	0.04
H_2O^+	6.87
H_2O^-	0.03
TiO_2	0.61
MnO	4.57
F	0.04
Less O for F	0.02
Total	99.95

Table 3.39. Analysis of Penn State Muscovite M-1 (%)

SiO_2	46.01
Al_2O_3	35.77
TiO_2	0.14
Fe_2O_3	0.83
FeO	0.72
MnO	0.04
MgO	0.33
CaO	0.12
BaO	0.008
Li_2O	0.03
Na_2O	0.76
K_2O	10.42
Rb_2O	0.134
Cs_2O	0.006
H_2O^+	4.31
H_2O^-	0.03
P_2O_5	0.12
F	0.38
Less O for F	0.16
Total	100.00

and the equivalent of 0.006% was found in the magnesium ammonium phosphate. Chloritoid is an example of a mica that contains little or no alkali metals.

Manganese was also determined on a separate sample after attack by nitric and hydrofluoric acids by the bismuthate method. Result was a value identical to that reported.

3.2.32. Muscovite and Other "White Micas"

Table 3.39 presents the analysis of a muscovite, M-I, prepared at the Pennsylvania State University with the intention of issuing it as a standard reference material. It was never so issued because it could not be made homogeneous at the laboratory subsampling level. This lack of homogeneity is characteristic of white micas. Examination under the microprobe often shows gross changes in composition across single crystals. For example, an optically pure white mica showed variations of 100% or more in potassium and magnesium contents within the same crystal.

An exception to the rule that white micas are intrinsically inhomogeneous is the Nancy Mica-Mg, which, with respect to potassium at least, is one of the few truly homogeneous micas of wide distribution.

Table 3.40 shows two complete analyses of the muscovite P-207, widely distributed among geochronologists before the necessity of examining standard reference materials for homogeneity was fully recognized. The first analysis, performed in the Denver laboratories of the USGS by E. L. Munson under the direction of L. C. Peck, differs from the second only in minor details. Most of the differences can be easily explained by the essential inhomogeneity of the material. For example, the high calcium content reported in the Penn State analysis could not be confirmed by spectrographic analysis of residual material from the same bottle. It is almost certain that most of the calcium reported originated in grains of calcite included in the sample analyzed.

The true K_2O content of the pure muscovite P-207 has been determined, using the microprobe, at 10.94%. Analyses of the material as distributed showed values ranging from 9.96% to 10.61%. Most of this variance was undoubtedly due to sample inhomogeneity.

The common assumption that differences between laboratories analyzing rock and mineral reference samples must be due to differences between analysts and methods must be revised. Most often, such differences merely demonstrate the inadequacy of methods of sample preparation and purification.

Table 3.40.   Two Analyses of the U.S. Geological Survey Muscovite P-207 (%)

	Analysis at Denver Laboratories of the USGS	Analysis at Penn State University
SiO_2	47.14	47.24
Al_2O_3	31.28	30.90
TiO_2	0.48	0.48
Fe_2O_3	3.98	3.91
FeO	0.79	0.91
MnO	0.11	0.12
MgO	0.94	0.93
CaO	0.09	0.18
BaO	—	0.01
Na_2O	0.57	0.67
K_2O	10.42[a]	10.22
Rb_2O	—	0.0846
Cs_2O	—	0.0006
H_2O^+	3.98	3.71
H_2O^-	0.02	0.31
P_2O_5	0.04	0.04
CO_2		0.08
F	0.34	0.34
O for F	0.14	0.14
Total	100.04	99.99

[a] No correction made for Rb and Cs.

Every geochemical analyst should try to avoid being drawn into "standards programs" and "round robins" promoted by persons, however prestigious, who have no understanding of sampling and subsampling problems. An enormous waste of time and talent (usually financed by government money) can be avoided if geochemical analysts will remain aware of such problems.

The two analyses of Table 3.40 show that competent classical analysts can agree fairly well in their analyses of even marginally homogeneous materials. A reason for this is that the competent classical analyst makes sure that the analysis represents the material he or she receives. The differences between the two analyses in the table are mainly due to differences between the two bottles delivered to the analysts.

REFERENCES

1. C. O. Ingamells, "A New Method for Ferrous Iron and Excess Oxygen in Rocks, Minerals, and Oxides," *Talanta*, **4**, 268–273 (1960).
2. F. Trusell and H. Diehl, "Efficiency of Chemical Desiccants," *Anal. Chem.* **35**, 674–677 (1963).
3. C. O. Ingamells and J. C. Engels, Preparation, Analysis, and Sampling Constants for a Biotite, National Bureau of Standards, Special Publication 422, Accuracy in Trace Analysis: Sampling, Sample Handling and Analysis, Proceedings of the Seventh IMR Symposium, October 7–11, 1974, Gaithersburg, Maryland (issued August 1976).
4. M. R. Heyes and J. Metcalfe, The Boron-Curcumin Complex in Trace Boron Determinations, United Kingdom Atomic Energy Authority. PG Report 251 (S), 1963.
5. G. Donnay and R. Allmann, "Si_3O_{10} Groups in the Crystal Structure of Ardennite, *Acta Crystallog.* **B24**(6), 845–855 (1968).
6. C. O. Ingamells and J. Gittins, "The Stoichiometry of Scapolite, *Can. Mineral.* **9**, 214–236 (1967).
7. S. A. Bilgrami and C. O. Ingamells, "Chemical Composition of Zhob Valley Chromites, West Pakistan," *Am. Mineral.* **45**, 576–590 (1960).
8. C. Frondel, "Two Yttrium Minerals: Spencite and Rowlandite," *Can. Mineral.* **6**(5), (1961).
9. E. F. Cruft, C. O. Ingamells, and J. Muysson, "Chemical Analysis and the Stoichiometry of Apatite," *Geochim. Cosmochim. Acta* **29**, 581–597 (1965).

CHAPTER

4

THE ELEMENTS

4.1. The Alkali Metals
    4.1.1. Primary Methods
           The J. L. Smith Separation
           Separation of Alkalies One from the Other
           Conversion of Chloroplatinates to Sulfates
           Flame Photometric Determination of Minor Alkalies
           Lead Carbonate Separation of Lithium
           Modified Berzelius Method for Isolating Alkali Metals
           Flame Photometry of Alkali Metals
           Preparation of Standard Solutions of Alkali Sulfates
           Potassium and Rubidium by Isotope Dilution Mass Spectrometry
           Ion Exchange Separation of Alkalies
           Gravimetric Determination of Sodium
           Rapid Gravimetric or Titrimetric Determination of Potassium
           Sodium Cobaltinitrite Reagent
    4.1.2. Secondary Methods
           Lithium Metaborate–Nitric Acid Solution
           Hydrofluoric–Sulfuric Acid Attack
           X-Ray Spectrometry
    4.1.3. Rapid Routine Methods for Specific Materials
4.2. Beryllium
    4.2.1. Analysis of Beryllium-Bearing Silicates
    4.2.2. Determination of Beryllium as Pyrophosphate
    4.2.3. Determination of Traces of Beryllium
    4.2.4. Colorimetric Determination of Beryllium Using Quinalizarin
    4.2.5. Fluorimetric Determination Using Morin
    4.2.6. Spectrographic Determination of Beryllium
    4.2.7. Determination of Beryllium by AA Spectrometry
4.3. Magnesium
    4.3.1. Primary Methods
    4.3.2. Rapid Methods
           Determination of Dolomite Content of Limestone
           Precipitation of Magnesium with Oxine
           Determination of Magnesium in Limestone or Dolomite
    4.3.3. Secondary Methods: X-Ray Fluorescence

4.4.   The Alkaline Earths
    4.4.1.   Scheme Recommended for Limestone
    4.4.2.   Primary Methods
            Analysis for Barium and Strontium in Silicates
            Conversion of Barium Chromate to Barium Sulfate
            Separation of Strontium
            Determination of Strontium
            Preparation of Standard Solutions of Alkaline Earth Nitrates
            Strontium (and Barium) by Isotope Dilution Mass Spectrometry
            Primary Analysis of Apatites and Phosphate Rocks
    4.4.3.   Secondary Methods
4.5.   Scandium, Yttrium, and the Rare Earths: Precipitation Reactions
4.6.   Titanium, Zirconium, and Hafnium
4.7.   Tantalum and Niobium
    4.7.1.   Determination of Trace Amounts in Rocks and Ores
    4.7.2.   Ion Exchange Separation
4.8.   Chromium
    4.8.1.   Titrimetric Determination of Chromium
    4.8.2.   Analysis of Chromite
            The Main Portion
            Determination of Manganese and Vanadium
4.9.   Molybdenum, Tungsten, and Rhenium
    4.9.1.   Analysis of Sulfide Precipitates Containing Molybdenum
    4.9.2.   Jones Reductor: Permanganate Method
    4.9.3.   Reduction of Molybdenum with Mercury
4.10.   Manganese
    4.10.1.   Titrimetric Determination of Manganese
    4.10.2.   Oxidation to Permanganate Using Bismuthate, Titration with Ferrous Iron
    4.10.3.   Oxidation to Permanganate by Persulfate, Titration with Arsenite
    4.10.4.   Oxidation to Mn(III), Titration with Ferrous Ion
    4.10.5.   Complexometric Titration of Manganese with EDTA
    4.10.6.   Colorimetric Determination of Manganese
    4.10.7.   Atomic Absorption Determination of Manganese
    4.10.8.   Sampling Considerations
    4.10.9.   Analysis of Sea Nodules and Other Manganese Minerals
    4.10.10.   General Procedure for Accurate Analysis of Manganese Mineral or Ore
4.11.   Iron
    4.11.1.   Determination of Ferrous Iron
    4.11.2.   Determination of Iron
4.12.   Nickel and Cobalt
    4.12.1.   Analysis of Limonitic and Garnieritic Ores
    4.12.2.   Analysis of Sulfide Ores

4.13.   Boron
      4.13.1.   Determination of Boron without Distillation
      4.13.2.   Colorimetric Determination of Boron Using Curcumin
      4.13.3.   Determination of Boron in Water
      4.13.4.   Other Colorimetric Methods for Boron
4.14.   Gallium, Indium, and Thallium
4.15.   Carbon and Carbon Dioxide
4.16.   Germanium
4.17.   Phosphorus
      4.17.1.   Phosphorus in Apatite and Phosphate Rock
      4.17.2.   Phosphorus in Tungsten-Bearing Materials
4.18.   Arsenic and Antimony
      4.18.1.   Decomposition of Samples
      4.18.2.   Distillation of Arsenic and Antimony
      4.18.3.   Determination of Arsenic by Heteropoly Molybdenum Blue Method
      4.18.4.   Determination of Antimony by Rhodamine B Method
4.19.   Selenium and Tellurium
      4.19.1.   Selenium
            Chemistry of Selenium
            Preparation of Standard Solutions of Selenium
            Determination of Selenium
      4.19.2.   Tellurium
4.20.   Oxygen Excess or Deficiency in Rocks
4.21.   Fluorine
4.22.   Hydrogen
      4.22.1.   Primary Methods: Moisture, H_2O^-, and Sampling Weight Baseline
      4.22.2.   Rapid Methods
      4.22.3.   Other Methods
4.23.   Noble Metals
      4.23.1.   Determination of Noble Metals by Fire Assay
      4.23.2.   Determination of Very Low Gold Contents in Sand or Stream Sediments
            Dissolution of gold by Aqua Regia
            Dissolution of Gold by Cyanide
4.24.   Uranium, Thorium, and Radium
      4.24.1.   Naturally Occurring Radioactive Isotopes
      4.24.2.   Decomposition of Uranium Ores
      4.24.3.   Separation of Uranium
      4.24.4.   Extraction of Uranium
      4.24.5.   Solution of Uranium in Sodium Fluoride
      4.24.6.   Decomposition of Thorium Ores
      4.24.7.   Separation of Thorium
      4.24.8.   Thorium Determination by Thorin Method
      4.24.9.   Decomposition of Ores for Radium Analysis

4.24.10.   Separation of Radium
References

4.1.   ALKALI METALS

Lithium is found in traces in almost all igneous rocks and in most natural waters. Lepidolite, spodumene, petalite, and amblygonite are the usual minerals containing major amounts of lithium:

Lepidolite, $(Li, K, Rb, Cs)_2, Al_2Si_3O_9, (F, OH)_2$
Spodumene, $LiAl(SiO_3)_2$
Petalite, $LiAlSi_4O_{10}$
Amblygonite, $AlPO_4 \cdot LiF$

In classical analysis, lithium may cause difficulties, especially when (as is often the case with lithium-bearing samples) fluorine and phosphorus are present. Retention of lithium in the ammonia group of elements is difficult to avoid, and a triple precipitation should be made when it is present in more than the usual traces. Phosphate should be removed with zirconium, either prior to the ammonia precipitation or a measured amount of zirconium should be added during the precipitation, and an appropriate correction made to the weight of ammonia group oxides. It is often convenient to determine lithium at the end of the main analysis in the filtrate from the oxalate precipitation. Before precipitating with phosphate to recover magnesium, an aliquot of the solution may be taken and treated with nitric acid to remove ammonium salts; then the solution is evaporated with sulfuric acid to obtain a neutral sulfate solution, which can be examined by flame photometry.

Because of the use of lithium isotope of mass 6 in nuclear technology [1], lithium reagent chemicals are of uncertain molecular weight when purchased because they may have been manufactured from lithium containing little ^6Li. Natural lithium is 7.4% ^6Li and 92.6% ^7Li, with an atomic weight of 6.94. Failure to take this into account may lead to appreciable error in lithium determination. Some manufacturers report the atomic weight of the lithium in their lithium chemicals.

Sodium and potassium are ubiquitous and should always be determined during any complete analysis.

Rubidium tends to occur with potassium. The most common silicate mineral containing major rubidium is lepidolite, which normally carries about 2% of Rb_2O, but most potassium minerals contain traces of rubidium.

A typical lepidolite analysis follows (analyst, R. B. Ellestad, University of Minnesota). This analysis is interesting because it was performed before the wide application of flame photometry in alkali metal determination; the potassium, rubidium, and cesium were separated chemically by fractional crystallization of their sulfates, chlorides, and chloroplatinates. Many years later, repetition of the analysis using modern instrumentation yielded almost identical values for all the alkalies.

SiO_2	49.19
Al_2O_3	24.81
Fe_2O_3	0.24 (total Fe as Fe_2O_3)
MnO	2.51
MgO	0.05
CaO	Trace
Li_2O	5.10
Na_2O	0.52
K_2O	10.25
Rb_2O	1.78
Cs_2O	0.19
H_2O	1.21
F	6.89
TiO_2	0.08
	102.82
Less O for F	2.90
	99.92

Cesium occurs in pollucite, $2Cs_2O \cdot 2Al_2O_3 \cdot 9SiO_2 \cdot H_2O$. It is present in most rocks at the part-per-million level and below. During analysis of lithium-bearing materials, cesium should be sought in the filtrate from the oxalate precipitation along with the lithium; a rubidium determination at this stage should be regarded with some suspicion because rubidium may have been introduced with reagents during the analysis.

4.1.1. Primary Methods

Satisfactory determination of the alkali metals by any method requires the use of reagents and distilled water that are essentially free from these elements. Blank runs should always be made, and the blanks must be small and reproducible. The necessary chemicals should be purchased in adequate quantity, thoroughly mixed before using, and examined for their alkali content. Distilled water should be tested by evaporating a large volume and using the flame photometer. Glassware, silica ware, and platinum must be thoroughly clean. Chromic acid cleaning solution should be prepared using chromic acid, CrO_3, not potassium dichromate. Each filter paper should be washed first with dilute acid and then with hot distilled water before use.

Lithium and cesium may be determined during the course of a systematic analysis, but the impossibility of obtaining reagents sufficiently free from sodium, potassium, and rubidium precludes this device for those elements; they are always determined on a separate sample by one or another of three basic primary methods

1. Gravimetrically, with flame photometric support, after separation by the J. L. Smith procedure.
2. Flame photometrically, after separation by the Berzelius procedure or a modification of it.
3. By isotope dilution mass spectrometry after chemical or ion exchange separation.

The J. L. Smith Separation

The most widely applicable and generally most reliable method for preliminary separation of the alkali metals from other constituents of a sample is still that of J. L. Smith. In this method, the finely ground sample is sintered with a mixture of calcium carbonate and ammonium chloride. The sinter is leached with water, ammonium carbonate is added to precipitate the calcium, and the alkalies are obtained in chloride solution, free from most other elements.

PROCEDURE. Transfer 0.5000 g air-dried or otherwise adequately prepared sample to a large, smooth agate mortar (one with a cavity 10 cm in diameter and 3 cm deep is a minimum size for convenient use), and grind slowly and carefully until no grit can be felt with the pestle. Note that pregrinding the sample may result in loss of the sampling weight baseline and is usually counterindicated. If it seems desirable to pregrind the material, as in the case of biotites or other micas, enough should be taken so that a baseline can be established by a determination of water, iron, or some other constituent.

To the finely ground sample in the mortar add 0.5 g ammonium chloride, and mix thoroughly with the sample by gentle rubbing with the pestle. The ammonium chloride is conveniently weighed out on the same weighing scoop or watch glass used to weigh the sample; it can then be used to recover the small amount of material adhering to the scoop. Add to the mixture in the mortar, a little at a time and mixing well with each addition, about four-fifths of 5 g calcium carbonate (specially prepared, alkali free).

Place a little of the remaining carbonate in the bottom of a J. L. Smith crucible (Fig. 4.1), and then add the contents of the mortar, transferring it as completely as possible using a flexible spatula. Rinse the mortar by rubbing it with small portions of the remaining calcium carbonate, and transfer the rinsings to the crucible. It is convenient to have a small aluminum block with a machined hole about 1 in. deep to act as a support for the crucible during this operation.

When all the calcium carbonate has been transferred to the crucible, cover it, and tap gently to pack the mixture. The charge should be compacted in the lower two-fifths of the crucible. Reserve the mortar, which may still retain a small proportion of the sample.

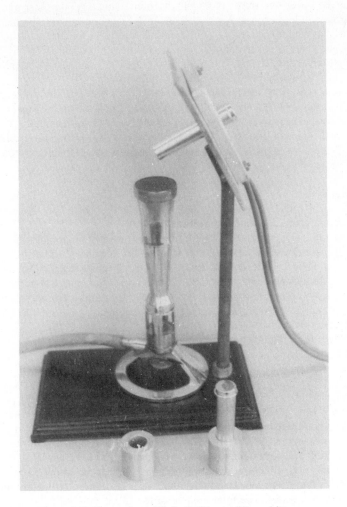

Figure 4.1. Illustration of J. L. Smith crucible and its use.

Support the crucible at an angle of about 45° in a special condensor (Fig. 4.1) consisting of an aluminum plate drilled, insulated, and plugged as shown through which coolant water can be passed. Alternatively, mount the crucible in a perforated asbestos board in such a way that the bottom half of the crucible can be heated to redness while the top remains relatively cool. Extensive investigation of this procedure has shown that, while potassium can be quantitatively retained without use of the water-cooled condensor, it is almost impossible to avoid losses of rubidium and especially cesium.

Heat the crucible with a very small flame at first (as with the crucible 3 in. above a $\frac{3}{4}$-in. flame) until the odor of ammonia is no longer detectable. Slowly increase the flame over a period of about an hour until the crucible is red hot over the lower two-thirds of its length; then, using a good Meker burner, heat to bright redness for an hour. It is essential that the crucible be heated very slowly at first because too rapid heating results in expulsion of ammonium chloride, with a corresponding loss of calcium chloride formed and hence in the efficiency of the flux. If heating is too rapid, the gases evolved (NH_3, CO_2, H_2O) may even blow the top off the crucible.

Alkali chlorides are appreciably volatile, and it is essential to keep the top of the crucible relatively cool. The condensor described assures that if a simple asbestos board is used, tests should be made to discover the heating conditions that will achieve attack of the sample without loss of alkali. A difficulty in using the condensor is that water originating in the sample may condense in the upper part of the crucible and run back into the heated zone, causing a burst of steam that may blow the top off the crucible; this can be avoided by delaying the water flow through the condensor block until the latter is hot to the touch, then allowing the whole assembly to cool before connecting the water lines and continuing the heating.

At the end of the heating period, allow the whole assembly to cool completely. Remove the crucible and stand it upright in its holder or in a small beaker. Take off the cover, and rinse out the small deposit of salts nearly always found inside it, collecting the washings in a 250-ml beaker. Add *cold* water to the crucible to about 1 cm from the top, and allow about an hour to convert calcium oxide to hydroxide. The small fusion cake should disintegrate to give a thick paste.

Wash the contents of the crucible into the 250-ml beaker with hot water using a total volume of about 100 ml. Reserve the crucible and lid if a second fusion and leach is contemplated.

Heat the lime suspension to boiling. Boil for several minutes and filter through an 11-cm S & S black ribbon paper that has been thoroughly prewashed with hot water, leaving as much as possible of the solid material in the beaker. Wash once by decantation, and wash the paper, especially around the upper edge, with numerous small portions of hot water. Catch the filtrate in a 600-ml beaker. To the residue in the 250-ml beaker, add a further 80–100 ml water and again boil, filter, and wash superficially by decantation. Repeat the operation a third time, finally transferring all the insolubles to the filter and washing with hot water. The total volume of the filtrate should be about 400 ml.

The purpose of the repeated boiling of the insoluble residue with water is to maintain a high pH so that magnesium will remain quantitatively in the precipitate. Magnesium hydroxide is insoluble in water saturated with calcium hydroxide but appreciably soluble in pure water.

Reserve the paper and precipitate for retreatment.

To the filtrate add 5 ml concentrated ammonia solution (prepared by dissolving pure NH_3 in water). Heat nearly to boiling, and add 5 ml saturated ammonium carbonate solution. Boil, allow to stand for several minutes to allow the precipitate to age and settle, and test the supernate with a drop of ammonium carbonate solution to be sure that calcium carbonate is all precipitated. Filter through an 11-cm S & S white ribbon paper, and catch the filtrate in a large (500-ml) silica dish. Wash thoroughly with *freshly boiled* hot water. Calcium carbonate is appreciably soluble in water containing carbon dioxide.

Put the two filter papers and their contents in the 250-ml beaker, and dry in the oven at 90–100 °C for several hours (overnight).

Evaporate the contents of the silica dish to dryness on the steam bath, covering the dish with a watch glass supported on a glass triangle. When the residue seems dry, remove the triangle, and continue heating, covered, on the steam bath for 1–2 h or more. Transfer the dish to a radiator or a sand bath and heat gently at first and then more strongly. Finally heat the whole surface of the dish and cover with a soft free flame to volatilize all ammonium salts. Care must be taken not to overheat; otherwise alkali salts (especially cesium) may be lost. Overheating is immediately detectable because minute amounts of alkali (especially sodium) give a flame color.

Transfer the contents of the two filter papers to the same agate mortar used to grind the sample. If the papers are handled carefully and were dried at 100 °C or less, they can be reused. Add to the residues in the mortar 0.5 g ammonium chloride, mix very thoroughly with the pestle, and transfer the mixture to the same Smith crucible that was used in the initial attack of the sample. Repeat the heating, leaching, filtration, precipitation with ammonium carbonate, and so on, exactly as before. The solution may be collected in the silica dish containing the major part of the alkalies; however, it is better to collect the second leach separately. This makes loss of alkalies less likely and also permits an estimation of the completeness of recovery.

Add to the alkali chlorides in the silica dish about 5 ml water, 2 drops of ammonia solution, and 2 drops of saturated ammonium carbonate solution. Digest on the steam bath for a few minutes, and filter through a 5-cm S & S blue ribbon or Whatman No. 44 paper, catching the filtrate in a 55-ml flat-bottom platinum dish. Wash the silica dish and the filter with a minimum of hot water. Cover the platinum dish with a silica glass cover raised on a silica glass triangle and evaporate to dryness on the steam bath. If these operations are carried out while the second fusion of the sample is in progress, the same silica dish, filter, and platinum dish may be used during its treatment. This makes complete recovery of the alkalies more certain and also helps diminish the introduction of extraneous alkalies.

When the alkali chlorides in the platinum dish are thoroughly dry, transfer the dish to an asbestos-covered wire gauze, and heat very slowly and cautiously with a small Bunsen flame, with the watch glass tightly covering the dish. Ammonium salts will sublime and collect on the cover. Remove the cover with a pair of platinum-tipped tongs, and heat it in a free flame to remove the salts. Put the cover back on the dish. Repeat the operation until no more ammonium salts are evolved. Finally pick up the dish with the tongs, leaving the cover in place, and heat very carefully in a small flame, keeping the dish in continuous motion to prevent local overheating. Great care is necessary in this operation to avoid loss of alkali chlorides, which are appreciably volatile below their melting points. If the dish is kept in a slow rotary motion, with the flame directed at the lip of the dish for an instant at each rotation, the heating can be discontinued at the instant a flame color is observed.

When ammonium salts have been removed, cool the dish, and add to it about 3 ml water, 1 drop of ammonia solution, and 1 drop of ammonium carbonate solution. Heat briefly on the steam bath to age the small precipitate of calcium carbonate. Filter through a 5.5-cm prewashed Whatman No. 44 paper, catching the filtrate in a weighed 55-ml platinum dish. Wash the paper thoroughly with a minimum of freshly boiled hot water. Evaporate the solution as before, dry, and heat over a small flame as before to remove ammonium salts. Weigh the dish plus alkali chlorides. Repeat the heating over a small flame until a constant weight is obtained.

Lithium chloride is appreciably hygroscopic, the other chlorides less so. During the final weighings of the chlorides, the heated dish should be allowed to cool in a desiccator over magnesium perchlorate. Evacuation of the desiccator is advisable prior to the final weighing. It is very important that the dish and desiccator attain the temperature of the balance room; this usually requires a cooling time of about an hour. Weighing must be done quickly, especially when lithium is present.

In the course of the above operations, it is usually possible to detect the presence of boron, molybdenum, and some other less common rock and mineral constituents. This is of considerable value during the analysis of unknown minerals and provides a good reason for determining the alkali metals before proceeding with the analysis proper. Boron gives a green flame during volatilization of ammonium salts from the watch glass. Molybdenum is almost always partly reduced by traces of organic matter and gives a blue color at one or another stage of the process. If boron is present, the mixed chlorides should be evaporated alternately with hydrochloric acid and methyl alcohol. If molybdenum is present, it may be removed with hydrogen sulfide or determined colorimetrically at an appropriate stage of the separation process.

During the heating of the chlorides, carbon from filter paper may cause a discoloration. This can usually be removed by careful and prolonged heating below the melting point of the chlorides; if excessive, it may be necessary to redissolve the chlorides and remove the carbon by filtration. If sulfate is present, it may be removed at this stage by adding a drop of 10% barium chloride solution.

If the J. L. Smith separation is being used for the first time, it is advisable to take known amounts of alkali chlorides and heat and weigh them repeatedly in order to obtain a feel for the operation. The testing of lithium and cesium chlorides in this way is especially instructive.

Any ammonium salts left with the alkalies, besides giving a high chloride weight, will cause error during the separation process.

Almost all rocks and minerals yield to the J. L. Smith attack, and recovery of sodium, potassium, rubidium, and cesium is 100% complete if the residue from the first fusion is reworked as described. A single fusion and leach will usually recover 95–98%. Since the only purpose in using this lengthy and rather difficult method is to obtain values of high accuracy on a few samples that can be used as calibrating standards, it seems obvious that every possible refinement should be employed. An alternative to the second fusion and leach is to thoroughly dry and mix the insolubles from the first fusion and leach, weigh them, fuse a portion with lithium metaborate, dissolve in nitric acid, and determine the alkali content by flame photometry. Application of this procedure to the insolubles from the *second* leach is also desirable, especially with unknown minerals. The author has never found more than 0.1 mg total alkalies in residues from the second fusion and leach except in the case of minerals containing much lithium. Lithium is incompletely recovered from minerals containing aluminum: Presumably it is partially retained in the solids as lithium aluminate.

Separation of Alkalies One from the Other

PROCEDURE. Wash the weighed chlorides into a porcelain or (better) a silica glass evaporating dish of 100 ml capacity with water. Dry, ignite, and weigh the platinum dish, and reserve it for the chloroplatinate precipitate. Use the empty weight to calculate the weight of chlorides; it should differ from the original weight by no more than 0.1 mg. An insoluble deposit on the dish or a cloudy solution of the chlorides may indicate the presence of silica or tungsten; the chloride solution should be filtered before proceeding.

To the solution in the evaporating dish add a measured volume of chloroplatinic acid solution calculated on the basis that all the chlorides are of sodium (or lithium, in the case of lithium minerals). This will ensure an excess. During this operation and the subsequent evaporation, ammonia must be absent; absorption of NH_3 from the air will result in precipitation of ammonium chloroplatinate and a false weight of the potassium group chloroplatinates.

Evaporate the solution on the steam bath *nearly* to dryness, stirring frequently to prevent local drying. Remove from the heat just before the yellow mass goes completely dry. Cool. Add about 0.5 ml of 80% (v/v) alcohol, conveniently from a small all-glass wash bottle, and triturate with a short, thick glass rod. Filter through a 5.5-cm Whatman No. 44 paper into a 50-ml beaker, retaining as much as possible of the precipitate in the dish. Repeat the trituration using 1-ml portions of 80% alcohol, and filter. Continue, keeping as much as possible of the residue in the dish, until the alcohol no longer shows a yellow color, indicating that the lithium and sodium chloroplatinates are all dissolved. Wash the paper free from soluble chloroplatinates with the 80% alcohol. The precipitate of potassium (rubidium, cesium) chloroplatinate has a golden color quite distinctly different from the orange color of the soluble sodium chloroplatinate.

The specific gravity of the alcoholic wash must be close to 0.85: It is prepared by putting exactly 40 ml water in a 200-ml graduate and filling to the 200-ml mark with *absolute* alcohol. Denatured alcohol should not be used. If there is doubt concerning the water content of the reagent alcohol, a hydrometer should be used to check the specific gravity of the diluted alcohol.

Evaporate the filtrate from the potassium chloroplatinate nearly to dryness on the steam bath. Dry the dish containing the potassium chloroplatinate and the filter paper until the odor of alcohol is no longer apparent. Wash the paper with hot water, catching the filtrate in the same platinum dish used to weigh the chlorides, and wash the chloroplatinates from the evaporating dish into it as well. Cover with a triangle and a watch glass, and evaporate to dryness on the steam bath. Transfer to an oven and heat at 135 °C for 1 h. Cool in a desiccator, and weigh potassium (rubidium, cesium) chloroplatinates.

During the drying, the watch glass should be left on the dish because decrepitation sometimes occurs. It is a good practice to cover the dish with a large watch glass, inverted, and to stand the dish on another watch glass. On removal from the oven, the watch glasses may be examined closely for specks of yellow-gold chloroplatinates.

Dissolve the chloroplatinates in hot water, transfer to a 100-ml beaker, and proceed with the removal of platinum and the conversion to sulfates as described below.

To obtain satisfactory results in the chloroplatinate separation, the reagent must be properly prepared. It should be checked before use for its sodium, potassium, and ammonium content, and a known weight of potassium chloride should be converted to chloroplatinate and weighed to be sure that the conversion is stoichiometric. The theoretical factor for conversion of chloroplatinate to oxide usually gives a weight of oxide about two parts in a thousand too low or too high; a discrepancy much greater than this should not be tolerated. Chloroplatinic acid is usually purchased in the form of a 10% solution. Its platinum content is best determined by precipitating with formate as in the following procedure and weighing.

Conversion of Chloroplatinates to Sulfates

PROCEDURE. Evaporate the solution of either sodium or potassium chloroplatinate to dryness on the steam bath. Do not heat longer than necessary. Add enough water to dissolve the precipitate, and heat to solution. Add 2–6 drops of ammonium formate solution prepared by neutralizing formic acid (90%) with ammonia. Heat, covered, until all the platinum is precipitated and the solution is water white. Then, but not before, add a single drop of 1:1 sulfuric acid, and evaporate to dryness on the steam bath.

It is most important to have all the chloroplatinate in solution before adding the formate; otherwise, recovery of alkalies will not be complete. If large amounts of rubidium or cesium are present, a relatively large volume of hot water may be required to put them in solution.

Digest the residue with 5–10 ml water, filter off the platinum, and wash very thoroughly with hot water, catching the filtrate in a weighed 55-ml flat-bottom platinum dish. Evaporate to dryness on the steam bath, and then heat gently over a small flame to remove ammonium salts, using a technique similar to that used in igniting the chlorides. It is not necessary to remove ammonium salts completely; their partial removal simplifies the following operation. Heating must be cautious because some of the alkalies may be present as hydroxides, which may volatilize if overheated.

Add sufficient sulfuric acid to convert all the alkalies to sulfates and provide a *slight* excess. It is well worth the trouble to calculate the addition.

Heat gradually on an aluminum hot plate; fume off excess acid, and slowly increase the temperature until bisulfates and pyrosulfates are decomposed. Finally ignite for 1 min in an oxidizing Meker flame. Weigh alkali sulfates. Repeat the ignition to constant weight.

With rubidium and cesium, there may be difficulty in removing excess SO_3: This may be overcome by adding to the partly ignited sulfates a drop of ammonium carbonate solution, drying, and igniting. Ignition in an open dish over a Meker burner should not be prolonged because of the appreciable volatility of the alkalies. The danger of loss by volatilization can be diminished by covering the platinum dish with a silica-glass cover during ignition: this not only increases the effectiveness of the heating by making a higher temperature possible inside the dish but also diminishes air circulation and therefore losses by sublimation.

It is of extreme importance to be sure that the Meker flame is oxidizing at all times. A reducing flame will reduce sulfates to sulfides, and losses by volatilization will occur.

Heating of alkali sulfates in a muffle furnace is not to be considered.

When the sulfates of both sodium and potassium have been weighed, they should be dissolved in a minimum of water. The solution should be filtered through a small paper into a beaker of suitable capacity (see below), and the paper should be washed with hot water very thoroughly. The paper should then be incinerated in the same dish that contained the sulfates, and the weight of dish plus the small amount of silica, platinum, and so on which is always present should be used in calculating the weight of sulfates.

Flame Photometric Determination of Minor Alkalies

Flame photometry is usually applied as a secondary technique: To use it in the primary mode, it is necessary to have the elements sought in pure solution so that there are no unknown interferences. The sulfate solutions prepared as described above meet this requirement.

In the usual case, flame photometric determination of lithium and potassium in the sodium sulfate fraction and of sodium, rubidium, and cesium in the potassium sulfate fraction yields highly accurate values for all the alkalies. With lithium-bearing minerals, flame photometric determination of lithium in the residues from the J. L. Smith separation is also required—or lithium must be determined on a separate portion of the sample.

The wide variety of flame photometers on the market makes it difficult to write a general procedure; however, certain essentials are common to all. Of particular consequence is the very large enhancement effect of potassium on the flame radiation of rubidium and cesium. This effect makes it essential that the potassium content of solutions in which rubidium and cesium are to be determined be closely controlled. The easiest way to do this is to dilute the solution of the weighed sulfates with water in such a way that an optimum concentration is achieved. Standard solutions of neutral sulfates may then be prepared to exactly match the unknowns.

Values obtained for lithium after the J. L. Smith separation are almost always low. Determination of lithium in the sodium fraction is nevertheless necessary so that the weight of sodium sulfate may be corrected. There is always a small amount of potassium in the sodium fraction and of sodium in the potassium fraction. Even with a lepidolite containing 7% Li_2O, less than a milligram of lithium was detected in the potassium sulfates.

Example. Recoveries of Li_2O in the J. L. Smith separation (double leach):

Sample	Li_2O Present	Li_2O Found
Biotite	0.26	0.22
Triphilite–lithiophilite	9.32	8.57
Riebeckite	0.44	0.40
Pegmatite	1.49	1.43
Lepidolite	5.37	5.17

Lead Carbonate Separation of Lithium

Lithium is almost always determined on a separate sample in rock and mineral analysis. If more than traces are present, it is necessary to determine it in the mixed chlorides or sulfates in order to correct a gravimetric sodium determination, but this value should never be reported, for it is almost certain to be low.

Of the many procedures devised for the separation of lithium prior to its primary determination, that using lead carbonate is the most reliable.

PROCEDURE. Weigh 0.2500 g (or more or less, depending on the sample) of the finely divided sample (preferably −200 mesh), into a 55-ml flat-bottom platinum dish. Cover with a silica watch glass, add a little water and 0.5 ml of 1:1 sulfuric acid. If CO_2 is evolved, heat gently with the dish covered until it is removed. If organic matter is present, add a drop of nitric acid. Cool, wash off and remove the cover, and add about 10 ml concentrated hydrofluoric acid. Heat slowly on an aluminum hot plate, stirring from time to time with a platinum rod. When the residue is nearly dry, increase the temperature until strong fumes of sulfuric acid appear. Cool, wash down with water, add a further 0.5 ml of 1:1 sulfuric acid, and again evaporate to fumes. If a clear solution is not obtained on the addition of water and gentle heating, it is probable that some undecomposed magnesium or calcium fluoride remains, and a third evaporation to fumes of sulfuric acid may be required.

Finally, evaporate and heat until free sulfuric acid is all removed.

Cool, take up the residue with water, and digest on the steam bath for some time to completely dissolve all soluble salts. Transfer solution and residue to a 200-ml high-form beaker, dilute to about 80 ml, cover, and heat to boiling, putting a piece of hardened filter paper under the end of the stirring rod to prevent bumping. Add a drop of methyl red indicator and then solid basic lead carbonate little by little to the boiling solution, keeping the cover on as much as possible to prevent loss, until the indicator color changes to pure yellow. Allow the precipitate to settle, and filter. Wash thoroughly with hot water. Collect the filtrate in a 250-ml beaker, and add about 50 mg lead carbonate to neutralize any remaining acid.

Evaporate the solution on the steam bath to 20–30 ml, and filter into a 50-ml volumetric flask. Cool, and dilute to the mark with water.

The solution contains substantially all the alkalies in the sample as well as some magnesium and a little lead. Attempts to determine sodium and potassium in this solution have lead to poor results, but lithium can be accurately determined if proper precautions are taken. Two methods are recommended:

1. Method of Bracketing Standards. Make a rough preliminary determination of lithium, sodium, and potassium in the solution from the lead carbonate separation using the flame photometer. Measure the magnesium content of the solution by AA spectrometry or by way of an EDTA titration of a small aliquot. Starting from pure solution with known concentrations of the several elements, prepare a mixed solution with slightly more and another with slightly less lithium than the estimate and with approximately the same concentration of magnesium, sodium, and potassium. Compare the signals generated by these solutions with that generated by the unknown.

2. Method of Standard Additions. Take aliquots of the unknown solution, and add to each a known amount of lithium. Dilute each to the same volume, and register the signals developed by these solutions. Plot added lithium against readout, and extrapolate to obtain the lithium content of the unknown.

It is probably best to use both methods when the concentration of lithium is high; in order to obtain results good to 1 part in 1000, the capabilities of the flame photometer must be extended to the limit. With lithium minerals, it is well to analyze the residues from the J. L. Smith separation for lithium using the lead carbonate–flame photometry method, and correct the weight of lithium–sodium sulfates. If the value so obtained agrees with the value obtained by the above procedure, there is some confidence in it.

By increasing the sample weight to 1 g, and decreasing the final solution volume to 25 ml, the lead carbonate–flame photometry method can determine as little as 0.0002% Li_2O with certainty.

Table 4.1 shows the weights and calculations involved in the determination of the alkali metals in a lepidolite.

Modified Berzelius Method for Isolating Alkali Metals

The J. L. Smith method for isolating the alkalies is usefully applied when the concentration of at least one of the alkali metals exceeds about 1%. With minerals containing less than this, time can be saved by using an HF attack and determining all the alkalies by flame photometry of their neutral sulfate solutions. The procedure is similar to that described for the separation of lithium; however, it is necessary to add magnesium to the initial mixture if all the sodium and potassium are to be recovered. If rubidium and cesium are to be determined, it may be necessary to start two portions for each sample, one of 50 mg for sodium and potassium and one of 500 mg for cesium and rubidium. The procedure that follows does not recover all the lithium in the sample.

PROCEDURE. Weigh 0.5000 g (for rubidium and cesium) and 0.0500 g (for sodium and potassium) of the suitably prepared sample into 55-ml flat-bottom platinum dishes. With most materials, even "pure" minerals, preliminary grinding to pass a 200-mesh screen is necessary if sampling error is to be avoided. Other sample weights may of course be used; but it is best to keep sample weight as small as practicable.

Cover the dish with a silica-glass cover, add a little water and 1 ml nitric acid, and warm gently to remove carbonate. Wash down with water and add 1 ml of 1:1 sulfuric acid and 10–15 ml hydrofluoric acid. Heat (uncovered) on

Table 4.1. Specimen Alkali Determination[a]

Chlorides	Chloroplatinates	Lithium and Sodium Sulfates	Potassium Sulfates
28.2594	28.3793	28.1907	28.2628
28.2590	28.3792	28.1903	28.3700
28.2586	28.0904	28.1904	28.3703
28.2586	0.2888	28.0907	28.3703
28.0902	−0.00099 Li	−0.0997	28.2635
0.1684	−0.00154 Na	−0.00453 Na	0.1068
−0.0780 K	−0.03096 Rb	−0.00030 K	−0.00026 Li
−0.00025 K	−0.00086 Cs	0.0949 Li$_2$SO$_4$	−0.00048 Na
−0.00040 Na	.2544 K$_2$PtCl$_6$		−0.01428 Rb
−0.00372 Na		=5.16 Li$_2$O	−0.00046 Cs
−0.01294 Rb	=9.86 K$_2$O	+0.01 in K$_2$SO$_4$	0.0913 K$_2$SO$_4$
−0.00042 Cs	−0.04 Blank		
0.0727 LiCl	−0.02 Blank	5.17 Li$_2$O	=9.87 K$_2$O
=5.12 Li$_2$O	+0.03 In Na$_2$SO$_4$		−0.04 Blank
	9.83 K$_2$O		−0.02 Blank
			+0.03 In Na$_2$SO$_4$
			9.84 K$_2$O

	Percent Oxide	Corresponding Weight of		
		Chloride	Sulfate	Chloroplatinate
Flame Determinations				
In sodium sulfates				
Li$_2$O	(5.06)	—	—	—
Na$_2$O	0.395	0.00372	0.00453	—
K$_2$O	0.032	0.00025	0.00030	—
In potassium sulfates				
Li$_2$O	0.014	0.00020	0.00026	0.00099
Na$_2$O	0.042	0.00040	0.00048	0.00154
K$_2$O	(9.8)			
Rb$_2$O	2.00	0.01294	0.01428	0.03096
Cs$_2$O	0.072	0.00042	0.00046	0.00086

Blanks

Fusion and leach: 0.07% Na$_2$O, 0.04% K$_2$O on 0.5 g

Chloroplatinic acid: 0.02% K$_2$O, 0.01% Na$_2$O/10 ml on 0.5-g sample

	Corrected Results	
Li$_2$O	5.17	(from J. L. Smith determination)
	5.37	(PbCO$_3$–flame photometer)
Na$_2$O	0.33	
K$_2$O	9.84	
Rb$_2$O	2.00	
Cs$_2$O	0.07	

[a] Lepidolite, Brown Derby Mine, Gunnison County, Colorado; 0.5000 g sample, undried. Moisture 0.04% (105°). Double leach.

327

an aluminum hot plate in such a way that no boiling occurs. Gradually increase plate temperature until fumes of sulfuric acid appear. Cool, dilute with 10–15 ml water, and warm gently until soluble salts are in solution, stirring with a platinum rod. If sample attack is incomplete, repeat the hydrofluoric acid addition and again evaporate to fumes and dilute.

Alternately evaporate to fumes and dilute until a clear solution is obtained; then add 2 ml of a magnesium sulfate solution (alkali free) containing 25 mg/ml MgO; or, if the magnesium content of the sample is known, add just enough magnesium sulfate to make a total of 50 mg MgO. Evaporate to fumes; then gradually increase plate temperature until excess acid has been removed and bisulfates and pyrosulfates are decomposed. Transfer the dish to a platinum triangle, and heat over a flame, cautiously at first and finally at the full heat of a good Meker burner for 1 min, *making certain that the flame is strongly oxidizing*. The dish should be covered with the silica-glass cover during this operation. If a reducing flame is used, sulfate will be reduced to sulfide and alkali metals will be lost by volatilization. Cool the residue, add 15–20 ml water, and digest on the steam bath for 30 min or more, stirring occasionally and adding more water if necessary. Up to this point the 50- and 500-mg samples are treated similarly.

Evaporate the suspension of the 50-mg sample to about 5 ml, and filter through a prewashed 5.5-cm Whatman No. 44 paper into a 50-ml volumetric flask. Wash the insoluble material thoroughly with hot water. Recovery of the alkali metals (except lithium) is at least 99% in most cases, but it may be wise to ignite the insoluble matter to remove paper, fuse with lithium metaborate, and dissolve the melt in dilute nitric acid; the resulting solution should show no sodium or potassium by flame photometry.

Filter the suspension of the 500-mg sample into a 150-ml beaker, breaking up the flakes of oxide with a rubber-tipped rod. Wash very thoroughly with hot water to a volume of 100 ml or more. Add about 50 mg precipitated calcium carbonate to the filtrate, and evaporate to a small volume on the steam bath. Filter through a small No. 44 paper into a 50-ml volumetric flask; wash with hot water, leaving enough room in the flask for later additions of potassium or sodium sulfate.

The pH of the final solutions should be about 6.0.

Examine the solutions with the flame photometer. Because of the wide variety of instruments available, a general procedure cannot be written; however, certain principles must be observed.

Flame Photometry of Alkali Metals

The sensitive lines available for flame photometry of the alkali metals are lithium, 671 nm; sodium, 589 nm, potassium, 766 and 770 nm; rubidium, 780 and 795 nm; cesium, 852 nm. If the available instrument cannot resolve the potassium and rubidium lines, accurate determination of these two elements is best accomplished using an AA spectrometer, although atomic absorption measurement of the alkalies is generally less precise and more subject to interferences than flame emission.

Emission from rubidium and cesium is enormously enhanced by potassium; accurate flame measurement of these metals requires that this enhancement be taken into account. In addition, most flame photometers do not completely resolve the potassium and rubidium lines. Solutions prepared as described above contain, in addition to the alkali metals, magnesium, calcium, and some manganese in the sample. These generally have a depressing effect on the emission from the alkalies (although calcium may lead to high values for sodium under some circumstances). Best results are obtained using a "cool" flame—air–natural gas or air–propane. In such a flame, emission from other elements is negligible; in hotter flames, atoms of, for example, calcium and magnesium may be excited, and their emission may not be excluded by the trivial monochromator usually used in flame photometers.

These difficulties are resolved in the following manner:

PROCEDURE. Take a small aliquot (e.g. 5 ml) of the solution of the 50-mg sample, dilute to 50 ml, and make a preliminary reading using pure sodium and potassium sulfate solutions to calibrate. If large amounts of lithium, rubidium, and cesium are present in the sample, they may be detectable at this stage. Prepare standard solutions containing calcium, magnesium, lithium, and so on to match the unknown solution as closely as possible, and arrive at a refined estimate of the sodium and potassium content of the sample.

To the solution of the 500-mg sample add enough standard sodium and potassium solution to bring the concentration of these elements to a standard value (usually 1000 ppm). Dilute to exactly 50 ml and mix. Make preliminary measurements of the rubidium and cesium content of the solution, and prepare bracketing standards containing all relevant elements. If the magnesium content of the samples is unknown, it may be desirable to determine it by AA spectrometry or by EDTA titration on an aliquot of the solution so that standard solutions containing exactly the right concentration of magnesium can be prepared.

Depression of potassium emission by magnesium is demonstrated by the following data, obtained using an air–natural gas flame. Solutions containing 1000 ppm K_2O and various concentrations of magnesium were flamed under identical conditions:

ppm MgO in Solution	Apparent ppm K_2O
500	1030
1000	1000 (standard)
1500	980
2000	960
5000	925

A major difficulty in determining the rarer alkalies by flame is that of obtaining pure salts, especially of rubidium and cesium. All standard solutions should be examined by any available method (mass spectrometry, emission spectroscopy, etc.) for impurities. The accurate determination of traces of rubidium in potassium by flame is a matter of extreme difficulty when no rubidium-free potassium salt is available. Small amounts of pure potassium salt may be prepared from "spec-pure," or reagent, potassium chloride by adding to a concentrated solution enough chloroplatinic acid to precipitate about one-tenth of the potassium. Rubidium and cesium chloroplatinates are less soluble in water than potassium chloroplatinate and are preferentially precipitated. Platinum can be removed from the purified solution using formate and the potassium converted to sulfate as described previously. Once a small amount of pure alkali sulfate has been prepared, it is a simple matter to find exactly how great is the contamination of the spec-pure, or reagent, salt in routine use and to correct all results accordingly.

Besides the generally greater freedom from interelement effects in flame emission (using a cool flame), an advantage of this technique over AA spectrometry is the very wide range of concentrations over which good results may be obtained. A further advantage lies in the possibility of using an internal standard; lithium is commonly used. It is worth noting that the excitation characteristics of lithium in the flame are quite different from those of potassium, rubidium, and cesium; this does not, as has been suggested, diminish the usefulness of lithium as an internal standard. The term *internal standard* is really a misnomer; the purpose of the lithium addition is to cancel out changes in atomization rate by ratioing the signal from the element sought to that from the constant concentration of lithium and to diminish the importance of accuracy in measuring aliquots. Thus, a solution containing a measured amount of added lithium may be diluted 1:10 or 1:20 with little change in the measured ratios.

If lithium internal standard is used and all precautions are taken, flame photometric determination of the other alkalies can be made precise and accurate to almost 1 part in 1000. Accuracy depends, of course, on the care with which standard solutions of the alkalies are prepared and on the availability of pure alkali salts.

Preparation of Standard Solutions of Alkali Sulfates

PROCEDURE. Weigh a suitable quantity of alkali sulfate that has been examined for its foreign alkali content, and dissolve in water. Two liters of sodium and potassium sulfate solutions containing 10,000 ppm Na_2O and K_2O, respectively, are advisable. Lithium, rubidium, and cesium solutions may be prepared in smaller quantity and lower concentration. The reagent sulfates cannot be accurately weighed; the solutions must be standardized after preparation.

Heat the sulfate solution for several hours near the boiling point. Cool, filter through a large Whatman No. 44 paper into a Pyrex bottle with a ring-seal stopper. Sterilize the solution by immersing in boiling water for several hours. Cool, and standardize as follows.

Using a calibrated pipette, transfer a volume of solution that contains about 0.25 g alkali sulfate to a 55-ml flat-bottom platinum dish. Add a drop of 1:1 sulfuric acid, and evaporate without boiling on an aluminum hot plate. Gradually increase plate temperature until all excess sulfuric acid is removed and acid sulfate and pyrosulfate have been decomposed. Then heat over a Meker burner, covering the dish with a silica-glass cover, for 1 min. Repeat the ignition to constant weight. The Meker flame *must be oxidizing*—that is, it must burn with excess air.

Wash the ignited sulfate from the dish through a small filter, and wash dish and filter thoroughly with hot water. Incinerate the paper in the dish, and weigh dish plus residue. The weight should be only slightly greater than the original weight of the dish alone. Calculate the concentration of alkali in the solution.

The sulfates of lithium and sodium are easily fusible in the Meker flame and should not be heated past the point at which fusion takes place. Rubidium and cesium sulfates fuse only with difficulty (melting points: Rb_2SO_4, 1060 °C; Cs_2SO_4, 1020 °C); they must be heated with the silica-glass cover in place, and if it is difficult to achieve a constant weight, the possibility that Rb_2O and Cs_2O are being volatilized should be investigated. The usual reason for such loss is failure to maintain oxidizing conditions in the flame.

Potassium and Rubidium by Isotope Dilution Mass Spectrometry

The technique of isotope dilution mass spectrometry has been extensively applied and refined by geochronologists and probably comes as close to being a perfected primary method as any routine method can be. The advantage of the method is the avoidance of a necessity to recover all the potassium or rubidium in the sample. At the start of the analysis, a known "spike" of a separated isotope of the element sought is added to the sample. After preliminary separations, the isotopic ratios are measured, and the concentration of the element of interest can be calculated.

Disadvantages of the method are the cost of the equipment, the limited availability of separated isotopes, and the consequent necessity to use small samples and risk introducing sampling error. In the rubidium–strontium age determination, sampling error is usually of little consequence because rubidium and strontium are determined and isotopically analyzed in the same sample; only if the sample is contaminated with a rubidium-bearing mineral of a different age is the sampling problem of consequence. Potassium–argon ages are usually estimated using separate samples for potassium and for argon determination, and sampling errors are sometimes gross. Methods in which potassium and radiogenic argon are determined on the same sample have not gained wide acceptance because of numerous procedural difficulties.

After spiking the sample with isotopes of rubidium, strontium, and/or potassium, the simplest method for obtaining appropriate solutions for mass spectrometric measurement is probably as follows:

PROCEDURE. Attack the spiked sample (usually 0.2 g or less of −200-mesh material) in a platinum dish with hydrofluoric, nitric, and sulfuric acids. Evaporate to fumes, and drive off excess acid. Leach the residue with water, and filter.

For potassium, the solution may be used without further treatment, after concentration by evaporation if necessary. For rubidium, it is necessary to make a separation from strontium, this is usually done by ion exchange on Dowex 50 resin.

This simplest form of the procedure is practically applicable only to low-potassium samples because of the cost of isotopic spike. With high-potassium samples, it is necessary to quantitatively separate the potassium, obtain it in a measured volume, and take a small aliquot for spiking. This makes the isotope dilution procedure for potassium less attractive: If all the potassium must be separated, it might as well be determined by a less expensive method.

Natural potassium isotopes have abundances ^{39}K, 93.1%; ^{40}K, 0.0119%; and ^{41}K, 6.9%. A spike enriched in ^{41}K, or in both ^{40}K and ^{41}K is used in isotope dilution analysis. The double spike permits estimation of the mass discrimination of the spectrometer, which is often appreciable—sometimes amounting to several percent in the potassium mass range. If a single spike is used, some other method for measuring discrimination must be applied. The only interfering element is calcium; ^{40}Ca makes up 96.9% of natural calcium: Most of the calcium is removed during the solution procedure outlined above. If too much calcium accompanies the potassium, ion exchange separation is necessary. Most often, use of a small volume of water (2–3 ml) during the leaching of the ignited residue followed by prompt filtration will lead to a low concentration of calcium in the solution applied to the filament of the mass spectrometer.

Ion Exchange Separation of Alkalies

In most K–Ar and Rb–Sr age determinattion, Dowex 50, 8× resin is used, usually in a column about 1 cm in diameter and 15 cm long. The spiked solution is loaded on the column in a small volume of hydrochloric acid solution of normality about 0.1. Elution of the metals with 0.5 N hydrochloric acid yields potassium and rubidium in the first effluent, calcium and strontium being held much more strongly by the resin.

Inorganic ion exchangers seem to be less frequently used; but they have been shown to be extremely effective in separating the alkali metals. Zirconium tungstate gives a clean separation of lithium, sodium, potassium, rubidium, and cesium in that order when they are loaded in 0.01

NH$_4$Cl solution and eluted with ammonium chloride in increasing concentrations (0.05 M for Li, 0.1 M for Na, 0.3 M for K, 0.75 M for Rb, 4.5 M for Cs).

Gravimetric Determination of Sodium

Sodium may be determined in a variety of materials by precipitation and weighing of sodium zinc uranyl acetate. The gravimetric factor is large— 1 mg sodium yields 0.067 g precipitate. With most samples, it is necessary to prepare a solution containing all the sodium, and take a small aliquot for the determination. The following general procedure may have to be modified to suit various sample compositions.

PROCEDURE. Prepare reagents as follows. Zinc uranyl acetate: Dissolve 10 g uranyl acetate $(UO_2)(C_2H_3O_2)_2\cdot2H_2O$ in 49 g water and 6 g of 30% (v/v) acetic acid, warming gently to hasten solution. Dissolve 30 g zinc acetate $Zn(C_2H_3O_2)_2\cdot3H_2O$ in 32 g water and 3 g of 30% (v/v) acetic acid, warming gently to hasten solution. Mix the two solutions, and if no precipitate appears, add about 1 mg sodium chloride dissolved in a drop of water. Stir, allow to stand for at least 24 h, and filter before using. Utilize the yellow precipitate to prepare an alcoholic wash solution. Shake or stir the precipitate with *absolute* alcohol to make a saturated solution. Allow to settle and decant. The alcoholic wash must be prepared within 24 h of use.

Transfer 0.1000 g properly prepared sample to a platinum dish and treat with a little water, 0.5 ml nitric acid, 0.5 ml of 72% perchloric acid, and 10–15 ml hydrofluoric acid. Evaporate to fumes of perchloric acid, cool, wash down the sides of the dish with water, add a further 0.5 ml perchloric acid, and evaporate slowly to dryness.

Cool, add no more than 5 ml water, and filter through a Whatman No. 44 paper that has been prewashed with dilute acid and water. Collect the filtrate in a small beaker (or in a volumetric flask if an aliquot is to be taken). Wash the platinum dish and the filter thoroughly with hot water. Evaporate the solution to 4–5 ml, cool, add 20 ml reagent, mix, and allow to stand for 30–40 min. Filter through a weighed fritted glass crucible of medium porosity. Wash once with 2 ml of the reagent, transferring the precipitate to the filter, then with five 1-ml portions of the alcoholic wash, and finally with five 1-ml portions of *anhydrous* ethyl ether. Allow to dry without heating, and weigh. The precipitate has the composition $(UO_2)_3ZnNa(C_2H_3O_2)_9\cdot6H_2O$. The gravimetric factor to convert from sodium zinc uranyl acetate to Na$_2$O is 0.02015. This factor is so large that the filtration step in the procedure can often be omitted; the small residue after the acid attack will seldom make any appreciable difference to the result.

Lithium precipitates with the sodium, so the procedure is inapplicable to lithium-bearing materials. The composition of lithium zinc uranyl acetate is somewhat uncertain, so that correction for lithium is dubious.

Rapid Gravimetric or Titrimetric Determination of Potassium

Potassium may be precipitated as $K_nNa_{(3-n)}Co(NO_2)_6 \cdot H_2O$. The precipitate may be weighed or the nitrite it contains may be titrated. Although the proportions of sodium and potassium in the precipitate are somewhat variable, n is very close to 2, and if a procedure such as the one to be described is closely followed, analysis of known samples will yield an empirical factor by means of which acceptable values may be obtained routinely. The method is useful when many samples must be analyzed for potassium on a routine basis. Rubidium and cesium are also precipitated; the method is not to be applied when appreciable concentrations of these elements are present.

PROCEDURE. Attack the sample (usually 1 g) with hydrofluoric and sulfuric acids, adding a little magnesium sulfate as described previously, and remove excess sulfuric acid by ignition of the residue. Complete removal of sulfuric acid is not necessary, and a Teflon dish may be used if platinum is not available.

Digest the residue with 30–40 ml hot water, and add 10 ml saturated (potassium-free) sodium acetate solution. Boil for 5–10 min, and filter through an 11-cm Whatman No. 44 paper. Wash thoroughly with 2% sodium acetate solution, measuring the volume of wash used if the sodium acetate contains appreciable potassium. Evaporate the filtrate to 15–20 ml, and add 10 ml cobaltinitrite reagent. Stir, and allow to stand at room temperature for several hours. Filter through a medium-porosity glass or porcelain filter crucible. Wash once with a few milliliters of cobaltinitrite reagent, then five times with 5-ml portions of a 1% solution of the reagent in water, and then twice with cold water.

Dry at 100 °C for 1 h, cool, and weigh $K_2NaCo(NO_2)_6$.

To finish the determination titrimetrically, transfer the yellow precipitate to a 400-ml beaker with a little water, and add 100 ml of 5% sodium hydroxide solution. Boil until the precipitate is completely decomposed. Cool the solution [which will contain a black precipitate of $Co(OH)_3$] to 50 °C, and add 1:4 sulfuric acid to neutralize the alkali and sufficient excess to make the solution 0.5 N in sulfuric acid. Titrate immediately with 0.1 N potassium permanganate. If the end point is uncertain, add an excess of permanganate and back titrate with 0.1 N ferrous ammonium sulfate solution.

The titration may be made somewhat cleaner by filtering out the cobaltic hydroxide before acidification and titration; in this case, the calculations must take care of the fact that the oxidizing effect of the Co^{3+} has been removed.

From weight of precipitate,

$$\text{Percent of } K = \frac{100nwK}{Ms}$$

From titration with $KMnO_4$,

$$\text{Percent of } K = \frac{NcnK}{110s} \quad [Co(OH)_3 \text{ not removed}]$$

$$= \frac{NcnK}{120s} \quad [Co(OH)_3 \quad \text{filtered out}]$$

In these equations, M is the molecular weight of $K_nNa_{(3-n)}Co(NO_2)_6 \cdot H_2O$, $= 421.99 + 16.10n$; w is the weight of the yellow precipitate; K is the atomic weight of potassium, $= 39.10$; N is the normality of permanganate solution; c is the volume of the permanganate solution; and s is the sample weight.

In principle, if the precipitate is both weighed and titrated, it is possible to determine a value for n during each determination. In practice, this is hardly worth the trouble, and results are not very satisfying. If this is done, the potassium content of the sample is found from

$$\text{Percent of } K = \frac{242.31}{s}(w - 0.03673Nc)$$

A major difficulty with this method is that of finding sodium acetate free from potassium: it is usually necessary to carry a blank determination through the method, being careful to use exactly the same amount of sodium acetate in every case.

Sodium Cobaltinitrite Reagent

Dissolve 80 g sodium nitrite in 100 ml water; for every 100 ml of this solution add 30 g $Co(NO_3)_2 \cdot 6H_2O$ and 10 ml glacial acetic acid. Allow to stand for 24 h or more, and decant from the yellow precipitate that is almost sure to appear. The reagent is reasonably stable if kept cool and in the dark.

The stoichiometry of the reactions involved in the above procedure is interesting, and development of the equations used in calculating potassium content makes an informative exercise; the overall reaction during solution of the precipitate and the titration of nitrite may be written as follows:

$$10K_nNa_{(3-n)}Co(NO_2)_6 + 5(3 + n)Na_2SO_4 + 22KMnO_4 + 28H_2SO_4 \rightarrow$$

$$10CoSO_4 + 60NaNO_3 + 22MnSO_4 + (11 + 5n)K_2SO_4 + 28H_2O$$

4.1.2. Secondary Methods

By far the largest number of determinations of the alkali metals are made by flame photometry without any prior separations. Interferences exist but can be overcome in adequate degree for almost any type of material. When rubidium or cesium are to be determined, AA spectrometry is preferred to flame emission spectrometry because few flame photometers contain a sufficiently large monochromator to resolve potassium and rubidium emission. For sodium and potassium, the use of lithium as an internal standard improves precision but not necessarily accuracy.

The use of a low-temperature flame eliminates most interferences in flame emission because few elements generate a flame signal at the temperature of an air–propane or air–natural gas flame. Band spectra from calcium and strontium are the most troublesome; their effect can be overcome by adding phosphate to the solution.

There are two chief methods of sample attack—solution in hydrofluoric and sulfuric or perchloric acids, usually in Teflon vessels with tight-fitting covers, and fusion with lithium metaborate followed by solution of the melt in dilute acid. As with all direct secondary methods, the details of the procedure must be tailored to the type of material being examined and to the type of readout device available. The following procedures are representative and are quite widely applicable.

Lithium Metaborate–Nitric Acid Solution

PROCEDURE. Weigh 0.2 g properly prepared sample and 0.7 g lithium metaborate into a porcelain crucible. Mix thoroughly with a small metal spatula (a glass rod is undesirable because static charges cause difficulty). Transfer the mixture to a graphite crucible that has been preignited and dusted with spectrographic graphite powder. Put the crucible in a muffle heated to 950 °C. After 15 min, remove the crucible from the muffle, swirl quickly to collect the melt into a single bead, and pour it into a plastic beaker containing 100 ml of 5% nitric acid and a magnetic stir bar. Stir to solution; this should take 5 min or less.

Dilute the solution appropriately, and measure the potassium–lithium ratio and/or the sodium–lithium ratio using the flame photometer. Volumes and dilutions are not critical; accuracy depends on the weighing of the sample and of the lithium metaborate. Consequently, the use of volumetric flasks is quite unnecessary.

Hydrofluoric–Sulfuric Acid Attack

PROCEDURE. Transfer the sample (0.1–0.5 g) to a Teflon beaker and add 20 ml hydrofluoric acid. Cover tightly, and digest, with occasional swirling, for an hour or more. Add 5–10 ml of 1:1 sulfuric acid, cover, and heat on a steam bath until decomposition is complete. Remove the cover, and evaporate, gradually increasing temperature until the sulfuric acid is fuming strongly. Cool, dilute cautiously with water, and add a measured volume of lithium sulfate solution so that the final concentration of lithium meets the requirements of the flame photometer. Dilute appropriately, and read the potassium–lithium or sodium–lithium ratio.

Both the above methods are best calibrated using analyzed standards similar in composition to the unknowns. Calibration using pure solutions of the alkalies is more easily done after the acid attack; however, use of the method in the primary mode can lead to appreciable error because of interference by other elements in the sample.

If the readout is attempted using an AA spectrometer, the most severe interferences are the alkali metals themselves; since atomic absorption is seldom used in the internal standard mode, the lithium addition is best omitted. Addition of a relatively large amount of cesium salt has been often suggested to minimize flame ionization effects, and this has proven effective; however, it is very difficult to obtain cesium salts free from potassium, and the large excess of cesium that is necessary may lead to an impossibly large blank.

The method of standard additions may be used, but this takes away the rapid character of the procedure. The standard addition device becomes invalid when the relation between signal and concentration is not linear, and when the blank is unknown—as it usually is in AA spectrometry—there is no easy way to obtain a just calibration.

X-Ray Spectrometry

X-ray spectrometry is always used as a secondary method. Availability of a suite of analyzed standards is a first requirement. Given such standards, X-ray determination of potassium using a crystal spectrometer is rapid, precise, and accurate. The major source of difficulty lies in the preparation of the sample, and the amount of attention given to this determines the excellence of the results. For potassium in silicates, interferences and interelement effects are very minor; they only become an important factor when sample preparation is meticulous and knowledgeable and highest accuracy is desired.

Determination of sodium by X-ray spectrometry is probably impractical.

Sample preparation techniques may be divided into three classes: direct dilution, fusion, and fusion with grinding. Of these, the fusion technique is undoubtedly best provided a suitable flux can be found, unacceptable losses by volatilization do not occur, and sensitivity is not seriously diminished. The fusion may be poured or pressed into a glass window or it may be ground to powder, mixed with a binder, and pressed into a pellet. Which procedure is best depends to a considerable extent on the nature of the sample.

Direct dilution involves mixing the prepared powder with a binder and pressing into a pellet. Advantages are that sensitivity is maintained at maximum, the procedure is relatively fast, and very little skill is required. This method is applicable in routine operations in which similar materials are examined and a large number of samples are processed. A large suite of analyzed samples is necessary for calibration. Mineralogical and particle size effects are gross.

The fastest way to examine a sample for its potassium, rubidium, and cesium content is simply to pour it into a cup with a mylar window on the bottom and measure the signals generated. Procedures of this kind are often very useful but, in the most favorable cases, are only marginally quantitative. An example of a situation in which this approach is of value is that in which a number of samples are to be dated by the rubidium–strontium method, and those with appreciable rubidium need to be selected. If the samples are all biotites, some known biotites are available, and all samples have been screened through the same mesh, results for potassium and rubidium by this fast method may be fairly quantitative. If there is a variety of minerals in the suite to be examined or if grain size varies, results are likely to be qualitative only.

Fusions are accomplished in graphite, vitreous carbon, or platinum alloy crucibles. The latter are formulated so that borate fusions do not wet the crucible. If the fusion is to be made into a glass window, it may be simply poured from the crucible onto a flat surface or it may be squeezed in a mold of some kind. Commercially available devices for fusing and molding are probably worthwhile, though home-made molds may be needed in special circumstances. Requirements are that the melt be thoroughly mixed, that the glass be cooled slowly to prevent shattering, and that there be a smooth flat surface for examination by the X-ray beam.

If the fusion is to be powdered and pressed into a pellet, it is first smashed in a hard-surfaced mortar, then further ground in a shaker mill, and finally mixed with binder either in the mill or in a mortar. While this is a longer procedure than that in which a window is poured, it is more generally applicable. Very few samples are completely attacked by any reasonable fusion method, and errors due to particles of unattacked mineral are minimized when the melt is crushed and ground to powder. Contamination from the grinding equipment may be a problem with elements such as cobalt, tungsten, and iron but is not a problem with the alkali metals.

The pelleting die should be of the kind that encloses the sample in a protective ring of backing. Some considerable effort has gone into the design of such dies, and optimum details of construction have been developed through trial and error. The surface against which the pellet is pressed is of supreme importance: With the lighter elements, the X-ray beam penetrates only a few micrometers into the surface—the analysis is of the surface of the pellet. In the die described by Fabbi (see Section 2.3.6), the pelleting surface consists of a "flashlight lens"—a disk of glass which is disposable. In determining elements like potassium, it is essential that these lenses be meticulously cleaned before use to remove the film of submicroscopic particles always present on new glass surfaces. If the pellets are pressed against a metal surface, this should be of hardened tool steel, tungsten carbide, or boron carbide. If softer materials are used, the surface quickly becomes roughened, particles of sample become imbedded in the metal surface, cross-contamination occurs, and unnecessary uncertainty results; for the light elements, every effort should be made to produce a mirror-smooth surface free from all blemishes and absolutely uncontaminated.

The press used to prepare the pellet should be capable of a pressure of at least 25 tons; procedure should be rigidly standardized—the pressure should be applied for exactly 1 m (or 30 s, or other fixed time), the same volume of backing material should be used in each pellet, and pellets should be placed in a desiccator as soon as they are prepared. It is advisable to use a vacuum desiccator containing anhydrous magnesium perchlorate, to evacuate the desiccator and to leave the prepared pellets therein for an hour or two before use.

If pellets of calibrating standards are carefully prepared and stored in a desiccating cabinet, they may have an almost indefinite life. The working surface of the pellets should never be touched after they have been prepared; they should be stored in pill boxes or other small individual containers with a ring of plastic so arranged that the pellet surface cannot come in contact with the lid of the box.

4.1.3. Rapid Routine Methods for Specific Materials

SODIUM AND POTASSIUM IN LIMESTONE. Weigh 0.5 g ammonium carbonate (alkali free) into a large agate mortar. Weigh out 5.00 g finely powdered sample (− 200 mesh) and add it in small portions to the mortar, thoroughly mixing with each addition. Transfer the mixture to a J. L. Smith crucible, and heat as described previously. Leach the ignited product with water, boil, and filter. Determine sodium and potassium in the filtrate using the flame photometer. Prepare a blank by taking 5 g alkali-free calcium carbonate through the procedure. Known amounts of alkali chlorides may be added to aliquots of the blank to provide calibrating standards.

SODIUM AND POTASSIUM IN MOLYBDENITE CONCENTRATE. Weigh 0.5 g molybdenite concentrate (-200 mesh) into a porcelain or platinum crucible. Roast at 600 °C for 2 h. Cool. Add 1 g lithium metaborate, and mix thoroughly with a metal spatula. Transfer to a graphite crucible that has been preignited and sprinkled with a little spectrographic-grade graphite powder. Put in a 950 °C muffle furnace for 15 min. Pour the melt into 100 ml of 4% nitric acid in a plastic beaker, and stir to solution. Dilute appropriately, and measure the sodium–lithium and potassium–lithium ratios using the flame photometer.

SODIUM AND POTASSIUM IN SULFIDE ORES CONTAINING COPPER. Transfer 0.1–0.5 g finely ground sample to a flat-bottom platinum dish. Add a little water, 10 ml hydrofluoric acid, 1–2 ml of 1:1 sulfuric acid, and 1–2 ml nitric acid. Heat without boiling until the sulfuric acid begins to fume. Cool. Take up the residue with water, transfer to a 100-ml beaker, dilute to about 80 ml, and add ammonia solution slowly with stirring until the acid is neutralized. Boil and filter into a 200-ml volumetric flask. Wash thoroughly with hot 1% ammonium sulfate solution. Prepare standard solutions containing ammonium sulfate and compare the unknown with these, using a flame photometer or an AA spectrometer.

RECOMMENDED PROCEDURE FOR ANALYSIS OF TRACE ALKALIES IN PERIDOTITES AND OTHER LOW-ALKALI ROCKS. Attack a 1-g sample with hydrofluoric, nitric, and sulfuric acids. Evaporate and dilute repeatedly until magnesium fluoride is decomposed and a clear solution is achieved. Evaporate to fumes, then to dryness, and ignite over an oxidizing Meker flame. Leach the residue with water, and filter to remove insolubles. The filtrate need not exceed 30–40 ml in volume. Dilute to 50 ml with water, and determine the magnesium content of the solution by AA spectrometry or by titrating an aliquot with EDTA, using Erichrome Black T as an indicator. If much calcium is present in the sample, this element should also be determined. Prepare standard solutions containing sodium, potassium, magnesium, and calcium and compare the unknowns to them using the flame photometer or the AA spectrometer.

A major difficulty in determining traces of alkalies (especially sodium) lies in the almost universal contamination of reagents. Acids sold in soda lime bottles always contain appreciable sodium. Salts such as magnesium sulfate should be examined for their alkali content, and solutions of them should be prepared in Pyrex bottles.

Trace Rubidium in Rocks. Govindaraju [2, 3] has devised an ingeniously simple method in which the sample is mounted on a threaded piece of metal and heated in the oxyacetylene flame of an AA spectrometer.

4.2. BERYLLIUM

Beryllium occurs in most rocks in the parts-per-million range. The most common mineral is beryl, $Be_3Al_2Si_6O_{18}$, commonly present in pegmatites, which are also likely to contain lithium minerals such as lepidolite. Analysis of beryllium-bearing rocks is likely to be complicated by the presence of fluorine, and beryllium minerals are apt to contain phosphate and fluoride in the form of contaminants. The usual traces of beryllium are most often determined by emission spectroscopy, using a dc arc source; high accuracy (± 1–2%) can be attained by skillful use of this technique if sampling and subsampling problems can be resolved. A number of colorimetric methods have been devised, most of which utilize organic dyestuffs; these are of doubtful composition or are difficult to obtain pure. Fluorimetric determination using morin is a long procedure that depends on the quality of the dye; some morin reagents contain relatively little of the beryllium-sensitive component. Of the organic colorimetric regents, quinalizarin (1,2,5,8-tetrahydroxyanthraquinone) probably remains the most reliable, but determination must be made by a mixed-color method or by a photometric titration and requires more than usual care and skill.

Isotope dilution mass spectrometry using a measured ^{10}Be addition may lead to useful measurement of traces of beryllium. Beryllium-10 has a half-life of about 10^6 years, so that there is no possibility of it occurring in natural samples. Beryllium-7 (half-life, 55 days) has been used to monitor the recovery of traces of beryllium during collection procedures from large samples.

During a classical analysis, beryllium behaves much like aluminum, except that its precipitation with the ammonia group is not quite complete at pH 6.5. A further difference is that beryllium sulfate is stable even above 1000 °C, so that the hydroxide should not be precipitated from sulfate solution if it is to be ignited to oxide and weighed.

The outstanding analytical characteristic of beryllium is its lack of an ability to form strong EDTA complexes, and this provides the simplest means for its separation. During the extended analysis of the ammonia group, beryllium follows aluminum through a sodium hydroxide separation and is eventually weighed as pyrophosphate with the $AlPO_4$. Fusion of this precipitate with pyrosulfate, addition of ammonium EDTA to complex the aluminum, pH adjustment, and addition of a little diammonium phosphate yields a precipitate of $BeNH_4PO_4$, which can be ignited to $Be_2P_2O_7$ and weighed.

Beryllium and aluminum may also be separated using oxine or tannin. These procedures are somewhat difficult and probably cannot be made to work reliably without first practicing with known mixtures. Both are

best applied when a little aluminum is to be separated from much beryllium. For the methods to be useful, phosphate and fluoride must be absent.

In the usual case where there is much more aluminum than beryllium,
fusion of the ignited mixed oxides with sodium carbonate followed by a
water leach leaves beryllium oxide undissolved. Phosphate must be absent; since most beryllium concentrates are likely to contain phosphate,
and the separation of aluminum and beryllium from phosphate is difficult,
the sodium carbonate separation is less attractive than the phosphate
separation.

4.2.1.   Analysis of Beryllium-Bearing Silicates

PROCEDURE. Determine phosphate and fluoride on separate portions of the
sample. Determine lithium and the other alkalies. (In what follows it is presumed that phosphate, fluoride, and lithium are all present).

Fuse 0.7 g of −200-mesh sample with 5 g sodium carbonate in the usual
way. Attack of beryl is likely to be slow, and a higher temperature than usual
may be required. Leach the melt with 50 ml water in a 300-ml platinum dish
and add 10 ml of 5% boric acid (more if fluorine exceeds 1–2% of the sample).
Acidify with hydrochloric acid and evaporate on the steam bath, uncovered,
to incipient dryness. Cool, add 5 ml methanol, and evaporate cautiously (overheating will cause the methanol to boil and spatter). Cool, add 5 ml concentrated hydrochloric acid, and evaporate. Repeat the alternate evaporations
with methanol and hydrochloric acid until boron (and fluorine) are all removed,
as evidenced by the lack of green color in a small colorless flame applied to
the drying salts. Proceed with the separation of silica in the usual manner.

If possible, avoid the use of sulfuric acid and pyrosulfate during the
solution of the residue from the silica determination; a sodium carbonate
fusion may be necessary, but most often the residue may be left in the
crucible and weighed with the ammonia group.

To the filtrate from the silica add a carefully measured volume of zirconium
oxychloride solution of known concentration, calculated to combine with all
the phosphate in the sample, and provide a 10-mg excess of ZrO_2. Proceed
with an ammonia precipitation as usual. If lithium is present, a triple precipitation is desirable.

To the filtrate from the ammonia precipitate add 1 ml of 3% tannin in water,
stir, digest for 15–20 minutes, and filter.

The precipitate contains a milligram or so of beryllium that escapes
precipitation with ammonia but may also contain rare earth metals, manganese, and elements such as cobalt, nickel, and copper. Depending on

its size and composition, it may be treated in a variety of ways. Usually it should be treated with nitric and perchloric acids to destroy paper and tannin and the small amount of beryllium recovered by ammonia precipitation at pH 7.5 in a small volume of solution and added to the main ammonia precipitate. Separation of calcium, magnesium, and so on from the main filtrate proceeds as usual.

After ignition and weighing, the ammonia precipitate should be fused for 2–3 h with 5 g sodium carbonate at a temperature just high enough to maintain a fluid melt. The Meker flame must be maintained in a fuel-lean condition. Fusion in a furnace is not desirable.

Leach the melt with 200–400 ml boiling water in a 600-ml alkali-resistant glass or silica glass beaker or (less desirable) a large Teflon dish.

Disintegration of the melt should be as complete as possible and may take an hour or more. It will save much trouble if the sodium carbonate fusion is released from the crucible by a brief reheating after the melt has cooled, as described in Section 2.4.12. After thorough scrubbing and rinsing, the platinum crucible should be ignited and weighed.

The precipitate in the sodium carbonate leach solution contains all the titanium, iron, rare earths, beryllium, and zirconium (including the zirconium added earlier) in the sample; the filtrate contains most of the aluminum and probably all the chromium (and uranium) in the sample. If chromium is present in small amount, it may be measured colorimetrically at this stage with no loss of solution. The two fractions of the ammonia group are now treated separately as follows:

1. SODIUM CARBONATE SOLUTION. Acidify with hydrochloric acid and boil to remove carbon dioxide. Double precipitate with ammonia in the usual way, recovering traces of beryllium with tannin. Ignite and weigh the aluminum fraction. Traces of beryllium are best determined in the ignited oxides spectrographically; if this technique is not available, the oxides should be fused with sodium carbonate and leached with water as before. Beryllium oxide may then be filtered off, washed thoroughly with water and weighed or put in solution and determined by AA spectroscopy.

2. SODIUM CARBONATE PRECIPITATE. Proceed in the usual manner with a pyrosulfate fusion, colorimetric measurement of titanium, removal of platinum, double or triple sodium hydroxide separation,† and determination of iron. Aci-

---

† Beryllium is not easily separated from iron and titanium using sodium hydroxide, and the iron-bearing solution should be examined for beryllium at some convenient stage by AA spectroscopy.

dify the sodium hydroxide solution with hydrochloric acid, double precipitate the beryllium with ammonia, adding tannin to recover the last traces, and ignite and weigh BeO plus a little Al_2O_3. The ignited oxides should be examined for sulfate, which will give a falsely high weight; this is conveniently done by X-ray fluorescence without loss. Spectrographic determination of aluminum in the oxides is acceptable, but it is best to separate the beryllium and weigh it as pyrophosphate.

4.2.2. Determination of Beryllium as Pyrophosphate

PROCEDURE. Fuse the impure beryllium oxide with potassium pyrosulfate (20 times the weight of oxides). Leach with water, add 5 ml sulfuric acid, and evaporate to fumes. Cool, add 50 ml water, and boil until soluble salts are in solution. Filter off any residue, ignite it, treat with hydrofluoric and sulfuric acids to remove any silica, fuse with a little pyrosulfate, leach with water, and return it to the main solution. Add concentrated ammonia dropwise with stirring to pH about 2.5 (indicator paper). Add a solution prepared by suspending 1–2 g EDTA in 20 ml water, stirring and adding concentrated ammonia until the acid is dissolved. Boil for a few minutes. Cool. Add 1 g diammonium phosphate dissolved in a little water and then enough saturated ammonium acetate solution to bring the pH to 5.2.

It is advisable to determine the volume of acetate needed by carrying a blank throughout the process and using a pH meter.

Heat the solution to boiling, and hold at steam bath temperature for an hour or more, stirring occasionally. Cool, allow to stand overnight, and filter through a 7-cm Whatman No. 44 paper. Wash with an ammonium acetate solution prepared by adding ammonia to a 1% solution of acetic acid until pH is 5.2 (pH meter). Burn off paper at a low temperature, and then ignite to 1000 °C in a muffle furnace. Cool and weigh $Be_2P_2O_7$ [4].

This procedure may be applied directly to the analysis of beryllium ores. The initial attack must be made by fusion with potassium fluoride to be sure of complete decomposition. Fluorine is then removed by fuming with sulfuric acid, and beryllium is precipitated as above. A double precipitation is necessary, and the final weighed product should be examined for contaminants, especially zirconium and sulfate, using the X-ray spectrometer.

4.2.3. Determination of Traces of Beryllium

Most often, the beryllium content of rock and mineral samples is so low that it may be ignored during the classical analysis. Traces are best determined spectrographically; however, calibration of instrumental techniques is best done using samples analyzed by primary methods so that accurate chemical methods may occasionally be needed. Because beryllium is frequently present in beryl—that is, as a major constituent of a minor mineral—sampling and subsampling problems must be considered in any careful determination of traces of beryllium in rocks. The use of large (10–20-g) samples of well-mixed material is desirable even though colorimetric, spectrographic, and AA spectrophotometric methods are sensitive enough to detect a few parts per million in much smaller samples. Whether the final readout is on an AA spectrometer or on a colorimeter is more or less irrelevant: the problem is in bringing a representative sample free from interfering elements up to the readout device.

Traces of beryllium are probably best separated from large quantitites of other elements by coprecipitation with an insoluble phosphate. Addition of EDTA inhibits precipitation of many other elements.

PROCEDURE. Transfer 10 g of −200-mesh sample to a 300-ml platinum dish. Add 5 ml water, 5 ml nitric acid, and 5 ml sulfuric acid, cover, and heat on the steam bath until any carbonate is removed. Remove the cover, and add 100 ml hydrofluoric acid. Evaporate at steam bath temperature with frequent stirring. Repeat the treatment with 100 ml hydrofluoric acid. Evaporate to near dryness, add 20 ml nitric acid, and evaporate. Repeat the nitric acid treatment and evaporation until it is judged that most of the fluorine and silica have been removed. Transfer the contents of the platinum dish to a 400-ml beaker using a minimum of water, and evaporate to strong fumes of sulfuric acid. (It may be advisable to take another beaker from the same lot, crush a portion of it, and examine it for beryllium with the spectroscope.)

Dilute the solution to 200 ml, add 1–5 g sodium EDTA, 1 ml phosphoric acid, and a few drops of methyl red indicator. If the sample contains no titanium, add 20–30 mg TiO_2 in the form of a sulfate solution. Heat to boiling, and add ammonia dropwise until the indicator changes color. Cool, stir in some paper pulp, and filter.† Drain the filter but do not wash. Transfer paper and precipitate to a 250-ml beaker, add 5 ml concentrated sulfuric acid, heat gently until the paper is thoroughly charred, and then add nitric acid dropwise while heating on a hot plate. Slowly increase hot plate temperature and add drops of nitric acid until organic matter is destroyed and the sulfuric acid is fuming strongly. Cool.

Dilute with water and boil until soluble salts are in solution. Examine any residue, and decide whether or not it should be discarded or treated further. Among minerals that may survive the preceding treatment are beryl (!), zircon, kyanite, chromite and other spinels, and some micas. These may be filtered off, washed with water, ignited, and fused with sodium peroxide in a zirconium crucible or sintered with peroxide in platinum at 480 °C. After solution in sulfuric acid, they may be returned to the main solution.

† It is advisable to add a little tannin to this filtrate, adjust pH to 7 with ammonia, digest, filter, and examine the precipitate for beryllium by AA spectrometry after solution in hydrochloric acid.

Determination of beryllium in this solution by AA spectrometry provides an attractive finish to the determination; however, the effects of phosphate and other ingredients of the solution are not well known, and a bad determination is possible. The method of standard additions should be used, and a nitrous oxide–acetylene flame is advised. The AA determination should be supported by a colorimetric determination.

All colorimetric methods for beryllium are subject to a number of interferences. Metals that give precipitates in sodium hydroxide solution must be removed, and this cannot be done effectively by adding sodium hydroxide to a solution because beryllium in trace amounts is coprecipitated to an intolerable extent. The mixture of oxides must be fused with sodium hydroxide and the melt leached with water; a difficulty in this procedure is that beryllium oxide is not reliably attacked by molten sodium hydroxide if it has been strongly ignited. A single sodium hydroxide separation does not lead to a clean separation of chromium, vanadium, uranium, scandium, and phosphate, all of which interfere in most colorimetric methods. Silicate may also interfere in the amounts dissolved from glassware. No general procedure can be written that will guarantee a final beryllium solution sufficiently pure for colorimetric determination, and the following should be used as a guide only.

PROCEDURE. To the impure sulfuric acid solution of the beryllium (e.g., that obtained after collection on titanium phosphate) add aqueous ammonia to pH 7.5 followed by 0.5 g tannin dissolved in a little water. Digest hot for a few minutes, cool, and filter. Wash with an ammonium sulfate–tannin solution as near in composition to the analytical solution as possible. Treat the paper and precipitate with nitric and perchloric acids and take to strong fumes of perchloric acid. Cool. A red or orange color indicates the presence of chromium. If chromium is present, heat to strong fumes and add small portions of sodium chloride (about 10 mg per addition) until all the chromium has been volatilized (as chromyl chloride).

Dilute the solution to about 10% in perchloric acid, and add sufficient zirconium oxychloride solution to precipitate phosphate, avoiding any large excess. Filter, and wash with 5% hydrochloric acid. Return the precipitate to the beaker with a minimum of water, add an equal volume of hydrochloric acid, and heat to boiling. Dilute to about 10% in hydrochloric acid and filter. Wash with 5% hydrochloric acid. Combine the filtrates. Reserve the residue for spectrographic or AA spectrometric examination of beryllium.

Again precipitate the beryllium and other metals with ammonia and tannin, filter, and wash. Burn off the paper at 450 °C in a gold crucible, and fuse the residue with a weighed amount of sodium hydroxide. Leach the melt with enough water to make a 5% solution of NaOH, filter, and wash with 5% sodium hydroxide solution. Dissolve the precipitate in hydrochloric acid and again precipitate with ammonia and tannin, burn off the paper, fuse with a weighed amount of sodium hydroxide, and leach with water. Filter. Reserve the residue.

Combine all residues, burn off paper, mix thoroughly, and examine for traces of beryllium by emission spectroscopy; alternatively, prepare a solution of the residues and look for beryllium by AA spectrometry. Determine beryllium in the sodium hydroxide solution (which should contain substantially all that in the sample).

4.2.4. Colorimetric Determination of Beryllium Using Quinalizarin

If the sodium hydroxide used to dissolve the beryllium and separate it from iron, titanium, and so on was weighed, and if the fusions were adequately protected from carbon dioxide in the air and in gas flames, beryllium will now be present in an alkaline solution containing no iron, little phosphate, and an acceptable concentration of carbonate. Colorimetric determination proceeds best in solutions 0.3–0.4 N in sodium hydroxide and containing 0.1–0.5 μg/ml beryllium. If a 10-g sample was used to begin with, and the beryllium has been collected in 50 ml of 5% sodium hydroxide solution, dilution of this solution to 150–200 ml will permit the determination of 1–5 ppm in the sample.

PROCEDURE. Prepare fresh a 0.05% solution of quinalizarin in 0.25 N sodium hydroxide. Filter. Also prepare a beryllium solution by adding standardized beryllium sulfate or nitrate solution to 5% sodium hydroxide and diluting to 0.3 N in sodium hydroxide; the beryllium concentration of this solution should be about 10 μg/ml.

To a 10-ml aliquot of the sample solution add 100–200 ml of 0.25 N sodium hydroxide solution. Add 0.05% quinalizarin from a microburette, observing the color change as the reagent is added. If no beryllium is present, the color change should be minimal. With beryllium, the solution will turn blue with the first drop of reagent. Continue adding reagent until the blue color is tinged with purple, indicating an excess of quinalizarin. Repeat the titration using the standard beryllium solution. In this way, a rough estimate of the beryllium content of the sample is made, and the effective concentration of the reagent is measured.

The determination is best finished in Nessler tubes. A volume of reagent calculated to be in excess of the beryllium is added to each of several tubes, a portion of the unknown solution is added to one of them, and a series of differing volumes of the standard beryllium solution added to the others. Each is then diluted to the mark with 0.25 N sodium hydroxide solution. The determination is made by visual comparison. Attempts to use a spectrophotometer, either in the conventional manner or during a colorimetric titration, have not proven very successful. The dye is not very stable in alkaline solution, and colors change fairly rapidly with time, so that direct observation of a series of solutions, each prepared at the same time, is preferred. Slight differences in hue are easily observed visually and provide warning that an interfering element may be present. Possible interfering elements that may have survived previous treatments are zinc, scandium, manganese, magnesium, and calcium. If the blue color obtained with the sample solution does not have the same hue as that obtained with the standard beryllium solution, a small addition of sodium EDTA or sodium cyanide may be added to both before addition of the reagent.

If difficulty is experienced with the colorimetric determination, the alkaline solution may be acidified with hydrochloric acid and the beryllium determined by AA spectrometry using the nitrous oxide flame. Calibration should be based on known beryllium solutions containing the same concentration of sodium salts as the unknown, and the effect of small amounts of phosphate and aluminum should be taken into account. A lanthanum addition is probably advisable.

4.2.5. Fluorimetric Determination Using Morin

Morin is a yellow dye obtained from a species of mulberry tree. Its composition is unknown; although the active ingredient is alleged to be 3,5,7,2′,4′-pentahydroxy flavone, the reagents as purchased vary widely in their behavior toward beryllium, some being very insensitive. Methods using morin for trace beryllium determination in rocks have been described in detail by Sandell [5] and Jeffery [6]. Probably the usefulness of these methods depends more on the availability of a high-quality reagent than on anything else. As with the colorimetric reagents, separation of the beryllium from interfering species is the major problem. The advantage offered by fluorimetric determination with morin is its much higher sensitivity, which diminishes the size of sample that must be taken. Sandell uses a 0.25-g sample of 100-mesh rock powder; since the beryllium in rocks is almost always present in beryl, the risk of gross subsampling error may invalidate any result, however well the determination is carried out. If Sandell's procedure is to be applied, 5-g sample of −100-mesh material should be taken, ground to pass a 200-mesh screen, and split 1:4. The resulting 1.25 g should be ground to pass a 270-mesh screen and mixed before removing the required 0.25-g subsample. Beryl does not grind readily, and it is essential that the material be passed 100% through the screens during preparation. There is some danger that particles of beryl may be trapped in the screen mesh and go unnoticed.

One 100-mesh grain of beryl weights about 10^{-5} g and contributes about 1 ppm beryllium to a 1-g sample, or 4 ppm to a 0.25-g sample.

The sampling constant K_s for beryllium in a rock containing 100-mesh beryl grains is about 10 kg; sampling error using 0.25-g subsamples is $\sqrt{K_s/w} = 200\%$!

4.2.6. Spectrographic Determination of Beryllium

Spectrographic determination by dc arc spectroscopy requires a preliminary fusion of at least 0.2 g of -200-mesh sample. Fusion with lithium or sodium metaborate or tetraborate decomposes beryl reliably. Crushing the fusion bead to -200 mesh yields a diluted and buffered sample that can be compared to synthetic samples prepared in a similar manner from spec-pure silica and other appropriate oxides. The high sensitivity of the beryllium 3130.42-Å line and its tendency to self-reversal make the dilution an advantage; in fact, higher dilutions may be necessary. Direct spectrographic determination without isoformation usually leads to poor results because of subsampling difficulties.

4.2.7. Determination of Beryllium by AA Spectrometry

Sensitivity for beryllium in an oxyacetylene flame is high; as little as 0.01 mg/l gives a useful signal at the 2349-Å line. Sulfate, silicate, phosphate, and aluminum interfere. It is desirable to have a relatively high concentration of potassium in solutions of both unknowns and standards. The commonly used lanthanum addition helps diminish interferences but does not seem to be as effective with beryllium as it is with some other elements. Instrument response is linear with concentration only at high dilutions, and readings should be repeated after a 1:5 dilution to be sure that self-absorption effects are under control.

4.3. MAGNESIUM

Magnesium occurs almost universally in rocks and is one of the 13 elements always reported in a "complete" rock analysis. It is a major constituent of many sedimentary rocks. The carbonate and oxide occur as dolomite and magnesite, respectively. Important silicate minerals are olivine, $(Mg,Fe)_2SiO_4$; serpentine, $3MgO \cdot 2SiO_2 \cdot 2H_2O$; and talc, $3MgO \cdot 4SiO_2 \cdot H_2O$. Huge amounts of magnesium occur in seawater, from which the metal is extracted commercially on a large scale. Large deposits of mixed salts of magnesium, potassium, and calcium are mined. They probably originated from the evaporation of ancient seas. Examples are carnallite, $MgCl_2 \cdot KCl \cdot 6H_2O$, and polyhalite, $MgSO_4 \cdot 2CaSO_4 \cdot K_2SO_4 \cdot 2H_2O$.

In the classical analysis, magnesium escapes precipitation with the acid group, the hydrogen sulfide group, the ammonia group, the persulfate group, and the oxalate group. It is then precipitated as magnesium ammonium phosphate in the filtrate from the calcium and ignited to magnesium pyrophosphate, $Mg_2P_2O_7$, which is an accurate weighing form. In very unusual circumstances, magnesium may contaminate the ammonia precipitate—usually in combination with phosphate. Such contamination can always be avoided by removing phosphate with zirconium before the ammonia precipitation and by making sure sufficient ammonium salts are present in the solution. With samples containing 40–50% MgO (e.g., dunite and peridotite), at least 15 g ammonium chloride should be present during the ammonia precipitation.

Possible contaminants of the magnesium phosphate precipitate are calcium and strontium, the rare earth metals, beryllium, and nickel. Of these, only calcium need always be considered; it is separated from the weighed magnesium pyrophosphate by solution in sulfuric acid and precipitation as calcium sulfate in 80% alcohol. Strontium accompanies the calcium.

Contamination of the magnesium precipitate with rare earths and beryllium is avoided by adding an excess of ammonia and tannin to the filtrate from the ammonia group and filtering before precipitating the magnesium.

Barium does not precipitate with the magnesium ammonium phosphate under the usual conditions of a rock or mineral analysis, at least when it is present at less than 3–4% BaO.

Nickel and perhaps some other elements may be present in small amounts in the magnesium phosphate if they make up several percent of the sample and are not adequately handled earlier in the analysis.

4.3.1. Primary Methods

There seem to be no generally applicable primary methods for determining magnesium that do not follow in essence the classical procedure. The invariable approach is to remove first the acid group of elements after suitable dissolution procedures and then the ammonia group and finally titrate the magnesium with EDTA or precipitate it with oxine or phosphate and determine it gravimetrically. The colorimetric determination of magnesium using thiazole yellow (titan yellow) is best calibrated using analyzed standards; use of this method in the primary mode yields poor results, partly because commercial titan yellow is a mixture of dyes that behave in unknown ways toward not only magnesium but also other elements. The isolation of the magnesium-active component of the commercial dye has been accomplished by King and Pruden [7], and its synthesis is described by King, Pruden, and James [8]. One may suppose that use of the purified dye may permit its application in the primary mode, but there remains insufficient information on the effect of other elements.

Shapiro and Brannock [9] determine magnesium in silicate rocks containing less than 2.4% MgO as follows:

PROCEDURE. Transfer 0.500 g of −200 mesh rock powder to a Teflon beaker of special design having a close-fitting lid. Add 15 ml solution B acid mixture (1-lb bottle of hydrofluoric acid, 165 ml concentrated sulfuric acid, 40 ml concentrated nitric acid, mixed in that order in a large polyethylene bottle cooled in a bath of ice water). Cover, and heat in a boiling water bath overnight. Remove the cover, and evaporate until acid fumes are no longer given off. Transfer the mixture to a 400-ml Vycor beaker using a minimum of water to police out the solids. Heat on a hot plate until excess sulfuric acid is removed. Add several drops of nitric acid and an equal volume of perchloric acid and heat to strong fumes of perchloric acid. Cool, and add about 225 ml water, 5 ml concentrated nitric acid, and 1 ml of 0.2% hydrazine sulfate solution (or more if necessary to reduce manganese to the divalent state). Dilute to 250 ml in a volumetric flask.

Transfer a 5.00-ml aliquot of the solution of the rock or mineral to a 100-ml volumetric flask. Add 5 ml polyvinyl alcohol solution (to 200 ml of a 0.1% solution of polyvinyl alcohol add 800 ml water, 5 ml of 1:1 sulfuric acid, 1.5 g $Al(NO_3)_3 \cdot 9H_2O$, 40 g hydroxylamine hydrochloride, stirring to complete solution after each addition). Dilute to about 70 ml with water and mix. Add 2 ml complexing solution (64 g potassium cyanide in a mixture of 400 ml triethanolamine and 600 ml water). Add 5 ml of 0.02% thiazole yellow, mix, and add 5 ml of 30% sodium hydroxide solution. Dilute to the mark, mix, and allow to stand for 20–30 min. Prepare blank and standard solutions in exactly the same manner using an acid mixture containing 20 ml nitric acid and 16 ml sulfuric acid per liter instead of the rock solution and adding suitable quantities of standard magnesium solution to each of several blanks. Measure absorbance at 545 nm.

This procedure is typical of several that have been devised to determine magnesium without prior separations: Such methods are useful when they are applied to large suites of samples all of approximately the same composition. Because of the differing compositions of commercial thiazole yellows, extensive modifications may be necessary depending on the behavior of a specific batch. Some batches of thiazole yellow do not perform well; others give quite acceptable results. The most efficient way to use methods of this kind is to forget about blanks and run a suite of known standards along with the samples, plot an empirical calibration curve, and read the magnesium contents of the samples from this curve.

With the development of AA spectrometry, the need for rapid chemical methods of this kind has almost disappeared. The same solution may be used, and calibration against analyzed standards is preferable to calibration against pure magnesium solutions. Since the development of fusion solution methods (e.g., the $LiBO_2$–HNO_3 solution method), the use of hydrofluoric acid to decompose silicates has diminished. Magnesium is among the long list of elements that may be determined by AA spectrometry in $LiBO_2$–HNO_3 solutions. Again, the method is best used in the secondary mode; however, the high dilutions required for AA magnesium determination (the best range is 0.2–2.0 ppm in solution) diminish any matrix effects to the point where calibration against pure magnesium solutions is feasible, especially if appropriate amounts of lithium metaborate are added to the standards, and acid concentrations are always nearly the same. Dilutions of standard solutions should of course be made with the same acid solution (usually 5% hydrochloric acid) as those of the samples to be analyzed.

4.3.2. Rapid Methods

Determination of Dolomite Content of Limestone

In many applications, the dolomite content of limestone and the calcium content of dolomite are of consequence. A rapid determination of the limestone–dolomite ratio is possible using differential thermal analysis. Sharma, Mukherjee, and Roy have reported successful use of this device in the analysis of some Indian limestones [10]. Dolomite in limestone as low as 0.3% can be detected, and this is an improvement on X-ray diffraction methods, which are probably not useful below a dolomite concentration of about 2%. The presence of dolomite results in a thermogram peak at about 800 °C, which diminishes as the dolomite content decreases. The use of a CO_2 atmosphere during development of the thermogram is desirable (see Rowland and Beck [11]) but not necessary if results are examined empirically. Probably the method must be locally calibrated since it may depend on a number of characteristics of the stone as well as on design characteristics of the apparatus used. Like other rapid methods, its usefulness is in the area of plant and product control, where immediate results are essential.

Precipitation of Magnesium with Oxine

In some cases, precipitation of magnesium with phosphate is undesirable; For example, when lithium must be determined in the filtrate. In such cases, 8-hydroxyquinoline (oxine) may be used to advantage.

PROCEDURE. To a solution from which the ammonia group of elements has been removed add aqueous ammonia to pH 10 (phenolphthalein indicator). Heat to near boiling, and add 2.5% oxine in 1 N acetic acid dropwise with stirring until a small excess is present, as shown by a yellow color in the supernatant liquid. The precipitate is bulky but coagulates rapidly and filters without difficulty.

Filter, and wash with hot water. Redissolve the precipitate in 20% hydro-chloric acid, heat to near boiling, and slowly add aqueous ammonia to pH 10. Add about 1 ml oxine reagent to be sure of an excess, filter through a weighed glass-frit crucible, wash with hot water, dry at 105 °C, and weigh $Mg(C_9H_6NO)_2 \cdot 2H_2O$. Return to the oven and heat at 160 °C for an hour. Weigh the anhydrous oxinate. The large gravimetric factor is an advantage when small amounts of magnesium are in question [$Mg:Mg(C_9H_6NO)_2 = 0.0778$].

Most of the oxine can be removed from the filtrate by simply evaporating in an open beaker on the steam bath. Any remaining oxine is destroyed or volatilized during the removal of ammonium salts with nitric acid prior to determining lithium by flame photometry.

Determination of Magnesium in Limestone or Dolomite

Prepare a solution of the stone as described under "Calcium." Transfer a 50-ml aliquot of the solution to a 250-ml beaker. Boil for a minute or two to remove carbon dioxide and add 3 ml of 15% ammonium chloride solution. Add a drop of aqueous methyl red indicator and then aqueous ammonia dropwise with stirring until the indicator just changes color. Drop a small ($\frac{1}{8}$-in.) square of nitrazine paper into the solution. If it shows a green or yellow color, add a single drop of ammonia. Stir, wash down the sides of the beaker with water, stir, add another small piece of nitrazine paper, and adjust with ammonia if necessary. Stir in a little paper pulp and filter at once through a 7- or 9-cm Whatman No. 40 paper. Wash with hot 2% ammonium chloride solution that has been adjusted to pH 6.5 by ammonia addition. Catch the filtrate in a 100-ml volumetric flask if further aliquots are to be removed for any purpose or if the sample is high in magnesium. Otherwise the titration of calcium plus magnesium may be carried out on the whole of the solution. If a 20-ml aliquot is taken (corresponding to a 0.1-g sample if the original sample was 1 g), proceed as follows: Add 0.05 g hydroxylamine hydrochloride, 5 ml of 15% ammonium chloride solution, 5 ml alkaline cyanide solution (40 g carbonate-free sodium hydroxide and 10 g potassium cyanide in 500 ml water), and enough eriochrome black T to give a good pink color. Titrate with the same EDTA solution used in the calcium determination to a pure blue. It is advisable to prepare a reagent blank solution with indicator added to act as a reference end point color. Subtract the titration due to calcium. Eriochrome black T indicator is best added in the form of a solid mixture of 1 part indicator to 20 parts potassium chloride. Solutions of this indicator are not stable.

4.3.3. Secondary Methods: X-Ray Fluorescence

Magnesium is the lightest element that can be usefully determined using ordinary X-ray equipment. A vacuum (or helium or hydrogen purged) instrument is necessary, and the sample must be pelleted or formed into a glass for maximum sensitivity. Since the fluorescent X-rays from magnesium are soft, readings are made on a very thin layer of material at the surface of the pellet. The effective sample weight is therefore only a few milligrams, and sampling problems are intractable. Maximum sensitivity is achieved when the sample is pressed directly into a pellet with a minimum of binder, but very fine grinding is essential. Fluxing with lithium metaborate or tetraborate (the latter for basic rocks) is usually desirable.

Calibration is always accomplished using analyzed standards; these should be treated in exactly the same manner as the unknowns. A difficulty arises here: Very often, standard rock samples are ground to pass a 200-mesh screen, while routine samples to be analyzed are more likely to have been prepared without screening. As a consequence, the same grinding and pelleting procedure may leave the unknowns coarser than the standards. With the light elements in general and with magnesium in particular, particle size has a large effect on the X-ray signals developed. Pellets prepared without fusion should be examined under the microscope: Few particles larger than 10 μm should be present, and standards and unknowns should have similar particle sizes.

Various shaker mills are used to prepare X-ray pellets. Very seldom do they provide a uniformly fine grind, and it is often necessary to hand grind the material after 5 or 10 min in the shaker mill. This is true not only for direct pressing but also when a fused sample is ground and pressed into a pellet.

X-ray methods are of great value in circumstances where a single product needs to be monitored. For example, in the cement and fertilizer industries the magnesium content of limestone is of interest and may be routinely determined very rapidly using energy-dispersive or wavelength-dispersive X-ray methods. The key to the successful application of such methods lies in exactly reproducing the sample preparation procedure for every sample and standard. Development of a suite of analyzed samples of the same material being examined, using primary methods, is essential. The use of purchased standards that originate from other raw material and are prepared by different procedures is likely to lead to biased results.

4.4   THE ALKALINE EARTHS

Calcium is the element of sedimentary rocks. It is found almost universally in other rock types in concentrations varying from major to trace. Many routine determinations of calcium are high because reagent acids are packed in soda lime glass containers and dissolve calcium from the glass. In accurate analysis, this contamination must be taken into account.

Strontium is commonly associated with calcium; many calcium minerals have their strontium counterparts, but these are generally rare. Strontium apatites are known but are of minimal importance.

Barium is more in character with the rare earth metals than with calcium and strontium, at least from the analytical point of view. It occurs most frequently in barite, $BaSO_4$, in contrast to calcium and strontium, which are most commonly found in carbonate rocks.

In classical analysis, calcium and strontium quantitatively escape precipitation with the acid group, the hydrogen sulfide group, and the ammonia group, except under a few special circumstances. For example, if tungsten is present in the sample, and it does not come down as tungstic acid or as tungsten sulfide, it may form calcium tungstate during precipitation of the ammonia group. If fluoride is not removed in the early stages of the analysis, calcium fluoride may precipitate with the ammonia group. Strontium is less apt to be caught in the ammonia group. Barium is almost never precipitated unless sulfate is present in the sample (or unless a sulfuric acid dehydration of silica is required, as in the analysis of titanium minerals).

In the analysis of sulfur-bearing materials, some barium will be precipitated as sulfate with the acid group (Si, Ta, Nb, W, etc.). This does not lead to any error in the silica determination because $BaSO_4$ is stable on ignition. In the usual procedure, where the residue from the HF determination of silica is simply added to the ammonia group, the presence of barium and sulfur in the sample will lead to error. If barium is present, the residue should be fused with pyrosulfate and added to the filtrate from the silica if no hydrogen sulfide separation is contemplated. The barium sulfate may then be recovered and weighed during the silica recovery from the ammonia group. If a hydrogen sulfide separation is to be used, the residue from the silica determination should be analyzed separately when it contains barium.

Strontium may partially accompany the barium, but most often it causes no difficulty. Special treatment is in order when strontium minerals are being analyzed.

In most cases, calcium, strontium, and barium will escape precipitation and will be quantitatively present in the filtrate from the ammonia precipitation. Manganese may then be separated after oxidation with persulfate, in which case most of the barium and part of the strontium will accompany the manganese. If this procedure is followed, barium and strontium should be determined on a separate portion of the sample. If much barium and strontium are present or if the available sample is limited, it is best to evaporate the filtrate from the ammonia precipitate with a large volume of nitric acid to destroy ammonium salts (if the dehydration of the silica was accomplished with nitric acid, hydrochloric acid must be added). Calcium, strontium, and barium may then be all precipitated with oxalate and carbonate and analyzed as a group by procedures to be described, or the barium may be recovered as sulfate before precipitating calcium and strontium.

If no attempt to determine barium and strontium is made, part of the strontium will be weighed with the calcium carbonate or oxide, and part will precipitate as sulfate during the calcium recovery from the magnesium pyrophosphate. Unless barium is present in substantial amounts (more than 1–2% of the sample), it will not be caught in any of the precipitates and will be present in the filtrate from the phosphate group. Recovery from this filtrate is usually not worth the trouble but may be necessary if the supply of sample is small or if some other element has to be determined at this stage (e.g., zinc, nickel, or copper). Statements to the effect that barium precipitates as phosphate with the magnesium during the classical analysis are incorrect.

One might suppose that determination of traces of calcium, strontium, and barium in the filtrate from the ammonia group by AA spectrometry would yield acceptable results, and this is indeed the case. However, it is necessary to first destroy ammonium salts if accurate results are to be obtained, and this is as much trouble as the conventional procedure. AA spectrometry should of course not be used for major constituents during an accurate analysis. Determination of strontium or barium in the phosphate filtrate by flame photometry or AA spectrometry is not feasible because of the gross interference of phosphate in the flame.

In some circumstances, the alkaline earths may be partially or wholly precipitated with the ammonia group if precautions are not taken. If phosphate, molybdate, tungstate, chromate, carbonate, or vanadate are present during the ammonia group separation, calcium phosphate, barium chromate, and so on may precipitate. In the analysis of phosphate rocks, it is essential that phosphate be removed before the ammonia group precipitation. This is best done with zirconium; if a carefully measured volume of zirconium oxychloride solution is used, the zirconium phosphate may be weighed together with the ammonia group of elements, and a valid weight of Fe_2O_3 + TiO_2 + Al_2O_3 + P_2O_5 + $\cdots$ may be obtained. The addition of an excess of ferric iron prior to precipitation with ammonia has been recommended; this expedient is a poor one because if enough iron is added to accomplish its purpose (a considerable excess is necessary), the precipitate is very bulky, difficult to wash, and otherwise inconvenient and unpleasant.

Interference from chromate is encountered in the analysis of chromite-bearing rocks when dehydration of the silica is done with perchloric acid. Barium chromate is insoluble at pH 6.5. Fortunately, chromites do not usually contain barium, and separation of the bulk of the chromium from the iron by precipitating with ammonia is feasible.

When molybdenum is present in more than traces, its separation with hydrogen sulfide prior to the ammonia precipitation is necessary to avoid the formation of calcium molybdate.

There does not seem to be any simple way to prevent tungsten from bringing down some calcium in the ammonia group; with tungsten-bearing materials, the presence of alkaline earths in the ammonia precipitate is always a possibility. It is probably best to take an aliquot of the pyrosulfate–sulfuric acid solution of the ammonia group and examine it by AA spectrometry when the possibility of alkaline earth contamination exists.

In analyzing sedimentary rocks (limestone, dolomite, shales, sandstones, phosphates rocks, etc.), the classical procedures need considerable modification. Organic matter is often present, sometimes to the extent of several percent. In limestones, the bulk of the material is calcium carbonate, and a meaningful analysis must report iron, aluminum, and so on to three figures instead of the usual two. In phosphates, rare earths are often present up to 1 or 2% of the total. Carbon and hydrogen may be present in several forms: organic carbon and hydrogen, carbonate carbon and hydroxyl hydrogen, and so on. Possibly the best way to begin a sedimentary rock analysis is to determine fluorine, phosphate, total hydrogen, and total carbon on separate portions of the sample. A spectrographic analysis will show the presence or absence of rare earths. Procedures can then be tailored to fit the composition of material.

4.4.1.   Scheme Recommended for Limestone

PROCEDURE. Determine total carbon, organic carbon, total hydrogen, and fluorine on separate portions of the sample. Weigh a 10-g and a 1-g sample into platinum dishes and heat slowly to 450 °C in a furnace. Hold the temperature at 450 °C for several hours. Cool and reweigh the 1-g sample. Increase the temperature of the furnace to 750 °C and heat both samples for a further 2–3 h. Reweigh the 1-g sample.

The 1-g sample is to be used for determining calcium and magnesium. Dissolve it in dilute hydrochloric acid, filter, wash with dilute acid, burn off the paper, and fuse the residue with sodium carbonate. Return the melt to the main solution. Dehydrate and filter off silica as usual and precipitate the ammonia group. Determine calcium, strontium, barium, and magnesium in the filtrate.

The 10-g sample is to be used for determining the minor constituents. Dissolve it in dilute hydrochloric acid in a large platinum dish. If fluorine is present, add 10 ml of 5% boric acid. Evaporate to dryness on the steam bath. Treat the residue alternately with 5 ml concentrated hydrochloric acid and 5 ml methanol (to remove boron), evaporating to dryness after each addition. Repeat until no green flame color can be perceived in a small colorless flame played on the surface of the dried material. Take up the residue with dilute hydrochloric acid and filter. Wash with dilute acid. Burn off the paper, fuse the residue with sodium carbonate, and proceed as for a silicate rock sample up to the point where silica has been collected and ignited. Reevaporate the main solution to recover any remaining silica, and combine this with the main part of the silica. Weigh, treat with hydrofluoric and sulfuric acids, and determine SiO_2 from loss in weight. Fuse the residue with pyrosulfate, leach with water, and combine all solutions before precipitating the ammonia group. If phosphate is present, add a measured amount of zirconium oxychloride. A triple ammonia precipitation is desirable. If rare earths are present, they should be recovered from the filtrate by addition of ammonia in large excess. The filtrate may be treated with ammonium persulfate to recover manganese and then discarded. Analysis of the ammonia group then proceeds as usual.

Results of a limestone analysis using this procedure are shown in Table 4.2.

4.4.2.   Primary Methods

Analysis for Barium and Strontium in Silicates

The following procedure is often used for the rapid determination of sulfur, chlorine, chromium, uranium, and some other elements in ores. It also provides a good separation of barium and strontium, which may then be determined gravimetrically or otherwise.

PROCEDURE. Weigh a 0.500 g sample into a 100-ml iron crucible, add 5 g sodium peroxide, mix thoroughly, and cover the charge with about a gram of sodium carbonate. Heat over a Bunsen or Tirrill burner in such a way that the inner cone of the flame forms a cool "button" on the bottom of the crucible. The crucible should be supported on a large fire-clay triangle, and fusion should take about 10 min. Pick up the crucible with the tongs, and swirl it in the flame for a few minutes, allowing the melt to become dull red. Cool. Pour 40 ml water into the crucible, and cover it at once with a watch glass. Have a pan of cold water handy, and heat the crucible cautiously on the steam bath; transfer at once to the cold water bath as soon as violent reaction begins. Finally, heat on the steam bath for an hour to decompose any remaining peroxide.

Filter through a double paper, a Whatman 541 (hardened) paper folded in-

Table 4.2. Complete Analysis of Limestone and Dolomite

	Weight (%) by GFS[a] Number			
	400	401	402	403
CaO	30.51	50.07	46.65	37.99
SrO	0.13	0.018	0.012	0.043
BaO	0.005	0.12	0.003	0.005
MgO	21.50	3.60	5.74	13.78
MnO	0.0052	0.011	0.019	0.018
SiO_2	0.075	2.09	2.63	1.81
Al_2O_3	0.029	0.22	0.53	0.43
TiO_2	0.004	0.020	0.032	0.026
Fe_2O_3 (total)	0.053	0.199	0.370	0.308
Na_2O	0.04	0.02	0.02	0.03
K_2O	0.01	0.06	0.16	0.14
CO_2 (acid solution)	47.36	43.28	42.82	44.56
C (organic)	0.02	0.03	0.31	0.28
S (total)	0.035	0.05	0.22	0.18
P_2O_5	0.000	0.035	0.048	0.044
H_2O^-	0.15	0.07	0.12	0.15
H_2O^+	0.10	0.16	0.29	0.20
H (organic)	0.00	0.00	0.03	0.02
Total	100.02	100.05	100.00	100.01
Ignition loss 950 °C	47.40	43.28	42.97	44.76
Total H as H_2O	0.30	0.25	0.64	0.54
Total C as CO_2	47.43	43.38	43.95	45.60
Acid-soluble FeO	0.03	0.10	0.12	0.14

Elements specifically sought spectrographically, but not detected (in all samples): Zn, Sn, Cr, Ni, V, Co, Mo, Be, Cu, Y, Yb, Zr, and Pb.

[a] GFS standard samples 400–403. G. Frederick Smith Chemical Company, Columbus, Ohio 43223. Analyses by C. O. Ingamells and N. H. Suhr, *Geochim. Cosmochim. Acta* **31**, 1347–1350 (1967).

side a Whatman No. 44 or an S & S blue ribbon paper. Wash crucible and filter with water and then with a solution containing 1% each of sodium hydroxide and sodium carbonate. Reserve the filtrate for determination of vanadium, uranium, chromium, sulfur, chlorine, and so on.

Wash the precipitate, which contains all the alkaline earths and rare earths in the sample, into a 250-ml beaker with water, transferring the solids as completely as possible. Put the papers in a platinum crucible and add several drops of nitric acid. Reserve the crucible and contents.

If the volume of water in the beaker exceeds 50 ml, evaporate to this volume on the steam bath. If the sample contains little or no calcium, add about 50 mg calcium carbonate at this stage. Add to the suspension of carbonates and oxides 5 ml of 1:1 sulfuric acid all at once while stirring. Warm on the steam bath until substantially all the iron is in solution. Add an equal volume of absolute alcohol, and allow to stand for several hours, stirring occasionally.

Filter through a large Whatman No. 44 paper, and wash with 50% (v/v) alcohol containing 2% sulfuric acid. Discard the filtrate. Transfer paper and precipitate to the crucible containing the original filter papers, and burn off paper in a furnace, bringing temperature up slowly to 450–500 °C.

Add 5 g sodium carbonate (and a little Na_2O_2 if carbon is present) to the mixed impure sulfates and mix thoroughly with a glass rod, breaking up any lumps. Fuse over a good Meker burner for at least 30 min at maximum temperature. Cool, and leach with water. Transfer the contents of the crucible to the original 250-ml beaker, and add 3 ml hydrochloric acid to the crucible. Leach the sodium carbonate melt on the steam bath until all soluble salts are in solution. Cool completely, and filter through a small S & S blue ribbon paper. Wash thoroughly with 2% sodium carbonate solution to remove all sulfate. After making sure the filtrate is perfectly clear, discard it.

Dissolve the precipitate in a little dilute hydrochloric acid using that which was added to the crucible earlier. Boil to remove carbon dioxide, add a little bromine water, and then add a slight excess of ammonia solution. Filter, and wash carefully, using a minimum of hot 2% ammonium chloride. Catch the filtrate in a 150-ml beaker. It is very necessary that (a) all carbonate be removed by boiling and (b) manganese be completely removed during this operation.

Discard the precipitate, acidify the filtrate with a little hydrochloric acid, and boil to remove all bromine. Add a very little *aqueous* methyl red indicator solution, and bring to the neutral point with diluted ammonia. Then add just enough diluted hydrochloric acid to restore the pink color of the indicator. Heat nearly to boiling, add 5 ml of 2 *M* ammonium acetate solution and then 5 ml of 10% ammonium dichromate solution. Stir, without striking the sides of the beaker with the rod, until precipitation of barium chromate begins. Cool, stirring occasionally. Allow to stand for several hours. Filter through a small Whatman No. 44 paper, transferring the precipitate quantitatively, and wash thoroughly with a minimum of cold water. Burn off the paper in a small weighed porcelain crucible, heat to constant weight at 500 °C, and weigh barium chromate $BaCrO_4$.

If more than 1% barium or strontium is present in the sample, the barium chromate should be reprecipitated before ignition and weighing. Dissolve the washed barium chromate in diluted colorless nitric acid, dilute to a suitable volume (100 ml for 0.1 g BaO), add ammonia until precipitation begins, heat nearly to boiling, add 5 ml of 2 M ammonium acetate solution per 100 ml solution, stir until precipitation is substantially complete, add a few drops of ammonium dichromate solution, cool, stand for 1–2 h, and filter. A porous porcelain crucible (Royal Berlin, medium porosity) may be used for the ignition, reducing the danger of decomposing chromate. If the ignition product is not pure yellow and cannot be improved by reignition, the barium chromate is probably contaminated with Cr_2O_3 and/or MnO_2 because of imperfect removal of manganese earlier in the procedure.

The barium chromate may also contain rare earth elements. It is advisable to look for these using the spectrograph. If present, the barium chromate should be converted to sulfate.

Conversion of Barium Chromate to Barium Sulfate

PROCEDURE. Transfer the ignited barium chromate to a 150-ml beaker using a minimum of hydrochloric acid in water. Dilute to about 50 ml, and add a drop of 30% hydrogen peroxide to reduce chromate. Evaporate nearly to dryness, add 1 ml hydrochloric acid, dilute to 50 ml, and filter through a small paper. Wash with water and discard any residue on the paper. Heat the filtrate to boiling, and add a single drop of 1:1 sulfuric acid. Boil for several minutes, and then add dropwise up to 2 ml of 1:1 sulfuric acid. Cool, allow to stand overnight, filter, wash with 1% sulfuric acid, and ignite and weigh $BaSO_4$.

Separation of Strontium

Strontium in the sample is quantitatively present in the filtrates from the barium chromate. It is precipitated with ammonium carbonate together with calcium. Calcium in the sample is not completely recovered. Strontium may be separated from calcium by precipitation of the nitrate in strong nitric acid solution, but this is a difficult procedure, and flame photometry or AA spectrometry are almost always used, except when strontium is a major constituent (>1–2%).

Determination of Strontium

PROCEDURE. To the filtrate from the barium chromate add a large excess (25 ml saturated solution per 10 ml) of ammonium carbonate. Stir, and allow to stand cold for several hours. Precipitation tends to be slow but is hastened by frequent stirring. Filter, and wash with 5% ammonium carbonate solution.

Dissolve the precipitate in dilute nitric acid, transfer the solution to a weighed platinum dish, evaporate to dryness, heat in an oven at 135 °C, and weigh calcium and strontium nitrates.

Dissolve the nitrates in water, filtering if necessary, and determine strontium by flame photometry or AA spectrometry, preparing bracketing standards containing both strontium and calcium to match the unknown solutions. Only a portion of the total calcium is found here; determining it is only necessary to compensate for its effect on the strontium determination. Having an accurate weight of the calcium and strontium nitrates permits a check on values obtained by the flame methods. Since the separations described have eliminated interfering ions (phosphate, silicate, sulfate, etc.), flame determinations can in this procedure be classed as primary, since calibration is accomplished using pure nitrate solutions.

Preparation of Standard Solutions of Alkaline Earth Nitrates

PROCEDURE. Obtain alkaline earth nitrates that are free from contaminants. It is especially difficult to find calcium nitrate containing no strontium; manufacturers' claims should not be accepted without verification. Weigh appropriate amounts and dissolve in water. Heat the solutions to boiling, and allow to stand for several days. Filter through Whatman No. 44 or S & S blue ribbon papers into volumetric flasks, and dilute to volume with distilled water. Store in properly cleaned Pyrex bottles.

Using calibrated pipettes, transfer appropriate volumes of the solutions to weighed 55-ml flat-bottom platinum dishes. Add a few drops of nitric acid, and evaporate to dryness on the steam bath. Add a few drops of water, and again evaporate to dryness. Transfer the dishes to an oven at 135 °C, and heat to constant weight at that temperature. Dissolve the weighed salts in water, filter through a small Whatman No. 44 paper, and wash thoroughly with hot water. Return the paper to the dish, burn it off at 450 °C, heat to 700–800 °C, and weigh. The residue should weigh no more than a few tenths of a milligram and probably consists mostly of silica. Use the weight of the dish plus residue in calculating the weight of nitrate obtained and hence the metal content of the standard solution.

Calcium in standard solutions may be determined by EDTA titration, but EDTA titration of barium and strontium is not entirely satisfactory. The titration is performed by adding sodium hydroxide solution to a diluted aliquot of the calcium solution to bring pH to about 11, adding murexide and titrating. The sodium hydroxide should be substantially carbonate free. Gravimetric calibration of strontium and barium solutions is preferred.

Strontium (and Barium) by Isotope Dilution Mass Spectrometry

In rubidium–strontium age determinations, the rubidium and strontium are determined on the same sample; sampling problems are thereby as a rule eliminated (only if the sample contains minerals of more than one Rb–Sr age is there any possibility of sampling difficulties; this is the main reason that Rb–Sr ages are more reproducible than K–Ar ages). In isotope dilution rubidium determination, slight strontium contamination of the sample placed on the mass spectrometer filament is of little consequence in the rubidium determination, but all rubidium must be removed before successful analysis of the strontium isotopes is possible. Most often, the separation is accomplished by ion exchange using Dowex 50 8× or 12× resin, usually 200–400 mesh. A shorter and often as effective procedure requires no ion exchange separation:

PROCEDURE. Treat a sample that contains about 250 μg rubidium with nitric, hydrofluoric, and perchloric acids in a platinum dish. For a 500-mg sample, use 1 ml nitric, 15 ml hydrofluoric, and 5 ml perchloric acids. The sample should first be wetted with a few milliliters of water, and nitric acid should be added first, with the dish covered in case carbonate is present. After gentle heating to remove carbon dioxide, the cover should be removed and washed off; then the hydrofluoric and perchloric acids should be added, and the mixture heated, without boiling, at gradually increasing temperatures until the perchloric acid fumes strongly. Repetitive dilution with water and evaporation to fumes of perchloric acid may be necessary, especially with samples containing much magnesium.

When a clear solution has been obtained, add an amount of ^{86}Sr spike solution, that will make the concentrations of ^{86}Sr and ^{87}Sr in the solution of comparable magnitude based on a preliminary rough determination of strontium in the sample (e.g., by flame photometry or AA spectrometry of a lithium metaborate–nitric acid solution as described below). Evaporate the solution to dryness. In some cases, it may be desirable to repeat the hydrofluoric–perchloric treatment to be sure of complete attack of the rock or mineral. Muscovites are especially resistant to this treatment, and unattacked muscovite flakes are difficult to observe. Leach the residue thoroughly with hot water.

For strontium determination, transfer a portion of the solution containing ^{86}Sr spike to a clean platinum dish and add several drops of concentrated sulfuric acid. Evaporate to dryness and heat to remove excess SO_3. Ignite the residue for 5–10 min over a Meker burner. Cool, add a few milliliters of water, and digest on the steam bath to dissolve alkali sulfates. Filter through a small Whatman No. 44 or S & S blue ribbon and wash thoroughly with hot water. Incinerate the paper in the dish, and add a little hydrochloric acid and a solution containing about 50 mg magnesium sulfate. Heat on the steam bath for a few minutes, and add several drops of sulfuric acid. Evaporate and ignite as before, and filter. Incinerate the residue, and treat it with 1–2 ml of 1:1 hydrochloric acid. Digest covered on the steam bath for 10–15 min, dilute with a little water, and filter. Evaporate the filtrate to near dryness, add a few drops of water, and transfer some of the solution to the filament of the mass spectrometer. If iron is present in unusually large amounts, it may have to be removed from the solution by ammonia precipitation; but a small amount of iron improves, rather than interferes with, the strontium analysis.

This time-saving procedure is not universally applicable; with samples containing much iron or much barium or with those of very low strontium content, the ion exchange separation is required.

In isotope dilution barium determination, ion exchange separations are seldom necessary, but it must be noted that reagent-grade perchloric acid often contains several hundred parts per million of barium: Careful blank determinations are essential.

One of the most common of calcium minerals is fluorapatite, $Ca_5(PO_4)_3F$, a major constituent of phosphate rock. Analysis of phosphate rocks and minerals is difficult when their composition is unknown at the start, and any general procedure needs modification, sometimes extensively, depending on the observed behavior during the analysis. Preliminary examination by emission spectrography and/or X-ray spectrometry will save much trouble. Rapid routine methods for phosphate rock analysis, while adequate for their intended purpose, should never be applied outside that purpose.

Primary Analysis of Apatites and Phosphate Rocks

PROCEDURE. Establish a sampling weight baseline by drying to constant weight at 105 °C. Determine total hydrogen by the Penfield method using a 0.5- to 1-g sample and adding 1 g finely ground quartz (-325 mesh) before fusion in a mixture of 5–10 g litharge, 1–2 g lead chromate, and 1 g copper oxide. The quartz addition is very necessary for complete decomposition of hydroxy-apatite, which may be present. Addition of a little calcium carbonate during the final stages of the fusion will help boil the last of the hydrogen from the phosphate melt, which stubbornly retains water even at 1000 °C. Such an addition is easily made during a Penfield determination but is nearly impossible if sample decomposition proceeds in a combustion train.

Determine fluorine on a separate sample, after fusion with sodium peroxide and steam distillation from sulfuric acid solution, by titration with thorium nitrate in 50% alcohol, or using a fluoride specific ion electrode [12].

Determine phosphate on a separate sample after fusion with sodium carbonate containing a little sodium peroxide and about 0.1 g added silica. Remove silica by nitric acid dehydration, precipitate with ammonium molybdate, convert to magnesium ammonium phosphate, and finaly weigh $Mg_2P_2O_7$.

Determine silica in a 1-g sample after fusion with sodium carbonate containing a little peroxide by hydrochloric acid dehydration of silica in the presence of aluminum chloride [13]. Separate the rare earths from the filtrate using oxalic acid, purify by ammonia precipitation at pH 10, weigh total rare earths plus thorium, and determine individual rare earths by emission spectroscopy.

Determine the alkali metals by standard procedures: Lithium determination is important because this element can sometimes cause difficulty by precipitating with the ammonia group elements.

Determine total carbon by combustion and acid-soluble carbon (carbonate) by solution in phosphoric acid. The evolved carbon dioxide may be weighed in an ascarite bulb or by a variety of other methods [14].

Examine the sample for radioactivity; this is usually due to uranium, but some thorium may be present.

Proceed with the main part of the analysis as follows, modifying procedures appropriately depending on the available information already obtained.

Weigh 1.0000 g into a weighed platinum crucible, dry at 105 °C, and reweigh. Add 0.2000 g finely powdered quartz (-325 mesh) of known purity (spec-pure silica is less desirable than quartz), 4 g sodium carbonate, and a small amount of sodium peroxide. How much oxidant to add will be determined by the difference between the total carbon and the carbonate determinations: If no oxidant is added, there is danger of reducing phosphate; if too much is added, attack on the crucible may be severe. Cover the crucible and fuse the mixture over a flame, gradually increasing temperature to the maximum of a good Meker burner. *It is essential that the flame be fuel lean at all times.* Cool the crucible on an aluminum plate: Reheat over the flame for 30–40 s just until the crucible becomes dull red. Cool, and transfer to a 300-ml platinum dish with 50 ml water. Put 5 ml concentrated hydrochloric acid in the crucible, cover it, and set aside. Allow the fusion cake to disintegrate; then add up to 20 ml hydrochloric acid from a pipette, keeping the dish covered. When effervescence has subsided, add the contents of the crucible, washing it out thoroughly. Reweigh the crucible.

Heat the solution of the sample on the steam bath until it is quiet, and then wash off and remove the cover. Evaporate *uncovered* to dryness. If much fluoride is present, add 10–20 ml hydrochloric acid and again evaporate to dryness. Proceed with a double dehydration of silica as in the analysis of silicates. Weigh impure SiO_2.

Determine the silica present by volatilization with hydrofluoric acid, but do not add sulfuric acid; use nitric and perchloric acids instead. From the weight of silica obtained, decide whether fluorine has been effectively removed (if much aluminum is present in the sample, it prevents the volatilization of fluorine). If there is a possibility of residual fluorine, add 50 ml perchloric acid to the solution in a 400-ml beaker, and boil until all hydrochloric acid is removed; then boil for a further 15 min until at least 5 ml of the added perchloric acid has been boiled away.

Make the solution from which silicon and fluorine have been removed 1–2 N in hydrochloric or perchloric acid either by dilution or acid addition, aiming at a volume of 200 ml, and add a solution of zirconium oxychloride of known concentration from a burette using a volume calculated to precipitate all the phosphate in the sample and provide an excess of about 10 mg ZrO_2. Digest at the boiling point for 20–30 min, and filter through a 12.5-cm Whatman No. 541 paper folded inside an S & S blue ribbon paper. Wash the bulky precipitate superficially with 5% hydrochloric acid, return it to the beaker using a minimum of water, add 20 ml concentrated hydrochloric acid, and digest on the steam bath for 0.5 to 1 h. Dilute to 200 ml, and repeat the filtration and washing using the same filter papers. A third treatment of the precipitate may be desirable to be sure of complete removal of calcium and other metals.

Ignite the precipitate of zirconium acid phosphate (after drying) in a large platinum crucible, weigh the zirconium pyrophosphate, fuse it with 10 g sodium carbonate, leach with water, filter, and wash with 2% sodium carbonate solution. Dissolve the precipitate in 1:20 sulfuric acid and examine it for TiO_2 using peroxide. Alternatively, the weighed zirconium pyrophosphate may be thoroughly mixed and then examined spectrographically. Titanium is usually the only contaminant if the washing of the precipitate was carefully done.

The residue in the platinum crucible used for the silica determination must be considered. It is usually small and may be fused with a little sodium carbonate, dissolved in dilute acid, and added to the main solution. However, in some circumstances, it must be treated separately. Tantalum and niobium are quantitatively retained in this residue and, if they are present, should be determined because there is no easier way of separating them. If the sample contains barite, some of the barium sulfate may be present in the residue and should not be added to the main solution. In this case, a pyrosulfate fusion is called for, barium being removed from the solution of the melt as the insoluble sulfate and titanium determined colorimetrically with peroxide. Bismuth phosphate is not very soluble in acids and may be found in the residue; in this case, fusion with sodium carbonate and leaching with water followed by acidification and treatment with H_2S may be necessary.

If the residue from the silica determination is small and does not contain barium, tantalum, or niobium, it may be left in the crucible and weighed together with the ammonia group of elements.

Evaporate the filtrates from the zirconium acid phosphate to remove excessive acid, and carry out a triple ammonia precipitation in the usual manner (if the precipitate is small, a double precipitation may suffice). If rare earths or beryllium are present, it is necessary to treat the filtrate from the ammonia group precipitation with a large excess of ammonia solution at nearly boiling temperature, cool to room temperature, stir in paper pulp, filter, and wash with 5% ammonium hydroxide. The precipitate is invariably contaminated by calcium and manganese and must be dissolved in acid and reprecipitated at least once.

Ignite and weigh the ammonia group oxides. Proceed with the analysis of the ammonia group in the usual manner. In principle, the total weight of ZrO_2 added can be subtracted from the total weight of oxides obtained, and one or another of the oxides can be determined by difference (usually aluminum); it is better, however, not to rely on this weight but to use it as a check on the total only.

If manganese is present in the sample, separate it by persulfate precipitation at pH 6.5. With high-manganese samples (e.g., triphylite), a triple precipitation is advisable to be sure that no lithium or calcium remains. If the sample contains little or no manganese, it is sometimes advantageous to add a known amount (about 10 mg) in the form of a manganese sulfate solution before the persulfate precipitation. This acts as a scavenger, removing small amounts of cerium, cobalt, rare earths, molybdenum, tungsten, and so on, which may have escaped previous separation. Most of any barium and strontium in the solution will also be removed.

The filtrate from the persulfate precipitation usually contains all the calcium, magnesium, and lithium in the sample. It may also contain most of the zinc and cadmium originally present and a little strontium. Perform a double oxalate precipitation, and finally weigh $CaCO_3$ and then CaO. Dissolve the calcium in nitric acid, and determine strontium contamination by flame methods.

Precipitate magnesium in the filtrate from the calcium with phosphate, and weigh $Mg_2P_2O_7$. Recover calcium and make corrections. (If lithium is to be determined, an aliquot of the filtrate from the calcium should be taken prior to the phosphate addition.)

Evaporate the filtrate from the magnesium to small volume on the steam bath and add 50–100 ml concentrated nitric acid. Have a cold water bath handy. Heat on the steam bath until reaction becomes violent, and then transfer the beaker to the cold water bath. Continue heating until ammonium salts are destroyed, evaporate to near dryness, and take up the residue with dilute hydrochloric acid. Leach thoroughly, and filter out insoluble material (mostly silica from the glassware). Precipitate zinc and cadmium as their ammonium phosphates, and ignite and weigh $(Zn,Cd)_2P_2O_7$. Separate cadmium with hydrogen sulfide and determine as sulfate.

Determine lithium in an aliquot of the filtrate from the calcium oxalate after evaporation with nitric and sulfuric acids (to remove ammonium salts and oxalate) by flame photometry in neutral sulfate solution using pure lithium and sodium sulfates to calibrate. Despite the high concentration of sodium, lithium can be determined with fair accuracy if it is present in percentage amounts in the sample. The standard sodium sulfate solution should be prepared from the sodium carbonate used in the dissolution of the sample in case it contains appreciable lithium. With many lithium phosphate minerals, it is possible to dispense with the sodium carbonate fusion at the beginning of the analysis; in this case, only incidental sodium is present in the final stages, and an excellent lithium determination is possible.

4.4.3. Secondary Methods

The alkaline earths are now most frequently determined routinely by AA spectrometry. This technique is rapid and precise to 1–2% of the amount present and, when properly applied, is more than adequate for most purposes. It cannot, however, be given the status of a primary technique unless it is applied to solutions from which all interferences have been removed. When solutions of rocks, ores, or minerals obtained by acid digestion (HF, HNO_3, and $HClO_4$) or fusion solution ($LiBO_2$–HNO_3,

Na_2O_2–HCl, etc.) are examined for their calcium, strontium, or barium content by AA spectrometry, it is best to similarly dissolve known samples of approximately the same composition and use these as calibrating standards. If this is done, results compare in accuracy with all but the most sophisticated primary procedures. In fact, there are instances where only thoroughly competent analysts can produce better values by primary methods. One should not forget, however, that without having the competent primary analyses to begin with, accuracy by secondary methods is not possible. The same remarks apply, of course, sometimes in greater degree, to X-ray fluorescence analysis, emission spectrometry, and other rapid techniques.

Signals generated by the alkaline earths in AA flame methods are grossly affected by silicate, phosphate, sulfate, and in fact almost all anions. These interference or matrix effects may be suppressed, but not entirely removed, by the addition of releasing agents, of which the most commonly used are lanthanum salts. Lanthanum in the flame in concentration much greater than those of the elements being determined diminishes the anion effects to a substantial degree.

4.5. SCANDIUM, YTTRIUM, AND THE RARE EARTHS: PRECIPITATION REACTIONS

The rare earth metals are not as rare as the name implies: Cerium, yttrium, and lanthanum are more abundant than "common" elements like arsenic, antimony, and mercury. Thulium, the rarest, is more abundant than gold or platinum. The rare earths are widely distributed and occur in traces in most rocks, especially in phosphates. Sphene and syenite may contain 1% or more.

During the classical analysis, the rare earths are precipitated—usually incompletely—with the ammonia group. For complete recovery, the filtrate from the ammonia group should be treated with excess carbonate-free ammonia and tannin.

Because the rare earths are separated from each other only with difficulty, they are normally isolated as a group and determined instrumentally, most often by emission spectroscopy.

Isotope dilution mass spectrometry permits measurement of lanthanum, cerium, neodymium, samarium, europium, gadolinium, dysprosium, erbium, ytterbium, and lutetium, but there are many difficulties: Scandium, yttrium, praseodymium, terbium, holmium, and thulium have only one isotope; this precludes their accurate determination by this technique.

AA spectrometry is not generally very satisfactory; sensitivities are not good, and there are many poorly understood interferences. X-ray spectrometry can give acceptable results for the lighter earths (usually after chemical separation and concentration) but fails with the minor heavy earths (Eu, Tm, Lu). ICP spectrometry gives satisfactory results [15–19].

Europium is of special interest because of its use as a red phosphor in cathode-ray tubes and because it is used as an indicator of oxidation potential in geologic processes. Europium is the only rare earth that can exhibit a stable valence of II in aqueous solution; in this condition, it acts like an alkaline earth (e.g., barium). A deficiency of europium in a geologic formation may indicate a reducing environment at some time in history. Samarium and ytterbium also form bivalent salts, but these are too unstable to be analytically useful.

Europium can be determined, when present in quantity, by reduction with amalgamated zinc in acid chloride solution followed by iodine titration. It may also be determined polarographically; the half-wave potential of $Eu^{3+} + e \rightarrow Eu^{2+}$ is about -0.7 V versus the standard calomel electrode.

The rare earths may be completely precipitated as hydroxides, oxalates, or fluorides. Precipitation of scandium is not complete; there seems to be no reliable way to quantitatively precipitate traces of scandium from solution. Scandium ammonium tartrate is nearly insoluble, and tartrate may be used to separate relatively large amounts from a number of elements, including the rare earths.

Precipitation as hydroxide requires a strongly basic solution (pH approaching 14) for complete precipitation of lanthanum, neodymium, and praseodymium. The presence of ferric iron improves recovery. The rare earths form weak carbonate complexes—Y(III) and Yb(III) are very soluble in potassium carbonate solution—so that carbonate concentration should be kept low. Peroxide complexes are also possible; thus, the leach of a peroxide fusion should be heated to remove all peroxide if rare earth recovery is to be complete.

Precipitation with ammonia requires that a large .excess be present, carbonate contamination be kept to a minimum, and concentration of ammonium salts be low. Precipitation should be from hot solution followed by cooling and long standing. Paper pulp should be added just before filtration, and the precipitate should be washed with 5% ammonia solution. Addition of tannin may be necessary for complete recovery.

Precipitation as oxalate requires a large excess of oxalic acid and a minimal concentration of other acid anions. Precipitation is not complete unless calcium is present or added. Calcium precipitates only slowly from

5% oxalic acid, and filtration of the rare earth oxalates should be delayed until some calcium oxalate appears, usually as gritty crystals that adhere to the glass. Scandium is not completely precipitated. Traces of yttrium and the heavier earths may not be precipitated if ammonium salts are present.

Precipitation as fluorides is usually employed during attack of minerals containing titanium, tantalum, or niobium. Treatment of these minerals or ores containing them with hydrofluoric acid, followed by dilution, will leave all the rare earths as insoluble fluorides and titanium, tantalum, niobium, iron, aluminum, and several other elements in solution. The mineral or ore should be warmed with concentrated HF until reaction is complete but should not be evaporated to dryness. After 1:1 dilution, filtration, and washing with dilute HF, paper and precipitate should be treated with sulfuric and nitric acids in platinum to destroy organic matter, and most of the sulfuric acid should be fumed off. The residue may then be sintered with sodium peroxide and the rare earths separated as hydroxides (see below).

Usually, the rare earths and yttrium may be separated from rocks, minerals, or residues by the following procedure or some modification thereof.

PROCEDURE. Transfer up to a 5- or 10-g sample to a large iron crucible and mix with 5–10 times its weight of sodium peroxide. If the sample contains little calcium, add 100 mg $CaCO_3$. Cover with a little sodium carbonate and heat slowly to incipient fusion. Cool, and leach with 100–500 ml water. Heat on the steam bath until peroxide is decomposed (several hours). Cool to room temperature and filter through a Whatman No. 541 paper folded inside a retentive paper. Wash with 3% sodium hydroxide containing a little sodium carbonate.

Using about 100 ml water, transfer the precipitate to a beaker. Treat it with 100 ml of 10% oxalic acid added all at once. Wash the filter with 5% oxalic acid and reserve it. Heat the oxalate solution on the steam bath (without stirring) until iron and titanium hydroxides are decomposed and dissolved. Cool.

Allow to stand with frequent stirring until gritty crystals of calcium oxalate can be detected sticking to the sides or bottom of the beaker.

Filter through the same papers as before, and wash with 5% oxalic acid. Burn off paper in a platinum dish or crucible at 450–500 °C.

From this point, procedure depends on the nature of the sample and the readout device to be used. Direct solution of the residue in dilute acid [with SO_2 addition to reduce Ce(IV)] may suffice if a solution technique (ICP or spark solution spectrometry) is intended. If dc arc spectroscopy is to be used, the rare earths had better be freed from calcium and other elements by ammonia precipitation. An internal standard element may be added during this operation.

PROCEDURE. Treat the residue containing the rare earths with 6 N hydrochloric acid and a little sulfuric acid, and heat on the steam bath until solution is complete. Evaporate to dryness, take up the chlorides in dilute HCl, and filter to remove silica, and so on. If the residue is large, it should be treated with HF and H_2SO_4, fused with a *little* pyrosulfate, and returned to the solution— or it may be examined separately for its rare earth content.

Remove all but 0.5–1 ml acid by evaporation, dilute appropriately with water, and add one-tenth the volume of concentrated carbonate-free ammonia. Allow to stand until cold, add paper pulp, and filter. Wash with 5% ammonia. Redissolve the hydroxides and reprecipitate as before. If more than a few milligrams of rare earths are present, ignite and weigh the oxides. Add radiation buffer and internal standard and analyze by dc arc spectroscopy.

Strontium carbonate and scandium oxide have been used as buffer and internal standard, respectively. Analytical line pairs are as follows:

Analytical Line	Internal Standard Line
Ce 4133.80	Sc 4023.68
Eu 4129.73	
Pr 4100.74	
Nd 4023.00	
Th 3469.92	
Tm 3462.20	Sc 3359.67
Ho 3456.00	
Er 3364.09	
Gd 3350.48	
Y 3330.88	
Tb 3324.40	
Dy 3319.88	
Sm 3306.37	
La 3303.11	
Lu 3281.74	
Yb 3192.87	

A major source of rare earths is monazite sand, a phosphate of the cerium metals (La, Ce, Pr, Nd, Sm) and thorium, usually containing grains of rutile, zircon, and various silicates. Analysis of monazite sand proceeds as follows:

PROCEDURE. Heat 10 g (or more) monazite sand in a silica dish with 20 ml (or more) concentrated sulfuric acid in a sand bath at temperatures up to 200–

250 °C (incipient fumes). When decomposition is complete, cool and add 300–400 ml water. Digest on the steam bath until solution is as complete as possible, and saturate the solution with hydrogen sulfide. Filter, and wash with cold 5% sulfuric acid saturated with H_2S; catch the filtrate in a 500- or 1000-ml volumetric flask. Cool and dilute to volume with water. Take aliquot portions of the solution for determination of rare earths, phosphate, and so on.

A large sample is necessary because of sampling consideratons. Unless the sand is very coarse, it is best to treat it without grinding. A preliminary separation of zircon, quartz, rutile, and so on is thereby accomplished.

To be sure of a representative subsample, the portion for analysis should be split out using a microsplitter.

Rare earths are separated from an aliquot (50–100 ml) of the sample solution by adding an equal volume of 10% oxalic acid, digesting on the steam bath (without stirring) until the precipitated oxalates become crystalline, and then allowing to cool with frequent stirring. The oxalates must be filtered off, ignited to oxides, dissolved in hydrochloric acid, and reprecipitated with oxalic acid. The double precipitation is necessary to remove phosphate.

The combined filtrates from the oxalates may be examined for iron, titanium, zirconium, and aluminum. These elements are separated from the oxalate solution by adding excess ammonia and tannin.

The precipitated oxalates may contain thorium and calcium. They should be ignited to oxides and dissolved in nitric acid. A double ammonia precipitation removes the calcium. Thorium can then be separated as peroxynitrate:

PROCEDURE. Dissolve the ignited oxides in 50 ml of 10% nitric acid, add ammonia to neutral, cool, and add one-tenth the volume of concentrated carbonate-free ammonia. Filter, redissolve in nitric acid, and reprecipitate the hydroxides. Again dissolve in nitric acid and evaporate the solution to dryness on the steam bath. Add water and reevaporate to remove all HNO_3, and dissolve the residue in 50–100 ml of 10% ammonium nitrate solution. Filter to remove any silica and wash with 10% ammonium nitrate. To the slightly warm solution add 30% hydrogen peroxide in slight excess. Filter, and wash with 10% ammonium nitrate containing a little peroxide. Dissolve the precipitate in dilute nitric acid and repeat the precipitation as before.

Boil the combined filtrates to remove peroxide and precipitate the rare earths with oxalic acid. Dissolve the thorium peroxynitrate in nitric acid and precipitate the thorium as oxalate. Ignite the oxalates to oxides and weigh.

Precipitation of the rare earth oxalates is best done by adding excess saturated oxalic acid all at once to the hot nearly neutral solution, digesting, without stirring, for an hour or more, and then allowing to cool with frequent stirring. Precipitation of thorium oxalate is best done by adding ammonium oxalate (5 g/100 ml) to the nearly neutral solution, and then adding 10 ml hydrochloric acid dropwise and slowly with stirring, interrupting the acid addition while a precipitate is forming.

The procedures given above are applicable to milligram amounts of the rare earths; traces will not be recovered. Precipitation of traces requires use of a carrier (calcium in oxalate precipitation, ferric iron in hydroxide precipitation). Scandium is never completely precipitated.

Scandium, in other than trace amounts, may be precipitated as the ammonium tartrate:

PROCEDURE. Evaporate a hydrochloric acid solution nearly to dryness on the steam bath, dilute to 50–100 ml, and pour the solution into 100 ml of 20% ammonium tartrate solution. Heat nearly to boiling; if any precipitate appears, add a little hydrochloric acid. Add dilute ammonia drop by drop to the boiling solution until an excess is present. Cool, filter, and wash with 2% ammonium tartrate solution.

Thorium, rare earths, zirconium, titanium, aluminum, and iron are not precipitated. Some yttrium may precipitate with the scandium.

Scandium acid thiocyanate is extracted into ether from 50% ammonium thiocyanate solution 0.5 N in hydrochloric acid (Fischer and Bock [20]; also see Sandell [21]). With equal volumes of ether and aqueous phase, three extractions will recover substantially all the scandium: Each extraction requires the addition of 5 ml of 2 N hydrochloric acid per 100 ml of solution to replace hydrogen ion extracted as HSCN.

Scandium may be recovered from the ether by evaporation to remove ether, then heating with hydrochloric and nitric acids. Other elements that may be partly extracted with the scandium include Fe(III), beryllium, aluminum, indium, titanium, and U(VI). Iron(III) may be extracted into ether from 6 N HCl prior to the thiocyanate addition or reduced to Fe(II), which does not extract.

4.6. TITANIUM, ZIRCONIUM, AND HAFNIUM

Titanium. Titanium is a very common element in the lithosphere (e.g., the average content of titanium is around 0.23% in granites, 0.90% in basalts, and 0.80% in syenites). Some titanium minerals are ilmenite ($TiFeO_3$), rutile (TiO_2), perovskite ($CaTiO_3$), pyrophanite ($MnTiO_3$), sphene, or titanite (SiO_2, TiO_2, CaO). Details concerning some of these minerals are given by Pascal [22].

Vinogradov [23] showed that considerable amounts of titanium occur in unweathered particles of clay, micas, amphibole, and lepidomelane. Bauxites are rich in titanium, zirconium, and hafnium.

Titanium has five naturally occurring isotopes: ^{48}Ti (73.94%), ^{46}Ti (7.93%), ^{47}Ti (7.28%), ^{49}Ti (5.51%), and ^{50}Ti (5.34%).

High strength and low density make titanium metal an economical and attractive material in advanced technology.

Separation of the ammonia group after filtration of silica results in precipitation of iron, aluminum, titanium, and phosphorus. [For example, two ammonia precipitations were performed followed first by ignition of oxides (see Section 3.1.4) and then by the recovery of small amounts of silica. The residue from the silica determination may contain some titanium and zirconium.] Titanium may be determined after combining the dissolved melt of fused residue from recovered silica with the main solution containing the ammonia group metals. In the primary method, titanium is determined colorimetrically with hydrogen peroxide.

PROCEDURE. To the solution of the ammonia group oxides (in a 100-ml volumetric flask or a larger flask if TiO_2 exceeds 1% of the sample), which should contain sulfuric acid between 0.7 and 2.5 M, add 1 ml of 30% hydrogen peroxide. Titanium(IV) gives an orange complex with hydrogen peroxide. Dilute to the mark, mix, and compare immediately with similarly prepared solutions of known titanium content using a Dubosq colorimeter.

Iron(III) may interfere. It could be complexed by adding as small quantities as possible of phosphoric acid, which complex ferric iron as ferric phosphate. However, phosphoric acid has a slight bleaching effect on the titanium that affects sensitivity.

It is important to use the same quantities of H_3PO_4 in prepared solutions of known titanium content. The limit of detection should be around 30 ppm titanium in the final solution.

Fusion or sintering with sodium peroxide is suitable for the decomposition of rocks containing titanium minerals as indicated by Dolezal [24].

PROCEDURE. Fuse about 1 g Na_2CO_3 in a platinum crucible and spread the melt over the crucible walls. This precaution, according to Harpham [25], allows the fusion to be carried out without damaging the platinum crucible. Then mix 0.5 g -200-mesh sample with 5 g sodium peroxide using a glass or platinum rod. Weigh 1 g sodium carbonate onto a piece of hard-surfaced paper. Rinse off the stirring rod in this. Transfer the sodium carbonate to the crucible, covering the peroxide sample mixture as completely as possible. Fuse for 5 min at 650 °C. After cooling rapidly, the melt is leached out with water and then dissolved with sulfuric acid. Dehydration of silica follows, and the ammonia group is separated as usual.

The following procedure is recommended for the analysis of sphene.

PROCEDURE. Fuse 0.5 g -200-mesh material with 3.5 g $LiBO_2$ in a large graphite crucible. Heat 20 min between 950 and 1000 °C. Pour the melt into 250 ml of 4% HNO_3 and stir to solution. Transfer, in stages, to a 250-ml beaker, adding 7 ml H_2SO_4. Evaporate to fumes. Add a little water and then 100 ml of 10% oxalic acid. Heat, stir, filter, and wash with 2% H_2SO_4. Put the residue in a platinum dish, burn off the paper at 500 °C, add sufficient 1:1 HCl to dissolve the precipitate, evaporate to dryness as in the usual silica determination. Add HCl, and filter. Wash with hot dilute HCl and then with water. Ignite the impure silica, weigh, and determine SiO_2 by evaporation with HF and H_2SO_4. Fuse the residue with $K_2S_2O_7$, leach with water, and add to the main solution, which contains the major constituents of sphene (Ca, Ti, Si) as well as appreciable Al, rare earths, etc..

Evaporate the solution to about 100 ml, heat to boiling and add ammonia slowly to pH 5. A gram of ammonium carbonate may be added to facilitate the precipitation of strontium. Allow to cool, stand cold until calcium oxalate starts to precipitate, and filter off Ca, Sr, rare earth oxalates. Dissolve the oxalates in a minimum of 1:1 HCl and reprecipitate from chloride solution with oxalic acid. (See Section 4–5). Finally ignite to carbonates at 500 °C.

Dissolve the rare earth carbonates (which contain some of the calcium in the sample) in 50 ml of 5% HCl and boil to remove carbon dioxide. Precipitate the rare earths by adding a large excess of carbonate-free ammonia to a hot solution, and allow to cool before filtering. Dissolve the precipitate in minimum dilute HCl and reprecipitate with ammonia. Ignite and weigh the rare earth oxides. Precipitate calcium in the filtrate as oxalate, ignite at 500 °C and weigh $CaCO_3$.

Half saturate the main solution containing the titanium as oxalate with NH_4Cl (about 40 grams). Heat to boiling, add 8 g of tannin dissolved in a little water. Cool, filter, and wash with a solution containing 30 g NH_4Cl, 10 g ammonium oxalate and 1 g tannin in 500 ml.

To the filtrate from the tannin precipitate add ammonia in 10% excess to precipitate Zr, Mg, Al as phosphates. Filter, wash with 5% ammonia, ignite in platinum at 500 °C., sinter with Na_2O_2, leach with water and filter. Determine Al_2O_3 in the filtrate as $AlPO_4$. Treat the precipitate with HCl, precipitate Zr as phosphate, weigh ZrP_2O_7. Precipitate Mg as usual and recover calcium from the weighed $Mg_2P_2O_7$.

Dry the tannin precipitate containing Ti, Ta, Nb and ignite at 500 °C. Fuse with pyrosulfate and recover SiO_2 as usual by fuming with sulfuric acid. Precipitate Ti, Ta, Nb with ammonia, ignite and weigh. Determine Ta, Nb in the residue spectrographically, by XRF, or by any other available method and correct the weight of TiO_2.

Zirconium and Hafnium. Zirconium may be found in granites and syenites in amounts ranging from 0.015 to 0.15% as well as in basic rocks such as basalts. Very often, hafnium is found with zirconium, and according to von Hevesy [26], the atomic ratio of zirconium to hafnium in minerals is around 100. For a few minerals, this ratio may be much smaller (e.g., alvite and thortveitite).

Some zirconium minerals are zircon ($ZrSiO_4$), baddelyite (ZrO_2), and malacon. Zirconium minerals are resistant to weathering and therefore may accumulate in sediments and soils.

Zirconium has five naturally occurring isotopes, and hafnium has six. The neutron capture cross section of these two elements is drastically different, very low for zirconium (e.g., makes this metal attractive in nuclear reactor technology) and very high for hafnium (e.g., may be used as an effective neutron-absorbing material). Outstanding mechanical properties combined with excellent resistance to corrosion by acids and alkalies may give to these metals an important place in chemical technology. Zirconium crucibles are used in laboratories with great success to perform fusions; however, in classical analysis, it may present serious disadvantages because forming peroxy compounds and also stable complexes such as phosphates, sulfates, and especially fluorides and oxalates.

In analytical chemistry, only the quadrivalent state is of interest for these metals. Compounds such as phosphate and arsenate are insoluble. Solutions of Zr(IV) polymerize strongly and often become colloidal, which may be prevented by addition of sulfuric acid [27]. For the analysis of a zircon sand Baud and Meurice [28] recommended the following procedure.

PROCEDURE. Fuse 1 g of -325-mesh sample with 10 g sodium hydroxide for 5 min. Leach the melt with warm water, dissolve with sulfuric acid, and evaporate on a sand bath until formation of white fumes. Add water and separate silica by filtration. To the clear filtrate add some hydrogen peroxide. Bring to a boil and precipitate with ammonia. Bring to a boil again, treat the hydrated oxide with nitric acid and hydrogen peroxide, filter, and add some 10% ammonium arsenate, boil, filter, dehydrate, and finally ignite at 1000 °C until constant weight of the ZrO_2 oxide.

For a good separation of, for example, Zr(IV) from Ti(IV), Fe(III), V(V), Th(IV), Mo(VI), W(VI), and Al(III), Charlot [29] proposes the following procedure:

PROCEDURE. To a 30-ml aliquot containing between 50 and 200 mg zirconium in molar HCl, add 15 ml concentrated hydrochloric acid and then 50 ml of a 15% mandelic acid solution (phenylglycolic acid). Let stand for 20 min at 85 °C. Filter. Wash with a hot washing solution made of 2% hydrochloric acid and 5% mandelic acid. For more details consult the literature [30–32]. Hafnium is following zirconium during the separation. Then, depending on the amount of zirconium present, a titration using EDTA [33] or a colorimetric determination using xylenol orange [34, 35] may be performed.

Hafnium is separated from zirconium by ion exchange. Then it may be determined separately by gravimetry as HfO_2 or by colorimetry with formation of the xylenol orange complex. Charlot [36] recommends a gravimetric determination of both zirconium and hafnium without separation, by weighing both elements first as oxides ZrO_2 + HfO_2 and second as selenites $Zr(SeO_3)_2$ + $Hf(SeO_3)_2$. Then solve the two equations for the two unknown values.

4.7 TANTALUM AND NIOBIUM

Tantalum and niobium (columbium) invariably occur together, mutually replacing each other in columbite–tantalite $(Mn,Fe)O·(Nb,Ta)_2O_5$ or microlite $Ca_2(Ta,Nb)_2O_7$. Microlite often contains uranium (probably as UO_2^{2+}) substituting for calcium and NaF substituting for CaO. Analysis of a microlite appears on Section 3.2.8. When niobium is the major element, the term *pyrochlore* is used.

Minerals and ores of tantalum and niobium are often very complex, with titanium, rare earths, tin, antimony, tungsten, zirconium, uranium, and vanadium likely to be present. Their accurate analysis presents numerous difficulties.

The solution chemistry of tantalum and niobium is marked by an extreme tendency to hydrolyze, yielding hydrated oxides or acids; these are generally insoluble in mineral acids (except HF) but form colloids or soluble heteropoly compounds (compare silicon). Hydrofluoric acid yields fluoride complexes, but these do not exist as simple ions in solution; there is always a degree of hydrolysis and polymerization. Potassium fluorotantalate K_2TaF_7 is not very soluble in dilute hydrofluoric acid; the corresponding niobium compound is much more soluble but tends to hydrolyze to K_2NbOF_5, which is moderately soluble in dilute hydrofluoric acid. Fluorotantalate and fluoro-oxyniobate behave very differently in ion exchange and cellulose columns, and this provides a means for separating the elements.

A useful method for attacking tantalum and niobium minerals involves fusion with sodium acid fluoride; however, the fluorides are appreciably volatile, and fusion temperature must be kept low. Sodium fluoride is preferred over potassium fluoride because the sodium fluoro compounds are soluble.

In the removal of silica by HF treatment from materials containing tantalum and niobium, an excess of sulfuric or nitric acid must be present, and evaporation must be slow and at low temperature if losses are to be avoided.

Fusions of tantalum and niobium minerals with sodium carbonate or hydroxide yield water-insoluble sodium salts. With potassium carbonate or hydroxide, solutions containing niobate and tantalate are obtained, but there is a strong tendency to hydrolyze, with partial precipitation of the oxyacids.

Fusions with alkali pyrosulfate attack most tantalum and niobium minerals; leaching the melt with water results in precipitation (usually incomplete) of the oxyacids. Extraction of the melt with oxalic or tartaric acids yields a solution in which the metals are loosely complexed; addition of free acid and boiling leads to hydrolysis of the tartrate complexes and (incomplete) precipitation of the oxyacids.

While hydrolysis leads to precipitation of the oxyacids from mineral acid solution, precipitation is never complete; colloids, loose complexes, and soluble heteropoly compounds are formed. The precipitation may be completed by the addition of tannin, which yields an insoluble sulfur yellow complex with tantalum and an insoluble bright vermilion complex with niobium. Titanium (brick red), tungsten (brown-orange), tin (off-white), and some other elements may precipitate. The iron and vanadium tannin complexes are dark blue to black.

Tannin precipitates tantalum, niobium, and titanium from weakly acid oxalate solution containing ammonium chloride. If an oxalic acid solution is treated with ammonia, ammonium chloride, and tannin, tantalum precipitates first, then titanium, and then niobium. By careful control of reagent addition, it is possible to achieve a partial separation into fractions enriched in each element. By repetitive treatment, a more or less complete separation is possible, but the development of other methods of separation has made this lengthy and difficult process obsolete. A preliminary partial separation using tannin may nevertheless be useful. For additional information see Dams and Hoste [37].

Tannin is a natural product obtained from Turkish or Chinese nut-galls and is inevitably variable in constitution. Each reagent batch should be checked for performance. Analysis and correction for ash content is necessary to avoid error. Attempts to use pure chemicals (e.g., digallic acid) to replace tannin have not been successful.

During the classical analysis, niobium and tantalum are rendered insoluble during the acid dehydration and weighed with the impure silica. If phosphates or fluorides are present, recovery is likely to be incomplete.

If tantalum and niobium are present in substantial amount, the residue from the silica determination should be treated separately from the main part of the ammonia group. How to analyze this residue depends on its composition: Besides minor amounts of iron, titanium, and other common elements usually present, it may contain tin (as unattacked cassiterite), tungsten, zirconium (as zircon or zirconium phosphate), and perhaps unattacked chromite. A general procedure follows. It may require modification in a particular case.

PROCEDURE. Fuse the residue from the silica determination with 1–5 g potassium pyrosulfate, and leach with a small volume (5–20 ml) of 5% sulfuric acid. Transfer to a beaker and add 50–100 ml of a 2% solution of tannin in 3% (v/v) sulfuric acid. Digest on the steam bath with occasional stirring for an hour or more and allow to cool overnight. Add paper pulp and filter. Wash with 3% sulfuric acid containing a little tannin. Ignite and weigh impure $(Nb,Ta)_2O_5$. Add nitric acid to the filtrate, evaporate to strong fumes of sulfuric acid to destroy tannin, dilute, and add the solution to the filtrate from the silica.

The weighed product may contain, besides tantalum and niobium, some titanium, germanium, and tungsten. It may also contain silica and other elements originating in the tannin. The simplest way to analyze this residue is to fuse it with pyrosulfate and cast the melt into a disc for X-ray analysis, calibrating by putting weighed amounts of pure oxide through an identical procedure. With large amounts of niobium and tantalum, the two elements should be partially separated using tannin; each fraction should be weighed and then examined by X-ray:

PROCEDURE. Fuse the oxides with 1–5 g potassium pyrosulfate and dissolve the melt in 50–500 ml of 3% ammonium oxalate solution. Heat to boiling and maintain a slow steady boil (put a piece of hardened filter paper under the end of the stirring rod). Slowly add freshly prepared 2% tannin solution and observe the colors produced.

If tantalum is present, a yellow color will develop (not necessarily a precipitate); further addition of tannin may produce a brown or red color. Tannin addition should be discontinued at this point. When a brown or red color appears, add 10–25 ml saturated ammonium chloride solution to the gently boiling solution. If there is an appreciable precipitate, add paper pulp, cool, filter, and wash with a solution containing ammonium oxalate, tannin, and ammonium chloride in approximately the same concentrations as the working solution.

Heat the filtrate to a gentle boil and add 1:10 ammonia solution drop-wise. When an appreciable precipitate has formed, cool, filter, and wash as before.

Again heat the filtrate to boiling and continue addition of dilute ammonia until precipitation is complete (pH ~ 5). A further addition of tannin may be necessary. Filter and wash as before.

Analyze the three precipitates as follows:

PROCEDURE. Burn off paper and tannin and ignite to oxides; weigh. Add 5 g (accurately weighed) of potassium pyrosulfate and fuse at the lowest effective temperature. Prepare a mold by placing a glass ring (3 mm high, 15 mm in diameter) on a microscope slide on a hot plate. Heat this assembly to about 350 °C. Pour part of the pyrosulfate melt into the ring and allow to cool.

Examine the pellet by X-ray. Prepare similar pellets using various quantities of pure oxides to match the unknown. From the weights of oxides and ratios of X-ray signals, calculate the concentration of oxides in the sample. During the tannin separation, the various elements that may be present are precipitated in approximately the order tantalum, titanium, niobium, tin, tungsten, vanadium, and iron. Complete separations are probably impossible, but the process can be made to yield fractions enriched in one or another of the elements.

Some experience is necessary for success, and it is advisable to prepare oxalate solutions with known amounts of, for example, tantalum and niobium and experiment with them before attempting analysis of an unknown. In analyzing a microlite, more than 90% of the tantalum and niobium were found in their two fractions, respectively.

The separation becomes increasingly difficult as the titanium content of the mixed oxides increases; The brick-red color of the titanium complex obscures the yellow and the bright vermilion of the tantalum and niobium complexes.

An advantage of the X-ray procedure is that all pyrosulfate fusions can be quantitatively recovered and retreated if fractionation of, for example, tantalum, titanium, and niobium turns out to be inadequate. They may also be used for colorimetric or titrimetric determination of the various elements.

4.7.1. Determination of Trace Amounts in Rocks and Ores

Trace analysis for niobium and tantalum is difficult. Direct determination (e.g., by X-ray or emission spectrography) yields poor results for a variety of reasons, especially for tantalum. With small samples, sampling errors are often gross. Solution techniques (AA and ICP spectrometry, spectrophotometry) suffer from the difficulty with which clean solutions of the elements can be prepared. For accurate results, preliminary concentration from a large sample is desirable. Details of the separation will differ according to the nature of the sample, but the following procedure may be used as a guide.

PROCEDURE. Treat 5–10 g of 100-mesh (or finer) sample with hydrochloric acid in a 200-ml platinum dish. Heat, covered, on the steam bath as long as any of the sample is dissolving, adding more acid if necessary; then evaporate to dryness to dehydrate soluble silica. Cool, add 0.5 ml concentrated hydrochloric acid and allow to stand for 5–10 min. Add 50 ml of a 5% solution of hydroxylamine hydrochloride, stir in some paper pulp, add 25 ml of 2% tannin solution, and digest on the steam bath for an hour. Cool, filter, and wash with 2% ammonium chloride containing a little tannin.

Burn off paper and tannin in the platinum dish, add 2–5 ml 1:1 sulfuric acid and 20 ml or more of hydrofluoric acid. Evaporate on the steam bath and then on a hot plate to fumes of sulfuric acid. Add 2–10 g potassium pyrosulfate and fuse at low temperature until only insoluble minterals (zircon, quartz) remain undissolved.

Cool, and leach the melt with 50–100 ml of 3% sulfuric acid. Transfer to a beaker. Add 20 ml of 2% tannin solution. Boil, and allow to cool overnight. Add paper pulp, filter, and wash with 2% sulfuric acid containing tannin. Burn off paper and tannin.

If the residue is small, it may be fused with 5 g pyrosulfate and the melt examined by X-ray as described above. Further separation is usually desirable:

PROCEDURE. Fuse the residue (which may contain calcium, magnesium, titanium, zirconium, manganese, iron, rare earths, and several other elements) with 2 g (or more) of potassium pyrosulfate. Cool, and leach the melt with 20–30 ml of 4% ammonium oxalate. Filter, and wash with water. Add dilute hydrochloric acid to a methyl red end point (pH ~ 5). Add 5 g ammonium chloride and 10 ml of 2% tannin solution. Digest on the steam bath for 1–2 h, cool, add paper pulp, and filter. Wash with 2% ammonium chloride containing tannin. Burn off paper and tannin in a silica crucible and weigh. The final product may contain besides Nb_2O_5 and Ta_2O_5, some TiO_2, SnO_2, WO_3, and SiO_2. It may be fused with pyrosulfate and examined by X-ray, dissolved in hydrofluoric acid for chromatographic separation of the niobium and tantalum, or treated for photometric determination, using thiocyanate for niobium (very satisfactory) and pyrogallol for tantalum (less satisfactory).

4.7.2. Ion Exchange Separation

From hydrochloric–hydrofluoric acid solution, tantalum and niobium are strongly absorbed on Dowex-1 8X resin. The following procedure is essentially that of Kallmann, Oberthin, and Liu [38].

PROCEDURE. Prepare a resin column 12 in. long and 1 in. in diameter containing Dowex-1, 100–200 mesh, washed with dilute nitric acid and screened to remove −200-mesh material and then washed several times with HF–HCl solution 1:250 ml HCl, 200 ml HF (concentrated reagents) in 1 l.

Decompose the sample or residue containing not more than 0.2 g of either tantalum or niobium using HF–HCl, sodium bisulfate fusion, or sodium peroxide fusion. Potassium salts may not be used because of the insolubility of K_2TaF_7. If sodium peroxide is used, the water leach of the melt must be neutralized with sulfuric acid before proceeding. Sodium pyrosulfate fusions ae leached with solution 1 without water addition. In every case, the final solution must contain the same concentrations of HF and HCl as solution 1.

Pass several column volumes of solution 1 through the column, and then add the cooled solution of the sample after removing insoluble minterals such as rare earth fluorides, cassiterite, or zircon by filtration.

Tantalum and niobium are absorbed on the first few centimeters of resin. Most other elements pass through the column. Tin and antimony may be partly retained, but titanium, tungsten, molybdenum, and zirconium are not. It may be convenient to wash them from the column with one column volume of solution 1 so that they may be determined in the effluent. Rare earths and thorium will have been precipitated as fluorides and mostly removed during the preliminary filtration.

Elute niobium from the column with 350 ml of a solution containing 140 g NH_4Cl and 40 ml/l concentrated HF. The flow rate should be about 125 ml/h.

Neutralize 350 ml NH_4Cl–HF solution with ammonia (to pH 6) and pass this through the column to elute the tantalum.

The niobium and tantalum are best separated from their respective solutions by precipitation with cupferron after complexing fluoride with excess boric acid (15 g) and adding 60 ml hydrochloric acid. If only small amounts are in question, addition of 5–10 mg zirconium (as oxychloride) may be added as a collector.

The ion exchange procedure is most useful when a large number of similar samples are to be analyzed. A number of columns can be set up, and the process can be semiautomated so that one technician can handle many samples. When one or two samples are in question and a complete analysis is required, the process is less attractive.

Niobium can be separated from titanium and tantalum using oxine. The mixed oxides are obtained in sodium acid sulfate solution containing tartrate and ammonium chloride. The pH is adjusted to 6.0 with ammonia, and niobium is precipitated by dropwise addition of pH 6 oxine solution. Titanium and tantalum remain in solution, from which they may be recovered with tannin. A double precipitation is advisable. A problem is the possibility of loss of niobium during ignition and the difficulty in washing the precipitate free of sodium salts.

Separation of niobium and tantalum from tungsten is difficult. The following procedures are essentially those of Powell, Schoeller, and John [39]. They are not useful when traces of niobium and tantalum are in question.

PROCEDURE. Fuse the mixed oxides with 5–10 g potassium carbonate. Cool, dissolve in 100–200 ml water, heat nearly to boiling, and slowly add a solution prepared by dissolving 4 g $MgSO_4 \cdot 7H_2O$, 10 g NH_4Cl, and 1 ml ammonia solution in 50 ml water. Filter and wash with 2% ammonium chloride. Recover tungsten from the filtrate with cinchonine and tannin.

Transfer the magnesium tantalate and niobate to a beaker with water. Add a few drops of hydrochloric acid, 5 g each of ammonium chloride and acetate, and 50 ml of 2% tannin solution. Digest on the steam bath for an hour, cool, filter, wash with 2% ammonium nitrate, ignite, and weigh.

The filtrate from the niobium and tantalum may contain a milligram or so of niobium, which may be recovered by the following procedure. Titanium, iron, and zirconium remain with the niobium and tantalum. When a little niobium and tantalum are to be separated from much tungsten, the insolubility of sodium niobate and tantalate is utilized.

PROCEDURE. Fuse the mixed oxides with sodium carbonate in platinum, adding a pellet of sodium hydroxide near the end of the fusion. Leach the melt with a sodium chloride solution containing 175 g/l NaCl. Volume should be as small as feasible (~25–30 ml). Allow the mixture to stand for several hours and filter through a small Whatman No. 44 paper using filter pulp. Wash with small portions of 175 g/l sodium chloride containing a little sodium hydroxide (1 g/l). Refilter if the filtrate is not perfectly clear.

Recover tungsten from the filtrate with cinchonine and tannin. Boil the paper and pulp containing tantalum and niobium with dilute hydrochloric acid, and add ammonium chloride and tannin to precipitate the earth acids.

4.8. CHROMIUM

In the small amounts in which chromium usually occurs in silicates, it is best determined colorimetrically using either the color of the chromate ion in alkaline solution or, for trace amounts, the almost specific reagent sym-diphenylcarbazide. In either case, the chromium is first oxidized by alkaline fusion with peroxide or by pouring the acid solution of Cr(III) containing hydrogen peroxide into an excess of alkali and filtering off the precipitated iron, titanium, and so on, after prolonged digestion on the steam bath and cooling.

Sodium carbonate fusions of chromium-bearing materials result in excessive attack on platinum crucibles if sufficient oxidizing agent is used to ensure complete oxidation and should be avoided if at all possible. For example, 0.7 g chromite is fused with 5 g Na_2CO_3 and 1 g Na_2O_2 over a Tirrill burner for 15 min and then over a Meker burner for 5 min, and the crucible may lose up to 18 mg platinum. Lesser amounts of peroxide result in less attack on the crucible, but the sample is incompletely decomposed in the fusion period given. Thus, any determination of chromium beginning with a sodium carbonate fusion and extraction of the melt with water is apt to be low unless the insoluble residue is retreated.

In procedures starting with sodium peroxide fusion, small amounts of iron (0.1–0.5 mg) always find their way into the filtrate; it should be evaporated to about half its volume, cooled thoroughly, and refiltered before attempting colorimetric determination of chromate. The diphenylcarbazide method has the advantage that traces of iron remaining in the solution are without effect. Despite the greater sensitivity of the reaction, the sample weight should remain the same as for the chromate method because of the danger that a smaller sample will be nonrepresentative. Aliquot parts of the final oxidized solution are taken for analysis.

PROCEDURE. Fuse 0.500 g of the sample with 5 g sodium peroxide in an iron crucible, holding the crucible in a pair of tongs and heating cautiously so that no spattering or local overheating occurs. The heating should be continued until the melt is fluid and quiet without it becoming more than dull red.

Note 1: If traces of chromium are in question, it is better to use a porcelain crucible for the fusion unless it can be established that the iron crucible contains no traces of chromium.

Cool, and transfer the crucible to a 400-ml beaker containing 50–75 ml water. Cover, and heat gently until the peroxide begins to dissolve rapidly. Then remove from the heat and let stand for a few minutes. Remove and wash off the crucible with water; clean it with a very little dilute hydrochloric acid using a policeman.

Add several small portions of sodium peroxide, or 1–2 drops of 30% H_2O_2, to the mixture, and digest on the steam bath for 2–3 h or boil hard for 10–15 min. Cool thoroughly, and filter through a hardened paper folded inside a retentive one (Whatman No. 41H inside an S & S blue ribbon). Wash with 1% sodium carbonate solution. Discard the precipitate.

Note 2: Retreatment of the precipitate may be worthwhile in some cases: About 1% of the chromium usually remains in the precipitate after thorough washing. For this reason, the procedure is not recommended for materials containing more than about 0.5% Cr_2O_3, though it is quite generally applicable to silicate rocks that contain less than this amount.

Evaporate the filtrate to about 70 ml, cool thoroughly, and filter through a retentive paper, conveniently the same one that was used before. If a new paper is used, it should be prewashed with 5% sodium hydroxide solution. Wash the paper with 1% sodium carbonate solution. Transfer the filtrate to a 100-ml Nessler tube. To two other similar tubes add 5 g sodium hydroxide and 2 g sodium carbonate dissolved in 80–90 ml water. To these add from a burette a standard solution of potassium dichromate until one is just slightly less and the other just slightly more highly colored than the unknown. From the amount of dichromate used, calculate the chromium present in the sample.

Reserve the three solutions for the determination of vanadium if this is required. If the chromium is to be determined with diphenylcarbazide, proceed as follows:

PROCEDURE. Prepare an alkaline solution containing the chromium as chromate (either as above or otherwise), and boil to be sure that all peroxide is removed. Cool, transfer to a volumetric flask, and dilute to volume. Transfer an aliquot containing about 0.05 mg Cr_2O_3 (not more than 0.1 mg) to a 100-ml volumetric flask containing 10 ml of 5% sulfuric acid. Add water to about 70 ml, then 10 ml of 0.05 percent sym-diphenylcarbazide in 5% acetic acid (see below). Dilute to the mark and mix. Measure the transmittance at 540 nm, and determine the chromium present from a calibration curve prepared using standard dichromate solution.

Note: The diphenylcarbazide reagent should be prepared fresh by dissolving 0.05 g in 5 ml glacial acetic acid, and then diluting to 100 ml with water and filtering.

Vanadium, when present in a relatively large amount compared to the chromium, may interfere: Molybdenum also interferes slightly. Neither element is likely to be present in silicates in sufficient amount to affect the determination.

4.8.1. Titrimetric Determination of Chromium

For larger amounts of chromium, a sulfuric acid solution containing it is prepared by acid attack or by treatment with a suitable flux, and the element is oxidized (with silver nitrate as a catalyst) by persulfate. The chromium is then reduced by adding a measured excess of ferrous ammonium sulfate, and the excess is titrated with standard permanganate. As applied to chromite, the procedure is as follows:

PROCEDURE. Transfer the solution of 0.5000 g of the sample obtained during the ferrous iron determination, for example, to an 800-ml beaker. Add 10 ml of 1:1 sulfuric acid, 4 ml nitric acid, 0.5–0.6 g silver nitrate dissolved in water, and 5 mg manganous sulfate if the sample does not contain appreciable MnO. Dilute to 400 ml and heat to boiling. Add to the boiling solution a freshly prepared 25% solution of ammonium persulfate until the chromium is all oxidized and the reddish or brownish color of oxidized manganese begins to appear.

Note 1: The manganese is added (if little or none is present in the sample) to provide an indication of the progress of the oxidation. It is otherwise difficult to decide when the reaction is complete. About 25–35 ml persulfate is normally required for complete oxidation of the chromium. It should be added a little at a time over a period of 5–15 min.

When the manganese color appears, discontinue the addition of persulfate, and boil for 10–15 min to completely decompose the excess. Then add a few drops of dilute hydrochloric acid and boil vigorously for 5 min to remove the chlorine generated. The manganese should be completely reduced, but only a little more than enough hydrochloric acid to accomplish this should be used.

Cool completely, and add an accurately weighed quantity of ferrous ammonium sulfate of known purity sufficient to reduce the chromate and provide a slight excess. Titrate the excess with standard permanganate solution, and correct the titration for a blank determined by oxidizing the proper amount of ferrous ammonium sulfate with standard dichromate in the presence of phosphoric, sulfuric, nitric, and hydrochloric acids and silver nitrate, adding excess permanganate, boiling to destroy the excess, cooling, and titrating with permanganate.

Note 2: It is important that the excess persulfate be completely destroyed before adding the hydrochloric acid to reduce manganese. Failure to do this leads to low results. Vanadium does not interfere in the titration despite the fact that it is reduced by ferrous iron because it is reoxidized by permanganate. The use of oxidation–reduction indicators such as diphenylamine or ferroin is not recommended despite the greater ease of end point detection because of the behavior of vanadium.

4.8.2. Analysis of Chromite

The constituents usually determined in the analysis of chromites are SiO_2, Al_2O_3, TiO_2, Fe_2O_3, FeO, MgO, CaO, MnO, H_2O, Cr_2O_3, V_2O_5, and sometimes NiO and ZnO. Of these, only H_2O and V_2O_5 can be determined in the usual manner.

Procedures involving fusion of chromite with sodium carbonate have been recommended, but these are not satisfactory; attack is often incomplete, and when an oxidizing agent (as nitrate or peroxide) is added, excessive amounts of platinum are dissolved from the crucible. Fusion with sodium peroxide (usually using an iron or nickel or zirconium crucible) is the most rapid means of obtaining complete solution, but this method of attack is limited in its application. Potassium pyrosulfate is capable of completely attacking chromite, but prolonged fusion is necessary and temperature must be controlled. A mixture of $K_2S_2O_7$ and Na_2CO_3 has also been recommended.

In the methods to be described, solution of the main portion is accomplished by heating with concentrated perchloric acid, and ferrous iron and chromium are determined on a single sample dissolved in mixed sulfuric and phosphoric acids in a special apparatus.

Fine grinding of the sample is essential: It is best to transfer 2–4 g of material to an agate mortar, grind briefly, and shake on a 300–325-mesh screen, repeating the operation until the whole sample has passed through. Excessive oxidation of ferrous iron does not occur during grinding if screening is frequent and thorough. Frequent screening also facilitates the operation because too fine a powder tends to plug the screen. Once the sample has all been reduced to 300 mesh, it should be thoroughly mixed in the mortar, and then by rolling on paper. If an accurate SiO_2 figure is required, a separate portion of the sample should be pulverized in a hardened steel mortar for this determination.

The Main Portion

PROCEDURE. Weigh 0.7000 g of the -300-mesh material into a 400-ml beaker. Add a little nitric acid and 50 ml of 72% $HClO_4$. Cover, and place a stirring rod in the beaker in such a position that acid condensing on the cover will run freely back into the solution without accumulating in drops. Heat on a plate, gently at first, gradually increasing the heat until finally the sample is all in solution and the condensing acid shows itself to be constantly boiling by the fact that the fumes within the beaker clear up and no sputter occurs when the condensate runs back into the solution. Continue boiling for 10–20 min after this stage is reached.

The nitric acid is added as a safety measure in case organic matter is present.

Dehydration of silicic acid and near complete oxidation of chromium cannot be accomplished merely by heating to fumes of perchloric acid; the acid must be boiled for 10 min or longer.

Place the hot beaker on a cold metal plate, and allow it to cool for 1–2 min, swirling once or twice; then cool quickly by plunging into a dish of cold water, swirling it.

The rapid cooling helps prevent reduction of chromate by peroxides present in the solution.

When cold, add 50 ml water, stir, heat rapidly to boiling, and boil for 1–2 min. Add 100 ml cold water and filter at once through a paper that has been washed with 25% perchloric acid. Scrub the beaker, transfer all the residue to the filter, and wash first with 2% perchloric acid and then with hot water.

The solution is boiled before filtration to remove chlorine and peroxides and then cooled by the addition of water before filtering to help prevent reduction of chromate by the paper. Washing the paper with perchloric acid before using it for the filtration serves to remove soluble organic material.

Transfer the paper and precipitate to a weighed platinum crucible, cover tightly, and heat until volatile matter has been removed. Uncover, burn off the paper, ignite to constant weight, and determine silica by volatilization with hydrofluoric acid as usual.

Because perchlorates may still remain in the paper, the preliminary heating should be done with the crucible tightly closed to prevent loss as a result of small explosions. Addition of a drop or two of nitric acid before burning off the paper may help to prevent such explosions.

No unattacked material should remain after the treatment with hydrofluoric acid; the residue will consist almost entirely of Cr_2O_3, a little of which seems to be rendered insoluble by the perchloric acid treatment.

Fuse the small residue with a little sodium carbonate, heating no more than is necessary to completely dissolve it and oxidize the chromium. Take up the melt with water, and add its solution to the main one. Heat to boiling, and boil for 5–10 min.

Precipitate with ammonia in the usual way, filter, using a double paper, and wash with 2% ammonium nitrate solution.

Carefully remove the inner paper containing the precipitate, wash it off into the precipitation beaker, refold the paper, and return it to the funnel. Dissolve the precipitate by adding to its water suspension 10 ml nitric acid and heating gently. When solution is complete, boil for a few minutes. Add about 5 ml in excess of ammonia and 1 drop of 30% hydrogen peroxide. Boil for 10 min to remove peroxide, make the solution just acid with nitric acid, and adjust the pH to 7.0–7.5 with ammonia. Heat to boiling and filter.

Repeat the precipitation a third time, macerating the inner paper and stirring it into the solution just before filtering through the same filter as before.

Because of the chromate present, manganese will unavoidably be precipitated with the R_2O_3, usually almost completely.

Many brands of reagent-grade hydrogen peroxide contain phosphate; and P_2O_5 thus introduced will be weighed with the ammonia precipitate: Thus, a minimum volume of low-phosphate peroxide should be used.

If the second filtrate is highly colored, a fourth precipitation should be considered. The filtrates from the separate ammonia precipitations should not be mixed at once; the progress of the separation may then be followed by observing the colors of the successive filtrates.

Ignite the precipitate to constant weight as usual, and weigh the ammonia group oxides. Treat the filtrate in the usual manner for the determination of calcium and magnesium. The oxalate precipitation may be omitted and calcium recovered from the ignited magnesium pyrophosphate.

Note: If the oxalate precipitation is omitted, the magnesium pyrophosphate must be examined for nickel, for this element precipitates quantitatively with a preponderance of magnesium in the absence of oxalate. The $Mg_2P_2O_7$ would also be examined for manganese as usual, though it is unlikely that much of it will escape precipitation with the ammonia group. A drop or two of hydrogen peroxide may be added to the ammonia solutions to ensure the nonprecipitation of chromium.

Fuse the ignited and weighed ammonia group precipitate as usual. Leach with water to which has been added 5 ml of 1:1 sulfuric acid. Evaporate or dilute to 80 ml, and add strong (50%) sodium hydroxide solution until the precipitating iron hydroxide redissolves on stirring only with difficulty. Heat 200 ml of 5% sodium hydroxide solution to boiling in a 400- or 600-ml beaker. Add 1 ml of 30% hydrogen peroxide to the nearly neutralized solution of the oxides, and pour quickly into the hot alkali solution. Cover at once to prevent loss by spattering, and then wash the contents of one beaker into the other completely. Digest the alkaline solution on the steam bath for an hour or more or until decomposition of peroxide is complete, stirring from time to time.

Allow the solution to cool completely, filter through a hardened (Whatman No. 41H) paper fitted inside a retentive paper (S & S blue ribbon). Wash thoroughly with 2% sodium hydroxide containing a little sulfate and then finally once or twice with small portions of cold water to remove some of the sodium salts.

Note: Before filtering, it is advisable to wash the papers with the alkaline wash solution to dissolve organic matter that may give a yellow color and invalidate the chromium determination that follows.

Evaporate the filtrate to less than 100 ml, refilter if necessary, transfer to a 100-ml Nessler tube, and determine chromium by colorimetric titration, adding standard chromate solution to standards in similar tubes containing the same amounts of alkali and salts. Reserve these standard solutions for the determination of vanadium (below).

Wash the precipitate containing the iron and titanium off the paper into the precipitation beaker, and add 10 ml of 1:1 sulfuric acid. Burn off the papers in a platinum crucible, fuse the residue with a minimum of pyrosulfate, and add the solution of this fusion to the main iron solution. Determine iron and titanium in the usual way.

Note: Because of the chromium, an unusually large amount of platinum is likely to be present: The H_2S separation should thus be carried out with particular care, or platinum (rhodium) may remain in sufficient amount to interfere with the titration of the iron. If the alkaline solution containing the vanadium and chromium had to be refiltered before colorimetric determination of the latter, the paper used should be ashed and the residue so obtained added after pyrosulfate fusion to the iron solution. A small amount of iron always remains in the alkaline solution in any case (usually about 0.2 or 0.3 mg): This is best determined colorimetrically (with o-phenanthroline) after vanadium has been determined.

Determine vanadium in the solution containing the residual chromium as follows: Make the solutions 1–2 N in sulfuric acid, and add a drop of peroxide. Boil until peroxide is decomposed and the chromium is all reduced to Cr(III). Determine vanadium as orange VO_2^{3+} by colorimetric titration using as standard solutions the same ones that were used for the chromium determination, reducing chromate with peroxide in the same way as in the unknown.

Alternatively, make the solution 5% in sulfuric acid, reduce the chromium with peroxide, boil, cool, and precipitate vanadium (and the small amount of iron that is almost always present) with cupferron [40]. Filter, burn off the paper and organic matter, fuse with a little pyrosulfate, dissolve in 1–2 N sulfuric acid, and determine vanadium with peroxide according to the directions of Sandell [41].

To obtain Al_2O_3, subtract V_2O_5, TiO_2, Fe_2O_3, Mn_3O_4, and Cr_2O_3 from the total R_2O_3.

Because there is some doubt about the oxidation state of chromium when it is ignited in an ammonia precipitate, the determination of aluminum as phosphate may be desirable. This avoids the necessity of obtaining an accurate weight of R_2O_3 and also makes unnecessary the determination of coprecipitated chromium and vanadium. The final aluminum residue ($AlPO_4$) should however be examined for chromium and a correction applied. Direct determination of alumina also removes uncertainties regarding the correction that should be made for manganese. On occasion, we have determined alumina by both methods—directly and by difference.

Determination of Manganese and Vanadium

PROCEDURE. Weigh 0.5000 g of the sample into an iron crucible and add about 5 g sodium peroxide. Mix, and fuse cautiously until a quiet melt is obtained. Leach with water, add a little peroxide, and digest on the steam bath for 1 h or until all the peroxide is decomposed. Cool thoroughly, filter, and wash with dilute sodium hydroxide solution containing a little carbonate.

Dissolve the precipitate in dilute nitric acid to which a drop of peroxide has been added, then add periodate, and oxidize the manganese. Add phosphoric acid and determine manganese colorimetrically in the usual way.

Make the filtrate 10% in sulfuric acid, add peroxide to reduce the chromium, boil off oxygen, cool, and precipitate vanadium with cupferron. Filter, burn off the paper and organic material, fuse with pyrosulfate, dissolve the melt in dilute sulfuric acid, and determine vanadium as peroxyvanadate by colorimetric titration.

4.9. MOLYBDENUM, TUNGSTEN, AND RHENIUM

Molybdenum resembles tungsten in that the metal has a high melting point while the oxide is volatile; MoO_3 cannot be heated in air above about 500 °C without loss. The disulfide, MoS_2, molybdenite, is the source of most molybdenum; some is extracted from powellite, $Ca(W,Mo)O_4$, and wulfenite, $PbMoO_4$.

During the classical analysis, most of the molybdenum is found in the filtrate from the acid group of elements if the dehydration is done with hydrochloric or sulfuric acid. With nitric and perchloric acids, more molybdenum remains with the acid group. In analyzing molybdenum-bearing rocks or minerals, separation of the acid hydrogen sulfide group is mandatory because molybdenum interferes badly in subsequent operations if it is not removed. Separation of molybdenum as sulfide, MoS_3, is rather difficult but can be quantitatively achieved if certain precautions are taken. If fluoride or phosphate are present in the sample, these must be removed in the early stages of the analysis (fluorine by the $H_3BO_3–CH_3OH$ treatment described elsewhere (see Section 3.1.6) and phosphate by precipitation with zirconium (see Section 3.1.12). The usual H_2S treatment needs some modification: Starting with a solution about $1N$ in hydrochloric acid, the hydrogen sulfide should be passed rapidly into the cold solution until precipitation is nearly complete before heating nearly to the boiling point and diluting to about $0.2–0.3 \ N$. During this procedure, some molybdenum is reduced, possibly to valence V, and remains in solution. The only effective way to recover this unprecipitated molybdenum is to boil the filtrate to remove hydrogen sulfide, add persulfate to oxidize the reduced molybdenum, boil to destroy the persulfate, and retreat the solution with hydrogen sulfide. If arsenic is prsent as arsenate, it may not be completely precipitated during these operations; the filtrate from the molybdenum recovery should again be boiled to remove hydrogen sulfide; then the arsenate may be reduced by sulfite and precipitated as sulfide. Many less troublesome procedures have been recommended, but in several trials, these have not proven satisfactory.

Tungsten is not precipitated by hydrogen sulfide in acid solution when present alone, but most or all of a small amount may be precipitated when large amounts of other elements, especially molybdenum, are present.

If the hydrogen sulfide separation is omitted or if all the molybdenum is not removed, difficulties may be expected during the rest of the analysis. Much of the molybdenum will usually precipitate with the ammonia group of elements as molybdate of calcium, barium, or lead or as basic iron molybdate. Some may escape precipitation; much of this will come down with the manganese if a persulfate separation is made. An indeterminate amount will go through the whole procedure and be lost.

Rhenium is one of the rarest of elements; it occurs almost exclusively in molybdenite, which may contain up to about a tenth of a percent of rhenium. Like molybdenum and tungsten, it is a high-melting metal that is easily oxidized to a volatile oxide. Rhenium is precipitable by hydrogen sulfide as Re_2S_7 but, during the ordinary analysis, is volatilized before reaching this stage. The volatility of rhenium compounds provides an easy means of separation. Rhenium is not lost during alkaline fusions if some simple precautions are taken.

Determination of rhenium in molybdenite or molybdenite-bearing ore proceeds as follows:

PROCEDURE. Treat a large sample (10–50 g) with 50–200 ml of 1:1 sulfuric acid, heat gently, and add nitric acid a little at a time until the MoS_2 is decomposed. Evaporate slowly until sulfuric acid just begins to fume, cool, add a little water, mix, and again evaporate to incipient fumes; repeat this operation until it is certain that all the nitric acid has been expelled. Add 10–20 ml phosphoric acid and transfer to a distilling flask fitted with a condenser using enough water to make the solution 75–80% in sulfuric acid. Heat the distilling flask to 170 °C, and hold at this temperature while bubbling constant boiling hydrochloric acid vapors through the solution. Generate the hydrochloric acid by boiling 6 N acid. About 200–300 ml distillate should be collected in an ice-cold receiver.

To the cold distillate add ammonia until the acid concentration is about 0.3 N, and saturate the cold solution with hydrogen sulfide. Filter, and wash with water containing hydrogen sulfide.

Note: To be sure of complete recovery, the filtrate should be treated with bromine in very slight excess and then again with hydrogen sulfide.

Dissolve the sulfide precipitate in 1:1 ammonium hydroxide, adding 30% hydrogen peroxide drop by drop until a clear solution is obtained. Evaporate the solution on the steam bath to remove much of the ammonia, make just acid with acetic acid, and add a 2% solution of nitron acetate. Cool in an ice bath to 4 °C or lower. Filter through a glass frit, wash with a minimum of ice cold 0.25% nitron acetate solution and then with 1–2 ml of ice-cold water. Dry at 105 °C and weigh nitron perrhenate, $C_{20}H_{16}N_4HReO_4$, which contains 33.1% rhenium.

Rhenium in the 6 N hydrochloric acid distillate may also be determined by AA spectrometry using an oxyacetylene flame and a wavelength setting of 2287 Å. Sensitivity is not high—the working solution should contain at least 20 ppm rhenium.

Colorimetric determination of rhenium is most often accomplished using thiocyanate; molybdenum, tungsten, and niobium also yield colored thiocyanate complexes.

Possibly the best colorimetric procedure for application to 6 N hydrochloric acid distillates is that using 2,4-diphenylthiosemicarbazide. The following procedure is that of Sandell [42].

PROCEDURE. To 5–10 ml perrhenate solution (6 N in HCl) containing up to 50 μg rhenium and no other metals except the alkalies, add 1.5–2 ml freshly prepared saturated reagent solution in methanol. Heat the solution in a 25-ml volumetric or other long-necked flask in a water bath at 80 ± 2 °C. After 20 min, cool in water and shake the solution with 25.0 ml chloroform for 2–3 min. Filter the chloroform extract through paper to remove water droplets and obtain absorbance at 510 nm.

Sandell describes several methods for separating and determining rhenium colorimetrically.

4.9.1.　Analysis of Sulfide Precipitates Containing Molybdenum

The acid hydrogen sulfide group of elements having been collected, it is usually best to obtain the metals in sulfuric acid solution as described by Hoffman and Lundell [43], and separate arsenic, germanium, rhenium, tin, antimony, and selenium by distillation as bromides. If the temperature of distillation is not permitted to rise above 200–205 °C, very little molybdenum will be found in the distillate. The sulfuric acid solution containing the nonvolatile metals may be treated in a variety of ways. Usually lead will be removed by coprecipitation with added strontium, but there is some danger of collecting some molyddenum with the lead sulfate. The latter should be examined for molybdenum, preferably by AA spectrometry. Small amounts of molybdenum in the sulfuric acid solution may be measured colorimetrically with thiocyanate.

When tungsten is present, the acid solution should be treated with 5 g or more of tartaric acid, made alkaline with sodium hydroxide, and heated on the steam bath for several minutes, so as to form the tungstate–tartrate complex and break up any heteropoly tungsten acids that may be present. The alkaline solution should be cooled and saturated with hydrogen sulfide and then made just acid with sulfuric acid while continuing the flow of H_2S for 15 min or more. The sulfide precipitate may then be treated as before with sulfuric and nitric acids and its analysis continued. The filtrate containing the tungsten can be freed from tartrate by evaporation with sulfuric and nitric acids and the tungsten precipitated with tannin and cinchonine.

Copper is separated from molybdenum by adding sodium thiosulfate to a boiling solution 3% in sulfuric acid; copper sulfide precipitates. Alternatively, 3 g granulated zinc may be added to a solution containing 5 ml sulfuric acid in 100 ml. Copper is deposited as metal.

The best gravimetric method for molybdenum is that in which it is precipitated and weighed as lead molybdate. Because the solution from which the molybdenum is to be precipitated invariably contains sulfate, care is necessary to prevent coprecipitation of lead sulfate. The solubility of lead sulfate is much increased in the presence of ammonium acetate, and if only a minimal excess of precipitant (lead acetate or nitrate) is used, the contamination can be avoided; nevertheless, a reprecipitation is advisable—or the ignited lead molybdate should be tested for its lead sulfate content.

PROCEDURE. Dilute the molybdate solution to about 300 ml (with less than 0.2 g molybdenum) and bring to pH 4 by addition of ammonia or acetic acid (methyl orange end point). Add 5 ml glacial acetic acid and 20–30 ml of 50% (w/v) ammonium acetate. From a burette add lead nitrate solution (40 g/l) drop by drop to the nearly boiling solution while stirring vigorously and constantly (a magnetic stirrer is advisable). From time to time stop the reagent addition and test a drop of supernatant liquid on a spot plate with a drop of freshly prepared 1% tannin solution. When no color reaction is observed, precipitation is complete. Add a 2-ml excess of precipitant and keep the solution at the boiling point for 15–20 min. Filter, and wash with hot 2% ammonium acetate until the washings test free of lead with sodium sulfide solution. If a reprecipitation is intended, washing may be omitted.

To reprecipitate the lead molybdate, dissolve the precipitate in 1:1 hydrochloric acid, dilute to about 200 ml, and add ammonia until the lead molybdate just begins to separate. Then add 25 ml of 50% ammonium acetate, heat to boiling, and add 1–2 ml lead nitrate solution. Digest at the boiling point for 15–20 min, and proceed as above.

Note: It may be desirable to test the filtrates from the lead molybdate with thiocyanate and stannous chloride to be sure all the molybdenum was recovered.

Molybdenum may also be precipitated and determined as oxinate, but the analyte must be free from all other metals except the alkalies.

PROCEDURE. Render the solution slightly acid to methyl red and add 5–10 ml of 15% ammonium acetate. Heat to boiling and add 5% 8-hydroxyquinoline slowly with stirring until present in slight excess (yellow supernatant liquid). Digest at the boiling point for 5–10 min, filter, and wash with hot water. A fritted glass crucible should be used. Dry to constant weight at 140 °C, and weigh $MoO_2(C_9H_6ON)_2$, which contains 23.05% Mo.

Note: The oxine procedure tends to give slightly high results, probably because small amounts of interfering elements are always present in practice.

In ore analysis, it is often of interest to determine not only total molybdenum but also "oxide moly" and MoS_2. The reason for this lies in the fact that the flotation procedures usually applied recover only the MoS_2— other molybdenum minerals finish in the tailings and are lost. The only way to distinguish analytically between the several forms in which molybdenum may be present is to rely on selective dissolution. Which procedure to apply depends on the mineralogy of the ore and the purpose of the analysis. Usually, a procedure must be tailored to the ore being examined. To apply a procedure that has proven satisfactory for one ore type to another ore type without investigation is to invite error. Some of the problems to be faced are listed:

1. Complete dissolution of a mineral in an ore depends on its liberation. Sample preparation must take the ore to a mesh size at which particles soluble in the leaching reagent are freed from gangue. In mining operations, which often require hundreds or thousands of "analyses," it is desirable to minimize sample preparation to the point where sampling errors are tolerable. If the minimum preparation does not liberate the soluble minerals, further grinding is necessary. How far such extra treatment must go can only be discovered by experiment or mineralogical examination of many samples.

2. Excess grinding must be avoided because there is always the possibility, especially in vigorous and rapid procedures, of oxidizing or otherwise altering the minerals in an ore.

3. Leaching efficiency depends on the size of the particles to be dissolved. Not only must the soluble mineral particles be liberated but they must also be reduced to a size that will lead to reasonably rapid solution rates. If leaching procedures are prolonged, there may be attack on those minerals that should not dissolve; this factor becomes very important when grain size is small.

Several methods are available for distinguishing between soluble and insoluble molybdenum and tungsten in ores. Some are more perfect than others; all need to be tested with the particular ore type being examined.

1. *Boiling Ore in Nonoxidizing Acids.* Molybdenite resists attack by boiling hydrochloric and dilute sulfuric acids. "Oxide moly" is determined colorimetrically in the filtrate, and "sulfide moly" is determined in the undissolved residue after attack by oxidizing acids.

2. *Attack with Sodium Hypochlorite.* If molybdenite is treated with strong sodium hypochlorite solution, and hydrochloric acid is added *slowly*, the MoS_2 is completely oxidized and dissolved. Presumably wulfenite ($PbMoO_4$) may not dissolve. The procedure is tricky: The reaction takes place only in a narrow pH range and generates free acid. Sodium hypochlorite reagents vary in their alkali content as well as in the amount of available chlorine they contain. The following general procedure is recommended: It may have to be altered depending on the ore and the character of the reagent hypochlorite.

PROCEDURE. Add 100–200 ml of 4% sodium hypochlorite solution to 1 g ore in a beaker, and introduce a magnetic stirring bar. Add several pellets of sodium hydroxide and stir vigorously. Add 1:1 hydrochloric acid dropwise and slowly until solution of the black MoS_2 begins, and then discontinue the acid additions. If the acid is added too fast, solution of the MoS_2 will not occur. Filter, wash with water, and examine the insoluble residue for the molybdenum it may contain. Mineralogical examination may be called for.

3. *Cold Acid Digestion.* Concentrated hydrochloric acid rapidly dissolves scheelite. Wolframite is dissolved much more slowly and, unless very finely divided, is not completely dissolved even after 12–18 h continuous agitation. Some varieties of wolframite dissolve more readily than others. Molybdenite is not attacked even when very finely divided. Wolfenite and powellite dissolve rapidly. The procedure may be useful in the examination of complex ores containing tungsten and molybdenum. Cassiterite (SnO_2) usually remains unattacked.

PROCEDURE. In a 500-ml Erlenmeyer flask, put a stirring bar that will sweep the whole bottom of the flask. Add the sample, which may weigh up to 5 g or more, and 200–250 ml *concentrated* hydrochloric acid. Place the flask on a magnetic stirring unit capable of turning the stirring bar at a slow and steady rate, and stir for an hour or more. Finely divided scheelite will go into solution in less than an hour. To dissolve wolframite, 18 h or more may be required even if the sample has been passed through a 200-mesh screen. In concentrated hydrochloric acid, tungstic acid does not precipitate, presumably because of the formation of the heteropoly paratungstate. During the digestion, the flask should be covered with a watch glass to prevent escape of hydrogen chloride. Temperature should not exceed 20 °C.

Filter through a glass fiber paper in a Büchner funnel, and wash with concentrated hydrochloric acid. The residue should contain all the molybdenite in the sample, much of the wolframite if digestion was abbreviated, and all of the cassiterite. The filtrate contains tungsten originating in scheelite, some of the tungsten from wolframite, all the molybdenum in wulfenite, molybdite, and powellite, and most of the iron originally present in hematite and magnetite.

Very often, the accurate primary analysis of complex materials benefits from separations of the kind described above. If a large sample (e.g., 10 g) is initially treated with, say, hydrochloric acid, and the insoluble portion is collected, ignited, and weighed, it may then be mixed and subsamples taken from it may be accurately analyzed. The soluble portion may be diluted to volume and aliquots taken for complete and accurate analysis. If the weight of insolubles and the aliquot of solubles are equivalent, the two analyses may be combined at any stage. For example, if a 10-g sample is treated with hydrochloric acid, and one-tenth of the insolubles and the solution are taken, the acid group may be separated from each and com-

bined in the same crucible for ignition, determination of silica, and separation of the earth acids (Nb, Ta). Such an expedient is valuable in the
analysis of complex sulfide ores that may contain molybdenum, tungsten,
copper, selenium, antimony, arsenic, and a host of other elements; besides, it may give a measure of soluble (oxide Mo) and insoluble (sulfide
Mo); a distinction between scheelite and wolframite; a measure of the
iron present in ilmenite (which usually does not dissolve in hydrochloric
acid) and in hematite; a distinction between rare earths present in silicates
and in soluble carbonate minerals; and a division of, for example, zinc
and cadmium between insoluble silicates and soluble sulfides.

A difficulty in such a procedure using hydrochloric acid as the leaching
agent lies in the nonavailability of a suitable filter. Glass fiber filters work
admirably, but the residue cannot be removed from them quantitatively.
It is necessary to analyze a filter and abandon efforts to determine the
elements it contains or to investigate the several filter types now available
for one that will contribute a minimum of difficulty. Polycarbonate (nu-
cleopore) filters show the most promise but are awkward to use. Ordinary
paper filters are not useful, being destroyed at once by concentrated hydrochloric acid. Even if the hydrochloric acid solution can be diluted to
6 N, when paper filters can tolerate the cold solution, enough organic
material is introduced from the paper to interfere in subsequent operations; hydroxide precipitations will leave iron in solution, complexed by
hydroxy-organic compounds. If paper is used, the resulting solution
should be evaporated to fumes with sulfuric acid and treated with nitric
acid to remove organics.

Contrary to popular belief, the sulfuric–nitric combination is more effective in removing organic materials than the perchloric–nitric
combination.

Titrimetric Methods. There seems to be no satisfactory volumetric
method for determining tungsten even after its separation from other elements. After separation as oxide, the WO_3 may be dissolved using a
known amount of sodium carbonate or hydroxide, and the excess titrated
with standard acid; however, if the oxide is pure enough to yield a good
result, it is much easier simply to weigh it.

Two useful redox methods for molybdenum are available: titration with
permanganate after reduction in a Jones reductor and reduction to Mo(V)
with mercury followed by titration with Ce(IV) using ferroin as indicator.
Titration of molybdenum with a standard solution of lead acetate using
tannin as an indicator yields fair results if conditions are carefully standardized; the procedure is essentially the same as in the precipitation of
lead molybdate for weighing and may be useful in situations where a rapid
control method is needed.

4.9.2.   Jones Reductor: Permanganate Method

PROCEDURE. Prepare a solution containing about 100 mg molybdenum in 5–10% sulfuric acid and free from organic matter, nitric acid, iron, chromium, arsenic, antimony, titanium, vanadium, niobium, tungsten, and uranium. Sulfur acids other than sulfuric must be absent. Nitric acid and thionic acids may be removed by strong fuming with sulfuric acid. Most other interferences may be removed by repeated precipitation with *excess* ammonia, adding excess iron if necessary to collect arsenic, antimony, and vanadium. Tungsten requires a hydrogen sulfide separation in tartrate solution and is therefore the most awkward interference; with small amounts of tungsten, it may be possible to make an empirical correction. Tungsten is reduced to an uncertain valence state in the reductor.

Since this procedure is most often applied during a systematic analysis of the hydrogen sulfide group, it is very necessary to be sure that paper and other organic materials are completely destroyed during the solution of the metals in sulfuric acid. The paper and sulfides should first be treated with concentrated sulfuric acid, any water additions being kept to a minimum. The paper will usually char and desintegrate within a few minutes; then the mixture should be heated gently and treated with nitric acid added a drop at a time just until the black char disappears. Heating should be continued, with dropwise additions of nitric acid only when char reappears, until the sulfuric acid is fuming strongly. The assumption that the solution is now free from organic material is false; after cooling, 2–3 ml nitric acid should be added, and evaporation to fumes repeated. If there is any trace of yellow or green in the solution, the nitric acid treatment should be again repeated. Finally, 2–3 drops of saturated potassium permanganate should be added and the solution evaporated to fumes until it is certain that all nitric acid has been expelled.

The solution is now ready for the removal of arsenic, selenium, and so on by volatilization as bromides [43], removal of lead as sulfate, and separation of molybdenum prior to its determination via reduction and titration. In some cases, when the metals forming volatile bromides are absent, the sulfuric acid solution may be treated with zinc to remove copper and the filtrate treated with ammonia to remove iron and titanium, both of which will be present only in small amounts. Removal of copper is best done by adding 3 g of 40-mesh granulated zinc to a cold solution containing 5 ml concentrated sulfuric acid in 100 ml and heating on the steam bath until the zinc is almost, but not quite, dissolved. The copper should be completely precipitated and may be determined by any appropriate method.

During the reduction of molybdenum in the Jones reductor, the effluent must be caught in a solution of ferric sulfate (free from ferrous iron or other reducing substances); contact with the air causes rapid oxidation of Mo(III).

4.9.3.   Reduction of Molybdenum with Mercury

In 3 N HCl, molybdenum is quantitatively reduced to Mo(V) by stirring or shaking with metallic mercury. It may then be titrated to Mo(VI) by ceric sulfate solution. Potassium permanganate or potassium dichromate solutions are not satisfactory as titrants. Ferroin is used as an internal indicator.

PROCEDURE. Evaporate the sulfuric acid solution containing about 100 mg molybdenum nearly to dryness and take up the residue in 25 ml concentrated hydrochloric acid. Dilute to 100 ml in a separatory flask, add 25 ml mercury and a stir bar, and stir vigorously over a magnetic stirring unit for 10–15 min. Remove the mercury and titrate the solution with $N/10$ ceric sulfate solution using ferroin as an indicator. Copper catalyzes the air oxidation of Mo(V) and should be absent. It may be desirable to pass carbon dioxide into the separatory flask during reduction and titration.

In the routine application of procedures such as the above, there must obviously be some short cuts: These must depend on the nature of the samples for which routine analysis is required. The more abbreviated the procedure, the less general its application. The important thing to recognize is that an abbreviated procedure designed to yield acceptable results for one type of material should never be applied to another type of material without investigation. The investigation should at least consist of putting a known sample through the procedure to uncover potential problems. For example, the following procedure has been applied with success to the determination of molybdenum in molybdenite concentrates:

PROCEDURE. Digest 0.2 g concentrate with aqua regia. Evaporate to dryness, and treat the residue with sodium hydroxide solution; acidify with hydrochloric acid and pour the mixture into boiling hot 10% sodium hydroxide solution. Boil for several minutes and filter. Wash with hot water. Precipitate molybdenum in the filtrate as lead molybdate, ignite, and weigh; or, after buffering the solution with ammonium acetate, titrate with lead nitrate solution using tannin as an external indicator. Note that after the titration, the determination can be finished gravimetrically, providing a control on the rapid, but inexact, titration.

Among elements that may interfere in this abbreviated procedure are tungsten, niobium and tantalum, and vanadium and aluminum, which may finish in the lead molybdate.

If the sample is given a preliminary leach with concentrated hydrochloric acid and the procedure is applied to the insoluble portion, the result will represent sulfide molybdenum only, and most of the interfering elements will have been removed. Gravimetric determination of sulfur in molybdenum and tungsten ores and concentrated by precipitation and weighing barium sulfate presents a difficulty because of the insolubility of barium molybdate and tungstate. The difficulty is overcome by adding

a large excess of mannitol, which forms a strong complex with molybdenum, or tartrate, which complexes tungsten.

4.10. MANGANESE

Manganese is almost universally present in rocks and minerals. It is an
essential minor constituent of steel, there being no substitute for it as a
deoxidizer and sulfur-controlling ingredient. More than 90% of commercial manganese is used in steel production. The most common manganese
minerals are pyrolusite, MnO_2, and psilomelane, a hydrated oxide of variable composition containing barium. Manganese carbonate, rhodochros-
ite, $MnCO_3$, occurs in sedimentary deposits, that at Butte, Montana, providing a large proportion of U.S. manganese. A huge, as-yet untapped
supply of manganese exists in manganese nodules on the ocean floor;
these contain nickel and cobalt as well as other elements and are mostly
hydrated manganese oxides in which the valence of the manganese is
variable. Manganese occurs as a minor constituent of many silicate minerals, including garnet (Section 3.2.22), lepidolite (Section 3.2.2), and
many others. Phosphate minerals frequently contain manganese [e.g., lithiophilite, $Li(Mn, Fe)PO_4$ (Section 3.2.21)]. Most ordinary rocks contain
less than 1% MnO, usually 0.1–0.3%.

The analytical chemistry of manganese is influenced by the several
valence states of the element in its compounds and solutions. The manganese–oxygen–water system is complex and greatly affected by a number of foreign constituents such as phosphate, fluoride, arsenate, and so
on. In hydrochloric acid solution, the stable manganese species is Mn^{2+}.
In phosphoric acid solution, stabilities of Mn^{2+} and Mn^{3+} are similar. In
sulfuric acid solution, Mn^{3+} disproportionates to Mn^{2+} and insoluble
MnO_2. In nearly neutral solution, Mn^{2+} easily oxidizes to insoluble
$MnO \cdot OH$ (the composition of the naturally occurring mineral manganite,
probably a major component of sea nodules). In alkaline solution, the
favored species is MnO_4^{2-}, manganate ion, but this may disproportionate
to MnO_4^-, permanganate, and MnO_2 or $MnO \cdot OH$. In acid solution, hydrogen peroxide immediately reduces all manganese species to Mn^{2+},
with the evolution of oxygen. When persulfate is added to an acid solution
of Mn^{2+}, oxidation occurs, usually to $MnO \cdot OH$ or MnO_2: In the presence
of silver ions, oxidation to permanganate is possible.

An important separation of manganese depends on its oxidation by
persulfate to insoluble $MnO \cdot OH$ in chloride solution at pH 6.5. This separation is interesting because it is thermodynamically impossible: The

precipitate is so slow to reach equilibrium with the solution that it is effectively taken out of the system.

In the classical scheme of analysis, very little manganese is normally found in the acid group of elements. During the sulfide precipitation, manganese normally remains in solution. If sufficient ammonium salts are present, manganese does not precipitate with the ammonia group unless (1) an oxidant is present (air oxygen, CrO_4^{2-}, VO_4^{3-}), (2) phosphate or arsenate are not sufficiently removed by Fe^{3+} or Al^{3+}, or (3) tungsten has escaped precipitation with the acid group. In the analysis of manganese minerals, especially phosphate minerals, the precaution of adding an excess of zirconium chloride should always be taken. If the zirconium addition is carefully measured, it does not invalidate the weight of the ammonia group oxides. It is also necessary to make the ammonia precipitation with care so as not to exceed pH 6.5: Solutions should be boiled thoroughly to remove air oxygen and chlorine before precipitation. Washing the ammonia group precipitate with ammonium nitrate as sometimes recommended is sure to lead to incomplete separation of manganese. With minerals containing several percent manganese, a triple ammonia precipitation is recommended.

Small amounts of manganese remaining in the ammonia group may easily be determined by treating the sulfuric acid solution of the pyrosulfate fusion with a little silver nitrate and ammonium persulfate, which converts the manganese to pink MnO_4^{2-}. This may be determined colorimetrically without any loss of solution using a Dubosq colorimeter. Silver in the solution can then be removed during the removal of platinum either by using hydrogen sulfide or metallic zinc, and the rest of the analysis can proceed normally.

If the filtrate from the ammonia precipitate is to be treated to recover traces of rare earth elements and beryllium, either using excess ammonia or using tannin, the precipitate may contain most of the manganese. In fact, if little manganese is present in the sample, addition of a measured amount may lead to more complete recovery of beryllium and the rare earths. The precipitate should be treated with sulfuric and nitric acids to destroy organic matter (paper and tannin), dissolved in dilute hydrochloric acid with the aid of a little hydrogen peroxide, and the beryllium and rare earths reprecipitated from small volume with ammonia. If much manganese is present, this operation may have to be repeated one or more times. It is important to remove all peroxide by boiling before attempting precipitation. Mn(II) hydroxide is soluble in alkaline solutions containing ammonium salts but very readily oxidizes to insoluble brown $MnO \cdot OH$ even merely on contact with the oxygen of the air.

Normally, the filtrate from the ammonia precipitation will contain substantially all the manganese in the sample. In rock analysis, no separation is necessary; in mineral analysis, especially when the sample contains appreciable manganese, it is best separated at this stage through oxidation to MnO·OH using persulfate.

> PROCEDURE. Evaporate the combined filtrates from a double or triple ammonia precipitation to 100–300 ml, the volume depending on the amount of ammonium salts present. Transfer to a 400-ml beaker, place a small piece of hardened filter paper under the end of the stirring rod to facilitate boiling, boil until all air oxygen and chlorine have been removed and methyl red indicator retains its red color in the boiling solution. Add dilute ammonia and freshly prepared saturated ammonium persulfate solution alternately in such a way that the pH of the solution remains between 5 and 7. Finally adjust pH to 6.5 (using nitrazine paper) and filter. Wash with ammonium chloride solution. Redissolve the precipitate and repeat the process: Details of the reprecipitation depend on the composition of sample. Further details of the procedure are given in Section 3.1.4.

The manganese precipitate may contain a number of elements that escaped complete precipitation with the ammonia group. Among these are barium, cobalt, nickel, tungsten, molybdenum, bismuth, vanadium, rare earths (especially cerium), uranium, and beryllium. Its further treatment depends on the sample and on the amount of manganese. In many cases, solution of the precipitate in dilute nitric acid containing peroxide followed by oxidation to permanganate with periodate will suffice; the permanganate can be measured colorimetrically.

Other procedures that are applicable when barium is present have been described in Section 3.1.15.

In rock analysis and in the analysis of minerals containing only a little manganese, the persulfate separation is normally omitted, and the manganese is collected with the magnesium and weighed as $Mn_2P_2O_7$ with the $Mg_2P_2O_7$ and then determined colorimetrically. A little manganese is likely to be found in the calcium oxalate precipitate: this is easily recognized by the brown color it imparts to the ignited CaO.

In the analysis of phosphate minerals and of rocks containing much phosphate, separation of manganese from the ammonia group elements is invariably imperfect unless phosphate is removed as zirconium phosphate before the ammonia precipitation. Fluoride has a similar effect; if all fluorine is not removed prior to the ammonia precipitation, manganese is easily oxidized to Mn^{3+}, is not reduced during the boiling with dilute hydrochloric acid, and precipitates at pH 6.5 as MnO·OH. Procedures for fluoride- and phosphate-bearing samples are described in Section 4.20.

In the analysis of chromium-bearing materials (e.g., impure chromite or chrome ores), it is often desirable to carry out the dehydration of the acid group using perchloric acid. This leaves chromium in the form of chromate, CrO_4^{2-}, and it is advantageous to precipitate the ammonia group without reducing the chromium to Cr^{3+}. Complete separation of chromium from the other ammonia group elements in this way is not possible; a reason for this is that any manganese in the sample is oxidized by the chromate to insoluble $MoO(OH)_2$ or $MnO \cdot OH$ during the precipitation, with corresponding reduction of chromium to Cr^{3+} and its precipitation as $Cr(OH)_3$. Thus, in the analysis of chromites and similar materials, provision for determination of manganese in the ammonia group is necessary.

For small amounts of manganese, oxidation to permanganate in the solution of the pyrosulfate fusion of the ammonia group oxides followed by colorimetric determination usually suffices. If much manganese is present, however, this procedure is undesirable because the condition of the manganese in the ignited oxides is in doubt. It is usually assumed to be present as Mn_3O_4, but its oxidation to manganate, MnO_4^{2-}, or MnO_3, is not unlikely. It is necessary to either ignore the weight of the ammonia group oxides and determine each of them separately and not by difference or to separate the manganese from the other oxides before weighing them. This separation is easily accomplished in the solution of the pyrosulfate fusion: In sulfate solution, manganese can be completely precipitated at pH 2 or lower by addition of ammonium persulfate and boiling. Solution of the precipitate in dilute sulfuric acid containing peroxide followed by reprecipitation leaves no weighable amounts of other elements in the precipitate, with the possible exception of titanium, niobium, tantalum, and tungsten, which are seldom present in sufficient amount to cause trouble.

PROCEDURE. To the sulfuric acid solution of the ammonia group oxides, obtained via pyrosulfate fusion, from which residual silica has been removed and with a volume of 100–150 ml add 1–2 g ammonium persulfate in the form of a saturated solution. Boil for a few minutes or heat on the steam bath for an hour or more. Higher oxides of manganese will precipitate. Filter, and wash with 1% sulfuric acid solution. Dissolve the precipitate in 5% sulfuric acid, adding just enough hydrogen peroxide to reduce the manganese. Heat to boiling, and add persulfate to reprecipitate the manganese. Filter, and wash with 1% sulfuric acid. Ignite and weigh Mn_3O_4 or Mn_2O_3. Ignition at 600 °C and below yields Mn_2O_3. Ignition at 900 °C yields Mn_3O_4. Alternatively, the manganese oxide may be dissolved and precipitated as $MnNH_4PO_4$ and then ignited to $Mn_2P_2O_7$ for weighing.

The filtrates from the manganese may be evaporated and treated by the usual procedures for titanium. A little manganese may remain in solution if the precipitation with persulfate took place in too acid a solution or if phosphate was present. In this case, it is advisable to add silver nitrate and persulfate before removing platinum with zinc, in the manner described above, and determine residual manganese colorimetrically. Precipitated manganese oxides are scavengers when formed in slightly acid solution; it is therefore better to put up with less than complete precipitation, determining the residual colorimetrically, than to raise the pH to the point where complete precipitation is certain. When much phosphate is present, complete precipitation is unlikely in any case.

During a rock analysis, it is usually easier to collect and weigh the manganese along with the magnesium, with which it precipitates as manganese ammonium phosphate and is ignited to $Mn_2P_2O_7$. The ignited phosphates are dissolved in a little sulfuric acid and taken to strong fumes. After filtration to remove, for example, silica and carbon, calcium is recovered as sulfate from 80% alcohol solution, the filtrate from the calcium sulfate is evaporated to remove alcohol, and manganese is oxidized to permanganate and determined colorimetrically. Details are given in Section 3.1.4.

In the analysis of rocks or minerals containing more than about 1% manganese, it should always be separated using persulfate or sulfide before precipitating calcium and magnesium. It is then best determined by precipitation as manganese ammonium phosphate and ignition to $Mn_2P_2O_7$. This method for determining manganese works extremely well if careful attention is paid to detail. It is very important that no oxidant be present (including air oxygen), since Mn^{2+} is easily oxidized to Mn^{3+} in the presence of phosphate, even in acid solutions.

PROCEDURE. To the hydrochloric acid solution of the manganese containing 200 mg or less of MnO add up to 15–20 g ammonium chloride (the amount depending on the manganese present and on the ammonium chloride that will be formed on neutralization). Dilute to 200–250 ml, add 1–2 g diammonium phosphate, and bring to a slow steady boil (put a piece of hardened filter paper under the end of the stirring rod). Add a few drops of methyl red indicator, and then slowly add aqueous ammonia until precipitation just begins. Continue the neutralization of the boiling solution by dropwise addition of 1:10 ammonia solution, making sure that the precipitate becomes crystalline before adding another drop, finally adding enough ammonia to discharge the red color of the indicator. Cool in a cold water bath, and stir in 5–10 ml concentrated ammonia solution. Wash down the sides of the beaker with 5% aqueous ammonia, and leave a layer of wash liquid on the surface.

After 4 h or more, filter, and wash with 5% ammonia solution. Dissolve the precipitate in dilute hydrochloric acid, and reprecipitate as before. The reprecipitation may be omitted if the first precipitate is flesh pink with no brown discoloration, and if alkali salts are not present. Burn off the paper, ignite, and weigh $Mn_2P_2O_7$. It is advisable to first dry the paper and precipitate in an oven at 60 °C, transfer as much as possible of the manganese ammonium phosphate from the paper to a weighed platinum crucible, and ignite the paper in another weighed crucible. If paper and precipitate are to be ignited together, the paper cone should be put in the crucible upside down so that most of the precipitate falls into the crucible. Paper should then be burned off at the lowest possible temperature (450 °C overnight in a muffle furnace) before ignition to 900–1000 °C.

4.10.1. Titrimetric Determination of Manganese

Several methods are available, each has its application. During a systematic analysis, gravimetric determination is preferred, so that the titrimetric methods are usually employed in separate determinative procedures applied to specific materials. An exception is the standardization of manganese solutions used in calibrating instrumental methods (AA spectrometry, spectrophotometry, etc.).

4.10.2. Oxidation to Permanganate Using Bismuthate, Titration with Ferrous Iron

PROCEDURE. Prepare a nitric acid solution of the sample, and boil to remove oxides of nitrogen, chlorine, and so on. Add water and/or nitric acid to produce a solution with about 0.5 mg/ml manganese, 15% by volume of nitric acid, and free from phosphate and fluoride. Cool the solution to about 10 °C, add a large excess of sodium bismuthate that has been tested for its oxidizing power, shake or stir rapidly for 1 min, dilute with an equal volume of cold (10 °C) water, and filter at once through glass fiber paper using a Büchner funnel. Wash with cold 5% nitric acid that has been boiled to remove nitrogen oxides. Washing should be slow and deliberate.

To the filtrate, add a weighed amount of ferrous ammonium sulfate that has been assayed for its ferrous iron content, and titrate the excess ferrous iron with $N/10$ potassium permanganate solution. The weight of ferrous salt should preferably lead to a 5–10-ml titration with the permanganate.

Important interferences in this procedure are cobalt, cerium, and chromium. The interference by chromium seems not to be due to its oxidation by the bismuthate, which would lead to high results for manganese, but to oxidation during the titration with permanganate, which leads to low results. If the reduction with ferrous salt and the back titration are quickly performed in cold solution, interference by a little chromium can be tolerated. Vanadium may cause difficulty in these circumstances because it is very slowly oxidized by permanganate in cold solution.

4.10.3. Oxidation to Permanganate by Persulfate, Titration with Arsenite

This procedure is especially convenient for the determination of small amounts of manganese when chromium, cobalt or antimony are present.

PROCEDURE. Prepare a nitric acid solution of the sample containing not more than about 5 mg manganese and boil to remove oxides of nitrogen. Dilute to about 10% by volume of nitric acid, cool, and add about 0.1 g silver nitrate. Prepare a nearly saturated solution of ammonium persulfate in cold water, and add 5 ml of this to the solution containing the manganese. Heat on the hot plate until the temperature of the solution reaches about 85 °C or until bubbles of oxygen (from decomposition of the persulfate) appear. The solution must not be allowed to boil. Cool at once in running cold water. When cold, titrate with an $N/50$ solution of sodium arsenite. If the titration is performed rapidly, the silver ion remaining in the solution causes no difficulty; however, the manganese is slowly reoxidized by residual persulfate, and slightly better results are possible if enough sodium chloride to react with the silver is added just before the titration.

With more than about 5 mg manganese/100 ml solution, the end point of the titration becomes indistinct: The reaction between arsenite and permanganate is not stoichiometric, manganese is not quantitatively reduced, and calibration of the arsenite solution must be empirical.

The procedure can be somewhat improved through use of mixed nitric, phosphoric, and sulfuric acids.

As applied to the determination of manganese in steel, the method is rapid and free from interference; however, calibration against standard steels of the same nature as the unknowns is necessary because of a rather vague end point (especially with chromium steels). The following procedure may be modified for use with a variety of materials (ferromanganese, manganese ores, slags, aluminum metal, manganese bronze, etc.).

PROCEDURE. Prepare an acid mixture containing 200 ml sulfuric acid, 250 ml phosphoric acid, and 500 ml nitric acid in 2 l. Dissolve 5.5 g silver nitrate in the mixture. Prepare a titrating solution by dissolving 8 g sodium carbonate and 2.4 g arsenic trioxide in boiling water and diluting to 2 l with water. Transfer a sample containing not much more than 10 mg manganese to a 500-ml Erlenmeyer flask and add 30–35 ml of acid mixture. Heat gently until solution is complete, and then boil to remove oxides of nitrogen. Add 100 ml hot water, heat to boiling, and add 10 ml freshly prepared ammonium persulfate solution (25 g in 85 ml water). Boil for 60–90 s. Cool in running cold water to less than 20 °C, add 75 ml ice-cold water, and titrate with arsenite to removal of the pink color of Mn(VII). Carry a known sample of a similar steel through the process, matching the end point of known and unknown. Calibrate empirically.

4.10.4. Oxidation to Mn(III), Titration with Ferrous Ion

In strong phosphoric acid solution containing pyrophosphate, manganese is readily oxidized to Mn(III) by almost any oxidant [except peroxide, which reduces Mn(III) to Mn(II)]. Nitric or perchloric acids are most convenient. If nitric acid is used, chromium is not oxidized, and the only common interfering element is vanadium; unless great care is taken, however, oxidation is not quite quantitative, and an empirical calibration is recommended. With perchloric acid as oxidant, chromium and cerium are quantitatively oxidized along with the vanadium and manganese. In principle, all four elements can be titrated electrometrically and separately determined in the same solution through careful control of pH and selection of titrant. In practice, there are easier ways. If only chromium and manganese are present, heating the phosphoric acid solution to about 350 °C will cause pyrolysis of chromate, its interference thereby removed. Some experience with the method is necessary for success in this operation, but it may be very useful when chromium and manganese both need to be determined. After titration of the chromium and manganese, evaporation and heating of the solution permits titration of manganese alone— or two samples may be carried through the procedure, one of which is carried through the extra heating step.

The same reaction ($Mn^{3+} + e^- \rightleftarrows Mn^{2+}$) may be utilized in reverse; pyrophosphate is added to the Mn^{2+} solution, and the manganese is oxidized to Mn^{3+} with standard permanganate, the end point being detected potentiometrically using a pH meter or similar device and a platinum electrode. This is an excellent procedure but leaves no possibility of repeating the titration in the same solution.

A strong phosphoric acid solution containing alkali pyrophosphate is a powerful solvent for siliceous materials; this leads to a very convenient procedure for direct determination of manganese, ferrous iron, and excess oxygen in rocks, minerals, and oxides. A difficulty lies in the fact that phosphoric acid, even of the highest quality, almost always contains reducing substances, probably organic, that must be removed if acceptable results are to be obtained.

PROCEDURE. Transfer 30 ml phosphoric acid to a 250-ml Erlenmeyer flask and add about 1 ml of $N/10$ dichromate solution. Heat on the hot plate, gradually rising the temperature until the yellow dichromate fades, being replaced by the yellow-green of Cr(III). (Some lots of phosphoric acid may require more than 1 ml of $N/10$ dichromate). Allow to cool, adding 10 ml water before the syrupy solution becomes too thick for easy mixing. Add a solution containing the manganese to be determined and then 5–10 ml nitric acid. Evaporate, controlling plate temperature so that evaporation takes place without boiling. When the volume has been reduced to about 40 ml, add 5–6 g sodium pyrophosphate, and continue heating until substantially all the nitric acid has been removed. Then add 1 ml of 1:1 nitric acid that has been freshly boiled to remove oxides of nitrogen. Continue heating for a few minutes (removal of all nitric acid is unnecessary and undesirable), cool to about 50 °C, add 10 ml cold water, mix thoroughly, cool to 10–15 °C, add 100–150 ml ice-cold water, and titrate with standard ferrous ammonium sulfate solution to nearly complete removal of the violet Mn(III) color. Add a drop of 1% diphenylamine in concentrated sulfuric acid and continue the titration to a permanent end point. The end point is so sharp that minute amounts of manganese can be titrated using $N/500$ ferrous solution; however, the solution is likely to contain some gelatinous silicic acid that slows the reaction; this is an advantage rather than otherwise.

The procedure may be adapted to the direct determination of manganese in a large number of materials:

PROCEDURE. Oxidize 30 ml phosphoric acid with dichromate as described in the previous procedure. When cooling, add only a few milliliters of water, just enough to prevent the acid from becoming too thick for convenience. Add sodium pyrophosphate (the pyrophosphate may be added before the oxidation with dichromate if it is suspected of containing organic matter, but this will increase attack of the flask, which is undesirable). Weigh a sample of, for example, silicate mineral, metal, and oxide that will not show more than about 40 mg manganese. Rotate the flask containing the syrupy phosphate mixture so that the mixture covers the wall of the flask, and then add the sample. Immediately disperse the sample by shaking or jiggling the flask; this is necessary to avoid clumping and consequent slow dissolution. Heat on the hot plate until solution of the sample is complete. Different materials require different heating periods and temperatures. Nitric acid may be added if necessary, but it is better to delay the addition of the nitric acid if solution occurs without it.

When solution is complete and the manganese is oxidized, continue as in the previous procedure.

The procedure may be modified to determine ferrous iron or excess oxygen in silicates and oxides. After oxidation of impurities in the phosphoric acid, about a gram of manganese sulfate is added followed by a measured volume of standard dichromate or permanganate. Dissolution of a sample containing ferrous iron leads to removal of an equivalent amount of Mn(III); the residual Mn(III) can then be titrated with ferrous iron as in the manganese procedure. If the sample contains excess oxygen, it is of course unnecessary to make the oxidant addition.

In many cases, oxidation of manganese to Mn(III) in phosphoric acid solution is best accomplished with perchloric acid. The sample may be dissolved in a mixture of nitric, perchloric, and phosphoric acids; heating to strong fumes of perchloric acid oxidizes chromium, vanadium, cerium, and manganese to valence VI, V, IV, and III, respectively. There is no need for the preliminary oxidation of impurities in the phosphoric acid. After cooling and dilution, titration with ferrous iron proceeds as usual. Evaporation of the titrated solution may follow: If all the perchloric acid is driven off and the solution heated to 350 °C to pyrolyze the Cr(VI), manganese (plus vanadium and cerium) may be titrated.

The behavior of vanadium in the procedures outlined above depends very strongly on the details of the procedure and on other elements in the sample. After titration with ferrous iron, vanadium is almost certainly in the V(IV) valence state. Presumably, manganese may be titrated in this solution with standard permanganate using a potentiometric end point. For details of the behavior of various elements as well as vanadium during the potentiometric titration of manganese with permanganate, reference should be made to the original work of Lingane and Karplus [44].

A useful property of the strong phosphoric acid–pyrophosphate mixture is its ability to dissolve glass fiber filter paper rapidly at relatively low temperatures. This is often very useful not only in manganese determination but also in analyses of a variety of precipitates containing an oxidized element. Barium may be indirectly determined by catching precipitated barium chromate on a glass filter, dissolving in phosphoric acid containing Mn(II), and titrating. Ce(IV) in precipitated rare earths may be quickly determined. Vanadium collected on ammonium phosphomolybdate or iron hydroxide can be determined in this way. The excess oxygen content of precipitated hydrated oxides of manganese can probably be found in no simpler manner.

For details on the determination of ferrous iron and excess oxygen, refer to the section on iron (Section 4.11) and oxygen (Section 4.22).

4.10.5. Complexometric Titration of Manganese with EDTA

EDTA titration of manganese is performed in alkaline solution (pH 8–10) and presents some difficulty because of the ease with which manganese is oxidized to insoluble black or brown hydroxy compounds at high pH. Ascorbic acid or hydroxylamine must be added to prevent air oxidation of the manganese. Addition of potassium cyanide masks cadmium, copper, cobalt, mercury, nickel, and zinc. Iron interference is difficult to overcome. Aluminum is imperfectly masked by triethanolamine. Possibly, the EDTA titration of manganese can be used to confirm the concentration of standard manganese solutions containing 0.1–1.0 mg/ml manganese. Otherwise, this method has limited usefulness. It is probably best to add an excess of EDTA to the alkaline manganese solution and back titrate with standard zinc solution using Solochrome Black T as indicator.

The manganic–EDTA complex is highly colored, and addition of excess EDTA to a neutral manganous solution followed by oxidation with sodium bismuthate in acetic acid solution yields a purplish color that can be measured photometrically.

Generally, EDTA titration is not the method of choice for manganese determination.

4.10.6. Colorimetric Determination of Manganese

Traces of manganese are usually determined colorimetrically or photometrically after oxidation to permanganate. The most convenient oxidant is periodic acid, and a nitric acid solution is employed. Oxidation is complete, and calibration against a standard permanganate solution is permissible:

PROCEDURE. Prepare a nitric acid solution containing 0.1–1.0 mg manganese, adjust the acid concentration to 10–15% in HNO_3, and add 0.5–1.0 periodic acid (H_5IO_6 or $HIO_4 \cdot 2H_2O$) dissolved in a little water or a similar amount of potassium periodate solid. Heat in a boiling water bath or boil until the manganese is oxidized, allow to cool, dilute to volume and measure transmittance at 522 nm.

If copper, nickel, cobalt, chromium, or other colored species are present, their effect may be canceled by adding a similar amount to the calibrating solutions, but usually it is best to separate the manganese from these metals by precipitation from acid solution with persulfate. Large amounts of chromium have an adverse effect because Cr(III) is only partially oxidized to chromate. Phosphate and fluoride may lead to low results because of imperfect oxidation to Mn(VII): Trivalent manganese is stabilized by phosphate and fluoride.

With very low concentrations of manganese, oxidation may proceed very slowly or not at all: The reason seems to lie in the nature of the reaction; it is autocatalytic and, once started, proceeds rapidly. Thus, if no color appears after an hour or so in the boiling water bath, a carefully measured addition of standard permanganate solution may initiate the reaction: The addition can be subtracted from the final result. Addition of a little silver nitrate may be advantageous.

If difficulty is experienced with the periodate oxidation in determining very low concentrations of manganese, persulfate in sulfuric–nitric–phosphoric acid solution may be used; the procedure is similar to that described for a titrimetric finish (Section 4.10.3). In using persulfate as an oxidant, especially when very little manganese is in question, it is essential to use a reagent that has been tested for its performance and to make a saturated solution in cold water, avoiding the addition of any solid particles of persulfate to the analyte. The reason for this seems to be that persulfate sometimes contains peroxy compounds that reduce oxidized manganese species immediately.

If chromium is present, photometric measurement is best made at 575 nm instead of 522 nm despite the lower sensitivity at the higher wavelength. If this device is used, standards and unknowns must be read without disturbing the wavelength setting, and the concentration–absorbance curve is likely to be nonlinear (becaue no absorption peak occurs at 575 nm). This is one case where a filter photometer is better than a spectrophotometer.

Certain elements, notably titanium, zirconium, columbium, and tantalum, seem to inhibit the oxidation of manganese by persulfate. This is of consequence when manganese is sought in the solution of the pyrosulfate fusion during a systematic analysis. In this case, it may be best to oxidize the manganese to $MnO \cdot OH$ or MnO_2 by omitting the silver nitrate addition, filtering, dissolving the precipitate in dilute nitric acid and peroxide, boiling to remove excess peroxide, generating the permanganate color with silver ion and persulfate, and returning the whole to the original solution after visual (Dubosq) or photometric measurement.

4.10.7. Atomic Absorption Determination of Manganese

Manganese can be determined in concentrations down to about 0.01 mg/l using an air-acetylene flame and the manganese line at 2795 Å. Silica in solution depresses the signal, and this appears to be the only major interference. For the determination of very low levels of manganese (as well as other elements) in effluent waters, seawater, and so on, extraction of manganese using oxine and chloroform may be used to achieve a concentration. A practical difficulty lies in the danger of contamination from the rather large amounts of reagents necessary.

Other extractants can be used; methyl isobutyl ketone extracts the cupferron complexes of manganese and other metals at pH about 7. Blank determinations should always be run through such procedures. In methods for rapid rock analysis, using lithium borate fusion or a hydrofluoric acid attack, calibration of AA manganese determinations should always be based on similar analyzed rocks taken through the analytical process: Limited experience with the lithium–fluoborate scheme suggested by Abbey [45] indicates that calibration against synthetic silicate solutions may be adequate, the fluoride present in the solution minimizing the oxide effects in the flame.

4.10.8. Sampling Considerations

Manganese in rocks commonly occurs as pyrolucite, MnO_2, and tends to occur generally in specific high-manganese minerals rather than as a minor constituent of common minerals such as feldspar and biotite. Consequently, sampling problems may be severe, especially if small subsamples are taken for analysis. While finely ground reference samples may yield reproducible results, analysis of unknown samples for manganese may present problems. If they are ground to pass, say, 100 mesh, sampleability for manganese may be poor; if ground much finer, manganese may be lost to the grinding equipment unless extreme precautions are taken. For example, the USGS granite G-1, which is supplied at about 80 mesh, yields manganese values between 0.018 and 0.032% MnO when 0.1-g samples of as-received material are taken. In exactly the same procedure applied to a large sample of G-1 ground to pass 200 mesh, values vary from 0.025 to 0.027% MnO. In the Nancy sample GR, 0.1-g samples of as-received material yield MnO values from 0.04 to 0.06%, while the Tanganyika tonalite T-1 run in the same way at the same time yields values from 0.095 to 0.097% MnO.

During the calibration of an emission spectrometric method for analyzing laterite ores, it has been observed that most low-manganese reference materials used in calibration yield erratic signals from subsample to subsample, while a few (e.g., biotite LP-6 Bio 40-60#) regularly yield the same signal within readout precision. Lateritic ores contain manganese in the form of hydrated higher oxides such as manganite, $MnO \cdot OH$, pyrolucite, and manganese silicates. Isolated particles of these minerals end up in the samples being analyzed and account for most of the manganese in the ore.

4.10.9.   Analysis of Sea Nodules and Other Manganese Minerals

Manganese nodules from the sea floor present some unusual analytical difficulties. Some of these difficulties are shared by manganese minerals in general.

The hydrogen content of sea nodules is high—up to about 20% H_2O in apparently dry powder. Establishment of a sampling weight baseline is difficult. In accurate analysis, it is best to determine the total hydrogen content of a sample weighed out at the same time as samples for other constituents.

Manganese deposits from seawater as $MnO \cdot OH$, $MnO(OH)_2$, and similar compounds; deposition begins on a piece of foreign mineral, often a shark's tooth or a broken fragment of a previously formed nodule. As a consequence, dried and crushed nodules may contain appreciable concentrations of almost any element, and these concentrations will differ from one nodule to another. It seems probable that the concentrations of, for example, copper, nickel, and cobalt in the deposited manganese oxide–hydroxide are very similar in nodules collected from the same locality and that observed compositional differences are due mostly to differences in the "seed" around which the nodule is formed.

Manganese ores such as psilomelane very often contain barium, phosphate, and fluoride. Since manganese nodules come from the sea, they may be expected to contain sodium chloride. Sometimes, however, no chlorine is found. The reason for this seems to be that during preliminary drying, chloride is oxidized to chlorine and escapes.

An excess oxygen determination is an essential part of any accurate analysis of manganese ores and minerals. This is best done by dissolving a sample in concentrated phosphoric acid containing Mn(II) and titrating the Mn(III) formed (see above).

4.10.10. General Procedure for Accurate Analysis of Manganese Mineral or Ore

PROCEDURE. Screen the sample through 115 mesh, with a minimum of grinding. Mix very thoroughly, and spread the powder on a piece of smooth paper; cover with filter paper, and allow to come to equilibrium with the laboratory atmosphere. Avoid overdrying. If the sample is obviously wet, place it in a desiccator for 1–2 days, and then bring to equilibrium with the laboratory atmosphere.

Again mix the sample, and split out as many portions as will be needed for the analysis. Weigh each portion. Determine total hydrogen on a 1-g (or 0.5-g) portion by the Penfield method.

Determine fluorine, chlorine, excess oxygen, and phosphate.

For the main part of the analysis, treat a 0.7–1.0-g portion with 1:1 hydrochloric acid in a silica dish, adding a minimum of hydrogen peroxide to reduce the manganese; do not heat the mixture. After several hours, dilute 1:5 and filter. Wash thoroughly with 5% hydrochloric acid. Return the filtrate to the silica dish, and add 10 ml saturated boric acid (if fluoride is present). Ignite the insoluble material at a low temperature to remove paper and other organic material, and fuse with a minimum of sodium carbonate. Leach the melt with water, and add to the main part of the sample in the silica dish. Evaporate, uncovered, to dryness, and continue with the separation and determination of silica by evaporation with HF and H_2SO_4 as usual. Fuse the small residue from the silica determination with a little sodium carbonate, leach with water, and add to the main solution.

By evaporation or dilution, make the solution 1 N in HCl, treat with hydrogen sulfide while heating to boiling, dilute to 0.2–0.3 N, and continue the hydrogen sulfide treatment while the solution is cooling to about 35 °C. Filter and wash with water containing H_2S.

Usually, the hydrogen sulfide precipitate will be small and composed mostly of lead and copper sulfides. It should be treated with concentrated sulfuric acid and nitric acid to remove paper. The metals can then be determined in a number of ways.

The filtrate from the sulfide precipitate should be heated on the steam bath, uncovered, to remove hydrogen sulfide as soon as practicable so as to limit the amount of sulfur precipitated by air oxidation. When most of the H_2S has been removed, bromine should be added to oxidize sulfur and iron and the excess bromine removed by boiling. A carefully performed ammonia precipitation, with the addition of a known weight of zirconium as oxychloride if phosphate is present, will leave most of the manganese and nickel in the filtrate, from which nickel may be removed with dimethylglyoxime and determined. Manganese may then be separated by oxidation with persulfate and determined by weighing $Mn_2P_2O_7$ as described above (p. 000). Separation and determination of calcium and magnesium proceeds as usual.

The weighed $Mn_2P_2O_7$ precipitate should be examined for cobalt and nickel after solution in hydrochloric acid.

The ammonia precipitate probably contains most of the cobalt in the sample. If an accurate weight of the ammonia group oxides is obtained and does not amount to more than 20–50 mg (the usual case), AA spectrometry may be used to advantage to determine minor constituents after determination of titanium with peroxide as usual.

Zinc, if present in the sample, will mostly escape precipitation; it may be recovered from the final filtrate from the magnesium determination, after removal of ammonium salts with nitric acid, either as phosphate or by ammonium sulfide precipitation. Smaller amounts of zinc are best determined by AA spectrometry (after separation from the phosphate in the solution!); larger amounts are determined weighing $Zn_2P_2O_7$. Titration with EDTA is a possible finish; probably the best indicator for this titration is xylenol orange at pH 5 in acetate-buffered solution. Eriochrome Black T has been used at pH 9–10 in ammonia buffer; obviously, this procedure should not be used unless other metals have been completely removed (especially magnesium or manganese).

Although manganese is completely precipitated by persulfate oxidation under the proper conditions, it is nevertheless wise to suspect the presence of this element in the magnesium, calcium, and zinc precipitates and to check them (after weighing) by oxidation with periodate in nitric acid solution followed by colorimetric determination.

4.11.　IRON

Iron occurs in rocks and minerals in both the ferrous and the ferric condition, and it is necessary to distinguish between these. Usually, total and ferrous iron is determined, and ferric iron is calculated by difference. The determination of ferrous iron is a relatively simple matter when no other reducing substances are present, and the sample is easily dissolved in a nonoxidizing acid or acid mixture. However, the presence of sulfides, organic matter, refractory minerals, and other elements in their lower valences complicates the matter. Thus, if a ferrous mineral contains tri-valent vanadium, there is no possible way of distinguishing this from divalent iron once the mineral has been put into solution: What one really determines in such cases is oxygen deficiency, and the reporting of it as ferrous oxide is really a convention, albeit a very reasonable one. Most rocks contain sulfides in greater or lesser amount; the effect these have on the determination of ferrous iron depends on their nature and partic-ularly on whether or not they are dissolved by the acid mixture employed. Thus, in the procedure most often used, pyrite is largely unattacked so that the iron it contains is reported finally as Fe_2O_3. This discrepancy can be corrected in reporting the analysis by determining sulfur and mak-ing an oxygen correction for it, provided pyrite is the only sulfur-bearing mineral present: Such a device is widely accepted and understood. Al-ternatively, the sulfur may be calculated to FeS_2 and the Fe_2O_3 figure reduced accordingly.

It is apparent that when more than one sulfide mineral is present in the sample, there are difficulties in properly reporting the analytical results. When sulfate is also present, the problem is even more difficult, and if the sample contains organic matter or refractory minerals such as chrom-ite as well, it is practically impossible to make any valid interpretation of the chemical results in themselves. All the analyst can do in many such cases is to report the results of determinations, giving total iron as Fe_2O_3, total sulfur as S, and so on, and leave the interpretation of these to others. Examples of materials for which a rational chemical analysis is almost impossible in these respects are oil shales, mineralized clays, and marl.

Despite these and other uncertainties, the determination of ferrous iron is an essential part of the regular rock and mineral analysis and, in most cases, gives a reasonably meaningful figure. The method chosen depends on the material, but the following is applicable to the great majority of silicate rocks and minerals.

4.11.1.   Determination of Ferrous Iron

PROCEDURE. Weigh into a large platinum crucible 0.5000 g of the rock powder. Add a few drops of water and 1–2 drops of 1:1 sulfuric acid (to dissolve car-bonate). Cover with a tightly fitted cover. After a few minutes, prepare a mixture of 5 ml freshly boiled and cooled water, 5 ml hydrofluoric acid, and 5 ml concentrated sulfuric acid by adding these reagents in the order named to a polyethelene beaker. Poor the hot acid mixture into the crucible containing the sample, cover at once, and place it on a triangle about 3 in. above a small flame. Heat the crucible with another flame until escaping steam indicates that the contents are boiling. Heat just at the boiling point for 5–10 min (longer with some samples).

Note: The crucible in which the sample is dissolved must be sufficiently large to permit boiling 15–20 ml of liquid without loss. The cover must fit tightly so that air does not enter the crucible.

The reaction mixture must be heated at once after adding the acid, bringing it to the boiling point as quickly as possible to expel air in the crucible before much of the sample has been attacked. The small flame used to keep the mixture at the boiling point during solution of the sample must be carefully regulated on the basis of actual trials.

While the sample is dissolving, add to 200 ml water in a 600-ml beaker 50 ml of 5% boric acid and 5 ml concentrated sulfuric acid. The water must be oxygen free and should be freshly boiled and cooled to be sure of this. Fill a burette with standard permanganate solution for the titration of the iron.

After 10 min heating, transfer the crucible quickly to the 600-ml beaker containing the boric–sulfuric acid mixture, tip it so that the contents mix with the solution, and titrate at once with the standard permanganate. Continue the titration until a faint pink blush colors the solution and persists for a few seconds despite stirring. From the volume of permanganate used calculate the percentage of ferrous oxide present in the sample.

Remove the crucible and cover, stir the solution, and look carefully for unattacked material. The nature of any residue may sometimes be determined by examining it with a hand lens through the bottom of the beaker. If the residue is black, the determination should be repeated using a more finely divided sample or increasing the time of heating; or the residue may be recovered and retreated by the same procedure.

Note: Pyrite, which remains as a brassy colored residue, is only slowly attacked by hydrofluoric and sulfuric acids but will slowly dissolve and consume permanganate if the titration is continued past the first indication of a transient pink blush in the well-stirred solution. Ferrous ion reacts almost instantly with permanganate in acid solution, so a fading end point indicates the presence of some other reducing substance. It is essential that boric acid be present in the titration mixture. It forms a strong complex with fluoride, and this has two effects, both necessary to the success of the determination. The boric acid prevents both the air oxidation of Fe(II) and the oxidation of Mn(II) by permanganate.

Ferrous iron in rocks and minerals may also be determined by the closed tube method. The sample, together with an excess of sulfuric acid (with or without hydrofluoric acid), is sealed in a heavy-walled glass tube in the absence of air and heated in a steel tube containing some volatile material to equalize the pressure on the glass. After decomposition is complete, the solution is transferred to a titration flask and titrated with permanganate. At first glance, it appears that this should be a better procedure than that described above.

However, it has the following disadvantages: The sulfides in the sample are wholly decomposed, and there is some doubt concerning the corrections to be applied on their account; any organic material in the sample reduces the sulfuric acid and causes a positive error in the FeO; and when it is necessary to use hydrofluoric acid to decompose the sample, much of this is consumed by reaction with the glass tube so that it is not wholly effective. In view of the fact that the method requires much more equipment and the expenditure of more effort and time, it is seldom used, except in special cases (e.g., in the determination of ferrous iron in chromite and other refractory minerals). Even with these, the indirect method of Seil [46] is preferred.

4.11.2 Determination of Iron

By far the most reliable and trouble-free method for the determination of iron in amounts exceeding a few milligrams is its potentiometric titration with potassium dichromate solution after reduction by stannous chloride in strong hydrochloric–sulfuric acid solution using a Hildebrand calomel electrode and a bright platinum electrode. A silver–silver chloride electrode may be substituted for the calomel.

The apparatus may be complex or simple. Excellent results can be obtained using home-made electrodes, a student potentiometer powered by a 1.5-V dry cell, and a null galvanometer and tapping key. The titration cell may also be connected to a sophisticated pH meter or a recorder or both. The use of a recorder saves a good deal of time because it is possible to trace a titration curve and determine very simply the inflections that represent the two end points.

The first end point must be approached in hot solution that contains about 30% by volume of concentrated hydrochloric acid and preferably about 5% by volume of sulfuric acid. As the solution cools, the end point becomes less sharp, and reaction rates become too slow for reliable end point detection. As usually performed, the subsequent titration of the iron takes place in an open beaker, and air oxidation occurs: It is necessary to titrate rapidly in order to minimize this source of error. Good results require that the operation be carried out with speed and assurance, thus, some practice is required.

The titration may be carried out in a flask fitted with a stopper carrying a CO_2 inlet tube, the necessary electrodes, or salt bridges leading to them and an opening to admit the burette tip. Such an arrangement is very useful when a determination of both ferrous and ferric iron is required on acid-soluble samples; however, the manipulative difficulties are greater than those involved in open-beaker titration, and it is not recommended for simple total iron determinations.

A very great advantage of the method is the possibility of repeating the determination without any complication. Often it is desirable to make a preliminary rough titration followed by a more careful one: During the second titration, the dichromate may be added very rapidly until almost all the iron is oxidized, thereby avoiding error due to air oxidation. The preliminary titration also has the effect of removing traces of oxidizable organic impurity in the solution (from previous operations involving filter paper, air-borne dust, etc.), which may lead to slightly high results.

The use of dichromate instead of permanganate or ceric salt is recommended because its solution may be prepared by direct weighing of $K_2Cr_2O_7$, which is easily obtained in high purity and because interferences are fewer when this oxidant is used. Both Mn(VII) and Ce(IV) solutions cannot be easily prepared in exact concentration; they must be standardized before use. There is therefore no possibility of a truly primary iron determination using them. Also, manganese and vanadium are completely without interference when dichromate is used, and this is not true with the other titrants mentioned.

The reaction of Cr(VI) with Fe(II) is not a simple reversible process; it is therefore necessary to approach the end point with caution, mixing the solution very thoroughly, and disregarding spurious potential jumps that sometimes occur. The best explanation of these is that Fe(II) reduces Cr(VI) to an unstable Cr(V) species with a higher oxidation potential, which may persist for an appreciable time because its disappearance must involve its simultaneous reduction and disproportionation; at very low Fe(II) concentrations, the kinetics of this process are slow. This phenomenon does not cause any serious difficulty. It is mentioned because it sometimes confuses an operator who is unaware of its cause.

The essential feature of the Hildebrand calomel electrode in this application is the salt bridge, which keeps the electrode proper from being heated by the solution. Obviously, many alternative arrangements are possible; however, a calomel electrode of the kind commonly used in pH measurements is not satisfactory. Limited experience with a commercial silver–silver chloride electrode indicates that it may be substituted for the Hildebrand electrode with little or no loss in performance. If expense is a consideration, there is no doubt that a Hildebrand electrode should be prepared and used. Glass vessels with suitably arranged stopcocks and a sealed-in platinum wire to make contact with the calomel–mercury paste can be purchased. However, a small bottle with a rubber stopper suitably drilled and a rubber tube with a pinch clamp to control the salt bridge will do as well. It is recommended that the potassium chloride electrolyte be $N/10$ rather than saturated.

The bright platinum electrode that completes the titration cell may consist of a short (1–2-in.) piece of pure platinum thermocouple wire sealed in a glass tube; contact to a copper wire leading to the potentiometer is made through a drop of mercury inside the tube. Manufactured electrodes are no more effective and are much more expensive, but they may, of course, be used. They have the rather serious disadvantage of being difficult to clean when inadvertently poisoned. Poisoning sometimes occurs during the analysis of samples containing metalloids such as arsenic. A wire electrode can readily be cleaned by alternate heating in a flame and treatment with hydrochloric acid. This is not true of all manufactured electrodes. The platinum electrode should be kept in a solution containing hydrochloric acid when not in use to avoid the need for frequent cleaning.

Care should be taken not to short the titration cell at any time. Before attempting any titrations, it is most advisable to determine the setting or the scale reading of the readout device that corresponds to each of the two end points. This will require the preparation of a titration curve. With a recorder, best results are obtained by setting up the titration cell using a solution containing a few milligrams of iron and the same concentrations of acids as will be used in titrations, dipping the tip of a burette containing dichromate into the solution, adding a *slight* excess of stannous chloride, heating to near the boiling point, and adding the oxidant at a slow constant rate while stirring steadily. The span and the range of the recorder should be adjusted so that the whole curve will be included without adjustment. External resistances or voltage dividers may be necessary for this, depending on the type of recorder. If no recorder is available, it will be necessary to plot the titration curve manually. This is somewhat difficult because of the necessity for keeping the solution hot during the first end point. However, the difficulty can be overcome by titrating to very near equivalence and permitting the ambient air to act as oxidant through the end point while heating the solution gently (keep it near 80 °C) on an electric plate—preferably a stir plate so that a magnetic stirring bar can be used. Just at the end point, the potential drifts rapidly as a result of air oxidation: On either side of the end point, such oxidation is too slow to be immediately perceptible.

It is common for the titration to give difficulty on a first trial. Some skill is required to reach the first end point rapidly before the solution cools and to do this with certainty. However, a very little practice makes the operation easy; and the highly satisfying reproducibility of the results makes the effort more than worthwhile.

It is very desirable to use a burette with an offset tip (a so-called titration burette) to minimize error due to heating of the titrant over the hot solution. Use of a hot plate during the titration—except in locating the end point as described—is not advised. A magnetic stirrer is a convenience but not a necessity. If magnetic stirring bars are used, they should be boiled and soaked for several hours in hydrochloric acid to be sure they are leak proof. They should also be treated after use because some have sufficient porosity to extract iron from the solutions in which they are used.

PROCEDURE. Heat the solution containing up to 200–300 mg iron, and which should contain 5 ml sulfuric acid and 30 ml hydrochloric acid in a volume of about 100 ml, nearly to boiling, and add stannous chloride solution slowly until the yellow color due to Fe(III) is discharged. Avoid a large excess.

Dip the platinum and calomel electrodes into the solution, and titrate as quickly as possible with N/10 dichromate solution to a potential corresponding to the first end point, as previously determined. This must be completed before the solution has cooled much below about 75 °C because at lower temperatures the reactions are too sluggish to be easily followed. At the end point, the measured potential will drift fairly rapidly without any dichromate addition because of air oxidation. After each small addition of dichromate, the direction of drift (after preliminary equilibration) may be observed. Before the end point, it will trend downward; at the end point, it will trend upward. If the solution is free from more than traces of colored ions, the yellow of ferric iron may just be detected at the end point. This is a useful guide for anyone attempting the operation for the first time.

in 30% hydrochloric acid at 80 °C, the minutest trace of Fe(III) gives a perceptible yellow color. It is possible to observe the sluggish nature of the reactions involved: A drop of dichromate will produce appreciable Fe(III) in the solution even when Sn(II) is still present. The latter slowly reduces the former, and the yellow color fades. The hotter the solution, the more rapid this reaction, and the easier it is to detect the equivalence point.

As soon as the excess stannous tin has been titrated, record the burette reading, and add dichromate as rapidly as possible until the second end point potential is approached. Finish the titration carefully, stirring very thoroughly with each addition of oxidant. Read the burette, and determine the volume of titrant used.

It is most advisable to heat the solution, evaporate it, and add more hydrochloric acid if necessary and retitrate.

In titrations of this kind, several very simple and subtle mistakes are frequently made. These deserve mention because they probably cause more trouble in practice than all the more academic problems put together. Most common is a failure to thoroughly wash down the walls of the titration vessel as the end point is approached. A minute droplet of stannous chloride reagent inadvertently left adhering to the upper part of the beaker through the first end point and then washed into the solution during the oxidation of the iron will produce a false result that never fails to puzzle the operator.

Another mistake involves the handling of standard solutions. Commonly, the solution is transferred from the reagent bottle into a beaker for easy pouring into the burette. This results in concentration of the solution by evaporation. Standard solutions should always be transferred directly from bottle to burette. The extent of concentration by evaporation of water from standard solutions can easily be demonstrated by placing a small beaker of solution on the balance pan and attempting to tare it. Solutions are commonly standardized to 1 part in 1000; in terms of evaporation loss, this corresponds to 0.1 ml in 100 ml, which can be lost in a minute or so of standing in an open beaker. Evaporation may also occur from the burette, especially if the upper portion is not clean, and droplets of solution adhere to the glass. This effect is very noticeable when the so-called microburettes are used with a funnel top. After droplets of solution have dried thereon, subsequent additions of reagent are concentrated. Since a burette of this type is read to the nearest 0.002 ml, that small volume evaporated in the process of filling the burette is significant. Steps must be taken to avoid this source of error if accurate values are to result from the work.

Burettes should always be calibrated and their draining characteristics established. In the iron determination, titrant is added as quickly as possible to minimize air oxidation. Time is necessary for the solution left on the burette walls to drain down.

When a satisfactory routine has been established, it is possible to reproduce results as closely as the burette can be read so that the final accuracy depends on the burette calibration.

Values by the method described are invariably about 5 parts in 1000 higher than those obtained using a silver reductor to reduce iron. Possibly, this is due to peroxide formation in the reductor [47]. The silver reductor may, however, also give high results under some circumstances: For example, if it becomes contaminated with platinum, it acquires the ability to reduce titanium to the trivalent state, and this, of course, affects the iron determination. A zinc reductor is inferior unless a volumetric determination of titanium is required, and then the whole character of the problem is changed.

For the use of the silver reductor in rock analysis, see Peck [48]. For the use of the zinc reductor and the zinc amalgam reductor, see Hillebrand et al. [49]. The latter is useful when determination of both titanium and iron is necessary.

When iron in a sample is much lower than about 2 wt. %, as good results are obtainable colorimetrically as by titration, and when iron is less than about 1%, colorimetric values are superior unless extreme precautions are taken in the titration to eliminate interferences and detect the end points exactly. Of all the colorimetric methods for iron, that using o-phenanthroline is probably most reliable. Dipyridyl and thioglycollic acid are two of a long list of colorimetric reagents for iron that may be used instead of o-phenanthroline. Examples of colorimetric reagents that are not usually as reliable but find occasional application are sulfosalicylic acid and alkali thiocyanate. These are useful when ferric iron must be determined and reduction to the ferrous condition is inexpedient for some reason.

A critical evaluation of the performance of several methods for determining iron in silicate rocks has been made by Mercy and Saunders [50].

It is interesting to note that values for total iron in standard silicate rocks obtained by competent analysts using a variety of methods differ very little. The wide variation sometimes quoted is due to the inclusion of data obtained in procedures that suffer from one or another fault or interference.

Iron may be determined colorimetrically to about 1 part in 1000 if a good spectrophotometer is available, all interferences are controlled, and calibration is carefully done. Thus, almost as good results can be obtained as in the titrimetric procedure. However, the labor involved is at least as great, and there is no advantage, as a rule, in the use of colorimetry in the determination of iron in the ammonia group precipitate except when less than about 1% is present in the sample and attainment of maximum accuracy through the use of differential measurement and other refinements of the spectrophotometric technique is unnecessary.

4.12. NICKEL AND COBALT

Nickel- and cobalt-bearing minerals are often encountered among lateritic accumulations above ultrabasic bedrocks in New Caledonia, Cuba, Philippines, Indonesia, Brazil, and so on. A major nickelferous mineral resulting from the weathering of peridotite is garnierite, $3SiO_2 \cdot 4(Mn, Ni)O \cdot 6H_2O$, which is mined at a large scale in New Caledonia. Nickel is also found with limonitic material, $(Ni,Fe)O(OH) \cdot nH_2O$, and provides under this form the main nickel ore reserves of the world. Nickel is also associated with sulfides in Ontario, Canada.

Major cobaltiferous minerals are asbolite and lithiophorite (associations with manganese), smaltite (association with arsenic), and skutterudite (association with nickel and arsenic).

Huge quantities of nickel and cobalt in association with manganese exist as nodules on the ocean floor where these elements occur as hydrated oxides. Nickel has five naturally occurring isotopes, while cobalt has only one. Nickel improves the resistance of stainless steel to corrosion and also the mechanical properties of many alloys. Cobalt is used to prepare hard alloys in the steel industry.

Nickel and cobalt belong to the zinc group in which are also found manganese, cadmium, and copper because their hydroxides are soluble in a mixture of ammonia and ammonium chloride by formation of complexes with NH_3. Nickel and cobalt have insoluble sulfides that precipitate in acetate buffer at pH 4.8, while manganese is staying in solution. This property may be used to effectively separate manganese from nickel and cobalt.

In the classical method of rock analysis, most of the nickel and some of the cobalt are determined in the filtrate obtained after precipitation of the ammonia group. Little nickel may be caught in the precipitate with most of the cobalt. Filtrates from the ammonia precipitate are treated with ammonium sulfide to precipitate manganese, nickel, cobalt, and zinc.

PROCEDURE. Transfer the filtrates from the ammonia precipitate to an Erlenmeyer flask. Add fresh colorless ammonium sulfide solution. From 5 to 25 ml ammonium sulfide solution may be used, depending on the elements present. Fill the flask to the neck with freshly boiled and cooled water, stopper, mix, and allow to stand overnight. Filter (S & S No. 589 white ribbon, preferably doubled) and wash with water containing a little ammonium chloride and ammonium sulfide. Redissolve the precipitate in hydrochloric acid and reprecipitate to be sure that no alkaline earths or magnesium remain. This precipitate will be redissolved to determine nickel using a dimethyldioxime precipitation and cobalt using a precipitation as potassium cobaltinitrite. Nickel and cobalt are determined in the ammonia group oxides after dissolution and determination of titanium and manganese. Check also filtrates from ammonium sulfide precipitations for traces of nickel and cobalt using an AA method.

One of the most reliable methods for determining nickel is the precipitation of its complex with dimethylglyoxime. The precipitate is so bulky that only small amounts of nickel can be handled conveniently. The reaction is practically specific. Cu^{2+}, Mn^{2+}, Zn^{2+}, Cd^{2+}, and Co^{2+} are separated if enough dimethylglyoxime is added to form all corresponding soluble complexes. Cu^{2+} and Co^{2+} slow down the precipitation of nickel. If too much Cu^{2+} is present, adding an excess of thiosulfate and performing the precipitation at pH 5.5 will overcome the interference. If large quantities of Mn^{2+} are present, the addition of sulfite prevents its oxidation by air. If the nickel content is low and the cobalt content is high, a double precipitation is necessary. In the presence of too much Zn^{2+}

and Cd^{2+}, addition of ammonia salts is beneficial in order to stabilize their complexes with NH_3.

PROCEDURE. Into an aliquot containing no more than 200 mg nickel, add 5 drops of concentrated nitric acid and 50 ml of a 50% tartaric acid solution. Adjust the volume to 300 ml and heat until near boiling. While stirring, slowly add 50 ml of a 1% dimethylglyoxime solution in concentrated ammonia. Let stand at room temperature for 3 h, stirring four times at regular intervals. Filter through a No. 597 S & S filter paper, and thoroughly rinse the beaker and filter paper with water. Remove the beaker containing the filtrate and set aside. Wash off the stem of the funnel and set the beaker from precipitation under the funnel. Dissolve the nickel precipitate using consecutive washes of hot 10% hydrochloric acid in water. Remove the filter paper and rinse the funnel with water.

Note: To ensure that all nickel is removed from the filter paper during the first precipitation, dribble a small amount of ammonia around the top of the filter paper. If even minute amounts of nickel remain, a red color will be noticed.

The filtrate from the first precipitation is set aside and an additional 10 ml of the 1% dimethylglyoxime solution is added. Mix and allow to stand overnight. Filter through No. 589 S & S filter paper (black ribbon). Wash the beaker and paper with water four times. If any traces of nickel are present, a red color is noticed. If nickel is present, leach into the original filtrate with consecutive washes of 10% hydrochloric acid in water. Proceed to the second precipitation.

Note: The purpose of the second precipitation is to minimize the effects of occlusion. The first precipitate normally carries down a small fraction of the total contaminant present in the original solvent. Thus, the solution containing the redissolved precipitate has a much lower contaminant concentration generating less occlusion during the second precipitation.

To the solution add 3 drops of concentrated nitric acid and 10 ml of a 50% tartaric acid solution. Stir and let stand 15 min. The precipitation of nickel is started by adjusting the pH to 7.2 with ammonium hydroxide. Add 1% dimethylglyoxime in ethyl alcohol until precipitation of nickel is complete.

Note: Very little dimethylglyoxime is required for complete precipitation of nickel (usually 5–10 ml).

Let stand 3 h at room temperature, and stir 2 times at regular intervals. Filter through a gooch crucible with vacuum; wash the beaker and crucible 10 times with a small amount of water.

Dry at 130 °C for 1 h. Then let stand inside a desiccator for 1 h, and weigh the crucible.

Note: Ten milliliters of dimethylglyoxime in ammonia can be added to the filtrate to check the complete precipitation of nickel.

Take the crucible that contains the nickel complex, and using a vacuum flask, dissolve the precipitate with consecutive washes of 10% hydrochloric acid and water. Verify complete removal by adding a few drops of ammonia.

If even a minute amount of nickel is left, a red color is noticed. Then dry the crucible at 130 °C until a constant weight is reached. Before each weighing operation, let the crucible stand inside the desiccator until it reaches room temperature.

Note: An empty crucible is carried through the drying and weighing steps and is used as a tare to correct for any deviations within the balance during the weighing steps.

Nickel is weighed as $Ni(C_4H_7N_2O_2)_2$.

Cobalt can be precipitated as potassium cobaltinitrite. Lingane [51] demonstrated that this method can produce highly accurate results. Quantities of cobalt from a few milligrams up to about 175 mg are determined with an error of less than 0.1 mg.

PROCEDURE. Obtain a nearly neutral solution containing no more than 175 mg cobalt. Bring the volume of solution to about 50 ml in a 250-ml beaker. Add 5 ml glacial acetic acid. Heat the solution to boiling and add 30 ml of boiling hot 50% potassium nitrite. To prevent spray loss due to the gas evolution, cover the beaker with a watch glass and add the potassium nitrite solution through the opening between the lip of the beaker and the watch glass. Allow the solution to stand at room temperature at least 1 h. Filter through a tared fritted glass or porcelain crucible. Wash the precipitate with 100 ml of 2% KNO_2 followed by five 10-ml portions of 80% ethanol and once with acetone. Dry in an oven at 110 °C and weigh.

4.12.1. Analysis of Limonitic and Garnieritic Ores

Screen the sample through 200 mesh with a minimum of grinding. This step is necessary because a small quantity of chromite may be present and cause difficulties during the fusion of the residue. Mix thoroughly, and dry at 110 °C until a constant weight is reached. Put the sample immediately in a desiccator to bring it back to room temperature. Again mix the sample, and split out as many portions as needed for the analysis. Weigh each portion. Determine total hydrogen on a 1-g portion by the Penfield method. Determine excess oxygen. For the main part of the analysis, treat a 2.0-g portion with 20 ml concentrated nitric acid and 20 ml concentrated perchloric acid in a 400-ml beaker. Heat slowly on a shaker hot plate. When the reaction subsides, increase heat until strong perchloric fumes are noticed. Continue until a low volume is attained. Remove the beaker from the hot plate and wash down with water. Add 10 ml concentrated hydrochloric acid, place on a hot plate, and bring to a boil. Filter through No. 597 S & S paper into a 600-ml beaker. Wash the beaker and paper with warm water; police the beaker and wash all

solids into filter paper. Transfer the paper containing the residue into a
25-ml platinum crucible. Ignite the insoluble material at a low temperature
to remove paper and other organic material. Fuse the residue following
the procedure and recommendations of Section 3.1.2, and conduct the
classical analysis.

4.12.2. Analysis of Sulfide Ores

Screen the sample through 150 mesh with a minimum of grinding. Mix
thoroughly, and dry at 110 °C until constant weight is reached. Put the
sample immediately in a desiccator to bring it back to room temperature.
Again mix the sample, and split out as many portions as needed for the
analysis. Weigh each portion. Determine total hydrogen on a 1-g portion
by the Penfield method. Determine excess oxygen.

For the main part of the analysis, treat a 5.0-g portion with 25 ml water
and 3 ml bromine in a 400-ml beaker. On most samples, it is best to put
the beaker into a cold water bath adding bromine slowly while mixing,
as initial reaction may be vigorous. Digest at room temperature for 4 h,
and then add 25 ml concentrated nitric acid. Boil gently until the excess
bromine is expelled. Filter through No. 597 S & S paper into a 600-ml
beaker. Wash beaker and paper with warm water; police beaker and wash
all solids into filter paper. Continue as indicated for limonitic and garnieritic ores.

4.13. BORON

Boron occurs as borax and boric acid and in tourmaline as well as in
several rare silicate minerals. Borax is mined in large quantities from
the salt deposits in dried-up lakes, notably in California, where there are
also large deposits of colemanite, $Ca_2B_6O_{11} \cdot 5H_2O$.

Tourmaline analyses are shown in Section 3.2.6. In the classical scheme
of analysis, boron of itself causes little difficulty except that some of it
may be brought down by the ammonia precipitate, causing a slightly high
value for any element (usually aluminum) determined by difference.
Boron contamination of the ammonia group oxides may occur when boric
acid is added to the sample to remove fluorine and is imperfectly removed
by methyl alcohol; such contamination is easily measured using the curcumin procedure of Heyes and Metcalfe [52].

The most useful analytical property of boron is its reaction with mannitol, in which the weak acid H_3BO_3 is converted to a strong acid, permitting a simple acid–base titration.

Boric acid is appreciably volatile, and acid solutions should not be boiled if their boron content is of consequence. Boric acid can be almost completely removed from perchloric acid solutions by evaporation.

By far the most effective method for determining traces of boron is that using curcumin; this reagent has been in use for a century or more, but only recently has its application been perfected, mostly as a result of work done at the United Kingdom Atomic Energy Administration, especially that of Heyes and Metcalfe.

Among many other colorimetric reagents for boron may be mentioned methylene blue, dianthrimide, carminic acid or carmine red (cochineal), quinalizarin, and several hydroxyanthraquinone dyes.

The outstanding separation of boron is its distillation as methyl borate, the optimum conditions for which were worked out by Chapin [53]. The description of Chapin's method as given by Hillebrand and Lundell [54] is as nearly perfect as any method description can be: Anyone following it exactly is sure to achieve an excellent result. Using boron-free reagents and glassware and calibrated burettes and carrying a blank determination through the procedure, boron values accurate to $\pm 0.01\%$ B_2O_3 are possible. The only serious interference comes from fluoride, and this can be overcome by making sure there is a large excess of aluminum present throughout the separation.

The method requires meticulous attention to detail and a thorough understanding of the many pitfalls to be avoided.

PROCEDURE. Weigh a sample of finely ground rock or mineral that contains not more than 100 mg B_2O_3. If the boron minerals are not water soluble (e.g., tourmaline), weigh a suitable quantity of anhydrous sodium carbonate to the nearest milligram, mix with the sample in a 50-ml platinum crucible, and fuse.

Fusion may have to be prolonged and the maximum temperature of a good Meker burner may be required to completely decompose some boron minerals. Not more than the minimum weight of flux necessary to decompose the sample should be used; normally, five times the weight of sample is sufficient. There is some possibility of losing boron during the fusion, and it is advisable to place a second platinum crucible on the lid of the fusion crucible and arrange a system to pass water through this crucible; this is easily done by mounting two tubes in a rubber stopper, one connected to a water pump and the other to a controlled water supply. Any volatilized boron will then be condensed on the lid of the fusion crucible. The condensing arrangement should be carefully set up—if anything goes wrong with it during the fusion, much inconvenience can ensue!

Cool the crucible, and stand it in a platinum dish or a low-boron glass beaker. Wash off the lid with a few drops—an absolute minimum—of water, adding the water to the crucible. Cover the beaker or dish containing the crucible with a soda lime glass cover (not Pyrex!), and using a pipette, add to the crucible exactly the amount of 1:1 hydrochloric acid necessary to react with the sodium carbonate flux.

The acid addition should be made slowly, and if it seems necessary to heat the mixture, this should be done on the steam bath. The solution must not be boiled, as this would lead to loss of boric acid. Careful tally of the volumes of water and acid added is necessary.

Set up a distilling apparatus consisting of a 500- or 1000-ml flask to deliver methyl alcohol vapor, a 250- or 500-ml working flask, a condenser with a condensing surface at least 40 cm long and 6–8 mm in diameter, and a receiving flask of 250–300 ml capacity fitted with a U-trap containing water to make sure no methyl borate escapes.

The whole apparatus should be made of low-boron (alkali-resistant) glass or of fused silica.

Transfer the sample (fused or unfused) to the working flask using an absolute minimum of water. Measure the volume of water used. Using a long-stem, wide-bore funnel, add 1 g anhydrous calcium chloride for each milliliter of water present, making sure that none of the calcium chloride adheres to the neck of the flask. If the sample contains fluorine and little aluminum, add also 0.5–1.0 g aluminum chloride.

Put 300 ml methyl alcohol in the first flask, and connect the apparatus using surgical-grade gum rubber connections. Put a boiling tube or boiling chips in the methyl alcohol flask. Take all precautions against spillage of methyl alcohol during the distillation. The methyl alcohol flask may be heated in a large water bath, and it should carry a safety tube, that is, a long glass tube reaching to the bottom of the flask and extending 50 cm above the stopper. The working flask should also be heated in a water bath, which may consist simply of a large porcelain dish or casserole. It is perhaps unnecessary to remind that no Pyrex or other borosilicate glass is permitted.

Heat the methyl alcohol flask until about 20 ml alcohol has accumulated in the working flask; then heat the water bath surrounding the latter with a small flame so adjusted that no further accumulation of alcohol in the working flask occurs. Continue until about 100 ml distillate has been collected. Add the contents of the water trap to the collection flask, and replace it with another. Continue the distillation.

After another 100 ml distillate has been collected, it is probable that all the boron has distilled, but there is no harm in collecting a further 50 ml or so in a third collecting flask to be sure of this. If most of the boron is not present in the first 100 ml distillate, this is an indication that insufficient calcium chloride was used or that the methyl alcohol contained water.

Measure the approximate boron content of each of the distillates as follows: To the distillate add 1 drop of 1% paranitrophenol in 75% ethanol, and titrate with 0.5 N sodium hydroxide (prepared and stored in a boron-free apparatus) to a rather diffuse end point. Add several drops of 0.5% phenolphthalein in 50% ethanol, and continue the titration until the pink basic form of the indicator just begins to appear.

A relatively strong alkali solution is used to limit the amount of water added to the distillate. The amount of alkali added between the two end points gives a (rather poor) approximation to the boron content of the sample.

Add to the solution twice as much 0.5 N sodium hydroxide as was added between the two end points, transfer the solution to a 500-ml flask using methyl alcohol to rinse the collection flask, and connect the flask to a condenser. Boil off the alcohol.

If the recommended amount of alkali was added, no boron will distill, but it does no harm to save the distillate and check it for any boron.

When the liquid containing the boron no longer boils in a boiling water bath, transfer it to a porcelain casserole, rinse the flask with a minimum of water, and heat over a small flame until the remaining alcohol has been boiled out. (The volume of solution at this stage should be less than 25 ml.)

Return the solution to the 500-ml flask, and add 1:1 hydrochloric acid in *small* drops just until the color of both indicators is discharged. Do not add more than just enough of the acid. Put a boiling tube in the flask, heat in a boiling water bath, attach a stopper with a tube running to a water pump, and boil under reduced pressure while the solution is cooling to room temperature.

The purpose of this treatment is to remove carbon dioxide, which will interfere in the subsequent titration.

Dilute the solution to 25 ml. Titrate the cold solution with 0.5 N sodium hydroxide until the yellow of paranitrophenol just appears. Bring back to colorless with 0.1 N hydrochloric acid and then to a faint yellow color with 0.1 N alkali. The end point is achieved when a small drop of 0.1 N hydrochloric acid discharges the yellow color completely.

Add 1 g mannitol, and titrate the boron with 0.1 N sodium hydroxide using a 10-ml microburette if the boron content of the sample is small. If the titration exceeds 1 or 2 ml, add a further 1 g mannitol and continue the titration to the end point, which is the faintest perceptible brown-pink.

When the mannitol is added, the yellow color of the paranitrophenol disappears. During titration it reappears and grows stronger. The appearance of the phenolphthalein end point is masked slightly by the yellow color of the paranitrophenol but is easily perceptible if a minimum of paranitrophenol was used.

In our experience, the second 100 ml distillate contains less than 1% of the boron in the sample provided water additions during the early stages of the analysis are carefully controlled. With samples containing minor amounts of boron, the increased titration blank following separate treatment of two distillates offsets the advantage of recovering the last traces of boron.

It is advisable to test the insoluble residues in the first distillation flask with curcumin to be sure that all boron-bearing minerals were completely decomposed during fusion or other treatment.

If the boron is separated by distillation and it turns out that there is too little for titrimetric determination, the distillates may be evaporated to dryness with calcium carbonate and the residue examined for boron with curcumin. Throughout all such procedures, borosilicate glass apparatus must be avoided.

4.13.1. Determination of Boron without Distillation

When a silicate rock or mineral is fused with sodium carbonate and the melt is leached with water, most of the boron is found in the solution; some, however, invariably remains in the residue. With the development of a rapid and accurate method for determining this residual boron, objections to this relatively simple approach have disappeared. Normally, 2–5% of the boron in the sample remains in the residue. Retreatment by sodium carbonate fusion is ineffective in recovering this remainder.

In mineral analysis, especially of tourmalines, when the supply of sample is limited, the residue from a sodium carbonate fusion and leach may be used for the determination of a number of elements that are quantitatively retained, for example, the rare earths and beryllium, titanium, zirconium, and in some circumstances tantalum and niobium. A very small portion of the residue is sufficient for the recovery of undissolved boron using curcumin.

PROCEDURE. Fuse a 0.5-g sample with 3–5 grams sodium carbonate. Cool. Release the melt from the crucible by brief reheating, and transfer it with a minimum of water to a boron-free glass beaker. Leach on the steam bath in 30–40 ml water until the melt is completely disintegrated. Filter, and wash the residue with hot water containing a little sodium carbonate. Collect the filtrate in a 300-ml flask, add about 1 ml more 1:1 hydrochloric acid than is required to neutralize the sodium carbonate, heat nearly to boiling (do not boil), and add an excess of dry precipitated calcium carbonate. Boil vigorously for 10–15 min. Filter, and wash with hot water that has been boiled to remove carbon dioxide. Filtration should be accomplished using a Büchner funnel and a pressure flask. When filtration is complete, remove the Büchner funnel and stopper the flask. With the flask still connected to the filter pump, heat the solution to boiling. Then allow it to cool with the pump still connected. Add phenolphthalein indicator, and titrate to a slight pink color with 0.1 N sodium hydroxide (free from carbonate). Add 1–2 g mannitol and titrate the boron with the 0.1 N sodium hydroxide.

More accurate results are obtained using a pH meter and a second-derivative method.

Accumulate all residues and determine their boron content using curcumin. Correct the titrimetric determination accordingly.

4.13.2. Colorimetric Determination of Boron Using Curcumin

A properly conducted determination of boron using curcumin is so robust (using Heyes and Metcalfe's adjective) that careful technique will determine the element to ±1% or better of the amount present from the sub-ppm level up to 10% or more in the sample. The color reaction is enormously sensitive, the boron–curcumin complex having a molar absorption coefficient of 180,000. Thus 0.1 μg boron can easily be detected in a volume of 100 ml, and the practical limit of detection is about 0.001 μg using refined procedures.

The known interferences in the boron–curcumin reaction are fluoride and nitrate. There seems to be a threshold below which fluoride does not interfere: If the fluoride content of the sample is less than the boron content, no interference is noted. Probably, aluminum helps prevent fluoride interference. In cases where little boron must be determined in the presence of much fluoride, the boron should be steam distilled with methanol, and a trap containing $AlCl_3$ in anhydrous methanol should be included in the distillation line. The absorbing reagent is prepared by dissolving aluminum metal in anhydrous methanol saturated with hydrogen chloride. Water must be excluded. (*Note:* aluminum dissolves slowly at first in the methanol–hydrogen chloride, but solution rate increases as the reaction proceeds; a cold water bath should be kept at hand!)

4.13.3. Determination of Boron in Water

Procedure. Transfer a volume of water containing not more than 5 μg boron to a gold (preferable), platinum, or plastic dish and add 1 pellet of sodium hydroxide (about 0.1 g). Evaporate on the steam bath to dryness (no more than 0.2–0.3 ml water should remain). Add 3 ml of 0.125% curcumin in anhydrous acetic acid, and stir with a platinum or plastic rod until solution is complete, warming gently if necessary. Cool. Add 3 ml of 1:1 sulfuric–acetic acid prepared by slowly adding 50 ml concentrated sulfuric acid to 50 ml anhydrous acetic acid (use an ice bath). Mix thoroughly, and allow to stand for 10–15 minutes at room temperature.

Prepare blanks using the same volume of boron-free water as the sample and also a much smaller (or larger) volume.

The purpose of the double blank is to make sure of a reliable reagent blank and to distinguish this from a blank introduced from apparatus or distilled water. In our experience, reagent blanks have been astonishingly small, and most reagent curcumin contains essentially no boron. If there is a large blank, a different reagent should be tried.

To the cold mixture add up to 30 ml anhydrous ethanol, and transfer the slurry to a 100-ml volumetric flask. Dilute to the mark with anhydrous alcohol, and allow the insoluble matter to settle (or use a centrifuge). Draw off a portion of the supernatant liquid and transfer to a 1-cm spectrophotometer cell. Read absorbance at 555 nm. Note that once the reaction mixture has been diluted with alcohol, it may be transferred to Pyrex ware without any concern for the boron content of the glass. Boron and curcumin do not react after the alcohol addition. The color is stable for several hours if water is carefully excluded from the system, and absolute alcohol is used. Denatured alcohol, 98% alcohol, and methanol or methylated spirits should not be used without first checking their performance.

Calibrating standards should be prepared by adding known amounts of boron to boron-free water and taking these through the whole procedure. The standard boron solution is best prepared by dissolving boric acid in water to make a solution with 1 mg/ml B_2O_3, diluting this solution 1:100 when needed, and using a microburette to dispense the diluted solution.

Note that the quantities of reagents given (3 ml) and the volume of alcohol (100 ml) must be adhered to. A lower dilution yields high values because the purple acid form of the dye is not completely removed, and it absorbs at about the same wavelength as the boron–curcumin complex.

In principle, boron in many rocks and minerals may be determined simply by fusing a 5–10-mg sample (observe sampling problems!) with 0.1 g sodium hydroxide in a gold crucible, leaching with water, evaporating, and proceeding as above. This is occasionally useful, especially in mineral analysis when only a small amount of material is available. Generally, however, sampling considerations lead to the necessity for a much larger sample: For tourmaline or tourmaline-bearing rocks, a 0.1–0.5-g sample may be fused with sodium hydroxide and leached with water in a platinum dish; a very small aliquot may then be taken and put through the procedure given above.

The boron–curcumin reaction may be made even more sensitive by dilution of the purple sulfuric–acetic acid solution containing boron and curcumin with water instead of alcohol. Thereby the unreacted curcumin and the boron–curcumin complex are precipitated. Filtration using a porcelain filter stick or similar device leaves the boron in the insoluble residue, which can be washed with water, dried, and extracted with a small volume of 0.1% acetic acid in anhydrous alcohol for colorimetric measurement.

Separation of boron by steam distillation with methanol from relatively large samples leads to the possibility of determining boron at the parts-per-billion level. It is only necessary to add an appropriate amount of sodium hydroxide to the collection flask, evaporate a suitable aliquot to dryness, and proceed as described above. It is probably unnecessary to again remark that boron-free glassware must be used throughout and that blank determinations should be taken through the whole process.

Numerous variations of the Heyes and Metcalfe procedure have been published. Many of these seem to have developed in attempts to use a substitute for anhydrous absolute alcohol. If available alcohol contains water, an appropriate use of acetic anhydride instead of acetic acid may be advantageous. Methyl alcohol may be substituted for ethanol; in our experience, results are not then as satisfying. Ethyl alcohol is denatured in various ways: Only by experiment is it possible to decide whether a particular formulation is effective. It is best to obtain pure anhydrous ethanol if at all possible. A difficulty worth mentioning arises if acetic anhydride is used: While the quantity of water in the reaction mixture must be kept at a minimum, some water is essential to the reaction. Use of an excess of acetic anhydride dehydrates the system to the point where the absorbance due to boron is seriously diminished. Additions of acetic anhydride should be carefully calculated to leave about 0.1 ml water in the reaction mixture.

4.13.4. Other Colorimetric Methods for Boron

Boron forms a triple complex with methylene blue and fluoride, and this appears to be a useful procedure, especially if one wishes to avoid the rather complicated process of separating boron and fluorine from samples containing both. The method has been applied to the determination of boron in steel by Bhargava and Hines [55], who suggest that it is less troublesome than that using curcumin; the latter is described by Harrison and Cobb [56].

Levinson [57] proposes dianthrimide as a reagent for determining boron in river water; his procedure is a modification of that of Rainwater and Thatcher [58]. It seems likely that procedures using dianthrimide and similar reagents will be replaced by curcumin procedures, if only because of the nuisance of working in concentrated sulfuric acid solution.

Carmine red (carminic acid, cochineal) is one of a number of dyes containing hydroxyanthraquinone (alizarin) that yield a blue color with boric acid in concentrated sulfuric acid. Fleet [59] describes a method for trace boron determination in rocks and minerals using carmine based on the Hatcher and Wilcox [60] method for water and biological materials. The rock or mineral sample is fused with potassium carbonate and the melt is leached with a mixture of hydrochloric acid, mannite, and Dowex 50 cation exchange resin. A small aliquot of the solution is treated with concentrated sulfuric acid and carmine red. Absorbance is measured at 585 nm against a blank determination.

4.14. GALLIUM, INDIUM, AND THALLIUM

These elements are highly dispersed in igneous rocks, and their geochemistry has been described by Shaw [61].

Concentration levels in ores and even in products are very low, and highly sensitive analytical methods must be used. Preliminary concentration from large samples may be required. As always with trace analysis, sampling considerations are important. Fortunately, these three metals may rarely form minerals of their own; they occur as minor constituents in minerals of other elements, and sampling characteristics will be those of the host minerals, which are themselves often sparsely distributed.

Gallium occurs universally in aluminum minerals and could be recovered during aluminum smelting; however, extensive modification of the process would be required. Most of the gallium probably ends up in the aluminum metal, where it has the effect of diminishing corrosion resistance. Gallium also occurs as a trace constituent of sulfides and becomes concentrated in residues from zinc refineries; much of it finishes in the zinc unless the process is modified to recover the gallium. According to Schoeller and Powell [62], flue dusts from some coals contain up to 1% gallium originating in sulfides, especially those containing germanium, and this is a major source of gallium in England. Price of gallium is extremely high, and its use is mostly limited to the semiconductor industry. Gallium has two naturally occurring isotopes, ^{69}Ga (60.5%) and ^{71}Ga (39.5%).

Because the properties of its hydroxide, Ga(III) may be included in the ammonia group. The chemical properties of Ga(III) are similar to the properties of Al(III) and Zn(II). The hydroxide precipitates as soon as pH 3 is reached and may redissolve around pH 9. If brought to a boil, the hydroxide may become much less soluble. $GaCl_3$ is volatile at 200 °C. Ga(III) in hydrochloric medium forms a purple complex with rhodamine B.

If occurring with sulfides, the gallium-bearing mineral may be decomposed using aqua regia.

PROCEDURE. Take a 10-g −200-mesh sample and digest it in aqua regia for 24 h; a hot plate may be used to aid the digestion process; however, the surface temperature must be kept around 100 °C due to the volatility of gallium at slightly higher temperature. Evaporate the sample to a paste, add 10 ml concentrated HCl, wash in 30–40 ml H_2O, and heat gently without reaching 100 °C. Filter, fuse the residue with $K_2S_2O_7$, leach the melt with water, and dissolve with minimum HCl and combine solution with original filtrate. Pour the filtrate into a separatory funnel. Rinse beaker with HCl and add to the separatory funnel until total HCl content is 6.5 M. According to Charlot [63], in the presence of Ti(III), substantial amounts of Sb(V), Au(III), Tl(III), and Fe(III) can be reduced to a valence state that will not interfere during the formation of the gallium complex with rhodamine B. Add 2 ml $TiCl_3$ at 15%. Let stand for 5 min, and extract twice with isopropyl ether. Wash both extracts with 1 ml of a 7 M HCl solution containing 2% $TiCl_3$. Bring extracts to dryness on a water bath, and then leach the residue with 5 ml of the same 7 M HCl, 2% $TiCl_3$ solution; heat gently to redissolve. Cool. Transfer 8 ml of a mixture of 3:1 chlorobenzene–carbon tetrachloride. Rinse the beaker with 1 ml of 7 M HCl, 2% $TiCl_3$ solution. Add 0.5 ml rhodamine B (0.5% in HCl, 6 N). Shake for 10 min. Filter the organic layer on glasswool and collect it in a 10-ml volumetric flask containing 1 ml alcohol. Wash the aqueous solution with 1 ml solvent. Add to the 10-ml volumetric flask. Bring to volume and measure at 562 nm.

For ores or minerals other than sulfide, the decomposition can be directly performed with $LiBO_2$, Na_2O_2, or Na_2CO_3 as described in Chapters 2 and 3.

Indium occurs in traces in zinc blends, tin ores, and many sulfides, especially of tin and antimony. It may be recovered during electrolytic refining of zinc, but quantities are relatively small. Indium is a relatively mobile element and may travel considerable distances from its original source [64].

Indium in ores at 1 ppm or even lower levels may be significant. At one time, it was thought that indium formed no natural mineral of its own; however, an indium mineral, roquesite, $CuInS_2$, has been recently reported in France [65], Japan [66], and New Brunswick, [67]. This mineral contains 47% indium.

According to Sutherland and Boorman [68], calcopyrite in the Mt. Pleasant (New Brunswick) ore contains 0.19% indium, sphalerite has 1.25% indium, and tetragonal stannite carries 2.1% indium.

Industrial releases of indium in western United States to the environment annually comprise about 28 tons to the air, 100 tons solid wastes to land, and 230 tons to land and/or water (chiefly to tailings ponds, which may dry up). Perhaps 30 tons indium are also contained in finished metal products, chiefly zinc oxide and slab zinc. Wastes disposed into water or land include red muds from bauxite milling (140 tons In), iron ore blast furnace slags and dusts (39 tons), Arizona copper mill tailings (33 tons), Arizona copper smelter slags (25 tons), urban waste (24 tons), collected coal fly ash (20 tons), lead–zinc mill tailings (14 tons), zinc smelting solid waste (7 tons), tin smelting solid waste, steel making dust, slags from lead smelting and slag fuming, and so on [69].

Indium has two naturally occurring isotopes, ^{113}In (4.33%) and ^{115}In (95.67%). New interest in indium arises from development of thin-film solar cells, with promise of eventually replacing silicon cells. Indium is used as an outstanding solder to metals and also in the semiconductor industry. Indium is a very attractive coating material to protect other metals against corrosion. Much of the work on concentrating, separating, and determining indium appears in the Russian literature, making library search somewhat difficult.

In(III) is a member of the ammonia group; however, as for Ga(III), its sulfide is not very soluble. The hydroxide precipitates as soon as pH 3 is reached and does not redissolve before pH 14.5 (gallium hydroxide redissolves as soon as pH 8.5 is reached).

Since the grade if indium ores is extremely low, leaching of large samples (e.g., 1–2 kg) with hydrochloric acid and an oxidant is necessary. This type of dissolution should attack most indium-bearing minerals. Concentration of the leachate will then yield a solution in which indium can be certainly determined by AA or X-ray spectrometry. Indium may be concentrated even further with zinc metal, which will precipitate it at the same time as other heavy metals. The precipitate may be redissolved with HNO_3 followed by boiling the solution for a short time.

Small amounts of indium can be determined colorimetrically using the 8-hydroxyquinoline method following the recommendations of Lacroix [70] and Sandell [71].

One of the most reliable ways to determine indium is by neutron activation.

Thallium occurrences have been described by de Albuquerque, Muysson, and Shaw [72].

There are a few rare thallium minerals such as crookesite, where thallium is associated with elements such as copper, silver, and selenium; lorandite, where thallium is associated with elements such as arsenic and sulfur; and urbaite, where thallium is associated with antimony, arsenic, and sulfur. Thallium has two naturally occurring isotopes, ^{203}Tl (29.5%) and ^{205}Tl (70.5%). All thallium compounds are extremely toxic.

According to Charlot, Tl(III) may be considered as a member of the ammonia group; however, Tl(I) resembles more the alkalies and also Ag(I) [73].

The most appropriate decomposition techniques for thallium minerals or thallium ores are acid digestions. Aqua regia may be more suitable for ores rich in pyrite; hydrochloric acid may be used for blende with a HNO_3 finish. In order to make these acid digestions successful, it is necessary that the sample be at least 100% -200 mesh. Several authors recommend to separate thallium from many other elements using zinc metal in acid medium near neutral. After the metallic residue is collected and redissolved, thallium can be effectively determined by a colorimetric method using rhodamine B [74].

4.15.   CARBON AND CARBON DIOXIDE

The determination of carbon in rocks is complicated by the fact that it may occur in at least three forms: carbonate, graphitic, and organic carbon. There are also a number of minerals that contain inorganic carbon, presumably in the same valence state as in carbonate, which do not easily give up carbon dioxide on boiling in acid. Scapolite is the best example. Carbon dioxide may also be present in the form of liquid or gaseous inclusions. If their matrix is not attacked during attempts at analyses, carbon is not completely recovered, and low results may be reported.

If graphite is present, a determination of total carbon by combustion and of acid-soluble carbonate gives graphite carbon by difference, provided all the carbonate is attacked by the acid. Graphite may also be determined by treating the sample with acid without the addition of any oxidant, filtering off the insolubles on an asbestos pad, and burning the latter in a combustion tube procedure and finally weighing the graphite as CO_2 in a weighing bulb.

Organic Matter Causes Difficulty. As a rule, all that can be done in an inorganic laboratory is to determine total carbon and calculate the weight percent of organic matter by subtracting carbonate carbon and using an empirical factor. This is commonly done in the analysis of soils. If carbonate, graphite, and organic matter are all present in the same sample, it may be possible to separate organic matter and graphite by treatment with a mild oxidant such as dilute nitric acid containing hydrogen peroxide; but uncertainties are great.

There are several alternatives to weighing the carbon dioxide obtained either by combustion or by acid treatment. It may be determined volu-

metrically by standard procedures of gas analysis, it may be absorbed in
alkali and determined by acid–base titration, or it may be measured con-
ductimetrically after absorption in water or by the techniques of gas chromatography. These alternatives are more attractive in microprocedures
because of their greater sensitivity. A well-standardized gravimetric procedure is probably more flexible and less given to difficulties than any
other, provided adequate sample is available.

Total Carbon. Total carbon in rocks is best done by combustion of a
relatively large sample (usually about 2 g) in a large combustion boat of
fire clay containing a bedding material such as alundum. A split combustion furnace is used with a silica combustion tube. Silica is used because it can endure considerable thermal shock. The split furnace makes
it possible to remove the hot tube without disturbing the train, so that it
is not necessary to wait for the whole furnace to cool between samples.
The use of fluxes to ensure complete destruction of the sample is almost
always necessary. Common fluxes are lead oxide, lead chromate, and
vanadium peroxide. The boat containing a mixture of flux and sample is
placed in the cold tube, and air is passed through the train until all extraneous CO_2 has been removed. The weighing bulbs are then placed in
position, and the tube is moved into the furnace and heated until reaction
is complete. A disadvantage of this *cold-start* method is that volatile organic matter is sometimes not burned; this difficulty may be overcome
by including a secondary furnace containing hot lead chromate, silver
oxide, or manganese oxide in the train.

In the *hot-start* method, the combustion tube is maintained at temperature continuously. The sample and flux are introduced directly into
the hot tube, which is stoppered at once. Oxygen, rather than air, is used
as the carrier gas.

It is advisable to cover the charge with a layer of inert material (alundum) to prevent premature ignition of samples containing combustible
matter. This provides the simplest and most rapid way of obtaining total
carbon, for if oxygen is used at high temperature, no volatile organic can
possibly escape combustion. The disadvantage of the method is the somewhat indeterminate blank obtained as a result of the introduction of a little
air each time the tube is opened. A possibility of some carbon monoxide
forming during combustion is easily avoided by temperature control and
by having a sufficiently large combustion tube.

Induction heating methods are rapidly replacing others in such procedures and have many advantages. However, the equipment is considerably more costly and more expensive to maintain and does not seem
to offer any very great advantages unless a very large number of samples

is involved. Induction furnaces should be considered primarily for their ability to determine sulfur rapidly and simply. If one is available, it may be readily converted to determine carbon.

A simple apparatus that does not require a combustion furnace will now be described. It will be found entirely adequate for the occasional total carbon determination and has the advantage that conversion to microprocedure is not difficult.

The *combustion tube* is a 9 × 1-in. Vycor test tube. Dried and CO_2-free air is drawn into it, exiting via a Vycor tube extending nearly to the bottom. The sample is contained in a porcelain boat and mixed with an oxidizing flux (V_2O_5 or $PbCrO_4$–PbO). Evolved gases are passed through a U-tube containing silver and manganese oxides prepared by heating silver permanganate and then through the same train as is used for the determination of acid-soluble carbonate (see below). The procedure for rocks is as follows:

PROCEDURE. Clean and ignite the Vycor tube and porcelain boat, and allow to cool. Weigh 0.5000 g powdered sample on a watch glass, and add to it 0.1–0.2 g powdered vanadium pentoxide or up to 1 g of 1:10 $PbCrO_4$–PbO or other suitable flux. Mix the sample and flux intimately on the watch glass with a small spatula. Mixing should be thorough. Vanadium pentoxide will be used when the carbon is largely carbonate and organic; lead oxide–chromate will be used when the sample is refractory (scapolite). Transfer the mixture to the small porcelain boat, and put it in the tube. Stopper, and connect the carbon train with a length of glass tubing replacing the weighing bulbs. Pass air through the apparatus for 10–15 min at a rate of 2–3 bubbles/s in the bubbler tube, corresponding to about 0.1 l/min.

When it is certain that all extraneous air has been driven from the train, put the weighed absorption bulbs in place, and heat the test tube, first with a small flame until water is driven off and no longer collects in the cool portions of the tube and then gradually increasing to the heat of a good Meker burner. Continue to heat with the Meker burner for 5–10 min, and then heat with an air–gas blast flame so that the boat and contents become red hot. Continue the blast until reaction is complete.

It is convenient to have the stopper rather than the combustion tube clamped in place. The boat may then be pushed into the tube with a suitably bent piece of wire and the tube fitted onto the stopper, being careful not to tip over the boat in the process. A sheet of bright aluminum sheet with a hole and slot so that it may be used as a heat shield and a similar small disk mounted on the gas exit tube are necessary to protect the rubber stopper from radiated heat.

One of the greatest difficulties in combustion carbon determinations is to control the behavior of sulfur. The silver–manganese oxide mixture obtained by heating silver permaganate is extremely effective: It not only oxidizes sulfur dioxide and fixes it as sulfate but it also stops the fume of polymerized SO_3 that frequently forms in combustion when temperature control is not adequate and can pass through most trapping devices.

In using the procedure, some difficulty may be encountered because of the expansion of gases during the preliminary heating. If gas development is too great, some carbon dioxide may back up into the ascarite in the purifying bulb and be lost. The correction of this difficulty is to heat the sample very slowly and carefully at first and to be prepared to increase the flow of incoming air as soon as back pressure develops.

> After reaction is complete and air has been drawn through the apparatus for about 10 min, remove the weighing bulbs, allow them to stand in the balance case until temperature equilibrium is reached, and weigh. A blank determination using ignited quartz should be carried through all the steps of the procedure. A normal blank amounts to about 0.2 mg.

Carbonate Carbon. The apparatus consists of 250-ml conical flask bearing a stopper with a reflux condenser and an addition funnel, the delivery end of which is bent upward on itself to prevent gases being trapped. Air enters the addition funnel through tubes containing anhydrone and ascarite. An additional U-tube with $CaCl_2$ may be used to prevent overloading of the desiccant. From the top of the reflux condenser, the gases are led to the train proper, which consists of tubes containing anhydrone, copper sulfate on pumice, anhydrone, followed by weighing tubes containing ascarite and anhydrone, a protective drying tube, and finally a bubbler tube so as to observe the rate of gas flow. A stopcock and a pinch clamp control gas flow. Many variations of this apparatus are in use. Possibly the best of these is that recommended by the American Society for Testing and Materials.

The copper sulfate on pumice is intended to absorb hydrogen sulfide, which may be evolved with the carbon dioxide.

To ensure gas tightness, the various connections are shellacked or varnished. Plugs of glass wool prevent the carrying over of solids from one tube to the next. Permanent connections are made with Tygon rather than rubber. Leaks may be detected by assembling the apparatus and applying suction at the exit end of the train with the stopcock in the addition tube closed. Leaks will cause high results if air is aspirated through the apparatus during a determination and low results if air or oxygen is forced through it. Errors from this source will be greater, relatively, if air is drawn through the apparatus, but the error is not appreciable if proper precautions are taken and connections are tight. Any rubber tubing should be thick walled and seamless.

As large a sample as can be conveniently handled without overloading the absorption bulbs should be used when there is no shortage of material. The effect of blanks is thereby minimized. When a small sample must be used, a blank using a similar weight of a CO_2-free sample should be put through the procedure both before and after the sample. Only in this way can one be certain of maximum accuracy: The two blanks should turn out exactly the same and should not exceed 0.1–0.2 mg CO_2.

PROCEDURE. Weigh 2.000 g rock powder (less if more than about 10% CO_2 is expected) into the reaction flask using a powder funnel and wash it to the bottom of the flask with about 25 ml cold water. Assemble the apparatus, with a piece of glass tubing replacing the absorption bulbs, and draw air through it for 15–20 min at about 0.1 l/min (2–3 bubbles/s in the bubble tube). Place the weighing tubes in the balance case to come to temperature equilibrium.

Weigh the bulbs against a similar one used as a tare, opening the stoppers for an instant just before weighing to equilize the pressure. When it is certain that all extraneous carbon dioxide has been removed from the train, close the stopcock on the addition funnel and then those on either side of the weighing tube position. Replace the glass tube with the weighing tubes. Remove the stopper from the addition tube, and add 25–30 ml of 1:1 hydrochloric or phosphoric acid. Restopper, and open the various stopcocks along the length of the train. When any immediate reaction has ceased, slowly heat the reaction flask to boiling, regulating the rate of aspiration to prevent evolved gases backing up into the inlet end of the train. Finally boil for 5 min, with air passing through the system at about 0.1 l/min, agitating the reaction flask from time to time to break up any clumps of sample that may form if soluble silicates are present. After boiling, remove the source of heat, and continue the slow passage of air through the apparatus until all carbon dioxide has reached the absorption tubes. It is well worthwhile to make a trial run using pure calcite and interrupting it in its final stages to weigh the absorption bulbs so as to discover how long a flushing period is necessary in practice.

When the transfer of carbon dioxide is complete, close the stopcocks, and remove the weighing tubes to the balance room to attain temperature equilibrium before weighing.

To lower the blank, water and acids free of carbon dioxide should be used. These are obtained by boiling vigorously and cooling rapidly just before use.

If an all-glass apparatus is used (no rubber stoppers), the procedure may be modified in various ways to take care of refractory samples (scapolite), samples containing organic carbon, and even certain carbides and sometimes graphite. The use of water to transfer the sample may be omitted, and mixtures of strongly oxidizing acids used. For scapolite, concentrated phosphoric acid prepared by evaporating the 85% reagent or by adding P_2O_5 to it and treating it with dichromate to remove organic

442   THE ELEMENTS

matter nearly always present in commercial phosphoric acid may be used.
Materials containing organic matter may be wet oxidized with chromic
and sulfuric acids. It is sometimes effective to use a solution of vanadium
pentoxide in strong sulfuric or phosphoric acid.

4.16. GERMANIUM

Most rocks and ores contain only minute concentrations of germanium.
Zinc blende may contain several parts per million; residues produced
during recovery of zinc provide a major source of commercial germanium.
It also occurs as germanite, a copper arsenic germanium sulfide, in South
African copper ores, from which it is recovered as a by-product.

Coal often contains appreciable germanium, which may be recovered
from fly ash and flue dusts in producer gas plants. Some low-grade coal
deposits have been mined for their germanium content, which may reach
100–200 ppm germanium. Germanium occurs in the organic fraction.

Analytically, an important property of germanium is the volatility of
its chloride and bromides. During a classical analysis, germanium is volatilized in the early stages.

The metal is a semiconductor and is widely used in electronics, light-
conducting glass fibers, and infrared optics (fiber optics, communication,
and night vision).

Germanium metal melts at 960 °C. The dioxide, GeO_2, is not at all
volatile at temperatures usually attained during analysis, but the monoxide, GeO, sublimes above about 700 °C; thus, ignition in the presence
of organic matter (e.g., paper, tannin, coal, and peat) must be cautious.
During ignition, chlorides must be absent to avoid loss of $GeCl_4$.

The sulfides, GeS and GeS_2, are also volatile above about 400 °C, and
one process for recovering germanium from concentrates utilizes this
property. GeS is red-brown to black and is soluble in alkali hydroxide
and sulfide solutions. GeS_2 is off white and precipitates from strong (6
N) hydrochloric or sulfuric acid solutions. Precipitation is not quantitative, and collection of traces of germanium as sulfide is unreliable even
when a collector (e.g., Hg) is added.

Precipitation as sulfide is prevented by oxalic or hydrofluoric acid
(compare tin), and this provides a means for separating germanium from
arsenic.

Tannin precipitates a white germanium complex from acid solution or
from nearly neutral oxalate solution (compare Nb, Ta).

The most common method for determining traces of germanium involves solution of the sample, distillation from 6 N hydrochloric acid

followed by colorimetric determination using phenylfluorone [75]. Samples containing sulfur, halogens, or organic matter must be treated very carefully to avoid loss by volatilization. It seems probable that many reported values for germanium are low because of this difficulty. Any ashing or roasting process must be carried out below 400 °C.

Rapid methods that avoid the distillation step (e.g., by direct extraction with HCl–CCl_4) should be tested for each material type before relying on results so obtained.

The final form of germanium after an effective ashing process is GeO_2, which exists in a soluble and an insoluble modification. The latter is unaffected by hydrochloric acid; to be sure of complete solution, the residue after ignition should be fused at high temperature with sodium carbonate.

Atomic absorption determination of germanium is not very satisfactory. Sensitivity is not high, and interferences are not well understood. Carbon furnace AA suffers from a memory effect, and formation of the volatile monoxide leads to erratic results. ICP spectrometry shows some promise as a determinative technique; the real difficulty lies in obtaining a clean solution without loss of germanium.

To obtain a solution from which germanium may be distilled, several procedures are available. For sulfide ores containing appreciable germanium, an acid attack is appropriate:

PROCEDURE. Treat 1 g of −150-mesh sulfide ore with water, nitric acid, and 1–2 ml sulfuric acid and heat on the steam bath until attack is complete, adding more nitric acid if necessary. Evaporate to fumes of sulfuric acid, transfer to a distilling flask with a minimum of water, and add an equal volume of concentrated hydrochloric acid. Distill *slowly* at the rate of about 1 ml/min, collecting the distillate in an ice-cold receiver containing enough ice water so that the end of the condenser dips beneath the surface.

Continue the distillation until half the volume has distilled. Add more hydrochloric acid to the distilling flask and repeat. Keep the two distillates separate.

Dilute the distillates to about 2 N in hydrochloric acid, using ice water, and slowly neutralize with ammonia, keeping the solution cold. Add 3 ml acetic acid and 5 g ammonium acetate; then add 5% tannin solution in excess. Heat on the steam bath for half an hour to an hour, add a little more tannin, cool, filter, and wash with 5% ammonium nitrate solution until *all* chloride is removed. Test the filtrate with silver nitrate to be certain of this. Transfer paper and precipitate to a platinum crucible and place in a 400 °C muffle furnace until all organic material is burned off, then ignite strongly and weigh GeO_2.

When germanium content is low (less than 5 mg in the sample), weighing as cinchonine germanomolybdate may be appropriate.

Procedure. Make the distillate containing the germanium 9 N in HCl, and extract with three or four 10-ml portions of carbon tetrachloride. Back extract the germanium with three or four 10-ml portions of water. To the cold water solution add 20 ml of 25% ammonium nitrate, 15–20 ml of 2% ammonium molybdate, and 3 ml nitric acid. To the clear solution add, with stirring, 10 ml of a cinchonine solution prepared by warming 0.25 g cinchonine and 1.5 ml nitric acid in 100 ml water. Stir occasionally during several hours. Filter through a porcelain-fritted crucible. Wash with a solution containing 7 ml nitric acid and 25 g/l ammonium nitrate. Dry at 150 °C and weigh $(C_{19}H_{22}ON_2)_4H_4GeMo_{12}O_{40}$. A blank determination should be carried through the procedure. The gravimetric factor is 0.0239 for germanium.

Zinc blende normally contains so little germanium that a large sample must be taken, even with a colorimetric finish:

Procedure. Moisten a 5–50-g sample in a porcelain or silica dish with water and add nitric acid slowly until decomposition is complete. Evaporate to dryness and heat slowly in a furnace until nitrates are decomposed. Transfer the residue to a distilling flask with water, add up to 300 ml hydrochloric acid, and distill. Determine the germanium colorimetrically with phenylfluorone as follows.

Transfer 10 ml of the aqueous solution to a 25-ml volumetric flask. Add 2 ml of 0.5% gum arabic and 6 ml of a phenylfluorone solution prepared by dissolving 0.05 g in 40 ml concentrated hydrochloric acid and 200 ml ethanol. Dilute to 25 ml with water. Mix. Hold at 25 °C in a water bath for 1 h, and measure transmittance at 510 nm against a reagent blank. If absorbance exceeds about 0.2 in 1-cm cells, repeat using a smaller aliquot. The method of standard additions may be used to advantage.

The phenylfluorone method has some peculiarities that need attention if bad results are to be avoided.

- During distillation, oxidizing agents must be absent; they generate chlorine, which leads to false, usually high, results due to oxidation of the reagent. If there is any chance of, for example, V(V) and Mn(III) in the material being distilled, sufficient ferrous sulfate should be added before distillation to reduce any oxidant. The use of chlorine or permanganate to maintain arsenic in the 5-valent condition to prevent its distillation is counterindicated when a colorimetric finish is intended.

- The pink-red-colored species is formed slowly in acid solution and is only very slightly soluble. It forms more rapidly in less acid solution, but the color is not as stable. Blank absorbance is affected by acidity and temperature and is appreciable; thus, acid concentration, reagent concentration, and temperature should be carefully controlled, and a standard addition procedure is recommended.

- Reagent phenylfluorone is not always of good quality. If results are unsatisfactory, a new batch of reagent should be tried. Recrystallization from alcohol containing a little hydrochloric acid may improve quality.

Determination of germanium in coal requires careful attention to ashing procedure. Ashing in an oxygen bomb (calorimeter) does not leave the germanium in condition for complete recovery because of formation of the insoluble modification of GeO_2 or of insoluble siliceous glasses. A peroxide bomb may be useful, but small sample size and the amount of peroxide required for combustion are disadvantages. The following procedure is probably best and is applicable with appropriate modification to materials other than coal.

PROCEDURE. Mix 0.5 g of -120-mesh material with 1.5–2 g sodium carbonate in a platinum crucible and cover with a further 1 g carbonate. Put the crucible in a cold muffle and bring the temperature slowly to 500 °C over a period of an hour or more. Cool, mix the contents with a platinum rod, and continue heating up to 600 °C for 1–2 h. Repeat the mixing and ignition at 600 °C until all carbon is gone; then fuse over a Meker burner for 15–20 minutes (oxidizing flame!) to be sure germanium is all solubilized. Leach the melt with water, transfer to a distilling flask, add hydrochloric acid, and distill.

Fusion with sodium peroxide may be advantageous for silicate samples containing more than traces of germanium (e.g., flue dust or fly ash) when a complete analysis is intended. If silica dehydration is carried out using nitric instead of hydrochloric acid, no germanium will be lost. A sulfide separation before precipitating the ammonia group from nitrate solution should be avoided unless much copper, cadmium, and so on are present. Recovery of germanium as sulfide from dilute nitric acid solution is imperfect; it should be delayed until after the pyrosulfate fusion, when the sulfate solution should be made 5 N in sulfuric acid, transferred to an Erlenmeyer flask, and saturated (cold) with H_2S. The flask should be stoppered and allowed to stand for 48 h or more before filtering and washing with 5 N H_2SO_4 saturated with H_2S. The precipitate may contain, besides germanium, some or all the copper, cadmium, silver, arsenic, antimony, platinum, selenium, tellurium, molybdenum, and tin in the sample. The filtrate should be boiled to remove H_2S and oxidized with bromine and/or persulfate before continuing with the analysis.

Unfortunately, a number of elements not considered members of the hydrogen sulfide group may be partially precipitated, presumably as mixed sulfides. Among these are iron, zinc, cobalt, nickel, gallium, indium, and thallium. If the sulfide precipitate is large, it is sure to contain an appreciable proportion of these elements. Also, recovery of arsenic and antimony is likely to be incomplete.

4.17. PHOSPHORUS

Phosphorus occurs as apatite in phosphate rock, which provides the commercial supply of the element and its compounds. Apatite is found in igneous and sedimentary rocks and in bones and teeth; thus, it may develop organically or inorganically and at low temperatures or high. The formula for apatite is $Ca_{10}(F,OH)_2(PO_4)_6$, with fluorapatite more stable and more common than hydroxyapatite. When calcium is precipitated as phosphate during an analytical process, the product may be a mixture of $Ca_2(PO_4)_2$ and hydroxyapatite. Any fluorine present will lead to the formation of fluorapatite.

There are several inorganic oxyacids of phosphorus:

hypophosphorous acid, H_3PO_2
phosphorous acid, H_3PO_3
hypophosphoric acid, $H_4P_2O_6$
metaphosphoric acid, $(HPO_3)_n$
orthophosphoric acid, H_3PO_4
pyrophosphoric acid, $H_4P_2O_7$

Ordinary reagent phosphoric acid usually contains small amounts of the other acids as well as ill-defined polymers. It may also contain appreciable organic matter; this is of importance when the acid is used in oxidation–reduction reactions. Complete removal of reducing substances from reagent phosphoric acid is a matter of some difficulty.

In the general analytical scheme, phosphorus is precipitated as iron and/or aluminum phosphate with the ammonia group of elements, *provided there is a large excess of these metals*. If much calcium is present, and especially if the sample contains fluorine, there is a good possibility of finding calcium (as apatite) in the ammonia precipitate. Other elements in the ammonia group may also precipitate as phosphate (beryllium, rare earths). With samples containing more than 1% or so of P_2O_5, the phosphate should be removed before precipitation of the ammonia group. In the analysis of phosphate minerals, the ordinary scheme of separation is not viable; special procedures are necessary.

Zirconium phosphate is insoluble in acids, and the addition of zirconium oxychloride to the filtrate from the acid group effectively removes phosphate from solution. The precipitate may be removed by filtration prior to further operations, or the zirconium may be added in exactly known amount (i.e., as a standardized solution of the oxychloride), in which case the weight of the ammonia group oxides may be corrected for the added ZrO_2 they contain. Use of iron as a collector of phosphorus has been recommended; this is not reliable except with relatively small amounts of phosphate.

Bismuth phosphate is not very soluble in acids; this fact is used during the gravimetric determination of bismuth as $BiPO_4$.

Whenever possible, a determination of phosphorus in an unknown sample should be made prior to other operations, usually on a separate sample. Additions of zirconium or other revisions of the analytical scheme may then be made on a rational basis.

Almost all metal phosphates, including zirconium pyrophosphate, are decomposed by fusion in sodium carbonate containing sodium peroxide. Leaching with water, filtration, and washing with sodium carbonate–hydroxide solution leaves the phosphate in the filtrate. If the precipitate is large, it should be retreated because it is difficult to achieve a complete separation in a single step.

There are two useful procedures for separating and determining phosphate: precipitation as ammonium phosphomolybdate and precipitation as magnesium ammonium phosphate. The phosphomolybdate forms in nitric acid solution containing ammonium nitrate. Chloride and sulfate retard or prevent precipitation, fluoride may be rendered harmless by boric acid addition, silicic acid leads to the formation of silicomolybdate and mixed silico- and phosphomolybdates, which may lead to high or low recoveries, depending on how much silica is precipitated and how much phosphate remains in the soluble yellow complex. A notable interfering element is tungsten, which tends to precipitate as tungstic acid or phosphotungstate. Zirconium (and possibly bismuth) may interfere in procedures in which the sample is treated with nitric and hydrofluoric acids, and insolubles are filtered off prior to precipitation with molybdate.

Of numerous colorimetric methods for determining phosphate, that using phosphomolybdenum blue is generally the most satisfactory. The yellow color of phosphomolybdate may also be used, and when tungsten is present, the orange vanadium–molybdenum–phosphate complex may be useful.

When small amounts of phosphate are to be precipitated with magnesium in ammonia solution, the addition of arsenate is useful: After collection of the mixed magnesium ammonium arsenate and phosphate, the arsenic can be removed by evaporation with hydrobromic acid or by precipitation with hydrogen sulfide, leaving the phosphorus ready for determination by any appropriate method.

While most phosphates and phosphate minerals are acid soluble, those containing thorium, zirconium, bismuth, tin, and titanium are not. Monazite, a rare earth thorium phosphate, is slowly soluble in hot concentrated sulfuric acid. In the complete analysis of rocks containing insoluble or slightly soluble phosphates, some phosphorus may precipitate with the acid group of elements and be counted as silica if sulfuric acid is used during the volatilization of SiF_4. It is best to use nitric acid (or perchloric) during the determination of silica in phosphate rocks and minerals.

Only occasionally is it desirable to determine phosphorus during the main part of analysis. When a small sample of a phosphate mineral needs an accurate analysis, the residue from a silica determination should be fused with sodium carbonate, dissolved in acid, and added to the main solution. Add an accurately measured volume of a standard solution of zirconium oxychloride sufficient to precipitate all the phosphate and an equal amount in excess. Evaporate to near dryness (repeatedly, with hydrochloric acid additions, if nitric acid is present), take up the residue with 1:1 hydrochloric acid, evaporate to about 20–25 ml, add 400–500 ml hot water, digest on the steam bath, add paper pulp, filter, and wash with 2% hydrochloric acid. Ignite the residue at low temperature to remove paper, fuse with sodium carbonate, leach with water, filter, and determine phosphorus in the filtrate by double precipitation with magnesium and ignition to $Mg_2P_2O_7$. The insoluble zirconium hydroxide should be examined for residual phosphorus (and for iron, titanium, tin, etc., as appropriate) before returning it to the main solution. If the added ZrO_2 is accurately known, its weight may be subtracted from the total weight of ammonia group oxides.

Usually, phosphorus will be determined on a separate portion of the sample prior to the main part of the analysis. How to proceed with this determination depends on the nature of the sample. Most often, the sample is attacked with an oxidizing acid or acid mixture or by an oxidizing fusion, silica is removed with hydrofluoric acid, boric acid is used to complex any remaining fluoride, and phosphate is precipitated with molybdate in ammonium nitrate–nitric acid solution and then converted to magnesium pyrophosphate for weighing. For small amounts of phosphorus, colorimetric methods are used. AA spectrometry is not very effective, except in special cases. Flame spectrometry, using the depression of calcium signals by phosphate, may find occasional use. X-ray spectrometry is only useful if well-analyzed calibrating standards similar in nature to the unknowns are available and is applied in routine work. Emission spectrographic or spectrometric determination is of limited value in rock and mineral analysis; most results are semiquantitative only.

4.17.1. Phosphorus in Apatite and Phosphate Rock

PROCEDURE. Sinter a 0.5000-g sample, with the addition of 0.1 g pure SiO_2 if little silica is present in the sample, with 1 g sodium carbonate and about 0.1 g sodium peroxide. A platinum or (less desirable) a zirconium crucible may be used. Leach the melt with water, add nitric acid, and dehydrate silica in a platinum or porcelain dish. Take up the residue in dilute nitric acid, filter, and wash with 2% nitric acid. Ignite the residue to remove paper, treat with hydrofluoric and a little perchloric acid, evaporate to dryness, and fuse the residue with a little sodium carbonate. Leach with water, and filter. Wash with 2% sodium carbonate solution. If the precipitate of hydrous oxides is large, it should be examined for phosphate before continuing.

Add the sodium carbonate solution to the main solution, add water and nitric acid to give 400–500 ml of a solution containing 30–50 ml nitric acid. Add 25–30 g ammonium nitrate and heat the solution to 90 °C. With stirring, add 25–30 g ammonium molybdate dissolved in water. Stir frequently until cold, and allow to stand for 8 h or more. Filter, and wash with cold 2% nitric acid.

Reagent ammonium molybdate does not have a reliable stoichiometry. If the reagent does not dissolve readily in water or if difficulties are experienced with the procedure described, ammonium molybdate reagent should be prepared as follows, and the procedure modified to take into account the more dilute reagent:

AMMONIUM MOLYBDATE REAGENT. Dissolve 200 g molybdic acid, MoO_3, in a mixture of 430 ml water and 150 ml concentrated ammonia solution. Filter, and add a solution containing 1300 ml water and 75 ml concentrated nitric acid.

Treatment of Ammonium Phosphomolybdate Precipitate. The composition of precipitated ammonium phosphomolybdate varies depending on the recipe used in preparing the reagent ammonium molybdate, numerous details of the procedure, and the nature of the sample being analyzed. In routine work, the precipitate may be collected in a porous-bottom glass or porcelain crucible, washed with dilute nitric acid and weighed. In this case, an empirical gravimetric factor determined by carrying known samples through the procedure should be used. The precipitate may also be dissolved in a measured volume of standard sodium hydroxide solution, and the excess alkali titrated with standard nitric acid. These rapid procedures are adequate if empirical factors are used and procedures are rigidly standardized. For small amounts of phosphorus, they are preferred; but for larger amounts (more than about 50 mg P_2O_5), the yellow precipitate should be converted to magnesium pyrophosphate for weighing.

CONVERSION OF AMMONIUM PHOSPHOMOLYBDATE TO MAGNESIUM PYROPHOSPHATE. Wash the yellow precipitate from the filter into a beaker, and wash the filter alternately with 25% ammonium hydroxide, water, and 5% hydrochloric acid. Use no more than enough ammonia to dissolve the precipitate completely. Make the solution just acid by adding first 1 g citric acid and then hydrochloric acid. Heat to near boiling (some ammonium phosphomolybdate may reprecipitate), and then add ammonia until the solution is just alkaline. If the solution is not perfectly clear at this stage, it should be filtered. The insoluble material may be zirconium or titanium phosphate; ignite it at low temperature, fuse with a little sodium carbonate, leach the melt with water, filter, and add the filtrate to the main solution. Dilute the solution with water so that it contains no less than 1 mg/ml P_2O_5.

Add 10–30 ml magnesia mixture (50 g $MgCl_2 \cdot 6H_2O$ plus 100 g NH_4Cl in 500 ml water; add 5–10 ml concentrated ammonia, allow to stand for 24 h or more, and filter; make just acid with hydrochloric acid for storage). Precipitate the phosphate by adding concentrated ammonia slowly with stirring, interrupting the addition with the precipitate is forming. If a few drops of methyl red are added to the solution, the ammonia may be added at a rate just sufficient to maintain the yellow alkaline form of the indicator.

When precipitation seems complete, add one-tenth the volume of the solution of concentrated ammonia solution, and allow to stand, stirring occasionally. Precipitation is hastened if each time the solution is stirred, the sides of the beaker are washed down with 5% ammonium hydroxide and the washings allowed to float on the solution.

Filter after 8 h or more, and wash with 5% ammonium hydroxide. If the precipitation was very carefully performed and the precipitate is heavily crystalline, it may be ignited to $Mg_2P_2O_7$ at this stage; usually, however, a second precipitation is desirable. If arsenic is present in the sample, or if arsenate was added to collect small amounts of phosphate, it may be removed before the second precipitation.

Dissolve the magnesium ammonium phosphate in a minimum of hot 5% hydrochloric acid. To remove arsenic, add 10 ml concentrated hydrobromic acid and evaporate to dryness on the steam bath. Take up the residue in 10 ml hydrochloric acid, dilute with water so that the solution contains not less than 1 mg/ml P_2O_5, and filter if necessary (any insoluble matter is likely to be extraneous silica). Wash with dilute hydrochloric acid. Add 1 ml magnesia mixture and then ammonia dropwise with stirring as before until the solution is 10% in concentrated ammonium hydroxide. Allow to stand for at least 4 h with occasional stirring. Filter, wash free from chlorides (test with silver nitrate solution) using 5% ammonium hydroxide, burn off paper at low temperature, and ignite to constant weight at 1000 °C.

After weighing the final product, the mass in the crucible should be carefully broken apart with a small spatula and examined for carbon. Often a perfectly white ignition product contains a core of unburned black material. The weight of this is usually small, but reignition at 1100 °C for a few minutes will usually remove it. Ignition of $Mg_2P_2O_7$ above 1000 °C should not be prolonged because P_2O_5 is slowly lost.

4.17.2. Phosphorus in Tungsten-Bearing Materials

The precipitation of phosphorus from nitric acid–ammonium nitrate solution as ammonium phosphomolybdate fails in the presence of tungsten, presumably because of the formation of phosphotungstic acid. With these materials, it is best to obtain the phosphorus and tungsten in concentrated hydrochloric acid solution, and then precipitate phosphorus as magnesium ammonium phosphate, adding arsenate to collect small amounts of phosphorus.

PROCEDURE. Transfer 0.5–5 g tungsten-bearing ore ground to pass a 200-mesh screen (finer grinding may be necessary with fine-grained wolframite ores) to a 500-ml Erlenmeyer flask (preferably with a wide mouth). Add 100–300 ml concentrated hydrochloric acid and a Teflon-coated or silica-glass magnetic stir bar. Cover the flask with a watch glass, and stir slowly (the stir bar should sweep the whole bottom of the flask, and stirring should be slow to limit abrasion of the flask) for 8 h or longer. Scheelite ores dissolve much faster than wolframite ores—stirring time depends on the nature of the sample.

Filter (by decantation as much as possible) through a glass fiber paper, and wash with concentrated hydrochloric acid. Filtration is likely to be difficult because of gelatinous silica: Every sample has its own characteristics, but allowing the precipitate to settle and decanting as much as possible will speed up the filtration. Most of the phosphorus will be found in the filtrate, but the precipitate on the glass fiber paper should be examined: Fuse paper and precipitate with sodium carbonate, leach with water, filter, and examine the filtrate for phosphorus. The filtrate contains silica, which should be removed by acid dehydration.

To the hydrochloric acid solution, or an aliquot of it, add 0.1 g sodium arsenate and, while stirring, enough potassium permanganate solution to oxidize any ferrous iron that may be present.

Add 1–10 g citric acid and 1–5 g magnesium chloride ($MgCl_2 \cdot 6H_2O$) to the hydrochloric acid solution or an aliquot of it.

In a 1000-ml beaker, put 500 ml of a solution containing enough ammonia to neutralize the hydrochloric acid and provide a 10% excess of ammonium hydroxide. Pour the hydrochloric acid solution into this solution. Stir occasionally, and allow to stand for 8 h or more.

Filter, and wash with 5% ammonium hydroxide.

Dissolve the precipitate in a minimum of hydrochloric acid, add 10 ml concentrated hydrobromic acid, and evaporate to dryness on the steam bath. Proceed with the separation and determination of phosphorus either through phosphomolybdate precipitation and weighing and titration or through reprecipitation as magnesium ammonium phosphate and ignition to $Mg_2P_2O_7$, depending on the amount of phosphorus present.

Other aliquots of the concentrated hydrochloric acid solution may be used for the determination of tungsten, rare earths, and so on.

Titrimetric Determination of Phosphorus. When alkalimetric titration of the ammonium phosphomolybdate precipitate is intended, it is convenient to filter through a pad of paper pulp. A paper pulp filter requires a little skill to prepare but saves much time in procedures of the kind to be described.

PROCEDURE. Filter the solution from which phosphorus has been precipitated as ammonium phosphomolybdate through a pulp filter, and wash very thoroughly with 2% potassium nitrate solution to remove acid. Suck the filter dry, transfer paper pulp and precipitate to the original flask or beaker, and add 20 ml or more of water. From a burette, add enough standard sodium hydroxide solution (usually 0.1 or 0.25 N) to dissolve the yellow precipitate and a 2–5 ml excess, stirring vigorously. Using phenolphthalein indicator, titrate the excess alkali with standard nitric acid solution. The titrants should be standardized against ammonium phosphomolybdate generated in the same procedure using analyzed standards or known samples salted with a known amount of phosphate. Theoretically, 23 mol NaOH react with 1 mol of the precipitate; in practice, the ideal formula, $(NH_4)_3PO_4 \cdot 12MoO_3$, is only approximately achieved.

The titration procedure is very convenient in cases where all the phosphorus in the sample is acid soluble, and there is a large proportion of insoluble material. In determining phosphorus in steels, irons, meteoritic material, low-grade limestones, and even some silicate rocks and minerals, treatment of the sample with nitric acid, oxidation with permanganate, reduction with peroxide, precipitation with molybdate, filtering, and titrating yields rapid and usually reliable results. As applied to steel, iron, or meteoritic material, procedure is as follows:

PROCEDURE. To a 0.5–2.0-g sample in a 250-ml Erlenmeyer flask add a little water and 25–30 ml of 1:1 nitric acid; have a bottle of cold water handy in case the reaction becomes violent. Boil gently for several minutes, and then add saturated potassium permanganate solution dropwise until there is a clear excess. Then add hydrogen peroxide to reduce the precipitated manganese oxides, and boil down to about 30 ml.

Without cooling, add 30 ml ammonium molybdate solution (200 g MoO_3 in 2 l, prepared as described above) and 25 ml of 1:1 nitric acid. Stopper the flask and shake for about a minute. Allow to stand for 20–30 min, and filter through a pulp filter, washing with 2% KNO_3. Titrate with sodium hydroxide and nitric acid as described above.

In some cases (e.g., graphitic iron), the insolubles may obscure the end point of the titration; it is then necessary to filter before precipitating the phosphorus.

Boiling with nitric acid converts all inorganic phosphorus compounds, and most organic phosphorus compounds to orthophosphate; with metallurgical products, phosphides may be present (e.g., Fe_3P in iron and steel). These are oxidized to phosphate. Some materials (e.g., ferrosilicon) may require the use of hydrofluoric and nitric acids for complete attack.

4.18. ARSENIC AND ANTIMONY

Many arsenic minerals are found with magmatic sulfides, and some are found as oxides. The most common minerals are realgar (As_2S_2), orpiment (As_2S_3), mispickel (FeAsS), and arsenates. Ardennite, an arsenic–manganese silicate mineral is studied in detail in Section 3.2.9.

Antimony is mainly occurring as stibnite (Sb_2S_3); many minerals are also found where antimony is associated with copper and sulfur.

Recently, interest for these elements has increased rapidly in geochemical exploration because of their reliability as pathfinders for gold deposits [76]. Today, many geochemical laboratories are in need of providing rapid and reliable determination of arsenic and antimony at very low levels. This is why the availability of homogeneous geochemical standards in which antimony and arsenic have been determined is becoming critical. Many rapid techniques using modern instrumentation are very dependent on these geochemical standards, and unfortunately few are available.

Arsenic has only one naturally occurring isotope, while antimony has two isotopes.

During the last decade, arsenic became an important element in the semiconductor industry. Industrial applications of antimony are still limited.

If sufficient iron or aluminum is present in the sample, arsenic will be caught in the ammonia group precipitate as aluminum or iron arsenate during the classical analysis. Antimony will also be quantitatively retained by the ammonia group. However, these elements will be determined more reliably after distillation as bromide from a separate sample, as explained below.

At very low pH, As(III) exists as AsO^+; from pH 1.5 to pH 9, we have arsenious acid. In acid medium, As(V) occurs mostly as arsenic acid H_3AsO_4. Arsenates behave as phosphates.

$AsCl_3$ boils around 130 °C; however, it starts to expel at lower temperatures. $SbCl_3$ distils at 225 °C; however, it starts to expel in appreciable amounts at 130 °C.

Most of their redox reactions are straightforward and quantitative and thus are often used in analytical chemistry.

4.18.1. Decomposition of Samples

A sodium carbonate fusion as described in Section 2.4.12 was successfully used for the analysis of ardennite. However, some authors pretend that sodium carbonate is not the most satisfactory flux.

According to Dolezal [24], an attractive way to decompose rock samples containing arsenic and antimony is by sulfoalkali fusion, called "freibergian decomposition." During fusion, arsenic and antimony convert to complex sulfides that are easily separated from insoluble sulfides by digesting the melt with warm water. The method requires very good skills.

Fusion with sodium hydroxide and small amounts of sodium peroxide is also very effective.

PROCEDURE. Fuse 5.0 g pure sodium hydroxide in a nickel crucible.
Note: Large amounts of water are released during the fusion, and this leads to procedural difficulties if not removed prior to introducing the sample.
After cooling the melt, add 1.0 g of −200 mesh rock powder, and then cover with 1.0 g sodium peroxide.
Note: The sodium peroxide is not a must; however, it will be beneficial to oxidize sulfides and organic matter if present.
Heat slowly over a small flame until the attack is complete; this may take at least 20 min.
Note: It is very difficult to determine completeness of attack, and this is a serious disadvantage of the method for many analysts. Many common silicate minerals are dissolved at once while some minor minerals may require up to 1 h or more low-red heat for complete attack.
Cool the melt and disintegrate it with minimum warm water. Heat it on a steam bath if necessary. Cool the solution; then acidify it slowly with hydrochloric acid.

4.18.2. Distillation of Arsenic and Antimony

Even though arsenic and antimony bromides have higher boiling points than corresponding chlorides, it has been found that distillation of these elements using HBr is easier than using HCl, especially for antimony.

If proper conditions are taken, arsenic and antimony are completely recovered by using a distillation with $HBr–HClO_4$ and CO_2 as a carrier. The distillation can be performed in less than 10 min at temperatures between 200 and 220 °C. Distill into 10 ml nitric acid. When distillation is complete, dilute distillate to 50 ml with water. An aliquot of this solution can be used to determine arsenic by the molybdenum blue method and antimony by the rhodamine B method.

The sample may be directly digested in the distillation apparatus as shown in Figure 4.2.

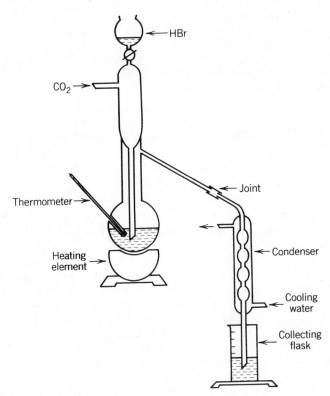

Figure 4.2. Illustration of distillation apparatus used for separation of arsenic and antimony.

4.18.3. Determination of Arsenic by Heteropoly Molybdenum Blue Method

In this method arsenic should be as As(V), and this is no problem since the distillation was performed with HBr and $HClO_4$. At this stage, all the arsenic should be present as As(V). Ammonium molybdate is added to form a heteropoly molybdiarsenate complex that is reduced by a reducing agent such as hydrazine sulfate to produce a very intense blue color. If present, germanium may be a source of interference. If handled carefully, germanium may be expelled during the distillation before arsenic [77]. The method has been described to a large extent by Sandell [5].

PROCEDURE. Take an aliquot of the distillate containing no more than 30 μg As, and evaporate the aliquot to dryness using a maximum temperature of 130 °C. Continue to heat at this temperature inside an oven for 1 h. Cool. Add 10 ml hydrazine sulfate–ammonium molybdate solution (e.g., dissolve 0.5 g ammonium molybdate in 50 ml 5 N H_2SO_4, dilute with water around 400 ml, add 5 ml of a 0.15% hydrazine sulfate solution, and dilute to 500 ml). Heat on a water bath for 20 min. Cool and dilute to 50 ml with the hydrazine sulfate–ammonium molybdate solution. Compare with similarly prepared solutions of known arsenic content using a Dubosq colorimeter or a spectrophotometer equipped with 2-cm cells at 700 nm.

4.18.4. Determination of Antimony by Rhodamine B Method

In this method, antimony should be as Sb(V), and this is no problem since the distillation was performed with HBr and $HClO_4$. At this stage, all the antimony should be present as Sb(V). However, any oxidizing agent may destroy the violet color of the rhodamine B chloro-antimonate. The method has been described to a large extent by Sandell [5].

PROCEDURE. Take an aliquot of the distillate containing no more than 50 μg Sb, and evaporate the aliquot to dryness using a maximum temperature of 130 °C. Continue to heat at this temperature inside an oven for 1 h. Cool. Add 20 ml of HCl, 6 N. Make sure no Sb(IV) is present [e.g., add 2 ml of a 1% sodium sulfite solution to reduce antimony to Sb(III), and then add 3 ml of a 4% ceric sulfate solution to reoxidize Sb(III) as rapidly as possible to Sb(V)]. Shake. Add about 12 drops of a 1% in 1 M HCl hydroxylamine hydrochloride solution to destroy the excess Ce(IV). Transfer into a separatory funnel. Add 4 ml of a 0.2% rhodamine B solution. After 1 min, extract with 25 ml benzene by shaking vigorously for 2 min. Separate phases. Read the absorbance at 565 nm.

4.19.   SELENIUM AND TELLURIUM

4.19.1.   Selenium

Selenium is found in small amounts in igneous rocks as selenide minerals such as lead selenide, mercuric selenide, bismuth selenide, cuprous and cupric selenides, and so on. Selenium is also found in sulfide ores in association with sulfur and sometimes silver. Small quantities of selenium leached out from soils or tailing ponds by streams have generated environmental concerns that make the accurate analysis of selenium an important necessity to many analytical facilities. As a poisonous trace in pure products (e.g., copper metal), selenium has been and remains a challenge to the metallurgist and also a constant preoccupation. It is also an element for which the literature contains many misleading statements regarding its behavior in solutions and during precipitation reactions. Its chemistry is mainly nonmetallic. The seemingly capricious behavior of selenium during experiments and analytical processes is due to a number of factors such as:

1.   The chemistry of selenium is not universally understood by analysts.
2.   The literature is replete with misstatements and contradictions.
3.   Thermodynamic calculations are often misleading; many selenium reactions are kinetically controlled.
4.   Analytical processes are subject to numerous interferences that are frequently ignored.
5.   No quick and easy determinative methods of high reliability exist.

Chemistry of Selenium

Elemental selenium, like sulfur, exists in several allotropic forms:

1.   The red amorphous selenium is precipitated by sulfur dioxide from cold selenite solutions containing hydrochloric acid. The precipitation is seldom complete; up to 1–2 mg/l remains unprecipitated even after long digestion periods, either in true solution or as a colloid.
2.   The red crystalline selenium is formed slowly from the amorphous variety in cold solution.
3.   The black crystalline "metallic" selenium is formed from the amor-

phous variety in suspension on heating near the boiling point of a hydrochloric acid solution. The red crystalline variety does not easily convert to the black variety under 100 °C.

Complete recovery of small amounts of precipitated selenium by filtration is difficult. If millipore filters are used, pore size must be 0.2 μm or less. If paper filters are used, it is necessary to add filter paper pulp even after long digestion in hot solution.

A standard method of distinguishing between Se(VI) and Se(IV) in solution involves reduction of Se(IV) to red amorphous Se(0) by sulfur dioxide in cold 4:1 hydrochloric acid. This is filtered out, and Se(VI) is then reduced by sulfur dioxide in hot solution and recovered. This process is only applicable with larger amounts of selenium because recovery of Se(IV) is incomplete, up to 1–2 mg/l remaining unprecipitated. Reduction of Se(VI) in hot HCl solution is slow; and heating must be continued for 30 min or more with sulfur dioxide addition to ensure complete recovery.

Precipitation of Se(IV) by sulfur dioxide in sulfuric acid solutions is very slow and usually incomplete. Se(VI) is not reduced by sulfur dioxide in sulfuric acid solution.

Attempts to verify these statements are often frustrated by the fact that reagent selenates are often highly contaminated with selenites. Some reagents labeled as metal selenate are almost entirely selenite. This may be one reason for the numerous false statements found in the literature. Elemental selenium dissolves in concentrated sulfuric acid to form a green solution of an addition compound $SeSO_3$. In commercial acid, selenium is always present in this form.

$$Se + SO_3 \rightleftarrows SeSO_3 \cdot$$

In this reaction, the valence of selenium is 0. The green color is not sufficiently intense to be useful in estimating concentration; it is bleached by tellurate, presumably because of formation of $SeTeO_3$. When both Se(0) and Se(VI) are present in acid solution, they may yield $SeSeO_3$; this may partly account for the seemingly capricious behavior of the element in sulfuric acid solutions. Dilution of the concentrated acid with water precipitates elemental selenium, traces of which remain in colloidal form and cannot be removed by filtration. If an analytical method for selenium in sulfuric acid is calibrated using SeO_2 dissolved in diluted selenium free acid, a false calibration is likely unless the Se(0) in the unknowns is oxidized to Se(IV) before processing. Oxidation is best accomplished by bromine; diluted nitric acid does not oxidize Se(0) reliably.

Se(0) may also form an addition compound $SeSO_2$ with sulfur dioxide, and this may explain the incomplete precipitation from cold sulfur dioxide saturated solutions.

Air oxidation of selenium to selenite may occur during filtration and washing unless some reductant is added to the wash solution.

When small amounts of selenium are to be precipitated by reduction to element, addition of a carrier ensures complete recovery. The most effective carrier is tellurium in the presence of copper; hypophosphorous acid and stannous chloride in 6 N hydrochloric acid are the most effective reductants. For complete recovery, a considerable excess of precipitant is needed, and the solution must be boiled for several minutes, or longer if Se(VI) is present. A 0.2-μm millipore filter is necessary for complete collection of the precipitate. Losses of selenium occur when hydrochloric acid solutions are boiled or evaporated. The reducing agent should be added before bringing the sample solution to a boil. Selenium (IV) is easily oxidized to Se(VI) in alkaline solution. Fusion with sodium peroxide and water leaching yields (Se(VI) in solution. Despite a high oxidation potential, Se(VI) is not easily reduced in acid solution; in most circumstances, it behaves like sulfate.

Sulfate and selenates are usually isomorphous and form mixed crystals; most selenates are somewhat more water soluble than the corresponding sulfates. Selenites are usually less soluble than selenates. Thus, liming sulfuric acid solution containing selenites and selenates will lead to precipitation of moderately insoluble calcium sulfate containing a part of the Se(VI) in mixed crystals and calcium selenite. Selenium remaining in solution will be in the form of soluble calcium selenate and selenite depending on the solubility product of $CaSeO_3$. On standing, the selenite in the alkaline mixture will slowly oxidize to soluble selenate. If any Se(0) is present, it will also slowly oxidize to soluble selenate.

Stable Form of Selenium in Solution. The stable form of selenium in aqueous solution is selenite, SeO_3^{2-}, analogous to sulfite. Oxidation to selenate, SeO_4^{2-}, analogous to sulfate, requires a strong oxidant.

Reduction of Se(VI). Thermodynamically, SeO_4^{2-}, is almost as powerful an oxidant as chromate; however, its reduction to selenite or to elemental selenium is difficult.

Reduction of Se(VI) in cold sulfuric acid solution by H_2S, SO_2, and so on does not normally occur. In hot solution reduction is unpredictable; it probably depends on catalytic reactions that are not understood. Halides, especially bromide, are known catalysts. The stability of Se(VI) and the strange behavior of Se(0) in sulfuric acid solutions has led to many difficulties and frustrations.

Hydrazine slowly reduces Se(VI) to Se(0) in hot solutions containing SO_2. Se(VI) pyrolyzes to Se(IV) at about 200 °C, the temperature depending on the matrix. In this respect, it resembles Cr(VI), which pyrolyzes to Cr(III) at about the same temperature. In a reducing environment, Se(VI) pyrolyzes to Se(0).

Separation of Se(0) and Se(IV) from Se(VI). Se(IV) is coprecipitated with ferric hydroxide over a wide pH range (4–6), probably as an addition compound $FeSeO_3OH$. Se(VI) is not coprecipitated. This provides a simple method for separating the two species prior to determination. Presumably any Se(0) will accompany the Se(IV). In using this separation, certain precautions are necessary. The acid solution should be sparged with CO_2 or nitrogen prior to addition of ammonia and ferric salt in order to remove the oxygen. An excess of ammonia must be avoided because Se(IV) is not reliably stable, and analysis should be performed as soon as possible after sample collection. If Se(0) is present, either colloidal or as $SeSO_3$, it is apt to deposit on the walls of the sample container; this will lead to spurious and nonreproducible assays. Addition of nitric acid to stabilize solutions containing Se(0) is not effective; in fact, this addition may favor precipitation of the selenium.

Separation of Se(0), Se(IV), and Se(VI) When the Three Species Are Present. In this case, it is probably best to collect three samples and treat them separately as follows:

1. For total selenium, add excess bromine, make 6 N in hydrochloric acid, and heat under reflux near the boiling point for 1 h or more. This will convert all selenium to Se(IV).
2. For Se(VI), add ferric chloride to the cold solution and precipitate with ammonia at pH 6. Analyze the filtrate after HCl addition and refluxing.
3. For Se(0) and Se(IV), add copper sulfate, make 6–8 N in HCl, and saturate the cold solution with sulfur dioxide. Add paper pulp and filter. Wash with cold water containing SO_2. Dissolve the precipitate in a mixture of hydrochloric acid and bromine.

Whole Bottle Technique. This technique is recommended for any liquid sample in which total selenium has to be determined. However, before receiving samples, the analyst should advise the sample submitter that a whole bottle technique will be used and that certain conditions should be met:

1. The bottle to be analyzed should contain an exact known amount of solution.
2. The sample should be acidified with 1 or 2 ml nitric acid.

Transfer the entire contents of a bottle to a beaker (400 ml). A blank bottle identical to the one that contains the sample should be submitted for the purpose of judging severity of contamination from sample containers. Add one drop of bromine to the bottle and about 10 ml water. Shake vigorously and add to beaker. Add 20 ml HNO_3 to the bottle. Shake vigorously and add to beaker. Add another 20 ml HNO_3 and shake, adding to beaker as before. Warm at just under boiling temperature for about 45 min or until Br_2 is volatilized. Transfer solution to a volumetric flask, either 250 or 100 ml size, and dilute with water to volume. Mix. Quantities of this "stabilized" sample may then be taken for analysis.

Decomposition of Solid Samples. A mixture of nitric acid and bromine may be effectively used on a very finely ground sample. A fusion may volatilize the selenium in the heating stages before the flux is even melted. Hydrochloric acid may be used as long as an excess of nitric acid is present since selenium may be volatilized from hot hydrochloric acid solutions. For many samples, the following method is recommended:

PROCEDURE. Transfer the sample to a distillation apparatus using no more than 25 ml water to wash sample into the apparatus. Add 5 ml bromine for each gram of sample and mix several times during at least a 2-h digestion period (some materials may require up to 16 h). Heat to 200–210 °C and distill into 10 ml HNO_3 while adding 15 ml HBr containing about 0.5 ml bromine, dropwise during the distillation. Dilute distillate, which should contain all selenium, to a 100-ml volumetric flask. For more details about distillation of selenium, see Hoffman and Lundell [78] and Walton [79].

Preparation of Standard Solutions of Selenium

SELENIUM (IV). Dissolve 1.405 g SeO_2 in water; dilute to 1 l. Take 100 ml and add 400 ml concentrated hydrochloric acid. Cool to 15–18 °C in a water bath (use a thermometer) and saturate with SO_2. Filter through a weighed porous porcelain crucible, wash with 4:1 HCl, then water, and then alcohol (no delay between washings). Dry at 100 °C and weigh. (It should weigh 0.1000 g.) Heat the filtrate to near boiling, add 1 g hydrazine sulfate, and pass SO_2 into the hot solution. The (black) precipitate represents Se(VI). If the reagent SeO_2 contains Se(VI), start with selenium metal: Dissolve 1.000 g selenium (black) in nitric acid (plus a little HCl if necessary). Evaporate to dryness on the steam bath. Dissolve in water and again evaporate to dryness. Repeat several times to remove all HNO_3. Dissolve in water, filter, wash with water, and dilute to 1 l.

SELENIUM(0). Prepare 0.1000 g red selenium as in 1. Dissolve by warming in 100 ml of 1% Na_2SO_3 solution. Dilute to 100 ml with water. This solution is not very stable; selenium oxidizes to Se(VI):

$$Na_2S^{+4}O_3 + Se^0 \rightarrow Na_2Se^0S^{+4}O_3 \tag{1}$$

$$\rightleftarrows Na_2S^0Se^{+4}O_3 \tag{2}$$

$$\rightleftarrows Na_2S^{-2}Se^{+6}O_3 \tag{3}$$

On acidification (1) should precipitate red Se(0). Species (3) is analogous to sodium thiosulfate, $Na_2S_2O_3$, with a selenium replacing one of the sulfur, and dilute acid should have no effect. The rates of the reactions (1), (2), and (3) are not known. Species (1) is probably light sensitive.

SELENIUM (VI). Fuse 1.405 g SeO_2 with 5 g Na_2O_2. Leach with 200 ml water and filter. Wash with water. Cool to 5–10 °C and slowly add dilute H_2SO_4 in amount calculated to neutralize the alkali added. Dilute to 1 l. Take 100 ml and precipitate and weigh Se(IV) as in (1). There is no guarantee that all the selenium is converted to Se(VI) during the fusion and no easy way to remove Se(IV) from the standard solution.

Determination of Selenium

Having obtained the several selenium species in solution as Se(IV), determination may be completed in several ways:

1. Hydride Generation Followed by Atomic Absorption Measurement: This method is not reliable for reasons explained below and is not recommended. The following discussion is necessary because many laboratories have a tendency to analyze selenium using evolution of SeH_2 followed by AA measurement.

In this method, Se(IV) is obtained in hydrochloric acid solution and converted to SeH_2 by addition of $NaBH_4$ (or other reductant). The gaseous SeH_2 is transferred to a hot tube using a carrier gas (argon), and absorption or fluorescence due to selenium atoms is measured. Interfering elements are numerous, almost invariably leading to low results. Pierce and Brown [80] group inorganic interferences as follows:

a. Little or no effect: Ca, Mg, K, Na, SO_4^{2-}, and Cl up to 300 mg/l.

b. Strong suppression: Cd, Co, Cu, Pb, Sr, Mo, Ni, Ag, Sn, V, Zn, MnO_4^-, VO_3^-, $S_2O_8^{2-}$, and NO_3^-. As little as 0.3 mg/l of Cd or Sr completely eliminated the Se signal.

c. Suppression controlled by changes in details of the apparatus and procedure: For example, standard solutions allowed to stand for 24 h in polyethylene cups lost 84% of the observed selenium.

In another paper by Pierce and Brown [81], further investigations showed that some elements (notably Ca) may actually increase the selenium signal. In this paper, Pierce and Brown included flameless AA in their evaluation and record as much as a 100% increase in flameless AA signal due to calcium. At the same selenium concentration, a 54% decrease in flameless AA signal is attributed to sulfate ion.

The Analytical Standard Committee of the (British) Chemical Society [82] investigated the determination of selenium in organic matter using a fluorimetric method as well as the hydride AA method. Important findings are that selenium must be present as Se(IV) for success and that sample preparation procedures are critical. Low results were not due to losses of selenium but to the fact that selenium was present (after sample decomposition) in a state in which it was not determined by the fluorimetric method.

Dennis, Moyers, and Wilson [83] determine selenium as selenide by differential pulse cathodic stripping voltammetry after conversion to SeH_2. They have investigated the SeH_2 borohydride generation process. They found that sulfite SO_3^{2-} has a distinct suppressing effect on SeH_2 generation. Also, as little as 15 µg sulfide caused serious interference. Interference of many metals, especially copper, nickel, cobalt, zinc, iron, aluminum, and lead, is reported to be severe. In the presence of 400 µg lead, for example, only 5% of 0.37 µg selenium was recovered.

2. Determination of Selenium and Arsenic by Coprecipitation with Tellurium Followed by X-ray Fluorescence Measurement

PROCEDURE. Use three aliquots of solution for each determination and transfer to 3–600-ml beakers (aliquots should contain 10 µg arsenic or selenium). Up to 100 ml or more may be used if arsenic or selenium is low. A blank should be carried through the procedure, and all reagents used for samples should be added to the blank. To all three beakers add 5 ml of 80 g/l copper solution, 10 mg tellurium, 10 ml HNO_3, and 20 ml $HClO_4$. To the second beaker add 20 µg arsenic and/or selenium. To the third beaker add 40 µg arsenic and/or selenium. Heat slowly at first and finally evaporate to heavy fumes of $HClO_4$. Continue heating until the volume has decreased to approximately 4 ml. Cool until just warm and add 100 ml water and 100 ml concentrated HCl and 2 drops of HF. Examine the solution carefully to determine if any insoluble material is present, if so, filter thru Whatman No. 541 paper and wash paper with 6 N HCl. While stirring briskly, add 20 ml of 50% H_3PO_2. A black precipitate should rapidly form. Heat to boiling and maintain temperature just under a boil for 5 min. Cool overnight and filter through a 25-mm, 0.25-μm millipore filter (the UGWP or equivalent is recommended). The apparatus should be adjusted so that the filter is level, and the millipore membrane is placed directly on the glass frit support with the dull side of the membrane up. Rinse the beaker with 6 N HCl containing 10% H_3PO_2. Wash down the sides of the filtering apparatus. Finally wash with water. On the final wash, stir up the precipitate and then allow it to settle undisturbed so that it is evenly distributed over the membrane. Transfer the membrane containing the precipitate to a damp paper towel inside a desiccator containing water. When ready for X-ray analysis, place the damp membrane with the precipitate side down onto a piece of mylar. Place the precipitate side down onto a round piece of glass and place it in a Fabbi die. Center the membrane and fill the chamber half with methyl cellulose. Press pellet to at least 30000 lb pressure for 30 s exactly. Release pressure and remove the pellet from the die. Place the pellet in a desiccator over $Mg(ClO_4)_2$ for at least 1 h. Determine selenium and arsenic using a wavelength dispersive X-ray fluorescence spectrometer for best results. This method is very reliable and highly recommended; however, it requires experience and good skill from the analyst.

3. Determination of Selenium by Colorimetry with 3,3'-Diaminobenzidine Extraction into Toluene: This procedure is lengthy and somewhat difficult, but it is accurate and reliable when properly performed [84].

4. Determination of Selenium by Fluorimetry: This method is extremely sensitive and may detect 0.01 ppm selenium when 2–5 g of sample has dissolved into the solution [85].

5. As a rapid method, the electrothermal atomization of selenium followed by AA measurement is very effective. This method is reviewed in detail in Chapter 5.

4.19.2. Tellurium

Tellurium is found around and with gold, lead, and silver ores. Some of the main tellurium minerals encountered are calaverite where tellurium is associated with gold; nagyagite, a complex association of tellurium with gold, antimony, lead, and sulfur; sylvanite, $(Au,Ag)Te_2$; and spiroffite and denningite in association with manganese and zinc as described in

Section 3.2.12. In the environment, tellurium may be easily reduced to the elemental state by organic compounds. Tellurium has eight naturally occurring isotopes.

Tellurides may be effectively decomposed by aqua regia; however, the sample preparation is critical and particle size should be 100% − 200 mesh. Decomposition can also be successful with a sodium peroxide fusion; however, it is not recommended. The decomposition described for selenium using nitric acid and bromine followed by a distillation recovers only very little tellurium. In fact, this method may be used to separate selenium from tellurium if using a mixture of HBr–HClO$_4$ [78]. Electrothermal atomization followed by an AA measurement is one of the most reliable means of determining tellurium and is described in Chapter 5 in detail.

4.20. OXYGEN EXCESS OR DEFICIENCY IN ROCKS

During the rock analysis, Fe(II) is determined after decomposition of the sample in acid and titration with a standard oxidant. The titration is usually assumed to be proportional to the Fe(II) content, and it is calculated to FeO. Fe$_2$O$_3$ is obtained after subtracting Fe(II) from a total iron determination.

This approach leads to false results in some instances as described by Ingamells [86]. The reason is simple: Iron is not the only multivalent element occurring in geochemical materials. Reporting a figure for *oxygen deficiency* or *oxygen excess* in the analysis of certain manganese ores seems far more logical.

The method described in this chapter depends on the stability in phosphoric acid–pyrophosphate mixtures of both Mn(III) and Mn(II). The sample is dissolved in a phosphoric acid mixture containing an excess Mn(II). If the material contains an excess oxygen, the resulting Mn(III) may be simply titrated with a standard ferrous solution. However, if the material presents an oxygen deficiency, a known excess of oxidant such as dichromate is added before dissolving the sample and the excess titrated with the standard ferrous solution.

This method is not applicable for materials containing sulfur and organic carbon or yielding peroxides on solution in acid. Some materials do not dissolve directly in phosphoric acid–pyrophosphate. As all methods, this method is not universal; nevertheless, it has proved to be very useful.

Necessary Reagents

- Ferrous ammonium sulphate: 0.01 N in 0.5% by volume sulfuric acid.
- Manganese sulfate: Dissolve 100 g $MnSO_4$, H_2O in 100 ml water and filter through glass fiber (no organics are permissible).
- Sodium pyrophosphate: Dissolve 200 g $Na_4P_2O_7$ in 200 ml water and filter through glass fiber.
- Phosphoric acid: reagent grade, 85% H_3PO_4.
- Potassium permanganate: 0.1 N, accurately standardized.
- Potassium dichromate: 0.1 N.
- Barium diphenylamine sulfonate: 0.2% in water.

The procedure starts with a preliminary treatment of the reagents to remove oxidizing and reducing impurities. This should be done just before digesting the sample to prevent the mixture to freeze into a glass on standing and generate serious difficulties.

PRELIMINARY TREATMENT OF REAGENTS. Prepare a reagent solution for each sample as follows: Add 20 ml H_3PO_4, 5 ml $Na_4P_2O_7$, and 2 ml of 0.1 N $K_2Cr_2O_7$ to a 250-ml wide-mouth conical flask made of thick glass. Mix and heat slowly, without boiling, gradually increasing the temperature until water is expelled and the excess chromate is destroyed (i.e., until yellow solution turns green). Cool until the flask can be handled, and then add 2 ml cold water. Mix, and cool to room temperature. Add 1.0 ml manganous sulfate solution, and rinse down the sides of the flask with a very small volume of water. Mix thoroughly and add a measured excess of standard (0.1 N) dichromate sufficient to oxidize the ferrous iron (or other oxidizable material) expected and provide a small excess. Again evaporate the solution to remove most of the extra water added with the standard oxidant.

SAMPLE PREPARATION. The sample weight should be chosen so that no more than 10 ml standard (0.1 N) oxidant need be used. If excess oxygen is to be determined, no more than 1.0 milliequivalents should be present. Most samples require grinding to -150 mesh or finer. If analyzing a spinel such as chromite, grinding the sample to -325 mesh is a must.

DIGESTION OF SAMPLES. Rotate flask on its side to cover inside wall with mixture, and then add the sample to the center of the flask. This acts as a washdown during the heating process. Cover with a glass paper to permit water to escape yet not permitting contaminants to enter. Heat without boiling, gradually increasing the temperature until attack is complete. When solution is complete, immediately cool until the flask can be handled, add 5 ml cold water, mix thoroughly, and cool in running water.

TITRATION. Dilute to about 20% in phosphoric acid, cool to room temperature, and titrate with 0.01 N ferrous ammonium sulfate solution, adding 2 drops of barium diphenylamine sulfonate indicator just before the end point. A blank determination should be performed.

Note: No organic paper nor deionized water should be used in this procedure to prevent large blanks and consequently erratic results.

4.21. FLUORINE

Fluorine is seldom present in silicate rocks in sufficient amount to interfere seriously with the determination of other constituents, but in material containing more than a few tenths of a percent, the element causes difficulties. Any not removed during preliminary operations may cause precipitation of the alkaline earths (as fluorides) in the ammonia precipitate or prevent complete recovery of the alumina. Extra pains must be taken with the determination of water and carbon, and a separate sample must be used for silica unless special procedures are followed, for the value for this element obtained in the course of the regular analysis of the main portion is almost certain to be low. With silicate rocks, it is usually safe to assume that fluorine will be completely expelled in the early stages of the analysis, causing error in the SiO_2 only. With minerals high in fluorine, its possible effect at every stage of the analysis must be considered. When silica in the sample is insufficient to accomplish complete expulsion of the fluorine, additional silicic acid may be added for the purpose, and the second evaporation with hydrochloric acid may be replaced by a perchloric acid dehydration, which is more effective in breaking up fluosilicates, fluoaluminates, and alkaline earth fluorides. The addition of extra silica does not ensure the expulsion of all fluorine. In such cases, silica is determined on a separate sample as described elsewhere (Section 3.1.3).

Until recently, fluorine and silica in fluorine-bearing minerals were best determined on the same portion of the sample (and this may still be desirable on occasion, particularly if the available material is limited in amount), the two elements being separated one from the other before determination. This separation, originally devised by Berzelius and modified by Hoffman and Lundell [87] and others, involves sodium carbonate fusion, leaching with water, filtration, precipitation of the silica by means of zinc oxide and nitric acid, followed by further operations designed to recover still unprecipitated silica and the fluorine remaining in the various precipitates. Silica is finally determined in the usual way (by acid dehydration and volatilization with HF), and fluorine is determined either by the lead chlorofluoride method or by thorium nitrate—alizarin S titration after distillation from perchloric acid solution.

When only a limited quantity of material is available for analysis, it is possible to incorporate the fluorine determination in the analysis of the main portion.

Shell [88] has shown that silica may be determined in fluorine-bearing minerals by dehydration with hydrochloric acid after alkali carbonate fusion in essentially the usual manner, provided a large excess of aluminum chloride is added to the solution of the melt before or during acidification. The aluminum presumably prevents loss of fluosilicic acid by forming AlF_6^{2-}. The same author and others have made an extensive study of the problems involved in fluorine determinations, and the following method is based on their work as well as that of Willard and Winter [89], Armstrong [90], Hoskins and Ferris [91], and others.

It should be noted that the method as here outlined is designed primarily for use with silicate rocks, and its application to other materials without modification is not always permissible. For example, phosphate rock will not yield all its fluorine under the conditions described; with materials high in fluorine, retreatment of the various residues and repeated distillation may be required. The method involves preliminary fusion with mixed alkali carbonates and zinc oxide followed by leaching with water, partial neutralization of the alkali with perchloric acid to render more complete the precipitation of silica and alumina, filtration and thorough washing with water, and finally steam distillation after evaporation of the filtrate and acidification with sulfuric acid. The distillate is concentrated by evaporation in the presence of excess alkali, and aliquot parts of it are titrated in buffered 50% ethanol with thorium nitrate using alizarin red S as indicator.

The use of sulfuric acid (instead of perchloric) in the steam distillation is perhaps open to criticism on the grounds that H_2SO_4 has a much greater effect on the titration than $HClO_4$. In fact, Shell advises against its use for this reason. Sulfuric acid has, however, certain advantages: It permits the use of mixed alkali carbonates, thereby lowering fusion temperature and making possible the use of large nickel crucibles (in which proper fusion cannot be accomplished over a Meker burner with zinc oxide and sodium carbonate alone) and it makes a higher distillation temperature possible (145 °C, instead of 135 °C), and this higher temperature has been shown to greatly increase the rate of recovery of the fluorine, thereby lessening both the time required and the volume of distillate that must be collected.

In a properly designed apparatus, with the exercise of reasonable care, the small amount of sulfate that finds its way into the distillate has a very minor effect on the subsequent titration under the conditions to be described. The extent of this effect can easily be ascertained, and if found to be serious, redesign of the apparatus or a slower rate of distillation may be called for.

Note: As an illustration of the applicability of the procedure to silicate rocks, two samples that showed 0.097 and 0.008% fluorine by the Berzelius method of separation followed by $HClO_4$ distillation gave 0.102 and 0.007%, respectively, using zinc oxide fusion with mixed carbonate followed by H_2SO_4 distillation. With standard samples (such as the Bureau of Standards Opal Glass No. 91), at least as good results are obtained with the latter method as with the former with considerably less trouble.

It should not be implied that sulfuric acid is always to be preferred to perchloric in this application. The latter has the outstanding advantage that it affects the titration very little even when considerable amounts distill over. It is undoubtedly to be preferred when direct distillations are made, particularly with materials containing calcium (as phosphate rock), when fluorine minerals are to be run, and when it is necessary to distill fluorine from the insoluble residues resulting from the leaching of carbonate fusions. However, with rocks, excellent results can be obtained in the usual case with sulfuric acid, and advantage is taken of this where possible. Among the many precautions that should be observed, the following deserve special mention: Alkaline solutions containing fluorine should be evaporated in platinum; evaporation in glass lead to low results. Operations carried out in glass should be made as short as possible. The use of low-boron glassware does not improve matters.

The conditions of distillation must be closely controlled: If too rapid, if carried out at too high a temperature, or if still design is poor, excessive sulfate will be found in the distillate and cause titration errors. Fine grinding of the sample (usually -115 mesh is sufficient) is usually necessary for complete decomposition under the conditions of fusion described. The water leaching of the carbonate–zinc oxide melt must be thorough. If preliminary leaching is done in the crucible and then the residue is twice boiled with water as described and washed thoroughly on the filter, very little fluorine will remain undissolved in the usual case. Among the exceptions to this rule is the case of phosphate rock, from which all the fluorine cannot be extracted by this means even by repeated fusion and leaching.

There must be enough silica present in the distillation mixture to form fluosilicic acid with the fluorine: With rocks, there is seldom any difficulty here, but with minerals, it may be necessary on occasion to add finely powdered quartz (not silicic acid!) before distillation, especially when fluorine is high. If this is not done, attack on the distilling flask with consequent introduction of boron may cause error. It may also prove difficult to complete the distillation of the fluorine.

The titration with thorium nitrate is really a semi-microprocedure. Its precision is best when the volume of titrating solution approaches 2.0 ml, and for maximum accuracy, aliquots of the proper size must be taken. It is quite necessary to prepare reference solutions as described to exactly detect the end point. The first sudden color change should be taken; titration to a full pink color is a mistake. The end point is not an easy one to observe; however, if the small beakers are placed on a white porcelain slab, with reference end points near by, and titration is done under white fluorescent light, little difficulty will be experienced after a few trials.

When the titration is carried out in 50% ethanol solution, the titration curve (milliliters of titrant vs. milligrams of fluorine) is a straight line down to the limit of detection. This is not true of titration in water solution. This factor is of particular importance when small amounts are in question. With the procedure to be described, it is easily possible to distinguish between 0.01 and 0.02% fluorine when ten-fiftieth of a 0.5-g sample is taken for titration.

A blank determination should be carried throughout the procedure and suitable corrections made. This is particularly important when a limited supply of material makes it necessary to use a 0.2-g sample, for in this case the blank may amount to 0.01 or 0.02% fluorine. A titration blank should also be determined; this will usually amount to 0.010–0.015 ml titrant.

The separation of fluorine is accomplished as follows:

PROCEDURE. Weigh 0.5000 g or less (0.2000 g if fluorine is known to be high or if the supply of sample is small) of the very finely ground material into a large nickel crucible. Weigh out 6–6.3 g fusion mixture (prepared by grinding together 70 g Na_2CO_3, 90 g K_2CO_3, and 30 g ZnO in an agate mortar and mixing thoroughly), mix the sample with about 5 g of this and cover with the remainder. Cover the crucible and heat at the full temperature of a good Meker burner for 30–35 min.

Note: The sample must be finely divided to ensure complete decomposition. In many instances, it is advisable to weigh out a little more than required, grind it as fine as possible in an agate mortar, and either determine the moisture content of this sample or dry it under the same conditions as the rest of the material taken for analysis before weighing out part of it for the fluorine determination. With biotites, grinding to pass a 115-mesh screen is advisable; this is not fine enough for most materials.

While a 0.5-g sample is desirable when fluorine is low, good results can be obtained with as little as 0.2 g. The smaller sample weight should always be used when fluorine is high. If the sample is high in fluorine and low in silica, it is advisable to add 0.3 g silicic acid to 0.2 g of the sample before fusion; finely powdered quartz (not silicic acid) should also be added to the distillation flask before introducing the sample solution.

Cool, add a little water to the crucible, and allow to stand for several hours, preferably overnight. Stand the crucible on a clean watch glass and cover the whole solution with a large beaker to prevent pickup of acids from the air.

Note: During the fusion, the molten carbonates tend to creep up the sides of the crucible, and some may even be found on the outside of the upper edge. During the leaching, the water solution sometimes wets the outside of the crucible; thus, it is placed on a clean watch glass and covered.

Wash the contents of the crucible into a 250-ml beaker. If the melt does not come free at once, place the crucible (after washing it off inside and out) on the hot plate for a few moments and loosen the solids by rubbing with a glass rod. A total of about 100 ml water should be used in transferring to the beaker.

Add to the crucible 10 ml of 1:10 perchloric acid, scrub briefly, and add the acid all at once to the carbonate solution. Wash out the crucible with a little water.

Note: The prolonged leaching should not be done in the beaker because this will cause introduction of boron from the glass. The purpose of the perchloric acid addition is to partly neutralize the alkali, leading to more complete precipitation of silica and alumina.

Care must be taken that the fusion and its water solution is protected from laboratory fumes, which may contain fluoride. Thus, the leaching should not be carried out at the same time as such operations as the volatilization of SiO_2 with HF. In the final titration, 2 μg fluorine may be a significant quantity.

Boil for 5 min, and filter at once through a large filter paper (S & S white ribbon, Whatman No. 40), and catch the filtrate in a 300-ml platinum dish. Wash with hot water. Return the contents of the filter to the beaker with 50 ml or so of water, and again boil. Filter through the same paper, and wash with hot water. Ignore the peptization that almost always occurs.

Evaporate the alkaline solution (in platinum) on the steam bath to the point of crystallization of sodium carbonate, stirring during the final stages to prevent the formation of large crystals, which may prove difficult to wash into the distillation flask.

Assemble the distilling apparatus, cleaning the distilling flask by shaking sodium hydroxide solution and coarse sharp sand in it and then washing with cleaning solution and water. Wash the evaporated sodium carbonate solution into the distilling flask using a minimum of water. Then add to the platinum dish 30 ml of 24 N sulfuric acid, and pour this slowly into the flask, a drop at a time at first, until evolution of CO_2 ceases. Wash out the dish with a little water.

Note: The additions are made through the side arm of the special Claisen-type flask by means of a curved funnel, the tip of which must reach well past the outlet tube so that none of the added material can run into the condenser. If the acid is added too quickly, an appreciable amount of sulfuric acid will be carried over with the rapidly evolved CO_2. This must be avoided, for excessive amounts of sulfate interfere in the subsequent titration. It is particularly important to clean the distilling flask with strong alkali when an ordinary rock sample (which may contain, say, 0.03% F) is to be run after a fluorine-bearing mineral. The gelatinous residue on the bottom of the flask is hard to remove and may still retain traces of fluorine. Sand is of great help in this cleaning operation. After using alkali, the flask should be washed thoroughly with an acid solution to remove possible soluble compounds of boron.

Wash out and remove the funnel in such a way that small amounts of sulfuric acid adhering to it will not enter the condenser. Stopper the side arm, and put a rubber policeman or the top of a medicine dropper over the end of the steam inlet tube.

Note: A minimum of water should be used in transferring the material to the distilling flask, since it must be removed by boiling before the steam distillation begins. It is best not to put the receiving flask in place until after the acidification of the solution in the distilling flask, for the carbon dioxide evolved would then fill it and make necessary the use of excessive amounts of alkali.

Put the receiving flask in place (a 200–300-ml conical flask is most convenient) and add to it several drops of normal sodium hydroxide (freshly prepared) and a little alcoholic phenolphthalein.

Heat the Claisen flask by connecting the heating coil and boil off water until the thermometer registers 135 °C. Then connect the steam generator and so regulate the current of steam and the heating of the Claisen flask so that the temperature of the air bath remains at 150–170 °C, that of the acid solution rises slowly to a maximum of 145 °C or slightly higher, and distillate collects at the rate of about 1 ml/min.

Continue until 200 ml or so of distillate has been collected, adding dilute sodium hydroxide as necessary to keep it alkaline to phenolphthalein. The distillation temperature should remain well above 140 °C for the greater part of the time. If this has not been the case, a larger volume should be collected, particularly if fluorine is likely to be high.

Note: With silicate rocks, there is usually not much point in collecting more than about 175 ml distillate. If the temperature rises above 145 °C, this will do little harm provided the higher temperature is not maintained. Temperature of 150 °C is permissible for only short periods, for at this temperature appreciable amounts of sulfuric acid volatilize. The aim should be to hold as closely to 145 °C as possible without exceeding this temperature. If there is any doubt concerning complete recovery, the receiver may be replaced with another when 175–200 ml has been collected and the distillation continued. Both fractions may then be titrated after evaporation as described below. If powdered quartz was not added originally, it is advisable to add a little at this stage to be sure there is sufficient silica to combine with the fluorine. Collection of a second or even third fraction of distillate is advisable when perchloric acid is used instead of sulfuric acid in the determination of fluorine on minerals (see below). With sulfuric acid, under the conditions described, very little fluorine will usually be found in the second 200 ml distillate.

When the distillation is judged to be complete, disconnect the apparatus, washing down the condenser tube into the receiver. Transfer the distillate to a large platinum dish (conveniently the same one that was used before), make certain that it is alkaline to phenolphthalein, and evaporate to a small volume for transfer to a 25- or 50-ml volumetric flask. If fluorine is known to be very low, the whole of the solution may be transferred to the titration beaker.

For phosphate rock and similar materials, the sample should be fused or sintered with sodium carbonate and peroxide and the fusion taken up with water and transferred to the distillation flask without any attempt to separate silica and alumina. If required, powdered quartz should be added to provide an excess of silica. With such materials and presumably others low in silica and alumina and high in calcium as well as with residues from sodium carbonate or other fusions that may require examination for fluorine, a perchloric acid distillation is to be preferred.

If perchloric acid is used, care must be taken that organic matter, (e.g., paper) has all been removed by gentle ignition (with the addition of alkali if there is any danger of loss of fluorine). If silica is low, finely powdered quartz should be added to prevent the solution of silica and boron from the flask.

In distillation with perchloric acid, the maximum temperature should be set at 145 °C, and 135–140 °C optimum. If fluorine is high, collection of at least 400 ml distillate is advisable, especially if the sample contains alumina. About 20 ml concentrated acid in excess of that required to neutralize the alkali used in the fusion or sinter is required. In other respects, the distillation is the same as that with sulfuric acid.

A specially designed apparatus has been described by Ingamells [92] that permits close control of the distillation and rapid and complete recovery of fluorine (see the apparatus in Figure 4.3).

TITRATION OF FLUORINE. The following solutions should be prepared:

Standard thorium nitrate: Weigh 1.38 g $Th(NO_3)_4 \cdot 4H_2O$ into a 500-ml volumetric flask, dissolve in water, add 0.2 ml nitric acid, and dilute to the mark with water. Standardize against pure sodium fluoride as described below.

Standard sodium fluoride: Dry pure NaF by heating it in a small platinum crucible at low red heat (below the melting point, 980 °C). Weigh exactly 0.1500 g, dissolve in water, and dilute to 200 ml in a volumetric flask. One milliliter of this solution contains 0.000339 g fluorine.

Sodium alizarin sulfonate indicator: Dissolve 50 mg in 100 ml water.

Buffer solution: Dissolve 9.5 g monochloracetic acid in water, and make up to 25 ml. To exactly half of this solution add a drop of phenolphthalein indicator and then 50% sodium hydroxide solution drop by drop until the indicator just changes color. Add the remainder of the acid solution and dilute to 50 ml. This solution must be prepared fresh: an old solution will cause poor end points and erroneous results.

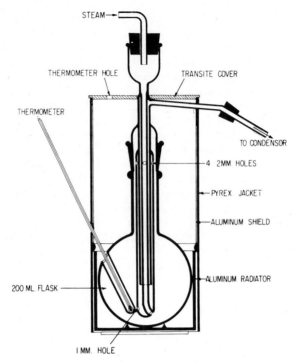

Figure 4.3. Illustration of special still for steam distillation of fluorine.

Diluted thorium nitrate: Dilute 2.0 ml of the standard thorium nitrate to 100 ml with water. Prepare fresh daily.

Standardize the thorium nitrate solution as follows:

PROCEDURE. Pipette several pairs of various quantities of the standard fluoride solution (as 0.5, 1.0, and 2.0 ml) into 50-ml beakers. Add water in an amount such that the total present at the end of the titration will be 15 ml including the volume of the titrant. For example, if 1 ml fluoride solution is taken and approximately 1 ml thorium solution will be required in the titration, add 13 ml water. Measure the water to the nearest milliliter. With unknowns, it will be necessary to make a preliminary titration to determine the volume of water to be added. It is important that the final concentration of alcohol in the solution be close to 50% by volume. Add 3 drops of indicator, 0.3 ml buffer solution, and 15 ml of 95% ethanol.

Using a microburette, titrate to the first indication of a permanent pink color, comparing the end points to standards made up as follows: Add to 15 ml ethanol in 50-ml beakers enough water to make a total of 15 ml, 3 drops of indicator, 0.3 ml buffer solution, and 0.0, 0.5, and 1.0 ml diluted thorium solution. If the color produced by 0.5 ml diluted thorium nitrate is taken as the end point, a titration blank of 0.01 ml should be subtracted from all titrations. Calculate the fluorine value of the standard thorium nitrate solution.

Determine the fluorine in distillate as follows:

PROCEDURE. Transfer the evaporated solution to a 50-ml volumetric flask (or a 25-ml flask if fluorine is low and the sample was small) and dilute to volume with water. For a preliminary determination, transfer 2 ml of this solution to a 50-ml beaker. Add 0.2 N hydrochloric acid drop by drop until the phenolphthalein is decolorized. Then add 3 drops of the alizarin indicator, and continue the addition of hydrochloric acid until the solution is pure yellow. Add 1 drop of 0.2 N hydrochloric acid in excess. Add 0.3 ml buffer solution, 13 ml water, and 15 ml of 95% ethanol.

Using a microburette, titrate to the first indication of a permanent pink color, comparing the end point to standards as was done in standardizing the titrant. From this approximate determination, calculate the proper aliquot to be used to give a titration of 1.5–2.0 ml. If fluorine is low, use a 10-ml aliquot. For samples very low in fluorine, further evaporation of the solution may be desirable.

The final titration should be made in duplicate, and the burette readings should check to 0.005 ml. The end point may give a little trouble at first but soon becomes easily recognized. The first sudden change in color, as seen against a well-lighted white background, should be taken. Reference end points should always be prepared as described.

Fluorine may also be determined either on distillate obtained as above or on filtrates from alkali carbonate fusions by the lead chlorofluoride method. This cannot be used with small amounts of fluorine but may be very useful when larger amounts are in question. As applied to filtrates from alkali carbonate–zinc oxide fusions, the procedure is as follows:

PROCEDURE. To the cold alkaline solution in a 400-ml beaker, which should contain from 0.02 to 0.05 g fluorine in a volume of 200–250 ml, add 0.3 g NaCl and then nitric acid (1:1) in amount calculated (from the weight of flux used) to slightly but definitely acidify the solution. Keep the beaker covered as much as possible during this and subsequent additions because of the large amount of CO_2 evolved. Add a few drops of methyl orange indicator and then 5% sodium hydroxide solution to just neutral to the indicator. Then add just 2 ml of 1:1 hydrochloric acid.

Add 5 g solid lead nitrate, let stand for several minutes, and then stir to complete solution. Add 5 g sodium acetate dissolved in a little water, and heat on the steam bath for 30 min, stirring from time to time. Cool, and allow to stand overnight.

Stir in a little paper pulp, allow precipitate and pulp to settle, and filter. Wash once with water, five times with saturated lead chlorofluoride solution, and again once with water. Return paper and precipitate to the beaker, add 100 ml of 5% nitric acid, and digest on the steam bath until all the PbClF is dissolved, stirring to macerate the paper. This will require 5–10 min.

Using a 5- or 10-ml pipette, add 5.00 or 10.00 ml standard silver nitrate solution made up so that 1 ml is equivalent to 0.005 g fluorine. Stir, and digest on the steam bath for 30 min, stirring frequently to coagulate the silver chloride and protecting the solution as much as possible from the light. Cool to room temperature in the dark, filter, and wash with very dilute (1:99) nitric acid to obtain all the unprecipitated silver in the filtrate.

Discard the precipitate, and add to the filtrate 2–3 ml ferric alum indicator. Titrate with standardized ammonium or potassium thiocyanate solution, the titer of which has been checked against the same silver nitrate solution used in the procedure.

If the silver and thiocyanate solutions are equivalent, and the former is made so that 1 ml = 1% fluorine on a 0.5-g sample.

Percent of fluorine = (milliliters of $AgNO_3$) − (milliliters of thiocyanate)

Prepare solutions as follows:

Lead chlorofluoride wash: Dissolve 10 g lead nitrate and add its solution to 100 ml of 1% sodium fluoride in 2% hydrochloric acid. Wash by decantation 8–10 times with water. Add 1 l water, and shake intermittently for some time. Filter.

Ferric alum indicator: Prepare a saturated solution of ferric alum and add sufficient colorless nitric acid to bleach the brown color of the water solution.

Silver nitrate, 1 ml of which should be equivalent to 1% fluorine on a 0.5-g sample: Dissolve 22.35 g $AgNO_3$ into 500 ml.

4.22. HYDROGEN

Hydrogen occurs in rocks and minerals in many forms. Determination of total hydrogen is an exercise of moderate difficulty. Dividing total hydrogen into its several components presents problems that cannot always be easily solved. The traditional rock or mineral analysis reports hydrogen as H_2O^- and H_2O^+, as "moisture" and "combined hydrogen." This convention is useful, but it must be remembered that H_2O^- is determined by measuring the loss in weight observed on heating a sample for 1–2 h at 100–110 °C, and H_2O^+ is the difference between H_2O^- and total hydrogen calculated to H_2O. Very often, such a procedure yields values that are not very meaningful. Loss in weight at 105 °C may not always be due only to loss of water; some minerals actually gain in weight when heated to 105 °C due to reaction with the ambient air; other constituents may volatilize. With some minerals (e.g., hydrous clay minerals), the loss in weight may depend on the humidity of the oven air, and with others (e.g., lateritic ores), the time of heating determines the extent of water loss. Moisture determinations on coal, lignite, and other organic or carbonaceous materials are always very empirical because slow oxidation occurs during the heating period, with volatilization of CO, CO_2, SO_2 and other gases.

Sedimentary rocks (limestone, marl, and shale) often contain carbon and sulfur compounds that are volatilized during moisture determination. Some rocks, especially those of radioactive nature, contain appreciable concentrations of absorbed or adsorbed or interstitial gases evolved and counted as moisture. Some volcanic rocks contain gaseous inclusions (usually CO_2), which may or may not be expelled on heating. Liquid inclusions containing water are common; they may or may not lose their water content on heating to 105 °C.

4.22.1. Primary Methods: Moisture, H_2O^-, and Sampling Weight Baseline

In any analysis, the sampling weight baseline must be controlled if accurate and meaningful results are to be obtained. Despite the apparent simplicity of the concept, it is sometimes very difficult in practice to establish a sampling weight baseline and even more difficult to relate the final analysis to the composition of the mass of material the analytical subsample represents. For example, lateritic ore as mined may contain 25% or more of water. If its actual water content is not measured, it is impossible to report accurately the metal content of the ore. Drying lateritic ores in the conventional manner (2 h at 105–110 °C) leaves a water content of up to 10–12%. The dried material is hygroscopic and may rapidly pick up 2–3% of water from the air, so if precautions are not taken, even the analysis of the dried laboratory sample may be in error. When a single constituent is of interest (e.g., nickel or cobalt), analysis for this constituent is not accurately reported unless the water content of the sample is also reported. With these ores, a statement that the sample was dried at 108 °C is insufficient because drying at 108 °C is not reproducible; how much water is lost depends on the time of drying, on the circulation of air in the oven, on local humidity, and even on how many samples are dried at the same time.

Calculations to dry weight is sometimes a useful expedient, but what is meant by dry weight must be carefully defined. With nonsiliceous laterites, which consist largely of goethite, $Fe_2O_3 \cdot H_2O$, ignition to 850–900 °C may establish a baseline, and the loss in weight gives a fair estimate of the water content of the analytical subsample (but not necessarily of the ore from which the sample was collected). With other materials (carbonate rocks, ferrous minerals, etc.), ignition loss may represent loss of carbon dioxide, hydrogen fluoride, chlorine, sulfur, and so on; if ferrous iron is oxidized, a weight gain may be observed. Metallic sulfides may either be converted to oxides or to sulfates, depending on sample composition.

Calculation to a different baseline from that actually employed during analysis is a simple matter that, however, seems to cause some difficulty, especially if only a partial analysis has been performed.

Example. A 2-kg sample of lateritic ore was received for nickel determination. The whole sample was transferred to a large pan and weighed to the nearest gram and then placed in an air circulating oven for 24 h, cooled, stirred, and allowed to come to equilibrium with the room air. It was crushed and screened to pass 10 mesh, and weighed. Loss in weight was 8.20%. The 10-mesh material was split four times, yielding a 250-g sample that was further reduced to 64 mesh. This was split twice, and the resulting 50-g sample was passed through a 150-mesh screen and submitted for nickel determination.

In the laboratory, several 2-g portions were split out and weighed so as to avoid difficulties due to loss or gain of moisture, and a Penfield water determination was performed at once. This showed 9.52% H_2O. Ignition of one of the other 1-g subsamples showed 9.03% ignition loss, indicating that some of the hydrogen was contained as hydroxyl in silicates or that an element (Co or Mn, perhaps) was oxidized during ignition.

The nickel content of one of the 1-g subsamples was found to be 2.22% NiO after fusion with peroxide, solution, and double precipitation with dimethylglyoxime. Leaching another of the 1-g subsamples with dilute acid yielded 2.03% nickel as NiO, indicating that part of the nickel occurred in acid-insoluble silicates.

The matter of interest is the nickel content of the original 2-kg sample and especially its acid-soluble nickel. The largest uncertainty in the data originates during the sample reduction to 150 mesh, during which some water may have been lost or gained, with proportionate increase or decrease in the nickel content; it is obviously important that the sample reduction be done knowledgeably.

Calculation of the NiO content of the original sample is carried out as follows:

Percentage of NiO in dried sample	2.22
Percentage of H_2O in dried sample	9.52
Percentage of other constituents	88.26 (by difference)
	100.00

To calculate to a water-free basis, the NiO value must be multiplied by a factor of $100/(88.26 + 2.22) = 1.1052$:

Percentage of NiO in water-free sample	2.455
Percentage of other constituents	97.545
	100.00

The water content of the original 2-kg sample is $8.20 + 9.52 = 17.72\%$; to find its nickel content, we must multiply by a factor of $100/(17.72 + 2.455 + 97.545) = 0.8495$:

Percentage of NiO in wet as-received sample	2.086	(1.91% acid soluble)
Percentage of H_2O in wet as-received sample	17.72	
Percentage of other constituents	82.864	
	100.00	

In this case, a Penfield water determination was probably unnecessary—the loss-on-ignition figure could just as well have been used; however, if need to reassay the sample should occur, some uncertainty is introduced because a loss on ignition is not generally reproducible. With sulfide ores in particular, there may be a large variation in loss-on-ignition determinations with slight changes in ignition conditions.

Although sampling and subsampling errors contribute a major proportion of the discrepancies between analysts and laboratories, a failure to pay sufficient attention to the sampling weight baseline is also a factor in analytical error. In the example given above, failure to follow a procedure such as that described and failure to report accurately the base on which the reported nickel value depends invalidate the analysis. The report should probably take the form

Percentage of NiO on sample as-received	
Total	2.09
Acid soluble	1.91
Percentage of NiO on water-free sample	
Total	2.45
Acid soluble	2.24

A common malpractice is to multiply the determined value on a dried sample by the percentage of loss on drying and subtract this from the determined value. In the example above, if this is done, one writes 2.22 × 17.72/100 = 0.3934 and 2.22 − 0.39 = 1.83% NiO in the wet sample. This is erroneous.

It is instructive to follow the weight change of a sample as temperature is increased. A thermogravimetric balance is convenient for this purpose, but most such balances accept so small a sample that sampling errors are likely to be large. Quite often there is an initial weight loss due to evolution of water; then at higher temperature, there is a further loss due to decomposition of carbonates followed by a gain in weight as ferrous iron is oxidized.

Example. A 2-g sample of the USGS basalt BCR-1 was heated at increasing temperatures, with the following results:

Heating Conditions	Loss or Gain in Weight (%)	
	Stepwise	Cumulative
Exposed to room air overnight	+0.040	+0.040
2 h at 110 °C	−0.450	−0.410
87 h at 110 °C	−0.015	−0.425
2 h at 150 °C	−0.440	−0.865
Exposed to room air 2 h	+0.465	−0.400
18 h at 150 °C	−0.545	−0.945
18 h at 450 °C	−0.195	−1.140
18 h at 750 °C	+0.270	−0.870
18 h at 1100 °C	+0.260	−0.610

Determination of total hydrogen as H_2O by the Penfield method showed 1.34% H_2O. It should be evident that a loss on ignition is no substitute for a total hydrogen determination in rock analysis.

In some cases, there may even be an increase in weight on ignition. One suboceanic basalt heated under the same conditions as BCR-1 in the above example, *gained* 1.02% in weight; its hydrogen content was 0.24% as H_2O.

Sulfide ores and rocks containing sulfides cannot be reliably dried by heating. An indeterminate proportion of the sulfur is fixed as sulfate, and iron may finish partly in the ferrous condition as magnetite.

Example. A 0.5-g sample of a sulfide ore (CAAS S-1) behaved as follows during heating:

Heating Conditions	Loss or Gain in Weight (%)	
	Stepwise	Cumulative
15 h at 105 °C	−0.08	−0.08
3.5 h at 160 °C	−0.04	−0.12
15 h at 425 °C	+2.72	+2.60
4 h at 590 °C	−4.94	−2.34
15 h at 775 °C	−4.28	−6.62
Repeat 15 h at 750 °C	0.00	−6.62
15 h at 975 °C	−0.30	−6.92
15 h at 1100 °C	−1.20	−8.12
15 h at 675 °C	+0.10	−7.92

A fair estimate of the H_2O content of this ore was made by determining sulfur in the residue after ignition and total sulfur in the unignited sample and assuming that iron finished as Fe_2O_3 in the ignited residue; such exercises are not usually worth the trouble and should not be substituted for a direct water determination.

CLASSICAL PROCEDURE. Air dry the prepared and weighed sample of rock, ore, or mineral by spreading the powder in a thin layer on albanene paper, covering with a second sheet of paper and allowing to remain in the weighing room for several hours. Reweigh. Mix thoroughly by rolling on the paper, and weigh out as many portions as are likely to be used for analysis into suitable crucibles, dishes, or beakers. Determine total water in one or two of the weighed portions by the Penfield method, the combustion tube method, or some other primary method. The total hydrogen determination should be carried out as soon as all the necessary samples have been weighed to avoid sampling baseline error. Obviously, for many or most materials, such a precaution is hardly necessary and can often be relaxed; however, with clay samples, vermiculites, limonitic rocks, chlorites, and even biotites and biotitic rocks, the precaution is necessary.

The Penfield tube method for total hydrogen determination is probably as good or better than any other in most cases. The modification here described was developed over a period of many years at the University of Minnesota, the Pennsylvania State University, and in the laboratories of the U.S. Geological Survey. The sample is mixed with a flux in the bulb of a Penfield tube and heated. Evolved water is collected in the neck of the tube, which is then pulled off in a flame. The portion of the tube containing the water is weighed, dried, and weighed again.

This process, simple in principle, requires attention to a large number of essential details, some of which are not very obvious. The sample must, of course, be completely attacked by the flux, and active fluxes such as sodium carbonate cannot be used in a glass tube. Many fluxes have been tried with variable success. The one that has proven most generally effective is flake litharge (lead oxide) containing lead chromate and copper oxide. Calcium carbonate may be added to provide gas evolution, which aids in liberation of the water. Sodium tungstate may be used; it must be prefused and a little excess of WO_3 should be added to neutralize any excess alkali. Lithium metaborate is worthless because even at 900 °C, it retains water. With phosphate rocks and minerals, addition of silica is necessary to evolve all the hydrogen.

PROCEDURE. Dry a Penfield tube thoroughly and use an aspirator to fill it with air at room temperature and humidity. Weigh the tube, add 0.5–1.0 g air-dried sample through a long-stem funnel, and reweigh to obtain an exact sample weight. Using a fresh funnel, add fluxes (usually 5–7 g PbO, 1 g PbCrO$_4$, 1 g CuO, and, for phosphate or carbonate rocks and minerals, 0.5 g quartz ground to pass a 325-mesh screen). Start an identical blank determination at the same time using all reagents but no sample (or a sample of low known hydrogen content; e.g., powdered quartz).

Mix the contents of the bulb thoroughly by rotating it, being careful that none of the charge is deposited on the upper portion of the bulb, where it may escape proper fusion. Stopper the tube with a small anhydrone bulb (to prevent "breathing" of atmospheric moisture), and then wrap a strip of towel saturated with ice water about the stem. Bend the tube at an angle of slightly more than 90° in a Meker flame so that it may be mounted in the apparatus shown in Figure 4.4, with the bulb supported in the heavy iron crucible as illustrated. Alternatively, the Penfield tube may be placed, without bending, in a specially constructed electric furnace such as that described by Peck [48]. The following description supposes that the apparatus of Figure 4.4 will be used.

Fill the ice bath with cold water and pack with crushed ice. After several minutes, remove the small drying tube, and deposit a small portion of dried calcium carbonate in the tube just above the bend, as shown, using a glass tube with a close-fitting glass rod. Stopper the tube with a capillary stopper, and heat the crucible and bulb with a large Meker-type blast burner, starting with a small flame and increasing the heat to a maximum over a period of about 20 min. Replenish the ice as necessary.

When fusion is complete, pull off the bulb with a pair of tongs without lessening the heat, and shape up the end of the tube containing the water, keeping it in the ice bath; then turn off the burner.

When cool, slide the wet towel to the bottom of the tube, and stand it at an angle in the ice bath. Remove the capillary stopper, and slowly insert a glass rod that closely fits the tube. Using another rod, push the first to the bottom of the tube. The purpose of the rods is to displace carbon dioxide; they

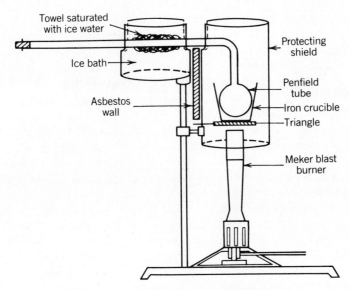

Figure 4.4. Illustration of mounted Penfield tube.

should fit closely so that no more than a fraction of a cubic centimeter of CO_2 can remain in the tube.

Restopper the tube containing the first glass rod, wipe it dry, and place it near the balance to come to temperature equilibrium. Weigh tube and rod, preferably against a similar tube and rod used as a tare. Make sure the tube is not statically charged—this can lead to a false weighing. Remove the rod, and touch it to a piece of pH paper; if pH is above 3–4, weighable amounts of contaminating acids are not present. A pH of 7–8 is normal, the traces of alkali coming from the glass or the reagents. Strongly alkaline water may indicate the presence of NH_3, which can be confirmed using Nessler solution. Ammonia has been found in the water generated from stilpnomelane, jarosite, shales, and some other materials.

If the water is strongly acid, the sample probably contains large amounts of chlorine, fluorine, or sulfur, and the determination should be repeated using larger amounts of lead chromate and calcium carbonate.

A normal blank determination yields 0.2–1.0 mg water. Part of this comes from the reagents, which can never be kept perfectly dry and part from the atmosphere. A zero blank is an indication of faulty technique unless atmospheric humidity is very low.

Different minerals require adjustment of the procedure depending on their composition. Amphiboles, for example, must be ground to pass a 200-mesh screen to ensure complete attack; during grinding, the mineral should be screened frequently so as to avoid overgrinding the fines. Biotites must not be ground any finer than is necessary because hydroxyl water is thereby lost. Sampling of hydrous minerals is a matter of extreme difficulty. As indicated above, the determination of total water in these materials affects the results for all other constituents and is therefore of some consequence even when the water content itself is not of interest. Thorough mixing of the air-dried sample just before taking a number of weighed portions, including one or more for total water, is essential.

An alternative to the Penfield method for total water determination is the combustion tube method most commonly used in organic analysis. The principle is essentially the same as in the Penfield method; the sample is fused with a flux, and the evolved water is collected and weighed. The method works well with materials that do not have a large moisture content; its overriding disadvantage is that there is no really practicable way to control the blank determination. If the sample is put in the combustion tube and dry carrier gas is drawn through to remove air moisture, some of the moisture in the sample will also be removed, and one of the main purposes of the determination, the establishment of a sampling weight baseline, will be lost. The best that one can do is to open the combustion tube and insert the sample in exactly the same manner each time and measure an empirical blank. This works better in dry climates than in damp. A microcombustion method described by Cremer, Elsheimer, and Escher [93] always gave a positive bias, which diminished as sample size increased. If 1–2-g samples are taken, the blank problem is not serious, and for samples that do not lose appreciable water in contact with dry gas, the combustion tube method is entirely adequate.

PROCEDURE. Construct a combustion train with the following components in sequence: a supply of dry gas (nitrogen or air or carbon dioxide), a combustion tube capable of being heated to 950–1000 °C, a calcium carbonate trap capable of being heated to 550 °C [decomposition temperature of $Ca(OH)_2$], a catalyst (mixed silver manganese oxide according to Körbl [94] is satisfactory) capable of being heated to 450 °C, a magnesium perchlorate weighing bulb, and a second magnesium perchlorate bulb to protect the first from the air.

Weigh the sample into a platinum or porcelain boat, and cover it with five times its weight of flux (the same flux used in the Penfield procedure). Place the boat in the combustion tube, and either heat it at once or first pass dry gas through the apparatus to remove the water introduced during loading. Which procedure to use depends on the nature of the sample. Weigh the absorption bulb, replace it in the train, and heat the tube and sample; the catalyst should be maintained at 450 °C. When fusion is complete, heat the calcium carbonate trap to 550 °C to release water trapped as $Ca(OH)_2$. Continue a slow flow of dry gas during the operation until it is certain that no water remains condensed within the tubes and connections. Usually 30–40 min after fusion is complete is sufficient if the train is compact and of good design. Long connecting tubes should obviously be avoided.

Weigh the absorption bulb containing the water. It is very necessary to tare the bulb against a similar bulb of comparable weight and to be very careful to dissipate static charges before weighing. If carbon dioxide is used as a carrier gas, the absorption bulb should be left in the train until just before weighing, and the stopper on the bulb should be closed immediately *after* removing it from the train and just before weighing. If nitrogen or air is used, the bulb should be momentarily opened just before weighing.

Good results by the Penfield method or the combustion tube method depend on a full understanding of the several pitfalls that exist and on a meticulous attention to detail. Either method is capable of an accuracy of ± 0.1 mg H_2O collected when all precautions are taken.

For the determination of total hydrogen in very small mineral samples, the microcoulometric method described by Cremer et al. is most satisfactory. It requires a modification of a commercially available apparatus, the essential component of which is a Keidel cell. The water evolved from a sample in microcombustion tube is purified by passage over calcium carbonate and a catalyst and introduced into the Keidel cell, where it reacts with phosphorus pentoxide. The product, phosphoric acid, is electrolyzed into H_2 and O_2, and the quantity of electricity required for this electrolysis is measured. Readout is in micrograms of H_2O, and up to 100 mg water can be registered. In practice, samples ranging from 10 to 50 mg are used. The major difficulties are in accurately weighing the small samples and transferring them to combustion boats without introducing extraneous water, in deciding on an appropriate blank, and in overcoming the unavoidable introduction of ambient air into the system during loading. Comparative results on the standard biotite LP-6 Bio 40–60# follow (Engels and Ingamells, 1977).

Total H_2O in split number 5-III-B-3: Penfield method, 1-g sample, 3.66% H_2O; microcombustion method, $\approx$50-mg samples, 3.58, 3.65, 3.60, 3.65, 3.61, 3.74, 3.75, 3.66, and 3.66. It is impossible to decide from these data whether the variance observed in the microcombustion values derives from sample inhomogeneity or from imprecision of the method or of the blank determination. However, a fault in the methods used to prepare LP-6 Bio 40–60# was revealed by analysis of another split, 9-II-C-2: microcoulometric determinations of total hydrogen on this split yielded the following values: 3.82, 3.77, 3.73, 3.87, 3.85, 3.80, 3.74, 3.79, 3.82, 3.73% H_2O.

Split 9-II-C-2 evidently contains about 0.2% more hydrogen as H_2O than split 5-III-B-3. The reason for this is as follows: Because of lack of oven space and other logistic reasons, LP-6 Bio 40–60# was dried in batches, with no attention to the time each batch was left in the drying oven. This was a mistake. It is now apparent that biotites held at 100 °C gradually oxidize; Fe(II) is slowly converted to Fe(III), and the associated hydroxyl is converted to H_2O. This otherwise carefully prepared standard biotite is therefore marred by reason of failure to control the last stage in its preparation, the drying process. We must note that a 0.2% uncertainty in water content leads to a 0.02% uncertainty in K_2O content and an uncertainty in potassium–argon age of about 2 my. This may serve to emphasize the importance of total hydrogen determinations and of the necessity to control the sampling weight baseline.

A microanalytical method that will probably be seldom used but can yield excellent results in skilled hands is the micro-Penfield method of Sandell [95]. It is generally similar to the macro-Penfield method described above but reduced in scale to handle samples of 20–30 mg. Like all micromethods, it suffers most from sampling problems and is applicable only to very pure minerals in special circumstances.

The only other primary method for total hydrogen that seems worth mentioning is that of Friedman and Smith [96], in which the sample is fused in an evacuated induction furnace. Evolved gases are passed over hot uranium metal, and the volume of hydrogen is eventually measured. This method, and others like it, are not easily applicable to materials that lose water at room temperature in a vacuum. An important consequence of the development of this method has been reported by Peck [48]; several amphiboles analyzed by the Penfield method yielded almost identical H_2O values using this totally different procedure. It had previously been alleged that Penfield H_2O values for amphiboles were invariably low because the "theoretical" formula of $(Ca, Mg, Si, etc.)(OH, F)_2$ was seldom achieved by analysis. The method of Friedman and Smith strongly supported the "discrepancy" and demonstrated the accuracy of the Penfield method.

4.22.2. Rapid Methods

Loss on Ignition. With many materials, a loss on ignition at a carefully selected temperature does indeed yield a fair estimate of water content. Used intelligently, a loss on ignition is a useful measurement rapidly acquired. Used blindly, it is an invitation to confusion. It may be useful to examine what happens to a variety of materials when they are heated.

Limestones and dolomites: These almost always contain organic matter, which will be oxidized and expelled at temperatures above 450 °C. Above 750 °C, most of the carbonate carbon will also be expelled. In limestones and dolomites of high quality, sulfur will be quantitatively retained on ignition at 750 °C. In low-grade carbonate rocks, some of the sulfur will volatilize. If high- and low-grade limestones are heated in the same furnace, those of low grade may evolve sulfur and those of high grade may absorb it, and results may be confusing.

Manganese nodules: These are formed in the ocean and always contain a "seed," which may be a shark's tooth, a pebble, a mineral crystal, almost anything. If the collector crushes a whole nodule and submits it for analysis, its compositon, which is basically $MnO \cdot OH$ plus H_2O plus, for example, adsorbed nickel or copper is complicated by the seed around which it formed. Obviously, manganese nodules as collected contain chloride from the seawater. If the crushed sample is ignited or even heated to 110 °C, most of the chloride will be lost via the reaction

$$MnO \cdot OH + Cl^- \rightarrow MnO + \tfrac{1}{2}Cl_2 \uparrow + OH^-$$

The water content of manganese nodules sometimes reaches 20%, even in air-dried material. Determinations of cobalt or nickel content that do not take this water content into account are likely to be qualitative only. A useful determination of heavy metal content must include a definitive statement of the sampling weight baseline.

Shales and other organic-bearing rocks: These contain organic material and also reduced iron compounds. Ignition leads to an increase in weight due to oxidation of iron and a decrease in weight due to burning of organic matter. The loss on heating to any temperature is not rationally interpretable without a complete knowledge of composition.

Limonitic ores often yield a useful loss on ignition that corresponds closely with their water content.

Granitic rocks containing little iron may yield a loss on ignition closely related to their water content, but mafic rocks usually do not.

Fluorine-bearing minerals (phlogopite micas, apatites, etc.) that also contain hydroxyl hydrogen are likely to lose HF or H_2SiF_6 on heating. It is practically impossible to drive off all hydrogen from phosphate rocks by simple heating.

PROCEDURE FOR CARBONATE ROCKS. Fill porcelain crucibles with powdered limestone or crushed calcite and heat to 750 °C in a muffle furnace. Cool, discard the contents, wash the crucibles in water, and dry them at 110 °C. (The purpose of this is to remove the glaze on the inner surfaces.)

Tare the crucibles in pairs, providing each with a closely fitting cover. It is convenient, but not necessary, that the working crucible of each pair be the slightly heavier of the two. Place the working crucible and cover on the left balance pan and its tare on the right. Add weights to the right pan to balance. Record the difference in weight.

Add 1–2 g air-dried sample to the working crucible, and add weights to the right pan to balance. Record the sample weight. Put both crucibles in the oven at 105 °C for 1 h. Transfer to a desiccator, cover the crucibles, and pump the air from the desiccator. Cool nearly to room temperature, and reweigh as expeditiously as possible.

Transfer crucible and tare to a muffle furnace set at 450 °C and roast at this temperature overnight. Transfer both crucibles to a desiccator, evacuate, and reweigh. Return the crucibles to the furnace, and increase temperature to 750–800 °C; hold at temperature for 2 h. Cool and reweigh as before. Report loss in weight at 105, 450, and 750 °C.

This procedure is usually applied when large numbers of samples have to be analyzed. Use of platinum crucibles would avoid the need for having a tare for each crucible; with porcelain crucibles, however, the effects of changing air density and the difficulty in cooling to exactly room temperature make taring mandatory. The crucibles must be covered to prevent absorption of water and carbon dioxide from the air. One must be careful not to heat low- and high-grade limestones in the same muffle furnace: Gases (especially SO_2) evolved from low-grade limestone may be absorbed by the high-grade limestone. If the muffle furnace has been used previously for roasting sulfide minerals at 450 °C, it should be set at 1100 °C overnight and then cooled before determining ignition loss on limestones; otherwise, sulfur compounds present in the furnace may be absorbed by the heated limestone. It is also necessary to be sure the laboratory atmosphere (which circulates through the muffle furnace) is substantially free from acid gases.

The purpose of the preliminary roast at 450 °C is chiefly to remove organic matter and oxidize sulfides and ferrous minerals; free circulation of air through the furnace must be provided. Most limestones will quantitatively retain all their sulfur under the conditions described; it may be convenient to make the sulfur determination on the ignited product. Retention of sulfur depends on there being a sufficient content of calcium or magnesium oxides in the product of ignition.

PROCEDURE FOR SULFIDE ORES AND MINERALS. Proceed as for carbonate rocks, except that if it is required to volatilize as much of the sulfur as possible, the first hour of heating at 450 °C should be done with the crucibles covered. If the samples contain molybdenite, MoS_2, ignition temperature should not exceed 650 °C, especially if the residue is to be used in a molybdenum determination.

PROCEDURE FOR PHOSPHATE ROCK AND MINERALS. It is advantageous to mix the samples with an equal weight of powdered quartz and to heat finally to 1100 °C.

Complete volatilization of hydrogen and fluorine from phosphates, especially apatite, is very difficult.

Rapid Modification of the Penfield Method. Shapiro and Brannock [97] describe a routine method for water determination in rocks, which is adequate for some purposes.

PROCEDURE. Mix 1 g rock powder and 2 g fused and crushed sodium tungstate in the bottom of an 18 × 250-mm Pyrex test tube. Mount the test tube horizontally in an ice bath to which some sodium chloride has been added, and stopper the tube with a loose-fitting cork. Weigh a 2 × 2-in. piece of filter paper in a weighing bottle; transfer the paper quickly to the test tube, replacing the cork at once to prevent condensation of water from the air. Heat the portion of the test tube containing flux and sample with a Meker burner for 5 min. Cool for 1 min, quickly remove the filter paper, and return it to the weighing bottle. Weigh the liberated water.

Methods Using Karl Fischer Reagent. Water may be titrated in anhydrous organic liquids using Karl Fischer reagent. Prior separation of the water is usually necessary, and then it may just as easily be weighed, except when only a little is present. When traces of moisture are of consequence, it may sometimes be measured quickly and accurately by simply adding a weighed portion of the sample to an appropriate organic liquid (e.g., methanol) and titrating the suspension with the reagent.

The Karl Fischer reagent consists of a mixture of sulfur dioxide, pyridine, and iodine in methanol. The reaction is complex but may be represented by the equation

$$H_2O + SO_2 + I_2 \rightarrow SO_3 + 2HI$$

The end point of the titration may be detected by observing the color change due to free iodine or electrometrically.

The Karl Fischer reagent is unstable and must be protected from the atmosphere. A two-reagent procedure may be preferable. In this procedure, the sample is suspended or dissolved in a methanol solution of SO_2 and pyridine, and the water is titrated with a methanol solution of iodine.

A closed-tube procedure may be used to separate water from a mineral sample prior to titration with Karl Fischer reagent:

PROCEDURE. Prepare a silica-glass tube about 8 mm in diameter and 20–25 cm long with one end closed. Introduce the sample, which may be no more than a few milligrams in weight, and add a suitable flux (sodium tungstate or lead oxide). In an oxy-hydrogen flame, pull a constriction in the tube 5–10 cm from the open end. Using a vacuum pump, remove air from the tube and seal it at the constriction using a small O_2–H_2 flame. Mount the tube horizontally with the constricted end in an ice bath, and heat the end containing the sample and flux in a Meker flame. Cool. Prepare a methanol solution, and titrate it with Karl Fischer reagent to remove water using an appropriate titration vessel. Cut off the end of the silica tube containing the water, and drop it into the prepared dry methanol solution. Titrate the water.

Calibrate the titrant by adding measured volumes of water to the titration vessel from a micrometer burette.

Generally, the weight of water that can conveniently be titrated is less than 0.1 g, and accurate results can be obtained with a few milligrams or even a few tenths of a milligram of water. The controlling factor is the success with which extraneous water (e.g., from the air) can be excluded from the titration apparatus.

A satisfactory titration apparatus can be improvised at low cost using an automatic burette to deliver the reagent, a beaker with a home-made plastic cover as a titration cell, and a pH meter as a readout device. If many titrations are required, it is probably a long-run saving to acquire a commercially available apparatus despite initial high cost.

4.22.3. Other Methods

Spectrographic measurement of water in minerals is possible, using the 3063.6-Å hydroxyl band head (Quesada and Dennen, 1968). Infrared spectra of minerals can be interpreted to obtain a rough estimate of water content. Necessary fine grinding or other pretreatment of the sample makes this of marginal value, except in special cases. In some cases, especially when the supply of sample is limited, it is possible to carry out the sodium carbonate fusion of the mineral inside an apparatus that will retain all the water evolved, which can then be measured; the operation is best carried out in a platinum boat inside a combustion tube, and the evolved water is collected in a magnesium perchlorate bulb and weighed. A difficulty in this procedure is the near impossibility of preventing spatter during the fusion.

4.23. NOBLE METALS

The importance of noble metals is rising very quickly in everyday life (e.g., silver in the photographic industry, platinum and palladium as catalysts, osmium in chemical processing, rhodium and ruthenium in petroleum processing, and gold for monetary status). This is without mentioning their historical value in jewelry and their growing importance in electronic applications. The economy of noble metals has been described in detail by Zysk [98].

Gold and platinum group metals are found in ultrabasic rocks such as peridotite, pyroxenite, harzburgite, and dunite. Gold, platinum, and palladium are found in alluvial products from weathering of these rocks such as laterite, serpentine, transported chromite, and so on. Platinum and palladium have been found in association with bismuth and tellurium and ruthenium and iridium with arsenic and sulfur. Silver is commonly found in association with copper and lead and sometimes with cobalt. Gold is found everywhere in the environment at a very low level (2 ppb). There are three important groups of gold deposits: alluvial deposits, where gold is liberated as native metal; vein deposits in association with quartz; or in association with sulfide minerals such as pyrite.

Before even thinking about analysis, precious metals present a formidable sampling challenge to the laboratory. From a practical standpoint, the sampling of precious metals is dominated by the fact that, often, they occur as pure metal and also the fact that their specific gravity is extremely high. Obtaining a representative sample is difficult for two reasons:

1. The constitution heterogeneity is very high since the content of the element of interest is very low (e.g., a 2-ppm gold deposit may be a good deposit).
2. The distribution heterogeneity is very high because of an odd specific gravity (e.g., segregation phenomena are likely to occur at each manipulation of the material).

Besides the annoying fact that there are still many philosophers that consider the theory of sampling impractical and preach that the only realistic way to perform good sampling is to follow empirical guidelines, which are the product of good experience and common sense, nobody with respectable intention will deny the outstanding statistical tool offered by Gy, Ingamells, Prigogine, and many others [99–101].

The objective of this discussion is not to be repetitious with what has been discussed in Chapter 1; however, it is critical for the reader to perceive the importance of sampling when dealing with noble metals.

Gy [102] has dealt very thoroughly with the problem of sampling noble metals and provides formulae and nomographs for estimating the optimum size of samples. Some practical examples are given in Chapter 6.

Because of their position in the periodic table, noble metals have a tendency to form covalent bonds. The analytical chemistry of these metals has been described by Beamish [103]. Noble metals are difficult to oxidize. They are oxidized by aqua regia to different extent. For example, gold, palladium, and platinum are easily attacked by aqua regia while rhodium is attacked only if finely divided as a powder and iridium is not attacked if pure. This has been used to separate noble metals from one another. After attack with aqua regia, gold, palladium, and platinum form soluble complexes with Cl^-; however, rhodium, ruthenium, silver, and iridium are insoluble. These insoluble products can be fused with litharge, and then lead and silver are dissolved by nitric acid while rhodium, ruthenium, and iridium are insoluble.

Gold forms stable complexes with Cl^-, Br^-, NH_3, CN^-, and SCN^-. Palladium and platinum also form stable complexes with CN^-. Some of the noble metals, particularly palladium, form complexes with dimethylglyoxime. The geochemistry of gold is complex and has been extensively investigated by Lingong and Swanson [104] and Boyle [105].

Many attempts have been made to develop analytical procedures for noble metals that will use acid decomposition of small samples, all of them failed for several reasons: The small sample cannot be representative; the precious metal is not properly liberated to be effectively digested by acids; when liberated, most of it has been lost during mechanical stress generated by the sampling procedure; and so on. For determination of noble metals, wisdom dictates that large samples have to be used if rapid success is expected.

So far, fire assay has been the most classic and successful approach for determining noble metals [106].

4.23.1. Determination of Noble Metals by Fire Assay

The normal fire assay separation of the noble metals (Ag, Au, Pt, Pd, Rh, Ir, Ru, and Os) proceeds as follows: The ore or product is mixed with an appropriate combination of reagents including lead oxide and fused. If the charge is properly compounded, about 20 g metallic lead are produced; this dissolves the precious metals and settles to the bottom of the fusion crucible. After cooling, the resulting lead "button" is removed, cleaned, and heated on a porous "cupel" made of bone ash and/or cement. The lead oxidizes to fusible PbO and is absorbed by the cupel, leaving a bead containing silver, gold, platinum, palladium, rhodium, and iridium. Osmium and ruthenium are volatilized; if these are sought, other methods for removing the lead are used.

While it is easy to come up with a general description of the fire assay separation, the details of the process are fraught with numerous difficulties. The flux composition must be matched to the ore or product being treated. The requirements are (1) reducing constituents must be just sufficient to release the proper amount of metallic lead; (2) flux viscosity must be sufficient to give the molten lead time to gather the precious metals before sinking to the bottom of the crucible; (3) sulfides must be completely oxidized to SO_2, which escapes; and (4) reaction must proceed slowly enough to permit complete attack of the sample before settling of the lead.

Cupellation cannot be employed when osmium, ruthenium, rhodium, or iridium are of interest; osmium and ruthenium volatilize and rhodium and iridium form oxides, insoluble in lead, that are partially lost to the cupel. In cupelling buttons containing platinum and palladium, silver must be added so that the silver bead contains no more than 3–4% platinum and palladium. Copper and nickel must be removed from the ore by acid leach.

For the assay of low-grade ores, Schoeller and Powel [107] take 10 assay tons, roast in a closed plumbago crucible, and leach with hydrochloric acid containing ammonium chloride. The residue is washed and dried and then subjected to crucible fusion. The purpose of the roasting and acid leach is to remove copper, nickel, and other base metals, which interfere in the fire assay. If silver is of interest, cupellation must be carefully controlled because of its appreciable volatility. Known weights of silver may be added and recovery monitored. There should always be at least three times as much silver as gold in the lead button, and a much higher proportion if, for example, platinum is present.

With ore containing sulfides, the problems are many. If fusion conditions are not exactly right, nickel–copper mattes may form, and these compete for the precious metals, leaving only a portion in the lead button. If oxidants are used to remove sulfide, osmium and ruthenium may be volatilized during fusion. Copper may finish in the lead button, and this leads to poor cupellation and losses of platinum and silver to the cupel.

Lead buttons containing osmium, ruthenium, rhodium, and iridium are dissolved in nitric acid. The solution is filtered and treated with a reducing agent. Platinum metals, gold, and silver are collected on a filter and analyzed by one or another of several (rather lengthy and difficult) classical procedures or by instrumental methods, for example, neutron activation analysis. AA or ICP spectrometry is most often used; difficulties with these techniques have not been fully resolved largely because complete ionic solution of the platinum metals, especially iridium, is not easy to achieve even given pure salts of the metals.

Once a detailed procedure for a specific ore or product has been successfully worked out, routine analyses of high accuracy are possible; however, such a procedure needs development for each ore type if accuracy is to be ensured. (See the section on the rapid method of determination of gold, palladium, and platinum in Chapter 5.)

4.23.2. Determination of Very Low Gold Contents in Sand or Stream Sediments

During geochemical exploration, extremely low gold contents are looked at, ranging from 1 ppb to only a few parts per million. Most of the investigated materials are sands or stream sediments. Several difficulties may arise:

1. Ordinary acid digestions are using a few grams subsamples, which are too small by three orders of magnitude or more.
2. Fire assay techniques are using 30-g subsamples, which are still too small by two orders of magnitude or more.
3. Fire assay shows gold contamination at the 1–10-ppb level, confusing the geochemist. This contamination cannot be controlled and is not reproducible; it comes from clay crucibles, PbO, silver inquart, dust from furnace, and so on.
4. The disseminated background gold should not be dissolved in order to better detect low trends.

At that stage, it becomes evident that an entirely new approach is needed. Quick calculations using Gy's principles (see Chapter 6) demonstrates that the use of a 5000-g sample of a −80-mesh material is necessary (e.g., this is a minimum weight). One way to dissolve large particles of free gold is to leach the sample with aqua regia or cyanide.

Dissolution of Gold by Aqua Regia

PROCEDURE. Roast a 5000-g sample for 48 h at a temperature between 450 and 500 °C. Cool, and transfer into a 10-l Pyrex jar. Add slowly 2500 ml aqua regia (HNO_3–HCl, 1:2). Allow to stand in dark hood 48 h or more. Draw off liquid as shown in Figure 4.5. Wash with distilled water. Evaporate if necessary. Bring to volume in a volumetric flask. Determine gold by atomic absorption using electrothermal atomization and standard addition techniques.

Dissolution of Gold by Cyanide

Cyanide leach of sand or stream sediments seems to be the most suitable approach since only the free gold or relatively coarse gold in the sample is dissolved [108]. The background gold finely disseminated in the silicate

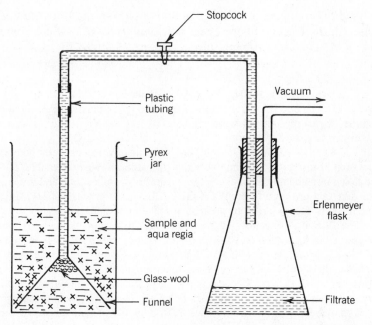

Figure 4.5. Illustration of apparatus used for the dissolution of gold by aqua regia in sand and stream sediments.

matrix is not dissolved by this technique or may be only a small part of it. The dissolution of gold may follow one of the three processes described below:

1. $4Au + 8CN^- + O_2 + 2H_2O \rightleftarrows 4Au(CN)_2^- + 4OH^-$

2. $2Au + 4CN^- + 2H_2O \qquad \rightleftarrows 2Au(CN)_2^- + 2OH^- + H_2$

3. $2Au + 4CN^- + H_2O + O \rightleftarrows 2Au(CN)_2^- + 2OH^-$

Reactions 1 and 3 have the same end products. Reaction 3 uses oxygen released by hydrogen peroxide. Reaction 2 liberates hydrogen and does not use oxygen. To dissolve a 150-μm gold particle may take up to 48 h. After digestion, gold is extracted using methyl tri-*n*-alkyl(C_8–C_{10}) ammonium chloride dissolved in diisobutylketone.

Following Groenewald [109], for a quaternary ammonium salt, the extraction reaction may be written as

$$(R_4NCN)_{org} + [Au(CN)_2^-]_{aq} \rightleftarrows [R_4NAu(CN)_2]_{org} + (CN^-)_{aq}$$

The equilibrium constant of this reaction is so high that most of the gold is found in the organic phase even with small ratio of organic volume to aqueous volume. The organic phase may be analyzed for gold and also silver either by AA or by ICP.

PROCEDURE. Roast a 5000-g −48-mesh sample for 48 h at a temperature between 450 and 500 °C. Cool, and transfer into a 10-l plastic jar. Add about 30 g lime. Measure 2920 ml water and add to the sample. Add 20 ml of 30% hydrogen peroxide. Stir until well mixed. Let stand for 12 h.

Add 60 ml of a 10% NaCN solution. Stir, and cover to prevent evaporation. Let stand for 4 days, stirring every 8 h. If evaporation is noticeable, it should be taken into account in final calculations. Stir one more time and immediately transfer to a pressure filter. Wash the jar. Discard the washes. Use this jar to catch the filtrate in order to prevent possible contamination.

The filtrate is checked for free cyanide. On a spot plate, add 1 drop of copper sulfate solution and 1 drop of ammonium sulfite solution. Add 2 drops of filtrate. If the copper sulfate and ammonium sulfite solution clears, there is at least 200 µg/ml free cyanide. The ammonium sulfite solution should be fresh (less than 1 week old). If there is not sufficient free cyanide left, return the solids to the pregnant solution, add 60 ml of 10% NaCN, and leach again for 4 days as before. If there is sufficient free cyanide left, proceed to the next steps.

Transfer 1000 ml filtrate to a separatory funnel and add 10 ml of a 1% methyl tri-n-alkyl(C_8–C_{10}) ammonium chloride dissolved in diisobutylketone. Shake for 1 min. Let phases separate for 20 min. Transfer the organic phase to a test tube and close it. Centrifuge the organic phase for 5 min. Determine gold and silver directly by AA or ICP by comparing sample to appropriate standards and blank.

4.24. URANIUM, THORIUM, AND RADIUM

It is not necessary to emphasize the importance of uranium and thorium since anything will be an understatement. These two elements are widely distributed in the environment. The presence of these naturally occurring radioactive elements may also generate some concerns, especially in mill products, residues, tailings, waste materials, effluents, and waters. Some of the decaying products such as ^{226}Ra, ^{224}Ra, ^{228}Ra, and ^{223}Ra are a matter of considerable public health interest. The occupational maximum permissible concentration for ^{226}Ra in drinking water is 10^{-7} µCi/ml, which is lower than that of any other naturally occurring radioisotope. For application to the general public, levels as low as 1% of this are of concern. In many areas, granites have relatively high uranium content. In nature, uranium exists as U(III), U(IV), and U(VI). The main uranium minerals are uranium oxides such as pitchblend [oxide of U(IV) and U(VI)], uranium silicates such as coffinite, uranium vanadates such as carnotite, uranium phosphates, uranium carbonates, and uranium arsenates. Uranium may be also found as sulfate, molybdate, tellurite, selenite, and niobiate [110]. The main thorium minerals are thorium oxides, such as thorianite, thorium silicates such as thorite, and thorium phosphates such as monazite.

4.24.1. Naturally Occurring Radioactive Isotopes

Uranium has three natural isotopes: ^{238}U (99.27%), ^{235}U (0.72%), and ^{234}U (0.0056%). Uranium-238 generates the main series of concern (^{230}Th, ^{226}Ra, ^{210}Pb, and ^{210}Po being the main isotopes causing health hazard). Uranium-235 generates the third series of concern (^{223}Ra being the main isotope causing health hazard). Thorium-232 is the main natural isotope of thorium and generates the second series of concern (^{232}Th, ^{228}Ra, and ^{224}Ra being the main isotopes causing health hazard).

4.24.2. Decomposition of Uranium Ores

A large quantity of silicate minerals may be decomposed by sintering, at low temperature, with sodium peroxide in platinum crucibles. In any case, the sample should be − 150 mesh. Resistant minerals such as chromite and cassiterite should be ground to − 325 mesh. The sample must be free of sulfides, and the temperature should not exceed 480 °C. Preliminary roasting of the sample is a good precaution.

PROCEDURE. Put the covered crucible in a cold muffle furnace, and bring the temperature slowly to 450–500 °C; then remove the lid and heat for 30 min. Overheating of the crucible during roasting must be avoided to prevent attack of the platinum by, for example, iron-bearing compounds and sulfides or arsenides. Cool the crucible containing the roasted sample.

Place the crucible on a clean porcelain plate. If a 1-g sample was used, weigh 4 g Na_2O_2. Mix thoroughly with a small clean platinum rod. Weigh 1 g Na_2O_2 onto a piece of hard-surfaced paper (Albanene). Rinse off the stirring rod in this peroxide. Transfer the Na_2O_2 to the crucible, covering the peroxide–sample mixture as completely as possible.

Cover, and transfer the crucible and contents to a furnace regulated at 200 °C for about 15 min. Then raise the temperature to 480 ± 5 °C for about 1 h. Cool the crucible for about 2 min. Put the crucible sideways in a 400-ml Pyrex beaker and place a watch glass on the beaker. Add, at once, 150 ml distilled water preheated at 60 °C and free of carbonates and CO_2. When the reaction is complete, wash the crucible twice in 25 ml water. Filter (tungsten, molybdenum, and vanadium remain in the filtrate). Rinse with diluted sodium hydroxide solution free of carbonates. Dissolve the precipitate with nitric acid, and make the final solution 1 N nitric.

If large quantities of iron are present, it may be necessary to complex uranium as a carbonate and separate it from these elements as discussed below. If vanadium is present, it is recommended to determine the uranium in the filtrate as well, since large quantity of vanadium may prevent complete precipitation of uranium. When uranium and vanadium are present together, their behavior becomes capricious.

4.24.3. Separation of Uranium

U(VI) is a member of the ammonia group; however, it presents some differences from other members. One of the main differences is the fact that some of the U(VI) complexes are very soluble. U(VI) gives stable carbonate complexes around pH 9 and provides an effective way to separate it from some other members of the ammonia group. Sandell [111] described how to separate uranium traces from elements such as Fe^{3+}, Ca^{2+}, Mg^{2+}, Mn^{2+}, Zr^{4+}, Co^{2+}, Ni^{2+}, as La^{3+} with a mixture of sodium and potassium carbonates. This has been successfully applied to the separation of a very small quantity of uranium in silicate rocks preparatory to a fluorimetric determination. Redissolution of the precipitate in hydrochloric acid followed by a second precipitation will lead to better uranium recovery [112].

Uranyl nitrate is effectively extracted by ethyl acetate in the presence of large quantities of aluminum nitrate. For many materials, this provides a quick separation of uranium from most interfering elements prior to fluorimetric determination.

The presence of tungsten in some ores may generate problems to the analyst. Returning to the carbonate dissolution procedure, once uranium is in solution and separated from the precipitate, the formation of uranyl nitrate must be established. Precipitation of tungsten during this conversion must be prevented. Tartaric acid is used to hold the tungsten in solution as a soluble complex, but it maintains a reducing medium. The addition of nitric acid in turn acts as an oxidant, freeing the uranium to form uranyl nitrate. However, not all uranium is released. The addition of hydrogen peroxide ensures the full recovery of uranium as uranyl nitrate. At the same time, tartaric acid is degraded to tartronic acid but still retains adequate complexing characteristics to prevent the precipitation of tungsten.

4.24.4. Extraction of Uranium

Many difficulties encountered in the determination of uranium have been overcome by the effective extraction of uranyl nitrate into ethyl acetate. Ions such as F^-, PO_4^{3-}, SO_4^{2-}, and Cl^- should be minimized to prevent difficulties.

PROCEDURE. Take an aliquot of solution and evaporate it to dryness. Dissolve in 10 ml aluminum nitrate solution.

Note: Dissolve 485 g aluminum nitrate, $Al(NO_3)_3,9H_2O$, in warm water. Heat if necessary. Bring to a final volume of 500 ml. If trace levels of uranium are being determined, it is necessary to purify this solution by shaking it with 200 ml ether.

Add 10 ml ethyl acetate, and shake vigorously for 3 min. Allow layers to separate for 10 min.

4.24.5. Solution of Uranium in Sodium Fluoride

The solution of uranium in sodium fluoride shows a green fluorescence that is easy to detect with the naked eye. The method is extremely sensitive (0.01 μg uranium in the fluoride pellet is the most appropriate working zone). Sodium fluoride should be mixed with sodium and potassium carbonate for best results.

PROCEDURE. Aliquot 1 ml of the ethyl acetate layer onto a platinum plate.

Note: Care should be taken to exclude the aqueous phase because the presence of the aluminum nitrate will give low results.

Dry under an infrared lamp. Flash platinum plate over a Bunsen burner to make sure organics are absent. Weigh 2.5 g flux into a platinum plate.

Note: Flux should be 40% potassium carbonate (anhydrous), 40% sodium carbonate (anhydrous), and 20% sodium fluoride. This flux should be thoroughly homogenized.

Fuse on a Claisse fluxer for 10 min; consistent timing is critical. Flux should melt within 90 s. Let cool while on fluxer for 3 min.

Note: After fusion, the pellet should be completely white with no unfused lumps. A brownish tinge indicates too high of a fusion temperature or too slow of a cooling rate or both [113]. Brownish spots may indicate the presence of organics. Lumps indicate that the fusion temperature is too low.

Invert the platinum plate on a desiccator tray, and separate plate and pellet. Place the pellet in a desiccator for 1 h before measurement is performed on the fluorimeter.

Note: Any substance containing over 10 ppm uranium should not be introduced in the fluorimeter to prevent contamination.

Submit blanks, blanks with standard additions, samples, and samples with two consecutive standard additions to the same procedure. Read the standard addition technique discussed in Section 2.3.7.

4.24.6. Decomposition of Thorium Ores

The decomposition of thorium ores is not as easy as primarily thought. Many authors agree that silica may generate some problems and retain some thorium if not handled properly [114–116]. The volatilization of most of the silica as SiF_4 is a necessary precaution.

PROCEDURE. Weigh 1 g of a -150-mesh sample in a platinum crucible. Roast sample if necessary at 500 °C for several hours to oxidize iron and remove sulfides if present. Cool, add a little water, and swirl until the sample is thoroughly wet. Using a pipette, add 10 ml of 1:1 nitric acid. Heat on the steam bath until any effervescence ceases. Use of a cover is necessary to prevent small losses. Introduce a platinum rod, and add 20 ml hydrofluoric acid. While heating on the steam bath, stir constantly until there is no danger that the sample will gelatinize. Heat, uncovered, as strongly as possible without any effervescence until most of the liquid has evaporated. Cool, wash down the sides of the dish with a little water, add 5 ml of 1:1 sulfuric acid and 20 ml hydrofluoric acid, mix thoroughly, cover with a Teflon cover, and heat on a steam bath overnight. If attack is incomplete, repeat the treatment with hydrofluoric acid. Evaporate slowly to fumes of sulfuric acid. Complete removal of fluoride may be a problem if much aluminum or magnesium is present. When removal of fluoride is judged complete, wash down the crucible with water, cover, and digest until all soluble sulfates are dissolved. Filter through a tight paper into an appropriate volumetric flask, and wash thoroughly with hot water. Scrub the crucible, and transfer its content completely to the filter. Dry the insolubles in the filter after placing it in the crucible. Burn the filter paper. Add 1.0 g sodium carbonate and fuse. Cool and dissolve the cake in minimum water, and add slowly 5 ml of 6 N sulfuric acid and then 3 ml hydrofluoric acid. Evaporate to dryness. Heat crucible until all water is gone; then heat to dull red for a few minutes. Cool, and dissolve the cake with nitric acid and combine this solution with the one already in the volumetric flask. Final solution should be around 3 N in nitric acid.

4.24.7. Separation of Thorium

Anion exchanger resin (e.g., Bio Rad 1-×8) permits the selective separation of thorium from a large number of other elements [117]. The reagent-grade resin is in the chloride form, which is converted to the nitrate form by washing with diluted nitric acid. The sample is run through the resin after the resin is free of chloride ions. While the thorium complex is strongly absorbed, other elements are eluted [118]. Once the thorium complex is absorbed, it is easily eluted with diluted hydrochloric acid.

PROCEDURE. Use an ion exchange column of about 10 mm diameter and 150 mm long as shown in Figure 4.6. Fill column two-thirds full with distilled water. Place a glass wool plug at bottom of column using a glass rod. Air should be absent, plug should be packed down completely, and any stray fibers will create a channel. Weigh out 8 g resin (BioRad Ag 1-x8 resin, 100–200 mesh, chloride form). In a 50-ml beaker, mix the resin with distilled water at a 1:2 ratio and introduce in the column. After this point, it is important not to let the resin become dry or receive any air at any time. Let the resin float slowly

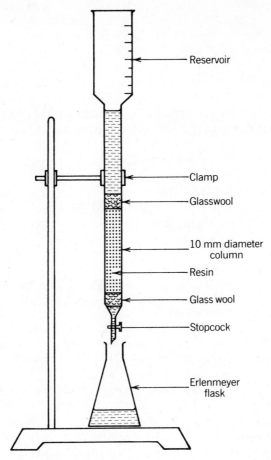

Figure 4.6. Illustration of the apparatus for the ion-exchange of thorium.

to the bottom to allow air bubbles to escape, if any. Open the stopcock to allow the slow release of the water (e.g., not faster than 1 drop/s) while adding the remainder of the resin–water mixture. A slight vacuum is necessary as the column fills with resin. After resin has settled, add another glass wool plug on top using a glass rod. The level of the solution should never go below the top of this glass wool plug. Charge column by passing 50 ml of 6 N nitric acid through it at a rate not faster than 1 drop/s.

Pass an aliquot of the digested sample at a rate not faster than 1 drop every 2 s (timing is critical). Rinse beaker with a 5 ml 6 N nitric acid. Rinse with 5 ml 6 N nitric acid. Pass 100 ml of 6 N nitric acid through the column. Empty contents of the 500-ml filtering flask, rinse it with distilled water, and return it to the apparatus. Pass 100 ml of 1.5 N hydrochloric acid through the column at a rate of 1 drop every 2 seconds. Pass 100 ml of 3 N hydrochloric acid through the column at the same rate. Transfer solution to a beaker. Add 5 ml concentrated nitric acid and 5 ml concentrated perchloric acid. Bring down to perchloric fume (this step gets rid of any organics coming from the resin). Cool, and transfer to an appropriate volume using 0.33 N hydrochloric acid.

Thorium could also be separated with rare earths using the oxalate precipitation as indicated in Section 4.5. However, platinum, titanium, and calcium may generate problems. Furthermore, the complete recovery of thorium is uncertain especially when present as a minor element [115].

4.24.8. Thorium Determination by Thorin Method

Thorin gives with Th(IV) a stable coloration around pH 1. Most interfering metals have been eluted by 6 N nitric acid. However, Ce(IV), Mo(VI), Sb(III), and W(VI) are still likely interferences. Tartaric acid can overcome problems with Mo(VI) and W(VI). Ascorbic acid can overcome problems with Ce(IV).

PROCEDURE. Transfer an aliquot of the thorium solution to a 10-ml volumetric flask. Add 1 ml of 0.1% thorin indicator (precise addition is critical). Bring to volume with 0.33 N hydrochloric acid, and read on spectrophotometer as soon as possible using 2-cm cells. Blank, standards, and spiked samples are treated in the same way. For best result, measure the absorbance between 475 and 650 nm using a spectrophotometer equipped with a scanner.

4.24.9. Decomposition of Ores for Radium Analysis

The presence of radium in this chapter may seem inappropriate since it is an alkaline earth. However, being strongly related to uranium and thorium in the environment makes it almost a necessity not to separate it from these elements in any discussion.

Radium-226 is the alpha emitter most commonly determined. Generally, it is too low in concentration to determine it directly by alpha spectrometry. Therefore, chemical separation is necessary to concentrate the isotope and eliminate matrix absorption problems. Most methods for ^{226}Ra employ barium sulfate as a collector during precipitation. This is precisely where difficulties may be encountered with decompositions of ores when sulfur and barium are present. Insoluble barium sulfate forms during acid digestions and recovery of radium may become very poor. Therefore, only a fusion procedure should be used for all solid samples to ensure that all radium and barium are in solution prior to selective precipitation.

Fuse 1 g − 150-mesh sample using 14 g sodium carbonate. To conduct the fusion properly, see the details in Section 2.4.12.

4.24.10. Separation of Radium

There are many ways to separate radium from other elements (e.g., ion exchangers, chromatography, solvent extractions, etc.); however, the separation that works best is the radium coprecipitation with barium sulfate. Goldin has written a remarkable document on the subject [119].

PROCEDURE. Dissolve the melt of the above fusion in warm water. Add 20 ml of 1 M citric acid. Add concentrated ammonium hydroxide until pH 8–9. Slowly add 10 ml lead nitrate solution (0.04 M), 3 ml barium nitrate solution (0.1 mg Ba/ml), and 5 ml ammonium sulfate solution (100 g/1). Adjust pH to 0.9–1.0 with 8 M sulfuric acid and continue stirring for 1 h. Filtter through HABP millipore filter and discard filtrate. Transfer filter to original beaker and rinse with distilled water. Add concentrated ammonium hydroxide to pH 8–9. Add 7 ml of 0.25 M EDTA and 5 ml ammonium sulfate solution. Heat to near boiling. Rinse and discard filter. Add glacial acetic acid to pH 4.8 and continue stirring for 1 h. Filter the barium–radium sulfate precipitate on a HABP millipore filter or equivalent. Mount filter on a metal planchet and count immediately using an alpha counting system. Radium isotopes are identified by the rate of growth of the daughter isotopes.

REFERENCES

1. S. Glasstone and R. H. Lovberg, *Controlled Thermonuclear Reactions,* Van Nostrand, Princeton, 1960, pp. 13–14.

2. K. Govindaraju, *Anal. Chem.* **40,** 24 (1968).

3. K. Govindaraju, *Appl. Spectry.* **20,** 302 (1966).

4. K. R. Krishnamoorthy and R. K. Iyer, "Homogeneous Precipitation of Beryllium by means of Trichloroacetic Acid Hydrolysis and Determination as Phosphate," *Analytica Chimica Acta,* **47,** 333–338 (1969).

5. E. B. Sandell, *Colorimetric Determination of Traces of Metals,* 3rd ed., Interscience, New York, 1959, pp. 319–324.

6. P. G. Jeffery, *Chemical Methods of Rock Analysis,* 2nd ed., Pergamon, Elmsford, N.Y., 1975.

7. King and Pruden, *Analyst* **92,** 83–90 (1967).

8. King, Pruden, and James, *Analyst* **92,** 695–697 (1967).

9. Shapiro and Brannock, Geol. Surv. Bull., 1144-A, 1962.

10. Sharma, Mukhertee, and Roy, *Technology* **2**(3), 140–142 (1965).

11. Rowland and Beck, *Am. Mineral.* **37,** 76 (1952).

12. R. A. Durst, Ion Selective Electrodes, National Bureau of Standards Special Publication 314, 1969, pp. 148–177, 208–210.

13. I. M. Kolthoff, E. B. Sandell, E. J. Meehan, and S. Bruckenstein, *Quantitative Chemical Analysis,* Macmillan, 4th ed., New York, 1969, pp. 653–662.

14. J. N. Latimer, W. E. Bush, L. J. Higgins, and R. S. Shay, *Handbook of Analytical Procedures,* United States Atomic Energy Commission, February 16, 1970, RMO-3008, pp. 161–170.

15. J. N. Walsh, F. Buckley, and J. Barker, "The Simultaneous Determination of the Rare-Earth Elements in Rocks Using Inductively Coupled Plasma Source Spectrometry," *Chem. Geology* **33,** 141–153 (1981).

16. R. Pribil and V. Vesely, "Determination of Rare Earths in The Presence of Phosphate," *Chem. Anal.* **56,** 23–24 (1967).

17. J. Kinnunen and B. Wennerstand, "Rapid Determination of The Rare Earths in Phosphate Rock," *Chem. Anal.* **56,** 24–25 (1967).

18. J. Kinnunen and O. Lindsjo, "Determination of Rare Earths in Phosphate Rock by Atomic Absorption Flame Photometry. I," *Chem. Anal.* **56,** 25–27 (1967).

19. J. Kinnunen and O. Lindsjo, "Determination of Rare Earths in Phosphate Rock by Atomic Absorption Flame Photometry. II," *Chem. Anal.* **56,** 76–78 (1967).

20. W. Fischer and R. Bock, *Z. Anorg. Chem.* **249,** 146–197 (1942).

21. E. B. Sandell, *Colorimetric Determination of Traces of Metals,* Interscience, New York, 1959.

22. P. Pascal, *Nouveau Traite de Chimie Minerale,* Vol. IX, Masson & Cie., Paris, 1963, p. 7.

23. A. P. Vinogradov, The Geochemistry of Rare and Dispersed Chemical Elements in Soils, Trans. from Russian published by Consultants Bureau, Inc., New York.

24. J. Dolezal, Decomposition Techniques in Inorganic Analysis, Pavel Povondra and Zdenek Sulcek, 1966 [American Elsevier, 1968].

25. E. W. Harpham, *Metallurgia* **52,** 93 (1955).

26. G. Von Hevesy and R. Hobbie, *Zeit. Anorg. Chem.* **212,** 142 (1933).

27. E. S. Pilkington and W. Wilson, *Anal. Chim. Acta* **33,** 577 (1965).

28. P. Baud, *Traite de Chimie Industrielle,* Vol. II, *Metalloids et Metaux,* Masson & Cie., Paris, 1951.

29. G. Charlot, *Chimie Analytique Quantitative,* Vol. II, *Methodes Selectionnees d'analyse Chimique des Elements,* Masson & Cie., Paris, 1974, p. 570.

30. E. Cerrai and C. Ctesta, *Anal. Chim. Acta* **26,** 204 (1962).

31. M. Rafiq, C. L. Rulfs, and P. J. Elving, *Talanta* **10,** 827 (1963).

32. G. Norwitz, *Anal. Chim. Acta* **35,** 491 (1966).

33. C. Sinha and S. D. Crupta, *Analyst* **92,** 558 (1967).

34. F. Culkin and J. P. Riley, "The Determination of Trace Elements of Zirconium Hafmium, Thorium and Cerium in Silicate Rocks," *Anal. Chim. Acta* **32,** 197–210 (1965).

35. P. M. Champion, P. Crowther, and D. M. Kemp, "Complex Formation of Zirconium and Hafnium with Xylenol Orange," *Anal. Chim. Acta* **36,** 413–421 (1966).

36. G. Charlot, *Chimie Analytique Quantitative,* Vol. II, *Methodes Selectionnees d'Analyse Chimique des Elements,* Masson & Cie., Paris, 1974, p. 428.

37. R. Dams and J. Hoste, "Separation and Determination of Tantalum and Niobium by Precipitation from Homogeneous Solution," *Talanta* **11,** 1599–1604 (1964).

38. Kallmann, Oberthin, and Liu, *Anal. Chem.* **34,** 609 (1962).

39. Powell, Schoeller, and John, *Analyst* **60,** 509 (1935).

40. H. J. Crump-Wiesner and W. C. Purdy, *Talanta* **16,** 124 (1969).

41. E. B. Sandell, *Colorimetric Determination of Traces of Metals,* 3rd ed., Interscience, New York, 1959, pp. 929–930.

42. E. B. Sandell, *Colorimetric Determination of Trace Metals,* 3rd ed., Interscience, New York, 1959, pp. 758–759.

43. J. I. Hoffman and G. E. F. Lundell, Volatilization of Metallic Compounds from Solutions in Perchloric or Sulfuric Acid, U.S. Dept of Commerce, National Bureau of Standards, Research Paper RP. 1198, Vol. 22, pp. 465–470, April 1939.

44. J. J. Lingane and R. Karplus, *Ind. Eng. Chem. Anal. Ed.,* **18,** 191 (1946).

45. S. Abbey, Geological Survey of Canada, Paper 70-23, 1970.

46. G. Seil, *Ind. Eng. Chem. Anal.* **15,** 189–192 (1943).

47. C. C. Miller and R. A. Chalmers, "Microanalysis of Silicate Rocks, Part 1, A Critical Investigation of the Use of the Silver Reductor in the Microvolumetric Determination of Iron, Especially in Silicate Rocks," *Chim. Analyt.* **42,** 501 (1960).

48. L. C. Peck, Systematic Analysis of Silicates—U.S. Geol. Surv., Bulletin 1170, 1964.

49. W. F. Hillebrand, G. E. F. Lundell, H. A. Bright, and J. I. Hoffman, *Applied Inorganic Analysis,* 2nd ed., Wiley, New York, 1953, pp. 108–112.

50. E. L. P. Mercy and M. J. Saunders, "Earth and Planetary Science Letters,

Precision and Accuracy in the Determination of Total Fe and Al in Silicate Rocks," **1,** 169 (1966).

51. J. J. Lingane, "Improved Gravimetric Determination of Cobalt as $K_3Co(NO_2)_6$," *Anal. Chim. Acta* **31,** 315–317 (1964).

52. M. R. Heyes and J. Metcalfe, The Boron-Curcumin Complex in Trace Boron Determinations, PG Report 251 (S), United Kingdom Atomic Energy Authority, 1963.

53. E. T. Wherry and W. H. Chapin, *J. Am. Chem. Soc.* **30,** 1687 (1908).

54. W. F. Hillebrand, G. E. F. Lundell, H. A. Bright, and J. I. Hoffman, *Applied Inorganic Analysis,* 2nd ed., Wiley, New York, 1953, pp. 752–753.

55. Bhargava and Hines, *Talanta* **17,** 61–66 (1970).

56. Harrison and Cobb, *Analyst* **91,** 576–581 (1966).

57. Levinson, *Water Res.* **5,** 41–42 (1971).

58. Thatcher, U.S. Geol. Surv., Water Supply Paper 1454, 1960.

59. Fleet, *Anal. Chem.* **39,** 253–255 (1967).

60. Hatcher and Wilcox, *Anal. Chem.* **22,** 567 (1950).

61. D. M. Shaw, *Physics and Chemistry of the Earth,* Vol. II, Pergamon, London, 1957, pp. 164–211.

62. W. R. Schoeller and A. R. Powell, *Analysis of Minerals and Ore of the Rarer Elements,* Hafner Publishing, New York, 1955, p. 75.

63. G. Charlot, *Chimie Analytique Quantitative,* Vol. II, Masson & Cie., Paris, 1974, pp. 424–425.

64. M. F. Kulikova, "Geochemistry of Gallium and Indium in the Oxidized Zone of Lead-Zinc Deposits in Soviet Central Asia," trans. from *Geokhimiya,* **10,** 1233–1245 (1966).

65. P. Picot and R. Pierrot, "La Roquesite, Premier Mineral d'Indium = Cu InS_2," *Bull. Soc. Franc. Mineral. Crist.* **86,** 7–14 (1963).

66. A. Kato and K. Shinohara, "The Occurrence of Roquesite from the Akenobe Mine, Hyogo Prefecture, Japan," *Mineral J. (Tokyo)* **5,** 276 (1968).

67. R. S. Boorman and D. Abbott, "Indium in Coexisting Minerals from the Mt. Pleasant Tin Deposit," *Can. Mineral.* **9,** 166–179 (1967).

68. J. K. Sutherland and R. S. Boorman, "A New Occurrence of Roquesite at Mt. Pleasant, New Brunswick," *Can. Mineral.* **54,** 1202–1203 (1969).

69. I. C. Smith, B. L. Carson, and F. Hoffmeister, *Trace Elements in the Environment,* Vol. 5, Ann Arbor Science Publishers, Ann Arbor, Mich., 1978.

70. S. Lacroix, "Treatment of the Equilibria Involved in the Extraction of the Hydroxyquinolates of Aluminum, Gallium, and Indium," *Anal. Chim. Acta* **1,** 260 (1947).

71. E. B. Sandell, *Colorimetric Determination of Traces of Metals,* 3rd ed., Interscience, New York, 1959, pp. 513–516.

72. C. A. R. de Albuquerque, J. R. Muysson, and D. M. Shaw, "Thallium in Basalts and Related Rocks," Paper presented at the International Geo-

chemical Congress, Moscow, U.S.S.R., July 20–25, 1971, *Chem. Geol.* **10,** 41–58 (1971).

73. G. Charlot, *Les Reactions Chimiques en Solution: L'Analyse Qualitative Minerale,* 6th ed., Masson & Cie., Paris, 1969, pp. 312–315.

74. A. P. Vinogradov and D. I. Ryabchikov, *Detection and Analysis of Rare Elements,* Israel Program for Scientific Translations, Jerusalem, 1962, pp. 241–242. Printed in Israel by S. Monson and translated from Russian by IPST staff.

75. C. L. Luke and M. E. Campbell, "Photometric Determination of Germanium with Phenylfuorone," *Anal. Chem.* **28**(8), 1273–1276 (1956).

76. R. W. Boyle, The Geochemistry of Gold and its Deposits, Geological Survey of Canada, Bulletin 280, 1979, pp. 143–147.

77. C. L. Luke and M. E. Campbell, *Anal. Chem.* **25,** 1589 (1953).

78. J. I. Hoffman and G. E. F. Lundell, Volatilization of Metallic Compounds from Solutions in Perchloric or Sulfuric Acid, U.S. Department of Commerce, National Bureau of Standards, Research Paper RP 1198, 1939.

79. H. F. Walton, "Separation by Distillation and Evaporation," in *Standard Methods of Chemical Analysis,* Ch. 8, p. 206, 1963. Frank J. Welcher., Ph.D editor D. Van Nostrand Company Inc. Princeton, NJ.

80. F. D. Pierce and H. R. Brown, *Anal. Chem.* **48,** 693 (1976).

81. F. D. Pierce and H. R. Brown, *Anal. Chem.* **49,** 1417 (1977).

82. Analytical Methods Committee, Chemical Society, *Analyst* **104,** 778 (1979).

83. B. L. Dennis, J. L. Moyers, and G. S. Wilson, *Anal. Chem.* **48,** 1611 (1976).

84. G. Charlot, *Chimie Analytique Quantitative,* Vol. II, *Methodes Selectionnees D'Analyse Chimique des Elements,* 6th ed., Masson & Cie., Paris, 1974, pp. 511–512.

85. C. A. Parker and L. G. Harvey, "Fluorimetric Determination of Sub-Microgram Amounts of Selenium," *Analyst* **86,** 54–62 (1961).

86. C. O. Ingamells, "A New Method for Ferrous Iron and Excess Oxygen in Rocks, Minerals, and Oxides," *Talanta* **4,** 268–273 (1960).

87. J. I. Hoffman and G. E. E. Lundell, "Determination of Fluorine and of Silica in Glasses and Enamels Containing Fluorine," *J. Res.* National Bureau of Standards, **3,** 581–595 (1929), Research Paper No. 110.

88. H. R. Shell and R. L. Craig, "Determination of Silica and Fluorine in Fluorsilicates," *Anal. Chem.* **26,** 996–1001 (1954).

89. H. H. Willard and O. B. Winter, "Volumetric Method for Determination of Fluorine," *Ind. Eng. Chem., Anal. Ed.* **5,** 7–10 (1933).

90. W. D. Armstrong, *J. Am. Chem. Soc.* **55,** 1741 (1933).

91. W. M. Hoskins and C. A. Ferris, *Ind. Eng. Chem., Analyt.* **8,** 6 (1936).

92. C. O. Ingamells, "The Application of an Improved Steam Distillation Apparatus to the Determination of Fluorine in Rocks and Minerals," *Talanta* **9,** 507–516 (1962).

93. M. Cremer, H. N. Elsheimer, and E. E. Escher, "Microcoulometric Measurement of Water in Minerals," *Anal. Chim. Acta* **60,** 183–192 (1972).

94. J. Körbl, *Chem. Listy* **49,** 858 (1955); **51,** 27 (1957).

95. E. B. Sandell, *Mikrochim. Acta* **38,** 487 (1951).

96. I. Friedman and R. L. Smith, *Geochim. Cosmochim. Acta* **15,** 218 (1959).

97. L. Shapiro and W. W. Brannock, "Rapid Determination of Water in Silicate Rocks," *Anal. Chem.* **27,** 560–564 (1955).

98. E. D. Zysk, *Precious Metals 1981. International Precious Metals Institute,* Pergamon Press, New York, 1981.

99. P. Gy, "Poids a Donner a un Echantillon, Abaques," *Rev. L'Industrie Mineral.* **38** (1956).

100. A. Prigogine, "Echantillonage et Analyse des Minerais Heterogenes a Faible Teneur., *Acad. Roy. Sc. Outre-Mer* **15** (1961).

101. A. Prigogine, "Echantillonage des Minerais Auriferes," *R.I.M.* **44,** 557–572 (1962).

102. P. Gy, *Sampling Nomogram: A Circular Calculator,* Minerais et Metaux, Paris, 1956.

103. F. E. Beamish, *Analytical Chemistry of the Noble Metals,* Pergamon Press, Oxford, 1966.

104. H. Ling Ong and V. E. Swanson, "Natural Organic Acids in the Transportation, Deposition, and Concentration of Gold," *Quarterly Colorado School of Mines,* **64**(1) (1969).

105. R. W. Boyle, The Geochemistry of Gold and Its Deposits, Geological Survey of Canada, Bulletin No. 280,

106. J. Haffty, L. B. Riley, and W. D. Goss, *A Manual on Fire Assaying and Determination of the Noble Metals in Geological Materials,* Geological Survey Bulletin No. 1445, U.S. Government Printing Office, Washington, D.C., 1977.

107. W. R. Schoeller and A. R. Powell, *Analysis of Minerals and Ores of the Rarer Elements,* Hafner Publishing, New York, 1955, pp. 325–394.

108. E. M. Hamilton, *Manual of Cyanidation,* McGraw-Hill, New York, 1920.

109. T. Groenewald, "Quantitative Determination of Gold in Solution by Solvent Extraction and Atomic Absorption Spectrometry," *Anal. Chem.* **41,** 1012 (1969).

110. E. A. Elevatorski, *Uranium Ores and Minerals,* Minobras, 1978. Library of Congress Catalog Card No.: 77-72238.

111. E. B. Sandell, *Colorimetric Determination of Traces of Metals,* 3rd ed., Interscience, New York, 1959, pp. 900–902.

112. P. Blanquet, *Anal. Chim. Acta* **16,** 44 (1957).

113. W. R. Schoeller and A. R. Powell, *The Analysis of Minerals and Ore of the Rarer Elements,* Hafner Publishing, New York, 1955, pp. 156–167.

114. E. B. Sandell, *Colorimetric Determination of Traces of Metals,* 3rd ed., Interscience, New York, 1959, p. 839.

115. S. Abbey, "Determination of Thorium in Rocks," *Anal. Chim. Acta* **30**, 176–187 (1964).

116. F. Culkin and J. P. Riley, "The Determination of Trace Amounts of Zirconium, Hafnium, Thorium and Cerium in Silicate Rocks," *Anal. Chim. Acta* **32**, 197–210 (1965).

117. J. Korkish and F. Tera, *Anal. Chem.* **33**, 1264 (1961).

118. J. S. Fritz and B. B. Garrlda, *Anal. Chem.* **34**, 1387 (1962).

119. A. S. Goldin, "Determination of Dissolved Radium," *Anal. Chem.* **33**(3), 406–409 (1961).

RAPID CHEMICAL METHODS

5.1. Introduction

5.2. Tailoring a Rapid Analytical Method

5.3. Determination of Molybdenum as Lead Molybdate in Molybdenite Concentrates

5.4. Molybdenum in Molybdenite Ores by Colorimetry

5.5. Molybdenum in Molybdenite Ores by Atomic Absorption

5.6. Rhenium in Ores by Extraction and Atomic Absorption

5.7. Determination of WO_3 in Scheelite and Wolframite Concentrates

5.8. Tungsten in Scheelite Ores by Colorimetry

5.9. Antimony, Arsenic, Selenium, and Tellurium in Ores and Geochemical Materials by Electrothermal Atomization and Atomic Absorption

5.10. Rapid Alkali Determination

5.11. Determination of Calcium in Fluorspar (Fluorite)

5.12. Determination of Calcium in Limestone or Dolomite

5.13. Determination of Calcium in Phosphate Rock

5.14. Rapid Analysis of Marl (Limestone)

5.15. Determination of Barium in Geologic Materials

5.16. Determination of Tantalum and Niobium by ICP

5.17. Rapid Determination of Basic Elements in Lateritic Ores by Atomic Absorption or ICP

5.18. Determination of Silicon Dioxide by Standard Addition and Atomic Absorption or ICP

5.19. Decomposition and Analysis of Chromium-rich Spinel

5.20. Determination of Silver in Ores by Atomic Absorption

5.21. X-Ray Sample Preparation

5.22. Determination of Mercury in Sulfide Residues

5.23. Rapid Methods of Determination in Environmental Analysis

    5.23.1. Instrumental Methods

    5.23.2. Conclusion

5.1. INTRODUCTION

Where complete instrumentation, including an emission spectrograph and X-ray fluorescence and neutron activation equipment, is available (e.g., chemical dissolution of samples generally not necessary), there may be little interest in "rapid" chemical methods of analysis except in specific instances where a special procedure is developed to provide a solution to a particular problem. However, complete instrumentation is by no means always available to everyone, and probably will not be in the foreseeable future, so that rapid chemical methods are frequently the only practical means of obtaining large numbers of analyses of geochemical samples.

Rapid procedures that prove so generally applicable and at the same time give such reliable results that they are used in preference to standard procedures have every right to replace the latter and become themselves the standard. Thus, it is no criticism of rapid methods to characterize them as methods that sacrifice accuracy and range of applicability for speed and simplicity of execution—it is only a matter of definition.

A short procedure involving only a few steps will often yield better results within its range of applicability than a much longer but more sophisticated one, especially in unskilled or semiskilled hands. There is often difficulty in deciding whether or not the procedure is applicable to a sample. For example, chromium may be very rapidly determined in certain rocks by fusing with sodium carbonate, leaching the melt with water, filtering, and measuring the intensity of the chromate ion color in solution, visually or with a photometer. If uranium is present, however, the method fails completely. Under some circumstances, iron and cerium interfere. Chromium present in minerals unattacked by molten sodium carbonate is not measured. As a further example, calcium and magnesium are commonly and successfully determined after suitable solution procedure by EDTA titration; but samples with strontium and the rare earths will yield erroneous results; there is no easy way to detect the error. The best rapid procedures are those designed to fill a particular need. The narrower the range of sample composition to be covered, the simpler and shorter the procedure can be made. It is not reasonable to expect rapid methods to be applicable, at least without knowledgeable modification, to the whole spectrum of geochemical samples and problems. As a rule, a great deal of effort and experiment guided by both the science and the art of the analytical chemist has been necessary to the development of successful rapid procedures. But their indiscriminate use must be avoided if they are to be reliable.

In what follows, an effort has been made to define the chief interferences and sources of error to be expected, but the only really reliable test of applicability is performance, and judicious use of standard samples similar to unknowns is strongly recommended as the surest way to avoid gross error. Unfortunately, not many geochemical standards are available. It seems unlikely, moreover, that a completely adequate selection

512 RAPID CHEMICAL METHODS

will ever exist. If at all possible, several (or, at the very least, one or two) of every suite of samples should be analyzed by comprehensive methods and used for control of the rapid procedures. In case lack of the necessary skills or equipment makes this impossible, it is wise to repeat critical determinations by different methods. Failure to obtain satisfactory agreement between results should lead to a search for interfering constituents. A checking device that is sometimes useful is to prepare and run synthetic standards made by blending known substances.

Dependence on standard samples is not a special weakness of rapid chemical methods—instrumental techniques are usually even more dependent on standards. In fact, most instruments do little more than compare samples known and unknown one to the other and often do not even present the possibility of preparing calibration curves directly from reagents of known purity, as is done in colorimetric or titrimetric procedures.

In general, carefully applied rapid chemical procedures are capable of producing results at least as good as those from the most sophisticated instrumentation; and the skilled technician working with equipment worth several hundred or at most a few thousand dollars is well able to compete, with respect to both quality and quantity of work accomplished, with highly instrumented facilities costing hundreds of thousands. It seems certain that the possibilities of shortening and speeding up rapid chemical procedures have by no means been exhausted, and the exercise of some ingenuity on the part of those using them can produce startling results.

It is a general characteristic of rapid procedures that directions must be followed exactly. During method development, a very great deal of time has often been spent in finding the optimum conditions for a reaction or process; and this is all wasted if careless and inaccurate manipulation is permitted. Reagents should be prepared exactly as specified, apparatus should be scrupulously clean, and procedures should be followed to the letter. Sometimes, of course, errors or ambiguities creep into the written procedures; but usually, if a procedure does not work, it is being misapplied or an error in solution preparation or manipulation has been inadvertently committed. Some procedures require more skill or more chemical sense than others: An initial failure may simply mean that the operator needs a little practice. Difficulties have a way of disappearing of themselves after a little perseverance.

Since rapid methods are apt to be used by those with a minimum of training in analytical chemistry, it seems reasonable to present them in as much detail as possible and to attempt definition of their shortcomings and failings so as to lessen the possibility of their misapplication.

Generally, rapid chemical procedures are not economically applied to the analysis of single samples: Considerable time is required for the preparation of special solutions and reagents. Special apparatus must be assembled for some determinations, and calibration procedures using standard samples or carefully prepared standard solutions are an essential preliminary undertaking. Only after numerous preparatory operations can one begin to take advantage of the rapidity of the determinative procedures.

Many ingenious devices have been proposed in the interest of speeding up the analysis of large numbers of samples. Some of the most practical of these are described by Shapiro and Brannock [1].

Other devices with obvious applicability are described in the catalogs of chemical apparatus dealers, and it is well worth the while of anyone attacking the problem of setting up a rapid analysis laboratory to look through these catalogs for gadgets that may lighten and speed the work. An operation that must be repeated innumerable times is the transfer of a definite volume of solution from one vessel to another. The use of ordinary transfer pipettes is tedious in the extreme, and except in cases where maximum volumetric accuracy is essential, serological pipettes with large tip openings and a rubber bulb of suitable size are preferable for dispensing reagents.

The chore of washing glassware can be much reduced by reserving flasks, pipettes, and so on for specific solutions and determinations. If this is done, a thorough rinsing after use is all that is required. Volumetric flasks can be stored completely full of distilled water, where they will usually remain free from grease through several determinations.

One part of the analysis that cannot be shortened without adverse effect on the results is the sample preparation. A smaller sample weight is commonly used than in a gravimetric analysis, and some considerable care is necessary to be sure that this is representative of the whole (see Chapter 1). It is particularly recommended that the sample, or a carefully prepared split of it, be ground to pass through a 200-mesh screen and thoroughly mixed in an agate mortar before weighing out portions for analysis. Grinding without screening is certain to lead to trouble (e.g., excess of large particles left and excess of insoluble material during the dissolution). Nylon monofilament screens are best, but an effective and inexpensive screen can be prepared from silk bolting cloth and an embroidery hoop. Metal screens are less desirable because of the unavoidable sample contamination that follows from their use, but they may be used when such contamination is of no moment.

Numerous schemes for the rapid chemical analysis of silicate rocks have been proposed, and reference to these is advised. One of the most widely used is that that of Shapiro and Brannock [1]. It is impossible to present a collection of procedures that will cover any and all cases, and that which follows makes no pretense at doing this.

5.2. TAILORING A RAPID ANALYTICAL METHOD

To set up a rapid analytical method can be a difficult task if no strategy is used. The knowledge of what has been exposed in the first four chapters of this book can provide for the analytical chemist enough fundamentals to prepare a typical flowsheet of what a new rapid method might be. Then through the iterative learning process from experiments, the method may shape up to its final form for a specific need. A recommended approach is illustrated in Figure 5.1. In this figure, six areas, which are each described by a diamond, are shown because they are, most of the time, responsible for analytical failure when overlooked. Experience has shown that they are frequently forgotten by the analyst. Each of these areas may completely ruin long and expensive efforts to achieve good accuracy and precision.

When a rapid method has been summarized into a flowsheet, it still needs to be written in detail following a standard format. A standard format is necessary to ensure that all the relevant information is available to the analyst. In the following proposed standard format 12 main sections have been listed:

Title of Method: Element in Material by Technique

1. *Scope:* A clear description of the applicability of the method is essential to the analyst.
2. *Summary:* Synopsis of the most important steps of the method.
3. *Interferences:* This section is too often overlooked, and it is a common mistake to put only a statement such as (1) Other elements ordinarily present in this material do not interfere in this method, (2) Provision is made in this procedure to correct for all interferences, or (3) The adjustment of the matrix with this reagent corrects for most potential interferences. Instead, it is essential to list all possible interferences and devices for overcoming them. The discussion may go a little beyond the defined domain of applicability in order to warn the analyst of potential shortcomings of the method under some conditions. In other words this discussion is placing the domain of applicability between identified limits. In this section, it is recommended to list other relevant references.
4. *Special reagents:* Enumerate and describe only reagents that are not normally found in an analytical laboratory. In this section, it is nec-

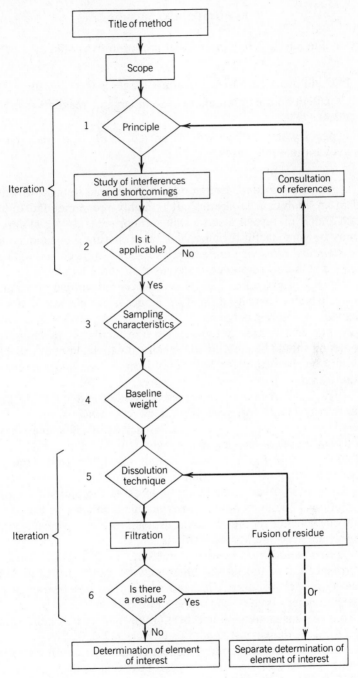

Figure 5.1. Illustration of recommended approach for tailoring rapid analytical method.

515

essary to emphasize a few points that are often overlooked:

- Indicate procedure to purify some reagents (see Section 2.2).
- Comment on the unreliability of some organic reagents (see Section 2.2.4).
- Type of water to be used (deionized or distilled) may be highly relevant in some cases.

5. Special apparatus: Enumerate only unusual items not ordinarily found in an analytical laboratory. It is highly recommended to show a diagram and/or picture of a setup using one or several special apparatus. This may be very useful for an analyst in preparing the same setup.

6. Standard reference materials: Provide references and data of recommended standard reference materials if they are available.

7. Sample preparation: Describe the required sample maximum particle size, which is usually critical for a successful digestion. If some of the elements of interest in the material have a strong tendency to segregate and generate distribution heterogeneity, the technique to minimize this phenomenon should be mentioned. Specify at which temperature the sample should be dried and for how long in order to choose a consistent baseline weight.

8. Procedure: A procedure should be written in detail to ensure successful transferability from one laboratory to another. A full description of every step is necessary. Sometimes too many shortcuts taken in written methods leaves the door open to too many minor changes; this may have a serious impact on the final result. Notes should be used frequently to clearly explain any potential pitfalls.

9. Method of calibration: Describe in detail the preparation of standard solutions. Recommend the most suitable technique to use (e.g., calibration curve using synthetic standards, calibration curve using standard reference materials, standard additions, etc.).

10. Instrumental parameters: Sometimes the procedure may be the same irrespective of the finish, which may be flame atomic absorption, furnace atomic absorption, ICP, and so on. The description of instrumental parameters is necessary, and these parameters may vary drastically from one instrument to another. Often, parameters need to be optimized for a particular instrument (see Section 2.3.7).

11. Calculations: Give mathematical equations that are clearly written with a glossary of all terms. Special attention shall be given to conversion factors, dilution factors, units, and so on.

12. Precision and accuracy: If following clear definitions, unnecessary confusion may be avoided:

- A measurement is unbiased when the mean (m) of the analytical error (AE) is equal to zero:

$$m(AE) = 0$$

- A measurement is biased when the mean of the analytical error is not equal to zero:

$$m(AE) \neq 0$$

- A measurement is accurate when the absolute value of the bias $|m(AE)|$ is not larger than a certain standard of accuracy $m_0(AE)$ regarded as acceptable,

$$|m(AE)| \leq m_0(AE)$$

Note: It is very important to make a distinction between an accurate measurement and an unbiased measurement. Even when carried out in an ideal way, an analysis is always biased. The bias is never strictly zero. It is often negligible or very small but always different from zero. Accordingly, we should speak of an accurate measurement but not an unbiased measurement. In fact, an unbiased measurement is a limiting case that is never encountered.

- A measurement is precise or reproducible when the variance of the analytical error $\sigma^2(AE)$ is not larger than a certain standard of precision $\sigma_0^2(AE)$ regarded as acceptable:

$$\sigma^2(AE) \leq \sigma_0^2(AE)$$

Depending on what is acceptable for $m_0(AE)$ and $\sigma_0^2(AE)$, several sections of the methods may have to be modified (e.g., sample weight, particle size, quality of SRM, number of points used for the calibration curve, standard additions, repetitive analyses, etc.).

In order to prevent repetition and provide maximum information within the scope of this book, only a limited number of the rapid chemical methods exposed in this chapter are written following the recommended standard format. These typical examples may be used as a guideline.

5.3. DETERMINATION OF MOLYBDENUM AS LEAD MOLYBDATE IN MOLYBDENITE CONCENTRATES

1. Scope: The purpose of this method is for the control determination of molybdenum in high-grade molybdenite concentrates as produced at mills. The method is designed to determine molybdenum with the best accuracy and precision possible and has been found extremely reliable via round-robin testing.

2. Summary: The molybdenum is dissolved into solution with nitric and perchloric acids and free from arsenic, antimony, phosphate, silicate, chromium, aluminum, iron, vanadium, and tungsten. The solution is buffered to pH 4.8 with a mixture of ammonium acetate and acetic acid.

The molybdate is precipitated as lead molybdate by the dropwise addition of a slight excess of lead nitrate to the nearly boiling solution. After settling hot for about 1 h, the lead molybdate is filtered, washed with a hot 2% ammonium acetate solution, ignited at 550 °C and weighed accurately.

It is essential to add the lead nitrate while stirring the solution vigorously at the same time. The solution should be hot and the reagent added dropwise and slowly and only in slight excess. These conditions allow gains in particle size, and the quantity of inclusions and occlusions is reduced. Completeness of precipitation may be monitored using tannin as an external indicator.

Lead nitrate is preferable to lead acetate as a precipitant because it can be accurately weighed. Reagent lead acetate may contain variable amounts of basic acetates, hydrates, and tetraacetate; thus, each batch should be analyzed for its lead content before using. Lead acetate is also more likely to contain objectionable impurities. Finally, when molybdenum ions are precipitated under proper conditions by a solution of lead nitrate, the precipitate is purer, more granular, and more readily filtered and washed than that obtained by adding a solution of lead acetate.

3. Interferences: Elements such as calcium, strontium, and barium may interfere by giving low results because of the formation of alkaline earth molybdates instead of lead molybdate; others such as arsenic, antimony, phosphate, silicate, chromium, aluminum, iron, vanadium, and tungsten lead to high results by contaminating the precipitate of lead molybdate. If too much sulfate is present and an excess of lead nitrate is used, the lead molybdate precipitate may contain small amounts of lead sulfate; a larger addition of ammonium acetate, which dissolves lead sulfate, usually overcomes this difficulty.

Interfering elements such as calcium, strontium, barium, and tungsten are removed by ammonia precipitation. If vanadium, arsenic, phosphate, and soluble silicate are present, small amounts of ferric iron are added. If lead or much calcium are present in the sample, some molybdenum may be lost, and the hydroxide precipitate should be examined for its molybdenum content by an atomic absorption, colorimetric, or ICP procedure.

Tungsten that is not removed during the ammonia precipitation is weighed as lead tungstate. If much tungsten is present in the sample, it is necessary to precipitate the molybdenum as sulfide from a tartaric–tartrate solution prior to its determination as lead molybdate.

4. Special reagents:

- The use of deionized water is not recommended: Organic impurities may be objectionable to the procedure in several ways. Use only good quality distilled water.
- Ammonium acetate solution (60%): Dissolve 600 g ammonium acetate in distilled water, dilute to one liter and filter through S & S black ribbon filter paper.
- Ammonium acetate wash solution: Dissolve 20 g ammonium acetate in distilled water and dilute to 1 l.
- Lead nitrate solution (2%): Dissolve 20 g lead nitrate in distilled water and dilute to 1 l.
- Methyl orange indicator (0.1%): Dissolve 0.1 g methyl orange in distilled water, filter, and dilute to 100 ml.
- Tannic acid indicator: Dissolve 1 g tannic acid in 300 ml distilled water.
- Ammonium chloride wash solution: Dissolve 20 g ammonium chloride in distilled water and dilute to one liter.

5. Special apparatus:

- Constant-rate burette, motor driven at a fixed delivery of about 5 ml/min.
- Magnetic stirrer, Teflon-covered stirring bars, and bar retriever.
- Shaker hot plate: The use of such a hot plate is recommended in order to perform the digestion effectively and prevent losses by spattering. Unfortunately, good shaker hot plates are difficult to find among known manufacturers.
- Number 1 porcelain crucibles.

- Use of Vycor glassware is recommended in order to avoid minor contamination from potential interfering elements. Vycor glass is practically pure SiO_2.

6. *Standard reference materials:* Analyze a standard reference material with each batch of samples and include the data obtained for the standard with the analytical report. Standard reference materials of molybdenite concentrates are best prepared by an in-house laboratory setup to prepare such standards as described in Chapter 7.

7. *Sample preparation:* Prior to the molybdenum analysis, it is recommended to prepare three different splits of each unknown sample. They must be dried and thoroughly deoiled. The bulk concentrate sample from the drying and deoiling are then individually blended in a closed mixing vial for 10 min before a sample is weighed out for the procedure.

8. PROCEDURE. Weigh a 0.25-g sample to the accuracy of the analytical balance (0.1 mg) and transfer it to a 600-ml beaker.

Note: MoS_2 sticks to most materials; therefore, it is recommended that the sample be weighed on a small piece of paper that will also be digested with the acids.

Add 25 ml nitric acid and 10 ml perchloric acid. Cover the beaker with a watch glass and heat on a shaker hot plate located in a fume hood equipped to handle perchloric acid fumes. Heat slowly at first for 15 min; then increase the heat and boil until the sample has been completely dissolved and the volume reduced to about 3 ml.

Note: The concentrate sample may be decomposed by a mixture of HNO_3 and $KClO_3$ instead of the HNO_3–$HClO_4$ mixture described in the text. However, for the sake of consistency, it is not recommended to choose one way or another and not to change.

After cooling the solutions slightly, add 15 ml hydrochloric acid, cover with a watch glass, and boil until all soluble salts are in solution.

Note: Overboiling at this stage will lead to low results. Remove the beaker from the hot plate as soon as the solution is clear.

Remove the solution from the hot plate. Rinse the watch glass and wash down the interior walls of the beaker with water, adjusting the final volume of the solution to about 50 ml. Add an excess of concentrated ammonia solution and heat to just boiling. Filter through a Whatman No. 40 paper to which a small amount of ashless paper pulp has been added.

Note: In many instances, the residue of unattacked material consisting of, for example, iron and aluminum hydroxides will contain no molybdenum; however, if MoS_2 is not completely liberated during sample preparation, or if lead, calcium, and certain other divalent elements are present, some molybdenum may remain undissolved. For these reasons, after precipitation with ammonia, it is necessary to investigate the residue.

Collect the filtrate in a 600-ml beaker. Wash the beaker and the precipitate eight times with the hot 2% ammonium chloride solutions.

Note: If the iron content of the sample is insufficient to precipitate phosphate, vanadate, and arsenate, an appropriate addition of ferric chloride solution should precede the ammonia precipitation. After the precipitation, a volume of at least 400 ml is necessary to prevent any occlusion of lead sulfate and other salts in the lead molybdate precipitate, leading to high results.

Dissolve the iron precipitate in hydrochloric acid and reprecipitate the iron with ammonium hydroxide. After filtration as before, the filtrates are combined for the lead molybdate precipitation step.

As a precaution, dry and ignite the paper containing the insolubles. Fuse the residue and precipitate using 1 g pure sodium peroxide in a pure iron crucible. Dissolve hydroxides in a 100-ml volumetric flask using hydrochloric acid. Make certain the final solution contains about 2% $AlCl_3$; then determine molybdenum by atomic absorption using a nitrous-acetylene flame, by a colorimetric method, or by an X-ray method. This loss of molybdenum is added to the final result obtained from the lead molybdate precipitation; if any is present, it is called the weight correction w_1.

Add 1 drop of methyl orange indicator to the filtrate and neutralize the excess ammonia with hydrochloric acid (orange to red end point), adding 2 drops in excess. Buffer the solution at pH 4.8 by adding 30 ml ammonium acetate solution and 10 ml glacial acetic acid. Place the covered beaker on the hot plate and heat the solution to 90 °C, maintaining this temperature until the precipitation of the lead molybdate.

Note: Do not boil or low results will be obtained.

Place the beaker containing the hot solution on a magnetic stirrer, introduce the stirring bar, and adjust the stirring speed to obtain good mixing action. Using the preset burette, add sufficient lead nitrate solution to precipitate the molybdenum as lead molybdate.

Note: During the lead molybdate precipitation, the addition of the lead nitrate must proceed under vigorous stirring of the hot solution, being careful about mechanical losses of droplets of solution. The reagent must be added dropwise and slowly and only in slight excess. If the precipitation is too rapid, salts such as ammonium chloride and lead sulfate may be occluded, causing high results.

The end point is reached when a drop of tannic acid indicator on a spot plate fails to show a brown coloration. Then add 5 ml precipitant in excess. With a magnetic retriever, remove the bar from the solution and rinse it with water.

Note: Do not place the retriever in the solution. Place it under the beaker and run it upward, pulling the stir bar to above the solution. Rinse off the bar and place it in a clean beaker and set aside for testing of molybdenum loss at a later stage.

Hold the solution containing the lead molybdate precipitate near the boiling point for 1 h.

Note: After the lead molybdate precipitation, the precipitate should be allowed to settle near the boiling point for at least 1 h. On digestion near 100 °C, the primary imperfect crystals are subjected to successive recrystallizations, and this greatly diminishes the quantity of occlusions.

Vacuum filter the solution through a Whatman No. 40 paper containing a small amount of ashless paper pulp. Carefully transfer the precipitate to the paper, scrubbing clean all traces adhering to the beaker. Wash the paper and precipitate four times with hot 2% ammonium acetate solution and then once with hot water.

Note: Excess washing with hot water will peptize the precipitate of lead molybdate and lead to low results. If much calcium (or strontium and barium) is present in the sample (above 1%), some molybdenum may precipitate as calcium molybdate. Also, if much sulfate is present, some lead sulfate may precipitate: dissolve the lead molybdate precipitate in 1:1 hydrochloric acid, dilute to 200 ml with water, boil, and add ammonia solution just to the beginning of precipitation. Clear the solution with a few drops of 1:1 hydrochloric acid. Heat to 90 °C, add 30 ml ammonium acetate solution and 3 ml 2% lead nitrate, and digest near the boiling point for 1 h. Filter and wash as explained for a single precipitation.

Transfer paper and precipitate to a No. 1 porcelain crucible and ignite in a muffle furnace at 550 °C.

Note: Lead molybdate is appreciably volatile above 600 °C.

When the paper is completely burned, remove the crucible from the furnace and cool inside a desiccator. After cooling to room temperature, transfer the ignited precipitate from the crucible to a balance pan and weigh.

Note: Weighing crucible plus precipitate is likely to introduce weighing errors unless an empty crucible is taken through all operations and used as a tare. If the ignited precipitate is transferred from the crucible to a balance pan, it is advisable to wash the empty crucible and the magnetic bar (earlier step) with 1:1 hydrochloric acid and determine the molybdenum losses as we did after the ammonia precipitation. This gives the weight correction w_2.

If tungsten is present, the procedure must include a hydrogen sulfide precipitation step for molybdenum. The same is true if wulfenite ($PbMoO_4$) or other lead minerals are present.

9. *Calculations:*

$$\text{Percent Mo} = \frac{AB + w_1 + w_2}{C} \times 100$$

where A = weight of $PbMoO_4$, g

$\quad\quad B$ = conversion factor from $PbMoO_4$ to Mo, 0.2613

$\quad\quad w_1$ = molybdenum loss from ammonia precipitations and insolubles

w_2 = molybdenum losses from transferring the $PbMoO_4$ precipitate from the porcelain crucible to the balance pan and from retrieving the magnetic stirring bar

C = sample weight, g

10. Precision and accuracy: It has been found experimentally via round-robin testing that the interlaboratory agreement precision with this method on a sample containing 50% Mo is 50 ± 0.20% (2σ).

The method provides unusually precise quantitative data allowing for very close control of a concentrate product. If precautions (see notes of procedure) are carefully followed, the accuracy is excellent.

5.4. MOLYBDENUM IN MOLYBDENITE ORES BY COLORIMETRY

1. Scope: The method is applicable to concentrations of molybdenum ranging from 0.01 to 20% in molybdenite ores. The method still performs very well in the concentration range between 0.005 and 60% molybdenum. The molybdenum determination includes molybdenum occurring as sulfide and as oxides.

2. Summary: The sample is dissolved with nitric, perchloric, and sulfuric acids. Stannous chloride reduces Mo(VI) rapidly to Mo(IV), which then disproportionates to Mo(V) and Mo(III). Only Mo(V) gives a colored thiocyanate complex. Accordingly, only half the potential color intensity is obtained. Addition of Fe(III) corrects this problem. Molybdenum is determined by measuring the absorbance of the amber molybdenum thiocyanate complex formed in butyl cellosolve–water solution at 470 nm.

3. Interferences: Interference from W(VI) may be eliminated by complexing it with ammonium citrate [2]. Rhenium and chromium can be separated from molybdenum by volatilizing them from a hydrochloric acid–perchloric acid medium. Moderate amounts of aluminum, arsenic, antimony, bismuth, beryllium, zinc, silica, and phosphate do not interfere. Interference from Ti(IV) may be eliminated by complexing it with fluoride. Fe(III), if present in small quantities, does not interfere because it is reduced by stannous chloride at the same time as molybdenum. Co(II) interferes if present in a quantity of half the amount of molybdenum.

If copper, titanium, iron, cobalt, nickel, platinum, palladium, rhodium, zirconium, thorium, uranium, niobium, tantalum, manganese, and cadmium are present in such quantities that interferences are expected [3], they should be separated from the molybdenum using a double precipitation with sodium hydroxide.

4. *Special reagents:*

- Molybdenum standard solution, 100 μg/ml Mo: Dissolve 0.150 g pure molybdenum trioxide in 10 ml of 0.4% (w/v) sodium hydroxide. Dilute the solution to 100 ml with water, and then make it slightly acid with 1:1 sulfuric acid. Transfer this solution to a 1-l volumetric flask, and dilute to volume with water.
- Daily molybdenum standard solution, 10 μg/ml Mo: Transfer 50 ml of the 100 μg/ml Mo standard solution to a 500-ml volumetric flask, and dilute to volume with water.
- Sodium thiocyanate solution, 5%: Dissolve 5.0 g reagent-grade sodium thiocyanate in water and dilute to 100 ml. Prepare solution daily. Filter the solution if necessary. This solution is more stable if kept in the dark.
- Stannous chloride solution, 35%: Dissolve 350 g stannous chloride in 250 ml hydrochloric acid. Heat the mixture until dissolution is complete. Pour the solution into 250 ml cold water, transfer to a 1-l flask and dilute to volume with 1:1 hydrochloric acid. Add 3 g metallic tin to keep the solution in the reduced state.
- Ferric chloride solution, 1% Fe: Dissolve 1.0 g pure iron wire in 50 ml of 1:2 hydrochloric acid. Oxidize the solution with a minimum amount of nitric acid, cool to room temperature, and dilute to 100 ml.
- Ethylene glycol monobutyl ether: packaged in glass.
- Hydrogen peroxide, 30%.

5. *Special apparatus:* Spectrophotometer capable of operating at a wavelength of 470 nm.

6. *Standard reference materials:* It is recommended to develop in-house standard reference materials using typical investigated ores and classical methods of analysis.

7. *Sample preparation:* Mineralogical investigation of the molybdenum ore under study is often critical in order to perform an appropriate sample preparation. If the molybdenite is occurring as coarse grains, the entire sample (e.g., core sample) should be reduced to only 6 mesh before splitting. This is even marginal in some cases, and it is essential that the whole sample be put through the screen, without loss of molybdenite particles, and thoroughly mixed before splitting. Further reduction to 100% −10-mesh should follow. Finally about 100 g of sample should be ground to 100% −100 mesh. If pulverized too finely, losses of MoS_2 material are very likely. Stage grinding during a very short time is the

most successful approach. Dry the sample at 105 °C to constant weight. During geochemical exploration, most of the time molybdenum is finely disseminated; thus, it is recommended to grind the sample to -200 mesh in order to obtain good recovery of molybdenum.

8. PROCEDURE. Weigh a 2.5-g sample and transfer it to a 250-ml beaker. Add 10 ml HNO_3, 5 ml $HClO_4$, and 10 ml H_2SO_4. Cover the beaker and digest the solution at boiling temperature for 20 min. Increase the heat until the nitric acid is removed and the sulfuric acid is refluxing.

Note: If the sample contains rhenium or chromium, they can be volatilized as the heptoxide and chromyl chloride by adding small increments of concentrated hydrochloric acid several times and evaporating the solution to fumes of perchloric acid each time.

Cool the sample, add 75 ml water, and heat gently to dissolve the soluble salts. Add 2 drops of 30% hydrogen peroxide.

Note: H_2O_2 assists in solubilizing manganese and molybdenum salts.

Boil about 5 min to get rid of excess hydrogen peroxide. Filter the solution through a dry Whatman No. 42 filter paper, catching the filtrate in a 250-ml volumetric flask. Carefully wash the beaker, paper filter, and insoluble eight times using hot water. Cool to room temperature, dilute to volume, and then mix thoroughly.

Note: If copper, titanium, iron, cobalt, nickel, platinum, palladium, rhodium, zirconium, thorium, uranium, niobium, tantalum, manganese, and cadmium are present in such quantities that interferences are expected, they should be separated from the molybdenum using a double precipitation with sodium hydroxide. The filtrate is collected in a 250-ml volumetric flask. If the sample contains appreciable tungsten, the solution may become cloudy or some tungsten trioxide may precipitate. Tungsten may be complexed using ammonium citrate.

Transfer an aliquot containing between 0.01 and 0.50 mg molybdenum, but no more than 60 ml, into a 100-ml volumetric flask. Dilute to 60 ml, and add 1 ml $FeCl_3$ solution and 7 ml sulfuric acid. Cool to room temperature.

Note: Some iron must be present in the solution during the formation of the Mo(V) thiocyanate complex when stannous chloride is used as the reducing agent. In the absence of iron, only half of the molybdenum present reacts with thiocyanate to form a colored complex.

Add 20 ml ethylene glycol monobutyl ether and mix the solution. Add 5 ml sodium thiocyanate solution and mix again.

Note: The deep red Fe(III) thiocyanate complex that forms at this stage is destroyed by stannous chloride during the reduction step. Potassium thiocyanate should not be substituted for sodium thiocyanate because of the relative insolubility of the potassium perchlorate that may occlude molybdenum during the complex formation.

Add 5 ml stannous chloride solution and immediately dilute to volume with water and mix thoroughly. Let stand for 20 min for full color development.

Note: The molybdenum–thiocyanate complex formed under these conditions is stable for at least 2 h.

Measure the absorbance of the solution at 470 nm.

9. *Method of calibration:* Prepare a blank and five calibrating standards made with 5-, 10-, 20-, 30-, 40-, and 50-ml aliquots of the daily standard solution into 100-ml volumetric flasks; then proceed as for unknown samples, starting when diluting to 60 ml and adding 1 ml FeCl$_3$ solution and 7 ml sulfuric acid.

10. *Instrumental parameters:* Usually the measurement of the absorbance is performed at 470 nm.

11. *Calculations:*

$$\text{Percent Mo} = \frac{ACE}{BWD} \times 100$$

where A = Mo in standard solution, g
B = absorption of corresponding standard solution
C = absorption of sample
W = weight of sample used, g
D = volume of sample aliquot, ml
E = sample dilution

12. *Precision and accuracy:* If the unknown samples are very different from one another, it is possible that interference phenomena become erratic, and in such a case, it is recommended to proceed to a double sodium hydroxide precipitation in order to improve precision and accuracy. Precision also depends on the good timing of each step. Accuracy depends mostly on how the digestion is conducted and on how meticulously the sample preparation is performed. Residues must be checked regularly for their molybdenum content using a fusion technique. This method should provide excellent precision and accuracy.

5.5. MOLYBDENUM IN MOLYBDENITE ORES BY ATOMIC ABSORPTION

1. *Scope:* The method is written for a molybdenum concentration range in the sample between 0.001 and 20% by weight. Molybdenum occurring as sulfide and molybdenum occurring as oxide are determined separately.

2. *Summary:* The nonsulfide molybdenum is dissolved using HCl. The solution is filtered and the filtrate is used for the determination of the HCl-soluble molybdenum. The residue is decomposed by a mixture of nitric and perchloric acids, and the resulting salts are dissolved by HCl.

This solution is used for the determination of molybdenum occurring as sulfide.

Molybdenum in both solutions is determined by atomic absorption at a wavelength of 313.3 nm in a dilute hydrochloric acid–aluminum chloride medium using a strongly reducing nitrous oxide–acetylene flame.

3. Interferences: When an air–acetylene flame is used, many interference phenomena in the flame are likely to occur. The use of a nitrous oxide–acetylene flame is recommended. When using a nitrous oxide–acetylene flame, most of the remaining interferences can be suppressed by adding an excess of a refractory element such as aluminum. The depression of the molybdenum response by high concentrations of iron in the air–acetylene flame is explained by incomplete volatilization. The depression is most significant in a fuel-rich air–acetylene flame and is not experienced at all in the much hotter nitrous oxide–acetylene flame.

4. Special reagents:

- Aluminum chloride, hexahydrate, 10 g/l: Dissolve 89.5 g $AlCl_3 \cdot 6H_2O$ in water, dilute to 1 l, and mix.
- Molybdenum standard solution, 1000 μg/ml: Weigh 1.5003 g high-purity MoO_3, transfer to a 1-l volumetric flask, and add 200 ml concentrated HCl. Warm to dissolve, cool to room temperature, and dilute to 1 l using water.

5. Special Apparatus: Atomic absorption spectrophotometer with a molybdenum hollow cathode source and a nitrous oxide burner.

6. Standard reference materials: It is recommended to develop in-house standard reference materials following guidelines described in Chap. 7. The mineralogical composition of SRMs should match that of unknown samples as closely as possible. It is important to remember that these rapid methods are tailored for a particular type of material.

7. Sample preparation: Molybdenite tends, sometimes, to occur in nuggets or in concentrated veins. When grinding the soft molybdenite into a homogeneous ore mixture, smearing may generate losses of molybdenite. In many cases, it is recommended to crush at least a 15-kg sample to −10 mesh, 3 kg to −24 mesh, and 300 g to −80 mesh. Proceed by stage grinding. During the last stage of grinding, some molybdenite may smear in the rotary mill. It can be removed by cleaning the mill with a small part of the processed sample. All grinding steps using the rotary mill should not exceed 15 s to avoid smearing. After each sampling stage, the sample should be thoroughly homogenized before splitting it. The weight of the final analytical subsample is dictated by results obtained for the sampling constant K_s. The study of this constant has been largely described in Chapter 1. The sample should be dried at 107 ± 5 °C until constant weight.

8. PROCEDURE. Weigh a 2.500-g sample and transfer it to a 400-ml beaker. Leach with 15 ml concentrated HCl and 35 ml water for 20 min on low heat. Swirl the contents of the beaker occasionally to prevent caking of the sample on the bottom of the beaker.

Filter through a fast-speed filter paper with paper pulp and wash with hot 2% HCl. The filtrate goes directly into a 100-ml volumetric flask. Add 10 ml aluminum chloride solution and dilute to the mark with water. Cool to room temperature and mix. This solution is used for the determination of nonsulfide molybdenum.

Transfer the paper containing the residue back into the 400-ml beaker. Add 20 ml HNO$_3$ and 15 ml HClO$_4$; place the beaker on a cold shaker hot plate and heat slowly until HClO$_4$ begins to fume. Turn up the heat and continue fuming HClO$_4$ for 10 min.

Note: When digesting a sample to perchloric fumes, splattering should be avoided or erratic results will be obtained.

Cool, and add 10 ml HCl, swirl the mixture to facilitate solution, and add 50 ml water. Bring the solution just under the boiling temperature. Transfer the contents of the beaker to a 250-ml volumetric flask, cool to room temperature, dilute to the mark with water, and mix.

Transfer an aliquot X of this solution into a 50-ml volumetric flask. Add 5 ml aluminum chloride solution and $Y \pm 0.02$ ml concentrated HCl.

Note: If $X = 10$ ml, $Y = 0.6$ ml in order to maintain the final solution at exactly 2% HCl; if $X = 5$ ml, $Y = 0.8$ ml; and so on.

Dilute to the mark with water, and mix. This solution is used for the determination of sulfide molybdenum.

9. *Method of calibration:* Aqueous solutions containing 1, 2, 3.5, 5, 7.5, and 10 µg/ml molybdenum in a matrix of 2% HCl and 1000 µg/ml aluminum are used to calibrate the AA spectrophotometer. Such standards are readily prepared from a 1000-µg/ml stock solution of molybdenum. When preparing calibration solutions, prepare a fresh 100-µg/ml molybdenum solution by pipetting 50 ml of the 1000-µg/ml solution to a 500-ml volumetric flask and diluting with water. Solutions for calibration are prepared as follows:

Concentration of Calibrating Standard (µg/ml)	Aliquot of 100 µg/ml Molybdenum Standard Solution (ml/l)	Quantity of HCl To Add Using Burette (ml)	Aluminum Solution Required (ml)
1	10	19.8	100 (10 g/l)
2	20	19.6	100 (10 g/l)
3.5	35	19.3	100 (10 g/l)

Concentration of Calibrating Standard (μg/ml)	Aliquot of 100 μg/ml Molybdenum Standard Solution (ml/l)	Quantity of HCl To Add Using Burette (ml)	Aluminum Solution Required (ml)
5	50	19.0	100 (10 g/l)
7.5	75	18.5	100 (10 g/l)
10	100	18.0	100 (10 g/l)

These calibration solutions cannot be used for more than 2 weeks.

10. *Instrumental parameters:* These parameters have been defined for a particular instrument by the authors; however, they may vary widely with other instrumentation. They may also vary from one molybdenum hollow cathode to another:

Wavelength	313.26 nm
Slit width	150 μm
Photomultiplier	R 106 UH
Integration	Auto
Scale expansion	1×
Nitrous oxide	11 SCFH
Acetylene	12 SCFH (fuel rich)
Lamp current	9 mA
Bandpass	0.5 nm
Aspiration rate	4.5 ml/min
Burner height	7 mm (varies widely; should be optimized for each instrument)

Figures 5.2–5.8 show how some of these parameters have been selected.

11. *Calculations:*

$$\text{Percent Mo} = \frac{ACD}{BW} \times 100$$

where A = molybdenum in the standard solution, g/ml
  B = absorption of corresponding standard solution
  C = absorption of sample
  W = weight of sample used, g
  D = total sample dilution

12. *Precision and accuracy:* Precision may greatly be affected if AA parameters are not optimized as shown in Figures 5.2–5.8.

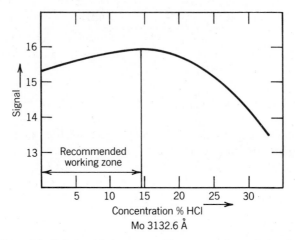

Figure 5.2. Relationship between acid concentration and signal.

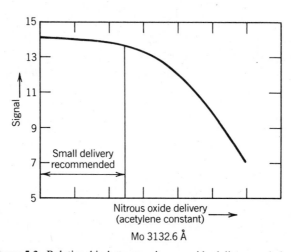

Figure 5.3. Relationship between nitrous oxide delivery and signal.

530

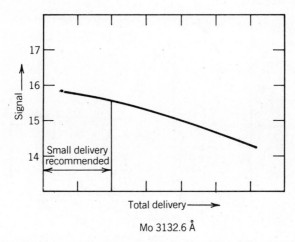

Mo 3132.6 Å

Figure 5.4. Relationship between total fuel and oxidant delivery and signal.

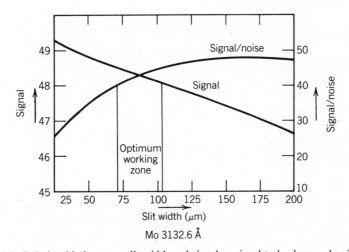

Mo 3132.6 Å

Figure 5.5. Relationship between slit width and signal or signal-to-background-noise ratio.

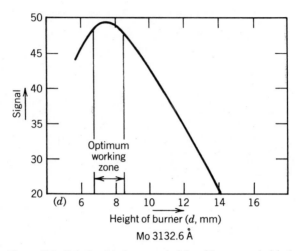

Figure 5.6. Relationship between height of burner and signal.

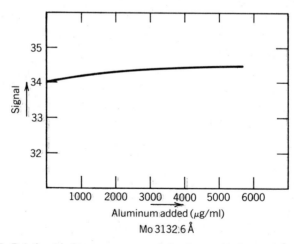

Figure 5.7. Relationship between amount of aluminum added to sample and signal.

532

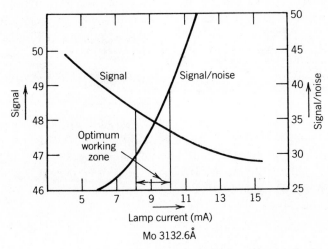

Figure 5.8. Relationship between lamp current and signal or signal-to-background-noise ratio.

Accuracy may be mostly affected by incomplete digestion of MoS_2 or by negligence of checking residues regularly for their molybdenum content using a fusion technique.

5.6 RHENIUM IN ORES BY EXTRACTION AND ATOMIC ABSORPTION

1. Scope: The method is applicable from one to several thousand ppm of rhenium in ores and particularly in molybdenum ores.

2. Summary: Sample is fused with KOH, KNO_3, and $NaClO_3$ in a nickel crucible. Melt is dissolved in water. Rhenium is extracted with 3% Aliquat 336 in MIBK. Rhenium is determined by atomic absorption at 2287.5 Å using a standard addition technique.

3. Interferences: The sample may be more rapidly digested with an acid digestion using $NaClO_3$ and HNO_3; however, it has been found that using a fusion is safer and provides generally better recovery. Using the 3% Aliquat 336 in MIBK extraction, no serious interferences are likely. Nevertheless, it is recommended to use a slightly more oxidizing $N_2O-C_2H_2$ flame condition than usually recommended by manufacturers. Erratic enhancement, which may be observed in unknown samples, is easily corrected by using systematically a standard addition technique. If the ratio of signal to background noise is too small at 2287.5 Å, use the 3460.5-Å line. It may be advantageous to increase the lamp current and decrease the spectral bandpass. Because optimum conditions may be difficult to reach for the analyst, it is recommended to use a strip chart recorder. The following documents are recommended: Bailey and Rankin [4] and Elliott, Stever, and Heady [5], the latter being a truly outstanding document.

4. Special reagents:

Rhenium stock solution: 1000 μg/ml rhenium. Dissolve 1g rhenium metal in about 20 ml HNO_3 and dilute to 1 l with water.

Aliquat 336 and methyl isobutyl ketone: Dissolve 30 ml Aliquat 336 in MIBK. Dilute to 1 l with MIBK.

Potassium hydroxide.

Potassium nitrate.

Sodium chlorate.

5. Special apparatus: Nickel crucibles, 75 ml capacity, and nickel covers for corresponding crucibles.

6. Standard reference materials: The National Bureau of Standards Molybdenum Concentrate Standard No. 333 may be used with this method. The certified value is 870 ppm rhenium.

7. Sample preparation: One sample must be ground to 100% −200 mesh. Samples should be split from larger bulk of material with precaution in order to minimize distribution heterogeneity (e.g., it is the only way to diminish segregation and grouping errors). It is recommended for the analyst to be familiar with the necessary steps to reduce properly a sample to a laboratory subsample (see Section 1.18).

All materials containing large quantities of MoS_2 should be deoiled and dried at 105 °C until constant weight is reached prior to weighing the analytical subsample.

8. PROCEDURE. Weigh from 0.5 to 5.0 g of sample, depending of the expected amount of rhenium present in the material, into a nickel crucible. Mix the sample with 8 g KOH, 4 g KNO_3, and 1g $NaClO_3$. Warm on a hot plate at approximately 450°C for half an hour, and then fuse over a Meker burner until there is complete digestion of the sample. Cool and dissolve the melt with warm water in a Teflon evaporating dish. Crucible and cover are rinsed with 4 ml hydrochloric acid, and this is added to the dissolving melt. Evaporate the solution slowly to about 40 ml and transfer it to a 100-ml volumetric flask. Add water until the volume of the solution is about 85 ml. Add exactly 10 ml of the 3% Aliquat 336 in MIBK solution. Shake for 1 min. Allow phases to separate. Add water to the mark. Shake again for 1 min. Allow phases to separate. Use the organic layer for the determination of rhenium by atomic absorption.

9. *Method of calibration:* Samples may be run against a standard curve in order to obtain a first estimate of the rhenium content of each sample. Then three equal aliquots of the unknown solution are transferred to an appropriate volumetric flask: one with the sample alone, one with the sample and a standard addition of the same order of magnitude as the estimate, and one with the sample and a standard addition twice the former addition. Apply the same procedure on the blank solution. Plot all the results on graph paper. Linearity should not be a problem up to 1000 μg/ml rhenium in the solution.

10. *Atomic absorption parameters:*

Wavelength: 2287.5 Å

Lamp current: 20 mA

Fuel: acetylene

Oxidant: nitrous oxide

Flame stoichiometry: slightly reducing; red cone no more than 1 cm high

Flow rate: Adjust nebulizing system so that the solution is aspirated at a rate of 3 ml/min.

If large numbers of rhenium determinations are planned, it is necessary for the analyst to optimize thoroughly all parameters following the recommendations of Section 2.3.7.

11. *Calculations:*

$$\text{Rhenium in sample (ppm)} = \frac{XV}{W}$$

where X = rhenium found for the unknown, μg/ml
   V = volume into which sample has been diluted, including all steps, ml
   W = weight of sample, g

12. *Precision and accuracy:* The precision of the method depends on the degree to which all the AA parameters have been optimized. In most cases, time spent optimizing these parameters is recovered by the satisfaction of working with a steady signal and minimum background noise. Accuracy should be excellent with the fusion method; comparison of results obtained for the SRM material and its recommended value should give an idea if the procedure has been well handled. Acid digestions may not give a good accuracy. Accuracy is also far better with the standard additions technique at the cost of slightly poorer precision.

5.7. DETERMINATION OF WO₃ IN SCHEELITE AND WOLFRAMITE CONCENTRATES

1. Scope: This method is intended for the control analysis of tungsten oxide in scheelite and wolframite concentrates as produced at the mines. The method is designed to accurately determine WO_3 in the range 10–80%

2. Summary: The sample is digested with concentrated hydrochloric acid and then with aqua regia. After evaporation of most of the acids and dilution, cinchonine is added to assure complete precipitation of tungsten (e.g., when tungstic acid is precipitated, some tungsten remains in solution, which can be recovered by precipitation with cinchonine; Tungstate can be precipitated by many organic compounds; cinchonine seems to be the best). The tungstic acid and cinchonine tungstate are filtered, dissolved, ignited, and weighed as impure tungsten oxide.

Any tungsten remaining in the insoluble residue is recovered by fusing the residue in a mixture of sodium carbonate and sodium peroxide. After leaching and acidifying the melt, tungsten is precipitated as before, dissolved, and combined with original redissolved tungsten precipitate.

Interfering elements are determined in the impure tungsten oxide and appropriate corrections are performed.

3. Interferences: Volatilization of silicon as SiF_4 is necessary in order to prevent high results.

In the presence of phosphate ions, complexes of cinchonine phosphotungstate will precipitate, necessitating a correction for this element.

If Ti(IV) is high, it may prevent the precipitation of tungsten by maintaining W(VI) in a colloidal state.

Mo(VI), Nb(V), Ta(V), and Ti(IV) may be partially precipitated, and corrections should be performed for these elements.

Other elements ordinarily present in scheelite and wolframite concentrates do not present interfering difficulties in this method.

4. Special reagents: The purity and concentration of the common chemical reagents used shall conform to Recommended Practice E50 (ASTM Reference):

Cinchonine, 100 g/l: Dissolve 100 g cinchonine in HCl (1:1), dilute to 1 l with HCl (1:1), and filter if necessary.

Cinchonine wash solution: Mix 5 ml HCl and 10 ml cinchonine solution and dilute to 500 ml with water.

Fusion flux: Mix 100 g sodium carbonate with 100 g sodium peroxide. Store in a well-capped polyethylene bottle.

FeCl₃–CuCl₂ solution: Dissolve 10 g $FeCl_3 \cdot 6H_2O$ in about 500 ml water and 100 ml HCl. Add 5 ml of 1000 μg/ml Cu^{2+} solution. Dilute to 1 l.

NaSCN solution: 100 g/l in water.

SnCl₂ solution, 150 g/l: Dissolve 30 g $SnCl_2$ in 50 ml HCl with gentle heat. After a clear solution is achieved, dilute to 200 ml with water. Prepare immediately before use.

SnCl₂ wash solution: Add 25 ml HCl and 25 ml SnCl₂ solution to 450 ml water.

Molybdenum standard, 0.1 g/l: Dissolve 0.1500 g MoO_3 by fuming with 10 ml H_2SO_4. Cool; then dilute with water to about 500 ml. Add 80 ml H_2SO_4 (1:1). Cool to room temperature. Transfer to a 1-l volumetric flask and dilute to volume with water.

Molybdenum standard, 0.01 g/l: Pipette a 50-ml aliquot of the 0.1-g/l standard solution into a 500-ml volumetric flask. Add 50 ml HCl and dilute with water to about 400 ml. Cool to room temperature; then dilute to volume with water.

WO₃ standard, 1 g/l: Weigh 1 g WO₃ into a nickel crucible. Add 10 g fusion flux and fuse. Dissolve in 50 ml of 10% NaOH solution and dilute to 1 l.

5. *Special apparatus:* Platinum dishes and nickel crucibles.

6. *Sample preparation:* The sample shall entirely pass a No. 200 (75-μm) sieve in order to be easily digested by acids. If the material is a wolframite, the sample shall entirely pass a No. 325 sieve.

Dry the sample to constant weight at 105 °C prior to analysis, cool it inside a desiccator, and keep it in there until ready to be weighed.

7. PROCEDURE. *Decomposition:* Weigh approximately 1 g of the sample to the nearest 0.0001 g into a 600-ml beaker (note 1). Add 350 ml HCl and cover the beaker with a watch glass (note 2). Place on a warm hot plate (not over 60 °C) or a steam bath for at least 3 h (notes 3 and 4). Use a glass rod to loosen any undissolved material from the bottom of the beaker at regular intervals during the digestion. Tilt the cover and evaporate to about 10 ml. Add 50 ml HCl and 25 ml HNO₃, close the cover, and warm on the hot plate until strong reaction ceases. Rinse down the cover and sides of the beaker with a little water; with the cover tilted, evaporate to about 10 ml.

Note 1: Since scheelite and wolframite minerals have a very high density, they are likely to segregate from other materials. In order to avoid introducing a large segregation error when weighing the sample, it is recommended to operate as follows: Empty the bag or vial containing the sample completely onto an adequate size sheet of paper. Do not try to homogenize the sample by hand since it is much more likely that segregation will take place rather than homogenization. Use a small spatula with a square extremity. Shape the sample roughly into a large square no more than $\frac{1}{4}$ in. thick. Divide the whole surface into nine (or more) equal subsurfaces and scoop out from the total thickness and from each subsurface a portion of the sample until the desired weight is obtained. The only way to deal effectively with segregation at this stage is to weigh a sample made of as many increments as possible.

Note 2: For scheelite concentrates, 100 ml HCl is enough for the digestion.

Note 3: Some methods recommend only 1 h digestion; however, allowing more time results in a more complete digestion.

Note 4: When decomposing the more refractory minerals such as wolframite, it has been found advantageous to cover the beaker with a sheet of polyethylene secured with a rubber band and to react the sample for 4 h on the steam bath.

Precipitation: Add 200 ml hot water and heat to a boil. Add 10 ml cinchonine solution while hot and allow to digest on a steam bath overnight (note 5). Filter through a 12.5-cm medium-porosity filter paper with ashless paper pulp and wash 10 times with the cinchonine wash solution. Save filtrate for end of filtration and insoluble step.

Note 5: Solution and recrystallization during digestion will improve the purity of the precipitate and also its filterability.

Filtration and insoluble: With a fine stream of water from a wash bottle, wash the precipitate back into the original beaker and evaporate to near dryness. Place a tared (55-ml) platinum dish under the original funnel. Dissolve precipitate with consecutive small amounts of warm concentrated NH_4OH and filter. Wash with small amounts of a warm solution made of 5% NH_4OH and 0.5% NH_4Cl. Evaporate solution to dryness and retain in platinum dish for WO_3 recovery from insolubles. Place the paper and residues in a nickel crucible. Ignite below 800 °C. Add 3 drops of H_2SO_4, and moisten the residues with HF and evaporate to dryness. Add 2.5 g Na_2O_2–Na_2CO_3 flux and fuse. Leach the fusion in precipitation water and acidify with nitric acid, combine with filtrate from precipitation step, dilute until about 200 ml, and add cinchonine; let stand overnight.

WO_3 recovery from insolubles: Dissolve precipitate and filter into original platinum dish as described in filtration and insoluble step.

Ignited tungstic oxide: Evaporate solution to dryness and ignite carefully over a small flame to remove any ammonium salts or organic matter (note 6).

Note 6: The temperature should not exceed 800 °C at any time or losses of WO_3 may occur.

Silica removal: Cool the platinum dish, and then moisten the tungstic oxide with water. Add 3 drops of H_2SO_4 and 5 ml HF and evaporate to dryness on the hot plate. Ignite to 800 °C. Cool in a desiccator and weigh. Examine the impure tungsten oxide for molybdenum, niobium, tantalum, titanium, and phosphorus by X-ray spectroscopy, gravimetry and colorimetry.

Determination of interfering elements by X-ray spectroscopy: Using appropriate synthetic standards in a WO_3 base, determine the percentage of

MoO$_3$, Ta$_2$O$_5$, Nb$_2$O$_5$, TiO$_2$, and P$_2$O$_5$ by X-ray spectroscopy. Alternatively, use the following chemical procedure.

Determination of interfering elements by gravimetry and colorimetry:

a. Add about 2.5 g Na$_2$CO$_3$ to weighed half of the ignited WO$_3$ from silica removal step and mix (note 7). (*Note 7:* Save the other half for determination of phosphorus in step c.) Cover with an additional 0.5 g Na$_2$CO$_3$. Fuse. Dissolve the fusion in hot water and heat; then filter through a 9-cm tight paper. Wash the dish and paper thoroughly with hot water and reserve the filtrate, place the paper and residue in a platinum crucible, and ignite. Add a little Na$_2$CO$_3$ and fuse again. Dissolve the fusion in water, filter, and wash thoroughly as before, combining the filtrate with the first filtrate. Transfer the residue and paper to a weighed platinum crucible. Ignite. Cool, add a few milliliters of HF and two drops of H$_2$SO$_4$ (1 + 1) to the residue, evaporate to dryness, and ignite again. Cool and weigh.

b. Transfer the combined filtrates from step a to a 500-ml volumetric flask, dilute to volume, and mix. Transfer a 50-ml aliquot of the solution to a 250-ml separatory funnel. Dilute to about 100 ml, and add 3 g tartaric acid and 20 ml H$_2$SO$_4$ (1 + 1). Add 10 ml FeCl$_3$–CuCl$_2$ solution and swirl. Add 10 ml NaSCN solution and swirl. After 5 min, add 10 ml SnCl$_2$ solution and swirl. Add 50 ml butylacetate to the funnel and shake for 1 min. Allow to settle; then drain off and discard the aqueous phase. Add 25 ml SnCl$_2$ wash solution and shake for 30 s. Allow to settle; then drain off the aqueous phase. Filter a portion of the organic layer through a dry filter paper; then measure its absorbance at 470 nm in 1-cm cells using butylacetate in the reference cell. Determine the milligrams of molybdenum in the aliquot by reference to a calibration curve prepared by adding 0, 5, 10, and 15 ml of 0.01 g/l molybdenum standard solution to a series of funnels containing 100 ml water, 3 grams tartaric acid, and 20 ml H$_2$SO$_4$ (1 ± 1), following the same photometric method.

c. Add about 2.5 g Na$_2$CO$_3$ to the other weighed half of the ignited WO$_3$ from silica removal step and mix. Cover with an additional 0.5 g Na$_2$CO$_3$. Fuse. Dissolve the fusion in hot water. Add HCl and magnesium chloride. Make strongly alkaline with NH$_4$OH, stir, and let stand overnight. Filter, wash, and ignite. Weigh as Mg$_2$P$_2$O$_7$, and calculate to P$_2$O$_5$.

8. Calculations: Sum the percentage of MoO$_3$, Ta$_2$O$_5$, Nb$_2$O$_5$, TiO$_2$, and P$_2$O$_5$ found by X-ray spectroscopy (see determination by X-ray spectroscopy); then multiply by the grams of WO$_3$ weighed in silica removal step and divide by 100 to obtain grams of impurities (A).

An alternative is to add the g carbonate-insoluble residues found in step a to the grams of molybdenum found in step b and to the grams of phosphorus found in step c. Multiply the total by 2 to obtain the grams of impurities (A). Impure WO$_3$: grams of WO$_3$ weighed in silica removal step (C). Then

$$\text{Percent WO}_3 = \frac{C - A}{\text{sample weight (g)}} \times 100$$

9. *Precision and accuracy:* It has been found experimentally that the interlaboratory agreement precision with this method on a sample containing about 60% WO_3 is 60 ± 0.30% (2σ). The method provides precise quantitative data allowing for a very effective control of a concentrate product.

If the precautions in the notes of this procedure are carefully considered, the accuracy should be excellent.

5.8.   TUNGSTEN IN SCHEELITE ORES BY COLORIMETRY

1.   *Scope:* The method is applicable to concentrations of WO_3 ranging from 0.1 to 60% in scheelite and powellite ores. Ores containing wolframite cannot be analyzed for WO_3 using this method unless some precautions are taken, which are described in items 7 and 8.

2. *Summary:* The sample is dissolved overnight in cold concentrated hydrochloric acid. W(VI) is reduced to W(V) with stannous chloride. W(V) is complexed with thiocyanate as yellow tungsten thiocyanate. Absorbance is measured at 400 nm.

3. *Interferences:* Nb(V) may generate some problems; however, it is easily complexed by oxalates [6]. Mo(VI) is interfering; fortunately, its color fades after half an hour [7], while the color of the tungsten complex is stable for at least 2 h. Fluorides and nitrates should be absent because they prevent the formation of the tungsten thiocyanate. When a large quantity of molybdenum is present, as with powellite or molybdite, it is recommended to use a scanning spectrophotometer. The cold temperature digestion is beneficial when large quantities of sulfides are present, such as molybdenite; also, cassiterite remains unattacked. A good document to read is the one by Fogg, Marriott, and Burns [8].

The thiocyanate procedure is not without weak points. For example, a lack of knowledge of the reaction mechanism leading to a necessity to rely on empirically devised recipes, which often work well with one type of material but not as well or not at all with another material. Another example is the tendency of thiocyanate to degrade to sulfur acids of unknown composition during the reduction of W(VI) to W(V), with consequent interference by metals that form sulfides and sulfosalts. Sometimes several colored thiocyanate complexes have unpredictable behavior during extraction into organic solvents. Finally, the tungsten may complex into heteropoly compounds (e.g., paratungstate, silicotungstate, etc.) from which it is removed with difficulty in acid solution. Some of these weak points can be overcome by addition of an organic radical to the reduced tungsten thiocyanate complex, as described by Donaldson [9].

4. Special reagents:

- Standard tungstic oxide solution: Dissolve 1.0 g pure tungsten oxide in 50 ml of 10% (w/v) sodium hydroxide. Dilue to 1 l. Store in plastic container. For daily use, dilute 25 ml of this mother solution into 500 ml with distilled water.
- Ferric chloride solution: Dissolve 10 g reagent-grade $FeCl_3$ in 100 ml water. Add 2 ml hydrochloric acid and filter.
- Stannous chloride solution: Dissolve 113 g reagent-grade $SnCl_2 \cdot 2H_2O$ in 250 ml hydrochloric acid. Heat to solution without boiling.
- Sodium thiocyanate solution: Dissolve 20 g reagent-grade NaSCN in 100 ml water.
- Reagent-grade ammonium oxalate.

5. Special apparatus:

- Spectrophotometer capable of operating at a wavelength of 400 nm.
- Large water bath capable of maintaining a constant temperature of 30 °C.
- Magnetic stirring bar covering most of the inside diameter of the flask used for the digestion.
- Magnetic stirring unit capable of turning the stirring bar at a slow and steady rate.

6. Standard reference materials: It is recommended to develop in-house standard reference materials, following the guidelines described in Chapter 7. The mineralogy of SRMs should match as closely as possible the mineralogy of samples to be analyzed. To use a wolframite ore SRM with scheelite-ore-like sample is looking for big trouble.

7. Sample preparation: The sample must be ground to −200 mesh to ensure the near complete liberation of the scheelite mineral from the gangue. In some cases, scheelite occurs as liberated tiny flakes that become electrically charged, especially in plastic containers and when the sample is very dry. This may generate an unsolvable segregation difficulty when the tungsten amount in the sample is very low. The use of very clean glass containers and large analytical subsamples is the only way to minimize this difficulty. In some rare cases, scheelite may be locked inside

acid-resistant minerals, and it is recommended to make a short mineralogical investigation of the scheelite occurrence. It may be necessary to
grind the sample finer or to fuse the residue for complementary tungsten
recovery. Wolframite ores may be successfully analyzed for tungsten
using this method if ground to -325 mesh.

8. Procedure. Weigh an appropriate analytical subsample into a volumetric flask and dilute to four-fifths with concentrated hydrochloric acid.

Note: If the sample contains less than 0.1% WO_3, weigh a 2.0-g sample into
a 200-ml volumetric flask; if WO_3 is between 0.1 and 0.5%, weigh a 1.0-g sample
into a 200-ml volumetric flask; if WO_3 is between 0.5 and 5% WO_3, weigh 1.0-
g into a 500-ml volumetric flask; and if WO_3 is between 5 and 20%, weigh a
0.5-g sample into a 1000-ml volumetric flask; and so on.

Carefully introduce a magnetic stirring bar into the flask and set on the
stirring plate.

Note: The length of the stirring bar should be nearly the diameter of the
flat portion of the bottom of the volumetric flask used.

Adjust speed to as slow as possible while keeping the sample in motion.
Allow to stir 9 h or overnight.

Note: In the case of wolframite ore, it is necessary to stir at least 18 h.

Remove stirring bar and rinse with concentrated hydrochloric acid. Dilute
to volume with concentrated hydrochloric acid and mix. Filter a portion of the
solution using a dry funnel and glass paper. From the filtered solution, take
an aliquot containing up to 0.50 mg WO_3 but with a volume of not more than
40 ml into a 100-ml volumetric flask. Add 1 ml of the $FeCl_3$ solution and enough
HCl to maintain a 40% v/v content.

Note: $FeCl_3$ helps to ascertain that all the tungsten is as W(VI).

Dilute to approximately 80 ml with water, mix, and cool. Add 5 ml stannous
chloride solution and mix. Add 5 ml sodium thiocyanate solution and mix. Add
0.5 g ammonium oxalate, dilute to volume with water, and mix thoroughly.
Place in the water bath set at 30 °C for 2 h. Measure absorbance at 400 nm.

9. Method of calibration: Prepare a blank and three calibrating standards made with 5-, 10-, and 20-ml aliquots of the daily standard solution
into 100-ml volumetric flasks, and then proceed as for unknown samples,
starting with the addition of $FeCl_3$.

10. Instrumental parameters: Usually the measurement of the absorbance is performed at 400 nm. Sandell [7] gives a good illustration of
the transmission curve of the tungsten–thiocyanate complex. When large
amounts of molybdenum are present, it may be advantageous to use a
scanning spectrophotometer.

11. Calculations:

$$\text{Percent } WO_3 = \frac{ACE}{BWD} \times 100$$

where A = WO$_3$ in standard solution, g
       B = aborption of corresponding standard solution
       C = absorption of sample
       W = weight of sample used, g
       D = volume of sample aliquot, ml
       E = sample dilution

12. Precision and accuracy: Precision depends mostly on the good timing of each step (e.g., all assays: blank, standards, and unknown shall be run in a very consistent manner). Good accuracy of the method depends mostly on how the digestion is handled and on how meticulously the sample preparation is performed. The method should provide excellent precision and accuracy.

5.9. ANTIMONY, ARSENIC, SELENIUM, AND TELLURIUM IN ORES AND GEOCHEMICAL MATERIALS BY ELECTROTHERMAL ATOMIZATION AND ATOMIC ABSORPTION

1. Scope: Electrothermal atomization inside an uncoated graphite furnace followed by AA measurement is used to measure the concentrations of antimony, arsenic, selenium, and tellurium after appropriate digestion of the ore samples. Depending on the material analyzed, detection limits range between 0.5 and 5 ppm of these elements in geochemical materials.

2. Summary: The sample is digested overnight at room temperature with Br$_2$. Then nitric acid is added with a little hydrochloric acid to complete dissolution of the minerals of interest and drive off the excess Br$_2$ by heating gently. As long as HNO$_3$ is present, a little HCl does no harm to the analyte.

After atomization inside the graphite furnace, arsenic is determined at 193.7 nm, selenium at 196.0 nm, tellurium at 214.3 nm, and antimony at 217.6 nm. For best results, an automatic deuterium background correction is used, and all results are displayed on a strip chart recorder.

3. Interferences: The most serious interference that may be encountered is high sulfate content in solution samples, which causes a severe suppression of atomization signals, possibly due to various losses when H$_2$SO$_4$ is volatilized [10]. The effect of other anions and cations are much less severe and can be overcome using the method of standard additions.

Another type of interference that occurs with some samples is background interference at the moment of atomization. Light scattering is one type of background interference, and it is caused by particles in the light path. This interference can be corrected by using a deuterium arc background correction system found on most modern AA spectrophotometers [11]. However, another type of background interference, molecular absorption falling within the spectral bandpass of the monochrometer, may not be accurately corrected for, leading to either positive or negative errors [12, 13].

Spectral interferences, although rare in AA spectroscopy, can occur when determining selenium in the presence of approximately 2000 or more times as much iron [14] or determining Sb in the presence of Pb or As in presence of Cd [15, 16]. All background interferences can be better eliminated through the use of Zeeman background correction [17–19].

Several things may be done to minimize the effect of background interference. Increasing the pyrolysis time and temperature may volatilize interfering substances before loss of analyte occurs. A smaller sample may be injected or greater dilution of sample solution may be performed. If interferences persist, a chemical separation of the analyte may be feasible.

Addition of nickel has been found beneficial: When the solutions have been properly prepared, a source of nickel is added to decrease the volatility of arsenic, selenium, tellurium, and antimony during the drying and pyrolysis stages that occur in the graphite furnace. It is thought that nickel selenide, arsenide, telluride, and antimonide are formed on the graphite surface [20]. These nickel compounds are stable up to about 1500 °C. Such temperatures are sufficiently high to allow most matrix salts to be completely burned off prior to atomization, which is the final rapid heating step. During the rapid heating step the nickel–analyte compound is destroyed, producing analyte atoms and a transient AA signal measured on a fast strip chart recorder.

The influence of nickel on the volatilization temperature of these elements has been studied by Welz, Akman, and Schlemmer [21]. It is also recommended to read the paper of Viets, O'Leary, and Clark [22].

   4.   *Special reagents:*

- Standard solutions: Prepare a 1000-ppm stock solution of each element (As, Se, Te, and Sb) by dissolving 1 g metal powder of each element in four different 1-l volumetric flasks using about 50 ml nitric acid. A little Br_2 may be added if the material is resistant to attack by nitric acid. However, before dilution to volume, the Br_2 should be driven off by gentle heating. Working solutions of 10 ppm are readily prepared by dilution of 10 ml stock solution to 1 l and adding 1% nitric acid to promote stability of the solutions.
- Nickel nitrate (reagent grade or better): prepare a 20-g/l solution by dissolving 99 g $Ni(NO_3)_2 \cdot 6H_2O$ in water and diluting to 1 l.

5. Special apparatus:

- Atomic absorption spectrophotometer, double beam, with background correction system.
- Graphite furnace for electrothermal atomization, compatible with the AA spectrophotometer.
- Strip chart recorder with fast response (e.g., full-scale response in less than 0.5 s).
- Hollow cathodes may be used; however, electrodeless discharge lamps (EDLs) give a better signal-to-background-noise ratio.

6. Standard reference materials: It is recommended to develop in-house standard reference materials following the guidelines described in Chapter 7.

The following two standard reference materials may be used with confidence:

CCRMP Lead Concentrate CPB-1	CCRMP Zinc Concentrate CZN-l
Sb; 0.36%	Sb; 0.052%
As; 0.056%	As; 0.026%
Se; 31 ppm	Se; 5 ppm
Te; 0.7 ppm	

7. Sample preparation: Significant, and often unrecognizeable, sampling problems may be present especially in liquid samples. All liquid samples should be treated by the whole bottle technique described in Section 4.19. Solid samples should be −200 mesh or finer and should be split from the larger bulk of material. Solid samples should also be dried at 105 °C until a constant weight is obtained prior to weighing the analytical subsample.

8. PROCEDURE. Weigh a 1-g sample and transfer to a 400-ml beaker. Add 10 ml cold water and 5 ml liquid Br_2 (be prepared to cool the sample to slow vigorous reaction if necessary). Let stand cold overnight. Add 10 ml HNO_3 and heat until Br_2 is completely expelled. Transfer to a 100-ml volumetric flask. Dilute to the mark and mix thoroughly. For some materials, this procedure may be modified as follows:

a. *Geochemical Material, Ores, Gypsum, and Some Sulfide Concentrates:* Weigh a 1-g sample and transfer to a 400-ml beaker. Add a few boiling chips. Add 15 ml HNO_3 and 15 ml HCl, and let stand at room temperature overnight with a watch glass on top of the beaker. Add 20 ml water. Bring the temperature to around 90 °C, and maintain it for 1 h, or until the solution is reduced to a volume of about 25 ml. Remove the beaker from the hot plate. Add 2 ml HCl. Cool and transfer to a 100-ml volumetric flask.

b.   *Tungsten Concentrates:* Use the distillation procedure described in Section 4.18.

9.   Method of calibration: At this point, the solution should be diluted to a volume giving optimum concentrations of the analyte that will allow the standard addition technique to perform in an area of good linearity. For example, if the estimated analyte concentration after dilution is 10 µg/ml, standards additions should be, respectively 10 and 20 µg/ml in order to generate a slope as close as possible to one that gives maximum precision.

Nickel should be added to the solutions until they all contain 200 µg/ml nickel.

10.   *Instrumental parameters:* Set up the AA spectrophotomer, graphite furnace, recorder, and EDL light source in accordance with the manufacturer's instructions. Use a temperature program that achieves a maximum pyrolysis temperature of 950 °C and a rapid atomization heating step to 2300 °C for selenium, arsenic, and tellurium. Antimony requires a temperature of 2500 °C. Sometimes it may be necessary to increase the pyrolysis temperature to as high as 1500 °C if significant background absorption is occurring at the atomization step. Record atomization peaks of all prepared solutions on a strip chart. The effect of the nickel blank can be very significant. Graphite cuvette memory problems caused by recombination of selenium with nickel on the graphite surface, following atomization, need to be carefully evaluated.

11.   *Calculations:* The concentration of analyte found from the plot of the results for the blank, sample, and samples with standard additions is multiplied by the dilution factor to yield the result in micrograms per gram.

12.   *Precision and accuracy:* Precision of one determination is about ±5% under optimum conditions. Precision may be better if averaging different measurements of the same solution. Precision is better with the EDL than with hollow cathodes.

Because of evidence that no analyte is lost in the graphite furnace during drying and pyrolisis, the accuracy of the measurement of analyte concentration is thought to be good. However, because complete decomposition of the sample is not made in the analysis of some materials, there still exists the possibility that some analyte may remain with the unattacked material. Therefore, for the highest accuracy, any residue of un-attacked material should be examined for analyte.

5.10. RAPID ALKALI DETERMINATIONS

Many procedures have been proposed for the preparation of solutions from rocks and minerals in which sodium and potassium may be quickly determined by flame photometry. Most of these are modifications of the Berzelius [23] procedure, in which the sample is attacked by hydrofluoric acid; fluorides are removed by fuming with sulfuric acid. Perchloric acid is used to advantage in some procedures. In the method described here, sulfuric and hydrofluoric acids are used, and the mixture of sulfates so obtained is ignited. The ignition converts the sulfates of iron and aluminum to oxides, which are insoluble in water and may be easily removed by filtration. This expedient was first suggested by Cantoni [24] and has been used by Abbey and Maxwell in the analysis of biotites for potassium [25]. Magnesium, calcium, and other elements accompany the alkalies. If magnesium is added to the sample at the start of the procedure, it makes the ignition proceed more smoothly and also plays the part of a radiation buffer in the flame, minimizing the effect of the extraneous elements in the final solution.

In cases where the alkali content of the sample is small, the procedure to be described is usually preferable to the much longer and more difficult J. L. Smith procedure because blanks can be reduced to an almost negligible amount with little trouble. The rapid procedure gives very satisfactory results in most cases where less than 2% by weight of either alkali oxide is present, and its application may be extended to include samples with up to 4% by weight or more, although considerable care is then necessary to attain fair accuracy. Flame procedures, when used alone, have the disadvantage of not providing an actual weight of either or both alkalies; thus, the whole success of the operation depends on the absence of elements that affect the flame, and no cross-check is possible. It is safe to say that the procedure that follows, and others like it, should never be applied to unusual or unknown materials unless there is some suitable way of checking its performance.

PROCEDURE. Transfer 0.05–0.50 g finely divided sample to a flat-bottomed platinum dish of 30–50 ml capacity. Moisten with a few drops of water, and add a solution of magnesium sulfate containing one-half the sample weight of MgO. Cover the dish, and add several drops of 1:1 sulfuric acid (about 0.3 ml of 1:1

acid for every 100 mg of sample). Heat gently on the steam bath to drive off any CO_2, cool, wash down and remove the cover, and add about 1 ml nitric acid and 5–15 ml hydrofluoric acid. Evaporate slowly on a hot plate or in a radiator, gradually increasing the temperature until the sulfuric acid fumes strongly. Cool, add about 20 ml water, and again evaporate and fume. Continue heating until all the excess sulfuric acid has been expelled. Heat the dry residue at the maximum temperature of the plate or radiator for half an hour or more, and then transfer to a triangle on a ring stand and heat with a free flame, cautiously at first, until no more SO_3 fumes appear. Then heat over a Meker burner, preferably on a platinum triangle. Ignition should not be prolonged because of the appreciable volatility of alkali salts. Cool. Add water to the dry residue, evaporate to dryness again, and repeat the ignition as before.

Reignition (the purpose of which is to remove the last traces of free acid) is not always necessary. If a large number of samples of similar nature are being run through the method, it would be well to check the pH of the solution resulting from a single ignition.

If only round-bottom platinum dishes are available, it is best to use a radiator rather than a hot plate. The same is true if evaporations are done in platinum crucibles.

Leach the ignited residue with water, digesting it on the steam bath for 2 h or more, and filter through a dense paper (Whatman No. 42) into a 50- or 100-ml volumetric flask. Wash thoroughly with hot water. Cool, and dilute to the mark.

Make preliminary readings for each of the alkalies, and prepare bracketing standards, adding to each standard the same amount of magnesium (as neutral sulfate) as was added to the sample at the beginning of the procedure. Then proceed exactly as described previously.

5.11. DETERMINATION OF CALCIUM IN FLUORSPAR (FLUORITE)

PROCEDURE. Weigh 0.500 g of − 100-mesh sample into a 400-ml beaker. Prepare a perchloric–boric acid mixture: To 70 ml water add 25 ml of 72% perchloric acid. Heat to 80 °C. Add solid boric acid with stirring (magnetic stir bar and hot plate) until no more dissolves. Allow the excess H_3BO_3 to settle. Keep the solution hot. To the sample add 20 ml of this solution. Swirl the beaker to be sure the sample is well dispersed. Heat gently below the boiling point with the beaker uncovered, gradually increasing the temperature until the perchloric acid begins to fume. Wash down the sides of the beaker with a minimum of water. Again evaporate, and continue fuming until salts begin to separate from the perchloric acid. Cool, and add 100 ml water, 2 g tartaric acid, and 3 ml hydrochloric acid. Dissolve 5 g oxalic acid in about 25 ml hot water, and add the solution to the sample. Put a stirring rod in the beaker, place a bit of hardened filter paper under the end of the rod, and heat the solution to incipient boiling. Slowly add ammonium hydroxide to the boiling solution until a drop of methyl red indicator solution colors the analyte yellow and the precipitate of calcium oxalate settles readily. Add a further 5 ml ammonium hydroxide and allow the solution to cool at room temperature. Filter through a Whatman No. 40 paper, and wash very thoroughly with a minimum volum of hot water. Return the filter to the beaker, and add 100 ml hot water and 20 ml of 1:1 sulfuric acid. Add a small crystal of manganese sulfate (1 mg is enough). Titrate immediately (before the solution cools) with $N/10$ potassium permanganate solution, stirring the solution vigorously and preventing the titrating solution from coming in direct contact with the pieces of filter paper floating in the solution. Calculate the calcium content of the sample on the basis of the reaction

$$5CaC_2O_4 + 2KMnO_4 + 8H_2SO_4 \rightarrow 5CaSO_4 + K_2SO_4 + 2MnSO_4 + 8H_2O + 10CO_2$$

In the analysis of fluorspar, it is often required that the calcium associated with carbonate, instead of fluoride, be determined. This may be done by determining the CO_2 content and subtracting its equivalent in $CaCO_3$ from the total calcium calculated to CaF_2. The carbonate calcium may also be determined by leaching the sample with 15 ml of 30% acetic acid at steam bath temperature for an hour, filtering, and washing with 30–40 ml water. The leachate may be analyzed for calcium, and the undissolved material may be subjected to the procedure given above after ignition to remove paper.

The calcium oxalate obtained in the above procedures may be ignited to carbonate or oxide and weighed, but routines based on the titration procedure are considerably faster if many samples are to be analyzed or if rapid turnaround time is important. If known samples of fluorspar are available, it is advantageous to use the method in the secondary mode, that is, calibrate the potassium permanganate against known fluorspar samples. Note that the procedure given provides a rapid silica determination: If required, magnesium may be determined on the filtrate from the calcium oxalate via an 8-hydroxyquinoline precipitation and titration or via a precipitation with phosphate. Magnesium may also be determined by AA spectrometry, as well as iron and other elements, in the filtrate from the calcium oxalate.

5.12. DETERMINATION OF CALCIUM IN LIMESTONE OR DOLOMITE

PROCEDURE. Prepare a porcelain crucible by filling it with calcium oxide and igniting at 750 °C for 1 h. Wash and dry the crucible. (The purpose of this is to remove the glaze, which would otherwise contaminate the samples; cru-

cibles so treated should be reserved for this purpose only.) Weigh 1.000 g of
−100-mesh sample into the crucible, dry overnight at 110 °C, cool in a desiccator, and weigh. Place the crucible, uncovered, in a cold furnace and bring
slowly to red heat to burn off organic matter and oxidize sulfur. Cover, and
hold at 750 °C for 3–4 h. Allow the crucible to cool in the furnace to dull red
heat before transferring to a desiccator. Evacuate the desiccator, and allow at
least 2 h before weighing.

For some purposes, the moisture and loss on ignition determinations may
be superfluous. The loss on ignition should usually not be reported as CO_2,
although it often gives a fair estimate: sulfur, organic matter, and water may
be included, and it is possible that all the CO_2 is not released at 750 °C.

Transfer the sample, ignited or not, to a 250-ml beaker, cover, and add a
little water and up to 10 ml concentrated hydrochloric acid. A few drops of
nitric acid may be added to help destroy organic matter.

When CO_2 has been removed, evaporate uncovered to dryness on the steam
bath. Transfer to a hot plate with a surface temperature of 200 °C or less, and
bake to remove excess acid and dehydrate the silica. When dry, cool, and add
to the 2.5 ml concentrated hydrochloric acid. Allow to stand for 10 min, add
25–30 ml water, and boil for 2–3 min. Filter at once through a 7-cm S & S
white ribbon or Whatman No. 40 paper, catching the filtrate in a 100-ml volumetric flask. Wash first with a little 5% hydrochloric acid and then with hot
water to a volume of about 90 ml.

Ignite the insoluble material at the same temperature as was used in the
ignition loss using the same crucible. Weigh the insoluble matter. It is usually
mostly silica. Whether it deserves further attention depends on the purpose
of the analysis.

Dilute the filtrate from the insoluble matter to 100 ml, and take aliquots for
the determination of calcium, magnesium, iron, manganese, and so on. For
calcium, direct titration with EDTA is usually possible; if much iron or aluminum are present, however, the end point may be obscured and a preliminary
separation is required. When this is the case, the procedure includes the determination of magnesium. When no separation is necessary, take a 10-ml
aliquot of the filtrate from the insoluble matter and transfer it to a 100-ml
beaker. Add a little water, 0.05 g hydroxylamine hydrochloride, 5 ml of a
solution containing 40 g carbonate-free sodium hydroxide and 10 g potassium
cyanide in 500 ml, and a small quantity of calcine indicator (any one of several
calcium-sensitive indicators may be used). Titrate with an EDTA solution containing 29.5 g ethylenediaminetetraacetic acid and 8.5 g/l carbonate-free sodium
hydroxide. Standardize the EDTA solution against a standard calcium solution
prepared by dissolving pure dried $CaCO_3$ in a minimum of hydrochloric acid
and diluting to volume. A 10-ml microburette should be used in the titration.
Alternatively, if only calcium is to be determined, a larger aliquot of the filtrate
from the insoluble matter and a 50-ml burette may be used.

A weakness of the EDTA titration of calcium is that strontium remains
undetected. Further, EDTA titration of strontium is not clear-cut; the end

point becomes indistinct, and there is a tendency to report high values. Barium has a similar effect. For example, EDTA titration of calcium in the GFS limestone 401 gave 50.25% CaO, as compared to 50.07% by carefully applied primary methods. This sample contains 0.03% SrO and 0.11% BaO. For many purposes, such differences are unimportant.

5.13. DETERMINATION OF CALCIUM IN PHOSPHATE ROCK

PROCEDURE. Weight 1.000 g of −100 mesh phosphate rock into a 200-ml flask and digest at the boiling point for 15–20 min in 30–35 ml of 1:1 hydrochloric acid. Cool, transfer to a 100-ml volumetric flask, and dilute to the mark with water. Mix, and filter through a dry S & S white ribbon or a Whatman No. 40 paper. Take exactly 50 ml of the filtrate and transfer to a 400-ml beaker. Add 10 ml of 1:1 sulfuric acid and 200 ml ethanol. Stir frequently for about an hour, and filter through a 12.5-cm Whatman No. 44 paper, covering the beaker and filter to prevent excessive evaporation of alcohol. Wash with 80% ethanol until the washings are free from phosphate. Dissolve the calcium sulfate in 500 ml water, dilute to exactly 500 ml, and take a 50-ml aliquot for EDTA titration. Alternatively, the calcium sulfate precipitate may be collected on a porous-bottom porcelain crucible, washed with 80% alcohol, heated to 600 °C, and weighed.

Attempts to determine calcium by EDTA titration in solutions containing phosphate have not been very successful. The presence of fluoride (usually present in phosphate rock) leads to slightly low results for calcium in most rapid procedures. The method described above for calcium in fluorite has been successfully applied to phosphate rocks. Calcite is a common contaminant of phosphate rock: It is usually estimated by a determination of acid-soluble carbon.

5.14. RAPID ANALYSIS OF MARL (LIMESTONE)

PROCEDURE. Put some powdered limestone in a new porcelain crucible and heat in the muffle at 800 °C for 2–3 h to remove the glaze, which would otherwise cause error by fusing with the samples. Wash the cooled crucible with water, scrubbing it thoroughly, ignite, and weigh.

Weigh 1.0000 g of 80-mesh and finer sample into the crucible, weigh crucible plus lid plus sample, and dry in the oven overnight or to constant weight at 110 °C. Cool in a desiccator and weigh. Calculate percent moisture.

Transfer the uncovered crucible to the cold muffle, and heat to redness. Cool to black heat, cover, and again heat to redness, holding the temperature at 800 °C for 3–4 h. Cool to black heat in the muffle, transfer to a desiccator, and weigh when cold. Calculate loss on ignition.

Transfer the contents of the crucible to a 250-ml beaker using first water and scrubbing thoroughly and then 2–3 drops of 1:1 HCl, again scrubbing thoroughly and washing with water to be sure that all the sample is transferred. The crucible should be reignited and reweighed at this stage to be sure that its weight has not changed appreciably. Save the weighed crucible for the silica determination.

To the ignited material in the 250-ml beaker add 10 ml concentrated HCl, stand for a few minutes, evaporate to dryness, and bake. Cool the residue, cover, and add 2.5 ml concentrated HCl. Stand 5–15 minutes; then add 25 ml water and heat to boiling. Boil 2–3 minutes, and filter at once, catching the filtrate in a 100-ml volumetric flask. Wash with 5% hydrochloric acid and then with hot water to a volume of about 90 ml, scrubbing the beaker and transferring all the insoluble matter to the filter.

Ignite the paper and precipitate in the same crucible that was used for the loss on ignition, and weigh siliceous material.

When the filtrate from the silica is at room temperature, dilute to exactly 100 ml, mix, and take out aliquots as follows:

1. 10 ml for Ca (optional, for low-Fe samples only) into a 200-ml flask,
2. 10 ml for Mn into a 150-ml beaker,
3. 10 ml for Fe into a 200-ml flask,
4. 50 ml for R_2O_3 into a 250-ml beaker (see below), and
5. 10 ml for Ca and Mg (optional, for samples low in iron and aluminum only) into a 200-ml flask.

These portions should all be measured out at the same time to prevent errors due to temperature effects.

Heat the 50-ml aliquot reserved for the R_2O_3 determination and boil for 1–2 min. Add 3 ml of 15% ammonium chloride and a little paper pulp, and precipitate at a boiling temperature with ammonia using a minimum of methyl red indicator and/or small pieces of nitrazine paper. Heat to boiling, filter into a 100-ml volumetric flask, scrub, and wash with hot 2% NH_4Cl to a volume of 90 ml.

When the filtrate is at room temperature, dilute to exactly 100 ml, and take out aliquot parts as follows:

6. 20 ml for Ca and Mg into a 200-ml flask,
7. 20 ml for Ca into a 200-ml flask, and
8. 50 ml for S into a 250 ml beaker.

After the silica determination is finished, transfer the R_2O_3 precipitate to the same weighed crucible, ignite, and weigh R_2O_3. This precipitate may be treated further if desired: The silica may be transferred to platinum and treated with HF and H_2SO_4, and the nonvolatile residue after ignition may be dissolved by carbonate or pyrosulfate fusion. Its solution should be diluted to volume and then half of it added to a similar solution of the R_2O_3, so that TiO_2 may be determined.

Treatment of Various Aliquots

1. Calcium (direct): To the 10-ml aliquot add 5 ml NaOH and KCN solution. If Mn is present, a little hydroxylamine hydrochloride should be added before the alkaline cyanide solution. Add a small scoop of calcein indicator, and titrate with versene.

2. Manganese: Evaporate the solution twice to dryness with nitric acid, and develop the permanganate color with persulfate or periodate as usual after adding the requisite amount of nitric acid, and titrate or measure the color photometrically.

3. Iron: Add several milligrams of salicylic acid and about 25 ml water. Add ammonia dropwise until the color changes from purple to brown; then restore the purple color to a maximum intensity by dropwise addition of 1:1 hydrochloric acid. A little sodium acetate facilitates this adjustment of pH. Titrate slowly with versene to complete removal of the purple color. If the final solution is more than just faintly yellow or the end point is poor, add more water or use a smaller aliquot.

4. R_2O_3 (see above).

5. To the 10-ml aliquot add a little hydroxylamine HCl and then 5 ml alkaline cyanide solution and a little calcein, and titrate to the calcium end point. Then add 5 ml of 15% NH_4Cl, allow to stand for 5 min, add eriochrome black T, and continue the titration to the magnesium end point. The addition of calcein may be omitted, and the Ca + Mg determined by titration with eriochrome black T. Some more ammonium chloride may have to be added to bring the pH to 10 for the titration of Mg or Mg + Ca.

6. To the 20-ml aliquot add a little $NH_2OH \cdot HCl$ and then 10 ml NaOH and KCN and a little calcein, and titrate with versene to the calcium end point; add 10 ml of 15% ammonium chloride, and let stand for 5–10 min; then add eriochrome black T and titrate to the magnesium end point. It may be necessary to dilute the solution somewhat before titration, particularly if magnesium is high; however, too great a dilution must be avoided.

7. To the 20 ml for calcium determination add 10 ml NaOH and KCN and a little calcein, and titrate to the end point with versene. If manganese is present, a little $NH_2OH \cdot HCl$ should be added before making the solution alkaline. The volume of solution should be kept as small as possible, but if magnesium is high, this element will interfere more in the more concentrated solution, so dilution to 75 ml or more may be necessary. If the volume is increased, more NaOH and KCN may be required. Use brilliant cresyl blue paper.

8.   Run sulfur gravimetrically as $BaSO_4$ as usual.

Versene solution: 29.5 g ethylenediamine tetraacetic acid plus 8.3 g NaOH in water, filter, dilute to 1000 ml. Standardize against pure $CaCO_3$ taken through the method.

Alkaline cyanide solution: 50 g NaOH in 50 ml H_2. Dilute to 100 ml in a stoppered cylinder. Let stand for Na_2CO_3 to settle. Draw off 80 ml into a polyethylene bottle, add 10 g KCN, and dilute to 500 ml.

Calcein: 25 g KCl, 2.5 g charcoal, and 0.25 g calcein ground together to pass a 100-mesh screen.

Eriochrome black T: Grind 0.25 g of the indicator powder with 25 g KCl.

5.15.   DETERMINATION OF BARIUM IN GEOLOGIC MATERIALS

Atomization generated by inductively coupled plasma provides an attractive detection limit of about 0.0004 μg/ml and virtually no spectral interference. However, the determination of barium in geologic materials may be difficult because of solubility problems. Mineral acids dissolve barium sulfate only slightly, and fusion using sodium peroxide is only partially successful. A fusion using sodium carbonate has been giving excellent results with a wide range of materials.

PROCEDURE.   Mix a 0.5-g sample with 3 g sodium carbonate in a graphite crucible and heat for 30 min at 950 °C.

Note: If the temperature exceeds 950 °C, the molten sample may percolate through the graphite crucible and be lost inside the furnace. However, the melting point of the mixture is close to 950 °C.

Pour the molten sample into 100 ml water in a polypropylene tri-pour beaker, stirring constantly. The small portion of the melt remaining inside the crucible is rinsed out with water after the crucible is cooled. Any residue of the melt that still remains is dissolved with hydrochloric acid after the filtration. Stir the solution for 20 min. Filter the solution and wash thoroughly with a 5% ammonium carbonate solution; the precipitate contains barium as a carbonate.

After diluting with water to 200 ml, the filtrate is tested for its barium content with the ICP spectrometer to verify that all the barium has been separated. The filter paper and the graphite crucible are washed with 10% hydrochloric acid. After filtering to remove graphite particles, the solution is diluted with 10% hydrochloric acid to a final volume of 50 ml and tested for its barium content with the ICP spectrometer.

Calibration and determination of barium may be accomplished with the following instrument conditions:

Incident power	1200 W
Relfected power	7 W
Concentric nebulizer flow	2.4 ml/min
Preflush time	30 s
Integration time	5 s
Number of integrations	3
Average profile position	

No background correction on peak or off-peak is required. Use of an SRM is recommended.

5.16. DETERMINATION OF TANTALUM AND NIOBIUM BY ICP

Determination of traces of tantalum and niobium from rocks or ores probably requires a preliminary separation. This may proceed as follows:

PROCEDURE. Digest 5 g of sample in a silica dish with 25 ml or more of hydrochloric acid for several hours on the steam bath. Evaporate to dryness and bake at 105 °C in an oven for an hour. Cool, add 1–2 ml hydrochloric acid and 50–100 ml water. Add 2–3 g hydroxylamine hydrochloride to the hot solution; then add filter paper pulp and 1 g tannin dissolved in 25 ml water. Heat on the steam bath for an hour, cool to 20–30 °C, and filter. Wash thoroughly with 2% ammonium chloride solution containing 1 g/l tannin. Discard filtrate. Burn off paper in a platinum dish. To the residue add 2–3 ml sulfuric acid and 20 ml (or more if necessary) hydrofluoric acid. Evaporate to fumes of sulfuric acid.

Fuse the residue with 5 g potassium bisulfate. Leach the cooled melt with 100 ml of 5% sulfuric acid and add 2 g tannin dissolved in 20 ml water. Digest hot for 2 h, and allow to cool overnight. Filter and wash with 2% sulfuric acid containing 1 g/l tannin.

Burn off paper and tannin in a platinum crucible. This residue may be soluble in hydrofluoric acid and thus yield a solution for ICP examination. The standard procedure is to fuse the residue with potassium bisulfate and dissolve the melt in saturated ammonium oxalate solution.

There will almost always be an insoluble residue of zircon, quartz, and so on after the ammonium oxalate treatment. Rare earth and calcium oxalates may precipitate. This residue should be examined for tantalum and niobium by dc arc spectroscopy because some tantalum minerals are resistant to both HF and pyrosulfate.

Note: Evaporation of hydrofluoric acid solution requires the addition of excess sulfuric acid to prevent volatilization of TaF_5 and NbF_5. Most of the HF should be removed at steam bath temperature before taking to fumes of sulfuric acid.

5.17.   RAPID DETERMINATION OF BASIC ELEMENTS IN LATERITIC ORES BY ATOMIC ABSORPTION OR ICP

1. Scope: This method enables the quick determination of the 12 major constituents of lateritic metalliferous deposits (e.g., silica, aluminum, magnesium, iron, manganese, cobalt, nickel, calcium, chromium, copper, zinc, and molybdenum). This list could be extended without difficulty for other basic elements; however, the purity of the sodium peroxide may become a problem.

The method is applicable to samples of unweathered substratum made of ultrabasic rocks such as peridotites, dunites, or harzburgites and also to serpentinites, garnierites, chlorites, and related minerals.

2. Summary: The sample is sintered at 480 °C with sodium peroxide in a zirconium crucible. The melt is digested with water and hydrochloric acid. After dilution and appropriate adjustment of the sodium and HCl matrix, the solution is ready for AA or ICP measurement of the elements of interest.

3. Interferences: Most of the proportional interferences (see Section 2.3.7) are corrected by using adequate flame conditions. As shown in item 10, these conditions may not always correspond with those giving the maximum sensitivity and those generally recommended by instrument manufacturers. Flame conditions are much more effective to correct these interferences than addition of lanthanum chloride, as some authorities suggested.

4. Special reagents:

- Sodium peroxide: reagent grade and finer than 20 mesh. This flux is effective only if very dry and fresh. The main impurities that may present a problem are calcium and potassium if the determination of these elements is attempted.
- Hydrogen peroxide: 30%.
- Standard solutions of the following elements:

  Nickel, cobalt, manganese, and copper: Dissolve 1.000 g metal in a minimum volume of 1:1 nitric acid and dilute to 1 l, which gives a concentration of 1000 μg/ml metal.

  Chromium: Dissolve 1.000 g chromium metal in 1:1 hydrochloric acid with gentle heating. Cool and dilute to 1 l, which gives a concentration of 1000 μg/ml chromium.

  Magnesium: Dissolve 1.000 g magnesium metal in 1:4 nitric acid. Dilute to 1 l, which gives a concentration of 1000 μg/ml magnesium.

Iron: Dissolve 1.000 g metal in 20 ml of 1:1 hydrochloric acid and dilute to 1 l, which gives a concentration of 1000 μg/ml iron.

Zinc: Dissolve 1.000 g zinc metal in 40 ml of 1:1 hydrochloric acid and dilute to 1 l, which give a concentration of 1000 μg/ml zinc.

Molybdenum: Dissolve 1.5003 g MoO_3 in 200 ml HCl. Warm if necessary, cool, and transfer to a 1-l volumetric flask, which gives a concentration of 1000 μg/ml molybdenum.

Aluminum: Dissolve 1.000 g metal in 20 ml hydrochloric acid with the addition of a trace of mercury salt to catalyze the reaction. Dilute to 1 l to give a 1000 μg/ml aluminum concentration.

Calcium: Dissolve 2.497 g dried calcium carbonate in a minimum volume of 1:4 nitric acid. Dilute to 1 l, which gives a concentration of 1000 μg/ml calcium.

Silica: Use finely ground quartz (-200 mesh), and follow the procedure described in item 8. Do a series of standard solutions containing 0, 10, 20, 40, 60, and 100 ppm SiO_2. Save standard solutions in polyethylene bottles for a maximum period of 2 weeks.

5. *Special apparatus:* 35 ml zirconium crucible, with zirconium cover.

6. *Standard reference materials:* It is recommended to develop in-house standard reference materials following the guidelines described in Chapter 7. At least two SRMs should be generated for each of the following ores: iron cap, limonite, weathered silicated substratum, and unweathered substratum. These SRMs will show large differences in their iron, magnesium, and silica contents.

7. *Sample preparation:* Because of the spinellike minerals present in these materials, it is important for all samples to be 100% -150 mesh. The material should be dried at 105 °C to constant weight and placed in a desiccator to cool until ready to be weighed. These materials may rapidly absorb air moisture; consequently, they are susceptible to large baseline weight inconsistencies if special precautions are not taken.

8. PROCEDURE. Transfer 4.5 g sodium peroxide to a zirconium crucible and then 1.000 g of sample on the top of Na_2O_2. Mix thoroughly with a glass rod. Weigh 0.5 g Na_2O_2 onto a piece of hard-surfaced paper (Albanene). Rinse off the stirring rod in this. Transfer the Na_2O_2 to the crucible, covering the peroxide–sample mixture as completely as possible. Cover, and transfer the crucible and contents to a furnace regulated at 200 °C for about 15 min. Then sinter at 480 ± 5 °C for 1 h.

Note: For the solid phase reaction to be most effective, it is important not to reach the fusion temperature. When sintering is complete, the crucible should cool for a maximum of 2 min. Put the crucible sideways in 400-ml Pyrex beaker. Add, at once, 150 ml distilled water preheated to 60 °C. When the reaction is complete, wash the crucible once with distilled water and twice with 20 ml concentrated hydrochloric acid: Heat near boiling until the solution is clear. Cool and add 2 drops of hydrogen peroxide.

Note: Addition of a small quantity of hydrogen peroxide has been found to keep silica in solution for a longer period of time.

When the solution is at room temperature, transfer to a 250-ml volumetric flask and dilute to the mark. This solution is called solution *A*.

9. *Method of calibration:* Trace elements such as copper, zinc, and molybdenum are determined using solution *A*. Because the matrix of this solution may not be very consistent, it is recommended to use standard additions for these elements.

Take a 25-ml aliquot of solution *A* into a 100-ml volumetric flask and dilute to volume. This gives solution *B*. The matrix is 0.5% Na_2O_2 and 4% HCl. This solution is used for the determination of nickel, cobalt, and calcium.

Take a 5-ml aliquot of solution *A* into a 100-ml volumetric flask and dilute to volume. This gives solution *C*. The matrix is 0.1% Na and 0.8% HCl. This solution is used for the determination of manganese, aluminum, chromium, and silica. It may also be used for nickel and cobalt.

Take a 5-ml aliquot of solution *C* into a 50-ml volumetric flask. Add 4 ml of 6 *N* HCl using a burette, and dilute to mark with water. This gives solution *D*. The matrix is 0.01% Na_2O_2 and 4.08% HCl. This solution is used for the determination of magnesium and iron.

Dilutions have been made in such a way that all measurements are performed in a zone of good linearity. A small curvature for manganese and iron may be corrected using a curve corrector.

10. *Instrumental parameters:*

Element	Wavelength	Optimum Working Range (ppm)	Fuel	Oxident	Stoichiometry of Flame
Ni	2320	1–15	C_2H_2	Air	Very lean
	3415	10–60	C_2H_2	Air	Very lean
Co	2407	1–15	C_2H_2	Air	Very lean
Mn	2795	1–15	C_2H_2	Air	Very lean
Fe	2483	1–15	C_2H_2	Air	Very lean
Cu	3247	1–10	C_2H_2	Air	Very lean
Zn	2139	0.1–2	C_2H_2	Air	Very lean

Element	Wavelength	Optimum Working Range (ppm)	Fuel	Oxident	Stoichiometry of Flame
SiO_2	2516	50–300	C_2H_2	N_2O	Very reducing
Mo	3133	10–60	C_2H_2	N_2O	Reducing
Mg	2852	0.1–0.5	C_2H_2	N_2O	Oxidizing
Cr	3579	1–15	C_2H_2	N_2O	Red cone
Al	3093	30–300	C_2H_2	N_2O	Red cone
Ca	4227	1–5	C_2H_2	N_2O	Red cone

11. Calculations: Absorption readouts should be adjusted in such a way that they give the percentage of the element directly without involving calculations.

12. Precision and accuracy: Precision is mostly affected by the stability of the spectrometer. An experienced technician should achieve a precision with a standard deviation not exceeding $\sigma = \pm 1\%$.

The main factor affecting the accuracy is the skill with which the sample digestion has been performed and the care brought to sample preparation (e.g., weight baseline, checking for undissolved residues, and loss of hard coarse material during sample preparation).

5.18. DETERMINATION OF SILICON DIOXIDE BY STANDARD ADDITION AND ATOMIC ABSORPTION OR ICP

1. Scope: The method is applicable for solutions containing from 2.0 to 1000 μg/ml SiO_2. Some types of solid samples can be analyzed for SiO_2 by this method provided that complete dissolution of the silica is possible as discussed in the interferences section.

2. Summary: The standard addition method is used to determine the instrumental response of the analyte in the sample matrix. The sample is dissolved by a mixture of HCl, H_2SO_4, and HF inside a polypropylene bottle. Excess HF is complexed using excess boric acid.

3. Interferences: Any type of precipitate that forms on dilution or addition of a standard will probably contain some SiO_2 and thus invalidate the results. It is recommended for the analyst to be extremely prudent about the validity of his results because the determination of silica using atomic absorption or ICP may seem attractive; however, many pitfalls are likely to occur.

At this stage, a short discussion on the nature of "soluble" silica seems necessary in order to attract the attention on potential difficulties.

The solubility of SiO_2 cannot usefully be considered within the conventional rubrics of solubility product, solubility in grams per liter, and so on. Silicic acid $H_2SiO_3 \leftrightharpoons H^+ + HSiO_3^-$ polymerizes more or less reversibly according to concentration in acid solution, in which it exists as a gel. There are no particles of SiO_2; instead a glasslike structure exists. Papers have been written on the structure of silicate solutions. Such structure, and the degree of polymerization of the silicic acid, are altered by simple agitation.

Precipitating agents exist that will remove soluble silica from solution. Organic gel-forming compounds (e.g., gelatin) of opposite charge to the silicic acid gel are used in analytical procedures. Zinc silicate is very insoluble at pH ~6, and zinc carbonate will precipitate silicic acid from its solution if the polymeric silicic acids can be depolymerized (e.g., by prolonged boiling in the presence of zinc ions).

Homogeneous artificial silicate minerals can be prepared from concentrated silicic acid solutions. Analysis of silicates is performed after solution of the whole sample in dilute acid.

Analytical difficulties in determining soluble silica are intractable. Colorimetric and AA methods are grossly affected by the degree of polymerization, which is not known and cannot be reproduced in calibrating standards. Reported values for soluble SiO_2 should be viewed with suspicion when they are generated by these methods.

Possibly the only certain ways to determine SiO_2 in solution are

a. to evaporate a large volume to dryness with acid, filter the dehydrated silicic acid, and weigh it (this is a lengthy and messy procedure, not suited to routine control work) and

b. to examine the solution by spark solution emission spectrometry.

The properties of silicate solutions have never been investigated with unqualified success. The literature on the subject is contradictory and vague and never deals with complex solutions containing large quantities of metals.

Iler quotes the work of Weitz et al. to the effect that polymeric silicic acids decompose completely to the monomeric acid on dilution, that is, the degree of polymerization depends on the concentration of silica. This may be true in pure silicic acid solutions if polymerization has not proceeded too far but is most certainly not true universally [26, 27].

It is suggested that the most likely way to remove silica from solutions is through the addition of zinc sulfate at some stage in the process where pH is about 6 and the solution is hot and agitated. Zirconium salts may possibly precipitate silicic acid from more acid solutions, but depolymerization is slower the more acid the solution. Another possibility is the addition of gelatin or related compounds to an acid solution.

    4.   Reagents: Quartz, −200 mesh; hydrochloric acid; sulfuric acid; hydrofluoric acid; and boric acid crystals.

    5.   Equipment and apparatus: Atomic absorption spectrophotometer.

    6.   Calibration and standardization: To prepare a standard calibrating solution accurately, weigh 0.5000 g pure quartz that has been ground to −200 mesh or finer. Transfer to a polypropylene bottle with airtight screw cap. Add 30 ml HCl, 3 ml H_2SO_4, and 8 ml HF to the bottle. Adjust the cap and heat in a drying oven set at 98 ± 3 °C for 1–5 h. Alternatively, the bottle may be placed in a beaker of boiling water for heating. Allow to cool before removing lid to check for complete dissolution. When dissolution is complete, add 75 ml of 2% boric acid solution to the bottle, rinsing lid with some of the 75 ml. Add an additional 300 ml of 2% boric acid to a 500-ml volumetric flask. (Prepare the boric acid solution by heating 20 g boric acid crystals in 600 ml water until dissolved, then diluting to 1 l. The boric acid complexes excess HF, rendering it inactive to glass and skin.) Transfer contents of bottle to the volumetric flask, and dilute to volume and mix. This solution is 1000 μg/ml as SiO_2 or 467.5 μg/ml as Si. Store in a polypropylene container. Make appropriate dilutions of this standard solution with 1.5% boric acid to obtain 25-, 50-, 125- and 500-μg/ml SiO_2 solutions.

    7.   Sample preparation: No preparation is required for solution samples. Solid samples must be dissolved by the method appropriate to the type of solid. For samples that are very alkaline or high in WO_3, some precautionary steps are needed as described later in step b.

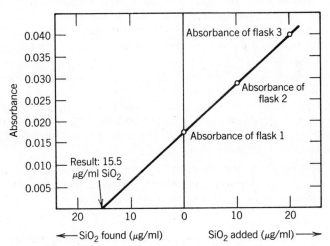

Figure 5.9. Illustration of plot of AA standard addition method for determination of silica.

8. PROCEDURE.

a.   Obtain estimates of SiO_2 concentration in samples. This can be done by running quick absorbance comparisons with the known concentration standard solutions prepared above.

b.   As shown in what follows, make appropriate aliquots of each sample into three plastic graduated or plastic volumetric flasks. Then make additions of the proper concentration standard, calibrating solutions to flasks 2 and 3 in the amounts indicated.

SiO_2 Estimate (ppm)	Sample Aliquot (to all 3 flasks) (ml)	Volume of Final Dilution (ml)	Concentration of Standard Calibration Solution Used for Additions (ppm)	Amount of Calibrating Solution Added (ml)	
				Flask 2	Flask 3
2–10	5	10	25	2 (5 ppm)	4 (10 ppm)
10–50	5	10	50	2 (10 ppm)	4 (20 ppm)
50–200	5	50	125	5 (12.5 ppm)	10 (25 ppm)
200–500	5	100	150	10 (15 ppm)	20 (30 ppm)
500–1000	5	250	500	5 (10 ppm)	10 (20 ppm)

The ppm values in parentheses are the final concentrations of the standard addition.

Solutions that are highly alkaline or high in tungsten will occasionally have precipitation problems when standards or boric acid are added. This can be prevented in most cases by adding some water to the flasks after aliquoting the sample but before making additions of the standard calibrating solution.

c.   Equivalent amounts of 1.5% boric acid must be present in all three flasks. Since the calibrating solution is 1.5% boric acid, amounts of boric acid solution equal to the total amount added as calibrating solution in flask 3 should be added to flasks 1 and 2. For example, if 10 ml standard calibrating solution is added to flask 3, 10 and 5 ml of 1.5% boric acid is added, respectively, to flasks 1 and 2.

d.   Dilute the three flasks or graduates to volume and mix.

e.   Make AA measurements on all three solutions. The silica signal is somewhat noisy at best, so instrument setup is critical.

The following instrumental parameters must be optimized with care.

Flame: fuel-rich nitrous oxide–acetylene with red cone to almost maximum height without flame turning to white or yellow and without excessive carbon buildup.

Burner height optimized for greatest signal and least noise: usually about 6 mm below center of light beam path.

Lamp current: 12 mA.

Bandpass: 0.3 nm.

Wavelength: 251.6 nm.

Nebulizer flow rate: 3.5–4 ml/min.

Background correction: necessary in most all cases.

f.   The three absorbance values are plotted on graph paper as shown in Figure 5.9, where the line intercepting the concentration axis is the concentration of SiO_2 in the first flask.

9.   Calculations: The value obtained from the plot for the concentration of SiO_2 in the first flask is multiplied by the dilution factor to yield the result in $\mu g/ml$ (ppm) SiO_2.

10.   Precision and accuracy: Greater precision and accuracy are achieved if the additions of SiO_2 to the second and third flasks are close to one and two times, respectively, the amount present in the first flask, so adjustment of the dilution and amounts of standard calibrating solution added may be advantageous in some cases. Accuracy is also greater if the line plotted through the three absorbance values is straight and parallel to a line plotted for absorbances of similar standard solutions containing no sample. It is recommended that the total concentration of SiO_2 in the third flask not exceed 60 $\mu g/ml$ so that the plotted line will remain linear. Results obtained with this procedure under optimum conditions are thought to be precise to ± 10% relative.

5.19.   DECOMPOSITION AND ANALYSIS OF CHROMIUM-RICH SPINEL

Spinels in general are not easy to decompose and often present serious problems in obtaining reliable analytical results [28].

The following procedure is particularly well adapted to the rapid analysis of a chromium-rich spinel (chromite), a mineral rich in chromium, iron, magnesium, and aluminum and may also contain small amounts of nickel, cobalt, manganese, calcium, and silica [29].

The rapid analysis of a chromite is an excellent exercise for the training of a young analyst because of good exposure to problems generated by sample preparation, sample solution, and simple chemical separations prior to instrumental measurements.

PREPARATION OF SAMPLE. The material for analysis is ground to pass a 325-mesh screen. This fine grinding does not appear to result in excessive oxidation of the ferrous iron. The grinding is done by hand in an agate mortar, and the material should be screened frequently to avoid doing excessive work on fine particles. This may be performed on a small 5-g 100-mesh sample. Contamination of the sample by silica from the mortar does not usually amount to more than about 0.1% of the total. After grinding, the sample should be thoroughly mixed because it may have been highly segregated during the screening operation. Dry the sample at 105 °C for 3 h before weighing the analytical subsample.

DECOMPOSITION OF SAMPLE. Protect the walls of a platinum crucible by fusing gently and swirling 2 g anhydrous sodium carbonate. Add 6 g sodium peroxide and then 0.5 g prepared chromite sample, and mix thoroughly. Cover the mixture with 4 g anhydrous sodium carbonate. Put in a furnace at 550 °C for 1 h (see Figure 5.10).

Note: If the temperature of the fusion exceeds 550 °C, some platinum is attacked and stays on the filter during the filtration as a grey deposit. This deposit is not attacked by a new fusion.

After cooling the crucible a few minutes, leach it with 100 ml distilled water. Do not leave the platinum crucible for too long in the water. Heat the solution

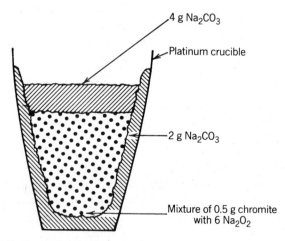

Figure 5.10. Protecting platinum crucible with Na_2CO_3 prior to Na_2O_2 fusion.

on a steam bath to remove the hydrogen peroxide. Filter on hard paper (e.g., filter through a No. 41 H paper placed inside a prewashed Whatman No. 42 paper). Wash with a cool dilute solution of NaOH and Na_2CO_3 (10 g of each per liter of distilled water). The solution is used to determine chromium and the precipitate to determine iron.

VOLUMETRIC DETERMINATION OF CHROMIUM. Acidify the filtrate with sulfuric acid until an orange color is obtained. Add an excess of solid $(NH_4)_2Fe(SO_4)_2$ weighed with precision. Cool to room temperature. Add 10 ml of a 1:5 sulfuric acid solution and 10 ml phosphoric acid (e.g., complexation of Fe^{3+} formed). Add 6 drops of 0.2% sodium diphenylamine sulfonate solution. Titrate the excess Fe^{2+} using $N/10$ potassium dichromate (1.733 g/l chromium).

Note: Elements such as V, Ba, Zr, Re, and Zn may interfere. The chromium determination may come out too low. To prevent this, it is necessary to precipitate the iron twice.

VOLUMETRIC DETERMINATION OF IRON. Dissolve the precipitate with HCl and look in the filter paper for unattacked chromite. Heat the solution to about 90 °C. Reduce all the iron using a 10% $SnCl_2$ solution (the stannous chloride is dissolved in HCl, 6 N). One drop of $SnCl_2$ in excess is the maximum that can be tolerated. Cool to room temperature. Add 10 ml of a 5% mercuric chloride solution, all at once, mix immediately, and add 10 ml of a 1:5 sulfuric acid solution, 10 ml phosphoric acid, and 6 drops of 0.2% sodium diphenylamine sulfonate solution. Titrate Fe^{2+} using $N/10$ potassium dichromate.

DETERMINATION OF OTHER ELEMENTS. For the determination of magnesium, aluminum, nickel, cobalt, and manganese, it is best to fuse another 0.5-g sample, as already described, after leaching the melt with 100 ml distilled water; hydroxides are dissolve with HCl. Heat near boiling until the solution is crystal clear. Dilute with distilled water to 200 ml in a volumetric flask. Run these elements by atomic absorption as described in the rapid analysis of lateritic material (Section 5.17).

5.20. DETERMINATION OF SILVER IN ORES BY ATOMIC ABSORPTION

1. Scope: The method is applicable from 0.2 to 100 ppm silver if flame atomization is used. Larger amounts may be determined by appropriate dilution into the working range of the standard solutions. The detection limit may be greatly improved if using electrothermal atomization inside a graphite furnace if the method is used to determine silver in geochemical materials.

   2.   Summary: The finely ground sample (-200 mesh) is decomposed by concentrated nitric acid. Any chloride ions present are complexed by a small amount of mercuric nitrate in order to prevent silver losses into the insoluble fraction. The final measurement may be performed either by flame atomic absorption, ICP, or graphite furnace atomic absorption. For the latter, a standard addition technique is used.

   3.   Interferences: When determining small amounts of silver by atomic absorption, background correction must be used to correct for light scattering and molecular absorptions, which can cause significant error. The loss of silver by formation of insoluble silver compounds is prevented through the use of mercuric nitrate.

   If the material is coarser than 200 mesh, it has been found experimentally via round-robin testing that the solution of silver minerals is incomplete. If the material is 100% -200 mesh, very few silver minerals will resist the nitric acid digestion. If it is found that some silver resists the digestion, it is recommended to fuse the residue with $LiBO_2$, followed by a solution of the melt using nitric acid only. If the residue is very large, it may be advantageous to reduce the sample to -325 mesh prior to digestion.

   4.   Special reagents: Mercuric nitrate solution: Dissolve 5 g $Hg(NO_3)_2 \cdot \frac{1}{2}H_2O$ in 10 ml nitric acid. Warm the solution if necessary and add water. Dilute to 100 ml.

   5.   Special apparatus: Atomic absorption spectrophotometer with automatic background correction, graphite furnace for electrothermal atomization, and strip chart recorder with fast response.

   6.   Standard reference materials: There are many standard samples with reliable silver values. Some of them are:

Sample	Silver Value (ppm)
Copper concentrate CCU-1 (Canmet)	139
Sulfide concentrate PTC (Canmet)	5.8
Zinc–lead–tin–silver ore KC-1 (Canmet)	1140
Zinc–tin–copper–lead ore MP-1 (Canmet)	59.5
Lead concentrate CPB-1 (Canmet)	626

   Many analyzed samples from various exploration projects should be available to the analyst. These analyzed samples should be prepared by a standard laboratory specially equipped to prepare SRMs. In Chapter 7, it is emphasized how important is this capability for a mining company.

   Most of the available standards should have been carefully analyzed for all their constituents, and the silver determination should have been performed at least by two entirely different procedures, including fire assay if possible.

7. *Sample preparation:* The sample should be ground and split in accordance with the sampling principles exposed in Chapter 1. It is assumed that the reader is familiar with all these principles.

As a rule of thumb, the sample should be ground to 100% -200 mesh.

When weighing the analytical subsample, precautions to minimize segregation errors should be taken. The following procedure is suggested:

- The sample must not exceed about 100 g; if bigger, it should be split down to a smaller size using an adequate splitter.
- Empty the vial or the bag containing the sample completely, above an adequate sheet of paper.
- Do not try to homogenize the sample at this stage, since any action may promote segregation.
- Use a small spatula with a square extremity.
- Shape roughly the sample into a large square less than $\frac{1}{4}$ in. thick.
- Divide the whole surface into a minimum of nine equal subsurfaces and scoop out from the total thickness and from each subsurface a little part of the sample until the desired weight is reached.

This procedure is far from being perfect; however, it is better than scooping out the analytical sample directly from the vial or the bag.

8. PROCEDURE. Transfer a 5-g sample into a 400-ml Erlenmeyer flask. Add 50 ml concentrated nitric acid, some boiling chips, and cover with a small watch glass. Heat on a shaker hot plate over low heat until the volume has been reduced to about 20 ml.

Note: The volume must not be reduced below 10 ml or insoluble silver compounds may form.

Remove from heat. Wash down the sides of the Erlenmeyer and the watch glass with water and dilute to about 50 ml. Add 3 drops of the mercuric nitrate solution. Heat again and boil gently for 5 min.

Transfer into a 100-ml volumetric flask. Dilute to volume with water and mix.

If the insoluble is going to be tested, filter the solution into the 100-ml volumetric flask and rinse with diluted nitric acid. Dry the filter, burn it at 600 °C, perform a fusion as described in Section 2.4.4 or 2.4.9, and then dissolve the melt in nitric acid instead of hydrochloric acid.

For better information, it is recommended to determine silver on both solutions separately.

9. *Method of calibration:* Flame atomic absorption: A blank and stan-

dards containing 0.05, 0.10, 0.25, 0.50, and 1.00 μg/ml silver in 10% HNO_3 are used to obtain the calibration curve.

Graphite furnace–atomic absorption: A blank and standards containing 0.02 and 0.04 μg/ml silver in 10% HNO_3 are made daily.

To all blanks and standards, 3 drops/100 ml solution of mercuric nitrate are added.

10. Instrumental parameters: For flame-atomic absorption, the important parameters are:

Light source	Hollow cathode
Wavelength	328.1 nm
Bandwidth	1.0
Burner head	Single-slot air–acetylene
Flame	Air–acetylene
Flame stoichiometry	Lean blue (very oxidizing)
Mode	Double-beam and background correction

Those for graphite furnace–atomic absorption are:

Light source	Hollow cathode
Wavelength	328.1 nm
Bandwidth	1.0
Mode	Single-beam and background correction
Readout	Peak height
Purge gas	Argon
Gas flow	30 cfh
Graphite tube	Uncoated and cylindrical

Indicated temperature, °C	125	125	400	400	1800	1800
Time, s	0	5	10	5	0	5

All measurements are recorded on a strip chart recorder with fast response.

11. Calculations:

$$\mu g/ml \text{ measured} \times \text{dilution factor} = \text{ppm found Ag}$$

$$\text{ppm found} \times 0.02917 = \text{oz/ton Ag}$$

$$\text{Dilution factor} = \frac{\text{volume of extract} \quad (ml)}{\text{sample weight} \quad (g)}$$

12. Precision and accuracy: The precision is reasonable. Flame atomic absorption $\sigma = \pm 3\%$ relative, and graphite furnace–atomic absorption $\sigma = \pm 7\%$ relative.

Accuracy depends mainly on the attention given to the determination of silver in the residue from the nitric acid digestion. Accuracy can be drastically affected in a sample with high halide content if no precautions are taken.

5.21.   X-RAY SAMPLE PREPARATION

In X-ray analysis, there are a number of methods of bringing solid samples up to the instrument. Because the technique usually requires many calibrating standards for quantitative results, it is desirable to prepare these in such a way that they can be reused repeatedly over a long period of time. If a bank of analyzed samples can be accumulated, these may repay the effort of preparing them many times over. There are some rather stringent requirements if such a bank of standards is to become useful:

1.   The prepared pellets or disks must be essentially permanent and must be stored and kept without damage or contamination.
2.   Very complete and careful records must be kept so that anyone can be sure of exactly repeating the preparation procedure on an unknown sample even several years later.
3.   There should be a limited number of rigidly defined preparation procedures; otherwise, the system becomes cumbersome and even unworkable.
4.   Adequate supplies of unprepared material should be available so that chemical analysis for any constituent can be performed as needed.

The following methods of preparation are recommended as the most universally applicable.

1.   *Direct Pressing.* Mix and grind 3 g of −325-mesh powder with 0.1 g chromatographic cellulose and press into a pellet with a methyl cellulose rim and backing. This procedure is required when low values (e.g., Co in laterite, W in Mo ores) are of interest. Matrix and particle size effects are large, and separate suites of standards are required for every sample type. Matrix correction schemes must be empirical, and preparation procedures must be rigidly adhered to.

2. *Fusion, Glass Window.* Mix 0.2 g of −200-mesh sample with 1.0 g of lithium metaborate and fuse in graphite or Pt-Au alloy crucible. Pour into a mold and cool slowly. The working surface must be optically flat. This procedure is most nearly ideal when loss of sensitivity due to dilution is not critical, fusion is complete, and any volatiles are not of consequences. Roasting prior to fusion may extend its usefulness.

3. *Fusion Plus Grinding and Pelleting.* Fuse a 0.5-g sample with 1.00 g $LiBO_2$ in graphite at 900 °C. Cool. Crush the resulting glass bead and grind to −200-mesh or finer. Proceed as in 1 above. This procedure is the most versatile and also the most time-consuming. It should be used with samples containing refractory minerals (wolframite, molybdenite, cassiterite, etc.) with which complete fusion is in doubt.

4. Other procedures may be devised to meet special circumstances: Examples are (a) fusion with sodium tetraborate when nonsilicates are to be analyzed (e.g., limonites), (b) collection of precipitates on millipore filters (e.g., As and Se on tellurium-copper), (c) use of sulfur as a binder (analysis of sulfide minerals), and (d) direct examination of powders (qualitative analysis, determination of sampling constants).

If several potential standards are available, each of these should be prepared for X-ray examination by one or more of the suggested procedures as a start toward accumulation of a bank of X-ray standards.

Standards should be preserved in a controlled atmosphere (desiccator cabinet) in such a way that their surfaces are kept uncontaminated by dust or by contact with their containers. The pill-box and curtain ring combination meets these requirements and permits easy labeling.

Each pellet should be examined under a binocular to be sure that particle size meets requirements (less than 10 μm for light elements) and that homogeneity is adequate. Pellets that show visible variations are useless and should be discarded.

The usefulness of the X-ray technique in a laboratory awaits a start of a standards preparation program.

5.22. DETERMINATION OF MERCURY IN SULFIDE RESIDUES

PROCEDURE. To the alkaline residue add water and 1:1 sulfuric acid in 10 ml excess. Use no more than enough water to control the reaction. Transfer to a separatory flask, add 0.1–0.2 g KBr and 1–2 ml liquid bromine. Cover the flask with a watch glass, and agitate with a Teflon stir bar for an hour or more.

Add 5 ml CCl_4, agitate, and separate. Repeat the CCl_4 extraction five to seven times, collecting the CCl_4–Br solution in a separatory funnel containing 5 ml 1:1 H_2SO_4 and 10–15 ml water. The purpose of this extraction is to remove all the sulfur (as S_2Br_2, etc.), while retaining the mercury in the aqueous phase. A small insoluble residue collects at the liquid–liquid boundary in this extraction. It contains mercury and must be retained in the separatory flask as much as possible.

Shake the separatory funnel containing the Br solution and dilute acid. Draw off and discard the CCl_4 layer. Wash the acid layer three times with 5-ml portions of CCl_4, and discard the washings. Add the acid layer to the main part of the solution.

Everything else that follows is unnecessary if a cold-tube AA is used.

The removal of sulfur must be complete: Otherwise, mercury will be lost during the next operation, in which bromide in the solution is oxidized to bromine and removed.

To the solution in the separatory flask, which should show a slight yellow color, add 5 ml CCl_4, agitate, and add 3% $KMnO_4$ solution slowly. When the pink color of MnO_4^- no longer disappears immediately on addition, remove the CCl_4 layer, collecting it in a separatory funnel containing 5 ml 1:1 sulfuric acid and 10–15 ml water. Add fresh CCl_4, and continue the addition of permanganate. Repeat until all bromide has been oxidized, and the resulting Br_2 has been removed by extraction. Do not add more than the smallest possible excess of $KMnO_4$. Collect all the portions of CCl_4 in the separatory funnel, and when the operation is complete, shake with the dilute acid and then discard. Return the acid washings to the main solution.

The solution should now show a slight brown color [due to $MnO(OH)_2$] and should be almost completely free from bromine. Add just enough diluted H_2O_2 to remove the brown color completely. Add about 0.2 g phenol and mix thoroughly for about 5 min.

If less than 2 mg mercury are present, add 5 ml $CHCl_3$, and titrate in the extraction flask with 0.1% dithizone in $CHCl_3$, removing the organic phase from time to time and replacing it with fresh $CHCl_3$. Titrate until the green color is not removed from the organic phase in 5 min violent agitation. Read the burette, add excess dithizone and agitate.

Reserve the $CHCl_3$ extracts for the colorimetric determination of Hg.

If more than about 2 mg mercury are present, agitate the solution with 5 ml $CHCl_3$, stand for 15–20 minutes, and run off the organic layer, together with any emulsion or solid particles that may be present, into a titration flask. Add 5 ml 1:1 sulfuric acid and 50–100 ml H_2O, and titrate with dithizone as above, finally adding excess, and reserve the organic phase. Only a small part of the total Hg will be found here.

Transfer the main part of the solution to a 200-ml volumetric flask, and titrate a suitable aliquot as above.

If only traces of mercury are present, it may not be possible to quantitatively recover it by dithizone titration because traces of bromine may still be present in the solution. In this case, add methyl red to the solution and then sodium hydroxide to neutral, and make just acid by adding a drop of 1:1 sulfuric acid. Add 2 ml M sodium acetate, and extract with dithizone in excess.

Determine mercury in the $CHCl_3$ solutions by the reversal technique, measuring the absorbance of the suitably diluted chloroform solution at 610 nm before and after shaking with 25% sulfuric acid containing bromide.

5.23. RAPID METHODS OF DETERMINATION IN ENVIRONMENTAL ANALYSIS

In several countries, environmental protection agencies are commissioned, among other things, to develop routine methods for analyzing industrial effluents for toxic constituents and are issued a number of "official" determinative methods. Results obtained by these methods are used to establish compliance or noncompliance with the law, which unequivocally establishes permissible concentration limits for a large number of toxins in effluent waters and wastes.

This may seem logical enough: There is a concentration limit for each regulated toxin. The actual concentration is determined; it is either in compliance or it is not.

There are, however, difficulties; these difficulties have not been properly addressed. As a consequence, frustration and confusion have developed, and large sums of money have been wasted. It is time to examine these difficulties and find ways of dealing with them.

One major difficulty lies in the analytical process itself. It is almost always possible for a competent and fully equipped analyst to determine the concentration of toxin X in a valid sample with a high degree of certainty, but not using rapid routine methods unless these have been designed for analysis of the specific sample type in question.

Rapid determinative methods are essential when large numbers of samples must be handled, but they are only applicable in circumstances and with sample types for which they are designed. The reason is that almost no determinative process is specific for the toxin sought. Given an unknown material in which toxin X must be determined, it is necessary to first separate X from its matrix, or at least from all constituents of the matrix which interfere in the readout process. This is impossible in a rapid routine.

Given a specific material, it is usually possible to devise a rapid direct method, often of high precision and reliability, for determining X, but this method may fail badly when applied to another material, or even a similar material from another source. A good example is the hydride generation AA method for selenium in water. With most ordinary potable waters, this method yields precise and reliable results. When applied indiscriminately to effluents or solutions from leach tests, results are often highly variable and absolutely erroneous.

Another example is the distillation–colorimetric method for total cyanide in water. False and nonreproducible results are caused by thiocyanate and by a variety of organic nitrogenous compounds commonly found in effluents.

A skilled and knowledgeable analyst can sometimes overcome interferences in a rapid method by changing details to suit the case in hand; however, most technicians are neither skilled nor knowledgeable and besides are instructed to follow the official recipe exactly.

Faced with clear evidence that an official method for determining X gives widely varying results, an exercise called the "round robin" is brought into play. In this exercise, samples of the material in question are sent to a number of laboratories for analysis. A coordinator evaluates the results and reaches conclusions.

Sometimes the results are concordant; this shows that all the laboratories followed the same recipe; it does not necessarily show that the method is interference free, that the average of all results is a good measure of the X concentration, or that the method will yield concordant results on all samples.

More often, results from the several participating laboratories are highly variable. There is no way to decide, or at least to persuade everyone, what is the cause of the variability. Some analysts may be incompetent or inadequately equipped. The method may be at fault. The samples may be labile or unstable. The shipping containers may have reacted with the samples. Contamination may be to blame. Each participant chooses his own reason.

Round robins in which a variety of determinative methods are applied by the several participants are especially likely to give variable results. The fact that different methods give different answers proves that some (or all) of the methods are faulty or incompetently applied. Among the various results, there may or may not be one that is very nearly correct.

Another difficulty lies in the fact that the extreme precautions against contamination necessary in trace element analysis are so expensive that few laboratories can afford them. Determination of traces of ubiquitous elements like lead, nickel, cadmium, and chromium cannot be successful unless every step in the collection, distribution, and analytical process is very carefully scrutinized and designed. Patterson and Settle [30] found

order-of-magnitude errors in routine lead determinations in biological materials and natural waters and showed that these errors are due to contamination from apparatus, reagents, and the laboratory atmosphere. Numerous other examples of contamination errors can be found in the
literature.

There is not much doubt that trace element round robins will yield
many results that are too high when common elements are determined.
Competent laboratories always run blank determinations; but when the
blank is an order of magnitude higher than the value sought, and is unavoidably variable, results will be imprecise and most always too high.

To subject data accumulated under the circumstances outlined above
to statistical manipulation and to draw conclusions therefrom is so obviously a mistake that one wonders why it is done. Since it *is* done, the
statistical process will be explained.

Any set of values may be averaged by adding them and dividing by
their number, N. This average is called the arithmetic mean, $\bar{x}$. The standard deviation s is a measure of the distribution of the data about the
average or mean. The square of the standard deviation, s^2, is called the
variance. It may be estimated in a number of ways, the most common
of which uses the formula $s^2 = \dfrac{(x_1 - \bar{x})^2 + (x_2 - \bar{x})^2 + \cdots (x_N - \bar{x})^2}{N - 1}$
where $x_1, x_2, \ldots x_N$ are the individual values.

Generally about 68% of the values will fall within one standard deviation from the mean, and one may follow the average, x, by an uncertainty
statement. If X is the unknown true value.

$$X = \bar{x} \pm s \quad (68\% \text{ confidence})$$

This means that we feel 68% certain that the true value, X, lies between
$\bar{x} - s$ and $\bar{x} + s$. If we would like to report the uncertainty in our estimate
of X with 95% confidence, we write

$$X = \bar{x} \pm 2s \quad (95\% \text{ confidence})$$

This statistical rubric applies to measurements subject to random errors;
its use in the treatment of nonrandom errors is a mistake, but that is what
is done (see Sections 1.15 and 6.1).

Having found an average of the round-robin data, a usual next step is
to eliminate "outliers." Most often, those data that lie outside two standard deviations from the mean are discarded; a new average and standard

deviation are then calculated and reported. Elimination of the outliers much improves the appearance of the data, and this is its only justification.

In a case where one laboratory uses an absolute method that overcomes all interferences and also carefully controls contamination problems and the rest use a rapid routine that does not take interferences into account and do not adequately control contamination, the average of the erroneous results so generated will be reported with a relatively small error statement. The laboratory generating the correct result is discredited, and the absolute method is judged in error and rejected; since it is likely to be more expensive and time-consuming than the rapid routine, this provides a further reason for its rejection.

It must be noted that a rapid routine method may yield results that are systematically either high or low. Selenium by the hydride generation–AA method yields, with samples containing Se(VI), Se(0), or any one of a number of interfering substances, results that may be an order of magnitude too low. For example, in an acetic acid extract of gypsum sludge, laboratories using the unmodified hydride generation–AA method found 0.28 and 0.061 ppm Se, while laboratories using more absolute methods found 1.1 and 1.2 ppm Se. Knowledgeable modification of the procedure yielded 1.34 ppm Se.

In considering this example, one may ask, "Why not simply go along with the hydride-generation method as written? It puts us in compliance, while our own results do not." Of course, the answer is that someone will sooner or later find their error, and the gain will be short-lived.

Cyanide determination by the distillation–colorimetric procedure yields, with samples containing thiocyanate or any one of a number of other nitrogen-containing compounds, results that may be an order of magnitude too high.

Note that we are not here concerned with small errors and uncertainties; we are talking about gross errors arising from the indiscriminate use of rapid routine methods of analysis and from the inability of most commercial laboratories to deal with contamination difficulties. To judge compliance or noncompliance and to enforce restrictive laws on the basis of results subject to gross uncertainties is an abomination. Regulations based on such an insecure foundation cannot be either fair or logical, especially if they lead to expensive plant or process modifications.

5.23.1.   Instrumental Methods

Of a number of instrumental trace element techniques, isotope dilution mass spectrometry (IDMS) and neutron activation analysis (NAA), properly applied, provide determinations of high reliability. There is little chance, however, that either can ever be used for routine analysis of environmental samples. Their function in this respect is to check on the reliability of lesser methods. In many cases, they have demonstrated that commonly used routines are grossly deficient. Both of these techniques require very expensive equipment and special skills; output is only a few samples per day.

In IDMS, samples are spiked with a separated isotope of the element in question; a spectrum of masses is obtained, and signals at various mass levels are ratioed. A major difficulty is the control of contamination. Sensitivities are high; as little as 10^{-8} g or less can be measured with certainty.

In NAA, samples are irradiated with neutrons from an atomic reactor or a neutron gun. The activated samples are allowed to decay for a period of time, which may range from seconds to months, depending on the elements sought; then their radioactivity is analyzed. Interpretation of the spectrum of emissions is complex. Sensitivity varies from element to element but is often much higher than that of the common routines (flame AA, for example).

Carbon furnace AA (CFAA) spectrometry shows promise as a highly sensitive routine method for many trace elements; however, interferences and matrix effects in this technique are not yet fully understood. Contamination is a major problem, as with IDMS. CFAA does not qualify as an absolute method, although it may do so as instrumentation improves and understanding increases.

Inductively coupled plasma spectrometry (ICPS) is a relatively new technique that shows promise as a rapid routine for determining traces of metals in solution. Like CFAA, it cannot qualify, in its present state of development, as an absolute method.

Other instrumental techniques may be described, but this would not add to the thesis here presented: The few absolute techniques capable of giving accurate answers are all time-consuming and expensive. They should be used to provide a base on which more practical control methods can depend. Without such a base, rapid routines will always be uncertain, and confusion will always be the rule. To apply statistical manipulations to data from rapid routines in the hope of extracting something useful from something erroneous is an exercise in futility.

5.23.2. Conclusion

Currently applied "official" or "approved" methods for determining toxins in waters and effluents are not very reliable and should not be used exclusively in determining noncompliance with regulations. Round robins do not provide useful information when the methods used are subject to interferences and matrix effects. Routine methods for determining specific toxins should be devised specifically for each sample type and should be investigated for validity using absolute methods of analysis. The misuse of statistical rubrics in data evaluation should be challenged (see Section 6.1).

REFERENCES

1. L. Shapiro, and W. W. Brannock, *Rapid Analysis of Silicate Rocks. A Contribution to Geochemistry*, Geological Survey Bulletin 1036-C, Revised from Circular 165, U.S. Government Printing Office, Washington, D.C., 1956.

2. G. Charlot, *Chimie Analytique Quantitative*, Vol. II, 6th ed., Masson & Cie., Paris, 1974, pp. 449–452.

3. E. B. Sandell, *Colorimetric Determination of Traces of Metals*, 3rd ed., Interscience, New York, 1959, pp. 644–650.

4. B. W. Bailey and J. M. Rankin, "Atomic Absorption Characteristics of Rhenium Resonance Lines Below 2500 Å," *Anal. Chem.* **43**(7), 936–937 (1971).

5. E. V. Elliot, K. R. Stever, and H. H. Heady, "Atomic Absorption Determination of Rhenium in Ores and Concentrates," *Atomic Absorption News.* **13**(5), 113–117 (1974).

6. B. McDuffie, W. R. Bandi, and L. M. Melnik, *Anal. Chem.* **31**, 1311 (1959).

7. E. B. Sandell, *Colorimetric Determination of Trace of Metals*, 3rd ed., Interscience, New York, 1959, pp. 886–889.

8. A. G. Fogg, D. R. Marriott, and D. Thorburn Burns, "The Spectrophotometric Determination of Tungsten with Thiocyanate," *Analyst* **95**, 854–861 (1970).

9. E. M. Donaldson, "Spectrophotometric Determination of Tungsten in Ores and Steel by Chloroform Extraction of the Tungsten–Thiocyanate–Diantipyrylmethane Complex," *Talanta* **22**, 837–841 (1975).

10. R. J. Clark and J. G. Viets, "Back Extraction of Trace Elements from Organometallic–Halide Extracts for Determination by Flameless Atomic Absorption. Spectrometry," *Anal. Chem.* **53** (1), (1981).

11. C. Hendrix-Jongerius and L. Degalan, "Practical Approach to Background Correction and Temperature Programming in Graphite Furnace Atomic Absorption Spectrometry," *Anal. Chem. Acta* **87**, 259–271 (1976).

12. R. A. Newstead, W. J. Price, and P. J. Whiteside, "Background Correction in Atomic Absorption Analysis," *Prog. Analyt. Atom. Spectrosc.* **1**, 267–298 (1978).

13. A. T. Zander, Factors Influencing Accuracy in Background, Corrected A.A.S., International Laboratory, Vol. 1, 1977.

14. D. C. Manning, "Spectral Interferences in Graphite Furnace Atomic Absorption Spectroscopy, I, The Determination of Selenium in An Iron Matrix," *Atomic Absorption Newsletter* **17**(5), September–October (1978).

15. J. J. Sotera, R. L. Stux, and H. L. Kah, "Significance of Spectral Overlap in Atomic Absorption Spectrometry," *Anal. Chem.* **51**(7), (1979).

16. R. J. Lovett, D. L. Welch, and M. L. Parsons, "On the Importance of Spectral Interferences in Atomic Absorption Spectroscopy," *Appl. Spectrosc.* **29**(6), (1975).

17. H. Koizumi, Various Applications of Zeeman Atomic Absorption Spectroscopy, Presented at 61st Canadian Chemical Conference and Exhibition, Winnipeg, Manitoba, June 1978.

18. A. T. Zander and T. C. O'Haver, Improved Accuracy in Background Corrected Atomic Absorption Spectroscopy, National Bureau of Standards, Special Publication 464, November 1977.

19. F. J. Fernandez, S. A. Myers, and W. Slavin, "Background Correction in Atomic Absorption Utilizing the Zeeman Effect," *Anal. Chem.* **52**, 741–746 (1980).

20. R. D. Ediger, "Atomic Absorption Analysis with Graphite Furnace Using Matrix Modification," *Atomic Absorption Newsletter,* **14**(5) (1975).

21. B. Welz, S. Akman, and G. Schlemmer, "Investigations of Interferences in Graphite Furnace Atomic Absorption Spectrometry Using a Dual Cavity Platform," Part I, "Influence of Nickel Chloride on the Determination of Antimony," *Analyst* **110**, 459–465 (1985).

22. J. G. Viets, R. M. O'Leary, and J. R. Clark, "Determination of Arsenic, Antimony, Bismuth, Cadmium, Copper, Lead, Molybdenum, Silver, and Zinc in Geological Materials by Atomic Absorption Spectrometry," *Analyst* **109**, 1589–1592 (1984).

23. J. J. Berzelius, *Pogg. Ann.* **1**, 169 (1824).

24. O. Cantoni, *Z. Anal. Chem.* **67**, 33–4 (1925).

25. S. Abbey and J. A. Maxwell, "Determination of Potassium in Micas, A Flame Photometric Study," *Chemistry in Canada,* September, pp. 37–41, 1960.

26. R. K. Iler, *The Chemistry of Silica. Solubility, Polymerization, Colloid, and Surface Properties, and Biochemistry,* Wiley, New York, 1979.

27. E. Weitz, F. Heinz, and M. Schuchard, *Chem. Ztg.* **74**, 256 (1950).

28. K. A. Rodgers, "The Decomposition and Analysis of Chrome Spinel—A Survey of Some Published Techniques," *Mineralog. Magazine,* **38**, 882–889 (1972).

29. S. A. Bilgrami and C. O. Ingamells, "Chemical Composition of Zhob Valley Chromites, West Pakistan," *Am. Mineral.* **45**, 576–590 (1960).

30. Patterson and Settle, NBS Special Publication 422, 1976.

BASIC CALCULATIONS AND RECOMMENDATIONS

6.1.   Fundamental Notions of Statistics
    6.1.1.   Definition of a Random Variable
    6.1.2.   Definition of a Probability Distribution
    6.1.3.   Construction of Histogram
    6.1.4.   Shape of a Distribution
    6.1.5.   Fitting a Distribution to a Model
    6.1.6.   Measuring Dispersion of Values
    6.1.7.   The Poisson Distribution
    6.1.8.   Conclusions
6.2.   Sampling
    6.2.1.   Construction of Sampling Nomographs
    6.2.2.   Construction of Sampling Nomographs Using Gy's Formula
    6.2.3.   Principles of Variography
    6.2.4.   Description of Heterogeneity of a Flowing Stream in Terms of a Variogram
    6.2.5.   Properties and Analysis of the Short-Range Term of the Variographic Function
    6.2.6.   Properties and Analysis of the Long-Range Term of the Variographic Function
    6.2.7.   Properties and Analysis of the Periodic Term of the Variographic Function
6.3.   Analytical Chemistry
    6.3.1.   A Review of Logarithms
    6.3.2.   Slow Reactions
    6.3.3.   Difference between Activity and Concentration
    6.3.4.   Cases Where Equilibrium Is Reached in Homogeneous Phase
            Transfer of Electrons
            Prediction of a Reaction: Strength of Oxidants and Reductants
            Transfer of Protons
            Prediction of a Reaction: Strength of Acids and Bases
            Transfer of Ions or Polar Molecules
            Prediction of a Reaction: Strength of Complexes
            Transfer of Electrons and Protons
            Prediction of a Reaction
            Transfer of Ions or Polar Molecules and Electrons
            Transfer of Ions and Protons

6.3.5.   Cases Where Equilibrium Is Reached in Heterogeneous Phase: Precipitation and Solubility Product
        Generalities
        Calculation of Solubility of Precipitate
        Solubility of Hydroxide as Function of pH
        Acidity of Cations
6.3.6.   Recommended References
6.4. X-Ray Matrix Corrections
6.5. Composition and Density of Slurries
6.5.1.   Use of *K* Lines and *L* Lines
6.5.2.   Use of Important Identities
6.5.3.   Discussion
6.6. Deconvolution of Energy-Dispersive X-Ray Using Poisson Probability Function
6.6.1.   Fitting a Peak to a Poisson Distribution
6.6.2.   Derivation of Equations
6.6.3.   Fitting a Single Poisson Distribution
6.6.4.   Procedure: Separation of EDX Peaks (Three Peaks)
6.7. Aerosols and Their Production
6.7.1.   Notes on Aerosols and Their Production
6.7.2.   Practical Problems in Atomizing Solutions for Flame Photometry
6.7.3.   Observations on Formation and Stability of Aerosols
6.7.4.   Proposal for Nebulizing System Producing More Stable Aerosol
6.8. Poisson Probability Distribution Applied to Geochemical Data
6.9. Fitting Geochemical Data to Poisson Distribution
6.10. Complex Distributions in Geochemical Exploration
6.11. Calculation of Mineral Formulas from Analyses
6.12. Reporting Halogens, Oxygen, and Sulfur in Rock or Mineral Analyses
References

Some of the content of this chapter is probably well known by the reader. However, the experience of authors proves that important fundamentals in statistics, sampling, and analytical chemistry are often long forgotten by the laboratory analyst. The problem may be even more acute if the analyst has been trained on the working bench and has never been acquainted with these fundamentals.

The objective given to this chapter is to provide tools to perform quick calculations in areas of sampling and analytical chemistry. Many of these tools may be acquired by short training and little practice. Another objective is to address the fact that in today's world, the use of statistics is often misleading because of the unawareness that many trace elements have a behavior that does not follow normally distributed laws. In many cases, the Poisson model is more valid, and in some cases it is the only one valid. The recognition of these facts could lead a wise analyst, chemist, geologist, or metallurgist to take appropriate preventive actions through good sampling practice and get the assurance that a phenomena will obey a normally distributed law. This is a necessary condition that may prevent useless round robins, endless discussions and meetings, unfortunate rejection of some outliers, and costly irresponsible programs.

The drama is the fact that the analyst is often a student who did not succeed in finishing college or someone from an unrelated field such as biology, geology, and many others; he or she is not an analyst. There are very few schools left around the world that produce talented and skilful analysts. Today's analyst has far more training in punching a personal computer keyboard than in the understanding of basic statistics and analytical principles absolutely necessary for doing a job effectively. Many people calculate a standard deviation with a pocket calculator and are unable to perform the same calculation by hand. There is a tendency to perform analytical chemistry by the "cookbook," following well established recipes. It would be a good idea for colleges or universities to revise their programs as far as analytical chemistry is concerned in order to produce the necessary skilful analysts of tomorrow.

6.1. FUNDAMENTAL NOTIONS OF STATISTICS

For anybody that directly or indirectly is making use of statistics, it is important to have some solid basic understanding of them. This is the first step that leads to a wise understanding of some limitations; then only can statistics become a powerful tool.

Statistics may become extremely complex rapidly, and many people are discouraged by the highly mathematical background of this discipline. Without going through academic mathematics and without becoming an expert, it is possible to know enough about basic statistics to solve the problems encountered in sampling and analytical chemistry. The following material is addressed to a reader that has not been seriously acquainted with statistics. Hopefully, this small introduction to statistics could be used by people that need to make simple use of them but do not have the resources, or background, or time to go into the mysterious depths of the subject.

6.1.1. Definition of a Random Variable

If a variable (i.e., determination of a particular element) can take one of many possible values, it is by definition a random variable. When an analytical subsample is collected from a jar, the analyst does not know which grade the element of interest will have; however, most of the possible grades are known. This is generally true for major and some minor constituents. We will see later that it is not true in the case of many trace elements.

6.1.2.   Definition of a Probability Distribution

All the possible values obtained after the determination of an element of interest are described by the probability distribution of its grade. The probability that the value x is between x_0 and $x_0 + dx$ located dx units away from x_0 may be written as

$$\text{Prob}\{x_0 < x < x_0 + dx\}$$

For clear understanding, it is recommended to draw a sketch each time it is possible, as shown in Figure 6.1.

On the sketch, it is clear that the product $f(x_0)dx$ is a probability of x between x_0 and $x_0 + dx$.

Note: A good statistician always draws a sketch. Things are much easier to understand on a sketch.

The probability that the value x is superior or equal to minus infinity may be written as

$$\text{Prob}\{x \geqslant -\infty\}$$

This probability is obviously equal to 100%. However, as shown in Figure 6.2, it is easier to read a sketch, on which it is clear that such a probability covers the entire probability distribution.

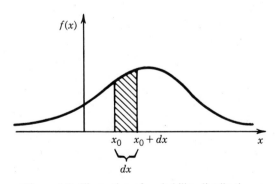

Figure 6.1. Illustration of probability distribution.

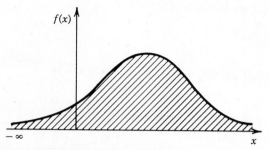

Figure 6.2. Illustration of probability being superior or equal to minus infinity. As a convention, the entire hatched area is 100% of one.

Note: When the histogram of a probability distribution is drawn, the statistical evaluation of the data is often obvious. Always put data on an histogram as soon as possible, before performing any statistical calculations. It is the most important thing to do.

6.1.3. Construction of Histogram

A histogram is simply a tabulation of the available data. It is the simplest statistical tool. How wide should each interval be is a matter of judgment, and as shown in Figure 6.3, it is clear that if the number of data is not large enough, some misleading conclusions could easily be addressed.

6.1.4. Shape of a Distribution

There are many types of distribution. In Figure 6.4, the distribution of gold grade in a sample is presented. This kind of distribution is obtained

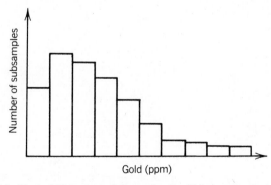

Figure 6.3. Typical histogram of the grade of gold in geochemical material.

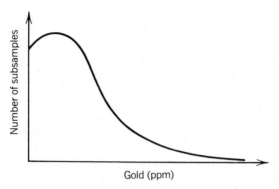

Figure 6.4. Skewed probability distribution often observed in gold deposits.

when the weight of analytical subsamples is too small to carry a representative number of gold flakes or tiny nuggets.

This type of distribution is frequently observed in the analysis of trace elements. A typical skewed distribution of this shape is the Poisson distribution already introduced in Section 1.15.

In Figure 6.5, the distribution of iron grade in a sample is illustrated. This type of distribution is obtained after the repetitive determination of iron in the sample, a typical distribution generally obtained with major constituents and some of the minor constituents that do not show noticeable distribution heterogeneity.

For many years, we have been accustomed to dealing with relatively high grade materials; even minor elements were in the range of 100–10,000 ppm. Naturally, the classical normal distribution (i.e., bell-shaped or

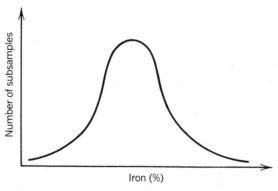

Figure 6.5. Probability distribution of the iron grade of a sample when many subsamples have been used to determine its iron content.

Gaussian distribution) became the only statistical tool we knew and used. In today's world, where trace elements in the range 0.1 ppb to 10 ppm are of great concern and where instrumental methods of analysis are using smaller and smaller subsamples, normally distributed phenomena are becoming scarce. This unfortunate fact is disturbing well-established habits.

6.1.5. Fitting a Distribution to a Model

There are three main parameters to define, as illustrated in Figures 6.6 and 6.7:

The mean $\bar{x}$, the average value of n determinations,

$$\bar{x} = \frac{1}{n} \sum_{i=1}^{n} x_i$$

The median γ, where 50% of the determination values x_i are above (or below) γ.

The mode Y, the most frequent value of the variable (i.e., determination of an element of interest).

During the analysis of trace elements in a sample, it may happen that a triplicate determination of a particular element shows three results grouped around the mode but nevertheless far from the true average. If

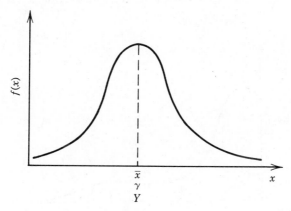

Figure 6.6. Position of mean $\bar{x}$, median γ, and mode Y in a symmetrical, bell-shaped Gaussian distribution.

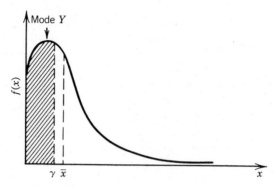

Figure 6.7. Typical position of mean $\bar{x}$, median γ, and mode Y in Poisson distribution.

more determinations could be performed, higher values may sometimes be obtained. It is a common practice to discard all the exceptionally high values because the failure of recognizing that the available data follows a strongly skewed distribution such as a Poisson distribution. These particular phenomena encountered in the analysis of trace elements have been pointed out by Jones and Beaven [1].

Sometimes, a distribution may show several modes, as seen in Figure 6.8 (see also the double Poisson distribution in Section 1.33).

6.1.6. Measuring Dispersion of Values

The most natural way to measure the dispersion of values is to compare all values x_i to their mean $\bar{x}$, which is done by measuring all the differences

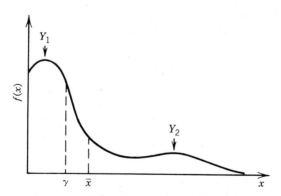

Figure 6.8. Illustration of bimodal distribution.

$x_i - \bar{x}$. However, the average of all $x_i - \bar{x}$ is obviously approaching zero, and it is more convenient to work with the square of these differences in order to avoid finishing with a value near zero. Then the square root of the average of all squared differences is taken. This approach is better than dealing with absolute values, which are more difficult to use in calculations.

By definition, the square root of the average of all squared differences between x_i and $\bar{x}$ is called the standard deviations $s*$:

$$s* = \left(\frac{\sum\limits_{i=1}^{n} (x_i - \bar{x})^2}{n} \right)^{1/2}$$

where n is the total number of calculated differences.

However, the estimated mean $\bar{x}$ in many cases is not a true mean, and it has an error of its own. Then it is necessary to also consider the distribution of the arithmetic mean, which gives

$$s** = \frac{s}{\sqrt{n}}$$

Then

$$s^2 = s^{*2} + s^{**2}$$

$$s^2 = \frac{\sum\limits_{i=1}^{n} (x_i - \bar{x})^2}{n} + \frac{s^2}{n}$$

$$s^2 - \frac{s^2}{n} = \frac{\sum\limits_{i=1}^{n} (x_i - \bar{x})^2}{n}$$

$$\frac{ns^2 - s^2}{n} = \frac{\sum\limits_{i=1}^{n} (x_i - \bar{x})^2}{n}$$

$$\frac{s^2(n - 1)}{n} = \frac{\sum\limits_{i=1}^{n} (x_i - \bar{x})^2}{n}$$

$$s^2 = \frac{\sum\limits_{i=1}^{n} (x_i - \bar{x})^2}{n} \cdot \frac{n}{n-1}$$

$$s = \left(\frac{\sum\limits_{i=1}^{n} (x_i - \bar{x})^2}{n-1} \right)^{1/2}$$

When using a large number of data (i.e., $n > 100$), the use of n or $n - 1$ does not make any difference.

When calculating the standard deviation s, it is recommended to write clearly all steps of the calculation in a table, as shown in the following practical example.

Percentage of Molybdenum Results Obtained During Analysis of Molybdenum Trioxide Concentrate	Deviation of x_i from Calculated Mean $\bar{x}$	$(x_i - \bar{x})^2$
61.24	+0.06	0.0036
61.10	−0.08	0.0064
61.35	+0.17	0.0289
61.25	+0.07	0.0049
61.00	−0.18	0.0324
61.20	+0.02	0.0004
61.25	+0.07	0.0049
61.23	+0.05	0.0025
61.23	+0.05	0.0025
61.20	+0.02	0.0004
61.10	−0.08	0.0064
61.11	−0.07	0.0049
61.02	−0.16	0.0256
61.42	+0.24	0.0576
61.22	+0.04	0.0016
61.24	+0.06	0.0036
61.15	−0.03	0.0009
61.09	−0.09	0.0081
61.17	−0.01	0.0001
61.00	−0.18	0.0324
61.20	+0.02	0.0004
61.16	−0.02	0.0004
Mean $\bar{x}$ = 61.18 Total: −0.03		0.2289

$$s = \sqrt{\frac{0.2289}{22 - 1}}$$
$$= 0.10$$

The result of this round-robin exercise gave an estimate of the molybdenum content of a molybdenum trioxide standard reference material with a standard deviation $s = 0.10$.

$$\text{Percentage of molybdenum} = 61.18 \pm 0.10$$

All these values could be arranged into a histogram as shown in Figure 6.9. In this example, it can be observed that the average is slightly toward the lower values (i.e., $\bar{x} = 60.18$) than the mode (i.e., $Y = 60.24$ or 60.25). It is important to note that the histogram probably gives more information than the plain calculation of the standard deviation. The method of determination was a gravimetric method. The best laboratories are grouping results between 61.20 and 61.29% Mo. Some laboratories show lower results because they are generating more molybdenum losses during the procedure. It turned out that the two isolated results in the range 61.30–61.49 were generated by the only two laboratories that have been checking the molybdenum content in the digestion residue, in the final empty crucible after weighing $PbMoO_4$, on the retrieved magnetic bar, and from the empty beaker in which the $PbMoO_4$ was precipitated. (See the rapid chemical method, Section 5.3.)

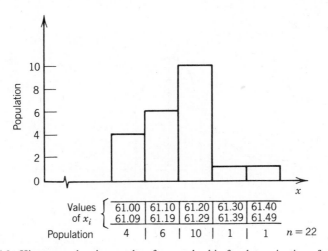

Values of x_i	61.00 61.09	61.10 61.19	61.20 61.29	61.30 61.39	61.40 61.49	
Population	4	6	10	1	1	$n = 22$

Figure 6.9. Histogram showing results of a round robin for determination of molybdenum in molybdenum trioxide concentrate.

Now the question is: Would it be wise to recommend the mean $\bar{x}$ of all data that is 60.18% molybdenum as the best value? Certainly not. In fact, the histogram is measuring to a full extent the seriousness of a big problem. Let us forget for a moment this histogram and back to the calculation of the plain standard deviation.

The standard deviation s says:

1. There are 68% chances that the molybdenum determination will fall between 60.08 and 60.28.
2. There are 95% chances that the molybdenum determination will fall between 59.98 and 60.38.
3. There are 99.7% chances that the molybdenum determination will fall between 59.88 and 60.48.

This reflects what would happen if the histogram had the normal bell shape shown in Figure 6.10. In this figure, the distribution is such that

$$\text{Prob}\{\bar{x} < x_i < \bar{x} + s\} = 34\%$$

$$\text{Prob}\{\bar{x} < x_i < \bar{x} + 2s\} = 47.5\%$$

$$\text{Prob}\{\bar{x} < x_i < \bar{x} + 3s\} = 49.9\%$$

The above example shows how reality may be different from the normally distributed ideal case. In analytical chemistry, sometimes things are not as random as one may think.

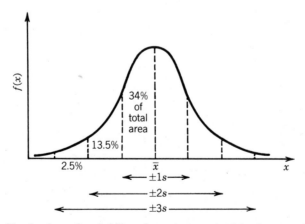

Figure 6.10. Usual values of probability of one datum point lying between the average $\bar{x}$ and one, two, and three standard deviations away from $\bar{x}$.

6.1.7. The Poisson Distribution

A binomial distribution applies when the mixture of minerals is made of two different components; these components are occurring as discrete grains and are roughly in equivalent proportions.

When dealing with trace elements, it may be advantageous to consider a limiting case of the binomial distribution where one constituent of interest occurs with a very small probability (i.e., where gold is occurring as very few random discrete nuggets in alluvial ores or even in unliberated gold ores after a comminution has been liberating the gold). Often these grains are made of pure mineral and may have a tremendous impact on the analytical result depending on whether they are present in the tiny final analytical subsample. Such a probability may be written as

$$\text{Prob}\{x = n\} = \frac{e^{-z}z^{n}}{n!}$$

where z is the hypothetical average number of pure mineral grains in the subsample. It can be observed in Figure 6.11 that as z is increasing, the mode of the distribution tends toward z. This observation gives some indications under which conditions the distribution tends to be normally distributed again.

The estimated standard deviation of a Poisson distribution is expressed as

$$s = \sqrt{z}$$

6.1.8. Conclusions

The determination of low-grade elements of interest and particularly trace elements should be preceded by a thorough mineralogical investigation to find out if there is a heterogeneity in composition and distribution that may be of great concern. The quick investigation may allow someone to take the decision of performing a complete sampling study of the material. Such preventive action leads to appropriate sampling procedures, good practice, and later to normally distributed data that is easy and inexpensive to interpret.

- When investigating data, always draw a histogram; the histogram may say it all without further calculations.
- If performing statistical calculations, be aware of their limitations, and be sure of the type of distribution you are dealing with: It may

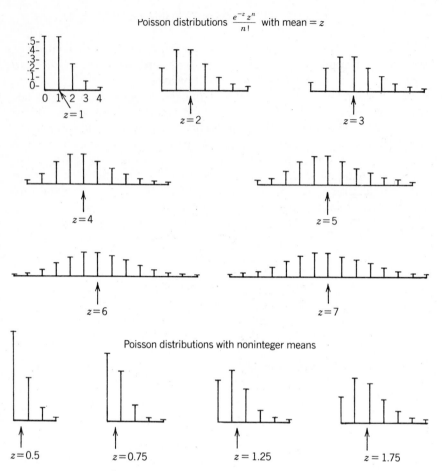

Figure 6.11. Different possible shapes for a Poisson distribution.

not be a normal distribution. This also implies that there are enough data. Never attempt statistical calculations with only a half dozen results.

- Unless the explanation of its origin is obvious, an outlier shall not be rejected; it may actually give extremely precious information.

If the reader wishes to improve his or her knowledge in statistics, the works of Cochran [2], Parratt [3], and David [4] are recommended.

6.2. SAMPLING

Gy [5] states, "The young metallurgist, miner, geologist, or chemist who begins his career after the University, realizes very soon that sampling problems are something to deal with constantly. Then he realizes, not without irritation, that his teachers did not provide him with enough information to deal comfortably with these unexpected difficulties. . . ."

Gy [6] also points out, "Mineral sampling is a statistical game that you can win by obtaining truly representative samples, but chances are you are losing." This book presents some aspects of sampling and a few of the necessary tools to deal with the difficulties attached to it. One of the most convenient tools is the sampling nomograph. It is essential to know how to construct such a nomograph because it provides a powerful tool to solve some complex problems in a matter of minutes. On the wall of an office it is a tool that shows facts.

6.2.1. Construction of Sampling Nomographs

The Simplest Case. Grains of the element of interest are liberated and are of the same size. From formula (1.5), we can calculate the necessary weight of the sample as follows:

$$W = \frac{K_s}{R^2}.$$

where R (%) is the relative uncertainty in a W-gram sample, and K_s is the sampling constant obtained by formula (1.7),

$$K_s = \frac{10^4 \times (K - L)(H - L)u^3 d_H}{K^2}$$

In the present example, $L = 0$ and $H = 100$; then,

$$K_s = \frac{10^6 \times Ku^3 d_H}{K^2}$$

If we are calculating the weight W necessary for a 10% sampling uncertainty with 68% confidence, we obtain

$$W = \frac{10^8 \times Ku^3 d_H}{K^2}$$

This may be simplified if we choose consistent units for K in the numerator and in the denominator. Here K is expressed as the overall grade in grams per ton. We may also write that

$$w = u^3 d_H$$

which is the weight of a single gold or silver particle. Then,

$$W\,(10\%) = \frac{10^8 \times w}{K(\mathrm{g})}$$

The nomograph in Figure 6.12 illustrates this equation for several values of K.

For gold or silver flakes with thickness 0.1 cm, their linear (screen) dimension u, in centimeters; their weight is

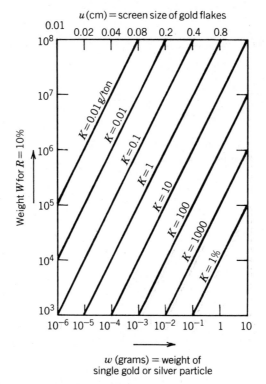

Figure 6.12. Nomograph indicating weight of sample to collect as a function of: (1) ore grade; (2) weight of largest gold or silver particles.

$$w = \tfrac{1}{10}u^3 d_H$$

where d_H is the density in grams per centimeter cubed. The nomograph in Figure 6.12 includes a scale converting w to the linear mesh size of gold grains of dimension $u \times u \times \tfrac{1}{10}u$ centimeters cubed.

The Real Case. Grains of the element of interest are liberated but of variable size. In this example, gold or silver is present as pure metal in completely liberated, randomly distributed grains of variable size. First, suppose there are two different sizes of weights w_1 and w_2. The weight necessary for a 10% sampling uncertainty with 68% confidence is

$$W(10\%) = \frac{10^8}{K^2}(K_1 w_1 + K_2 w_2)$$

where $K = K_1 + K_2$ is the overall grade in grams per ton, K_1 is that portion of the grade due to particles of weight w_1, and K_2 is that portion of the grade due to particles of weight w_2.

Suppose now that $K_1 = K_2 = \tfrac{1}{2}K$; then the above equation becomes

$$W(10\%) = \frac{10^8 \times (w_1 + w_2)}{K}$$

If the two grain sizes are very different, the necessary value for $W(10\%)$ depends heavily on the larger grains.

Figure 6.13 shows $W(10\%)$ as a function of w_2 when $w_1 = 10^{-4}$ g for a material in which $K_1 = K_2 = \tfrac{1}{2}K$ for different values of K. In any real case, there is a distribution of sizes, which may vary from one deposit to another. One may write

$$W(10\%) = \frac{10^8 \times w_c}{K}\left(\frac{K_1}{K}\frac{w_1}{w_c} + \frac{K_2}{K}\frac{w_2}{w_c} + \frac{K_3}{K_c}\frac{w_3}{w_c} + \cdots + \frac{K_n}{K}\frac{w_n}{w_c}\right)$$

and describe the dimensionless quantity in parentheses as a size distribution factor. This is the device used by Gy [5]. The size distribution factor is determined from actual experience with the type of ore being examined. The weight w_c is a concensus weight calculated from a particle shape factor and densities. The particle shape factor is also found from experience with the type of ore being examined [5].

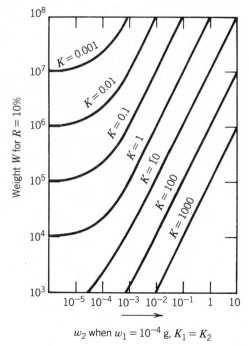

Figure 6.13. Nomograph indicating weight of sample to collect as a function of: (1) w_2 when $w_1 = 10^{-4}$ g; (2) ore grade when $K_1 = K_2$.

For alluvial gold deposits,

$$W_c \left(\frac{K_1}{K} \frac{w_1}{w_c} + \cdots \right) \approx \frac{d_H \cdot u_{max}^3}{20}$$

where u_{max} is the linear dimension of the largest flakes in the deposit. This leads to the expression

$$W (10\%) \approx \frac{10^8}{K} u_{max}^3 \quad \text{(with } u_{max} \text{ in centimeters)}$$

Figure 6.14 plots this relationship for various values of K. Values obtained using this nomograph are in close agreement with values obtained by Gy in materials where gold is completely liberated (alluvial deposits) or not liberated at an early stage of the sampling procedure but liberated later, when the particle size is reduced.

Notes: The sample weight for native silver is approximately half that for gold, other factors being the same. The sample weight for 95% confidence is four times greater than for 68% confidence; all calculations are based on the 68% confidence level.

The nomographs are intended to apply only to alluvial deposits in which the gold is 100% liberated.

6.2.2. Construction of Sampling Nomographs Using Gy's Formula

A very useful nomograph is the one that may be drawn from the Gy general equation (1.3) (Chapter 1),

$$S_R^2 = G \frac{u^3}{W}$$

In order to put as much information as possible on the nomograph, it is convenient to draw it using logarithmic coordinates; consequently, take

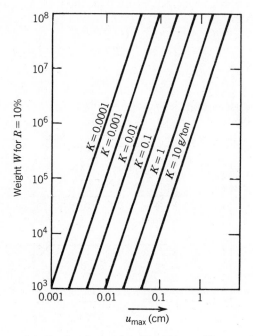

Figure 6.14. Nomograph indicating the weight of sample to collect as a function of: (1) ore grade; (2) linear dimension of largest flakes in deposit.

the logarithm of Gy's equation,

$$\log S_R^2 = \log G + 3 \log u - \log W$$

To simplify this equation, $\log S_R^2 = Y$, and $\log W = x$. Then,

$$Y = \log G + 3 \log u - x$$

Since G is a constant for a given particle size u, the derivative Y' of this equation is $Y' = -1$. It is thus easy to conclude that each line representing the sampling error as a function of the sample weight for a given particle size will have a slope of -1 on the graph. Figure 6.15 shows a typical example of a nomograph.

The nomograph may be used to optimize sampling procedures as shown in Figure 6.16, where a maximum fundamental sampling error of ±5% was initially chosen for each splitting and comminution stage. The example shows a route x-A-B-C-D-E-F-G to follow, where none of the cycles generated an error above this limit. This route determines the exact sampling procedure to be followed. From this, the total fundamental error [5] may be calculated as

$$S_{R,\text{total}}^2 = [f(A) - f(x)] + [f(C) - f(B)] + [f(E) - f(D)] + [f(G) - f(F)]$$

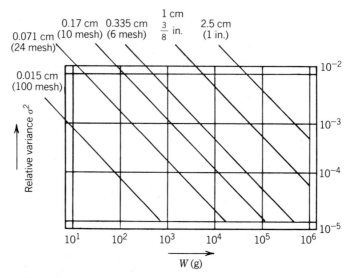

Figure 6.15. Example of nomograph obtained for estimation of low-level SiO_2 in pure petroleum coke.

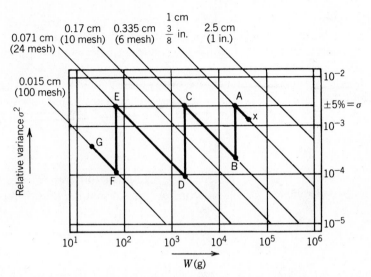

Figure 6.16. Illustration of ideal sampling procedure used for estimation of low-level SiO₂ in a pure petroleum coke; chosen maximum fundamental sampling error, σ = ±5%.

For more information on sampling nomographs concerning the Gy sampling theory, refer to the literature [7, 8].

6.2.3. Principles of Variography

In recent years, especially in the field of geochemical exploration and mining evaluation, variography became more and more important. Its success is due to the work of Krige [9, 10], Matheron [11, 12], and Gy [13] and many followers. The technique described in this section is extremely effective in the study of the continuity of a parameter of interest (i.e., assay grade, grain size distribution, etc.) in materials such as mill feeds, tailings, concentrate shipments, wastes of environmental concern, and so on. It is also a powerful tool in processes using kilns, furnaces, autoclaves, and so on, where process control is critical.

It was important to present the principles of variography in this book because they provide an effective means of estimating all the components of the overall estimation error s^2_{total},

$$s^2_{\text{total}} = s^2_{\text{analytical}} + s^2_{\text{sampling}} + s^2_{\text{process trends}}$$

A single experiment may be performed to study each of these terms; it is called *a variographic experiment*. Generally, the best area to perform a variographic experiment is along a flowing stream. A variographic experiment is the measurement of a continuity index as a function of distance or time along the flowing stream (i.e., conveyor belt, rotary kiln, melted metal, stream sediments, etc.).

The most normal way to compare two values, that is, two-assay grades $Z(x)$ and $Z(x + h)$ at two points x and $x + h$ that are h units away from x, is to consider the difference d between them [4]:

$$d = Z(x) - Z(x + h)$$

Since we are not really interested in the sign of the difference but rather in the absolute value, we should consider the absolute value of the difference d:

$$|d| = |Z(x) - Z(x + h)|$$

However, this value, which expresses the dissimilarity between two individual points, is only a beginning. We actually need the average difference between two points h units apart. Thus, we should consider the value

$$|\bar{d}| = \frac{\sum_{i=1}^{N} |Z(x_i) - Z(x_i + h)|}{N(h)}$$

for all possible points x_i and $x_i + h$.

Since absolute values are not easy to deal with in calculus, we shall consider the average squared differences. We define this value as d', also called the variographic function $2Vf(h)$; thus,

$$2Vf(h) = \bar{d}' = \frac{1}{N(h)} \sum_{i=1}^{N(h)} [Z(x_i) - Z(x_i + h)]^2$$

Note that $d' \neq d$.

For mathematical convenience [11, 12], we use the semivariographic function $Vf(h)$; thus,

$$Vf(h) = \frac{1}{2N(h)} \sum_{i=1}^{N(h)} [Z(x_i) - Z(x_i + h)]^2$$

where $N(h)$ is the total number of pairs separated by a distance h.

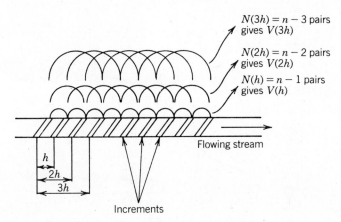

Figure 6.17. Schematic computation of variogram using pairs of samples from increments a given distance h apart.

Finally, to make things even easier, it is recommended to work with dimensionless relative values:

$$V_R f(h) = \frac{1}{2N(h)\bar{x}^2} \sum_{i=1}^{N(h)} [Z(x_i) - Z(x_i + h)]^2$$

As expected, the differences between values at greater distances apart are found to increase as h increases, and thus $Vf(h)$ increases with h; this is true for all variograms except the truly random ones. As an example, take n samples at intervals of h seconds regularly distributed along a flowing stream, as shown in Figure 6.17. This will give $n - 1$ pairs to compute $V(h)$, $n - 2$ pairs to compute $V(2h)$, $n - 3$ pairs to compute $V(3h)$, and so on.

The variographic function can thus be calculated as shown in Figure 6.18. In a variographic experiment time series are considered to characterize overall deviations where the order of data is highly relevant as opposed to statistical populations where the order of data is irrelevant. This point is illustrated in Figure 6.19.

6.2.4. Description of Heterogeneity of a Flowing Stream in Terms of a Variogram

Let $f(t)$ be any nonspecified time function. The study of a large number of such functions leads to the conclusion that in many cases, they can be

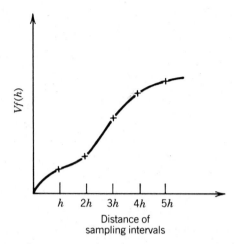

Figure 6.18. Typical variogram from flowing stream. As expected, the difference between values at greater distances apart are found to increase. This is generally true for all variograms except the truly random ones.

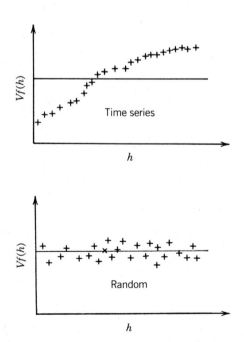

Figure 6.19. Example of typical variogram (top) and variogram of the same data distributed at random (bottom).

602

broken up into a sum of four terms:

$$f(t) = f_K + f_1(t) + f_2(t) + f_3(t)$$

where f_K = a constant describing the average properties of $f(t)$; in this case, f_K is the average content of the element of interest

        $f_1(t)$ = a short-range term, quasi-random, which represents the random nature of sampling

        $f_2(t)$ = a long-range term, nonrandom, which represents changes observed in mineral deposits, along streams, or in metallurgical processes

        $f_3(t)$ = a periodic term or periodic fluctuation, introduced by processing, handling, or an attempt to keep a variable between two known limits

The variographic function $f(t)$ is thus the sum of three terms, the variographic function of the constant term f_K being obviously zero. In the nomenclature of the variogram,

$$V(f) = Vf_1(h) + Vf_2(h) + Vf_3(h)$$

which is the sum of the variographic functions $f_1(t)$, $f_2(t)$, and $f_3(t)$, respectively. Here $Vf(h)$ is expressed as a variance.

6.2.5. Properties and Analysis of the Short-Range Term of the Variographic Function

The short-range variations of $f(t)$ result mainly from the particulate structure of the material being investigated. Relative to the A and B constants introduced in Section 1.3, this term reflects a situation where

$$\frac{A}{W} \neq 0 \quad \text{and} \quad \frac{B}{N} = 0$$

In this case B is related to large-scale inhomogeneity, or process trend. The short-range term may be the sum of several other terms, all of random nature. For example, the same analytical subsample can be analyzed about 30 times, and the variographic function $Vf_1(h)$ obtained reflects the small-scale inhomogeneity of this particular subsample and the reproducibility of the analytical procedure.

The short-range term may be studied another way. For example, one increment of a material is collected at very short intervals along a stream,

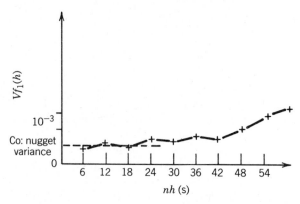

Figure 6.20. Example of variogram showing short-range term (feed to a flotation plant): h = 6 s (0.1 min). This variogram is essentially flat throughout the interval $0 < h < 30$ s.

conveyor belt, wire, or any one-dimensional object. At least 30 of these increments are collected, prepared, and analyzed separately. This time the variographic function obtained reflects the small-scale inhomogeneity of the running material, the sampling and subsampling errors, and the analytical errors.

These two experiments may be summarized by a typical example obtained in a molybdenum flotation plant, as illustrated in Figure 6.20.

The objective of this experiment is to evaluate how all the n increments of a given sample shall be composited knowing that if all increments are analyzed, all variances introduced by the small-scale inhomogeneity of the running stream, the sampling and subsampling operations, and the analytical method will be divided by n. However, if all the increments are composited into one sample, only the variance introduced by the small-scale inhomogeneity of the running stream is divided by n.

6.2.6. Properties and Analysis of the Long-Range Term of the Variographic Function

The long-range variation of $Vf(h)$ represents a continuous trend, all discontinuities being taken into account by the short-range term $Vf_1(h)$. In order to study this term, it is recommended to collect at least 60 increments at longer intervals (i.e., for the short-term, $h = 6$ s, then for the long-term, $h = 60$ s).

An example of the variogram obtained after performing such an experiment is illustrated in Figure 6.21. As far as sampling is concerned, the interesting part of this variogram is the value, or approached value, of the slope with which it is increasing. This gives a valid indication of how often increments should be collected in order to minimize the total variance introduced between each increment.

It is frequent, after such an experiment, to conclude that the time (or space) between each increment is too long (or too large). It is recommended to check if the long-range term is cyclic.

6.2.7. Properties and Analysis of the Periodic Term of the Variographic Function

The periodic term of the variographic function $Vf(h)$ is represented by the third term, denoted as $Vf_3(h)$, and is the result of periodic quality fluctuation changes caused by process-related operations, such as reclaiming from an ore pile, intermittent discharge from a process stream, regular adjustments of parameters between specified limits, and so on. The periodic phenomenon does not need to be strictly sinusoidal for its effects to be detrimental to systematic sampling.

This term may be studied in the long-range term already exposed and in a very long range term that consists of collecting about 80 increments at very long intervals, such as one order of magnitude longer than the one collected to study the long-range term (i.e., for the long term, $h = 60$ s; then for the very long term, $h = 600$ s).

Figure 6.22 shows a periodic variogram obtained in a flotation plant. First we draw two curves C_1 and C_2 enveloping the line C. Note that C_1 and C_2 can be superimposed by translation parallel to the ordinate axis.

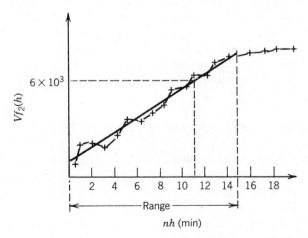

Figure 6.21. Variogram of long-range term. Example of feed material to flotation plant: $h = 1$ min.

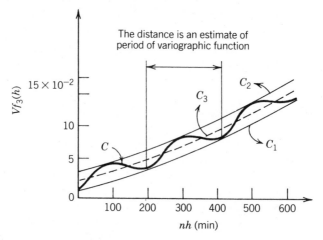

Figure 6.22. Variogram showing periodic term (feed to a flotation plant): $h = 20$ min.

The curve C_1 represents the first two terms of the variographic function $Vf_1 + Vf_2(h)$. The curve C_1 is tangent to the curve C for all values of $h = KT$, where K is an integer and T is the period of the variographic function. Here C_2 is a graphical representation of $Vf_1 + Vf_2(h) + 2Vf_3$, and it should be tangent to C for all values of $h = \frac{1}{2}(2K + 1)T$, and C_3 is a curve around which the periodic phenomenon occurs. The distance of translation of C_3 to C_2 represents the amplitude of the periodic function. The distance of translation of C_1 to C_2 represents the contribution of the periodic term to the variographic function.

When dealing with cyclic phenomena, the first and most desirable action is to implement a random stratified sample selection scheme. It is the only feasible solution to the problem of obtaining suitable samples. This matter has been extensively studied by Gy [14].

6.3. ANALYTICAL CHEMISTRY

The theory concerning the calculation of solubility products and equilibrium and stability constants has been solidly developed by many authors. The objective of this chapter is to emphasize practice and to give to the analyst a means of thinking and reacting quickly in the face of small difficulties, sometimes at the cost of approximations. One of the most powerful ways to acquire these skills is to practice using logarithms. The modern analyst is incapable of using the most valuable principles of analytical chemistry in an effective manner unless he or she feels extremely comfortable with the manipulation of logarithms.

6.3.1. A Review of Logarithms

Theorem. For each positive real number x, there is a unique real number n such that

$$a^n = x$$

Definition. If x is any positive real number, the unique exponent n such that $a^n = x$ is called the logarithm of x with base a and is denoted by $\log_a x$, $n = \log_a x$ if and only if $a^n = x$. Logarithms with base 10 are useful in analytical chemistry. It is customary to refer to such logarithms as common logarithms and to use the symbol $\log a$ as an abbreviation for $\log_{10} a$.

The following formulae are very important to remember:

$$\log(ab) = \log a + \log b$$

$$\log \frac{a}{b} = \log a - \log b$$

$$\log a^n = \log a + \log a + \cdots + \log a = n \log a$$

$$\log \frac{1}{a} = \log 1 - \log a = \operatorname{colog} a$$

In order to perform quick calculations, it is only necessary to memorize the logarithm of the first five prime numbers:

$$\log 1 = 0$$
$$\log 2 = 0.30$$
$$\log 3 = 0.48$$
$$\log 5 = 0.70$$
$$\log 7 = 0.85$$

All other logarithms are quickly calculated by using the above formulae.

Examples:

$$\log 4 = 2 \log 2 = 0.60$$

$$\log 6 = \log 2 \times 3 = 0.30 + 0.48 = 0.78$$

$$\log 8 = \log 2 \times 2 \times 2 = 0.30 + 0.30 + 0.30 = 0.90$$

$$\log 9 = 2 \log 3 = 0.96$$

With little practice, mental arithmetic can be performed for all these calculations. How to write a number as a power of 10? For example,

$$a = 10^b$$

Approach: Take the logarithm of the above expression,

$$\log a = b \log 10 = b$$

Then,

$$a = 10^{\log a}$$

from which the following rule can be stated:

In order to write a number as a power of ten, calculate the logarithm of this number and put it as an exponent above ten

---

Examples

1.
$$2 = 10^x$$

$$\log 2 = x \log 10 = x$$

Then

$$2 = 10^{0.30}$$

2.
$$12 = 10^x$$

$$\log 12 = x \log 10 = x$$

$$\log 12 = \log (2 \times 2 \times 3) = 0.30 + 0.30 + 0.48$$
$$= 1.08$$

Then

$$12 = 10^{1.08}$$

3.
$$34 = 10^x$$

$$\log 34 = \log(17 \times 2) = \log 17 + \log 2$$

$\log 17$ = average between $\log 16$ and $\log 18$, which is a good enough approximation in most cases

$$\log 16 = 4 \log 2 = 1.20$$

$$\log 18 = \log 2 + 2 \log 3 = 0.30 + 0.96 = 1.26$$

Then

$$\log 17 = \tfrac{1}{2}(1.20 + 1.26) = 1.23$$

$$\log 34 = 0.30 + 1.23 = 1.53$$

Then

$$34 = 10^{1.53}$$

This is all that is needed to know about logarithms for a good understanding of the following material and for applying it in a very effective way.

6.3.2. Slow Reactions

In analytical chemistry, we usually are concerned with reversible reactions in which homogeneous equilibrium is rapidly established:

$$A + B \rightleftarrows C + D$$

When equilibrium is reached, the velocity of the reaction between A and B is equal to the velocity of the reaction between C and D.

The speed of the reactions is expressed by the law of mass action, first clearly stated by Guldberg and Waage

---

velocity of reaction between A and B:

$$v_1 = K_1 \,|\, A \,\|\, B \,|$$

velocity of reaction between C and D:

$$v_2 = K_2 \, |\, C \, \| \, D \,|$$

Note 1: In all the following material the symbol $|\ |$ indicates a concentration of a particular compound in gram-formula-weight per liter, also called the molar concentration.

Note 2: K_1 and K_2 are constants.

When

$$v_1 = v_2$$

$$K_1 \, |\, A \, \| \, B \,| = K_2 \, |\, C \, \| \, D \,|$$

or

$$\frac{|\, C \, \| \, D \,|}{|\, A \, \| \, B \,|} = \frac{K_1}{K_2} = K$$

If we consider the reaction between $2A$ and B,

$$2A + B \rightleftarrows C + D \quad \text{or} \quad A + A + B \rightleftarrows C + D$$

$$\frac{|\, C \, \| \, D \,|}{|\, A \, \| \, A \, \| \, B \,|} = \frac{|\, C \, \| \, D \,|}{|\, A \,|^2 |\, B \,|} = K$$

The equilibrium constant for the general reversible reaction,

$$aA + bB + \cdots \rightleftarrows mM + nN + \cdots$$

is

$$\frac{|\, M \,|^m |\, N \,|^n \cdots}{|\, A \,|^a |\, B \,|^b \cdots} = K$$

Convention. K is such that compounds on the right of the reaction are put as the numerator in the relation. Applications of this law are very useful to predict chemical phenomena. However, reactions are sometimes very slow and predictions made with this law are not valid.

In the short following study, we will consider two cases:

1. The case where equilibrium is reached.
2. The case where equilibrium is not reached.

6.3.3. Difference between Activity and Concentration

One of the problems encountered in analytical chemistry is to estimate activities of solutions, which is the real term of interest that should be taken into account in the law of mass action. The only value that is easily accessible is the concentration. In an ideal solution, such as a very dilute solution, the activity is usually equal to the concentration; however, many solutions are not ideal.

In reality, the activity can be related to the concentration by an activity factor f in such a way that

$$\text{Activity} = f \times |\text{ concentration }|$$

If solutions are very diluted, usually $10^{-2} M$, $f \approx 1$, and we may assume in this case that activity and concentration are equal.

If the concentrations are superior to $10^{-2} M$, calculations made without the necessary precautions bring false results. If necessary, f can be experimentally determined, which often is the most effective way of determining it because its determination by calculation can be complicated and uncertain.

In Section 6.3.4, we assume that the activity is roughly equal to the concentration. In various cases, this approximation will not seriously affect the reasoning; however, it is a limitation to keep in mind. The following calculations are strictly semiquantitative, and in some cases, where concentrations are high, predictions will have to be verified or corrected according to results from quick experiments.

6.3.4. Cases Where Equilibrium Is Reached in Homogeneous Phase

Transfer of Electrons

The Oxidation Number. An element has an oxidation number N^+ when the corresponding elementary ion has an n^+ charge, in other words, when the element lost n electrons. In the case of a ferrous iron salt, the iron has the oxidation number II^+. In potassium permanganate, $KMnO_4$, we have K^{I+}, $4O^{II-}$, and Mn^{VII+}.

> If the oxidation number is positive, it cannot be greater than the number of electrons on the outer shell.
>
> If the oxidation number is negative, it cannot be smaller than the number of electrons necessary to saturate the outer shell.

Convention. Arabic numbers indicate ion charge: For example, Fe^{3+}, $MnO_4{}^-$, $SO_4{}^{2-}$, and so on.

Roman numbers indicate oxidation number, irrespective of the formula of the compound: Cu^{II+} in $CuCl^+$.

Definitions. A compound, ion, or molecule that fixes an electron and becomes a reductant is called an *oxidant*. A compound, ion, or molecule that liberates an electron and becomes an oxidant is called a *reductant*.

$$\text{oxidant} + ne^- \rightleftarrows \text{reductant}$$

Example.

$$Cu^{2+} + e^- \rightleftarrows Cu^+$$

> *Rule:* For an oxidant to fix electrons, it is necessary to have a reductant that is a donor of electrons.

Example.

$$2Fe^{3+} + 2e^- \rightleftarrows 2Fe^{2+}$$
$$\text{and} \quad Sn^{2+} - 2e^- \rightleftarrows Sn^{4+}$$
$$\overline{\text{Total: } 2Fe^{3+} + Sn^{2+} \rightleftarrows 2Fe^{2+} + Sn^{4+}}$$

Prediction of a Reaction: Strength of Oxidants and Reductants

If we consider the red–ox reaction,

$$2Ce^{4+} + Sn^{2+} \underset{2}{\overset{1}{\rightleftarrows}} 2Ce^{3+} + Sn^{4+}$$

the problem is to know if this reaction goes toward direction 1 or toward direction 2. In this case, the reaction will have a strong tendency to go toward direction 1 because Ce^{4+} oxidizes Sn^{2+} and Sn^{4+} cannot oxidize Ce^{3+}.

> *Rule:* To the strongest oxidant corresponds the weakest reductant, and vice versa

Electrochemical Reactions. Consider a solution with the following red–ox system:

$$a(\text{Ox}) + ne^- \rightleftarrows b(\text{Red})$$

When a platinum wire is introduced in the solution, because of continuous electron exchange with compounds in solution, it takes a potential E of equilibria equal to

$$E = E_0 + \frac{RT}{nF} \ln \frac{(\text{Ox})^a}{(\text{Red})^b}$$

where
E_0 = constant characteristic of red–ox system
R = gas constant
T = absolute temperature
n = number of electron involved in reaction
F = the Faraday (96.49 coulombs)
$\ln$ = Neperian logarithm
(Ox) and (Red) = activities of oxidant and reductant, respectively

After simplication, under normal conditons, we write

$$E = E_0 + \frac{0.06}{n} \log \frac{(\text{Ox})^a}{(\text{Red})^b}$$

If we apply to the electrode a potential $E' > E$, the equilibria is broken, and some of the reductant is going to oxidize; therefore the ratio $(\text{Ox})^a/(\text{Red})^b$ will increase until $E = E'$.

> By forcing a potential to the electrode, we start a red–ox reaction

Example. Assume that we are mixing a ferric salt Fe^{3+} with a stannous salt Sn^{2+}. The red–ox potentials are given by

$$E_{Fe} = 0.77 + 0.06 \log \frac{(Fe^{3+})}{(Fe^{2+})}$$

$$E_{Sn} = 0.13 + \frac{0.06}{2} \log \frac{(Sn^{4+})}{(Sn^{2+})}$$

The ferric solution has a higher potential than the stannous solution; then

$$2Fe^{3+} + Sn^{2+} \rightarrow 2Fe^{2+} + Sn^{4+}$$

Consequently, the ratio $(Fe^{3+})/(Fe^{2+})$ is diminishing and so does E_{Fe}. The ratio $(Sn^{4+})/(Sn^{2+})$ is increasing, and so does E_{Sn}. The reaction is stopping when $E_{Fe} = E_{Sn}$.

Notion of Apparent Normal Potential. For the accurate determination of solution potentials, it is necessary to take into account the activity of ions, as already mentioned earlier. However, the only values easy of access are the concentrations. To circumvent this difficulty, we rewrite the Nernst equation as

$$E = E_0 + \frac{0.06}{n} \log \frac{f^a_{Ox} \mid Ox \mid^a}{f^b_{Red} \mid Red \mid^b}$$

or

$$E = E_0 + \underbrace{\frac{0.06}{n} \log \frac{f^a_{Ox}}{f^b_{Red}}}_{E_R} + \frac{0.06}{n} \log \frac{\mid Ox \mid^a}{\mid Red \mid^b}$$

We see that it is convenient to include the entire factor f into E_0 and work with what is defined as the apparent normal potential E_R. It is simply a new constant that has to be defined each time for a particular new medium.

Then we work with concentrations using the following equation:

$$E = E_R + \frac{0.06}{n} \log \frac{\mid Ox \mid^a}{\mid Red \mid^b}$$

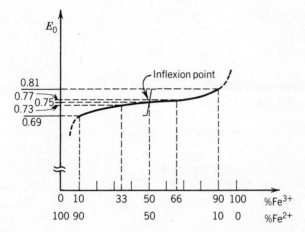

Figure 6.23. Illustration of variation of potential of a Fe^{3+}–Fe^{2+} buffer solution as a function of concentrations of Fe^{3+} and Fe^{2+} ions.

As a first approximation, when $|\,Ox\,|$ and $|\,Red\,|$ vary, E_{apparent} varies also but not very quickly and to some extent may be considered as constant and can be experimentally determined.

Calculation of Potential of Buffer Solution
Definition. We call a red–ox buffer solution a solution containing the oxidant and the reductant of a same couple in equivalent quantities.

It is easy to demonstrate that such solutions have the property of fixing the potential at a determined value.

Example. Consider the couple Fe^{3+}/Fe^{2+} with $E_R = 0.75$ V, where

$$Fe^{3+} + e^- \rightleftarrows Fe^{2+}$$

$$E = 0.75 + 0.06 \log \frac{|\,Fe^{3+}\,|}{|\,Fe^{2+}\,|}$$

Figure 6.23 is a graph that gives the potential of the solution as a function of the respective concentrations of Fe^{3+} and Fe^{2+} ions. If we have 50% Fe^{3+} and 50% Fe^{2+}, the ratio $|\,Fe^{3+}\,|/|\,Fe^{2+}\,| = 1$, and

$$E = E_R = 0.75$$

Now suppose we have 66% Fe^{3+} and 33% Fe^{2+}; then the ratio $|\,Fe^{3+}\,|/|\,Fe^{2+}\,| = 2$, and

$$E = 0.75 + 0.06 \log 2 = 0.77$$

If we had 33% Fe^{3+} and 66% Fe^{2+},

$$E = 0.75 - 0.06 \log 2 = 0.73$$

Finally, suppose we have 90% Fe^{3+} and 9% Fe^{2+}; then the ratio $|Fe^{3+}|/|Fe^{2+}| = 10$, and

$$E = 0.75 + 0.06 \log 10 = 0.81$$

For $|Fe^{3+}|/|Fe^{2+}| = 1/10$, we have

$$E = 0.75 - 0.06 \log 10 = 0.69$$

For 50% oxidant, there is an inflexion point on the curve; at this concentration, the buffer effect is the strongest. These buffer solutions have important practical applications.

Mixture of Several Oxidants or of Several Reductants. When in a solution we have simultaneously several oxidants such as Ce^{4+} and Fe^{3+}, no reaction is occurring. However, if we start to reduce the solution with a reductant, it is the strongest oxidant that starts reacting; in this case, it would be Ce^{4+}. The situation is identical for a mixture of two reductants; that is, the strongest reductant will oxidize first.

Calculation of Potential of Mixture of Oxidant and Reductant of Two Different Couples at Equivalent Concentrations. At the equivalent point of concentrations we have, for couple 1,

$$E = E_1 + \frac{0.06}{n} \log \frac{|Ox|^a}{|Red|^b}$$

For the couple 2, we have

$$E = E_2 + \frac{0.06}{n} \log \frac{|Ox|^a}{|Red|^b}$$

At this point, all concentrations are equivalent; then

$$2E = E_1 + E_2 + \frac{0.06}{n} \log 1$$

$$E = \tfrac{1}{2}(E_1 + E_2)$$

Example. Reduction of a mixture of Ce^{4+} and Fe^{3+} by Cr^{2+} (E_0 of $Cr^{3+}/$ $Cr^{2+} = -0.40$ V). First, Cr^{2+} reduces the strongest oxidant Ce^{4+}, and at point B, as shown in Figure 6.24 all the Ce^{4+} has been reduced and a mixture of Ce^{3+}, Fe^{3+}, and Cr^{3+} is left. Curve $A–B$ corresponds to the buffer solution Ce^{4+}/Ce^{3+} with an inflexion point at $E = E_0$ of $Ce^{4+}/$ $Ce^{3+} = 1.70$ V. After B, the Cr^{2+} reduces Fe^{3+}, which is the only oxidant left, and curve $B–C$ corresponds to the buffer solution Fe^{3+}/Fe^{2+} with an inflexion point at $E = E_0$ of $Fe^{3+}/Fe^{2+} = 0.75$ V. After C, we have a mixture of Ce^{3+}, Fe^{2+}, Cr^{3+}, and Cr^{2+} in excess, and the reaction stops. In B, we have two oxidants Fe^{3+} and Cr^{3+} and one reductant Ce^{3+}; however, the strongest oxidant is Fe^{3+}; the potential at this point is

$$E_B = \tfrac{1}{2}(1.70 + 0.75) = 1.22 \text{ V}$$

In C, we have two reductants Ce^{3+} and Fe^{2+} and one oxidant Cr^{3+}; however, the strongest reductant is Fe^{2+}; the potential at this point is

$$E_C = \tfrac{1}{2}(0.75 - 0.40) = 0.18 \text{ V}$$

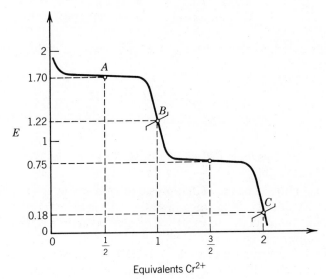

Figure 6.24. Reduction of Ce^{4+} and Fe^{3+} by Cr^{2+}.

Transfer of Protons

Acids and Bases. An ionized compound or molecule capable, when in solution, of donating an hydrogen ion, H^+, or proton is an *acid*. A *base* is an ionized compound or molecule capable, when in solution, of fixing a hydrogen ion or proton and forming an acid. This has been described by the theory of Bronsted:

$$\text{Acid} \rightleftarrows \text{base} + H^+$$

In solution, a strong acid such as HCl behaves as follows:

$$HCl \rightarrow H^+ + Cl^-$$

and a weak acid such as HCN as

$$HCN \rightleftarrows H^+ + CN^-$$

It is important to note that the ion NH_4^+ is a weak acid:

$$NH_4^+ \rightleftarrows H^+ + NH_3$$

Some hydrated cations behave as acids:

$$Al^{3+}(\text{aqueous}) \rightleftarrows 3H^+ + Al(OH)_3$$

Polyacids give ions that behave at the same time as acids and bases; for example,

$$H_2CO_3 \rightleftarrows HCO_3^- + H^+$$

$$HCO_3^- \rightleftarrows CO_3^{2-} + H^+$$

The ion HCO_3^- is called an amphiprotic ion.

Notion of Indifferent Ion. In the presence of a strong acid such as HCl, which is completely dissociated in water, Cl^- is an extremely weak base since it has no tendency to fix H^+ ions. For this reason, we call the Cl^- ion an indifferent ion.

Similarly, a strong base such as NaOH is completely dissociated in water; consequently, Na^+ is an extremely weak acid and is also an indifferent ion.

Prediction of a Reaction: Strength of Acids and Bases

Reaction between Two Acid–base Systems. Consider the acid–base system A_1–B_1 and the other acid–base system A_2–B_2. First, we have the equilibrium

$$A_1 \rightleftarrows B_1 + H^+ \qquad (6.1)$$

If we had B_2, it fixes some H^+ ions,

$$B_2 + H^+ \rightleftarrows A_2 \qquad (6.2)$$

Consequently, H^+ ions will be shared between the two systems A_1–B_1 and A_2–B_2. In some cases, all the H^+ of reaction (6.1) will be fixed by the B_2 base. The H^+ ions cannot exist free in solution, and they are attracted by the solvent. In this case, we may consider water as a base:

$$H^+ + H_2O \rightleftarrows H_2O^-$$

If we put a strong acid in water, all H^+ are fixed by water. If a weak acid is put in water, few H^+ are fixed by water.

Water also behaves as an acid:

$$H_2O \rightleftarrows OH^- + H^+$$

In this case, it may react with bases such as NaOH.

Notion of pH. Pure water may behave as an amphiprotic ion:

$$H_2O \rightleftarrows OH^- + H^+ \qquad (6.3)$$

$$H_2O + H^+ \rightleftarrows H_3O^+ \qquad (6.4)$$

We may apply the law of action mass to these reactions and particularly to reaction (6.3),

$$\frac{|OH^-\,\|\,H^+\,|}{|H_2O\,|} = \text{constant } K'$$

However, $|H_2O|$ is also a constant: We have pure water, and in the denominator we have only water. We would like to know its concentration in molecule gram/liter:

$$| H_2O | = \frac{1000}{18} \approx 55.5 \text{ molecule g/l}$$

Then, the constant $| H_2O |$ may pass with the constant K' since the product of two constants is another constant.

$$| OH^- | \| H^+ | = K = 10^{-14.2} \approx 10^{-14}$$

> The product $| OH^- | \| H^+ |$ is called the ionic product of water.

In pure water, we have $| OH^- | = | H^+ |$; then

$$| H^+ |^2 = 10^{-14}$$
$$| H^+ | = 10^{-7} \quad \text{(neutral solution)}$$

An acid solution is a solution where an excess of ion H^+ is added:

$$| H^+ | > 10^{-7} \quad \text{in acid solution}$$
$$| H^+ | < 10^{-7} \quad \text{in basic solution}$$

At his stage, it appears more convenient to use the notion of pH.

> *Definition:* The pH is the cologarithm of the concentration of the H^+ ions, keeping in mind that to be exact we should use activities.

$$pH = \text{colog} | H^+ | = -\log | H^+ | = \log \frac{1}{| H^+ |}$$

For a neutral solution, we have

$$pH = \log \frac{1}{10^{-7}} = \log 10^7 = 7 \log 10 = 7$$

$$pH < 7 \quad \text{in acid solution}$$
$$pH > 7 \quad \text{in basic solution}$$

pH of Strong Acid and pH of Strong Base

Example 1:   we have a 1 M HCl solution, and we want to calculate the pH. We know that all the HCl is dissociated:

$$\underset{\substack{1\,M\,\text{or}\,1\text{H}^+ \\ \text{g/l}}}{HCl \to Cl^- + \quad H^+}$$

Then

$$pH = -\log 1 = 0$$

This calculation was valid only because HCl is completely dissociated, being a strong acid.

$$\boxed{pH \text{ of strong acid} = -\log C}$$

where C is the concentration of H^+ ions in grams per liter. The strong acids for which this method applies are HCl, HBr, HI, HNO_3, $HClO_4$, and the first acidity of H_2SO_4.

Example 2:   we have a solution 1 M NaOH and want to calculate the pH. All the NaOH is dissociated,

$$NaOH \to \underset{\substack{\text{indifferent} \\ \text{ion}}}{Na^+} + \underset{\substack{1\,M\,\text{or}\,1\text{OH}^- \\ \text{g/l}}}{OH^-}$$

We also know that

$$|\,OH^-\,\|\,H^+\,| = 10^{-14}$$

$$1 \times |\,H^+\,| = 10^{-14}$$

Then

$$pH = \text{colog } 10^{-4} = 14$$

Now suppose we want to calculate the pH of an $N/10$ NaOH solution:

$$\tfrac{1}{10} \times |\,H^+\,| = 10^{-14}$$

$$pH = -\log(10 \times 10^{-14})$$

$$pH = -\log 10 - \log 10^{-14}$$

$$pH = 14 - 1 = 13$$

$$\text{pH of strong base} = 14 + \log C$$

The strong bases in which this method applies are NaOH, KOH, LiOH, and $Ba(OH)_2$.

Example 3: We have an $M/25$ perchloric acid solution and would like to calculate the pH:

$$HClO_4 \rightarrow ClO_4^- + \underset{1/25H^+ \text{ g/l}}{H^+}$$

$$|H^+| = \tfrac{1}{25}$$

$$pH = -\log C = -\log \tfrac{1}{25} = -(\log 1 - \log 25)$$

$$pH = \log 25 = \log 5^2 = 2 \log 5$$

$$pH = 2 \times 0.70 = 1.40$$

Notion of pK. When we are working with a weak acid or a weak base, it becomes necessary to consider the law of mass action:

$$A \rightleftarrows B + H^+$$

$$\frac{|B \| H^+|}{|A|} = K$$

In this case, the dissociation of the acid A is a function of K. For an acid that is very much dissociated, there is a large quantity of base B and H^+ ions in the solution; then the value of K is high.

For a relatively strong acid, the value of K is high.
For a weak acid, the value of K is small.

Then, if we know the value of K, we also know the strength of the acid, or its aptitude to provide H^+ ions. For many classical acid–base systems, the value of the strength of the acid is defined by measuring the value of K.

Example. For $HF-F^-$, we have $K = 10^{-3.2}$. For reasons used in defining pH, we may now define the pK, which greatly simplifies the calculations:

$$K_{HF} = 10^{-3.2}$$

$$pK = colog\ K$$

$$pK_{HF} = 3.2$$

As a consequence, a strong acid has a smaller pK than a weak acid

---

Example.

HF is a stronger acid than HCN
HF/F$^-$ has a pK = 3.2
HCN/CN$^-$ has a pK = 9.2

Following is a table showing some common acids and the value of their pK. This table may be useful to the analyst to predict some reactions:

Acid	Value of pK
$H_2C_2O_4$, oxalic acid	1.2, 4.1
H_3PO_4, phosphoric acid	2.1, 7.1, 12
HF, hydrofluoric acid	3.2
HCH_3CO_2, acetic acid	4.8
H_2S, hydrogen sulfide	7.1, 15
H_2CO_3, carbonic acid	6.4, 10.3
NH_4^+, ammonium ion	9.2
HBO_2, boric acid	9.2
H_3AsO_3, arsenious acid	9.2
H_3AsO_4, arsenic acid	2.2, 7.0, 11.5
HCN, cyanic acid	9.3
$H_4P_2O_7$, pyrophosphoric acid	1.0, 2.5, 6.1, 8.5
H_2SO_4, second acidity of sulfuric acid	1.9
H_3PO_2, hypophosphorous acid	1.0
EDTA	2.0, 2.7, 6.2, 10.3
Tartaric acid	3.0, 4.4

pH of Buffer Solution. Suppose we introduce into a solution an acid A at the concentration x and a base B at the concentration y. Here A and B are also from a same acid–base system. We have

$$A \underset{x}{\overset{}{\rightleftarrows}} B + H^+_{}$$

where x and y change until the law of mass action is respected:

$$\frac{|\, B\, |\, |\, H^+\, |}{|\, A\, |} = K$$

From this equation, we may isolate $|\, H^+\, |$,

$$|\, H^+\, | = K \frac{|\, A\, |}{|\, B\, |}$$

We know it is convenient to work with the colog of such an equation:

$$pH = pK + \log \frac{|\, B\, |}{|\, A\, |}$$

This important formula is useful for the calculation of the pH of a mixture of acid and base from a same system. Such a mixture is called a buffer solution.

Example 1:   We have a 0.1 M HCH_3CO_2 and 0.2 M $NaCH_3CO_2$ solution and want to calculate the pH of this solution:

$$HCH_3CO_2 \rightleftarrows CH_3CO_2^- + H^+$$

$$pH = 4.8 + \log \frac{2/10}{1/10}$$

$$pH = 4.8 + \log \frac{2}{1}$$

$$pH = 4.8 + 0.3 = 5.1$$

Example 2:   We have a 1.5 M HHC_3CO_2 and 1.5 M $NaCH_3CO_2$ solution and want to calculate the pH of this solution:

$$pH = 4.8 + \log 1$$

$$pH = pK = 4.8$$

> In the equimolar mixture of an acid with its corresponding base, we have pH $=$ pK.

As illustrated in Figure 6.25, at the extremities of the curve, pH cannot be calculated because activities are becoming very strong.

> If a buffer solution, if we had small quantities of a strong acid or strong base, the variation of the pH is very small: We buffered the solution around the pK of the acid–base system.

pH of Weak Acid. If we put into a solution one weak acid at the concentration C, what is the pH of such a solution?

$$A \rightleftarrows B + H^+$$

According to the law of mass action

$$\frac{|\,B\,\|\,H^+\,|}{|\,A\,|} = K \qquad (6.5)$$

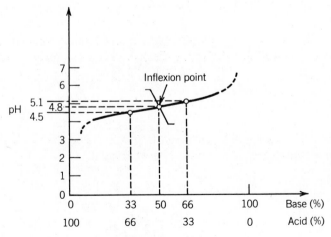

Figure 6.25. Illustration of variation of pH of a HCH_3CO_2–$CH_3CO_2^-$ buffer solution as a function of concentrations of HCH_3CO_2 and $CH_3CO_2^-$.

Two important things are happening:

1. Since the acid is weak, it is not very much dissociated in solution; then at the equilibrium, we have

$$|A| \approx C \quad \text{(quantity of } H^+ \text{ being very small)} \qquad (6.6)$$

2. The dissociation of the acid corresponds to the introduction of H^+ and B, both at the same concentration:

$$|B| = |H^+| \qquad (6.7)$$

Putting (6.6) and (6.7) into (6.5), we obtain

$$\frac{|H^+ \parallel H^+|}{C} = K$$

$$|H^+|^2 = KC$$

$$2 \operatorname{colog}|H^+| = \operatorname{colog} K + \operatorname{colog} C$$

$$2\,\text{pH} = \text{p}K - \log C$$

$$\boxed{\text{pH of weak acid} = \tfrac{1}{2}\text{p}K - \tfrac{1}{2}\log C}$$

Example. We have a 0.1 M NH$_4$Cl solution and want to calculate the pH. It is a salt of a strong acid and a weak base; accordingly, it is totally ionized:

$$NH_4Cl \rightarrow Cl^- + NH_4{}^+$$

However, Cl^- is an indifferent ion, the result being that we have only $NH_4{}^+$ in water to consider, and it is a weak acid:

$$NH_4{}^+ \rightleftarrows NH_3 + H^+$$

$$\text{pH} = \frac{9.2}{2} - \frac{1}{2}\log\frac{1}{10}$$

$$\text{pH} = 5.1$$

pH of Weak Base. Suppose we put NH_3 in water; then practically all NH_3 will stay free, and only a very small quantity will fix some H^+ ions from the water:

$$NH_3 + H^+ \rightleftarrows NH_4{}^+$$

$$B + H^+ \rightleftarrows A$$

The H^+ ions fixed by B are coming from the water; then in order to form one A, it was necessary to take one H^+ from the water. Their concentrations are equal, and we write

$$|A| = |OH^-| \tag{6.8}$$

According to the mass action law, we write

$$\frac{|B\|H^+|}{|A|} = K \tag{6.9}$$

At the beginning, the concentration of the base was C,

$$|B| = C \tag{6.10}$$

Practically, this concentration does not change when introducing the base into water; then putting (6.8) and (6.10) into (6.9),

$$\frac{C|H^+|}{|OH^-|} = K$$

$$\frac{C|H^+|^2}{10^{-14}} = K$$

$$|H^+|^2 = \frac{K \times 10^{-14}}{C}$$

$$2 \operatorname{colog}|H^+| = \operatorname{colog} K + 14 - \operatorname{colog} C$$

$$\boxed{\text{pH of weak base} = \tfrac{1}{2}pK + 7 + \tfrac{1}{2}\log C}$$

Example. We have a 0.1 M NaCN solution and want to calculate the pH.

We know that all the NaCN is dissociated,

$$NaCN \rightarrow CN^- + Na^+$$

However, Na^+ is an indifferent ion, and we are dealing with the system

$$CN^- + H^+ \rightleftarrows HCN \qquad pK = 9.2$$
$$pH = \tfrac{1}{2} \times 9.2 + 7 + \log \tfrac{1}{10} = 11.1$$

pH of Mixture of Acid with Base from Two Different Systems but Existing in Equivalent Quantities. Suppose we have a mixture of A_1 with B_2: A_1 belongs to the system A_1–B_1, with

$$A_1 \rightleftarrows B_1 + H^+ \qquad pK = pK_1 \tag{6.11}$$

B_2 belongs to the system B_2–A_2, with

$$B_2 + H^+ \rightleftarrows A_2 \qquad pK = pK_2 \tag{6.12}$$

To some extent, and depending on their respective pK, the acid A_1 will react on the base B_2 with formation of base B_1 and acid A_2:
For (6.11),

$$pH = pK_1 + \log \frac{|B_1|}{|A_1|}$$

For (6.12),

$$pH = pK_2 + \log \frac{|B_2|}{|A_2|}$$

At the equilibrium, these two pH's are equal; then

$$2pH = pK_1 + pK_2 + \log \frac{|B_1| \, |B_2|}{|A_1| \, |A_2|}$$

If the concentration of A_1 and B_2 are equivalent, $\log(|B_1\|B_2|/|A_1\|A_2|)$ will be near zero:

$$pH = \tfrac{1}{2}(pK_1 + pK_2)$$

It is important to note that, in a first approximation, the pH given by this formula does not depend on concentrations. Of course, this formula is valid only when A_1 and B_2 are in nearly equivalent quantities. However, the ratio of these concentrations may vary between $\frac{1}{3}$ and 3 without generating noticeable errors.

Example. We have a NH_4F solution and want to calculate the pH. NH_4F is a neutral salt formed from a weak acid combined with a weak base,

$$NH_4F \rightleftarrows \quad \overbrace{F^-} \quad + \quad \overbrace{NH_4^+}$$

$$pK = 3.2 \qquad pK = 9.2$$
$$\text{concentration} = C_1 \quad \text{concentration} = C_2$$

Obviously, C_1 and C_2 are equal; then

$$pH = 1.6 + 4.6 = 6.2$$

Transfer of Ions or Polar Molecules

Complexes contain a central atom or ion, usually a metal, that is chemically bonded to various ions or molecules. All metal ions in aqueous solution may be regarded as complex ions since they are combined with water molecules.

> When describing a complex compound, the central metal ion is called the acceptor, and the attached groups are known as donor groups or ligands

In addition to the presence of simple ions and molecules in the surroundings of the metal ion, larger organic molecules may combine to form a special class of compounds known as chelates. Associations such as NO_3^-, ClO_4^-, NH_3, and NH_4^+ are called primary complexes. Another kind of complex exists that is less stable and corresponds to the theory of Alfred Werner. In these complexes, there is a central atom, in general a cation, around which there are molecules, or primary complexes, or ions fixed by auxiliary bonds, which do not follow simple rules.

Examples. Experience proves that around a copper atom, four NH_3 molecules may be fixed:

$$Cu(NH_3)_4^{2+}$$

NH_3 is a primary complex that can be explained by ordinary covalent bonds. However, bonds between copper and the four NH_3 molecules cannot be explained by any of the common bonds (i.e., covalent, ionic, etc.). Around the copper atom, we can start to fix one NH_3:

$$Cu(NH_3)^{2+}$$

then a second NH_3: $Cu(NH_3)_2^{2+}$

then a third NH_3: $Cu(NH_3)_3^{2+}$

then a fourth NH_3: $Cu(NH_3)_4^{2+}$

There is an upper limit to the number of molecules of NH_3 that may be fixed around the copper atom: This limit defines the coordination number.

Generally (but not always) the coordination number is equal to twice the maximum valence of the central element.

Examples. Ag^+ has a coordination number equal to 2:

$$Ag(CN)_2^{-}$$

Cu^{2+} has a coordination number equal to 4:

$$Cu(NH_3)_4^{2+}$$

Zn^{2+} has a coordination number equal to 4:

$$Zn(NH_3)_4^{2+}$$

Fe^{2+} and Fe^{3+} have a coordination number equal to 6:

$$Fe(CN)_6^{4-} \qquad Fe(CN)_6^{3-}$$

When an element has two possible valences such as iron, the coordination number is calculated by using the maximum valence. The coordination number of iron is always 6, even if it is Fe^{2+}.

> In a complex:
>
> 1. The total charge of the complex is equal to the total charge of its components.
> 2. For the calculation of the coordination number, a molecule counts for one, a monovalent ion counts for one, a bivalent ion counts for two.
> 3. Physical and chemical properties are very different from those of its components.

Examples. For Al^{3+}, $n = 6$ and we may encounter complexes such as AlF^{2+}, AlF_2^+, AlF_3, AlF_4^-, and so on. It is interesting to note that the neutral complex AlF_3 is also the insoluble compound. Al^{3+} may form complexes with oxalic ions $C_2O_4^{2-}$:

$$Al(C_2O_4)^+ \qquad Al(C_2O_4)_2^- \qquad Al(C_2O_4)_3^{3-}$$

Chelates. Consider a molecule of oxine, also called 8-hydroxyquinoline, where N has three covalent bonds and a free pair of electrons:

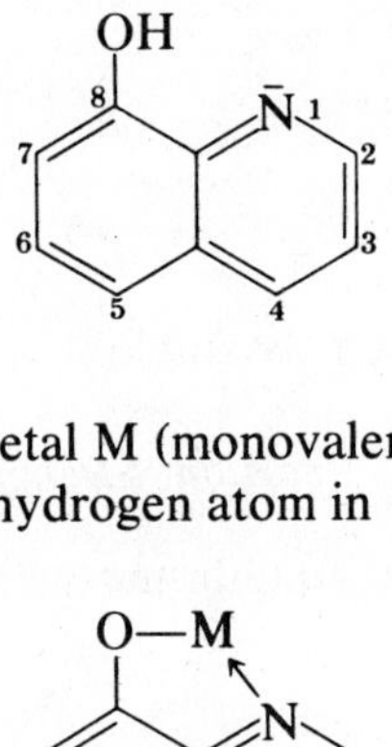

Let's take the cation of a metal M (monovalent); under some conditions, it will take the place of the hydrogen atom in $-OH$ and form an oxinate,

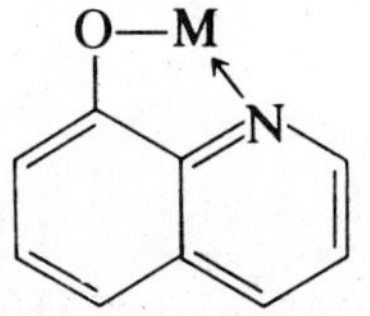

The free pair of electrons around N is very near the atom M and by attraction with it is forming a secondary complex. Then a five-sided cycle is closed. This new molecule is very stable and gives extremely insoluble compounds.

Example. $Al^{3+} + 3 \text{ oxinates} \rightarrow Al(ox)_3 \downarrow$

This yellow-green oxinate is used to titrate traces of aluminum.
   Consider another molecule, such as *o*-phenantroline,

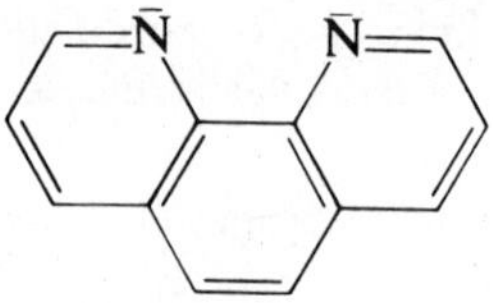

With some cations, the free pairs of electrons on both N will get closer
to each other and may react with formation of a cycle with five sides.

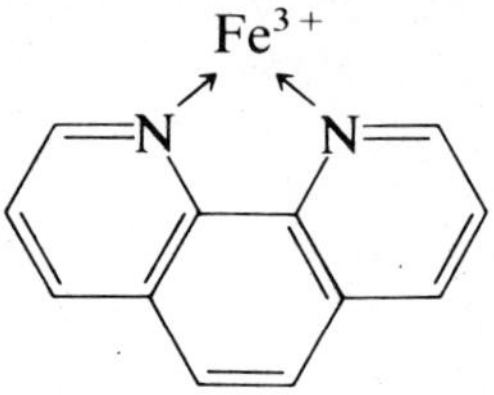

Fe^{3+}

dark red

Fe^{2+}

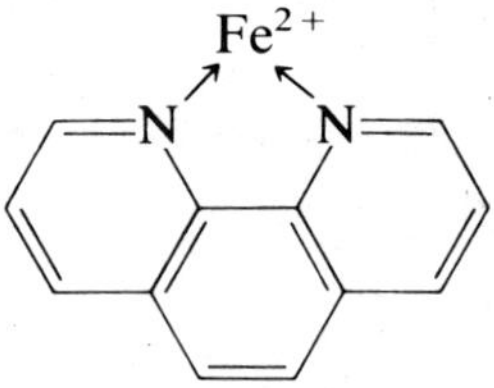

pale blue

These compounds are extremely stable.

Prediction of a Reaction: Strength of Complexes

In general, a complex is in equilibrium with the ions or molecules that
originates it.

Example.

$$HgI_2 + 2KI \rightleftarrows (HgI_4)K_2$$

$$Hg^{2+} + 4I^- \rightleftarrows (HgI_4)^{2-}$$

To this equilibrium, we may apply the mass action law:

$$\frac{|Hg^{2+}||I^-|^4}{|(HgI_4)^{2-}|} = K$$

> If K is very small, the complex is very stable (i.e., ferrocyanide).
> If K is very large, the complex is not stable (i.e., ferrous acetate).

Definition. I^- is called a particle x, Hg^{2+} is called an acceptor A of the particle x, and $(HgI_4)^{2-}$ is called a donor D of the particle x:

$$D \rightleftarrows A + x$$

Examples.

1. Suppose x is H^+; we have

$$Acid \rightleftarrows base + H^+$$

2. Suppose x is e^-; we have

$$Reductant \rightleftarrows oxidant + e^-$$

If we apply the mass action law,

$$\frac{|A \parallel x|}{|D|} = K$$

then

$$|x| = K \frac{|D|}{|A|}$$

And we may define a px as we defined a pH:

$$px = -\log |x|$$

> $$px = pK + \log \frac{|A|}{|D|}$$

Notion of Polycomplex, Amphoterization, and Disproportionation. The ion Cu^{2+} has a coordination number of 4; sometimes it is possible to fix five

NH_3. It is done by action of liquid NH_3 on the Cu^{2+} ion:

$$Cu(NH_3)_5{}^{2+} \rightleftarrows Cu(NH_3)_4{}^{2+} + NH_3 \quad \text{with } pK_5 = -0.5$$

$$Cu(NH_3)_4{}^{2+} \rightleftarrows Cu(NH_3)_3{}^{2+} + NH_3 \quad \text{with } pK_4 = 2$$

$$Cu(NH_3)_3{}^{2+} \rightleftarrows Cu(NH_3)_2{}^{2+} + NH_3 \quad \text{with } pK_3 = 2.7$$

$$Cu(NH_3)_2{}^{2+} \rightleftarrows Cu(NH_3)^{2+} + NH_3 \quad \text{with } pK_2 = 3$$

$$Cu(NH_3)^{2+} \rightleftarrows Cu^{2+} \qquad\quad + NH_3 \quad \text{with } pK_1 = 3.3$$

When an ion can fix several particles, it is a polycomplex, and generally we have

$$pK_5 < pK_4 < pK_3 < pK_2 < pK_1$$

Often, these pK constants are not known because they are too close to each other, and we know only the overall constant,

$$pK_{overall} = pK_1 + pK_2 + pK_3 + pK_4 + pK_5$$

In order to differentiate quantitatively two different complexes, it is necessary for their pK to be different by at least six units. In the case of copper and NH_3, the complexes are mixed together, and therefore it is difficult to differentiate them from one another.

Any intermediary ion that may be a donor or an acceptor at the same time is called an amphoter [i.e., $Cu(NH_3)_2{}^{2+}$]. Some amphoters are not capable of staying in solution and they generate reactions that

1. may increase the number of fixed particles or
2. may decrease the number of fixed particles.

Example. $3C_2O_4{}^{2-}$ particles have a total of six negative charges, which corresponds to the coordination number of aluminum. However, the ion $Al(C_2O_4{}^{2-})^+$ does not exist. If we put in solution the exact quantities of Al^{3+} and $C_2O_4{}^{2-}$ that give $Al(C_2O_4{}^{2-})^+$, the following reactions take place:

1. One reaction that increases the number of $C_2O_4{}^{2-}$ particles fixed,

$$Al(C_2O_4{}^{2-})^+ + C_2O_4{}^{2-} \rightleftarrows Al(C_2O_4{}^{2-})_2{}^-$$

2. At the same time, one reaction that decreases the number of $C_2O_4^{2-}$ particles fixed,

$$Al(C_2O_4^{2-})^+ \rightleftarrows Al^{3+} + C_2O_4^{2-}$$

These two reactions occur simultaneously as follows:

$$2Al(C_2O_4^{2-})^+ \rightleftarrows Al^{3+} + Al(C_2O_4^{2-})_2^-$$

The $Al(C_2O_4^{2-})^+$ does not exist in solution; it disproportionates into Al^{3+} and $Al(C_2O_4^{2-})_2^-$ ions.

This may have interesting applications, such as knowing that two oxalate ions are necessary in order to complex one aluminum ion.

Destruction of Complexes by Dilution. It is not difficult to demonstrate that when a colored complex is diluted, not only the color is diluted but also the complex is destroyed to some extent.

Suppose that at the beginning of the reaction the donor D is at the concentration C. The idea is to find out the concentrations of all the constituents (i.e., donor D, acceptor A, and particle x) at the equilibrium as a function of C and α, a dissociation factor.

If one particle is exchanged, by definition, we have

$$|A| = |x|$$

where $|x|$, the dissociated part associated with A, is equal to $C\alpha$. At the equilibrium, the concentration of D is

$$|D| = C(1 - \alpha)$$

Then, at equilibrium, we have

$$\underset{C(1-\alpha)}{D} \rightleftarrows \underset{C\alpha}{A} + \underset{C\alpha}{x}$$

Then, according to the mass action law,

$$\frac{C\alpha C\alpha}{C(1 - \alpha)} = K$$

After simplification,

$$\frac{\alpha^2}{1-\alpha} = \frac{K}{C}$$

If the concentration C is diminishing, that is, if a dilution is performed, since K is a constant, the ratio K/C is increasing. If K/C increases, the ratio $\alpha^2/(1-\alpha)$ increases in the same way. To increase $\alpha^2/(1-\alpha)$, only α needs to increase.

Conclusion: When a complex is diluted, the concentration is diminishing but the dissociation coefficient is increasing, which indicates that the complex is destroyed to some extent.

This property of complexes can have serious consequences during colorimetric procedures if precautions are not taken.

Transfer of Electrons and Protons

If we take the example of MnO_4^-, which is a strong oxidizing agent, it becomes more complicated: We are dealing with the particle MnO_4^- in which manganese is as Mn^{VII+}. It is also known that in acid medium, Mn^{VII+} reduces to Mn^{2+}; however, Mn^{7+} does not exist in an aqueous solution. As a general guideline, we may

1. write the equation as $Mn^{VII+} + 5e^- \rightleftarrows Mn^{2+}$;
2. rewrite this equation using existing particles, $MnO_4^- + 5e^- \rightleftarrows Mn^{2+}$; and
3. balance this new form of the equation, $MnO_4^- + 5e^- + 8H^+ \rightleftarrows Mn^{2+} + 4H_2O$.

Example. Reduction of dichromate into Cr^{3+}: We know that we are dealing with $Cr_2O_7^{2-}$ and Cr^{3+}. The First step:

$$Cr^{VI+} + 3e^- \rightleftarrows Cr^{3+}$$

Second step:

$$Cr_2O_7^{2-} + 6e^- \rightleftarrows 2Cr^{3+}$$

Third step:

$$Cr_2O_7^{2-} + 6e^- + 14H^+ \rightleftarrows 2Cr^{3+} + 7H_2O$$

Example. Reduction of Se^{IV+} to Se^0: We know that we are dealing with SeO_3^{2-} and Se^0. First step:

$$Se^{IV+} + 4e^- \rightleftarrows Se^0$$

Second step:

$$SeO_3^{2-} + 4e^- \rightleftarrows Se^0$$

Third step:

$$SeO_3^{2-} + 4e^- + 6H^+ \rightleftarrows Se^0 \downarrow + 3H_2O$$

In practice, this reduction is difficult to perform, and the reaction is slow (see Section 4.19).

Prediction of a Reaction

Reduction of V^{V+} to V^{IV+}:

$$V^{V+} + e^- \rightleftarrows V^{IV+}$$

$$VO_2^+ + e^- \rightleftarrows VO^{2+}$$

$$VO_2^+ + e^- + 2H^+ \rightleftarrows VO^{2+} + H_2O$$

The advantage of this type of reaction is that we can calculate the potential as a function of the pH.

$$E = f(pH)$$

If we apply the Nernst equation to be above reaction,

$$E = E_0' + \frac{0.06}{1} \log \frac{|VO_2^+ \| H^+|^2}{|VO^{2+} \| H_2O|}$$

The concentration of H_2O is a constant and we may include it with E_0':

$$E = E_0' + 0.06 \log \frac{|VO_2^+|}{|VO^{2+}|} + 0.06 \log |H^+|^2 - 0.06 \log |H_2O|$$

$$E = E_0 + 0.06 \log \frac{|VO_2^+|}{|VO^{2+}|} + 0.06 \log |H^+|^2$$

or

$$E = E_0 + 0.06 \log \frac{|VO_2^+|}{|VO^{2+}|} - 0.12 \text{ pH}$$

Most of the time we make an assumption that $|VO_2^+|/|VO^{2+}| \approx 1$. Then

$$E = E_0 - 0.12 \text{ pH} \quad \text{with } E_0 = 1.0 \text{ V}$$

For pH = 0:

$$E = 1.0 - 0 = 1.0 \text{ V}$$

For pH = 10:

$$E = 1.0 - 1.2 = -0.2 \text{ V}$$

Now suppose the medium is very oxidizing:

1. $|VO_2^+|/|VO^{2+}| = 10$. Then

$$E = 1.0 + 0.06 - 0.12 \text{ pH} = 1.06 - 0.12 \text{ pH}$$

2. $|VO_2^+|/|VO^{2+}| = 100$. Then

$$E = 1.0 + 0.12 - 0.12 \text{ pH} = 1.12 - 0.12 \text{ pH}$$

The curve with $|VO_2^+|/|VO^{2+}| = 1$ separates the two areas, reducing and oxidizing, as indicated in Figure 6.26.

The advantage of these curves is to show if the considered ions are reducing or oxidizing at a particular pH.

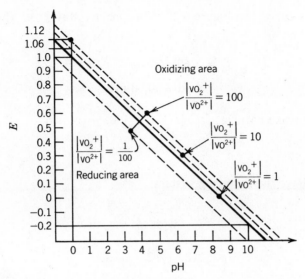

Figure 6.26. Potential as a function of pH in the Red–Ox system VO_2^+–VO^{2+}.

Exercise. Can MnO_4^- oxidize Cl^- and under which conditions? First we have the following system:

$$MnO_4^- + 4H^+ + 3e^- \rightleftarrows MnO_2 + 2H_2O \quad \text{with } E_0 = 1.69 \text{ V}$$

Then

$$E = E_0' + \frac{0.06}{3} \log \frac{\mid MnO_4^- \parallel H^+ \mid^4}{\mid MnO_2 \parallel H_2O \mid^2}$$

Since H_2O is a constant, we include it with E_0':

$$E = E_0 + 0.02 \log \frac{\mid MnO_4^- \mid}{\mid MnO_2 \mid} + 0.02 \log \mid H^+ \mid^4$$

$$E_0 + 0.02 \log \frac{\mid MnO_4^- \mid}{\mid MnO_2 \mid} - 0.08 \text{ pH}$$

If the convention of having $\mid MnO_4^- \mid / \mid MnO_2 \mid = 1$ is made, we obtain

$$E = 1.69 - 0.08 \text{ pH}$$

This gives curve 1 in Figure 6.27. If we make $| MnO_4^- |/| MnO_2 | = 10$, we obtain

$$E = 1.69 + 0.02 - 0.08 \text{ pH}$$
$$1.71 - 0.08 \text{ pH}$$

and if we make $| MnO_4^- |/| MnO_2 | = 100$, we obtain

$$E = 1.73 - 0.08 \text{ pH}$$

We also have the following system:

$$\frac{Cl_2}{2Cl^-} \text{ with } E_0 = 1.40 \text{ V}$$

$$2Cl^- - 2e^- \rightleftarrows Cl_2$$

Cl^- is an indifferent ion since it comes from HCl, which is a strong acid.

$$E = 1.40 + \frac{0.06}{2} \log \frac{| Cl_2 |}{| Cl^- |^2}$$

If we make $| Cl^2|/| Cl^- |^2 = 1$, we obtain $E = 1.40$ V irrespective of the pH; then curve 2 in Figure 6.27 is represented by a straight horizontal line.

For the reaction between the two systems to be quantitative, the difference of potential between the oxidant and the reductant should be at least 0.2 V

Following are the conclusions we can draw from a graph such as the one shown in Figure 6.27.

- In acid medium, MnO_4^- oxidizes Cl^- quantitatively if the pH < 3.7.
- Around pH 3.7, both curves 1 and 2 are near each other; then MnO_4^- will oxidize Cl^- only partially.
- If the pH > 3.7, MnO_4^- cannot oxidize Cl^-, and theoretically Cl_2 could oxidize Mn^{IV+}.

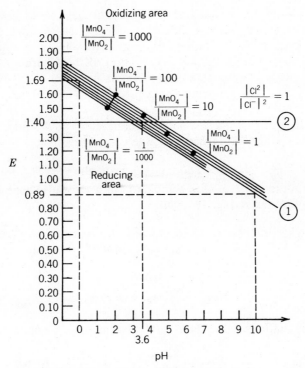

Figure 6.27. Illustration of conditions under which MnO_4^- may oxidize Cl^-

Transfer of Ions or Polar Molecules and Electrons

If some Fe^{3+} and Fe^{2+} ions are introduced into a solution, the potential of the solution is given by

$$E = E_0 + 0.06 \log \frac{|\,Fe^{3+}\,|}{|\,Fe^{2+}\,|}$$

Then, when SCN^- ions are added, Fe^{3+} is complexed, and the concentration of Fe^{2+} is not changing.

Conclusion: When $|\,Fe^{3+}\,|$ is diminishing after being complexed by SCN^-, the potential E is also diminishing and the solution is becoming reductive.

Rule: If we add an ion complexing the oxidant of a red–ox system, the solution becomes more reductive, and vice versa

---

However, when the complexing agent acts on both ions (i.e., reductant and oxidant), it may become difficult to predict the resulting potential.

Example 1:   If we are adding CN^- ions into a solution containing Co^{3+} and Co^{2+}, both complexes $Co(CN)_6{}^{3-}$ and $Co(CN)_6{}^{4-}$ are formed. However, the $Co(CN)_6{}^{3-}$ complex is far more stable that the $Co(CN)_6{}^{4-}$ complex, and consequently the ratio $|\,Co^{3+}\,|/|\,Co^{2+}\,|$ is becoming very small and the solution more reductive.

Example 2:   If we have a solution containing Fe^{3+} and Co^{2+} and if we want to characterize these ions, we can add SCN^-:

a.   With the Fe^{3+}, a red complex is formed.
b.   With the Co^{2+}, a blue complex is formed that is completely covered by the red ferric complex.

In order to see the blue complex of Co^{2+}, it is necessary to complex first Fe^{3+} with F^-. This technique has many applications in analytical chemistry.

Transfer of Ions and Protons

The pH may influence the stability of a complex. If we have the systems

$$BaC_2O_4 \rightleftarrows Ba^{2+} + C_2O_4{}^{2-} \quad Kc$$

$$C_2O_4{}^{2-} + H^+ \rightleftarrows C_2O_4H^- \quad Ka$$

we can demonstrate that the pH, by affecting the stability of BaC_2O_4, has an effect on the concentration of the Ba^{2+} ions. Calculate the concentration of the Ba^{2+} ions. In order to do this, suppose the complex is not dissociated very much, which allows consideration of its concentration constant:

$$|\,BaC_2O_4\,| = C$$

$$|\,Ba^{2+}\,| = |\,C_2O_4{}^{2-}\,| + |\,C_2O_4H^-\,|$$

We have also, from the law of action mass,

$$Kc = \frac{|\,Ba^{2+}\,\|\,C_2O_4{}^{2-}\,|}{|\,BaC_2O_4\,|}$$

and

$$Ka = \frac{|\,C_2O_4{}^{2-}\,\|\,H^+\,|}{|\,C_2O_4H^-\,|}$$

from which we are able to write

$$| C_2O_4H^- | = \frac{| C_2O_4^{2-} \| H^+ |}{Ka}$$

and

$$| Ba^{2+} | = | C_2O_4^{2-} | + \frac{| C_2O_4^{2-} \| H^+ |}{Ka}$$

$$| Ba^{2+} | = | C_2O_4^{2-} | \left(1 + \frac{| H^+ |}{Ka}\right)$$

$| C_2O_4^{2-} |$ is an unknown. Let us calculate it:

$$| C_2O_4^{2-} | = \frac{CKc}{| Ba^{2+} |}$$

Then

$$| Ba^{2+} | = \frac{CKc}{| Ba^{2+} |} \left(1 + \frac{| H^+ |}{Ka}\right)$$

or

$$\boxed{| Ba^{2+} | = \sqrt{CKc \left(1 + \frac{| H^+ |}{Ka}\right)}}$$

Discussion: If $|H^+| = Ka$, then pH = pK, and

$$| Ba^{2+} | = \sqrt{2CKc}$$

If $| H^+ | \ll Ka$, then pH > pK, and

$$| Ba^{2+} | = \sqrt{CKc}$$

If $| H^+ | \gg Ka$, then pH < pK, and

$$| Ba^{2+} | = \sqrt{C \frac{Kc}{Ka} | H^+ |}$$

6.3.5.   Cases Where Equilibrium Is Reached in Heterogeneous Phase: Precipitation and Solubility Product

Generalities

If we introduce some NaCl in a solution and add some $AgNO_3$, we obtain an AgCl precipitate. NaCl and $AgNO_3$ are completely dissociated:

$$NaCl \rightarrow \boxed{Cl^-} + Na^+$$
$$AgNO_3 \rightarrow NO_3^- + \boxed{Ag^+}$$
$$Cl^- + Ag^+ \rightleftarrows AgCl \downarrow$$

To this system we may apply the law of mass action: If we consider an excess of AgCl in solution, this excess is not in equilibrium; however, some AgCl is still in solution and is not dissociated, and we may write

$$Cl^- + Ag^+ \rightleftarrows \underset{\text{dissolved}}{AgCl} \rightleftarrows AgCl \downarrow$$

The concentration of AgCl dissolved is a constant equal to its maximum solubility in water under normal conditions of pressure and temperature. It is a constant as long as the precipitated AgCl is in excess.

Constant Concentration. Let us introduce in a beaker a solution of dissolved AgCl and then some AgCl crystals: AgCl is dissolving with a certain speed v_1, but the crystal is forming also with another speed v_2. The equilibrium is reached when v_1 is equal to v_2. This statistical equilibrium may be very easily disturbed (i.e., by a current of water that is increasing v_1 or a saturated AgCl solution that is increasing v_2):

$$\frac{|\, Cl^- \,\|\, Ag^+ \,|}{|\, AgCl \text{ dissolved} \,|} = K$$

If AgCl is in excess, $|\, AgCl \text{ dissolved} \,|$ is a constant, and

$$|\, Cl^- \,\|\, Ag^+ \,| = s = \text{solubility product of AgCl}$$

In general,

$$aA + bB \rightleftarrows cC \text{ (dissolved)}$$

$$\frac{|\, A \,|^a |\, B \,|^b}{|\, C \,|^c} = K$$

since $|C|^c$ is a constant if C is in excess,

$$|A|^a|B|^b = s$$

Examples of current solubility products are:

$$\text{AgCl:} \quad |Ag^+ \parallel Cl^-| = s = 10^{-10}$$

$$\text{Ag}_2\text{CrO}_4: \quad |Ag^+|^2|CrO_4^{2-}| = s = 10^{-12}$$

$$\text{Al(OH)}_3: \quad |Al^{3+} \parallel OH^-|^3 = s = 10^{-33}$$

$$\text{HgS:} \quad |Hg^{2+} \parallel S^{2-}| = s = 10^{-53}$$

Calculation of Solubility of Precipitate

Definition. The solubility of a precipitate is the concentration expressed in ion-gram per liter of one of the ions going into solution.

Examples.

1. In the case of AgCl, we call the solubility of AgCl the concentration in grams per liter of one of the ions going into solution:

$$\text{Solubility of } Ag^+ = S_{Ag^+} = |Ag^+|$$

We know that

$$|Ag^+ \parallel Cl^-| = 10^{-10}$$

In this case

$$|Ag^+| = |Cl^-|$$

Then

$$|Ag^+|^2 = 10^{-10}$$

or

$$|Ag^+| = 10^{-5}$$

2. In the case of Ag_2CrO_4,

$$| CrO_4{}^{2-} \| Ag^+ |^2 = 10^{-12}$$

$$| CrO_4{}^{2-} | = \frac{| Ag^+ |}{2}$$

$$\frac{| Ag^+ |^3}{2} = 10^{-12}$$

$$| Ag^+ |^3 = 2 \times 10^{-12} = 10^{-11.7}$$

$$| Ag^+ |^3 = 10^{-11.7}$$

$$| Ag^+ | = 10^{-3.9}$$

Remark: Do not confuse solubility with solubility product.

Solubility of Hydroxide as Function of pH

Let us consider an insoluble monohydroxide that is able to redissolve either in acid or alkaline medium. In other words, we may say that it is an amphoteric hydroxide:

Acid medium: $\begin{cases} MOH \rightleftarrows M^+ + OH^- & \text{which gives } | M^+ \| OH^- | = s_1 \\ MOH + H^+ \rightleftarrows M^+ & \text{aqueous} \end{cases}$

Alkaline medium: $\begin{cases} MOH + OH \rightleftarrows MO^- + H_2 \\ MOH \rightleftarrows MO^- + H^+ & \text{which gives } | MO^- \| H^+ | = s_2 \end{cases}$

There is a small quantity of MOH that is soluble and in equilibrium with MO^- and H^+. If we call M the dissolved metal, we have

$$S_M = | M^+ | + | MO^- | + \underset{\substack{\text{dissolved fraction} \\ \text{that is very small}}}{| MOH |}$$

Then

$$S_M \approx | M^+ | + | MO^- |$$

---

or

$$S_M = \frac{s_1}{|OH^-|} + \frac{s_2}{|H^+|}$$

We can change $|OH^-|$ by $10^{-14}/|H^+|$,

$$S_M = \frac{s_1 + |H^+|}{10^{-14}} + \frac{s_2}{|H^+|}$$

To simplify the equations,

$$\frac{s_1}{10^{-14}} = K_1 \quad \text{and} \quad s_2 = K_2$$

$$S_M = K_1 |H^+| + \frac{K_2}{|H^+|}$$

or

$$S_M = K_1 \times 10^{-pH} + K_2 \times 10^{pH}$$

S_M being the solubility of the hydroxide as a function of pH.

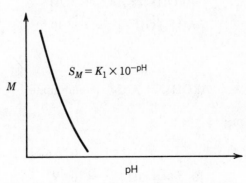

$$S_M = K_1 \times 10^{-pH}$$

Figure 6.28. In acid medium, if the pH is increasing, the solubility of a hydroxide is decreasing.

Discussion of K_1 and K_2. Since s_1 and s_2 are roughly of the same order of magnitude, we have

$$K_1 > K_2$$

If $|\,H^+\,|$ is high, the second term becomes negligible; then

$$S_M \approx K_1 \times 10^{-\text{pH}}$$

and we have a decreasing exponential as shown in Figure 6.28. In alkaline medium, we have

$$S_M \approx K_2 \times 10^{\text{pH}}$$

and we have an increasing exponential as shown in Figure 6.29.

Acidity of Cations

Rule: Less the solubility of a hydroxide, more the acidity of the corresponding cation

Example. We start with a solution containing 10^{-2} mol Al^{3+} and we want to know at which pH the $Al(OH)_3$ hydroxide is precipitating.

$$Al(OH)_3 \rightleftarrows Al^{3+} + 3OH^-$$
$$|\,Al^{3+}\,\|\,OH^-\,|^3 = s = 10^{-33}$$

In fact,

$$Al(OH)_3 \rightleftarrows AlO_2^- \quad (\text{aluminate})$$

We started from

$$|\,Al^{3+}\,| = 10^{-2}\ \text{mol}$$
$$10^{-2} \times |\,OH^-\,|^3 = 10^{-33}$$
$$|\,OH^-\,| = 10^{-10.3}$$

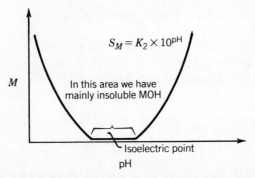

Figure 6.29. In alkaline medium, if the pH is increasing, the solubility of a hydroxide is increasing.

But

$$| H^+ \| OH^- | = 10^{-14}$$

Then

$$| H^+ | = \frac{10^{-14}}{10^{-10.3}}$$

$$| H^+ | = 10^{-3.7}$$

$$pH = 3.7$$

There are several important groups of hydroxides:

1. Very acid group (i.e., hydroxide precipitating under pH 1): W^{6+}, Mo^{6+}, Si^{4+}, Ce^{4+}, Ti^{4+}, Nb^{5+}, and Ta^{5+}.
2. Hydroxides precipitating between pH 1 and pH 2: V^{5+} and Zr^{4+}.
3. Hydroxides precipitating between pH 2 and pH 4: Th^{4+}, Fe^{3+}, Ga^{3+}, V^{4+}, and Al^{3+}.
4. Hydroxides precipitating between pH 4 and pH 8: U^{4+}, Cu^{2+}, Zn^{2+}, Ni^{2+}, Co^{2+}, and Mn^{2+}.
5. Hydroxides precipitating above pH 8: hydroxides of alkaline earths.

6.3.6. Recommended References

The following documents are recommended highly to the reader who wishes to extend his knowledge of analytical chemistry: (1) Charlot [15, 16, 18–20] and (2) Machtinger and Rosset [17].

6.4. X-RAY MATRIX CORRECTIONS

Absorption and enhancement effects need to be taken into account in accurate XRF analysis. The following empirical correction scheme is based on the work of Heinrich and Rasberry and others [21]. It is especially useful when the mineralogy of interfering elements is not known and a suite of similar samples is to be analyzed after direct pressing or partial fusion. The following symbols are used:

$$S = \text{observed counts for the element of interest, } X, \text{ background corrected and deconvoluted}$$

$$A, B, C, \cdots = \text{observed counts for the interfering elements } A, B, C, \text{ and so on}$$

$$C_1 = \text{true (unknown) counts for the element } X$$

$$K_1, K_2, K_3, \cdots = X \text{ contents of calibrating standards}$$

$$f = \text{a factor to convert corrected counts to \%: K (\%)} = fC$$

Generally,

$$S = C_1(1 + aA)(1 + bB)(1 + cC) \cdots$$

If we have $N + 1$ calibrating standards and N interfering elements, we may in principle determine N factors $a, b, c,$ and so on. With one interfering element and two standards,

$$S_1 = C_1(1 + aA_1)$$

$$S_2 = C_2(1 + aA_2)$$

$$\frac{C_1}{C_2} = \frac{K_1}{K_2}$$

Solving these three equations gives

$$a = \frac{K_1 S_2 - K_2 S_1}{A_2 K_2 S_1 - A_1 K_1 S_2}$$

Example. Two copper standards with 0.2 and 0.8% Cu contain differing iron concentrations. Copper counts S_1 and S_2 are 100 and 300, respectively. Counts A_1 and A_2 for ion are 2000 and 10000:

$$a = \frac{0.2 \times 300 - 0.8 \times 100}{10000 \times 0.8 \times 100 - 2000 \times 0.2 \times 300} = -2.94 \times 10^{-5}$$

The true copper counts C_1 and C_2 are

$$C_1 = \frac{S_1}{1 + aA_1} = \frac{100}{0.941} = 106$$

$$C_2 = \frac{S_2}{1 + aA_2} = \frac{300}{0.706} = 424$$

The factor $f = 0.2/106 = 0.8/424 = 0.00188$

Given S, A for an unknown sample,

$$\%Cu = \frac{fS}{1 + aA}$$

For example, if a sample gives 250 counts for copper and 5000 counts for iron, $S = 250$, $A = 5000$, and

$$\%Cu = \frac{0.00188 \times 250}{1 - 2.94 \times 5000 \times 10^{-5}}$$

$$= 0.55\%$$

When there are two interfering elements, three known samples are required:

$$S_1 = C_1(1 + aA_1)(1 + bB_1)$$
$$S_2 = C_2(1 + aA_2)(1 + bB_2)$$

$$S_3 = C_3(1 + aA_3)(1 + bB_3)$$

$$\frac{C_1}{C_2} = \frac{K_1}{K_2} \qquad \frac{C_1}{C_3} = \frac{K_1}{K_3}$$

or, approximately,

$$\frac{S_1K_2}{S_2K_1} = \frac{1 + aA_1 + bB_1}{1 + aA_2 + bB_2}$$

$$\frac{S_1K_3}{S_3K_1} = \frac{1 + aA_1 + bB_1}{1 + aA_3 + bB_3}$$

These equations may be solved simultaneously for the factors a and b. Then, for an unknown sample,

$$\%X = fS/(1 + aA)(1 + bB)$$

With more than two interferences, solving for $a, b, c, \cdots$ becomes difficult and also often yields unsatisfactory factors. It is better to find a standard in which one interfering element predominates, thereby obtaining a clean value for one of the factors.

For example, with laterite ores varying from limonite to garnierite and containing varying amounts of manganese and chromium, we may wish to correct Ni readings for Fe, Mn, and Cr:

$$S_{Ni} = C_{Ni}(1 + a\text{Fe})(1 + b\text{Mn})(1 + c\text{Cr})$$

We may find a by examining two garnierites having different Ni contents and little Mn and Cr and then b using samples with little Cr and different Mn contents. Any two chromium standards will then give a value for c. Reiteration may be necessary to discover the best values for a, b, c.

Obviously, there must be complete confidence in the calibrating standards and in the sample preparation process.

In dealing with a specific material type (e.g., laterite ore), a single factor may often be adequate. This follows because a second interfering element is often related to the first; for example, laterites with much iron are likely to contain little aluminum. Although attributed to iron, a single factor carries an automatic compensation for aluminum or other related elements.

A factor may deal with mineralogical effects. For example, in laterites containing chromite, the correction to iron due to chromium reflects as much the mineralogy as the interelement effect. For reasons such as this, use of theoretical absorption coefficients is not valid when heterogeneous materials are being examined.

6.5. COMPOSITION AND DENSITY OF SLURRIES

In principle, one should be able to calculate the density of a slurry (i.e., the proportion of solids) as well as the X content of the solids from two X-ray signals each generated from the element of interest, X.

How well this can be done in practice depends on the precision with which the two measurements can be made.

6.5.1. Use of K Lines and L Lines

If a sample containing X is excited by X-rays of appropriate energy, K electrons and L electrons will generate fluorescent X-ray spectral lines. The K lines are of higher energy than the L lines and will be absorbed more strongly by the matrix. One may write

$$I_L = kI(1 - \alpha M) \tag{6.13}$$

$$I_K = I(1 - \beta M) \tag{6.14}$$

where I_L is the measured intensity of the L line, I_K is the measured intensity of the K line, k is a factor less than 1 relating the intensities of the two lines, α and β are absorption coefficients, and M is the concentration of the matrix. For any element with readable K and L lines (e.g., Mo), k, α, and β are determinable constants for a particular excitation system. I_L and I_K are simultaneously measurable quantities that can be continually monitored.

Solving (6.13) and (6.14) for I and M gives

$$M = \frac{kI_K - I_L}{kI_K\alpha - I_L\beta)} \qquad I = \frac{I_K}{1 - \beta M}$$

The calculated intensity I is proportional to the concentration of X in the slurry, and the calculated value M is proportional to the concentration of solids in the slurry.

6.5.2. Use of Important Identities

Various seldom used identities relate volume proportions, weight proportions, and compositions. If

$$X = \text{wt. \% of } X \text{ in slurry}$$

$$X_H = \text{wt. \% of } X \text{ in solids}$$

$$X_L = \text{wt. \% of } X \text{ in liquid}$$

$$p_w = \text{wt. \% of liquid in slurry}$$

$$q_w = \text{wt. \% of solids in slurry}$$

$$p_v = \text{vol. \% of liquid in slurry}$$

$$q_v = \text{vol. \% of solids in slurry}$$

$$d_L = \text{density of liquid}$$

$$d_H = \text{density of solids}$$

$$d = \text{density of slurry}$$

then we have the following identities:

$$p_w = 1 - q_w = \frac{X_H - X}{X_H - X_L}$$

$$q_w = 1 - p_w = \frac{X - X_L}{X_H - X_L}$$

$$d = \frac{d_H d_L}{d_H p_w + d_L q_w}$$

$$= \frac{d_H d_L (X_H - X_L)}{d_H (X_H - X) + d_L (X - X_L)} \tag{6.15}$$

Of the quantities in these equations, we know the density of the solids, d_H, the density of the liquid, d_L, and X-ray measurements give a measure of X and X_H.

If the X content of the liquid is zero,

$$X = aI \tag{6.16}$$

$$X_H = 100 \times \frac{bI}{M} \tag{6.17}$$

where a and b are experimentally determined factors relating calculated X-ray intensities to concentration in weight percent. (*Note:* If the X content of the liquid is not zero and is variable, X-ray measurements alone do not give useful information).

Given the concentrations X and X_H from (6.16) and (6.17), the density of the slurry may be calculated from (6.15) with $X_L = 0$:

$$d = \frac{d_H d_L X_H}{d_H X_H - d_H X + d_L X}$$

One may also calculate

$$q_w = \frac{X}{X_H} \qquad q_v = \frac{d_L X}{d_H X_H - d_H X + d_L X}$$

6.5.3. Discussion

The suggested device resembles the other currently used to monitor slurry density, that is, measurement of scattered radiation and its attenuation. The difficulty is that the scattered radiation is not related to X content, and relationships such as those in Section 6.5.1 do not exist.

Whether or not the suggested device is in practice feasible is a question to put to the instrument's manufacturer.

6.6. DECONVOLUTION OF ENERGY-DISPERSIVE X-RAY PEAKS USING POISSON PROBABILITY FUNCTION

Because energy-dispersive X-ray analyzers have poor resolution, a simple method for deconvoluting two, three, or more overlapping peaks is presented in this section. The method was presented for the first time by Ingamells and Fox [22].

6.6.1. Fitting a Peak to a Poisson Distribution

Measure peak height A. Call the position of the peak channel $Z - \frac{1}{2}$. Find position to the left of the peak where signal is $\frac{1}{2}A$. Call the position of this channel $Z - x$, and call the signal in this channel B.

Measure the signal in channel C to the right of the peak channel the same distance from the peak as channel B is to the left. Call the position of this channel $Z + x - 1$.

If the signals in channels B and C are identical, the distribution is more appropriately treated using the Gaussian approximation. If the signal in channel C is less than the signal in channel B, the roles of the two channels should be reversed, that is, the channels should be numbered from right to left instead of from left to right. It is here assumed that the signal in channel C is greater than the signal in channel B. We have the following relationships:

$$\frac{A}{B} = Z^{x-1/2} \frac{(Z-x)!}{(Z-\frac{1}{2})!} = 2$$

$$\approx Z^x Z^{-1/2}(Z-x)! \left(\frac{e}{Z-\frac{1}{2}}\right)^{Z-1/2} \left(\frac{3}{\pi(6Z-2)}\right)^{1/2} \quad (6.18)$$

where

$$Z^x(Z-x)! \approx 2 \left(\frac{2Z-1}{2e}\right)^Z \sqrt{\frac{2\pi e Z(6Z-2)}{6Z-3}}$$

$$\frac{A}{C} = R = 2 \left[\left(1+\frac{x-1}{Z}\right)\left(1+\frac{x-2}{Z}\right) \cdots \quad \text{for } 2x-1 \text{ terms}\right] \quad (6.19)$$

Find values for x and Z that satisfy (6.18) and (6.19). Calculate $A' = e^{-Z}Z^Z/Z!$

Divide the abscissa distance between channels $Z - x$ and $Z + x - 1$ into $2x - 1$ equal parts, yielding x channels numbered $Z - x, Z - x + 1, \cdots, Z, Z + 1, \cdots, Z + x - 1$.

Find a factor f that will convert the signal in channel Z to equal A'. Multiply the signal in all other channels by this same factor. Calculate the corresponding Poisson distribution from the Poisson formula,

$$P = \frac{e^{-Z}Z^n}{n!}$$

and compare with the manipulated experimental distribution.

Since, in practice, Z and x are whole numbers, tables of values may be prepared to provide solutions to (6.18) and (6.19).

Develop a factor F and a constant Q to convert experimental abscissa readings (e.g., minutes and chart divisions) to channel numbers. The original chart is then completely described by the parameters Z, f, F, and Q.

6.6.2. Derivation of Equations

A characteristic of the Poisson distribution is that the probabilities

$$P_Z = \frac{e^{-Z}Z^Z}{Z!} \quad \text{and} \quad P_{(Z-1)} = \frac{e^{-Z}Z^{(Z-1)}}{(Z-1)!}$$

are the same when Z is an integer. In fitting to a curve, the peak probability lies between P_Z and $P_{(Z-1)}$ and may be approximately represented by

$$P_{(Z-1/2)} = \frac{e^{-Z}Z^{(Z-1/2)}}{(Z-\frac{1}{2})!}$$

A value for $(Z-\frac{1}{2})!$ may be calculated using the gamma function or an improved Stirling approximation:

$$n! \approx \left(\frac{n}{e}\right)^n \sqrt{\frac{\pi(6n+1)}{3}}$$

The Poisson probabilities at the peak channel and the channel in which the signal is half the peak signal are then in the ratio

$$\frac{P_{(Z-1/2)}}{P_{(Z-x)}} = 2 = \frac{e^{-Z}Z^{(Z-1/2)}/(Z-\frac{1}{2})!}{e^{-Z}Z^{(Z-x)}/(Z-x)!}$$

The Poisson probabilities at the peak channel and the equidistant opposite channel are in the ratio

$$\frac{P_{(Z-1/2)}}{P_{(Z+x-1)}} = \frac{A}{C} = R = \frac{e^{-Z}Z^{(Z-1/2)}/(Z-\frac{1}{2})!}{e^{-Z}Z^{(Z+x-1)}/(Z+x-1)!}$$

Substituting $(Z-1/2)! = Z^{(x-1/2)}(Z-x)!/2$ from (6.18),

$$R = 2Z^{(1-2x)}\frac{(Z+x-1)!}{(Z-x)!}$$

which is the same as (6.19).

Selection of Z and x must be guided by equations (6.18) and (6.19): In principle, these equations are soluble simultaneously for Z and x; in practice, this does not seem to be possible. It is necessary to resort to trial and error methods or to use a table of corresponding values.

With highly skewed distributions, the approximation of the peak value by $P_{(Z-1/2)}$ loses validity, as also does the Stirling approximation of $n!$. In such cases, formulae using the gamma function must be applied.

The ordinary Stirling approximation, $n! \approx (n/e)^n\sqrt{2\pi n}$ should not be used.

6.6.3. Fitting a Single Poisson Distribution

Normalize total corrected counts under the peak to 1.0000. Call the normalized counts at the peak value (in the peak channel) P_Z. Find two channels $Z - w$ and $Z + z$ in which the normalized counts are $P_Z/2$. The problem is to assign a channel number Z to the peak channel: We have

$$P_Z = 2P_{Z-w} = 2P_{Z+z} = \frac{Z^Z}{e^Z Z!} = \frac{2Z^{(Z-w)}}{e^Z(Z-w)!} = \frac{2Z^{(Z+z)}}{e^Z(Z+z)!} \quad (6.20)$$

We must select Z such that equation (6.20) is true. From (6.20), $(Z + z)!$ $= 2Z^z Z!$ and $(Z - w)! = 2Z^{-w}Z!$ or $(Z + 1)(Z + 2) \cdots (Z + z) = 2Z^z$ and $Z(Z - 1)(Z - 2) \cdots (Z - w) = Z^w/2$, or $(1 + 1/Z)(1 + 2/Z) \cdots (1 + z/Z) = 2$ and $(1 - 1/Z)(1 - 2/Z) \cdots (1 - w/Z) = 1/2$, or $\ln(1 + 1/Z) + \ln(1 + 2/Z) + \cdots + \ln(1 + z/Z) = \ln 2 = 0.69315$ and $\ln(1 - 1/Z) + \ln(1 - 2/Z) + \cdots + \ln(1 - w/Z) = \ln(1/2) = -0.69315$.

There do not seem to be explicit solutions, but we may take values for Z and z or Z and w, and prepare tables of values. With $Z = 50$, and $z = 1, 2, \cdots$:

$Z = 50$:

z	$\ln(1 + 1/Z) \cdots$	w	$\ln(1 - 1/Z) \cdots$
1	0.0198	1	-0.0202
2	0.0590	2	-0.0610
3	0.1173	3	-0.1229
4	0.1943	4	-0.2063
5	0.2896	5	-0.3116
6	0.4029	6	-0.4395
7	0.5339	7	-0.5903
8	0.6823	8	-0.7647
9	0.8479	9	-0.9631

Thus Z may be taken to be 50 when z is close to 8, and w lies between 7 and 8:

$Z = 40$:

7	0.6599	6	-0.5560
8	0.8422	7	-0.7484

$Z = 30$:

| 6 | 0.6543 | 5 | -0.5337 |
| 7 | 0.8640 | 6 | -0.7568 |

6.6.4. Procedure: Separation of EDX Peaks (Three Peaks)

Select a window wide enough to include several background channels.

Number the channels in the window consecutively; thus, 1, 2, 3, $\cdots$, z, $\cdots$, y, $\cdots$, x, $\cdots$, where z, y, x are the channel numbers at the peak energies for each of three elements.

Select Z, Y, X such that $Z - Y = z - y$, $Z - X = z - x$. To begin with, a value for Z must be arbitrarily selected; $Z = 40$ is probably a good preliminary selection.

Determine the average number of background counts per channel, and subtract background counts from the counts in each channel.

Total all background-corrected counts within the window. Normalize all corrected counts to total 1.00000.

Draw a regression curve through all the points on a plot of channel number versus number of counts. Read the number of normalized counts in channel numbers Z, Y, X from the regression curve. Call the values obtained P_Z, P_Y, P_X, respectively.

Calculate

$$A = \frac{0.977205}{\sqrt{6Z + 1}} \qquad B = Ae^{Z-Y}\left(\frac{Y}{Z}\right)^Z \qquad C = Ae^{Z-X}\left(\frac{X}{Z}\right)^Z$$

$$E = \frac{0.977205}{\sqrt{6Y + 1}} \qquad D = Ee^{Y-Z}\left(\frac{Z}{Y}\right)^Y \qquad F = Ee^{Y-X}\left(\frac{X}{Y}\right)^Y$$

$$J = \frac{0.977205}{\sqrt{6X + 1}} \qquad G = Je^{X-Z}\left(\frac{Z}{X}\right)^X \qquad H = Je^{X-Y}\left(\frac{Y}{X}\right)^X$$

$$a = \frac{P_Z(EJ - FH) + P_Y(CH - BJ) + P_X(BF - CE)}{A(EJ - FH) + D(CH - BJ) + G(BF - CE)}$$

$$b = \frac{P_Z(FG - DJ) + P_Y(AJ - CG) + P_X(CD - AF)}{B(FG - DJ) + E(AJ - CG) + H(CD - AF)}$$

$$c = \frac{P_Z(DH - EG) + P_Y(BG - AH) + P_X(AE - BD)}{C(DH - EG) + F(BG - AH) + J(AE - BD)}$$

Note that the denominators in each of these fractions are the same.

Find the sum $a + b + c$. If $a + b + c$ is less than 1.0000, use a higher value for Z, and repeat the above calculation. If the sum $a + b + c$ is greater than 1.0000, use a lower value for Z, always keeping $Z - Y = z - y$, $Z - X = z - x$.

Plot the two selected values for Z against the two derived values for $a + b + c$. Draw a straight line between the points, and pick off a value for Z that will yield $a + b + c = 1.00000$. Calculate final values for Z, Y, X and final values for a, b, c.

Multiply the total number of background-corrected counts by the factors a, b, c to obtain the counts due to each of the three elements, respectively.

To construct a calculated histogram for comparison with the actual histogram, list the Poisson distributions for those values of Z, Y, X that yield $a + b + c = 1.0000$, and multiply each distribution through by the corresponding factors a, b, c. This gives the calculated distributions normalized to total counts equal to 1.0000. Multiply each factored Poisson probability by the total background-corrected counts within the window, and add the average background counts to each.

If necessary, revise the estimates of P_Z, P_Y, P_X and repeat.

Notes: The equations given above are based on an improved Stirling approximation for the factorial function:

$$n! \approx \left(\frac{n}{e}\right)^n \sqrt{\frac{\pi(6n + 1)}{3}}$$

which is good to 1 part in 10,000 at $n = 20$ and improves as n increases. The chief advantage in using the approximation is the ability to utilize final values for Z, X, Y which are not integral without having to employ the gamma function. The usual Stirling approximation

$$n! \approx \left(\frac{n}{e}\right)^n \sqrt{2\pi n}$$

is not good enough to yield fully satisfactory results.

6.7. AEROSOLS AND THEIR PRODUCTION

6.7.1. Notes on Aerosols and Their Production

It can be shown that in an aerosol with droplets *all the same size,* there is a droplet size that is thermodynamically stable. Such stable aerosols are commonly produced in smokestacks, volcano emissions, and other inadvertancies such as smog [23].

The stable droplet size depends on (1) the surface tension of the dispersed liquid, (2) the concentration of the dispersed phase, and (3) the physical characteristics of the carrying phase.

In principle, it should be possible to disperse a liquid into droplets within a carrying gas in such a way that the resulting aerosol is indefinitely stable. This has not been accomplished in controlled laboratory experiments; it is commonly approximated in Nature and in effluent gases from industrial installations.

If a device (e.g., a supersonic atomizer) can be fabricated so that the droplets it produces are of controlled size, preparation of a stable aerosol should be possible. Such an aerosol would be ideal in flame atomic absorption, argon arc, ICP or other analysis depending on the dispersion of a liquid sample in a gas.

Atomizers of the types generally used in flame atomic absorption and ICP analysis generate a spectrum of droplet sizes. During the life of the aerosol, these droplets collide and coalesce; they may also change size because of evaporation during their short lifetime or because they attach moisture in the ambient gas to themselves. When two droplets collide, they form a single droplet, unless both are of the critical stable size. The consequence is as follows: a generated aerosol immediately begins to deteriorate; most of its droplets collide with other droplets and eventually most droplets become so large that they are precipitated on the walls of the container. A small population of droplets which happen to attain the critical stable size remains. In a randomly generated population of small droplets, only a very few will end up in a stable condition.

The theory of droplet coalescence is well developed and relatively simple. Given an aerosol with a known weight of dispersed phase per unit volume of aerosol, the half-life of a droplet is

$$t_{1/2} = \frac{k}{n_0} \text{ seconds}$$

where k is a constant: $k = \lambda/(k'T)$; λ is viscosity of the medium, k' is Boltzmann constant, T is temperature in °K, and n_0 is the number of droplets per centimeter cubed at time t_0. For H_2O droplets in argon, $k \approx 3 \times 10^9$ s.

Only droplets with a critical stable size are immune to this theory, so long as all are less than about 10^{-3} cm in diameter. Larger droplets are removed from the system by impingement on surfaces in any real atomizing device. It must be noted that the stable droplet size depends on the concentration of the dispersed phase; thus removal of large droplets alters the concentration and therefore the stable droplet size.

6.7.2.   Practical Problems in Atomizing Solutions for Flame Photometry

1. The atomizer must generate particles (droplets smaller than 10^{-4} cm in diameter. This means small tubes and high carrier gas pressure, at least 1 atmosphere over ambient.

2. Organ tube vibrations must be avoided: No tube should increase in cross section in flow direction. No sharp edges or corners.

3. Resonance vibrations must be avoided: The more asymmetric the system, the better.

4. Aerosol should have a residence time of at least 1–2 s before reaching excitation stage.

5. Atomizer must be arranged so that no droplets run into it from atomizing chamber.

6. There must be a whirlwind motion, to remove droplets larger than 10^{-3} cm.

7. Positioning of the two atomizer tubes is very critical. The angle between them must be slightly less than 90 °C: The tube ends should be bevelled; machining should be good on a micro scale, and they should either be very rigidly held to prevent any reed-type vibration, or very unrigidly held to eliminate high-frequency vibration.

8. Flow of the aerosol into the excitation area should be quiet and free from vibrations of either the organ tube or the resonance variety.

9. The atomizer should be mounted off center; even slightly off-center makes a large improvement.

10. Provision for frequent or continuous rinsing of the atomizer assembly is essential for continuous operation. This is partly contradictory to item 5, and there must be some kind of alternation between reading and washing cycles.

11. There must be no dead spaces in the gas flow where aerosol can accumulate and be released in "bursts".

12. A curved surface at 60 ° to the atomizer blast is effective in removing large (10^{-3}) droplets. If the droplets are fired at a flat surface, they bounce off and are not removed.

13. Sample inlet must be a capillary tube: otherwise time between samples is long.

14. Rate of flow of sample into the atomizer is not very critical; it is the proportion of the sample which is reduced to droplets smaller than 10^{-4} cm which matters: The smaller the solution flow rate, the larger the proportion of fine droplets.
15. Solutions must be free of solid particles.
16. Expansion of the carrier gas causes cooling, and salts may crystallize.

6.7.3. Observations on Formation and Stability of Aerosols

Given an aerosol with a known weight of liquid made up of n_0 droplets of a given size, the half-life of a droplet is

$$ t_{1/2} = \frac{k}{n_0} \text{ seconds} $$

where k is a constant: for H_2O droplets in argon, $k \approx 3 \times 10^9$ s. This applies only to droplets smaller than 10^{-3} cm in diameter; particles much larger than 10^{-3} cm are removed by centrifugal forces.

If the aerosol at time t_0 contains 0.1 ml liquid per liter, the half-life of droplets depends on their size:

Radius (cm)	$t_{1/2}$ (s)
10^{-6}	10^{-5}
10^{-5}	10^{-2}
10^{-4}	10
10^{-3}	Removed by centrifugal forces

It is evident that the atomizer must produce droplets of radius smaller than 10^{-4} cm to be effective; it is also evident that finer droplets will rapidly coalesce, and the useful aerosol contains droplets of about 10^{-4} cm in diameter.

After formation of an aerosol, it is advantageous to delay its excitation for at least 1–2 s before exciting it, so that it will stabilize.

6.7.4. Proposal for Nebulizing System Producing More Stable Aerosol

The apparatus showed in Figure 6.30 may present considerable advantages on existing devices currently used in atomic absorption and ICP. When an inductively coupled plasma is used as a source in emission spec-

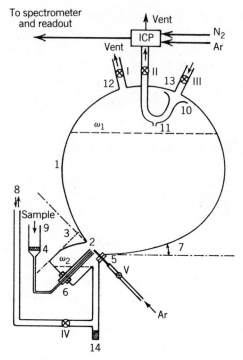

Figure 6.30. Illustration of nebulizing system producing stable aerosol.

trometry, the quality of the aerosol is critical for the control of physical interferences that may be generated inside the plasma torch. This has been studied by various authors [24, 25]. The apparatus is described by the following items numbered in Figure 6.30.

1. Bulb with a capacity between 1 and 4 l.
2. Atomizer: The atomizer assembly must not be under any dripping surfaces. For details see Figure 6.31.
3. A cone-shaped baffle is necessary.
4. The solution should be free of any solid particles; it is recommended to filter the solution.
5. Adjustable gland.
6. Adjustable gland, out of center.
7. The bulb should be distorted in such a way that no trap forms inside.
8. Reservoir for distilled water.

9. Small funnel to introduce the sample.
10. At this place the tube is flared.
11. Small hole for drainage.
12. Check valve.
13. Annular space same diameter as the flared tube.
14. Drain.

Procedure for Using Apparatus

1. The bulb is full of argon gas. Valves I and V are opened. Valves II, III, and IV are closed. The sample is introduced to the atomizer for a period of about 30 s.
2. Valves I and V are closed. The aerosol is ready to be sent to the ICP torch.
3. Open valves II and IV; then read the signal. The signal should be collected until the internal standard reaches a predetermined level. Water level comes around w_1.
4. Close valves II, and open valve III to run the water out of the bulb. Water level comes around w_2.
5. Close valves III and IV. Open valves I and V.
6. Rinse the atomizer.
7. Repeat steps 1–6 for a new sample.

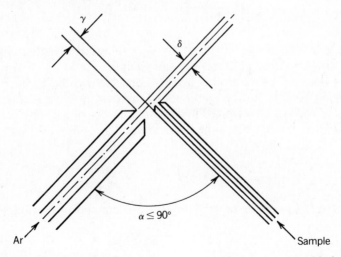

Figure 6.31. Illustration of Penn State atomizer: α, γ, and δ dimensions are critical and must be optimized for each atomizer.

Remarks

1. A piston may be used instead of water to displace the aerosol. The advantage would be a more rapid displacement of the aerosol; however, the disadvantage would be no washing possible.
2. Use of a supersonic nebulizer. The advantage is that no atomizer is necessary. The disadvantages are numerous: many breakdowns, inadequate control, and theory not understood.
3. A heater may be installed inside the bulb to evaporate the aerosol droplets, with also a condensor to precipitate the water. The advantage would be a gain in sensitivity; the disadvantage is to complicate the apparatus.
4. No tube should increase in cross section in the flow direction.
5. Any irregularity in the bulb shape is good but no sharp edges or corners should be present.

6.8. POISSON PROBABILITY DISTRIBUTION APPLIED TO GEOCHEMICAL DATA

Among mathematical devices that have been underutilized is the Poisson distribution function,

$$P_n = \frac{e^{-Z}Z^n}{n!} \qquad (6.21)$$

which gives the probability P_n of finding n things in a single sample when the average number of things per sample is Z.

The Poisson distribution is usually considered to be a discrete distribution, in contrast to the continuous Gaussian, or normal, distribution. It may however be converted to a continuous distribution by substituting $n! = n\Gamma(n)$, where $\Gamma(n)$ is the gamma function,

$$\Gamma(n) = \int_0^\infty e^{-x}x^{n-1}\,dx \qquad (6.22)$$

In practice, $n!$ may be replaced by the Stirling approximation,

$$n! = e^{-n}n^n\sqrt{2\pi n} \qquad (6.23)$$

which becomes more exact as n increases. In some applications, a closer approximation

$$n! = e^{-n}n^n\sqrt{\tfrac{1}{3}[\pi(6n + 1)]}$$

is desirable. Various devices are available to improve the validity of these approximations. We note that

$$P_0 = e^{-Z}$$

$$P_1 = \frac{e^{-Z}Z}{1!} = \frac{P_0 Z}{1}$$

$$P_2 = \frac{e^{-Z}Z^2}{2!} = \frac{P_1 Z}{2}$$

$$P_3 = \frac{e^{-Z}Z^3}{3!} = \frac{P_2 Z}{3}$$

$$\vdots$$

$$P_{14} = \frac{e^{-Z}Z^{14}}{14!} = \frac{P_{13} Z}{14}$$

Thus, for example, if one would like to find $P_{0.5}$ when $Z = 2$, one may begin (with the help of the computer, of course) by calculating, say,

$$P_{14.5} = \frac{e^{-2}2^{14.5}}{(14.5)!}$$

$$\simeq \frac{e^{-2}2^{14.5}}{e^{-14.5}\,14.5^{14.5}\sqrt{2\pi \times 14.5}}$$

Then

$$P_{13.5} = \frac{14.5 P_{14.5}}{2}$$

$$P_{12.5} = \frac{13.5 P_{13.5}}{2}$$

$$\vdots$$

$$P_{0.5} = \frac{1.5 P_{1.5}}{2}$$

Using the Poisson distribution as a substitute for the Gaussian, which it approaches as Z becomes large, it is convenient to use a factor f such that

$$h = fP_{Z-0.5} = \frac{fe^{-Z}Z^{(Z-0.5)}}{(Z - 0.5)!}$$

where h is the peak height of the Gaussian curve.

This device has been used in resolving the photometric spectrum of the thiocyanate complexes of tungsten, molybdenum and rhenium, and also in the deconvolution of energy-dispersive X-ray peaks in a rapid method for determining tungsten in the presence of copper and zinc [22].

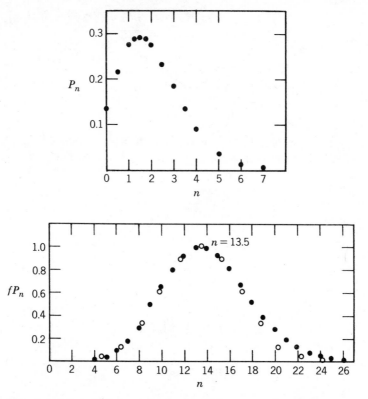

Figure 6.32. Comparison of Poisson distribution with $Z = 14$ with Gaussian distribution factored to match Poisson distribution at $n = 13.5$ and $n = 8.2$: ●, Poisson distribution for $Z = 14$; ○, Gaussian distribution.

Before the advent of the computer, such operations were inpracticable. This probably accounts for existing lack of interest and failure to develop the Poisson distribution into a useful tool.

Many real data sets fit the skew exhibited by the Poisson distribution, and a value for Z can be chosen to match this skew. Only when Z approaches 100 or more does skew in the Poisson distribution become unimportant in most applications.

The distribution represented by the approximation in equation (6.23), may be called pseudo-Poisson. In Figure 6.32, the Poisson distribution with $Z = 14$ is compared with the Gaussian distribution factored to match the Poisson at $n = 13.50$ and $n = 8.20$.

Also shown is the distribution with $Z = 2$. This may represent, for example, the distribution of assay values from an ore containing nuggets of ore mineral in which the probability of finding no nuggets in a random sample is 0.135 and the average number of nuggets is 2.

Evaluation of such skewed data should not be attempted using Gaussian ("normal") statistics!

6.9. FITTING GEOCHEMICAL DATA TO POISSON DISTRIBUTION

To discover if and how data fit a Poisson distribution, estimation of two essential parameters is required:

C = histogram interval, or the contribution of a single equant grain of a minor mineral to the overall X content of a mixture

Z = a parameter determining the shape of the distribution curve

The weights of samples must, of course, be known!

Use of a computer may find a best fit simply by trial and error, but if K, the overall concentration of X, L, the concentration of X in gangue, and S, the standard deviation in the data, are known, then

$$Z = \frac{(K - L)^2}{S^2} \quad \text{and} \quad C = \frac{S^2}{K - L}$$

(Refer to Chapter 1.)

6.10.　COMPLEX DISTRIBUTIONS IN GEOCHEMICAL EXPLORATION

Real data seldom fit any simple model perfectly. In examining exploration data, a product of two or more distributions may be required to achieve a fit. An example is the double Poisson distribution. A more complex distribution is used in evaluating diamond-bearing deposits [26]. This device may also be applied to alluvial gold deposits.

To fit a double Poisson distribution, four parameters must be estimated, Z, z, C, and c (see Section 1.33). Procedure is then as follows: Calculate $f = C/c = z/Z$ from equation (1.68) and π_0 from equation (1.65), using best estimates of L and C or c and thence preliminary estimates of Z and z from equation (1.69).

Substitute these estimates in equation (1.65), thereby obtaining the proportion π_0 of gangue values. Averaging these lowest values gives a new estimate to L. Reiteration, with a new estimate of C or c leads to a best fit.

Figure 6.33 shows some idealized double Poisson distributions developed during examination of data from a molybdenum deposit.

Figure 6.34 demonstrates a double Poisson distribution when $z = Z = 1$.

Figure 6.35 shows the results of applying a double Poisson distribution to exploration tungsten data.

Computer Requirements for Skewed Data Manipulation

1. Code data with coordinates and store (up to 10,000 data) (temporary storage); for example:

X East	Y North	Z Elevation	Au (ppm)	Ag (ppm)
1751.544	1558.886	1024.896	3.055	10.250
1756.505	1571.443	1057.047	0.139	7.750

may be stored as

00	00	00	03055	01025
05	13	32	00139	00775

Change any zero data to 00001

2. Arrange in order east, north, and elevation and store (permanent storage). Erase temporary storage.

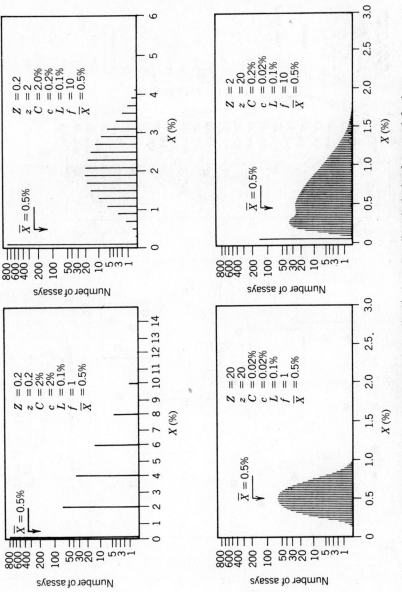

Figure 6.33. Some idealized Poisson distributions. Reading clockwise from top left, these may represent the progress of an exploration of an ore body: (1) large nugget effect, large reduction error; (2) reduction error partially corrected; (3) reduction error further corrected, larger field samples; (4) much larger field samples, minimum reduction error. As sampling errors are corrected, distribution approaches normality.

671

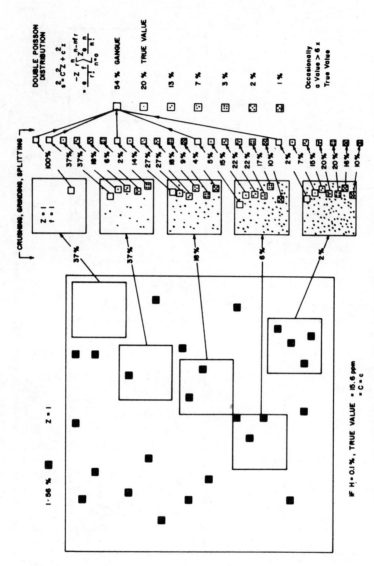

Figure 6.34. Reduction of samples, each of which contains, on average, one equant grain of ore mineral, by a process whereby each subsample contains, on average, one reduced grain of ore mineral.

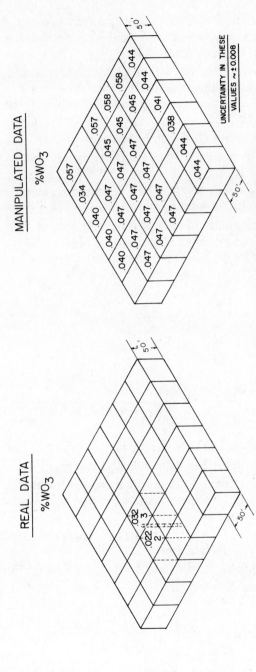

Figure 6.35. Original and manipulated grades for 50-ft blocks within 400-ft block of tungsten ore.

673

3. Given coordinates XYZ, pull N data for a relevant element (e.g., Au) within these coordinates and store for quick access (up to 500 data). Call this data set 1.

4. Given coordinates xyz within XYZ, pull data within xyz and find average, $\bar{y}$. Increment xyz and repeat until all xyz within XYZ have been averaged. Store averages and the number Q^1 of data in each xyz. Call this data set 2.

5. Given a number Q, flag those xyz with $Q \neq Q^1$.

6. Find the average K_1, standard deviation s_1, and harmonic mean h_1 of data set 1.

7. Find the average K_2, standard deviation s_2, and harmonic mean h_2 of data set 2, omitting those data for which $Q^1 \neq Q$.

8. Provide an ability to choose other methods of grouping; for example, given a number Q, pull groups of Q from data set 1 without reference to xyz but in predetermined order.

9. Calculate

$$\frac{A}{w} = \frac{s_1^2 - s_2^2}{1 - 1/Q}$$

$$B = s_1^2 - \frac{A}{w}$$

$$L = \frac{Q h_1 (K_1 - h_2) - h_2 (K_1 - h_1)}{Q(K_1 - h_2) - (K_1 - h_1)}$$

Print and store K_1, K_2, s_1, s_2, h_1, h_2, A/w, B, L, N.

10. Display data set 2 in diamonds [27] and print.

11. Arrange XYZ data set 1 in ascending order for relevant constituent (e.g., Au) and calculate running average and standard deviation. Print and store.

12. Count n data for which running average is less than L. Calculate $\pi_0 = n/N$. Print and store.

13. Given a number f, calculate $Z = (e^{-f} - 1)/\ln \pi_0$. Print and store.

14. Calculate a distribution

$$\pi_r = \sum P_n \cdot P_r = \sum \frac{e^{-Z} Z^n}{n!} \cdot \frac{e^{-nf}(nf)^r}{r!} = \frac{e^{-Z} f^r}{r!} \sum \frac{Z^n e^{-nf} n^r}{n!}$$

with $r = 0, 1, 2$ etc. and display a histogram (see Figure 6.36).

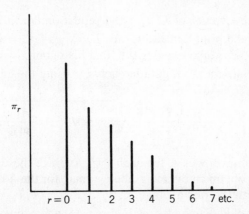

Figure 6.36. Histogram of a typical Poisson distribution.

15. Given an increment c, change histogram abscissa as in Figure 6.37.

16. Prepare a histogram of the XYZ data in the following way (see Figure 6.37).

    a. Note the number n of the datum with running average that equals or just exceeds L. $\pi_0 = n/N$. The assay of this datum gives x_0, the x value of the first histogram bar.

    b. Add c, $2c$, $3c$, . . . to this value to get $x_1, x_2, x_3, \cdots$.

    c. Count the number of data in each group to obtain $\pi_1, \pi_2, \cdots$.

    d. Overlay the histogram of step 15 and observe fit (or lack thereof).

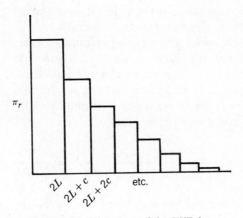

Figure 6.37. Histogram of the XYZ data.

17. Enter new values of L, f, c and repeat until a fit is achieved.

18. Report and print final values for L, π_0, c, f, $C = cf$, Z, $z = Zf$, K, A/w, B. Make copy of the final histogram.

19. Manipulate the XYZ data in each xyz using the equation

$$K_i = \frac{y_i + L}{2} + \frac{1}{2}\sqrt{(y_i - L)^2 + \frac{2A(y_i - L)}{w(K - L)}}$$

Generate averages $\overline{K}$ for each xyz. Average these averages over XYZ to obtain an overall grade estimate for the XYZ block. Compare with the raw average K.

20. Display $\overline{K}$ in diamonds (see Figure 6.35) and compare with display of raw data from step 10. Print.

21. Pull another XYZ block and repeat.

22. Repeat for another element (e.g., Ag), starting at step 3.

6.11. CALCULATION OF MINERAL FORMULAE FROM ANALYSES

Calculation of a mineral formula from an analysis may serve several purposes. It may be used by the crystallographer in determining mineral structure, by the analyst to estimate the correctness of his work or to discover whether impurities are likely to be present, and by the originator of the sample for any one of a variety of reasons having to do with the purpose of his investigation.

The general process of calculation is illustrated in Table 6.1 with biotite PSU-5-110 as an example. This sample was selected because of its unusual purity and freedom from extraneous minerals.

Referring to the table: The weight percentage of the various oxides and nonmetals is listed in column 1. Each is divided by the corresponding molecular weight to obtain a number of moles (column 2) and the number of cations (column 3) and anions (column 4). In the case of biotite it is known that the anions in the mineral formula should ideally total 12 (or 24); a factor f is therefore calculated:

$$f = \frac{12}{\text{sum of the values in column 4}} = \frac{12}{2.7189} = 4.4135$$

The cations in column 3 are multiplied by this factor (column 5), and +

Table 6.1. Biotite PSU 5-110-Formula Calculation

Analysis (wt. %)	(1) Molecular Weight (M)		(2) Moles (% ÷ M)	(3) Cations (C)	(4) Anions (A)	(5) Cations $\times \dfrac{12}{2.7189}$ C	(6) Charges × Valence
SiO_2	38.63	60.06	0.6432	0.6432	1.2864	2.838	+11.356
						0.161	0.483
Al_2O_3	13.08	101.94	0.1283	0.2566	0.3849	3.000	11.839
						0.972	+ 2.916
TiO_2	1.55	79.90	0.0194	0.0194	0.0388	0.086	0.344
Cr_2O_3	0.23	152.02	0.0015	0.0030	0.0045	0.013	0.039
Fe_2O_3	2.50	159.70	0.0157	0.0314	0.0471	0.139	0.417
FeO	8.75	71.84	0.1218	0.1218	0.1218	0.538	1.076
NiO	0.02	74.69	0.0003	0.0003	0.0003	0.001	0.002
MnO	0.14	70.73	0.0020	0.0020	0.0020	0.009	0.018
MgO	19.94	40.32	0.4945	0.4945	0.4945	2.183	4.366
						3.941	+ 9.178
CaO	0.18	56.08	0.0032	0.0032	0.0032	0.014	0.028
SrO	0.005	103.63	0.0001	0.0001	0.0001	0.001	0.002
BaO	0.45	153.36	0.0029	0.0029	0.0029	0.013	0.026
Na_2O	0.26	61.99	0.0042	0.0084	0.0042	0.037	0.037
K_2O	10.00	94.19	0.1062	0.2124	0.1062	0.937	0.937
Rb_2O	0.03	186.96	0.0002	0.0004	0.0002	0.002	0.002
						1.004	1.032
H_2O^+	3.52	18.02	0.1953	0.3906	0.1953		− 1.724
H_2O^-	0.30	18.02	0.0166	0.0332	0.0166		0.147
F_2	0.30	38.00	0.0079	(0.0158)	0.0078		0.070
							− 1.941
P_2O_5	.06	141.96	0.0004	0.0008	0.0020	0.004	+ 0.02
				Σ_{anions}	2.7189		

Conversion factor $f = \dfrac{12}{2.7189}$

Formula: $(SiAl)_3{}^{+11.84}(Al,Ti,Fe^{2+},Fe^{3+},Cr,Ni,Mn)_{3.94}{}^{+9.18}P_{0.004}{}^{+0.02}$
$(Ca,Sr,Ba,Na,K,Rb)_{1.00}{}^{+1.03}(OH,F)_{1.94}{}^{-1.94}O_{10.06}{}^{-20.13}$

and − charges are calculated from the normalized values in columns 5 and 6 and the valences of the various elements and groups.

In biotite, it is known that hydrogen is largely present as hydroxyl ion (OH^-), and the calculation assumes this in finding the factor f. In some other minerals (e.g., vermiculite) hydrogen may be present as H^+, H_3O^+, or as loosely bound H_2O. In such cases, the calculation must be modified, and with unknown minerals some trial and error may be necessary, especially to augment an X-ray crystallographic study or other investigation of mineral structure and stoichiometry.

Normalization need not always be relative to anions as in the example. The fibrous amphibole of Table 3.9, which contains no alumina, may be normalized to Si = 8.000, whereby the formula calculates to

$$Si_{8.000}(Fe, Mn, Mg, Ca, Li, Na, K)_{7.00}(OH, F)_{1.00}O_{23.00}$$

Unfortunately, few mineral samples submitted for analysis are pure enough to yield clear-cut formulae. Sometimes, a single contaminant is present; it may be identified and "removed" during calculation. For example, scapolites may be normalized to $(Ca − C − S) = 5.00$; contaminating calcite (common in scapolites) is thereby "removed" from the analysis [28].

Many mineral crystals consist of a solid solution of two end-members, and different samples may contain different proportions of these. Examples are meionite and marialite in scapolite, sodium and potassium feldspars, magnesian and ferrous olivine, and so on. Sometimes, single crystals of a mineral vary from one end to the other; variation is hardly detectable chemically—only with the electron microprobe can it be certainly observed.

A general formula for a crystalline mineral in which there are two chemical species in solid solution, and in which isomorphous substitution occurs, may be written

$$(1 − m)(A_aB_bC_c \cdots Z_z) + m(A_{a'}B_{b'}C_{c'} \cdots Z_{z'})$$

or

$$A_{a(1-m)+a'm}B_{b(1-m)+b'm}C_{c(1-m)+c'm} \cdots Z_{z(1-m)+z'm}$$

In the process of normalization, one of the subscripts, or an arithmetic combination of two or more of them, is placed equal to a whole number, the value of which is often known from physical data. For convenience in calculation, this number may be made equal to unity: a factor will then convert the empirical formula to correspond with the physical data. If the element Z is used for normalization, that is, N_Z is made equal to unity, the formula becomes

$$A_{\frac{a(1-m)+a'm}{z(1-m)+z'm}}B_{\frac{b(1-m)+b'm}{z(1-m)+z'm}}C_{\frac{c(1-m)+c'm}{z(1-m)+z'm}} \cdots Z \qquad (6.24)$$

If the numbers a, b, c, $\cdots$ and a', b', c', $\cdots$ are to be dealt with as constants, it is often necessary to let the letters A, B, C, $\cdots$, Z stand for groups or combinations of elements selected on the basis of a knowledge of crystal structure, ionic radii, electronic charge, electronegativity, and so on. Preliminary assumptions may have to be made and tested for validity. One assumption which cannot be avoided is that the subscripts z and z' of the normalizing element or group of elements (though not necessarily equal) are indeed constants. For silicates, oxygen (or O + OH + F + Cl) is often used, despite the fact that it is never determined directly, but is calculated from assumed stoichiometric relationships.

From formula (6.24), an expression for m may be obtained using any one of the elements, A, for example. If N_A is the number of atoms of the element A in the formula normalized to $N_Z = 1.0$,

$$m = \frac{N_A z - a}{a' - N_A z' + N_A z - a} \qquad (6.25)$$

For any other element W, m may now be eliminated, and

$$N_W = N_A \left(\frac{w'z - wz'}{a'z - az'} \right) + \left(\frac{a'w - aw'}{a'z - az'} \right) \qquad (6.26)$$

If the normalizing element Z is present in only one end member of the isomorphous series, $z = 0$, $Z' = 1$, and

$$N_W = \frac{w}{a} N_A + w' - \frac{w}{a} a' \qquad (6.27)$$

This is a linear equation in N_A and N_W, with slope S_W and intercept I_W, where

$$S_W = \frac{w}{a} \quad \text{and} \quad I_W = w' - S_W a' \qquad (6.28)$$

Table 6.2. Two-Mineral Analyses[a]

	I				II		
	Oxide (wt. %)		Calculate to $K = 1.0$		Oxide (wt. %)		Calculate to $K = 1.0$
SiO_2	64.57	Si	3.899	SiO_2	66.28	Si	10.708
Al_2O_3	18.95	Al	1.349	Al_2O_3	19.42	Al	3.698
TiO_2	0.04	Ti	0.002	TiO_2	0.03	Ti	0.004
Fe_2O_3	0.11	Fe	0.005	Fe_2O_3	0.26	Fe	0.032
FeO	0.02	Fe	0.001	FeO	0.04	Fe	0.006
MgO	0.01	Mg	0.001	MgO	0.04	Mg	0.010
			0.001				3.750
			1.358				
CaO	0.31	Ca	0.020	CaO	0.29	Ca	0.050
SrO	0.14	Sr	0.005	SrO	0.36	Sr	0.034
BaO	0.48	Ba	0.011	BaO	0.26	Ba	0.017
Na_2O	2.16	Na	0.252	Na_2O	7.94	Na	2.487
			0.288				2.588
K_2O	12.98	K	1.00	K_2O	4.85	K	1.000
Rb_2O	0.04	Rb	0.001	Rb_2O	0.004	Rb	0.000
H_2O^-	0.01			H_2O^-	0.00		
H_2O^+	0.09	H	0.004	H_2O^+	0.04	H	0.005
		O	10.507			O	28.877
Total	99.91				99.81		

[a] Analyst: E. H. Oslund.

Using the above relationships, it is possible to calculate the stoichiometry of both end members of an isomorphous series from any two analyses. This can be done graphically, by plotting the number of atoms of various elements or groups of elements against the number of atoms of the m-determining element (N_A), after normalization to $N_Z = 1.0$, where Z is an element which occurs in only one end member. The various slopes and intercepts obtained are substituted in equation (6.28), which are then solved simultaneously for $a, b, c, \cdots$ and $a', d', c', \cdots$.

The calculation may be done arithmetically, for

$$S_W = \frac{N_W(\text{II}) - N_W(\text{I})}{N_A(\text{IX}) - N_A(\text{I})} = \frac{w}{a}$$

and

$$I_W = N_W(\text{II}) - S_W N_W(\text{I}) = w' - \frac{w}{a} a' \tag{6.29}$$

in which W refers to any element or group of elements, and I and II refer to the two analyses being considered.

A similar process may be used when an analysis is needed of two intimately mixed minerals which cannot be easily separated one from the other. Analysis of two fractions, enriched in one mineral and the other respectively, will serve the purpose as well as that of the completely separated materials. The process is illustrated using a known mineral as a simple example. Analyses I and II in Table 6.2 are of two samples of a mineral that may reasonably be assumed to have a sodium component and a potassiuam component. The formula for this mineral may be written

$$m(K_{z'}Na_{j'}Al_{b'}Si_{a'}O_{c'}) + (1 - m)(K_zNa_jAl_bSi_aO_c). \qquad (6.30)$$

If we assume that $z = 0$, we may use K as the normalizing element. and apply equations (6.29): With Si as the m-determining element, we obtain

$$S_O = \frac{28.877 - 10.507}{10.708 - 3.899} = 2.69 = \frac{c}{a}$$

$$I_O = 28.877 - 2.70 \times 10.708 = -0.01 = c' - 2.69a'$$

$$S_{Na} = \frac{2.487 - 0.252}{10.708 - 3.899} = 0.328 = \frac{j}{a}.$$

$$I_{Na} = 2.487 - 0.328 \times 10.708 = -1.025 = j' - 0.328a'$$

$$S_{Al} = \frac{3.698 - 1.349}{10.708 - 3.899} = 0.345 = \frac{b}{a}$$

$$I_{Al} = 3.698 - 0.345 \times 10.708 = 0.004 = b' - 0.345a'$$

If c is taken to be equal to c', these equations yield the following values for the subscripts in formula (6.30):

$$m(K_{1.00}Al_{1.08}Si_{3.13}O_{8.40}) + (1 - m)(Na_{1.02}Al_{1.08}Si_{3.12}O_{8.40}).$$

Multiplying through by the factor 8.00/8.40 gives

$$m(K_{0.95}Al_{1.03}Si_{2.98}O_{8.00}) + (1 - m)(Na_{0.97}Al_{1.03}Si_{2.97}O_{8.00}).$$

If Na + Ca + Sr + Ba is used instead of Na, and Al + Ti + Fe + Mg is used instead of Al, and the result is normalized to $N_O = 8.00$, the following is obtained:

$$m(K_{0.98}Al^0_{1.04}Si_{2.98}O_{8.00}) + (1 - m)(Na^0_{1.01}Al^0_{1.04}Si_{2.97}O_{8.00}).$$

A similar process has been used by Ingamells and Gittins (loc. cit.) to resolve scapolite analyses into meionite and marialite.

6.12. REPORTING HALOGENS, OXYGEN, AND SULFUR IN ROCK OR MINERAL ANALYSES

Conventionally, all elements in a silicate analysis are reported as oxides, except the halogens (fluorine, chlorine, etc.), which are reported as elements. A correction to the total is then applied to compensate for the fact that too much oxygen has been reported. Since it takes two atoms of halogen to replace one atom of oxygen, the correction is found from

$$\text{O for (F, Cl, }\cdots) = \frac{\text{atomic weight of oxygen}}{2 \times \text{atomic weight of halogen}} \times (\%\text{F, Cl, }\cdots)$$

That is, $\frac{16}{38} = 0.421$ for fluorine and $\frac{16}{71} = 0.225$ for chlorine.

The reason for this procedure is that there is usually no way to identify the element or elements with which the halogen is actually associated.

In the case of sulfur, when knowledge of its valence state is lacking, it is usually reported as SO_3, although in rocks it may well be present as sulfide—or even, as in some antarctic rocks, as free sulfur. If this is known to be the case, one or another of the metals (usually iron) should be calculated to an equivalent amount of sulfide, and the total sulfur content reported separately as percentage of sulfur.

Sulfide minerals present a problem. Often they are partially oxidized during preparation for analysis, and fall short of perfect sulfide stoichimetry. Contaminating oxides complicate the mineral calculation. A sulfide such as MnS_2 that has been partially oxidized may be treated with a nonoxidizing acid in a carbon dioxide atmosphere, yielding H_2S, free sulfur, and sulfate or sulfite, with Mn^{2+} in solution. Volatilized hydrogen sulfide may be trapped in alkaline solution, oxidized to sulfate and converted to $BaSO_4$ for weighing. Free sulfur may be removed from solution by filtration and determined by combustion and titration, and sulfate may be determined as $BaSO_4$ (after titration of any sulfite present). The reported analysis may appear as follows:

MnS	68.38	87.00	0.7860	0.7860	1.000
S	25.33	32.07	0.7898	—	1.005
MnO	2.95	70.93	0.0416	—	.053
SO₃	3.30	80.00	0.0412	—	.052

Formula: 95% MnS_2, 5% $MnSO_4$.

It has been suggested from time to time that analysts should report elemental concentrations only, for that is what, for the most part, they determine. This eminently reasonable suggestion suffers from the fact that there exists no easy way to determine oxygen content with precision and accuracy. Neutron activation analysis has been used, as well as complex procedures using halogen fluorides; these are unlikely to become available on a routine basis.

The solution to this problem has been to determine ferrous iron, FeO, in silicate analysis. The procedure is really a way of measuring the amount of oxygen necessary to convert all the elements in the sample to arbitrary valence states—Mn^{2+}, Fe^{3+}, V^{5+}, and so on. It should on this account be called an oxygen deficiency determination. There exist many materials containing no iron that will yield a value for ferrous iron in normal procedures—the uranium mineral pitchblende, for example. There are also

Table 6.3. Analysis of PSU 61-1437

Analysis	Mol. wt.	Moles	Cations	Anions	Cations $(\times f)$	Charges
SiO_2	0.18	(omitted; introduced during grinding in agate)				
Al_2O_3	16.01	101.94	0.1571	0.3142	0.4713	0.6164	+1.8492
Cr_2O_3	42.97	152.02	0.2827	0.5654	0.8488	1.1092	+3.3276
TiO_2	0.70	79.90	0.0088	0.0088	0.0176	0.0173	+0.0682
Fe_2O_3	8.91	159.70	0.0558	0.1116	0.1674	0.2190	+0.6570
FeO	21.78	71.84	0.3032	0.3032	0.3032	0.5949	+1.1898
MgO	8.69	40.32	0.2155	0.2155	0.2155	0.4228	+0.8456
CaO	0.04	56.08	0.0007	0.0007	0.0007	0.0014	+0.0028
MnO	0.24	70.93	0.0034	0.0034	0.0034	0.0067	+0.0134
V_2O_5	0.32	181.90	0.0018	0.0036	0.0090	0.0071	+0.0355
NiO	0.14	74.69	0.0019	0.0019	0.0019	0.0037	+0.0074
Total	99.98				2.0388		+7.997

M_2O_3 trioxides: $1.1092 + 0.6164 + 0.2190 = 1.9446$
MO_2 dioxides: $0.0173 = 0.0173$
M_2O_5 pentoxides: $0.0071 = 0.0071$
MO monoxides: $0.5949 + 0.4228 + 0.0014 + 0.0067 + 0.0037 = 1.0295$

Formula: $M^{+3}_{1.94}M^{+4}_{0.02}M^{+5}_{0.01}M^{+2}_{1.02}O^{-8}_{4.00}$

If vanadium and titanium are present in the trivalent state, instead of the valence states assumed, the formula becomes

$$M^{+3}_2M^{+2}_1O^{-8}_4$$

minerals that contain elements in higher valence states (e.g., pyrolucite) occurring in rocks; a ferrous iron determination in such materials does not have much meaning. If the Mn^{3+} or Mn^{4+} content exceeds the Fe^{2+} content, the ferrous iron convention fails entirely. In such a case an excess oxygen determination is in order. In the analysis of an apatite (Table 3.32), the measured oxygen deficiency is greater than that required to convert all the Fe_2O_3 to FeO. In Denningite (Table 3.16), the excess oxygen is probably due to the presence of either Te^{6+} or Mn^{3+}. In chromites (Table 3.17), although oxygen deficiency is reported as FeO, there is evidence that vanadium and titanium are present as V^{+3} and Ti^{+3}, respectively. Calculation of a mineral formula from the analysis of PSU 61-1437 illustrates this; (Table 6.3). If the oxygen deficiency were distributed among the vanadium and titanium cations, the calculated formula would approach the theoretical

$$M_{2.00}{}^{3+} M_{1.00}{}^{2+} O_{4.00}{}^{8-}$$

Oxygen excess or deficiency is of importance in the analysis of ferrites and other semiconducting materials. These frequently contain manganese and iron, and their semiconducting properties are due to various combinations of Fe^{2+}, Fe^{3+}, Mn^{2+}, and Mn^{3+}. All four of these species may be present in a bulk ferrite sample. All an ordinary analysis can report is an overall excess or deficiency over or below that required by Fe^{3+} and Mn^{2+}. Vanadium and titanium may be present in various valence states, complicating the issue.

Consider the ferrite of Table 3.30. Iron is reported as Fe_2O_3 (68.63%) and FeO (3.02%). The material contains 18.08% of manganese reported as MnO. There is no guarantee that some of the manganese is not present as Mn^{3+}. A better reporting of this analysis would show all the iron as Fe_2O_3, all the manganese as MnO, and a value for oxygen deficiency of 0.336%.

REFERENCES

1. M. P. Jones and C. H. J. Beaven, *Sampling of Non-Gaussian Mineralogical Distributions. Geological, Mining and Metallurgical Sampling,* M. J. Jones, ed., Proceedings of meetings organized by the Institution of Mining and Metallurgy and held in London in January 1972, September 1972, and July 1973, pp. 12–19, The Institution of Mining and Metallurgy, 1971.
2. W. G. Cochran, *Sampling Techniques,* Wiley, New York, 1977.

3. L. G. Parratt, *Probability and Experimental Errors in Science*, Wiley, New York, 1961.

4. M. David, *Geostatistical Ore Reserve Estimation—Developments in Geomathematics*, 2, Elsevier, New York, 1977.

5. P. M. Gy, *L'echantillonnage des Minerais en Vrac*, Vol. 1, Societe de l'Industrie Minerale, Saint-Etienne, France, January 15, 1967.

6. P. M. Gy, "Sampling or Gambling?" *Coal Mining Process.* 18(9), 62–69 (1981).

7. P. M. Gy, "Poids a donner a un echantillon, Abaques," *Revue l'Industrie Minerale* 38, (1956).

8. P. M. Gy, "Nomogramme d'echantillonnage," Sampling Nomogram, Probenahme Nomogramm, Minerais et Metaux, Paris, 1956; German edition published by Gesellshaft Deutscher Metallhütten und Berglente.

9. D. G. Krige, "The Role of Mathematical Statistics in Improved Ore Valuation Techniques in South African Gold Mines," in *Topics in Mathematical Geology*, Consultants Bureau, New York, 1970.

10. D. G. Krige, *A Review of Development of Geostatistics in South Africa.* Advanced Geostatistics in the Mining Industry, Reidel, Dordrecht, Netherlands, 1976, pp. 279–294.

11. G. Matheron, "Principles of Geostatistics," *Econ. Geol.* 58, 1246–1266 (1963).

12. G. Matheron, *Random Functions and Their Applications in Geology. Geostatistics: A Colloquium*, Plenum, New York, pp. 79–87.

13. P. Gy, "Variography—International Symposium of Applied Statistics, Golden, Colorado—April 1964," *Colorado School of Mines Quarterly* 59, 693–711 (1964).

14. P. Gy, *Sampling of Particulate Materials—Theory and Practice*, 2nd ed., Elsevier, Amsterdam, 1982.

15. G. Charlot, *Cours de chimie analytique generale—Solutions aqueuses et non aqueuses*, Masson & Cie., Paris, 1967.

16. G. Charlot, *Cours de chimie analytique generale—Methodes electrochimiques et absorptiometriques, chromatographie*, Masson & Cie., Paris, 1971.

17. M. Machtinger and R. Rosset, *Cours de chimie analytique generale*, Masson & Cie., Paris, 1972.

18. G. Charlot, *Les reactions chimiques en solution—L'analyse qualitative minerale*, Masson & Cie., Paris, 1969.

19. G. Charlot, *Chimie analytique quantitative—Methode chimiques et physicochimiques*, Masson & Cie., Paris, 1969.

20. G. Charlot, *Chimie analytique quantitative—Methodes selectionnees d'analyse chimique des elements*, Masson & Cie., Paris, 1974.

21. S. D. Rasberry and K. F. J. Heinrich, "Calibration for Interelement Effects in X-ray Fluorescence Analysis," *Anal. Chem.* **46**(1), 81–89 (1974).

22. C. O. Ingamells and J. J. Fox, "Deconvolution of Energy-Dispersive X-ray Peaks Using the Poisson Probability Function," *X-Ray Spectrom.* **8**(2), 79–84 (1979).

23. H.-P. Aubauer, "Two-Phase Systems with Dispersed Particles of a Stable Size," Acta Metallurgica **20**, 165–180 (1972)

24. J. Farino and R. F. Browner, "Surface Tension Effects on Aerosol Properties in Atomic Spectrometry," *Anal. Chem.* **56**, 2709–2714 (1984).

25. D. D. Smith and R. F. Browner, "Influence of Aerosol Drop Size on Signals and Interferences in Flame Atomic Absorption Spectrometry," *Anal. Chem.* **56**, 2702–2708 (1984).

26. H. S. Sichel, Statistical Valuation of Diamondiferous Deposits, Proc. 10th APCOM Symp. South African Institute of Mining and Metallurgy, Johannesburg, 1972, pp. 17–25.

27. C. O. Ingamells, "Evaluation of Skewed Exploration Data—The Nugget Effect," *Geochim. Cosmochim. Acta* **45**, 1209–1216 (1981).

28. C. O. Ingamells and J. Gittins, *Can. Mineralog.* **9**, Part 2, 214–236 (1967).

CHAPTER

7

GUIDELINES FOR PREPARATION OF GEOCHEMICAL STANDARDS

7.1. Introduction
7.2. Purpose of Geochemical Standards
    7.2.1. Calibration
    7.2.2. Control
7.3. Criticism of Common Practices
7.4. Commercially Available Standard Reference Materials Are Not Always the Solution
7.5. Preparation of Geochemical Standards
7.6. Trace Elements in Reference Materials
7.7. Guidelines for Conducting Preparation of Standard Reference Material
    7.7.1. Nature of SRM and Its Applications
    7.7.2. Yield
    7.7.3. Separation, Purification, and Packaging
    7.7.4. Investigation of Inhomogeneity
    7.7.5. Typical Sampling Procedure for Preparing SRM of Ore
7.8. Selection of Analytical Methods and Relevant Data
7.9. Setting up Standards Laboratory
    7.9.1. Standard Reference Materials
    7.9.2. Complete Primary Analysis
    7.9.3. Analyzed Minerals for Microprobe Standards
    7.9.4. Preparation and Storage of Solutions
    7.9.5. Preparation of Synthetic Mixtures
    7.9.6. Analysis of Unusual Materials
    7.9.7. Investigation of new Analytical Techniques
7.10. Microprobe Standards
7.11. Use of Standards in Calibration
7.12. Standards for K–Ar Dating
7.13. Sources of Information
References

7.1. INTRODUCTION

Mining exploration usually involves collection and assay of a large number of samples. Data are evaluated, and a decision to proceed beyond the exploration stage may follow. In these days of diminishing reserves, increasingly marginal deposits must be investigated. Assay values must continuously meet higher standards of accuracy as the difference between a feasible and nonfeasible grade narrows. As the overall grade estimate becomes more critical in estimating feasibility, the number of assay samples increases and can only be handled through the use of rapid instrumental determinative methods. Some of these methods were described in Chapter 5.

Rapid methods, often highly precise, must be calibrated; in practice, this can only be done effectively through the use of standards reference materials similar in composition, mineralogy, and particle size distribution to the unknowns. Preparation of these SRMs is a first essential step in mine evaluation.

If these calibrating standards are deficient, introduction of bias into a set of exploration data may lead to abandonment of a viable property or the exploitation of one that contains too little of the mineral of interest to be profitable. Accurate primary analysis of selected samples is an absolute must.

Should exploration data appear favorable, a small exploratory mining test may follow. Metallurgists will bench test the ore and develop an appropriate extractive process. A pilot plant may then be built to test and modify this process under operating plant conditions. Preliminary mining and construction and operation of the pilot plant may cost several million dollars. During bench tests and piloting, a large number of samples are developed: concentrates, middlings, tails. Analytical problems increase in difficulty, and sample volume becomes overwhelming in the absence of well-calibrated instrumental methods. Again, calibration requires the primary analysis of selected samples if instrumental data are to be accurate.

7.2. PURPOSE OF GEOCHEMICAL STANDARDS

7.2.1. Calibration

Instrumental methods of analysis depend heavily on accurately analyzed standards. Preparation of a useful standard reference material requires the following:

1. A complete primary analysis.
2. Reliable uncertainty statements and sampling constants or other measures of inhomogeneity.

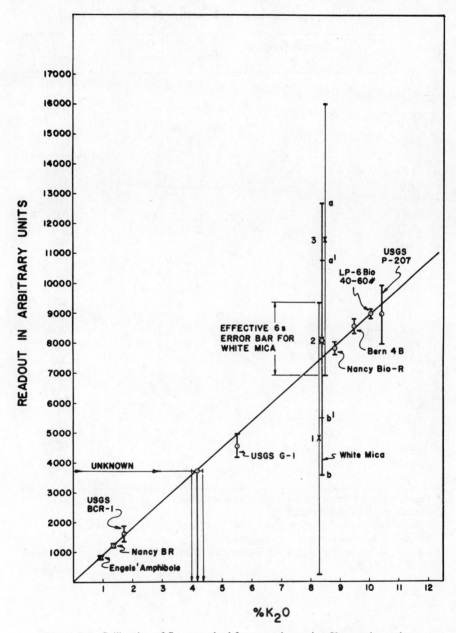

Figure 7.1. Calibration of flame method for potassium using 50-mg subsamples.

689

Table 7.1.   Composition and Homogeneity of LP-6 Bio # 40-60

	Sample Weight (g)	Weight (%)[a]	K_s (g)
SiO_2 [b]	0.7	38.33 ± 0.1	0.001
Al_2O_3 [b]	0.7	15.18 ± 0.05	0.001
TiO_2 [b]	0.7	1.67 ± 0.01	0.05
Fe_2O_3 [c]	3.7[d]	2.25	—
FeO [b]	0.5	8.55	K_s for FeO probably high
Total Fe_2O_3	—	11.75 ± 0.02	0.005
Cr_2O_3	—	0.06 ± 0.02	—
V_2O_5	—	0.02	.—
MnO [b]	—	0.11 ± 0.01	0.01
NiO	—	0.05 ± 0.01	—
CoO	—	0.01	—
ZnO	—	0.01	—
MgO [b]	0.7	19.32 ± 0.05	0.005
CaO [b]	0.7	0.21 ± 0.04	2.
BaO	—	0.22	0.01
Li_2O	—	0.005	—
Na_2O [c]	—	0.09 ± 0.02	1.0
K_2O [c]	10.0[e]	10.03 ± 0.02	0.005
Rb_2O [c]	—	0.025	—
Cs_2O	—	0.0006	—
H_2O^+ [b]	1.0	3.53 ± 0.2	—
H_2O^- [b]	0.7	0.13	—
F	—	0.26	—
P_2O_5 [b]	—	0.01 ± 0.02	Very high
CO_2 [b]	2.0	0.00	—
Less O for F		0.11	
Total		99.96	

[a] The ± is the maximum range expected between bottles.
[b] Determination by primary methods.
[c] Determination by secondary methods, PSU 5-110 used as calibrating standard.
[d] Average of 37 results using 0.1-g subsamples.
[e] Average of 100 results using 0.1-g subsamples. Primary determination by V.C. Smith showed 10.03% K_2O.

690

Table 7.2.   Trace Elements in a Nickel–Copper Sulfide Ore and Concentrate from Same Ore (ppm)

Element	Control Standards		Ore	Concentrate
	Found	Recommended		
As	750	420	41	26
Te	1200	1260	60	105
Bi	475	420	30	105
Pb	305	240	55	565
Zn	325	310	110	960

3.   An established sampling weight baseline.

4.   Known mineralogy.

5.   A sufficient supply.

Figure 7.1 shows a calibration curve for a flame photometric method for potassium determination using standard reference materials for which accurate analyses and inhomogeneity data are available. The sampling constants are used to construct error bars on the readout value for each standard [1]. Table 7.1 shows the certificate of analysis for one of these standard reference materials.

7.2.2.   Control

In devising routine control methods of analysis, chemical or instrumental, it is obviously desirable to make the method as rapid and as simple as possible, consistent with adequate accuracy and precision. The most reliable way to check the performance of such methods is to apply them to similar samples that have been rigorously analyzed by primary methods. Table 7.2 gives some trace element determinations on nickel–copper sulfide ores and concentrates with supporting determinations on standard reference materials. Without these supporting data, there would be much less confidence in the reported values.

7.3.   CRITICISM OF COMMON PRACTICES

Table 7.3 shows the analysis of one of several thousand field samples collected during the exploration of a tungsten–molybdenum ore body and the ranges of heavy-metal values found in these samples. After several

Table 7.3.   Typical Ore Sample

		Traces:
SiO_2	84.00	Cu
Al_2O_3	5.84	Nb
TiO_2	0.07	Ta
Fe_2O_3	3.20	Sb
CaO	1.08	Sn
WO_3	0.72 (range—0 to 3%)	Se
MoS_2	0.36 (range—0 to 3%)	Na
Bi_2O_3	0.48 (range—0 to 2%)	K
As_2O_3	0.40 (range—0 to 5%)	Ba
ZnO	0.32 (range—0 to 5%)	etc.
S not MoS_2	0.47	
Total	99.63 (includes P_2O_5, H_2O, 2.15% F)	

of these samples had been analyzed by primary methods, the calibration of a rapid X-ray method for molybdenum and tungsten was not difficult. Tables 7.4 and 7.5 show the results of an attempt to establish the molybdenum and tungsten contents of the calibrating standards by the "round-robin" process. It should be evident that this exercise was not very useful.

Table 7.4.   Round Robin for Molybdenum[a]

Laboratory	Molybdenum (%)
1	0.122
2	0.168
3	0.15
4	0.365
5	0.405
6	0.15
7	0.12
8	0.14
9	0.242
10	0.11
	Average, 0.197
	Range, 0.11–0.405
	$s/\sqrt{N}$, 0.033
	Shall we accept
	0.197 ± 0.033
	as a standard
	value?

[a] Example: One of 25 samples circulated among 10 laboratories.

Table 7.5. Round Robin: Tungsten as WO_3

Sample	Lab 1	Lab 2	Lab 3	Lab 4
1	0.355	0.220	0.37	0.354
2	0.293	0.220	0.25	0.237
3	0.980	1.28	0.90	0.952
4	0.180	0.100	0.07	0.16
5	2.34	3.36	1.87	2.0
6	0.090	0.120	0.13	0.11

Table 7.6. Concentrates

	Mo-1	Mo-2	Bi	As-1	As-2	CaF_2	WO_3
WO_3	0.40	0.28	0.46	0.74	1.28	0.22	2.81
Bi	4.42	3.02	4.82	0.42	1.36	0.058	—
MoS_2	14.6	37.8	0.55	0.77	2.69	0.056	—
Cu	2.00	0.96	2.54	0.094	0.28	0.010	—
Pb	2.24	1.86	2.38	0.13	0.42	0.016	—
Zn	8.56	3.84	13.0	0.18	5.68	0.054	—
As	7.59	2.21	10.7	5.4	16.5	—	1.16
Fe	11.00	5.56	13.6	9.22	15.0	1.52	9.24
Ca	1.74	2.05	1.52	3.05	3.12	44.1	1.28
F	—	—	—	—	—	40.1	—

Table 7.7. Tails

WO_3	0.11	0.37	0.47
Bi	4.52	0.016	0.068
MoS_2	1.96	0.016	0.080
Cu	2.24	0.004	0.017
Pb	3.02	0.008	0.028
Zn	10.5	0.030	0.14
As	6.06	0.12	1.12
Fe	10.7	3.16	6.84
Ca	1.52	2.36	2.84

Table 7.8. Division of MoS$_2$

MoS$_2$ content of ore	0.302%
MoS$_2$ content of tailings	0.010%
Recovery	96.7%
Weight of ore treated daily	40,000 tons
Weight of MoS$_2$ recovered daily	117 tons
Weight of MoS$_2$ in tails daily	4 tons
Dollar value unrecovered MoS$_2$	$40,000
Value unrecovered MoS$_2$ per year	$14,600,000

It should be noted that calibration of an energy-dispersive X-ray method for determining tungsten and molybdenum also requires that the standards be analyzed for other elements because of potential severe matrix and interference effects.

Once bench testing and piloting begins, the complexity of samples increases. Tables 7.6 and 7.7 show typical analyses of products from the ore of Table 7.3. Analysis of samples such as these by the methods commonly used by participants in round-robin exercises is nearly impossible. "Dissolve it and run it by AA" does not meet the requirements.

Analysis of "tails" is very important. Although MoS$_2$, for example, may be only 0.01% in the tailings, this may be very significant. Table 7.8 approximates the situation at an operating mine that produces tailings with 0.01% MoS$_2$ and shows losses amounting to $14 million a year.

7.4. COMMERCIALLY AVAILABLE STANDARD REFERENCE MATERIALS ARE NOT ALWAYS THE SOLUTION

In this framework, commercially available standard reference materials (SRMs) are not very helpful, for several reasons:

1. Widely distributed SRMs are usually much more finely ground than the samples to be analyzed. This affects rapid X-ray methods, which examine samples without isoformation, and also rapid chemical methods, which rely on dissolution of the critical mineral away from gangue or interfering minerals.

2. Mineralogy may be different. This also affects rapid X-ray and chemical methods. A procedure that quickly extracts all the tungsten from a −200-mesh scheelite ore is worthless when applied to a −100-mesh wolframite ore.

3.   Most SRMs are not supplied with complete analyses. Even if the certificate value for one or two elements can be confidently accepted, lack of information on interfering elements severely limits the usefulness of the SRMs. Samples free from copper and zinc, for example, are almost worthless in calibrating an EDX method for tungsten when unknowns contain these elements.

4.   Seldom does an SRM certificate carry a convincing and useful statement concerning degree of homogeneity. In some cases, the application (misapplication?) of statistics gives some evidence that each bottle is the same for certified elements within acceptable limits; but no measure of sampleability at the subsample weight demanded by the instrumental method to be calibrated: for example, a standard precious-metal ore, provided in 2-kg lots, may indeed contain the advertized 3 ppm gold in each lot, but there is no assurance that each gram—or each assay ton—will contain just 3 ppm gold, and there is no way an analyst can calculate, from the data supplied, what sample weight must be taken to achieve a desired degree of precision and accuracy.

5.   Most often, the certificate value is an adjusted average of numerous determinations using different methods and different sample weights. Those few submitted values obtained by skillfully applied primary methods are averaged in with hundreds obtained by rapid slop methods. Little consideration is given to subsampling variance. The consequence is a wide spread of values, the causes of which are obscure and the manipulated average of which inspires little confidence.

6.   Those SRMs that do exist are mostly simple rocks: Almost no effort has been made to issue completely analyzed complex ores and products of the kind that must be routinely dealt with in mining and metallurgical laboratories.

7.5.   PREPARATION OF GEOCHEMICAL STANDARDS

When an analytical laboratory is faced with the necessity for generating hundreds—sometimes thousands—of determinations per day, it is very desirable to avoid unnecessary effort and at the same time to keep accuracy at an adequate level. Of the three stages in an analysis—sample preparation, analysis, and data transfer—the last two are automated by various means; sample preparation remains the operation requiring time and labor. Grinding and screening are especially time-consuming, and it is fortunate that methods exist to minimize these operations. Table 7.9 shows the calculated sample weights necessary for 1% sampling uncertainty (68% confidence) for molybdenite ores and products at various

Table 7.9. Sample Preparation

MoS$_2$ Content (%)	Sample Weight for 1% error (g) By Mesh Size[a]			
	No. 10	No. 35	No. 100	No. 200
20.0	1,100	18	0.8	0.12
1.0	22,000	360	16	2.4
0.1	220,000	3,600	160	24
0.01	2,200,000	36,000	1,600	240
0.001	22,000,000	360,000	16,000	2,400

[a] Effective mesh size of MoS$_2$ particles in the sample.

mesh sizes. These weights are the sampling constants K_s for these mesh sizes. If a higher sampling uncertainty is acceptable, the necessary sample weight may be calculated from [2]

$$w = \frac{K_s}{R^2}$$

where R is the acceptable uncertainty. For example, if a 10% sampling uncertainty is acceptable, only $K_s/100$ grams need be taken.

In any well-designed exploration, the degree to which each field sample represents the surrounding ore is known approximately: With highly segregated ore bodies, the field sample may carry an uncertainty of 30% or more. In such a case, uncertainty introduced during reduction, subsampling, and analysis may reach 10% (one-third of the field sampling uncertainty) without affecting data evaluation. Reduction and subsampling procedures should be developed with this in mind so as to minimize effort. For example, suppose one has to determine MoS$_2$ in several hundred field samples for which 10% precision is adequate; examination of Table 7.10 shows that it will be sufficient to split out 80 g at the 35-mesh level and reduce it to pass 65 mesh if 1-g subsamples are to be taken for analysis.

All those samples selected for development into calibrating standards should be prepared in the same manner as the samples to be analyzed; finer grinding leaves them in a less useful condition because their characteristics are then not the same as the samples to be analyzed. A rapid technique using minimally ground and screened material should be calibrated against standards similarly ground and screened. Especially with X-ray techniques without isoformation, failure to observe this necessity may lead to error.

Sampling constants can be estimated by simple experiment. Table 7.10 shows the X-ray signals developed by molybdenum in 12 untreated 35-mesh samples. It is only necessary to fill 12 cups in an automatic sample changer and push a button; a slave computer collects data and calculates the variance and the sampling constant. The amount of grinding and screening necessary for the desired sampling precision can then be calculated.

In applying the experimental method for finding sampling constants, it is essential to observe one precaution: if samples are too small, there is a possibility that all of them may contain no grain of ore mineral; variance will be low (as will signals), and a spurious result will follow. Table 7.11 shows minimum sample weights for molybdenite ores and products containing 0.3, 0.01, and 0.001% MoS_2. Testing, for example, 1-g samples

Table 7.10. Determination of
Sampling Constant: $K_s = R^2w =$
$10^4(K - L)(H - L)u^3d/K^2$

35-Mesh Samples Observed X-ray Counts
7,500
13,500
8,050
9,150
14,600
11,350
6,200
15,400
7,450
12,350
11,450
10,900
Standard deviation, 3,005
Average, 10,658
Coefficient of variation, 28 = R (%)
Sampling constant K_s [a]
795 grams at 35#
100 grams at 65#
12 grams at 150#
1.5 grams at 270#
Approximate MoS_2 content, 0.1%

[a] Effective X-ray sample weight, 1 g.

Table 7.11. Minimum Sample Weight (g)[a]

Mesh Size	u (cm)	w_{min}		
		$K = 0.3$	$K = 0.01$	$K = 0.001$
10	0.168	7.6	227	2270
35	0.042	0.12	3.7	37
100	0.0149	0.006	0.17	1.7
200	0.0074	0.001	0.02	0.2

[a] At w_{min}, a 75% chance for 1 grain in any sample. $w_{min} = H/K(u^3d)$: for MoS_2, $H = 100$, $d = 4.8$.

of 35-mesh ore with 0.3% MoS_2 will yield meaningful results; the same test applied to 1-g samples of 10-mesh ore with 0.01% MoS_2 will not yield meaningful results.

7.6. TRACE ELEMENTS IN REFERENCE MATERIALS

One of the most discouraging—and disappointing—aspects of current programs for developing SRMs is the almost universal failure to pay attention to the sampling characteristics of the rock powder, despite the work of numerous authors who have dealt with this matter [2, 3–14].

One flake of gold $80 \times 80 \times 20$ μm weighs about 3 μg and will just pass a 200-mesh screen. If this flake exists in a 1-g sample of -200-mesh rock powder, this sample will contain 3 ppm gold. If one takes successive 1-g portions of a rock powder containing 3 ppm gold in the form of 200-mesh flakes, about 37 in every 100 samples will show no gold, about the same number will show 3 ppm, about 18 will contain two flakes of gold and show 6 ppm, and a few samples will show 9 or even 12 or more ppm gold. If 10-mg samples are taken for spectrographic analysis, most will show no gold; an occasional sample that contains a flake of gold will show 300 ppm gold. The sampling constant K_s is 10,000g.

If one takes an assay ton of this ore, the uncertainty in the assay is $R = \sqrt{10,000/29} = 19\%$ with 68% confidence, or 56% with 98% confidence. In a round-robin exercise in which the participants use sample weights ranging from 0.1 g to 90 g or more, reported values for gold will vary from close to zero up to 10 or 20 ppm gold, even if no participant falls into analytical difficulties. Given enough of these widely scattered values, the average may well approach the true value; but it should be evident that the material is of little use as an SRM unless it is made clear to the user what uncertainty may be expected for any sample weight.

While this is an extreme example, it should demonstrate several more or less obvious conclusions:

1. No material should be circulated in a round-robin exercise unless its sampling characteristics for critical constituents have been first estimated. It is not constructive for numerous investigators to expend a total of hundreds of hours of work generating numbers of no practical usefulness.

2. In reporting preferred or certificate values for an SRM, a sampling constant for at least the most important constituents should be given so that the user will know how large a sample must be taken to achieve a desired uncertainty.

3. Statistical exercises in which values derived from samples of different weight are manipulated are invalid unless those weights are taken into account.

4. Reporting evidence that each lot of an SRM has the same X content within acceptable limits is not sufficient as a statement of homogeneity. Homogeneity is relative; it depends on sample or subsample weight.

5. Grinding a potential SRM to an impalpable powder without screening invites disaster. It would be better to supply a coarser powder together with a homogeneity statement; then the user could grind the material further if that suited the purpose. Very finely ground reference samples are not suitable for calibrating procedures that are to be applied to real-world samples, for reasons stated earlier in this chapter.

A general conclusion is that the current efforts to produce geochemical reference materials lack sophistication and need extensive alteration if their astronomic expense is to be justified.

7.7. GUIDELINES FOR PREPARATION OF STANDARD REFERENCE MATERIAL

7.7.1. Nature of SRM and Its Applications

Scrupulous consideration must be given to the intended uses of an SRM. For example, a chromite SRM containing 200 ppm nickel and 50 ppm cobalt cannot, under any circumstances, be used to calibrate a procedure that is intended to be used on a limonite ore containing the same amounts of nickel and cobalt occurring in lithiophorite.

Improper use of an SRM in procedures for which it is not intended or certified leads to erroneous results, confusion, and waste of time and effort.

7.7.2. Yield

The preparation of an SRM is a very expensive endeavor, and it would be very unfortunate to find out that after all the time and money spent on its preparation and analysis, not enough material is left to satisfy all the potential users.

A thorough evaluation of quantities needed for the homogeneity tests and preliminary analyses must be performed to ensure that a useful supply remains for future distribution.

Some very valuable SRMs should not be used unnecessarily in routine work for which they were not originally intended.

7.7.3. Separation, Purification, and Packaging

Depending on the uses intended for the SRM, it may be necessary to proceed to a complete separation of the different phases of the material chosen to prepare the SRM. If large quantities are required, it is unlikely that the SRM will be prepared using a single crystal; therefore, the material would have to be crushed until the liberation size of the mineral of interest is reached. This could be achieved by microscopic examination of thin sections. The selection of the method of separation is influenced by grain size requirements or by the alteration of mineral structure during mechanical processing. Heavy liquid separation is convenient. However, it is not always recommended, especially if the SRM is intended to be used in mass spectrometry or if large amounts of material have to be separated.

Packaging is critical, all vials should be cleaned by scrubbing with a brush in soap and water to remove the submicroscopic impurities that are always present on new glass. They are treated with concentrated sulfuric acid containing chromic acid, rinsed with water, washed with dilute hydrochloric acid containing hydrogen peroxide, rinsed with distilled water, and oven dried.

The use of plastic containers is not recommended due to possible segregation introduced by adhesion phenomena. The adhesion or the stickiness between some particles and a plastic surface should be an important consideration. Thus, surface roughness and porosity of the wall surface of the containers receiving the SRM are important parameters. Historically, glass has been proven satisfactory.

An experienced analyst will immediately determine if an SRM has been given proper attention during packaging; it is not uncommon, for instance, to find that fine particles are adhering to the part of the glass vial that is always difficult to wash thoroughly underneath the shoulder before the neck of the container. It is recommended to use a container with a large neck diameter or of the same diameter as the container itself.

7.7.4. Investigation of Inhomogeneity

Investigation of inhomogeneity of geochemical materials at various sampling levels for each constituent of interest is a must. Engels and Ingamells proposed several guidelines that gave satisfaction in many instances [1, 8, 11]. An investigation of shape, particle size distribution, and mineralogical and liberation factors as proposed by Gy could also be performed as explained in Section 1.1.

It is not acceptable for an SRM to be certified for the content of its elements of interest without including with the certificate values a measure of the uncertainty between containers, including sampling weight baseline, and a calculation of the K_s sampling constant or its Gy equivalent.

Thorough consideration of these sampling constants is necessary before obtaining a consensus average after statistical treatment of all reported values. It is clear that if no attention is given to sampling constants, before selecting subsample weights for analysis, all the efforts provided to prepare the SRM can be lost and end up in a disappointing failure.

7.7.5. Typical Sampling Procedure for Preparing SRM of Ore

Figure 7.2 shows a typical sampling procedure used for the preparation of an SRM. Such a procedure may vary widely depending on the objectives and nature of the material; however, there are certain steps that often generate problems. Attention is emphasized on these problems in a series of notes.

Note 1. Attempt to homogenize material by coning and quartering is a very poor approach for several reasons:

1. Coning is promoting segregation, certainly not homogenization. Segregation is introduced by the sifting of free-flowing particles, a mechanism depending on different angles of repose of various minerals, the fluidization mechanism of fine particles, and so on.

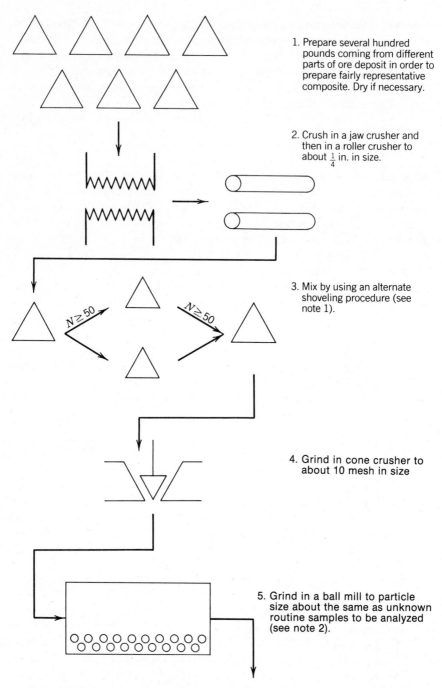

1. Prepare several hundred pounds coming from different parts of ore deposit in order to prepare fairly representative composite. Dry if necessary.

2. Crush in a jaw crusher and then in a roller crusher to about $\frac{1}{4}$ in. in size.

3. Mix by using an alternate shoveling procedure (see note 1).

4. Grind in cone crusher to about 10 mesh in size

5. Grind in a ball mill to particle size about the same as unknown routine samples to be analyzed (see note 2).

Figure 7.2. Illustration of typical sampling procedure for the preparation of an SRM.

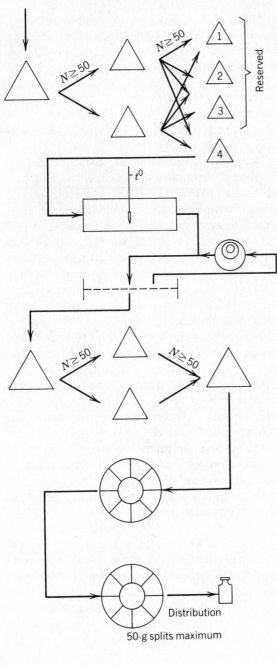

6. Divide sample into four batches using an alternate shoveling procedure (see note 3). Reserve three splits, choose one at random for the following.

7. Complete drying of the sample at an acceptable sampling baseline temperature (see note 4).

8. Screen sample at size opening about the same as unknown routine samples to be analyzed (see note 5). The upper fraction being recycled after passing through rotary mill.

9. Mix by using an alternate shoveling procedure (see note 1).

10. Batch is divided into about 16 portions of equal size using spinning riffler (see note 6).

11. Same operation is repeated for each portion using a same type but smaller spinning riffler until size for distribution is reached (see note 7).

12. As SRM arrives, the user should split it thoroughly see note 8).

2. Quartering is equivalent to making a composite sample of two increments which is in contradiction with equation (1.8) of (Section 1.3) where it is clearly recommended that N should be large in order to minimize segregation (i.e., $N \geqslant 50$).

Pierre Gy goes one step further and states that an alternate shoveling technique (i.e., faster and cheaper) is always more effective to reduce segregation than coning and quartering. Coning and quartering is still very popular more by tradition than logic.

Note 2. A very finely ground SRM, less than 200 mesh for example, should not be used for calibrating procedures if the unknown samples are not going to be -200 mesh as well. This is a very common pitfall, and in the real world of geochemical routine analysis, samples are rarely finer than 90% -100 mesh. Failure to pay attention to this may generate bias during X-ray analysis as well as unnoticed problems with incomplete sample digestion. It is common for a laboratory to perform perfectly well with the SRM and totally fail with unknown samples made of the same material. The conclusion is that an SRM should not only match the unknown samples in content of elements of interest, in mineralogy, but it should also match them in its particle size distribution.

Note 3. Make sure that N is larger than 50 at all times for any newly generated subsample. It may be necessary to adjust the size of the shovel accordingly.

Note 4. Sampling weight baseline is a source of apparent inhomogeneity, which is often ignored and the cause of bias. The user of an SRM should not forget that water occurs in several forms: absorbed water (moisture), adsorbed water (moisture), water of crystallization, and combined water (hydroxyl). Of course, an appropriate correcting factor could be applied after a determination of H_2O^+ and H_2O^-; experience proves that these determinations are often forgotten.

Note 5. In order to avoid large accidental inhomogeneity problems that could be generated by large particles finding their way in the SRM, sooner or later in the analytical subsample, an SRM should be thoroughly screened with a sieve of known size. The choice of the size opening of the sieve should match the maximum particle size of the unknown samples of the users. The conclusion is that a homogeneous -200-mesh SRM may not be a good SRM for some users.

Note 6. For the same reasons mentioned in Note 1, splitting the SRM with a chute riffler having only five riffles going to one side and five to the other or with popular table samplers is not recommended. The objective is to raise N to a high value to prevent segregation errors, and the only way to perform this is to use a spinning riffler, which is fed by a

Figure 7.3. Illustration of typical small riffle splitter to be used to split an SRM.

mass feed hopper followed by a small vibratory conveyor. The use of the hopper and the vibratory conveyor could be objected to since they are generating segregation: The hopper should be small and fed several times accordingly and the conveyor should be as short as possible. When using these devices, the fact that N can be very high (i.e., several hundred) largely overcomes the small disadvantages indicated above [15].

Note 7.   The size of splits for distribution should not be too large. Experience proves that splits larger than 50 g are not recommended. There is no guarantee that a larger split will be in turn properly split for further use.

Note 8.   The SRM received by a user should be thoroughly split before distribution with a small riffle splitter as described in Figure 7.3.

7.8.   SELECTION OF ANALYTICAL METHODS AND RELEVANT DATA

As Abbey [16] stated, "The reliability of a result depends more on who produced it than on how it was done. There is not such a thing as a bad method, only bad analysts who fail to allow for its limitations." This statement, which reflects many years of experience, know how, and a long commitment to the cause of SRMs, is very true. The reader will find it beneficial to refer to some of the other work of Abbey [17, 18]. A knowledgeable and well-experienced analyst will not use an analytical method that was not intended to be used with a specific material. A method should be tailored for a particular material (it is a rule of thumb already mentioned in Section 5.2) and could be the object of lengthy testing work.

If there is not a "bad" method, there is, however, a difference between a good rapid method and a primary method. In most cases for a particular material, a method will exceed the others in accuracy and precision, and according to our definition mentioned in the introduction of Chapter 3, such a method is called a primary method and should be used during classical analysis. It is unlikely that a rapid method could become a primary method because its objectives are different. A rapid method is tailored to a particular material and may only work for this material; while a primary method is generally lengthier, inappropriate for routine work, and involves the state of the art in analytical chemistry. The main purpose of a primary method is to analyze SRMs.

An example may illustrate our point better: Suppose that during the analysis of a MoS_2 concentrate intended to be used as an SRM, molybdenum was determined by gravimetry and also by X-ray. The results obtained from both methods cannot be averaged; the reason is that the X-ray method, which may be very precise, can only be as accurate as the SRMs of the same material used to calibrate the machine. Then, by definition, the X-ray method is a secondary method and always will remain as such. An SRM cannot and should not be certified by averaging determinations generated on one hand by primary methods and on the other hand by secondary methods. For this reason, during the preparation and analysis of an SRM, reported data should always be accompanied by a complete description of the method used. The originator of the SRM has responsibility to only select reported values than can show respect to basic requirements irrespective of the fact that some values generated by an unknown method may look good.

A blind statistical evaluation of data based on the fact that values falling in the most likely range, or the most frequent range of values that "will enjoy privileges," is as a very dangerous approach, particularly in the case of trace elements. Such an approach fails in several instances:

1. It tries to force the estimated average toward the mode, and there are no apparent reasons why in the case of trace elements the true average should be close to the mode unless a calculation of the sampling constant K_s proves otherwise.

2. It depraves the very valuable solitary efforts by competent people of being rewarded in obtaining the only reliable result.

3. It assumes that constitution and distribution heterogeneities introduce negligible errors, a very optimistic assumption.

In the analysis of trace elements, an outlier cannot be rejected unless the facts show where the method went wrong. The knowledge of the K_s constant is a very effective means of eliminating the outliers and it helps to segregate meaningful outliers from true outliers.

7.9. SETTING UP STANDARDS LABORATORY

In many instances, because the availability of relevant SRMs on the market is often unlikely, it is feasible for a company to set up its own standards laboratory; it is indeed a very highly recommended step; however, it cannot be done without taking many precautions. Such a facility would have a very heavy responsibility to fulfill the following objectives:

1. The preparation, storage, and monitoring of SRMs of ores, concentrates, middlings, tailings, products, etc.

2. Analysis of SRMs by primary methods carefully selected.

3. Preparation and primary analysis of pure minerals for microprobe standards.

4. Preparation and storage of standard solutions.

5. Preparation of synthetic mixtures of known composition for specific purposes.

6. Analysis of unusual materials and development of methods for routine and classical analysis of these materials.

7. Investigation and tailoring of new analytical techniques.

7.9.1 Standard Reference Materials

Given an adequate suite of analyzed standards, rapid instrumental methods of analysis can be quickly and accurately evaluated and calibrated. A complete primary analysis is lengthy and expensive and should not be attempted unless there is an adequate supply of the analyzed material. For a company's internal use, at least 10–20 kg of carefully prepared and thoroughly mixed rock or 1–2 kg of mineral powder should be prepared and split into numerous fractions, each of which has exactly the same composition with respect to relevant elements. Tests for degree of homogeneity must precede analysis so that it is certain that the analysis of one fraction represents all the others.

7.9.2. Complete Primary Analysis

A complete and accurate analysis of an ore or concentrate is a lengthy operation, requiring special skills and equipment. Typically, a well-equipped and competently staffed laboratory can produce about eight complete analyses per month. An example of such an analysis is given in Table 7.12. It would be desirable for a company to have complete analyses of this sort for all major ores and products. This would be of special benefit in new explorations, where unsuspected concentrations of valuable elements or nuisance constituents might be of real consequence.

7.9.3. Analyzed Minerals for Microprobe Standards

The selection, separation, purification, and analysis of naturally occurring minerals for use as microprobe standards is a time-consuming, expensive, and difficult operation. Nevertheless, if the microprobe is to be maximally useful, analyzed mineral standards are an absolute necessity. A competently staffed and well-equipped standards laboratory can meet this need.

There is no commercial source of mineral standards for use in calibrating microprobe analyses.

Ingamells and Engels indicated the difficulty of producing an analyzed mineral standard LP-6 Bio 40-60# [19].

7.9.4. Preparation and Storage of Solutions

In AA or ICP analyses, calibration is effected using standard solutions. Often commercial reagents contain appreciable trace contaminants (e.g., cobalt in nickel salts and tungsten in molybdates).

When trace element AA or ICP determinations are based on blank determinations using such reagents, errors are likely. A function of the standards laboratory will be to generate solutions of known concentration and free from undesirable trace impurities. Murphy investigated the role of the analytical blank in accurate trace analysis [20].

Some solutions cannot be accurately prepared from metal salts because the metal content is not exactly calculable from the formula—especially with hydrates (e.g., $Na_2MoO_4 \cdot 2H_2O$ and $Co(NO_3)_2 \cdot 6H_2O$). Solutions must be analyzed for their metal content after preparation.

Table 7.12. Analysis of Canadian Sulfide Ore[a]

SiO_2	34.84
Al_2O_3	9.73
TiO_2	0.79
ZrO_2	0.015
SnO_2	0.0004
V_2O_5	0.037
Cr_2O_3	0.044
NiO	1.92
CuO	1.06
BeO	0.0003
CoO	0.065
PbO	0.026
ZnO	0.039
As_2O_3	0.055
Fe_2O_3	33.54 (total Fe as Fe_2O_3)
MnO	0.13
MgO	3.84
CaO	3.97
Na_2O	1.04
K_2O	0.61
Rb_2O	0.0042
BaO	0.019
SrO	0.011
H_2O^+	3.04
Ignition loss	5.15[b]
$H{\leftrightarrow}O^-$	0.07
P_2O_5	0.10
S	12.03[c]
F	0.02
Rare earths + ThO_2	0.01
Total	100.16
Correction for F	0.01
Correction for free Fe	0.11
Total	100.04

[a] Analysis by Ingamells and Suhr.
[b] Total loss on ignition less H_2O^+, $H{\leftrightarrow}O^-$, and As_2O_3 plus SO_3 in residue.
[c] Total sulfur as S not included in total.

709

7.9.5. Preparation of Synthetic Mixtures

In many cases, adequate natural materials suitable for development into SRMs are unavailable. In such cases, mixtures of known composition need to be prepared to use as calibrating standards. For example, a potassium analytical problem was resolved through mixing of small quantities of analyzed potassium minerals with pure molybdenum oxide. Such mixtures are difficult to homogenize and require special facilities for checking their composition and homogeneity.

7.9.6. Analysis of Unusual Materials

In the extractive metallurgy of ores, fractions containing rare and valuable minerals (e.g., heavy sands) are accumulated. These present severe analytical and sampling problems that need to be resolved if exploitation of these fractions is contemplated. Only when the chemical and mineralogical compositions of a particular separate have been defined can rapid control routines be devised. Application of determinative methods to complex unknown materials are apt to yield spurious results. A function of the standards laboratory is to examine such materials, analyze them completely, and suggest routine methods for determining constituents of interest.

7.9.7. Investigation of New Analytical Techniques

New analytical techniques (e.g., ICP spectrometry, photoacoustic spectrometry, and light scattering) cannot be adequately evaluated and utilized without trial and development to fit specific needs. The inductively coupled plasma source affords an example: It is becoming an increasingly attractive device that promises to resolve a number of awkward problems. The standards laboratory should be able to develop this device into an on-stream monitor for leaching, solvent extraction, or many other processes.

It is doubtful that the operating cost of a standards laboratory can be charged to specific projects except in isolated cases. Corporate funding would seem to be necessary. The generated standard reference materials will have a long life—probably several years—and may be used in controlling analytical work for future projects centered anywhere within the company. The full usefulness of the laboratory will probably not develop for at least two years after it begins operations and will depend on development and maintenance of technical competence, adequate equipment, and a good record-keeping and storage system.

Once an SRM or pure mineral has been properly analyzed and certified, it becomes a permanent asset that should be carefully housed and maintained. The initial cost is high but will be recovered many times over by developing an ability to quickly and confidently calibrate cost-effective instrumental routines.

It must be recognized that the standards laboratory will have to be staffed by competent analysts paid in accordance with their skills. Initially, these skills will have to be developed to suit the special nature of the work.

7.10. MICROPROBE STANDARDS

The electron microprobe is widely used in the analysis of minerals and has the enormous advantage, from the point of view of the crystallographer, that clean individual grains can be examined even without separation from gangue or matrix.

Despite the development of primary methods of calibration using pure metals or synthetic crystals, a need for analyzed natural minerals remains, if not for direct calibration, to check the sometimes complex calculations and corrections required when operating in the primary mode.

Very few mineral separates are sufficiently pure and "homogeneous" to serve well as microprobe standards. The USGS has a massive compilation of minerals that have been used or suggested as microprobe standards. This compilation is in the form of an open file report that can be consulted and used by anyone. (Inquiries may be addressed to the U.S. Geological Survey, Reston, Virginia, 22092.)

The compilation rates minerals according to method of analysis, summation, formula unit, homogeneity, and problems encountered during preparation and analysis. Ratings are in the form of a number from 0 to 14. Materials with scores of 10 or better are usable as known unknowns; those with a rating from 4 to 9 can be used as general-purpose standards. Those with ratings 10–14 are few and far between, and many are exhausted, or exist only in carefully guarded private collections (e.g., that of the authors).

An important source of information on microprobe standards is the journal *Geostandards Newsletter*. Back issues carry reports from a number of authors and institutions and give sources from which small samples of analyzed minerals can be obtained.

These authors posess a number of analyzed minerals that have been widely distributed.

Among the minerals of which there is a plentiful supply and are likely to be available for some time may be included the Nancy phlogopite mica-Mg, available from K. Govindaraju, CRPG, C.O. no. 1, 54500 Van-doeuvre-Nancy, France, the Penn State orthoclase Or-1 (Table 3.7) and the Penn State pyroxene Px-1 (Table 3.27), available in milligram quantities from the authors or from the Mineral Constitution laboratories, State College, Pennsylvania.

Engels' amphibole (Table 3.9) is available in small quantities from the authors. It is used primarily as K–Ar dating standard. Its rating on the USGS microprobe scale is 4, reflecting inverse variation of count rates for iron and magnesium and an occasional "rogue" grain. Sigma ratios for sodium, calcium, aluminum, and silicon are good to excellent.

Many available mineral geostandards are not usable as microprobe standards. A mineral to be used as a microprobe standard must have a sampling constant $K_s < 0.000$ g for the constituent of interest. Some samples may be usable for some elements and not for others. For example, various fractions of LP-6 Bio 40-60# show inappreciable differences in iron, aluminum, maganese, and silicon but unacceptable differences in potassium, magnesium, and calcium. This sample would rate low on the USGS scale but has the advantage of being available and of well-established composition.

7.11. USE OF STANDARDS IN CALIBRATION

Geostandards (and this includes locally produced standards) should only be used to calibrate an analytical procedure if

1. sampling constants have been estimated and
2. the standards perform in the same way as the unknowns to be examined during the procedure.

1. Sample constants are used to construct error bars on the readout value for each standard (see Figures 7.1 and 7.4). It is best to draw $6s$ error bars, calculated from

$$\pm E = 6s = 6\frac{Ry}{100} = \frac{6}{100}y\sqrt{\frac{K_s}{w}}$$

where E is the half-length of an error bar that contains $99+\%$ of a large number of readout values from w-gram subsamples, y is a single readout value, R is the relative subsampling error in percentage of the amount of

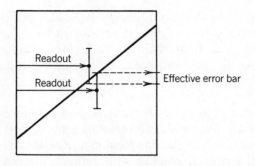

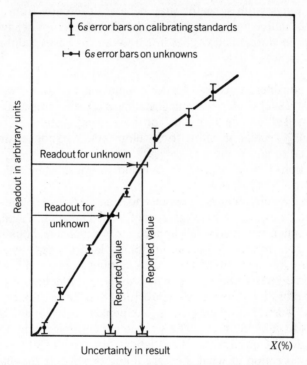

Figure 7.4. Typical curve obtained when calibrating flame photometer.

the constituent of interest present, and s is the standard deviation expected due to the inhomogeneity of the standard material measured by the sampling constant K_s. This principle has been developed by Ingamells and Switzer [10] and Engels and Ingamells [1], and its application will remove much of the uncertainty from the calibration process.

2. It is often essential that calibrating standards have physical characteristics similar to those of the unknowns to be examined. In a direct-

pressing X-ray fluorescence procedure, grain size effects are of paramount importance; calibrating standards should be prepared in just the same way as the unknowns. This excludes many widely circulated and well analyzed finely powdered geostandards from usefulness in the real world, where fine grinding is out of the question when hundreds of samples must be treated in as short a time as possible.

Mineralogical composition of standards and unknowns are often required to be similar. In a rapid fusion solution procedure, if the constituent of interest is contained on the one hand in a mineral easily attacked and on the other in a mineral attacked with difficulty, trouble is likely. For example, tungsten in scheelite may behave differently from tungsten in wolframite: native gold and gold in telluride ores do not behave the same in many procedures; and copper in carbonates is easily soluble, while in sulfides it is not.

These considerations lead to the conclusion that there will probably never exist a supply of geochemical standards that will meet all calibration needs. It will always be necessary for those needing standard ores, rocks, or minerals to make an effort to develop them "in-house" to suit their own special requirements.

In Section 1.26, the need to develop in-house standards has been emphasized.

The first requirement is the ability to analyze selected samples of the material to be examined by primary methods. This requirement is seldom met, and much confusion and unnecessary expensive floundering have resulted. This is especially true in mining laboratories, where there is seldom any recognition of the truths outlined above. Large expenditures—sometimes running into many millions of dollars—go into collecting exploration samples that are turned over to minimally trained "analysts" sometimes equipped with expensive instruments but without the means to calibrate them. The situation is confounded by the fact that field samples are often reduced to "laboratory samples" by laborers who have no perception of what they are doing or why. If the material they are hired to crush, split, and blend contains interesting-looking crystals, these are removed, to be displayed on the mantelpiece. If the prescribed procedure involves the troublesome process of putting the material through a screen, the few stubborn particles remaining after a suitable effort are simply discarded.

In one real example, a sampler was fired because he took too long to prepare −200-mesh material for analysis. His successor received high praise for tripling the output of the Sample Preparation Department. His success was due to the clever procedure he devised: He simply took the

$\frac{1}{2}$-in. crushed field sample supplied and shook it on the screen until he had the required 50 g of -200 mesh material. Only occasionally was extensive crushing or grinding needed!

Having acquired a suite of analyzed standards, the next step is to estimate sampling constants for each constituent of interest in each standard. This is done by repetitive determination using the analytical procedure devised especially for the material to be examined and using minimal sample weights. It is unnecessary to convert instrument readings to concentrations because the sampling constant K_s can be calculated directly from instrument readings. A minimum of eight—and preferably more—readings should be accumulated for each constituent of interest in each potential standard.

The cost of such a program may be high, but it is miniscule compared to the cost of collecting and treating the exploration samples and fades into insignificance when compared to the cost of either abandonment of a viable property or exploitation of a nonviable deposit.

When standards have been analyzed and sampling constants estimated, calibration curves, with error bars, can be drawn, and the analytical procedure can be put into operation. Knowing the sampling characteristics of the material being examined, rational uncertainties can be attributed to each result. These must be taken into account during data evaluation.

Data evaluation should begin as soon as any results become available. A watch should be kept for gross faults in exploration procedure. Field sample size, initial reduction procedures, subsampling, and analytical procedures should be continually scrutinized for their adequacy. New calibration standards may need development. A tight, knowledgeable, and continuing control can, in the long run, save enormous amounts of money and avoid disastrous mistakes.

One should note that properly prepared and well-analyzed standards are the key to the effective use of the numerous rapid instrumental determinative methods now available to geologists, exploration geologists, mining engineers, and everyone who needs rapid analyses of many samples. An inability to generate or obtain adequate standards can founder an otherwise useful operation.

Figure 7.4 shows the type of curve that might be obtained in calibrating a flame-photometric method for an alkali metal, X. Error bars have been exaggerated to demonstrate the principle. Note that the curve passes through all the error bars.

In cases where a calibrating standard has a relatively poor sampling constant, and this constant is known, repetitive determinations may establish a good calibration point, *especially if chance supplies two widely different readout signals*! The overlap of the two error bars gives a much shorter error bar. It is extremely unlikely that the two disparate readings will be more than $6s$ apart.

A calibration point generated by the average of two readings will carry an error bar of $6s/2$, which may be longer than the overlap error bar illustrated in the inset of Figure 7.4.

If, as is usual in rapid determinative procedures, the calibration curve is computerized, the distinction between the $6s/2$ error bar and the overlap error bar can be made during programming. Sometimes, several determinations on a poorly sampleable standard can yield a strong calibration point.

7.12.   STANDARDS FOR K–Ar DATING

The potassium–argon and rubidium–strontium methods for determining the geologic age of rocks and minerals was developed during the 1950s, largely through the efforts of Nier [21] and Goldich [22] at the University of Minnesota. It soon became apparent that carefully prepared and analyzed standards were desirable, so that the many geochronologists newly attracted to this interesting field could compare techniques and results. One of the first geochronological standards prepared and distributed in quantity was the biotite B-3203 from the Massachusetts Institute of Technology. Much time and effort was expended by numerous laboratories worldwide in an attempt to establish firm K_2O and Ar* values for this material. Symposia were held to discuss the annoying variability in results reported by analysts of undisputed competence.

Other organizations developed standards for worldwide circulation, and discordant results remained a puzzle that was not resolved until it became recognized that these standards were inhomogeneous with respect to potassium and argon. At the University of Minnesota and at Penn State University, the practice was to reduce as-received samples to pass a 100-mesh screen before analyses for potassium even though grinding biotite to pass a 100-mesh screen is something of a chore. As a consequence, results from these laboratories showed a consistency that other laboratories could not match. At that time, no one recognized that the fine grinding essentially removed the inhomogeneity problem; it was used at Minnesota and Penn State because the method applied—J. L. Smith fusion and classical analysis—demanded it, not because there was any awareness of the sampling problem. Only a few others used the classical approach, notably L. Peck's USGS laboratory, and these agreed substantially with the results from Minnesota and Penn State.

Sampling constants for many of the K–Ar biotite and other geochron-

Table 7.13.   Sampling Constants for Potassium

Sample	K₂O (wt. %)	Sampling Constants K_s^*
Nancy phlogopite	10.18	0.000
PSU biotite 5-110	10.00	0.000
LP-6 Bio 40-60#	10.03	0.005
Bern biotite 4B	9.47	0.012
Nancy biotite Bio-R	8.80	0.012
USGS muscovite P-207	10.3	0.17
MIT biotite B-3203	9.05	0.19
Bern muscovite 4 M	10.40	0.02
PSU orthoclase Or-1	14.92	0.000

ological standards have now been determined, and the source of difficulties experienced in the early days of K–Ar dating has been recognized.

The standard LP-6 Bio 40-60# was prepared at the USGS laboratories in Menlo Park in an effort to provide the geochronological community with a reliable standard biotite for calibration and control of K–Ar dating procedures. Selection of this material followed lengthy examination of about 20 possible biotite-bearing rocks in California and Washington State. The biotite–pyroxenite LP-6 was selected because microscopic examination showed richness in inclusion-free biotite, and the rock was available in quantity.

Preparation of this 99.9% pure biotite has been described in the proceedings of the Seventh National Bureau of Standard Materials Research Symposium [19].

Results of this effort were somewhat disappointing in that the sampling constants for potassium and argon were not zero as was hoped, but 0.005 and 0.05 g, respectively. The extensive work done on this material did, however, serve to illuminate some of the previously unrecognized sampling difficulties that have plagued K–Ar geochronology from its infancy.

Sampling constants for several of the most widely known standards are listed in Table 7.13.

Despite its imperfections, LP-6 Bio 40-60# has served well. Hopefully, the lessons it teaches may lead to the development of a truly homogeneous biotite for use as a standard.

7.13. SOURCES OF INFORMATION

Among the most important sources of information on the availability and character of geostandards are the South African National Institute for

Metallurgy, where numerous widely distributed ore and rock standards have originated, publications by the Geological Surveys of Canada, the United States, the Soviet Union, Japan, and many other countries, and especially the publishers of *Geostandards Newsletter,* the CRPG in Nancy, France. Special note is made of the contributions of Abbey of the Canadian Geological Survey, among which the most recent appears as Appendix I in "Rock and Mineral Analysis" by Johnson and Maxwell [23]. In this article, Abbey reports usable values for numerous trace elements and major constituents in most available geochemical standards. Among Abbey's subheadings are "Nature of the Raw Data," "Derivation of Usable Values," "The Select Laboratory Method," "Verification of Usable Values," and "Presentation of the Data" and a description of the Canadian Certified Reference Materials Project. This article also contains a wealth of information concerning sources of geochemical standards and an extensive bibliography. It can be described as 58 pages of concentrated information of interest to anyone using, producing, or otherwise interested in geostandards [16–18].

REFERENCES

1. J. C. Engels and C. O. Ingamells, "Geostandards, A New Approach to Their Production and Use," *Geostandards Newsl.* **1,** 51–60 (1977).

2. C. O. Ingamells and P. Switzer, "A Proposed Sampling Constant for Use in Geochemical Analysis," *Talanta* **20,** 547–568 (1973).

3. A. W. Kleeman, "Sampling Error in the Chemical Analysis of Rocks," *J. Geol. Soc. Australia* **14,** 43–48 (1967).

4. F. E. Wickman, "The Amount of Material Needed for a Trace Element Analysis," *Ark. Mineral. Geol.* **3**(6), 131–139 (1962).

5. P. Gy, *Sampling Nomogram, A Circular Calculator,* Minerais et Metaux, Paris, 1956.

6. P. Gy, The Sampling of Broken Ores, A Review of Principles and Practice, Opening lecture of a meeting organized by the Institution of Mining and Metallurgy, July 2, 1973, London, pp. 194–205, Geological Mining and Metallurgical Sampling.

7. P. Gy, The Sampling of Particulate Materials, A General Theory, Opening lecture of the symposium "Sampling Practices in the Mineral Industries," Australasian Institute of Mining and Metallurgy, Melbourne, Australia, September 16–17, 1976, pp. 17–33.

8. P. Gy, Contribution a l'Etude de l'Heterogeneite d'un Lot de Matiere Morcelee, Thesis, Nancy, France, 1972.

9. J. C. Engels, "Effect of Sample Purity on Discordant Mineral Ages Found in K–Ar Dating," *J. Geol.* **79,** 609–616 (1971).

10. A. D. Wilson, "The Sampling of Rock Powders for Chemical Analysis," *Analyst* **89,** 18–30 (1964).

11. C. O. Ingamells, "Control of Geochemical Error Through Sampling and Subsampling Diagrams," *Geochim. Cosmochim. Acta* **38,** 1225–1237 (1974).

12. H. Jaffrezic, B. Villemant, and J. L. Joron, "Dosage des Elements en Traces dans les Roches: Estimation de l'Erreur Due a leur Repartition Statistique dans les Poudres d'Echantillon," *Geostandards Newsl.* **2**(1), 57–60 (1978).

13. F. W. Quackenbush and R. C. Rund, "The Continuing Problem of Sampling," *J.O.A.O.C.* **50,** 997–1006 (1967).

14. J. Visman, "A General Sampling Theory," *Mat. Res. Stds.* **9**(11), 8–13 (1969).

15. T. Allen and A. A. Khan, "Critical Evaluation of Powder Sampling Procedures," *Chemical Engineer,* May, 1970.

16. S. Abbey, Reliability in the Analysis of Rocks and Minerals, Presented at the symposium on "Reliability in Analytical Chemistry," at the 1980 Federation of Analytical Chemistry and Spectroscopy Societies (FACSS) Meeting 1981.

17. S. Abbey, U.S. Geological Survey Standards, A Critical Study of Published Analytical Data. Canadian Spectroscopy, V.15, 10-16, 1970.

18. S. Abbey, Studies in "Standard Samples" of Silicate Rocks and Minerals 1969–1982, Geological Survey of Canada, Paper 83-15, 1983.

19. C. O. Ingamells and J. C. Engels, *Preparation, Analysis, and Sampling Constants for a Biotite,* National Bureau of Standards Special Publication 422, Proceedings of the 7th IMR Symposium, held on October 7–11, 1974, Gaithersburg, Maryland, pp. 401–419 (issued in 1976).

20. T. J. Murphy, The Role of the Analytical Blank in Accurate Trace Analysis, National Bureau of Standards Special Publication 422, Proceedings of the 7th IMR Symposium Held on October 7–11, 1974, Gaithersburg, Maryland, p. 509–537 (issued in 1976).

21. A. O. Nier, "A Redetermination of the Relative Abundances of the Isotopes of Carbon, Nitrogen, Oxygen, Argon, and Potassium," *Phys. Rev.* **77,** 789–793 (1950).

22. S. S. Goldich, C. O. Ingamells, N. H. Suhr, and D. H. Anderson, "Analyses of Silicate Rock and Mineral Standards,"*Can. J. Earth Sci.* **4,** 747–755 (1967).

23. W. M. Johnson and J. A. Maxwell, "Rock and Mineral Analysis," *Chem. Anal. Ser.* **27,** Wiley. (1981).

INDEX

Absorbers, heavy, 142
Accuracy, 191, 517
  X-ray methods, 142
Acid, pH of strong, 621
Acid decomposition, noble metals, 492
Acid group, members, 233
Acids, 114, 618
  strength, 619
Actinolite, 294
Activity, 611
Aerosol, 138
  formation, 663
  production, 660
  stability, 663
Age:
  derived, of low-potassium mineral, 71
  true, of low-potassium mineral, 71
Age determination, 68
  rubidium-strontium, 362
Air buoyancy, 89, 90, 95
Air density, determination, 99
Alfred Werner theory, 629
Alizarin, 434, 467
Alkalies:
  flame photometric determination, 323
  ion exchange separation, 332
  rapid determinations, 339
  separation, 321
Alkali feldspars, 265
Alkali metals, 314
  analysis of traces in peridotites and other
    low-alkali rocks, 340
  Berzelius isolation method, 326
  determination, 123
  flame photometry, 328
  primary methods, 315
  rapid routine methods, 339
  secondary methods, 336
  X-ray spectrometry, 337
Alkali minerals, acid attack, 336

Alkaline earth nitrates, preparation of
    standard solutions, 362
Alkaline earths, 355
  primary methods, 358
  secondary methods, 367
Alkali sulfates, preparation of standard
    solutions, 330
Aluminon, 116
  purification, 116
Alvite, 376
Ammonia, 115
Ammonia group, 434
  separation and analysis, 206
Ammonia group oxides, 249, 447
  first precipitation, 209
  ignition, 212
  second precipitation, 209
Ammonium salts, fusion, 185
Ammonium sulfide group, 241
Amphibole, 233, 268
Amphibolite, 10
Amphiprotic ion, 618, 619
Amphoterization, 633
Analysis, 191
  complete primary analysis, 708
  unusual materials, 710
  work sheet, 260, 261, 262
Analytical chemistry, calculations, 606
Analytical error, 22, 50
Analytical subsample, 41
Analytical techniques, investigation, 710
Analytical variance, 23
Analyzed minerals, 708
Andalusite, 297
Antimony, 238, 453
  decomposition of samples, 454
  determination of Rhodamine B method,
    456
  distillation, 455
  in ores and geochemical materials, 543

Apatite(s), 301, 446
    primary analysis, 364
Aperiodic principle, 90
Ardennite, 274, 453
Argon, radiogenic, 70
Arsenate, 247
Arsenic, 237, 453
    decomposition of samples, 454
    determination:
        of heteropoly molybdenum blue method,
            456
        by tellurium coprecipitation, 463
    distillation, 455
    in ores and geochemical materials, 543
Arsenic mineral, 274
Asbolite, 422
Assay values, distribution, 2
Atomic absorption, 148
    acid concentration, 162
    atomization, 150
    atomization problems, 662
    background noise, 151
    burner position, 158
    calibration curve, 160
    choice of wavelength, 162
    concentration of other elements, 162
    detection limit, 152
    determination of basic elements in laterites,
        556
    drift, 151
    flame optimization, 156
    flame type, 153
    infinite dilution, 167
    interferences, 162
    internal standard, 167
    lamp current, 154
    limit of linearity, 160
    nebulization, 149
    principle, 148
    ratio of signal to background noise, 152
    reproducibility, 158
    slit width, 156
    standard additions, 166
    stoichiometry, 157
    total delivery of flame, 157
Atomic fluorescence, emission, 124
Atomization, problems, 662
Atomizer, 662, 663, 664, 665
Atoms, free, 138
Attack, incomplete, 231

Automatic pipetting, 105
Average, weighted, 49

Baddelyite, 376
Balance:
    analytical, 87
    calibration of weights, 93, 96
    constant-load device, 91
    equality of beam arms, 92
    location, 96
    temperature and humidity effects, 95
    ultimate weighing, 95
    vibrations, 96
Barite, 355
Barium, 254
    contamination, 77
    determination in geologic materials,
        554
    isotope dilution mass spectrometry,
        362
    in silicates, 358
Barium chromate, conversion to barium
    sulfate, 361
Base, 618
    pH of strong, 621
    strength, 619
Beakers, 106
Beryl, 225
Berrylium, 341
    determination:
        by AA spectrometry, 349
        as pyrophosphate, 344
        using Quinalizarin, 347
        of traces, 344
    flourimetric determination using Morin,
        348
    spectrographic determination, 349
Beryllium-bearing silicates, analysis, 342
Berzelius method, isolation of alkali metals,
    326
Bias, 517
Biotite, 256
    analysis of chlorine-bearing biotite, 264
    analysis of MIT Biotite B-3203, 263
    analysis of PSU 5-110, 259
    lithium, 324
Bismuth, 236
Bismuth phosphate, 447
Blending, 74
Borax, 426

Boron, 426
  colorimetric determination using Curcumin, 431
  colorimetric methods, 433
  determination:
    without distillation, 430
    in water, 431
    distillation as methyl borate, 427
    separation, 427
Bronsted theory, 618
Buffer solution:
  pH, 623
  potential, 615
Burette, drainage, 102

Cadmium, 237
Calcium, determination, 218
  in fluorspar, 548
  in limestone or dolomite, 549
  in phosphate rock, 551
Calcopyrite, 435
Calibration curve, flame photometer, 713
CaO, correction for manganese content, 220
Carbon, 437
  carbonate, 440
  total, 438
Carbonate rocks, loss of ignition, 487
Carbon dioxide, 437
Carbon furnace AA, 576
Carmine red, 433
Carnallite, 350
Cassiterite, 184, 199, 232
Catapleite, 275
Cations, acidity, 648
Cerium, 368
Cesium, 315
Chapin's method, 427
Chelates, 629, 631
Chemical microscopy, 127
Chemical reactions, slow reactions, 609
Chlorite, 556
Chloritoid, 306
Chloroplatinates, conversion to sulfates, 322
Chondrites, 227
Chromite, 32, 184, 199, 225, 232, 279
  analysis, 386
  determination of vanadium, 390
  manganese determination, 390
  noble metals, 491
  sampling, 16

Chromium, 383
  determination, 280
  titrimetric determination, 385
Chromium-rich spinel, decomposition, 563
Classical analysis, 191
  of typical minerals, 256
Classical silicate analysis, 192
Cleaning of glassware, 77
Cobalt, 246, 422
  analysis of sulfide ores, 426
  assays in lateritic ore, 19
Cobalt data, histogram, 20
Cobaltinitrite, sodium, 335
Colemanite, 426
Colorimeter, Dubosq, 122
Colorimetric methods, difficulties, 635, 636
Columbite, tantalite, 377
Columbium, 183
Combustion tube, 439
Complexes, 629
  destruction, 635
  dilution, 635
  strength, 632
Concentration, 611
  constant, 644
Conductivity test, 114
Contamination, 75
  cleaning equipment, 79
  Clerici solution, 241
  cross-contamination, 79
  crushing equipment, 78
  detergents, 77
  dish-washing compounds, 77
  glass containers, 77
  glassware, 106
  mixing cloths, 77
  origins, 75
  papers, 77
  screens, 79
  sodium chloride, 76
  tramp iron, 76
  wiping cloths, 77
Coordination number, 630, 631
Copper, 237
Crookesite, 436
Crucible:
  gold and silver, 110, 178
  graphite, 172
    and vitrified carbon, 111
  J. L. Smith, 317

Crucible (*Continued*)
  platinum, 96, 106, 178
  porcelain, 88, 111
  silica and glass, 112
  zirconium, nickel, and iron, 110, 178
Curcumin, 430, 431, 432, 433
Cyanide determination, 575

Data, evaluation, 53
Decomposition, Freibergian decomposition,
    454
Denningite, 276, 464
Density:
  of rocks and minerals, determination,
    100
  of slurries, 653, 654
Desiccants, 113
Desiccators, 113
Determination, 191
Dichromate, iron determination, 418
Dilutors, 105
Dimethylglyoxime, 423, 424
Diopside, 297
Diphenylcarbazide, 384, 385
Direct-reading spectrometers, 129
Dispersion of values, 586
Disproportionation, 633
Dissolution procedures, 170
  incomplete, 231
  perchloric-hydrofluoric, 177
  phosphoric acid, 181
  sinter, 183
  sodium hydroxide, 180
  sodium peroxide, 178, 184
  sodium and potassium pyrosulfate, 181
  sulfuric-hydrofluoric, 175
Distilled water, 113
Distribution:
  double Poisson, 63
  fitting to a model, 585
  Gaussian, 23, 67
  normal, 23
  Poisson, 24
  shape, 583
Dithizone, 571
Dolomite, 349
  complete analysis, 359
  determination:
    of calcium, 549
    of magnesium, 353
  manganese, 305

Dolomite content in limestone, determination,
    352
Drilling, diamond, 69
Dunite, 350, 556
Dysprosium, 368

EDX peaks, separation, 659
Electrochemical reactions, 613
Electrode:
  Hildebrand, 217, 418
  pH and glass, 125
  silver–silver chloride, 417
Electron microprobe analyses, 34
Electrons:
  free, 138
  transfer, 611
Electrons and ions, transfer, 641
Electrons and protrons, transfer, 636
Emission, flame, 123
Emission spectrometry, 128
Emission spectroscopy, 127
Energy-dispersive X-ray, deconvolution of
    peaks, 655
Environmental analysis, rapid methods,
    572
Erbium, 368
Eriochrome black T, 354
Error(s):
  drainage, 104
  evaporation, 103
  nonrandom, 574
  weighing, 88
Europium, 368
Exploration, optimum and minimum field
    sample weights, 47
Exploration cost, minimizing, 50

Ferrite, manganese zinc, 298
Ferrous iron, 273
  determination, 415
Field sample, 41
Fire assay, noble metals, 492
Flame:
  oxidizing, 196
  strongly oxidizing, 200
Flame emission, 123
Flame optimization, 156
Flame photometer, calibration, 713
Flame photometry, atomization problems,
    662
Flame type, 153

Fluorapatite, 364, 446
Fluorides:
  precipitation, 370
  removal, 176
Fluorine, 467
  determination of calcium, 548
  steam distillation, 468, 469, 470, 471, 472
    still, 474
  titration, 473
Fluorophlogopite, 266
Fluorspar, silica content, 231
Fluxes, 116
  lithium metaborate, 116
  potassium carbonate, 122
  potassium pyrosulfate, 121
  procedures using lithium metaborate, 119
  purification, 118
  sodium carbonate, 120
  sodium peroxide, 121
Free atoms and electrons, 138
Fusion, 196, 197
  ammonium salts, 185
  in bombs, 179
  in iron, 179
  lithium metaborate, 170
  in platinum, 179
  sodium carbonate, 186
  sodium hydroxide, 180
  sodium peroxide, 178
  sodium and potassium pyrosulfate, 181
  tantalum and niobium minerals, 378
Fusion oxidants, nitrates, 196

Gadolinium, 368
Gallium, 434
Gallium-bearing mineral, decomposition,
  435
Gangue concentration, determination, 32
Garnet, 290
Garnierite, 422, 556
Garnieritic ores, 425
Gaussian distribution, 23, 67
Gaussian model, 67
Geoanalytical proficiency, 191
Geochemical analyst, duty, 50, 51, 52, 53
Geochemical data, fitting to Poisson
  distribution, 669
Geochemical exploration, complex
  distributions, 670
Geochemical standards, 511, 687
  calibration, 688

control, 691
criticism of common practices, 691
preparation, 695
Geochemistry, Poisson probability, 666
Geochronology, 68
Geostatistics, 6, 46, 50, 54, 599
  nugget effect, 6
  variograms, 6, 600
  zone of influence, 6
Germanite, 442
Germanium, 238, 442
  in coal, 445
  in sulfide ores, acid attack, 443
Glass:
  alkali-resistant, 428
  care for new glassware, 77
  silica, 112
  Vycor, 113
Glassware:
  contamination, 106
  washing, 513
Gold, 491
  dissolution:
    by aqua regia, 494
    by cyanide, 494
  low contents in sand or stream sediments,
    494
  sampling, 594
  sampling nomograph, 595
  stable complexes, 492
Gold ores, sampling, 48
Gold ware, 110
Grains, average number, 15
Grain size, effective, 27
Granite, (USGS) G-1, 15, 24
Graphite, determination, 437
Graphite crucibles, 111
Guldberg and Waage law, 609

Hafnium, 373, 376
  separation from zirconium, 377
Halogens, 682
Harmonic mean, 54
Harzburgite, 556
Heavy absorbers, 117, 119
Heterogeneity of flowing stream, 601
Heterogeneous phase, 644
Histograms, 19
  construction, 583
H_2O^-, determination, 194
Holmium, 368

Homogeneity, 25
  LP-6 Bio # 40–60, 690
Homogeneity constant, 5
Homogeneous phase, 611
H_2S group, elements, 236
Hydride generation, selenium, 462
Hydrobiotite, 301
Hydrofluoric acid attack, alkalies, 326
Hydrogen, 476
  primary methods, 477
Hydrogen determination, rapid methods, 486
Hydrogen sulfide group, separation, 234
Hydroxide, solubility, 646
Hydroxyapatite, 446

ICP:
  atomization problems, 662
  computer scan, 139, 140
  configuration, 137
  determination of tantalum and niobium, 555
Ilmenite, 199, 225, 373
Inclusions, 25
Increments, 37
Indium, 434
Inductively coupled plasma spectrometry,
    135, 576
  limitations, 576
Industrial effluents, analysis, 572
Influences, 138
Information on standards, sources, 717
Inhomogeneity, 61
Insoluble compounds, 631
Instrumental methods, 575
Instrumentation, 122
Interferences, 162
  additive, 163
  classical analysis, 224
  internal standard, 167
  matrix effect, 163
  nonspecific, 163
  proportional, 163
  saturaton, 164
  solutions, 163
  standard additions, 166
  suppression:
    of additive interferences, 165
    of proportional interferences, 166
Internal reference, 131
Internal standard, 330
  rare earths, 371

Internal standard method, 167
Internal standard principle, 129
Ion(s):
  acceptor, 629
  donor, 629
  indifferent, 618
  transfer, 629
Ion exchange separation, alkalies, 332
Ionic product of water, 620
Ions and electrons, transfer, 641
Ions and protons, transfer, 642
Iridium, 491
Iron, 414
Iron crucibles, 110
Iron determination, 76, 217, 417
Isotope dilution mass spectrometry, 129, 341,
    575
  potassium and rubidium, 331
Isotopes:
  lithium, 314
  potassium, 332
  zirconium and hafnium, 376

Jarosites, 214
Jones reductor, 398

K-Ar dating, standards, 716
Karl Fisher methods, 489
K_2O, determination in a mixture, 11
Kyanite, 225

Laboratory sample, preparation, 81
Laboratory subsamples, 26
Lanthanum, 368
Laterite:
  noble metals, 491
  rapid determination of basic elements, 556
Laterite ore(s), 21
  water determination, 478, 479
  X-ray analysis, 652
Law of mass action, 609
Lead, 236
Lepidolite, 258, 314, 315
  analysis of purified lepidolite PSU 60-1252,
    265
  determination of alkalies in lepidolite
    60-1252, 266
  lithium, 324
Leucophosphite, 287
Liberation factor, 3

Ligand, 629
Limestone:
  complete analysis, 359
  determination:
    of calcium, 549
    of dolomite content, 352
    of magnesium, 353
  rapid analysis, 551
  recommended scheme for analysis of
    alkaline earths, 358
  sodium and potassium, 339
Limonitic ores, analysis, 425
Lithiophilite, 400
Lithiophorite, 422
Lithium, 314
  internal standard, 131
  lead carbonate separation, 324
Lithium metaborate, 116, 142, 170
  procedures, 119
  purification, 118
Lithium metaborate fusion, nitric acid
    solution, 336
Lithium tetraborate, 143
Logarithms, review, 607
Lognormal model, 67
Lorandite, 436
Loss on ignition, 194, 486
  carbonate rocks, 487
Lutetium, 368

Magnesite, 349
Magnesium, 349
  determination, 220
    in limestone or dilomite, 353
  precipitation with oxine, 353
  primary methods, 350
  rapid methods, 352
  secondary methods, 354
  X-ray fluorescence, 354
Magnetite, 199, 225
Malachite, 32
Malacon, 376
Manganese, 243, 400
  accurate analysis, 413
  analysis of sea nodules, 412
  atomic absorption determination, 411
  colorimetric determination, 410
  determination, 220
  dolomite, 305
  oxidation:

  to Mn(III), 407
  to permanganate using bismuthate,
    405
  to permanganate by persulfate, 406
  sampling considerations, 411
  titration:
    with arsenite, 406
    with EDTA, 409
    with ferrous iron, 405
  titrimetric determination, 405
Mannitol, 429
Marl, rapid analysis, 551
Matrix corrections, 8
  X-ray, 650
Mean, 585
  harmonic, 54
Median, 585
Mercury, 236
  determination in sulfide residues, 570
Meteorites, 227
Methylene blue, 433
Micas (white), 308
Microchemical work, 95
Microline, 199
Microlite, 272, 377
Microprobe standards, 708, 711
Microscope, 126
Microscopy, chemical, 127
Microtitration, 101
Mineral:
  density, 9
  proportion, 9
Mineral formula, calculation, 676
Mineralogical composition factor, 3
Mispickel, 453
Mixing, 74
Mode, 585
Model:
  Gaussian, 67
  Poisson, 37
  two-mineral, 35
Moisture determination, 477
Molybdenite, 390
  sodium and potassium, 340
Molybdenum, 238, 390
  colorimetric determination, 523
  determination:
    by atomic absorption, 526
    as lead molybdate, 518
  histograms, 61

Molybdenum (*Continued*)
  manipulation of data points, 60
  reduction with mercury, 398
Molybdenum Blue (heteropoly) method,
    arsenic, 456
Molybdenum ore:
  analysis, 393
  sampling, 18
Molybdenum sulfide, 227
Molybdenum titration, permanganate method,
    398
Monazite, 371
Morin, 348
Mortar, steel, 78
Most probable result, plot *vs.* sample weight,
    36
Most probable value, 37
Mud, density, 12
Muffle furnace, 197
Muscovite, 308

Nagyagite, 464
Nebulizers, ICP, 137
Nebulizing system, production of stable
    aerosol, 663, 664
Neodymium, 368
Nepheline, 281
Nernst equation, 614
Neutron activation analysis, 575
Nickel, 242, 422
  analysis of sulfide ores, 426
Nickel crucibles, 110
Niobium, 377
  determination:
    by ICP, 555
    of trace amounts, 380
  ion exchange separation, 381
Niobium minerals, fusion, 378
Noble metals, 239, 490
  acid decomposition, 492
  alluvial deposits, 491
  constitution heterogeneity, 491
  determination, 492
  distribution heterogeneity, 491
  fire assay, 492
  sampling, 491, 594, 595, 596
Nomographs, construction, 593, 597
Normal distribution, 23
Nugget effect, 18, 47, 50

Olivine, 282
O-phenanthroline, 422
Optimum sample weight, 52
Organic matter, difficulty during analysis of
    carbon, 437
Organic reagents, purity, 115
Orthoclase, 13
Outliers, 67
Oxidant(s), 612
  mixture, 616
Oxidation number, 611
Oxygen, 682
Oxygen deficiency, 273, 465
Oxygen excess, 465

Palladium, 242, 491
Particle size, 12
Particle size distribution factor, 3
Pegmatite, 341
  lithium, 324
Pelletizing die, 144
Penfield method:
  rapid modification, 489
  water determination, 479
Penfield tube, illustration, 483
Penfield tube method, hydrogen determination,
    481
Perchloric acid, evaporation and precautions,
    178
Periodotite, 350, 556
  alkali metals, 340
  weathering, 422
Perovskite, 373
Petalite, 314
Petrographic microscope, 126
pH:
  buffer solution, 523
  mixture of acid with base, 628
  notion, 619
  weak acid, 625
  weak base, 627
Phenylfluorone method, 444
pH meter, 125
Phosphate, 247
  contamination, 77
Phosphate rock:
  analysis, 291, 292, 293
  determination of calcium, 551
  primary analysis, 364

silica analysis, 231
Phosphomolybdate (ammonium):
  conversion to magnesium pyrophosphate,
    450
  treatment of precipitate, 449
Phosphorus, 446
  in apatite, 449
  in phosphate rock, 449
  titrimetric determination, 452
  in tungsten-bearing materials, 451
Photographic plate, 127
Pipette(s):
  drainage, 102
  lambda, 105
pK:
  common acids and bases, 623
  notion, 622
Plasma, 136, 138
Plasma torch, ICP, 136
Platinum, 491
  ammines, 240
  base metal contamination, 197
Platinum crucible:
  loss of weight, 204
  protection, 108, 564
  treatment, 200
Platinum ware, 106
Point of diminishing returns, sampling of a
    mountain, 53
Poisson distribution, 591
  double, 63
  fitting single Poisson distribution, 658
Poisson formula, 17
Poisson model, 37, 64
Poisson probability, 666
  X-ray, 655
Poisson statistics, 23
Polar molecules, 629
Pollucite, 315
Polycomplex, 633
Polyhalite, 350
Porcelain ware, 111
Potassium, 314
  calibration curves, 132, 133, 134, 135
  carbonate, 122
  isotope dilution mass spectrometry,
    331
  in limestone, 339
  in molybdenite, 340

pyrosulfate, 121
  rapid gravimetric determination, 334
  rapid titrimetric determination, 334
  sampling constants, 717
  in sulfide ore containing copper, 340
Potassium-argon dating, 71
Potassium feldspar, analysis, 267
Potential:
  apparent, normal, 614
  calculation in mixture of oxidant and
    reductant, 616
Powellite, 390
Praseodymium, 368
Precipitation, 644
Precision, 191, 517
  X-ray methods, 142
Preparation, laboratory sample, 81
Probability distribution, 582
Procedures, standard format, 514
Proportions:
  by volume, 9
  by weight, 9
Protons, transfer, 618
Protons and electrons, transfer, 636
Protons and ions, transfer, 642
Psilomelane, 400, 413
Pyrite, 415
Pyrochlore, 272, 377
Pyrolusite, 400
Pyrophanite, 373
Pyroxenes, 294

Radioactive isotopes, naturally occurring, 497
Radium, 496
  analysis in ores, 502
  in ores, decomposition of samples, 502
  separation, 503
Random variable, 581
Rapid analytical method, tailoring, 514
Rapid chemical methods, 510
Rare earths, 241, 368
Reaction:
  electrochemical, 613
  prediction, 619, 632, 637
  between two acid-base systems,
    619
Reagents, organic, 115
Realgar, 453
Recorders, strip chart recorder, 126

Red-ox buffer, 615
Reductant, 612
  mixture, 616
  zinc and silver, 421
Reference materials, trace elements, 698
Relative deviation, 25
Relative variogram, 601
Releasing agent:
  aluminum, 165
  lanthanum, 165, 368
  strontium, 165
Reproducibility, 517
Residues, spectrographic examination, 245
Result, most probable, 37
Rhenium, 390
  determination by extraction and atomic
    absorption, 533
Rhodamine B, 435, 437, 455, 456
Rhodium crucibles, 109
Riebeckite, lithium, 324
Roquesite, 435
Round robin, 573
  problems, 692, 693
Rubidium, 314
  isotope dilution mass spectrometry, 331
  traces in rocks, 340
Rubidium-strontium, age determinations, 362
Ruthenium, 491
Rutile, 373

Samarium, 368
Sample, 8
  effect of variable weight, 49
  laboratory, 7
  reduction, 26, 672
Sample preparation, X-ray, 569
Sample reduction, molybdenum ore, 28, 29
Sample weight:
  minimum, 45
  optimum, 52
  optimum field sample weight, 54
Sampling:
  analytical care (Gy's quote), 43
  calculations, 593
  gold ores, 48
  gold and silver, 594
  hard facts, 48
  herring, 2
  long-range fluctuations, 603
  periodic fluctuations, 605

  for potassium-argon dating, 70
  purposes, 68
  short-range fluctuations, 603
  subtilities, 68, 69
Sampling campaign, planning, 44
Sampling constant, 3
  determination, 697
  Gy, 3
  Ingamells and Switzer, 4
  K_s, 24
  Visman, 30
Sampling diagrams, 35
  construction, 38
  molybdenum, 40
  usefulness, 41
Sampling methods, in the field, 69
Sampling nomographs, construction, 593, 597
Sampling problem, emission spectroscopy,
    128
Sampling slide rule, Gy, 30, 41
Sampling theory(ies):
  Gy's fundamental error, 3
  relationships, 5
  Visman, 5
Sampling weight baseline, 477
Scandium, 368
  extraction, 373
Scapolite, 275
Scheelite, 451, 536
Screen tests, 81
Sea nodules, analysis, 412
Segregated mixtures, sampling diagrams, 37
Segregated orebodies, 63
Segregation, 7, 31, 37, 53, 61
Segregation constant, 5
Separation, J. L. Smith, 316
Selenium, 239, 457
  chemistry, 457
  colorimetric determination, 464
  decomposition of solid samples, 461
  determination, 462
    by tellurium coprecipitation, 463
  electrothermal atomization, 464
  fluorimetric determination, 464
  hydride generation, 462
  in ores and geochemical materials, 543
  preparation of standard solutions, 461
  in solution, stable form, 459
  whole bottle technique, 460
Selenium (VI), reduction, 459

Selenium (0) and selenium (IV), separation
    from selenium (VI), 460
Selenium (0), selenium (IV), and selenium
    (VI), separation, 460
Serpentine, 349
    noble metals, 491
Serpentinite, 556
Shape factor, 3
Silica:
    dehydration, 198, 199
    determination, 201
        by atomic absorption, 559
    filtration, 202
    recovery of small amounts, 213
    second dehydration, 203
    vitreous, 112
Silica determination, interfering elements, 206
Silica glass, 112
Silica separation, interferences, 228
Silicate analysis, beryllium-bearing silicates,
    342
Silicates, basic and actual analysis, 193
Silicon dioxide, determination by atomic
    absorption, 559
Sillimanite, 199, 225
Silver, 236
    determination by atomic absorption, 565
    nomograph, 595
    sampling, 594
Silver ware, 110
Single grain, contribution, 12
Sinter:
    sodium carbonate, 183
    sodium peroxide, 184
SiO2, evaporation, 205
Size, effective linear mesh size, 12, 27
Skewed data, manipulation, 54, 670
Skutterudite, 422
Slurries, composition and density, 653
Smaltite, 422
Smith, J. L.:
    method, 185
    separation, 316
Sodalite, 270
Sodium, 314
    contamination, 77
    gravimetric determination, 333
    in limestone, 339
    in molybdenite, 340
    in sulfide ore containing copper, 340

Sodium carbonate, 120
    fusion, 186
    sinter, 183
Sodium peroxide, 121
    sinter, 184
Solubility of precipitate, calculation, 645
Solubility product, 644
Solution procedures, 185
Solutions of metals, X-ray methods, 143
Specimen, 69
    atomic absorption, 148
    atomic fluorescence, 124
    emission, 128
    ICP, 135
        advantages, 141
        analytical characteristics, 138
        chemical interferences, 138
        disadvantages, 141
        interferences, 138
        minimizing interferences, 139
        physical description, 136
        spectral interferences, 139
Spectroscopy, emission, 127
Spencite, 285
Sphalerite, 435
Sphene, 283, 373
Spinel, 199, 225
    decomposition, 563
Spiroffite, 276, 464
Spodumene, 314
SRM, 707
    investigation of inhomogeneity, 701
    selection:
        of analytical methods, 705
        of relevant data, 705
    typical sampling procedure, 701
Standard deviation, estimated, 21
Standard addition method, 130, 131, 166
Standard format, analytical methods, 514
Standard laboratory, 707
Standard reference material:
    applications, 699
    commercially available, 694
    guidelines, 699
    nature, 699
    separation, purification, and packaging,
        700
Standards:
    calibration, 8, 712
    geochemical, 688

Standards (*Continued*)
  K-Ar dating, 716
  sources of information, 717
Standard samples, 512
Standard solutions, preparation and storage,
    708
Stannite, 435
Statistical models, fitting, 67
Statistics, fundamental notions, 581
Stibnite, 453
Still, distillation of fluorine, 474
Stirling approximation, 657, 658, 660
Stoichiometry, 153, 157
Stream sediments:
  determination of gold by cyanide dissolution,
    494
  gold, 494
Strontium, 254
  determination, 361
  isotope dilution mass spectrometry, 362
  separation, 361
  in silicates, 358
Subsample(s), 8
  average, 15
Subsampling diagram, amphibole-orthoclase
    mixture, 39
Sulfate, 248
Sulfide containing molybdenum, analysis,
    393
Sulfide ore containing copper, sodium and
    potassium, 340
Sulfur, 682
Sylvanite, 464
Synthetic mixtures, preparation, 710

Talc, 349
Tannin, 378, 379, 442
Tantalate, potassium fluoro-tantalate,
    377
Tantalum, 183, 377
  determination:
    by ICP, 555
    of trace amounts, 380
  ion exchange separation, 381
Tantalum minerals, fusion, 378
Teflon vessels, 175
Tellurium, 239, 457, 464
  in ores and geochemical materials, 543
Tellurium minerals, 276
Terbium, 368

Thalenite, 285
Thallium, 240, 434
Thiazole yellow, 350
Thiocyanate procedure, 540
Thorin method, thorium, 502
Thorium, 496
  determination by Thorin method, 502
  separation, 500
Thorium nitrate, standardization, 474
Thorium ores, decomposition, 500
Thortveitite, 376
Thulium, 368
Tin, 238
Titan yellow, 350
Titanite, 373
Titanium, 373
  contamination, 77
  standard solution, 215
Tourmaline, 225, 269, 426
  analysis, 230
Toxic constituents, analysis, 572
Toxin(s), 572
  in waters and effluents, 576
Trace element content, variation, 68
Trace elements:
  environment, 574
  in reference materials, 698
Triphylite, 289
Triphylite-lithiophilite, 289
  lithium, 324
Tungsten, 390
  colorimetric determination, 540
  determination in scheelite and wolframite,
    536
  volumetric method, 397
Tungsten-bearing materials, phosphorus, 451
Tungsten ore, 41
  sample reduction scheme, 42
Tyler standard screen scale, 81

Ulvospinel, 285
Uncertainty, 41
Uranium, 253, 496
  extraction, 499
  separation, 498
  solution in sodium fluoride, 499
Uranium ores, decomposition, 497

Vanadium, 253
Variation, coefficient, 25

Variographic experiment, 599
  long-range term, 604
  periodic term, 605
  short-range term, 603
Variography, principles, 599
Variograms, 6
Vermiculite, 301
Vitreous silica, 112
Vitrified carbon crucibles, 111
Volumetric apparatus, 100
Volumetric flask, 100
Vycor combustion tube, 439
Vycor glass, 113
Vycor glassware, 520

Wastes, analysis, 572
Water:
  distilled, 113
  ionic product, 620
  in minerals, spectrographic measurement, 490
Water determination:
  Karl Fisher methods, 489
  Penfield method, 479
Water effluents, analysis, 572
Weak acid and base, pH, 625
Weighing:
  single-deflection, 91
  single-swing, 91
  ultimate, 95
Weight baseline, sampling, 477
Whole bottle technique, 460

Wolframite, 451, 536
Wulfenite, 390

X-ray:
  fitting peak to Poisson distribution, 655
  K lines, 653
  L lines, 653
  matrix corrections, 650
  pelletizing die, 145, 146, 147
  Poisson probability, 655
  sample preparation, 569
X-ray energy dispersive, deconvolution, 655
X-ray fluorescence, 141
  determination of magnesium, 354
X-ray methods, 141
  pelletizing die, 144
X-ray peaks, separation, 659
X-ray spectrometry, 142
  alkali metals, 337
  pellet preparation, 143

Ytterbium, 368
Yttrium, 368

Zinc, 246
Zinc blende, 444
Zinc group, 423
Zircon, 199, 225, 233, 376
Zirconium, 373, 376
Zirconium crucibles, 110
Zirconium phosphate, 447